# A Practical Guide to MR-Linac

Indra J. Das • Filippo Alongi
Poonam Yadav • Bharat B. Mittal
Editors

# A Practical Guide to MR-Linac

## Technical Innovation and Clinical Implication

Springer

*Editors*
Indra J. Das
Department of Radiation Oncology
Northwestern Memorial Hospital,
Northwestern University Feinberg
School of Medicine
Chicago, IL, USA

Poonam Yadav
Department of Radiation Oncology
Northwestern Memorial Hospital,
Northwestern University Feinberg
School of Medicine
Chicago, IL, USA

Filippo Alongi
Department of Advanced Radiation
Oncology
Ospedale Sacro Cuore Don Calabria
Negrar di Valpolicella, Verona, Italy

Bharat B. Mittal
Department of Radiation Oncology
Northwestern Memorial Hospital,
Northwestern University Feinberg
School of Medicine
Chicago, IL, USA

ISBN 978-3-031-48164-2 ISBN 978-3-031-48165-9 (eBook)
https://doi.org/10.1007/978-3-031-48165-9



This Springer imprint is published by the registered company Springer Nature Switzerland AG
The registered company address is: Gewerbestrasse 11, 6330 Cham, Switzerland

Paper in this product is recyclable

*My personal dedication goes to my late grandfather Jagdeo Das who provided me dream of aspiration through education. Additionally, I extend my loving gratitude to my family (Sununta-my wife, Avanindra-son, Anita-daughter, and twin grandchildren Avan and Aska who were born during inception of this book). Their unwavering support and patience have been invaluable, allowing me the space and time on this book.—**Indra J Das***

*Special thanks to all my patients. All cases, their feelings, the fear, the trust are in my mind. All efforts to cure cancer remain to me as experience for improving the next.—**Filippo Alongi***

*A great many individuals deserve credit and recognition from whom I have the opportunity to learn and grow. Especially, the two who enriched my career in academia are Dr. Bhudatt R. Paliwal, whose deep knowledge for therapeutic physics helped me find innate delight in work and modesty, and Dr. Bharat B Mittal, whose unprecedented guidance turned work into an adventure.—**Poonam Yadav***

*I'm grateful to my patients for their trust in me.—**Bharat B Mittal***

# Preface

Over the years, significant strides have been made in the field of radiation oncology, with cutting-edge innovations driving improved precision treatment that is improving the patient outcomes. One recent revolutionary advancement that is transforming the landscape of radiation treatment is MR-Linac, an amalgamation of magnetic resonance imaging (MRI) and linear accelerator (Linac) technology, which has joined forces to create a paradigm shift in cancer care. This novel approach offers an unprecedented level of precision, enabling real-time visualization of the tumor and surrounding healthy tissues during treatment delivery. Unlike conventional radiotherapy, where imaging is performed separately from treatment, MR-Linac technology offers the unprecedented ability to capture high-resolution images of the tumor and surrounding healthy tissues immediately before and even during the radiation treatment process. This real-time imaging (MR-guided Radiation therapy, MRgRT) capability has the potential to reshape the way treatment can be planned and executed.

The utilization of imaging to optimize precision and accuracy throughout the entire process of radiation therapy delivery has been called "image-guided radiation therapy" (IGRT). By comparing current images to those obtained during simulation procedures, the patient's positioning and/or the geometry of the radiation beams may be adjusted to more precisely address the radiation dose to the target. Radiation therapy has long been guided by X-ray images; in fact, historically, the portal films or, later on, electronic megavoltage images (obtained with electronic portal imaging devices, EPID) represented an early type of IGRT. Portal films and EPID images are two-dimensional evaluations that, lacking 3-D visualization of soft tissue, provide limited information with a poor image quality. The EPID images can indicate reasonably well the location of the isocenter and of the field margins relatively to the bony landmarks. However, the treatment target could be a mobile soft tissue mass within the body, and patient repositioning based on bony landmarks alone is subject to significant uncertainties. Due to these inaccuracies, it is inevitable to include a larger volume of normal tissue in the planning target volume to ensure target coverage.

Different types of IGRT modalities are now available. The target location may be identified by several methods for localization of surrogates, including implanted fiducial markers, external surface markers, or anatomical features, by planar imaging, fluoroscopy, kV or MV computed tomography (CT), ultrasound and X-ray imaging, electromagnetic localization, or optical surface imaging. Previously, for precise treatment, weekly portal film was developed and changes were made. However, in the late 1990s, the concept of adaptive therapy was developed. Based on changes seen in imaging of the treatment area, treatment shifts were made in the next fraction. This led to adaptive therapy using daily CT simulation, CT-on rail, and cone beam CT. In most of these efforts, imaging had time delays from several days up to a week. However, a real-time adaptive approach is needed where tumors can be visualized before treatment. The recent introduction of MR-Linac systems into clinical practice can be interpreted as the latest step in this IGRT sequence, opening new perspectives in radiation treatment delivery. The on-board MR imaging allows clinicians to improve the quality of image guidance compared to conventional CT-based IGRT, by increasing the soft tissue contrast visualization of the anatomy without the addition of further radiation doses. Thus, the advent of these hybrid machines may represent a game-changer for the radiation oncology community.

During the development of this book, we collaborated with leading experts, researchers, and clinicians who have been at the forefront of adopting and refining MR-Linac technology, especially MRgRT. Their invaluable insights provide a comprehensive overview of the challenges and opportunities that lie ahead in harnessing the full potential of this transformative technology and are presented in various chapters in this book. Additionally, we examine the evolving clinical evidence surrounding MR-Linac technology, including its applications in various cancer types and treatment scenarios.

Furthermore, this book endeavors to address potential concerns and limitations of MR-Linac technology. As with any innovative medical technology, it is vital to assess the feasibility, safety, training requirements, and cost-effectiveness of implementation. Our exploration offers a balanced perspective on the practical challenges faced by medical institutions and the ongoing efforts to overcome them.

In conclusion, MR-Linac technology is a testament to human ingenuity and our unwavering commitment to advancing patient care. As this transformative technology continues to mature, we are poised on the precipice of a new era in radiation treatment, one where personalized, precise, and real-time therapies become the norm rather than the exception. While the promise of MR-guided radiation therapy is undeniable, we strive to offer a well-rounded perspective on the potential obstacles faced by healthcare institutions and the ongoing efforts to overcome them. It is expected that this book will support the dissemination of essential MR-Linac clinical and technical knowledge.

Chicago, IL, USA Indra J. Das
Negrar di Valpolicella, Italy Filippo Alogi
Chicago, IL, USA Poonam Yadav
Chicago, IL, USA Bharat B. Mittal

# Acknowledgment

A book like this *A Practical Guide to MR-Linac: Technical Innovation and Clinical Implication* requires the support, encouragement, and contributions from institution, collaborators, and many authors. Our gratitude extends to them for their expertise in the field of MRI who shared their knowledge and passion that is reflected in respective chapters.

# Contents


1 **Introduction to MR-Linac** . . . . . . . . . . 1
Bharat B. Mittal, Indra J. Das, Poonam Yadav, and Filippo Alongi

2 **Role of MRI in Radiation Oncology** . . . . . . . . . . 5
Indra J. Das, Poonam Yadav, Filippo Alongi, and Bharat B. Mittal

3 **Clinical Necessity and Patient Selection in MR-Linac** . . . . . . . . 21
Poonam Yadav, Filippo Alongi, Bharat B. Mittal, and Indra J. Das

4 **Education and Training in MR-Based Radiation Treatment** . . . 35
Luca Nicosia, Riccardo Filippo Borgese, Kerryn Brown, and Filippo Alongi

5 **An Introduction to MR Sequences Used in Oncology with Application to MR-Linac** . . . . . . . . . . 45
Hersh Sagreiya, Karthik Sundaram, Jooyoung (James) Sohn, and Poonam Yadav

6 **Motion Management and Tracking** . . . . . . . . . . 69
Elizabeth Huynh, Eenas A. Omari, Sara N. Lim, and Raymond H. Mak

7 **sCT and Dose Calculation** . . . . . . . . . . 89
Emilie Alvarez-Andres, Fernanda Villegas, Anaïs Barateau, and Charlotte Robert

8 **Treatment Planning Considerations for an MR-Linac** . . . . . . . . 123
Justine M. Cunningham, Jennifer L. Dolan, Kate Aldridge, and Ergys Subashi

9 **The First Low-Field MRI-Guided Radiation Therapy Hybrid Integrated System: MRIdian** . . . . . . . . . . 149
Poonam Yadav, Jooyoung (James) Sohn, and Ahtesham U. Khan

10 **Elekta Unity System** . . . . . . . . . . 175
Jochem W. H. Wolthaus, Eenas A. Omari, Xinfeng Chen, and Bram van Asselen

**11 The Alberta Rotating Biplanar Linac-MR, a.k.a., Aurora-RT™** . . . . . 193
B. Gino Fallone, Satyapal Rathee, Nicola de Zanche, Eugene Yip, Keith Wachowicz, and Jihyun Yun

**12 Clinical Applications of MR-Linac in Oligometastatic Disease** . . . . . 217
Katie N. Lee, Martijn Intven, Filippo Alongi, and Mai Anh Huynh

**13 Management of Pancreatic Cancer with MRI-Guided Adaptive Radiotherapy** . . . . . 243
Ritchell van Dams, Sylvia S. W. Ng, John Hayes, and Tarita O. Thomas

**14 The Role of MR-Guided Radiation Therapy in the Management of Liver Malignancies** . . . . . 257
Nicki Karimi-Mostowfi, Luca Boldrini, Cihan Gani, and Ann Raldow

**15 Magnetic Resonance Imaging-Guided Radiation Therapy: Advances and Opportunities for Thoracic Malignancies** . . . . . 273
Julius Weng, Jinzhong Yang, Raymond H. Mak, Mohamed Abazeed, and Percy Lee

**16 MRI-Guided Radiation Therapy for Prostate Cancer: Less Is More When Technology and Evidence Intersect** . . . . . 285
Luca Valle, Rosalyne Westley, Filippo Alongi, Sean Sachdev, and Amar Kishan

**17 Breast Radiotherapy Using MR-Linac** . . . . . 307
Chiara De-Colle, Anna M. Kirby, Desiree van den Bongard, and Maureen Groot Koerkamp

**18 Considerations for Using MR Linac for the Treatment of Patients with Gynecologic Cancer: A Practical Guide and Early Clinical Experience** . . . . . 323
Lorraine Portelance, Ina Jürgenliemk-Schulz, Kyle R. Padgett, Rosa Patricia Castillo, and Astrid van Lier

**19 MR-Guided Radiotherapy in the Pediatric and Adolescent Patient** . . . . . 355
Margaret M. Kozak, Sean Mahase, Bryan Traughber, Mitchell Machtay, and John Buatti

**20 MR-Linac-Guided Adaptive Radiotherapy for Brain Tumors** . . . . . 375
Tugce Kutuk, Jay Detsky, Arjun Sahgal, and Rupesh Kotecha

**21 Treatment of Head and Neck Cancers with MR-Linac** . . . . . 395
Julia Khriguian, Laila Gharzai, Jolien Heukelom, Brigid McDonald, and Clifton D. Fuller

**22 The Australian MRI-Linac: A 10-Year Journey** . . . . . . . . . . . . . . 425
Gary Liney

**23 Low-Field MR Imaging** . . . . . . . . . . . . . . . . . . . . . . . . . . . . . . . . . . 433
Yingli Yang, Shaolei Li, Daniel A. Low, Zhimin Li, and Peng Hu

**24 Image-Based Biomarkers in Magnetic Resonance-Guided Radiotherapy (MRgRT)** . . . . . . . . . . . . . . . . . . . . . . . . . . . . . . . . . 459
Andrea D'Aviero, Haidy Nasief, Charlotte Robert, and Davide Cusumano

**25 Epilogue and Outlook** . . . . . . . . . . . . . . . . . . . . . . . . . . . . . . . . . . . 469
Filippo Alongi, Bharat B. Mittal, Indra J. Das, and Poonam Yadav

**Index** . . . . . . . . . . . . . . . . . . . . . . . . . . . . . . . . . . . . . . . . . . . . . . . 473

# Editors and Contributors

## About the Editors

**Indra J. Das, PhD** is Director of Medical Physics, Professor and Vice Chair of Physics at Northwestern Memorial Hospital and Northwestern University Feinberg School of Medicine in Chicago. He is internationally acclaimed medical physicist with expertise in radiation dosimetry, treatment planning, small fields, proton beam, and MR-Linac. His profile can be viewed at: https://www.feinberg.northwestern.edu/sites/rad-oncology/faculty/profile.html?xid=48210.

**Filippo Alongi, MD** is Head of Advanced Radiation Oncology Department at IRCCS Sacro Cuore Don Calabria Cancer Care Center of Negrar in Verona, Italy and full Professor at the University of Brescia, Italy. He has been involved with 1.5 T MR-Linac over 7 years and has written extensively. He is highly visible in European societies and on editorial board of many journals. His profile can be viewed at:

**Poonam Yadav, PhD** is Co-director of MR-Linac program, Chief of Clinical Physics and Associate Professor at Northwestern Memorial Hospital and Northwestern University Feinberg School of Medicine in Chicago. She has been active with MR-Linac over its inception and has published extensively on this technology. She is also expert on radiation dosimetry and treatment planning. Her profile can be viewed at: https://www.feinberg.northwestern.edu/sites/rad-oncology/faculty/profile.html?xid=52112.

**Bharat B. Mittal, MD** is Willian N. Brand Professor and Chair of Radiation Oncology at Northwestern University Feinberg School of Medicine in Chicago. He specializes primarily in Head and Neck and Lymphoma. He has keen interest in MR-Linac who made it possible to acquire the technology at Northwestern Memorial Hospital. He had been the chair of the department for nearly 30 years and have published extensively on every aspect of the radiation oncology. His profile can be viewed at: https://www.feinberg.northwestern.edu/sites/rad-oncology/faculty/profile.html?xid=12553.

## Contributors

**Mohamed Abazeed** Department of Radiation Oncology, Northwestern University Feinberg School of Medicine, Chicago, IL, USA

**Kate Aldridge** Department of Radiation Oncology, Henry Ford Health, Detroit, MI, USA

**Filippo Alongi** University of Brescia and Advanced Radiation Oncology Department, IRCCS Sacro Cuore Don Calabria Hospital, Negrar di Valpolicella, Italy

Università di Brescia, Direttore del Dipartimento di Radioterapia Oncologica Avanzata, Brescia, Italy

Department of Advanced Radiation Oncology, IRCCS Sacro Cuore Don Calabria, Cancer Care Center, Negrar di Valpolicella, Italy

University of Brescia, Brescia, Italy

Advanced Radiation Oncology Department, Cancer Care Center, IRCCS Sacro Cuore Don Calabria Hospital, Negrar di Valpolicella, Italy

Department of Advanced Radiation Oncology, IRCCS Ospedale Sacro Cuore Don Calabria, Negrar di Valpolicella, Italy

Advanced Radiation Oncology Department, IRCCS Sacro Cuore Don Calabria Hospital, Negrar di Valpolicella, Italy

**Emilie Alvarez-Andres** Department of Radiotherapy and Radiation Oncology, Faculty of Medicine and University Hospital Carl Gustav Carus of Technische Universität Dresden, Dresden, Germany

OncoRay—National Center for Radiation Research in Oncology, Faculty of Medicine and University Hospital Carl Gustav Carus, Technische Universität Dresden, Dresden, Germany

Helmholtz-Zentrum Dresden-Rossendorf, Dresden, Germany

**Bram van Asselen** Department of Radiotherapy, University Medical Center Utrecht, Utrecht, the Netherlands

**Anaïs Barateau** Univ Rennes, CLCC Eugène Marquis, Inserm, LTSI - UMR 1099, F-35000 , Rennes, France

**Luca Boldrini** Department of Bioimaging, Radiation Oncology and Hematology, Fondazione Policlinico Universitario "Agostino Gemelli" IRCCS, Rome, Italy

**Desiree van den Bongard** Department of Radiation Oncology, Amsterdam University Medical Centers, Cancer Center Amsterdam, Cancer Treatment and Quality of Life, Cancer Biology and Immunology, Amsterdam, The Netherlands

**Riccardo Filippo Borgese** Department of Advanced Radiation Oncology, IRCCS Sacro Cuore Don Calabria, Cancer Care Center, Negrar di Valpolicella, Italy

**Kerryn Brown** Department of Radiation Oncology, ONJ Centre, Austin Hospital, Heidelberg, VIC, Australia

**John Buatti** Department of Radiation Oncology, University of Iowa Hospitals and Clinics, Iowa City, IA, USA

**Rosa Patricia Castillo** Radiology Department, Miller School of Medicine, University of Miami, Miami, FL, USA

**Xinfeng Chen** Department of Radiation Oncology, Medical College of Wisconsin, Milwaukee, WI, USA

**Justine M. Cunningham** Department of Radiation Oncology, Henry Ford Health, Detroit, MI, USA

**Davide Cusumano** Fondazione Policlinico Universitario "Agostino Gemelli" IRCCS, Rome, Italy

UO Fisica Medica e Radioprotezione, Mater Olbia Hospital, Olbia, Italy

**Andrea D'Aviero** UO Radioterapia Oncologica, Mater Olbia Hospital, Olbia, Italy

**Ritchell van Dams** Department of Radiation Oncology, Dana-Farber Brigham Cancer Center, Harvard Medical School, Boston, MA, USA

**Indra J. Das** Department of Radiation Oncology, Northwestern Memorial Hospital, Northwestern University Feinberg School of Medicine, Chicago, IL, USA

**Chiara De-Colle** Department of Radiation Oncology, University Hospital and Medical Faculty, Eberhard Karls University Tübingen, Tübingen, Germany

**Jay Detsky** Department of Radiation Oncology, Odette Cancer Centre, Sunnybrook Health Sciences Centre, University of Toronto, Toronto, ON, Canada

**Jennifer L. Dolan** Department of Radiation Oncology, Henry Ford Health, Detroit, MI, USA

**B. Gino Fallone** Department of Medical Physics, Cross Cancer Institute, Edmonton, AB, Canada

Department of Oncology, Medical Physics Division, University of Alberta, Edmonton, AB, Canada

MagnetTx Oncology Solutions, Edmonton, AB, Canada

**Clifton D. Fuller** Department of Radiation Oncology, The University of Texas MD Anderson Cancer Center, Houston, TX, USA

**Cihan Gani** Department of Radiation Oncology, University Hospital Tübingen, Tübungen, Germany

German Cancer Research Center (DKFZ) Heidelberg and German Consortium for Translational Cancer Research (DKTK), Partner Site Tübingen, Tübingen, Germany

**Laila Gharzai** Department of Radiation Oncology, Northwestern Memorial Hospital, Northwestern University Feinberg School of Medicine, Chicago, IL, USA

**John Hayes** Department of Radiation Oncology, Northwestern Memorial Hospital, Northwestern University Feinberg School of Medicine, Chicago, IL, USA

**Jolien Heukelom** Department of Radiation Oncology (Maastro), GROW School for Oncology and Reproduction, Maastricht University Medical Centre, Maastricht, the Netherlands

**Peng Hu** School of Biomedical Engineering, ShanghaiTech University, Shanghai, China

**Elizabeth Huynh** Department of Physics and Engineering, London Regional Cancer Program, London Health Sciences Centre, London, ON, Canada

**Mai Anh Huynh** Harvard Radiation Oncology Program, Brigham and Women's Hospital/Dana Farber Cancer Center/Massachusetts General Hospital/Boston Children's Hospital, Boston, MA, USA

Department of Radiation Oncology, Dana-Farber/Brigham and Women's Cancer Center and Harvard Medical School, Boston, MA, USA

**Martijn Intven** Division of Imaging and Oncology, Department of Radiotherapy, University Medical Centre Utrecht, Utrecht, the Netherlands

**Ina Jürgenliemk-Schulz** Department of Radiation Oncology, University Medical Center Utrecht, University Utrecht, Utrecht, the Netherlands

**Nicki Karimi-Mostowfi** Lewis Katz School of Medicine at Temple University, Philadelphia, PA, USA

**Ahtesham U. Khan** Department of Radiation Oncology, Northwestern University Feinberg School of Medicine, Chicago, IL, USA

**Julia Khriguian** Department of Radiation Oncology, The University of Texas MD Anderson Cancer Center, Houston, TX, USA

**Anna M. Kirby** Department of Radiotherapy, Royal Marsden NHS Foundation Trust and Institute of Cancer Research, Sutton, United Kingdom

**Amar Kishan** Department of Radiation Oncology, David Geffen School of Medicine, University of California Los Angeles, Los Angeles, CA, USA

**Maureen Groot Koerkamp** Department of Radiation Oncology, University Medical Centre Utrecht, Utrecht, the Netherlands

**Rupesh Kotecha** Department of Radiation Oncology, Miami Cancer Institute, Baptist Health South Florida, Miami, FL, USA

**Margaret M. Kozak** Department of Radiation Oncology, University of Iowa Hospitals and Clinics, Iowa City, IA, USA

Department of Radiation Oncology, UT Southwestern Medical Center, Dallas, TX, USA

**Tugce Kutuk** Department of Radiation Oncology, Miami Cancer Institute, Baptist Health South Florida, Miami, FL, USA

**Katie N. Lee** Harvard Radiation Oncology Program, Brigham and Women's Hospital/Dana Farber Cancer Center/Massachusetts General Hospital/Boston Children's Hospital, Boston, MA, USA

Department of Radiation Oncology, Dana-Farber/Brigham and Women's Cancer Center and Harvard Medical School, Boston, MA, USA

**Percy Lee** Department of Radiation Oncology, City of Hope National Medical Center, Los Angeles, CA, USA

**Astrid van Lier** Department of Radiation Oncology, University Medical Center Utrecht, University Utrecht, Utrecht, the Netherlands

**Sara N. Lim** Department of Radiation Oncology, Northwestern Memorial Hospital, Northwestern University Feinberg School of Medicine, Chicago, IL, USA

**Gary Liney** Sydney, NSW, Australia

**Shaolei Li** Institute for Medical Imaging Technology, Ruijin Hospital, Shanghai Jiaotong University School of Medicine, Shanghai, China

Department of Radiology, Ruijin Hospital, Shanghai Jiaotong University, School of Medicine, Shanghai, China

**Zhimin Li** Department of Radiology, Northwestern University, Chicago, IL, USA

**Daniel A. Low** Department of Radiation Oncology, University of California, Los Angeles, CA, USA

**Mitchell Machtay** Department of Radiation Oncology, Penn State Cancer Institute, Hershey, PA, USA

**Sean Mahase** Department of Radiation Oncology, Northwestern University Feinberg School of Medicine, Hershey, PA, USA

**Raymond H. Mak** Department of Radiation Oncology, Brigham and Women's Hospital, Dana-Farber Cancer Institute, Boston, MA, USA

**Brigid McDonald** Department of Radiation Oncology, The University of Texas MD Anderson Cancer Center, Houston, TX, USA

**Bharat B. Mittal** Department of Radiation Oncology, Northwestern Memorial Hospital, Northwestern University Feinberg School of Medicine, Chicago, IL, USA

**Haidy Nasief** Department of Radiation Oncology, Medical College of Wisconsin, Milwaukee, WI, USA

**Sylvia S. W. Ng** Department of Radiation Oncology, Sunnybrook Odette Cancer Centre, University of Toronto, Toronto, ON, Canada

**Luca Nicosia** Department of Advanced Radiation Oncology, IRCCS Sacro Cuore Don Calabria, Cancer Care Center, Negrar di Valpolicella, Italy

**Eenas A. Omari** Department of Radiation Oncology, Medical College of Wisconsin, Milwaukee, WI, USA

**Kyle R. Padgett** Radiation Oncology Department, Sylvester Comprehensive Cancer Center, Miller School of Medicine, University of Miami, Miami, FL, USA

**Lorraine Portelance** Radiation Oncology Department, Sylvester Comprehensive Cancer Center, Miller School of Medicine, University of Miami, Miami, FL, USA

**Ann Raldow** Department of Radiation Oncology, University of California, Los Angeles, Los Angeles, CA, USA

**Satyapal Rathee** Department of Medical Physics, Cross Cancer Institute, Edmonton, AB, Canada

Department of Oncology, Medical Physics Division, University of Alberta, Edmonton, AB, Canada

**Charlotte Robert** UMR 1030 Radiothérapie Moléculaire et Innovation Thérapeutique, ImmunoRadAI, Université Paris-Saclay, Institut Gustave Roussy, Inserm, Villejuif, France

Department of Radiation Oncology, Gustave Roussy, Villejuif, France

**Sean Sachdev** Department of Radiation Oncology, Northwestern University Feinberg School of Medicine, Chicago, IL, USA

**Hersh Sagreiya** Division of Abdominal Imaging, Department of Radiology, Hospital of the University of Pennsylvania, Philadelphia, PA, USA

**Arjun Sahgal** Department of Radiation Oncology, Odette Cancer Centre, Sunnybrook Health Sciences Centre, University of Toronto, Toronto, ON, Canada

**Jooyoung (James) Sohn** Department of Radiation Oncology, Northwestern University Feinberg School of Medicine, Chicago, IL, USA

**Ergys Subashi** Department of Medical Physics, Memorial Sloan Kettering Cancer Center, New York, NY, USA

**Karthik Sundaram** Division of Abdominal Imaging, Department of Radiology, Hospital of the University of Pennsylvania, Philadelphia, PA, USA

**Tarita O. Thomas** Department of Radiation Oncology, Northwestern Memorial Hospital, Northwestern University Feinberg School of Medicine, Chicago, IL, USA

**Bryan Traughber** Department of Radiation Oncology, Penn State Cancer Institute, Hershey, PA, USA

**Luca Valle** Department of Radiation Oncology, David Geffen School of Medicine, University of California Los Angeles, Los Angeles, CA, USA

Veterans Affairs Greater Los Angeles Healthcare System, Los Angeles, CA, USA

**Fernanda Villegas** Radiotherapy Physics and Engineering, Medical Radiation Physics and Nuclear Medicine, Karolinska University Hospital, Stockholm, Sweden

Department of Oncology-Pathology, Karolinska Institute, Stockholm, Sweden

**Keith Wachowicz** Department of Medical Physics, Cross Cancer Institute, Edmonton, AB, Canada

Department of Oncology, Medical Physics Division, University of Alberta, Edmonton, AB, Canada

**Julius Weng** Division of Radiation Oncology, The University of Texas MD Anderson Cancer Center, Houston, Texas, USA

**Rosalyne Westley** Clinical Oncology Academic Urology Unit, Institute of Cancer Research, The Royal Marsden Hospital, London, UK

**Jochem W. H. Wolthaus** Department of Radiotherapy, University Medical Center Utrecht, Utrecht, the Netherlands

**Poonam Yadav** Department of Radiation Oncology, Northwestern Memorial Hospital, Northwestern University Feinberg School of Medicine, Chicago, IL, USA

**Jinzhong Yang** Division of Radiation Oncology, The University of Texas MD Anderson Cancer Center, Houston, TX, USA

**Yingli Yang** Institute for Medical Imaging Technology, Ruijin Hospital, Shanghai Jiaotong University School of Medicine, Shanghai, China

Department of Radiology, Ruijin Hospital, Shanghai Jiaotong University, School of Medicine, Shanghai, China

**Eugene Yip** Department of Medical Physics, Cross Cancer Institute, Edmonton, AB, Canada

Department of Oncology, Medical Physics Division, University of Alberta, Edmonton, AB, Canada

**Jihyun Yun** Department of Medical Physics, Cross Cancer Institute, Edmonton, AB, Canada

Department of Oncology, Medical Physics Division, University of Alberta, Edmonton, AB, Canada

**Nicola de Zanche** Department of Medical Physics, Cross Cancer Institute, Edmonton, AB, Canada

Department of Oncology, Medical Physics Division, University of Alberta, Edmonton, AB, Canada

# Introduction to MR-Linac

# 1

Bharat B. Mittal, Indra J. Das, Poonam Yadav,
and Filippo Alongi

## Historical Perspective

Radiation therapy is an integral part of cancer treatment where a multimodality approach has proven to have the superior outcome. The advances in radiation oncology are intertwined with innovations in engineering, radiological and physical sciences, and computer technology, along with a better understanding of cellular interactions, the effects of radiation on biological tissues, and the roles genetics and epigenetics play in determining radiation response. The discoveries made over the past several decades have been instrumental in the development of precision radiotherapy through advances in technology and cancer biology.

The use of radiation for the treatment of malignant tumors has experienced a technological revolution since the discovery of X-rays by the German physicist Wilhelm Conrad Röntgen on November 8, 1895 [1], and the subsequent discovery of radioactive elements by Henri Becquerel in 1896 and Marie and Pierre Currie in 1898.

Within a year after the discovery of X-rays, Emil Herman Grubbe, a chemist and homeopath from Chicago, claimed to have used X-rays to treat a patient with breast cancer. However, the first verifiable priority claim can be credited to two general practitioners, Tage Sjögren and Tor Stenbeck, who reported two successful cancer treatments at the Swedish Society of Medicine in 1899 [2]. Since then, advances in radiologic sciences, both diagnostic and therapeutic, combined with innovations in computer science, have played a critical role in the development of sophisticated technology to deliver radiation for cancer treatment.

X-Ray tube technology to image in two dimensions (2-D) evolved into three-dimensional (3-D) imaging with the invention of computed tomography (CT) scans in 1972 by the British engineer Godfrey Hounsfield and physicist Alan Cermak of Tufts University [3, 4]. CT scanners became widely available for diagnostic purposes in 1980. In the 1990s, the use of CT scans was the first step in 3-D image reconstruction via a CT simulator [5], and this spurred advances in radiation planning and delivery which evolved into 4-D planning and delivery, taking tumor motion into account. This was the first step towards adaptive radiotherapy.

In 1969 Raymond Vahan Damadian, an American physician, hypothesized that benign

B. B. Mittal (✉) · I. J. Das · P. Yadav
Department of Radiation Oncology, Northwestern Memorial Hospital, Northwestern University Feinberg School of Medicine, Chicago, IL, USA
e-mail: b-mittal@northwestern.edu; indra.das@nm.org; poonam.yadav@northwestern.edu

F. Alongi
University of Brescia and Director of Advanced Radiation Oncology Department, IRCCS Sacro Cuore Don Calabria Hospital, Negrar-Verona, Italy


I. J. Das et al. (eds.), *A Practical Guide to MR-Linac*, https://doi.org/10.1007/978-3-031-48165-9_1

cells had longer relaxation times than malignant cells [6]. Further work by Paul Christian Lauterbur, an American chemist, and Sir Peter Mansfield, an English physicist, resulted in the invention of the magnetic resonance imaging (MRI) scanner, for which they received Nobel Prize in Medicine in 2003. In 1977, MRI was first used to scan the whole body.

Superficial X-ray technology to treat cancer gave way to the use of deeply penetrative radiation from radioactive elements like cobalt-60 or high-energy X-rays produced by linear accelerators. The cobalt-60 teletherapy machine was invented in the 1950s, and in 1951 physicist Harold Johns, from the University of Saskatchewan, became the first researcher to treat a patient with radiation emitted by cobalt-60. In parallel, linear accelerator technology evolved. In 1928, R. Wideroe built the first research accelerator that emitted positive ions at about 50 keV. In 1952, Henry Kaplan and Edward Ginzton began building a medical linear accelerator. In 1953, the first patient was treated using a medical linear accelerator in London. In 1989, the first hospital-based particle therapy center became operational. Advances in hardware, electronics, and computer technology quickly resulted in intensity-modulated radiotherapy and image-guided radiotherapy [7, 8]. A detailed historical perspective has been provided by Rammohan et al. [9].

The next evolution in adaptive radiotherapy was the integration of online MRI with external beam radiotherapy [10–12]. This became possible once the technical problem of interference between the magnetic field (which distorted the accelerator) and the metals in the linear accelerator (which distorted the magnetic field) was solved by active shielding of the magnet. The advantages of MRI over CT-based images include better soft tissue definition, the ability to monitor tumor motion, and real-time treatment adaptation taking interfraction and intrafraction changes in the shape and location of tumors and surrounding organs into account. This technology will eventually lead to MRI-based physiologic changes and biomarkers that could potentially advance precision radiotherapy and outcome assessment using image texture (radiomics). Integrating MRI technology with linear accelerators is a paradigm shift in delivering radiation based upon the anatomic and physiologic changes in a tumor to optimize therapeutic ratio.

## MR-Guided Radiation Therapy

The first clinically active MR-guided radiotherapy machine was installed at Washington University in St. Louis, and the first patient was treated in January 2014. The machine consisted of three cobalt-60 sources mounted on a ring integrated with a 0.35-Tesla magnet with the same isocenter [13]. It included an adaptive treatment planning system to efficiently modify the treatment plan and deliver radiation based upon MRI information. The next development was integrating MRI with linear accelerators (MR-Linac). In North America, the first MR-Linac consisting of a 0.35-T magnet with a 6-MV photon linear accelerator received FDA approval in 2017 and was installed at Henri Ford Hospital in Michigan. The first patient was treated in July 2017. Simultaneously, a prototype system combining high-field MRI with a linear accelerator was being developed in Europe. The first such system was installed at UMC Utrecht with the first patient with spine metastases treated in May 2017. This MR-Linac consisted of a 1.5-T magnet integrated with a ring-based gantry including a 7-MV standing wave linear accelerator [14]. In the last 5 years, there has been an exponential growth in the number of patients treated using MR-Linac worldwide. There are two other types of MR-Linac units under development, one in Canada and the other in Sydney, Australia, that have been described in the literature [12]. It is expected that MR-Linac technology will create a paradigm shift in the management of cancer that will result in improved therapeutic ratios, tumor dose escalation, and improved tumor control, representing a step towards personalized radiotherapy.

## Outline of the Book

This book consists of 25 chapters divided into three sections. The first 11 chapters are devoted to the various aspects of MR-Linac. Chapters 1–8 describe clinical necessities, education and training, imaging sequences, motions and tracking, synthetic CT, and treatment planning. The next three chapters, 9–11, are devoted to in-depth coverage of three systems which are commercially available, i.e., the ViewRay MRIdian system, the Elekta Unity system, and the Alberta Aurora Magneton system.

Chapters 12–21 cover disease-specific clinical implementation of MR-Linac without bias towards any specific vendor. The treatments covered in this book include common types and sites of cancer, such as oligometastatic, pancreas, liver, lungs, prostate, breast and gynecological tumors, pediatrics, brain, and head and neck. The chapters on clinical sites have common outlines, e.g., an overview of disease, need for MRI, contouring, dose-volume histogram (DVH) constraints, dose escalation, online adaptation, evaluation, and outcomes. Clinical chapters provide unbiased views of the most common technologies (Unity and MRIdian) wherever applicable.

At the end, the book covers the Sydney project MR-Linac (Chap. 22). There is significant interest in low-field imaging for various disease sites, which is covered in Chap. 23. Chapter 24 covers artificial intelligence, radiomics, texture analysis, and biomarkers in the context of MRI. The final chapter provides an epilogue on the future of MR technology in the management of cancer.

## References

1. Roentgen W. Ueber eine neue Art von Strahlen. In: Sitzungsberichte der Physikalisch-Medizinischen Gesellschaft zu Wuerzburg, vol. 9; 1895. p. 132–41.
2. Mould RF. The early years of radiotherapy with emphasis on x-ray and radium apparatus. Br J Radiol. 1995;68:567–82.
3. Hounsfield GN. Computerized transverse axial scanning (tomography): part I. Description of system. Br J Radiol. 1973;46:1016–22.
4. Hounsfield GN. Nobel award address. Computed medical imaging. Med Phys. 1980;7:283–90.
5. Coia LR, Schultheiss TE, Hanks GE. A practical guide to CT simulation. Madison, WI: Advanced Medical Publishing; 1995.
6. Damadian R. Tumor detection by nuclear magnetic resonance. Science. 1971;171:1151–5.
7. Keall PJ, Sawant A, Berbeco RI, Booth JT, et al. AAPM Task Group 264: the safe clinical implementation of MLC tracking in radiotherapy. Med Phys. 2021;48:e44–64.
8. Das IJ, Sanfilippo NJ, Fogliata A, Luca CL. Intensity modulated radiation therapy: a clinical overview. Bristol: IOP Publishing; 2020.
9. Rammohan N, Randall JW, Yadav P. History of technological advancements towards MR-Linac: the future of image-guided radiotherapy. J Clin Med. 2022;11:4730.
10. Hall WA, Paulson E, Li XA, Erickson B, et al. Magnetic resonance linear accelerator technology and adaptive radiation therapy: an overview for clinicians. CA Cancer J Clin. 2022;72:34–56.
11. Das IJ, Yadav P, Mittal BB. Emergence of MR-Linac in radiation oncology: successes and challenges of riding on the MRgRT bandwagon. J Clin Med. 2022;11:5136.
12. Das IJ, McGee KP, Tyagi N, Wang H. Role and future of MRI in radiation oncology. Br J Radiol. 2019;92:20180505.
13. Mutic S, Dempsey JF. The ViewRay system: magnetic resonance-guided and controlled radiotherapy. Semin Radiat Oncol. 2014;24:196–9.
14. Raaymakers BW, Jürgenliemk-Schulz IM, Bol GH, et al. First patients treated with a 1.5 T MRI-Linac: clinical proof of concept of a high-precision, high-field MRI guided radiotherapy treatment. Phys Med Biol. 2017;62(23):L41–50. Published 2017 Nov 14. https://doi.org/10.1088/1361-6560/aa9517.

# 2 Role of MRI in Radiation Oncology

Indra J. Das, Poonam Yadav, Filippo Alongi,
and Bharat B. Mittal

## Introduction

The evolution of radiation oncology over 125 years has been surprisingly very positive, with the ultimate goal of achieving improved survival and reduced normal tissue toxicity. More and more patients are living longer, as shown in the American Cancer Society (ACS) report that cancer mortality has been dropping since 1990 in every disease site [1]. This success story is due to the multimodality approach to cancer management, in which radiation therapy plays an important role.

Radiation does not distinguish between tumor and normal tissues; hence, it is the user who has to choose the proper tools with which to apply radiation. Multiple tools (planning and treatment) are used to optimize radiation so that it is focused on the tumor. The goal in radiation therapy is to maximize the dose to the tumor while minimizing the dose to normal tissues, also termed organs at risk (OARs). In this context, the location of the tumor is a critical determinant in planning the radiation dose. Optimal treatment planning is achieved by selecting adequate imaging modalities to discern the tumor and OARs and by adding appropriate margins to optimally irradiate tumors. In early 2000, image-guided radiotherapy (IGRT) led the way in imaging and treating patients in a way that avoids radiation exposure to OARs. Imaging became a cornerstone of radiation therapy for volume delineation from pre-treatment verification to post-treatment evaluation.

In the early days of radiation therapy, it was mandatory to do weekly portal filming to verify patient setup and significant anatomical changes [2]. Eventually it was realized that a week is too long and a better system should be adapted, which was the beginning of the era of adaptive radiation therapy (ART) [3–5]. Standard clinical practice involved compensating for changes or anatomical shifts in the following day's treatment. Additionally, adaptive treatment was undertaken with the help of daily computed tomography (CT) scans [6, 7]. This led to CT-on-rail [8, 9] and then cone-beam CT [10–12] for localization, but in most cases, there was still a time lag. Therefore, a technology with real-time visualization and adaptation was needed. Leaping forward to the present day, this is possible with magnetic resonance-guided linear accelerator (MR-Linac) radiation therapy, which provides visualization of tumors, minimizes registration

I. J. Das (✉) · P. Yadav · B. B. Mittal
Department of Radiation Oncology, Northwestern University Feinberg School of Medicine, Chicago, IL, USA
e-mail: indra.das@nm.org; poonam.yadav@northwestern.edu; b-mittal@northwestern.edu

F. Alongi
Università di Brescia, Direttore del Dipartimento di Radioterapia Oncologica Avanzata, Brescia, Italy


I. J. Das et al. (eds.), *A Practical Guide to MR-Linac*, https://doi.org/10.1007/978-3-031-48165-9_2

errors, and adapts treatment plans in real time in a method called MR-guided radiotherapy (MRgRT).

Since the advent of the X-ray, the question of delineation of organs and tumor has dominated, and a balanced view is needed. In radiation oncology, magnetic resonance imaging (MRI) is commonly used as a secondary imaging modality to identify and define the extent of soft tissue tumors. MRI is a key imaging modality utilized in radiation oncology to guide treatment planning and delivery. The images produced by MRI allow radiation oncologists to accurately target the tumor site, sparing healthy tissues and organs from unnecessary radiation exposure. In addition, MRI provides important information about the response of tumors to treatment, helping clinicians determine whether additional treatment is necessary. MRI is a noninvasive imaging technique that uses a magnetic field and radio waves to produce detailed images of the internal structures of the body. This imaging modality is particularly important in radiation oncology because it provides high-quality, three-dimensional images of soft tissue that are not available with other imaging modalities such as CT or X-rays.

One of the advantages of MRI in radiation oncology is its ability to differentiate between tissue types based on their relaxation time in a magnetic field. MRI tends to provide tissue characterization for anatomical and disease type. This makes it possible to differentiate between normal and abnormal tissue and help delineate tumor boundaries. This is achieved by selecting imaging sequences (see Chap. 5) by tuning the magnetic field and radiofrequency waves used for imaging to provide the optimal view of the tumor and surrounding tissues. The diagnostic capabilities are appealing, as it does not require ionization radiation exposure. This accurate imaging is critical for treatment planning, helping clinicians to determine the optimal radiation dose and delivery method to achieve the best possible outcomes for patients.

MRI also plays an important role in IGRT, a highly precise radiation delivery technique that uses real-time imaging to accurately target the tumor. During IGRT, MR images are used to visualize the position of the tumor, accounting for any movement or changes in the patient's anatomy during treatment. This precise targeting of the tumor site minimizes damage to surrounding healthy tissues and organs, reducing treatment-related side effects.

MRI is not a perfect system and has many difficulties. One of the main problems with MRI is that the images of air and bones are alike, and differentiation is often difficult, especially when there is an air/bone interface. The MR signal is also not unique, as it varies with MR sequences that are dependent on the magnetic field strength. Unlike CT data where the Hounsfield unit (measure of tissue attenuation) is only dependent on the kilovoltage of the machine [13], the MR signal is multifactorial and requires a good understanding of the MR theory. In this context, radiation treatment planning that is dependent on the electron density cannot be easily performed with MR images. This problem is being addressed using synthetic CT (sCT), which is discussed in greater detail in Chap. 7.

The use of sCT is growing and is an active area of research. Most of the initial approaches have used an atlas- or bulk density-based Dixon method, but now they are based on machine learning and artificial intelligence (AI) as described in various references [14–23]. In the bulk model, air, water, fat, and soft tissue and bone densities are mapped with the MR signal and a correlation is used to create a sCT [14, 15]. In general, soft tissue CT numbers do not change much against water, and hence, the dose calculation using the Dixon method usually provides satisfactory results [24, 25].

This chapter deals with MR-only approaches and provides various aspect of CT and MR-Linac technology in comparison.

## Margin

Difficulty in delivering a tumoricidal dose is compounded by uncertainty in location and position, and hence, the concept of margin is introduced. It is also due to the notion that recurrences could be avoided if large margins are given to the target. A certain amount of margin is required to cover a tumor for radiation treatment; however, it is at the expense of an extra dose to the normal tissues. The International Commission on Radiation Units and Measurements, in publications ICRU-50 and ICRU-62 [26, 27], provided international guidelines for defining the target, gross tumor volume (GTV), clinical target volume (CTV), planning target volume (PTV), and OARs. Whereas the GTV is defined as gross tumor volume, the CTV is based on subclinical disease, and the PTV accounts for setup and organ motion. The latter two volumes have margins with a significantly greater amount of irradiated tissue, which in turn may lead to more treatment-related complications.

The simple calculation of volume and expansion is shown in Fig. 2.1. This provides a measure of the extra volume treated for a hypothetical spherical tumor with radiation. The extra volume is computed based on the PTV margin. Margin is a safety net to make sure that the entire target is treated and reduces the probability of missed tumor volumes that could produce recurrence. For small margins, the relationship for extra volume irradiated is a nearly linear function (Fig. 2.1), but as margin increases, the functional form becomes steeper. For example, for a 3-cm-radius tumor, a margin of 5 mm versus 2 mm adds 2.8-fold volume, whereas a 10-mm margin adds a 6.4-fold volume, thus increasing the probability of the complication rate.

The question is how to reduce the margin while minimizing the cost, e.g., not covering the target volume adequately and thus allowing recurrence. This is a complex question and requires understanding of the available imaging tools. Current radiographic pre-post imaging is not suitable for soft tissue contrast, and visibility is relatively poor; thus, discerning the tumor versus OARs is often difficult and the margin is hard to reduce. Additionally, the most current tools are passive and cannot be used in real time, which is needed due to intra- and inter-fraction target motion and more so with organs which are susceptible to physiological changes such as breathing and gas-

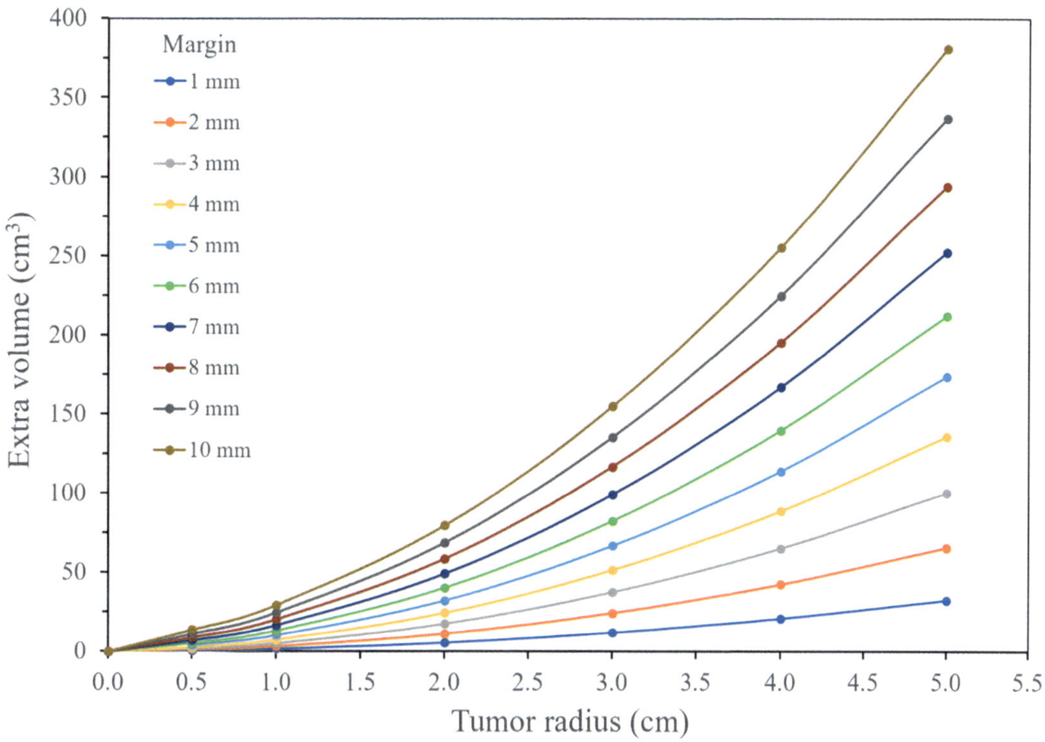


**Fig. 2.1** Extra volume irradiated for a given tumor size and irradiation margin

tric and intestinal movements. ART and IGRT have evolved over 25 years with the help of fiducial markers and better imaging tools, but still lack real-time implementations. Better imaging and gating could help reduce the margin, which is where MRgRT can play important role [28].

Figure 2.1 is theoretical in nature, but has been proven in clinical practice. To manage interfractional setup error, large margins are advocated. For example, in prostate cancer, CTV to PTV expansion varies from 5 mm to 15 mm depending upon the institution. This margin could be significantly reduced, as shown in the MOMENTUM trial [29] using MR-Linac. Christiansen et al. [30] analyzed the data using a normal tissue complication probability (NTCP) model for standard (IMRT) and adaptive MR-Linac treatment with a small margin (CTV to PTV expansion; were 3 mm R-L, 4 mm S-I, and 5 mm A-P with additional 2-mm uniform margin and 5, 7, and 12 mm in intermediate-grade prostate cancer. In the standard arm the total margins from CTV were 12 mm, 14 mm, and 19 mm in R-L, AP, and SI directions, respectively). It was shown that the incidence of complications, including pain, incontinence, hematuria, rectal bleeding, proctitis, and GI toxicity, was significantly lower in the MR-Linac arm. Additionally, margin reduction was so effective in reducing toxicity that the MIRAGE trial for prostate cancer was closed much earlier than planned, as shown by Kishan et al. [31] where grade 2 toxicity was significantly lower with a 2-mm margin with MR-Linac compared to a 4 mm with CT. Alongi et al. [32] analyzed the data from 100 prostate cancer patients treated with MR-Linac for quality of life (QOL). They evaluated a 5-mm uniform margin from the CTV and showed a significant reduction in toxicity that was favorable for QOL.

The margin reduction related to ART also has clear dosimetric advantages in other disease sites such as the liver, lung, pancreas, and lymph nodes [33]. Michalet et al. [34] showed that for the treatment of the adrenal gland, the use of ART with a smaller margin provided superior PTV dose coverage and nearly 50% reduction in duodenum and stomach dose, thus reducing normal tissue complications and improving overall survival. Pancreatic cancer, which has a universally poor prognosis due to critical structures in the vicinity, has been shown to have promising results when treated with adaptive MR-Linac that provides real-time imaging with a reduced margin [35, 36].

## Multimodality Imaging

X-Rays, MRI, positron emission tomography (PET), and ultrasound are the basic imaging tools that are used in clinical practice; here we compare and contrast imaging via X-rays and MRI.

### X-Ray Imaging

The most common imaging method is X-ray, acquired in either planar (radiograph) or 3D (CT). Planar images have been used since the beginning of X-ray radiography, but CT data became available with the development of the CT scanner in 1972 [37, 38]. A revolutionary change took place with the advent of digitally reconstructed radiographs (DRRs) [39]. This allowed the visualization of the PTV and OARs on a reconstructed image based on axial CT data. In 1990, the voxel processor called a CT simulator became popular. It had three components: a CT scanner, virtual simulation, and laser alignment [40]. The CT simulator created a paradigm shift in the way radiation oncologists view and delineate tumors and normal structures. However, CT images are based on the attenuation properties of the tissue and do not provide soft tissue contrast. Other imaging modalities are thus required for real-time visualization of non-bone structures. Additionally, X-rays are ionized radiation that adds an extra dose to the entire body. A few additional features of X-irradiation should be considered:

- Dose escalation is needed for ablative treatment.
- In general, the higher the dose, the higher the probability of complications.

- Visualization in real time is needed to provide confidence in beam delivery.
- Reduce margin to avoid OAR dose and related toxicities.

## Superiority of MRI

With the advent of MRI in 1971 [41], various types of images became popular as described by Bitar et al. [42] and in Chap. 5 of this book. MRI is based on the nuclear magnetic resonance of water molecules in the different tissues of the body and their relaxation time in a magnetic field. It is used for clinical-level diagnosis and for providing anatomical details. MRI contrast comes from the differing water content in different tissues. Also, with extrinsic additives such as gadolinium chelates or iron oxide particles that can change the relaxation properties of the mobile protons of the water, additional contrast can be created, which is exploited clinically. Another advantage of MRI is that one can create different images based on various relaxation times in terms of T1 and T2, as has been described by Das et al. [43]. Additionally, MRI parameters and sequencing that are useful in radiation oncology have been described by Jacob et al. [44]. Soft tissue characterization by MRI is so superior that it is considered the gold standard imaging modality for brain tumors and all soft tissue such as the pancreas, liver, prostate, etc., which are discussed in other chapters of this book. In a recent article, Srinivasan et al. [45] described the usefulness of MRI in visualizing brain, prostate, and gastrointestinal sites.

A few examples of the superiority of MRI are shown in the following figures. Figure 2.2 shows CT and MRI data for a prostate cancer patient in the axial and sagittal plane. The prostate is diffi-

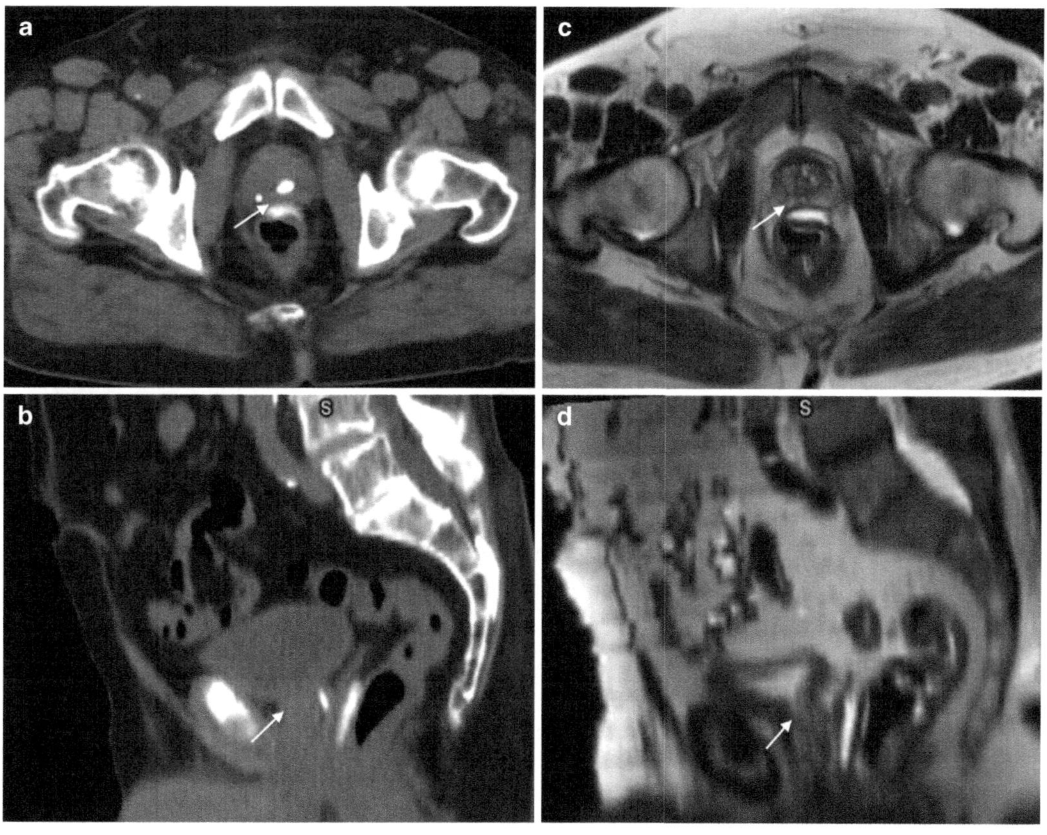


**Fig. 2.2** Axial and sagittal images of a prostate cancer patient with CT (**a**, **b**) and MRI (**c**, **d**). Please note the superior soft tissue contrast in MRI, where visibility of prostate is much better (arrow indicates the prostate gland)

cult to delineate in CT images, whereas it is relatively easy in MRI due to soft tissue contrast. When the magnetic field is increased, the soft tissue visualization is even better. Another example of soft tissue contrast and tissue visualization is shown for a case of liver lesion (Fig. 2.3) with CT and MRI. Once again, the visibility of the lesion is better in the MR images. Figure 2.4 shows a

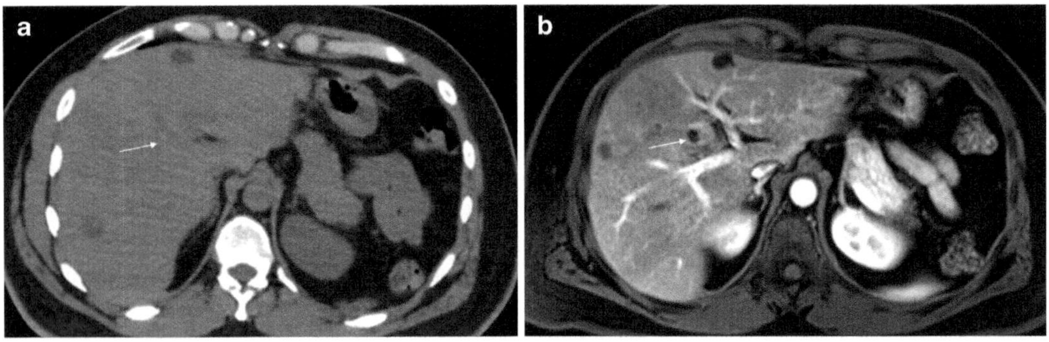


**Fig. 2.3** Imaging of a liver cancer patient by CT (**a**) and MRI (**b**). The arrows show the location of the lesion

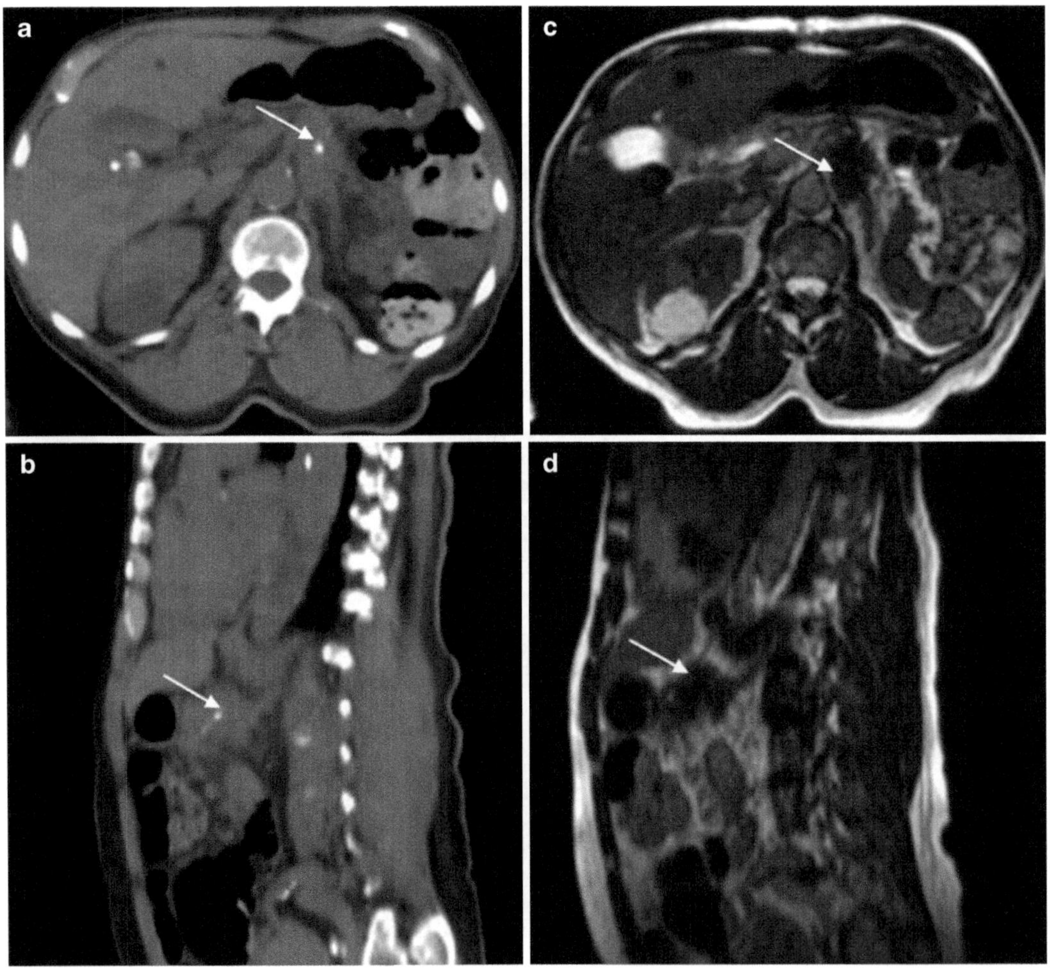


**Fig. 2.4** A case of pancreatic cancer; (**a**, **b**) CT with clips and (**c**, **d**) MR images. Note that the lesion is not visible in the CT image (arrows), whereas it is clearly visible in the MR images

case of pancreatic cancer. CT and MRI images are shown in the axial and sagittal planes and indicate a superior soft tissue contrast with MR images that can be used for visualization and adaptive therapy which is not possible with CT images and will be even worse in cone-beam CT images.

## High-Z Artifacts

Prosthetic devices are common in the elderly population in whom the probability of cancer incidence is also higher. Most prosthetic devices are made of high-Z metal and have very high attenuation [46–50]. This creates problems in CT images due to photon starvation and streaking artifacts (Fig. 2.5). It is very hard to delineate any structure with these artifacts, as the CT numbers are distorted and normal anatomic boundaries are lost, leading to poor dose calculation [51–55]. MR images can improve the artifacts; however, there is a certain amount of distortion depending on the prosthesis that have blooming effect: MR effect as shown in Fig. 2.6. Additionally, dose calculations from MR images are still in infancy with prosthetic devices due to the uncertainty around electron density.

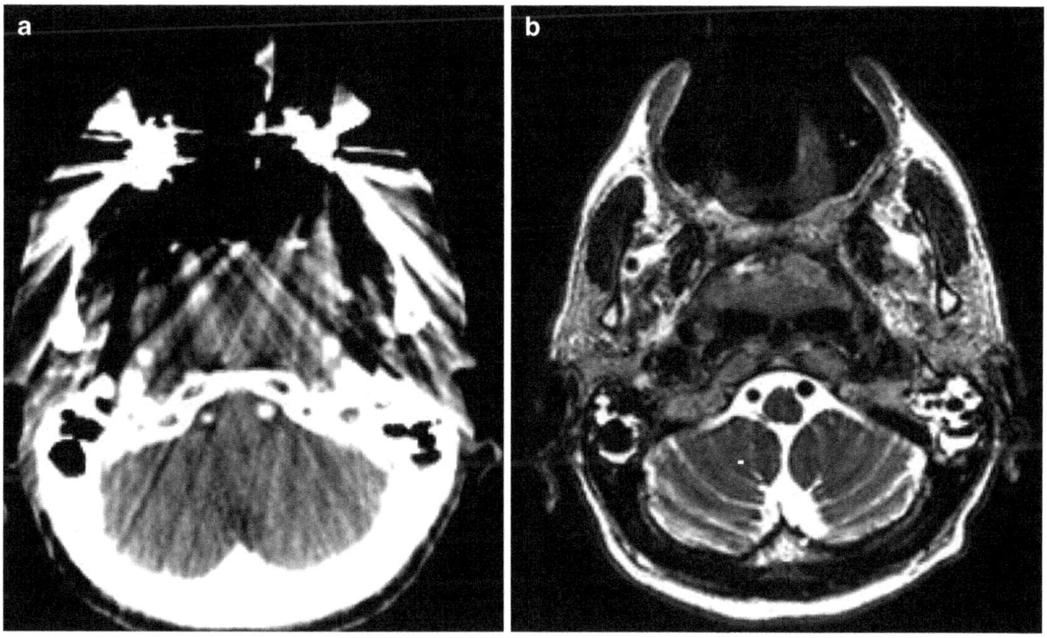


**Fig. 2.5** Image artifacts due to a dental filling. (**a**) CT, (**b**) T2-weighted MRI (adapted from Das et al. [56])

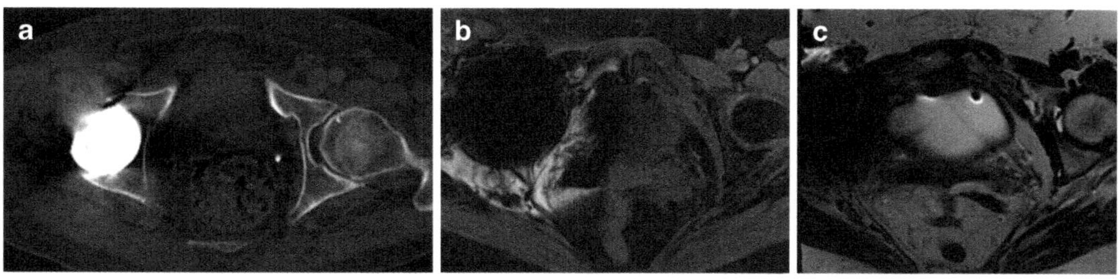


**Fig. 2.6** Images of hip prosthesis by (**a**) CT and (**b**) T1-weighted and (**c**) T2-weighted MRI

## MR-Only Approach

With the wide array of MR sequences suitable for characterizing tissues and measuring tumor volume and treatment response, it is generally agreed that MR offers a high signal-to-noise ratio and contrast, with submillimeter resolution. This contributes to better flexibility and greater utility in MR images over CT that leads to improved target delineation, which is a critical component in radiation oncology [57, 58]. Tenhunen et al. [59] showed that MR can be used by itself for treatment planning in 92% of prostate cancer patients, and only 8% patients needed supplementary CT data.

With better soft tissue contrast and MRgRT, it is obvious that an MR-only option for patient treatment could be chosen. There are many publications advocating an MR-only philosophy [40, 60–62]. Due to its appeal, many institutions are looking into installing an MR simulator, similar in many ways to a CT simulator [40], that can provide DRRs. Early attempts at MR-based DRRs [63, 64] were relatively inferior but gave way to the concept which is being explored with modern technology in conjunction with sCT (see Chap. 7 in this book). The rosy picture of the MR-only approach is not shared by everyone, as indicated in a point/counterpoint article by Bayouth et al. [65]. However, there is adequate sentiment that an MR-based approach will provide superior treatment outcomes, as it can mitigate inter- and intra-fractional motion in real time [58].

## MR Simulation

Treatment simulation is a prerequisite for the radiation therapy process and has been used since the first simulator was devised in 1966 in Philadelphia [66]. Currently, every radiation oncology department has a CT simulator that consists of a CT scanner, a virtual simulator for DRR, and a laser marking device; this setup has been described in detail by AAPM Radiation Therapy Committee Task Group No. 66 [67]. Now there is increasing use of MR images in radiation therapy to supplement CT data, mainly for improved soft tissue definition in the treatment of the prostate, brain, liver, and pancreas. However, MR images require fusion and registration, processes that create additional uncertainty in volume delineation. As we move to reduce margin and rely on high-contrast and high-resolution MR images, the need for MR simulation has become a priority in radiation oncology. In this context the MR simulator was introduced for the treatment of prostate cancer at Fox Chase Cancer Center with a 0.23T magnet [68]. Similar efforts were also carried out by Kapanen et al. [69] to have MR-only treatment planning for prostate cancer.

There has been significant momentum towards the use of MR simulators, which has been described by AAPM Task Group No. 284 [70]. Owing to the superiority of MR images over CT, many institutions have adopted diagnostic MR images in radiation oncology planning, but due to time constraints, immobilization requirements, and selection of sequences in radiation oncology, the necessity for a dedicated MR simulator system was realized. Glide-Hurst et al. [71] described an open 1.0T magnet for MR simulation that has very similar features to that of a CT simulator such as a flatbed, a laser device, and incorporation of immobilization devices. Moore-Palhares et al. [57] described their first 5 years of experience with an MR simulator. Many vendors are now providing MR simulators. Tyagi et al. [72] described the Philips MR simulator in detail. Significant difficulties still exist in terms of full implementation of MR simulators in clinical practice, such as contouring, body size, bloom artifacts, sCT issues, the air-bone interface, and the training of staff [72–74].

Figure 2.7 provides images of two MR simulators of different design and magnetic field strength. The basic components of immobilization devices, laser, and flatbed are similar to that of a CT simulator. MR simulators have evolved from 0.23T to 3T, so institutions can tailor their needs for MR simulators depending on budget and functionality. One caveat is that MR simulators cannot be used in lieu of CT simulators, since MR simulation is only capable of covering

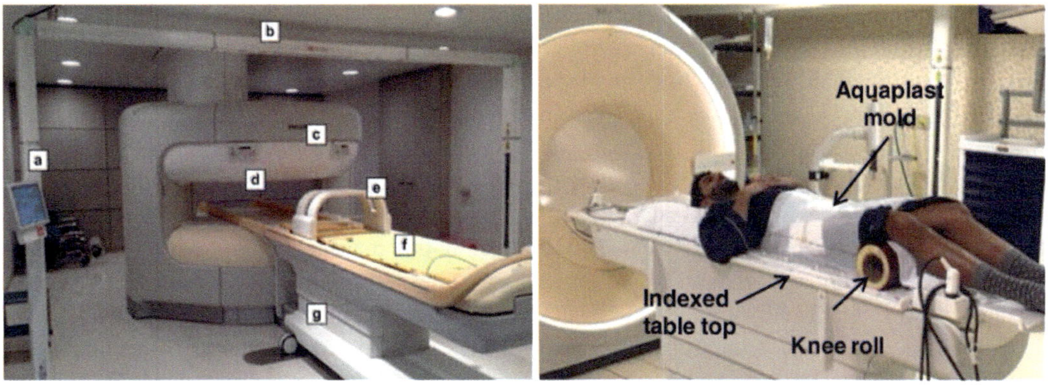


**Fig. 2.7** MR simulators. (Left) Philips open high-field MR simulator with (**a**, **b**) a laser system, (**c**) vertical field magnet, (**d**) MR laser, (**e**) body coil, (**f**) MR-compatible flatbed, and (**g**) motor drive. Adapted from Glide-Hurst et al. [71]. (Right panel) 3T Philips Ingenia scanner. Note similar components to that of a CT simulator. Adapted from Tyagi et al. [72]

approximately 1/3 of the total patient population with soft tissues, as shown by Moore-Palhares et al. [57].

## Contouring

Target volume and OAR delineation is a prerequisite for radiation treatment planning. Accurate imaging with tissue contrast is required for tissue segmentation and contouring. As we move towards an MR-only approach for treatment planning, it is imperative that volume delineation be accurate. However, there is a large amount of literature demonstrating that variability among planners and modalities is significant. Since MRI provides better soft tissue contrast, it was used in prostate cancer volume delineation as early as 1990. However, the volumes drawn from CT and MRI are not identical, as shown by several investigators who advocate a multimodality imaging approach [75–77]. In general, for prostate, CT provides a 27% larger volume than MRI. The target volume variability between CT and MRI is also observed in other disease sites including the brain, lung, breast, and head and neck, as shown in the literature [78–87]. This is primarily due to the better soft tissue contrast in MRI, and thus, volume delineation can be done accurately. In brain tumors, Weltens et al. [84] showed that tumor volume variability among clinicians is significant, even for regularly shaped tumors. It was noted that on average GTV by MRI was 69.7 $cm^3$ compared to 59.5 $cm^3$ by CT (a 17% difference, p value <0.05). For the pancreas and liver, the variability between CT and MRI was 5–15 $cm^3$ and 20–60 $cm^3$, respectively [88].

In the era of MR-Linac and MR simulation, it is expected that contour delineation variability will be minimized, as anatomical structures can be seen with a high degree of confidence. These developments will lead us into a new era of autocontouring, which is discussed in the following section.

## Autocontouring and Segmentation

Segmentation and contouring have been highly variable in most disease sites, and this is an issue for the radiation oncology community to resolve [78–84, 89–93]. Thanks to advances in technology with machine learning and AI, most OARs can be efficiently contoured with a high degree of accuracy, as shown by numerous publications [94–99]. Virtually, all radiation planning vendors are attempting to enter this market and provide autosegmentation software. We expect that within 5 years, autocontouring will be available on most CT simulators and treatment planning systems.

Automated image segmentation is essential in MR-Linac treatment, especially for adaptive therapy, in whose rapid workflow time is of the essence. MR-based segmentation, especially in MR-Linac, has challenges due to the choice of magnetic field and imaging sequence. Fast et al. [100] compared MR-based manual and autocontouring in locally advanced lung cancer. It was shown that interobserver variability in target delineation was significantly small when evaluated in terms of the Dice similarity Index and Hausdorff distance. For autocontouring, Dice similarity was 0.95 and Hausdorff distance <3 mm with $p < 0.05$. Such encouraging findings will lead to early adaptation of autocontouring. Liang et al. [101] provided a novel algorithm that yields high accuracy in delineating the liver, the kidney, and the spinal canal on T1-weighted MRI acquired on-board of radiation therapy platforms.

## Role of Ciné Imaging

One of the advantages of the MR signal is that it carries motion information that can be used for any dynamic imaging. This has been exploited in cardiac and many other sites, where motion is an important aspect [102–104]. To complement four-dimensional CT in liver cancer, Akino et al. [105] proposed orthogonal ciné imaging for respiratory motion. They successfully presented data for internal target volume (ITV) and motion management. Cine MRI is a special MR sequence with very fast data acquisition (milliseconds) to look for motion that can be viewed in any orthogonal plane. Cine MRI has been used mainly for cardiac motion for 20 years as shown by Ho et al. [106]. Multiple images that are acquired in rapid succession are strung together like in cinema to give the visual perception of motion. These images are called frames. Cine imaging is now exploited in MRIdian view ray system for MRgRT. In the MRIdian system, 4 frames/sec or 8 frames/sec are available that can be selected by the user for motion visualization. The utility of cine MRI images and their transportability to any other system is very limited, as this cannot be handled by most non-cine treatment planning systems. MR-Linac vendors are also not prepared to transfer these images outside their domain. Kim et al. [107] provided an intuitional approach in managing the cine images especially for four-dimensional MRI use in contouring, evaluation, and motion management.

At the present time, use of cine MR images in MR-Linac is more for motion management, which is discussed in Chap. 6 of this book. One example of cine MRI is also shown in Fig. 2.8, where the sagittal images show the tracking structure in red and the corresponding expansion volume in yellow. When these two structures are in phase, the beam is turned on automatically in a ViewRay system and manually in an Elekta Unity system. Details of motion management in the context of adaptive therapy have been described by Winkel et al. [108] for the Unity system.

## Dose Calculation

Dose calculation is dependent on the electron density, which is acquired from the CT number. However, when the MRI-only approach is used and inhomogeneity correction is needed, a totally different approach is required. This is performed using sCT, which is discussed in Chap. 7 of this book. For soft tissues with minimal high-density material like the rectum and prostate, the difference in dose between the MR-only approach and CT is minimal. For the lung and rectum, Wang et al. [24, 25] showed that with simplified sCT using the Dixon approach, the dose differences are <3%. However, more advanced methods are needed to create accurate sCT data for accurate dose calculations in the MR-only approach. This is discussed in detail in Chaps. 7 and 8 in this book.

## MR Education and Training

Apart from the above issues, the MR-only approach to radiation therapy requires specific

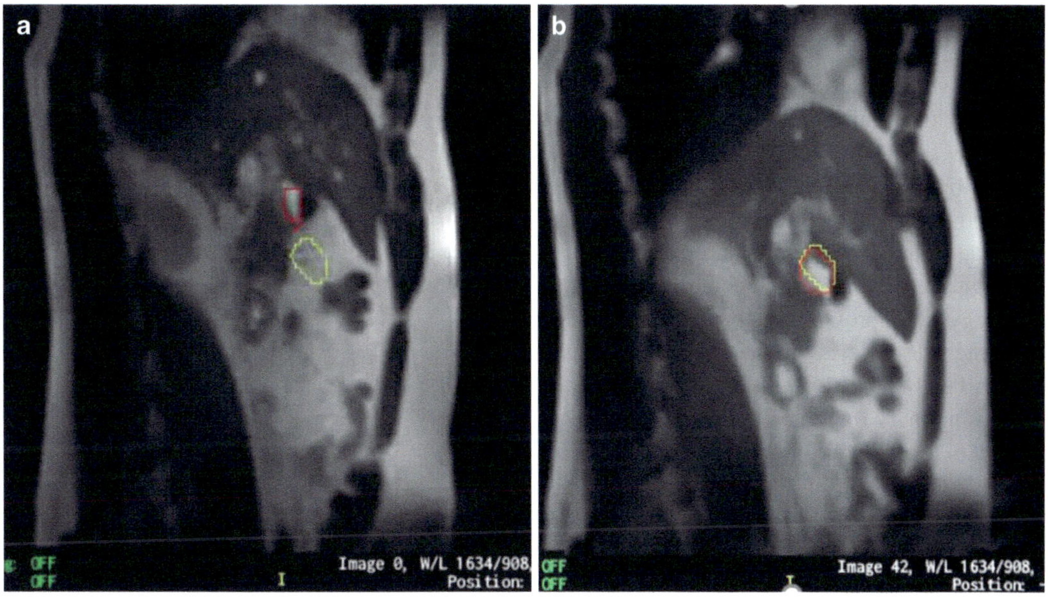


**Fig. 2.8** (**a**) Cine image of a pancreas under treatment on a ViewRay MRIdian system using 4 frames/sec. The red contour is the tracking stucture (a subset of the lesion) being monitored for treatment. The yellow contour is the expansion structure usually kept uniformaly to 3 mm. (**b**) When red and yellow structures are in phase based on cine MR images, the beam is turned on

training and education of staff including physicians, physicists, dosimetrists, therapists, and others. Medical residents in the field of radiology have a poor perception of MRI training, as shown by Wald et al. [109]. As radiation oncology becomes its own discipline and splits from radiology, MR training in radiation oncology is relatively scant. Education and training were a significant barrier to the early adoption of MR simulators, as stated by Moore-Palhares et al. [57]. This topic is addressed in detail in Chap. 4 of this book, but a few facts are worth noting here.

Despite widespread use and exponential growth of MRI in radiation oncology, didactic education and training for radiation oncology staff is lacking. The problem is that expertise in MR falls largely within radiology and its translation to radiation oncology is very slow. This fact was felt by the AAPM, which decided to form an ad hoc committee to explore the issue and provide guidance (see McGee et al. [110]). Singer et al. [111] described survey data showing that 61% of the radiology residents did not feel comfortable with MRI, due to lack of resident training. It is clear that MR simulation and MR-Linac will require robust training to improve knowledge and understanding in this area.

The role of MRI is growing globally, and global training and involvement in MR is urgently sought [112]. Upon the formation of an MR consortium (Unity and ViewRay), Hall et al. [113] provided the views of members that training in MR is urgently needed. They cited lack of training as one of the negative points in MR-Linac implementation. Cahoon et al. [114] provided views from Australia on the training of technologists in MR, especially MR simulation for radiation oncology. It was felt that most training is given by the vendors, whereas training would be more effective if provided by clinicians. The lack of comprehensive education in basic MR physics, MR safety, image acquisition, image interpretation, and knowledge of MR anatomy could be a long-term problem if not resolved by improved training mechanisms.

## Summary

MRI is a valuable tool in radiation oncology, providing critical information to guide treatment planning and delivery. Its high-quality, three-dimensional images allow for precise identification and delineation of tumor boundaries, aiding in the accurate delivery of radiation therapy. MRI has significantly advanced the field of radiation oncology by improving treatment outcomes and reducing side effects with the aid of MR simulation and MR-Linac. Integration of education and interdisciplinary training should allow this technology to reach its full potential.

## References

1. ACS. American Cancer Society. Cancer facts and figures 2023. https://www.cancer.org/content/dam/cancer-org/research/cancer-facts-and-statistics/annual-cancer-facts-and-figures/2023/2023-cancer-facts-and-figures.pdf. 2023.
2. Byhardt RW, Cox JD, Hornburg LG. Weekly localization films and detection of field placement errors. Int J Radiat Oncol Biol Phys. 1978;4:881–7.
3. Yan D, Vicini F, Wong J, Martinez A. Adaptive radiation therapy. Phys Med Biol. 1997;42:123–32.
4. Yan D, Lockman D, Brabbins D, Tyburski L, et al. An off-line strategy for constructing a patient-specific planning target volume in adaptive treatment process for prostate cancer. Int J Radiat Oncol Biol Phys. 2000;48:289–302.
5. Jaffray DA. Image-guided radiotherapy: from current concept to future perspectives. Nat Rev Clin Oncol. 2012;9:688–99.
6. Lattanzi J, McNeeley S, Pinover W, Horwitz E, et al. A comparison of daily CT localization to a daily ultrasound-based system in prostate cancer. Int J Radiat Oncol Biol Phys. 1999;43:719–25.
7. Lattanzi J, McNeely S, Hanlon A, Das I, et al. Daily CT localization for correcting portal errors in the treatment of prostate cancer. Int J Radiat Oncol Biol Phys. 1998;41:1079–86.
8. Wong JR, Grimm L, Uematsu M, Oren R, et al. Image-guided radiotherapy for prostate cancer by CT-linear accelerator combination: prostate movements and dosimetric considerations. Int J Radiat Oncol Biol Phys. 2005;61:561–9.
9. Wong JR, Gao Z, Uematsu M, Merrick S, et al. Interfractional prostate shifts: review of 1870 computed tomography (CT) scans obtained during image-guided radiotherapy using CT-on-rails for the treatment of prostate cancer. Int J Radiat Oncol Biol Phys. 2008;72:1396–401.
10. Gayou O, Miften M. Commissioning and clinical implementation of a mega-voltage cone beam CT system for treatment localization. Med Phys. 2007;34:3183–92.
11. Gayou O, Miften M. Comparison of mega-voltage cone-beam computed tomography prostate localization with online ultrasound and fiducial markers methods. Med Phys. 2007;35:531–8.
12. Nielsen M, Bertelsen A, Westberg J, Jensen HR, et al. Cone beam CT evaluation of patient set-up accuracy as a QA tool. Acta Oncol. 2009;48:271–6.
13. Das IJ, Cheng CW, Cao M, Johnstone PAS. CT imaging parameters for inhomogeneity correction in radiation treatment planning. J Med Phys. 2016;41:1–11.
14. Cusumano D, Lenkowicz J, Votta C, Boldrini L, et al. A deep learning approach to generate synthetic CT in low field MR-guided adaptive radiotherapy for abdominal and pelvic cases. Radiother Oncol. 2020;153:205–12.
15. Cusumano D, Placidi L, Teodoli S, Boldrini L, et al. On the accuracy of bulk synthetic CT for MR-guided online adaptive radiotherapy. Radiol Med. 2020;125:157–64.
16. Farjam R, Tyagi N, Deasy JO, Hunt MA. Dosimetric evaluation of an atlas-based synthetic CT generation approach for MR-only radiotherapy of pelvis anatomy. J Appl Clin Med Phys. 2019;20:101–9.
17. Han X. MR-based synthetic CT generation using a deep convolutional neural network method. Med Phys. 2017;44:1408–19.
18. Kim J, Garbarino K, Schultz L, Levin K, et al. Dosimetric evaluation of synthetic CT relative to bulk density assignment-based magnetic resonance-only approaches for prostate radiotherapy. Radiat Oncol. 2015;10:239.
19. Lei Y, Harms J, Wang T, Tian S, et al. MRI-based synthetic CT generation using semantic random forest with iterative refinement. Phys Med Biol. 2019;64:085001.
20. Tyagi N, Fontenla S, Zhang J, Cloutier M, et al. Dosimetric and workflow evaluation of first commercial synthetic CT software for clinical use in pelvis. Phys Med Biol. 2017;62:2961–75.
21. Spadea MF, Maspero M, Zaffino P, Seco J. Deep learning based synthetic-CT generation in radiotherapy and PET: a review. Med Phys. 2021;48:6537–66.
22. Spadea MF, Pileggi G, Zaffino P, Salome P, et al. Deep convolution neural network (DCNN) multiplane approach to synthetic CT generation from MR images-application in brain proton therapy. Int J Radiat Oncol Biol Phys. 2019;105:495–503.
23. Tang B, Wu F, Fu Y, Wang X, et al. Dosimetric evaluation of synthetic CT image generated using a neural network for MR-only brain radiotherapy. J Appl Clin Med Phys. 2021;22:55–62.
24. Wang H, Chandarana H, Block KT, Vahle T, et al. Dosimetric evaluation of synthetic CT for magnetic resonance-only based radiotherapy planning of lung cancer. Radiat Oncol. 2017;12:108.

25. Wang H, Du K, Qu J, Chandarana H, et al. Dosimetric evaluation of magnetic resonance generated synthetic CT for radiation treatment of rectal cancer. PLoS One. 2018;13:e019088.
26. ICRU Report 50. Prescribing, recording, and reporting photon beam therapy. Bethesda, MD: International Commission on Radiation Units and Measurements; 1993.
27. ICRU Report 62. Prescribing, recording, and reporting photon beam therapy (supplement to ICRU report 50). International Commission on Radiation Units and Measurements: Bethesda, MD; 1999.
28. ICRU Report 97. MRI-guided radiation therapy using MRI-linear accelerators. Bethesda, MD: International Commission on Radiation Units and Measurements; 2022.
29. de Mol van Otterloo SR, Christodouleas JP, Blezer ELA, Akhiat H, et al. The MOMENTUM study: an international registry for the evidence-based introduction of MR-guided adaptive therapy. Front Oncol. 2020;10:1328.
30. Christiansen RL, Dysager L, Hansen CR, Jensen HR, et al. Online adaptive radiotherapy potentially reduces toxicity for high-risk prostate cancer treatment. Radiother Oncol. 2022;167:165–71.
31. Kishan AU, Ma TM, Lamb JM, Casado M, et al. Magnetic resonance imaging-guided vs computed tomography-guided stereotactic body radiotherapy for prostate cancer: the MIRAGE randomized clinical trial. JAMA Oncol. 2023;9:365–73.
32. Alongi F, Rigo M, Figlia V, Nicosia L, et al. 1.5T MR-guided daily-adaptive SBRT for prostate cancer: preliminary report of toxicity and quality of life of the first 100 patients. J Pers Med. 2022;12:1982.
33. Nierer L, Eze C, da Silva MV, Braun J, et al. Dosimetric benefit of MR-guided online adaptive radiotherapy in different tumor entities: liver, lung, abdominal lymph nodes, pancreas and prostate. Radiat Oncol. 2022;17:53.
34. Michalet M, Bettaïeb O, Khalfi S, Ghorbel A, et al. Stereotactic MR-guided radiotherapy for adrenal gland metastases: first clinical results. J Clin Med. 2022;12:291.
35. Chuong MD, Bryant J, Mittauer KE, Hall M, et al. Ablative 5-fraction stereotactic magnetic resonance-guided radiation therapy with on-table adaptive replanning and elective nodal irradiation for inoperable pancreas cancer. Pract Radiat Oncol. 2021;11:134–47.
36. Chuong MD, Herrera R, Kaiser A, Rubens M, et al. Induction chemotherapy and ablative stereotactic magnetic resonance image-guided adaptive radiation therapy for inoperable pancreas cancer. Front Oncol. 2022;12:888462.
37. Hounsfield GN. Computerized transverse axial scanning (tomography): part I. Description of system. Br J Radiol. 1973;46:1016–22.
38. Hounsfield GN. Nobel Award address. Computed medical imaging. Med Phys. 1980;7:283–90.
39. Das IJ, McGee KP, Desobrey GE. The digitally reconstructed radiograph. In: Coia LR, Schultheiss TE, Hanks GE, editors. A practical guide to CT simulation. Madison, WI: Advanced Medical Publishing; 1995. p. 39–50.
40. Coia LR, Schultheiss TE, Hanks GE. A practical guide to CT simulation. Madison, WI: Advanced Medical Publishing; 1995.
41. Damadian R. Tumor detection by nuclear magnetic resonance. Science. 1971;171:1151–5.
42. Bitar R, Leung G, Perng R, Tadros S, et al. MR pulse sequences: what every radiologist wants to know but is afraid to ask. Radiographics. 2006;26:513–37.
43. Das IJ, Sagreiya H, Yadav P, Allen BD. Basics of MR imaging for the radiation oncologist. In: Ozyar E, Onal C, Hackett SL, editors. MR Linac radiotherapy, a new personalized treatment approach. London: Academic Press; 2022. p. 5–32.
44. Jacobs MA, Ibrahim TS, Ouwerkerk R. AAPM/RSNA physics tutorials for residents: MR imaging: brief overview and emerging applications. Radiographics. 2007;27:1213–29.
45. Srinivasan S, Dasgupta A, Chatterjee A, Baheti A, et al. The promise of magnetic resonance imaging in radiation oncology practice in the management of brain, prostate, and GI malignancies. JCO Glob Oncol. 2022;8:e2100366.
46. Baissalov R, Sandison GA, Donnelly BJ, Saliken JC, et al. Suppression of high-density artefacts in x-ray CT images using temporal digital subtraction with application to cryotherapy. Phys Med Biol. 2000;45:N53–9.
47. Wei J, Sandison GA, Hsi WC, Ringor M, et al. Dosimetric impact of a CT metal artifact suppression algorithm for proton, electron and photon therapies. Phys Med Biol. 2006;51:5183–97.
48. Lewis M, Reid K, Toms AP. Reducing the effects of metal artefact using high keV monoenergetic reconstruction of dual energy CT (DECT) in hip replacements. Skelet Radiol. 2013;42:275–82.
49. Lell MM, Meyer E, Kuefner MA, May MS, et al. Normalized metal artifact reduction in head and neck computed tomography. Investig Radiol. 2012;47:415–21.
50. Andersson KM, Nowik P, Persliden J, Thunberg P, et al. Metal artefact reduction in CT imaging of hip prostheses—an evaluation of commercial techniques provided by four vendors. Br J Radiol. 2015;88:20140473.
51. Ojala J, Kapanen M, Sipila P, Hyodynmaa S, et al. The accuracy of Acuros XB algorithm for radiation beams traversing a metallic hip implant - comparison with measurements and Monte Carlo calculations. J Appl Clin Med Phys. 2014;15:162–76.
52. Higashigaito K, Angst F, Runge VM, Alkadhi H, et al. Metal artifact reduction in pelvic computed tomography with hip prostheses: comparison of virtual monoenergetic extrapolations from dual-energy computed tomography and an iterative metal artifact

reduction algorithm in a phantom study. Investig Radiol. 2015;50:828–34.
53. Mullins JP, Grams MP, Herman MG, Brinkmann DH, et al. Treatment planning for metals using an extended CT number scale. J Appl Clin Med Phys. 2016;17:179–88.
54. Carrasco P, Jornet N, Duch MA, Panettieri V, et al. Comparison of dose calculation algorithms in slab phantoms with cortical bone equivalent heterogeneities. Med Phys. 2007;34:3323–33.
55. Yadav P, Chang SX, Cheng CW, DesRosiers CM, et al. Dosimetric evaluation of high-Z inhomogeneity used for hip prosthesis: a multi-institutional collaborative study. Phys Med. 2022;95:148–55.
56. Das IJ, McGee KP, Tyagi N, Wang H. Role and future of MRI in radiation oncology. Br J Radiol. 2019;92:20180505.
57. Moore-Palhares D, Ho L, Lu L, Chugh B, et al. Clinical implementation of magnetic resonance imaging simulation for radiation oncology planning: 5 year experience. Radiat Oncol. 2023;18:27.
58. Otazo R, Lambin P, Pignol JP, Ladd ME, et al. MRI-guided radiation therapy: an emerging paradigm in adaptive radiation oncology. Radiology. 2021;298:248–60.
59. Tenhunen M, Korhonen J, Kapanen M, Seppala T, et al. MRI-only based radiation therapy of prostate cancer: workflow and early clinical experience. Acta Oncol. 2018;28:1–6.
60. Pollard JM, Wen Z, Sadagopan R, Wang J, et al. The future of image-guided radiotherapy will be MR guided. Br J Radiol. 2017;90:20160667.
61. Chandarana H, Wang H, Tijssen RHN, Das IJ. Emerging role of MRI in radiation therapy. J Magn Reson Imaging. 2018;48:1468–78.
62. Nousiainen K, Santurio GV, Lundahl N, Cronholm R, et al. Evaluation of MRI-only based online adaptive radiotherapy of abdominal region on MR-linac. J Appl Clin Med Phys. 2023;24:e13838.
63. Ramsey CR, Oliver AL. Magnetic resonance imaging based digitally reconstructed radiographs, virtual simulation, and three-dimensional treatment planning for brain neoplasms. Med Phys. 1998;25:1928–34.
64. Ramsey CR, Arwood D, Scaperoth D, Oliver AL. Clinical application of digitally-reconstructed radiographs generated from magnetic resonance imaging for intracranial lesions. Int J Radiat Oncol Biol Phys. 1999;45:797–802.
65. Bayouth JE, Low DA, Zaidi H. MRI-linac systems will replace conventional IGRT systems within 15 years. Med Phys. 2019;46:3753–6.
66. Kramer S, Kusner D, Gunn WG. Clinical experience with the Jefferson Hospital radiotherapy simulator. Radiology. 1966;87:134–6.
67. Mutic S, Palta JR, Butker EK, Das IJ, et al. Quality assurance for computed-tomography simulators and the computed-tomography-simulation process: report of the AAPM Radiation Therapy Committee Task Group No. 66. Med Phys. 2003;30:2762–92.
68. Mah D, Steckner M, Palacio E, Mitra R, et al. Characteristics and quality assurance of a dedicated open 0.23 T MRI for radiation therapy simulation. Med Phys. 2002;29:2541–7.
69. Kapanen M, Collan J, Beule A, Seppälä T, et al. Commissioning of MRI-only based treatment planning procedure for external beam radiotherapy of prostate. Magn Reson Med. 2013;70:127–35.
70. Glide-Hurst CK, Paulson ES, McGee K, Tyagi N, et al. Task group 284 report: magnetic resonance imaging simulation in radiotherapy: considerations for clinical implementation, optimization, and quality assurance. Med Phys. 2021;48:e636–70.
71. Glide-Hurst CK, Wen N, Hearshen D, Kim J, et al. Initial clinical experience with a radiation oncology dedicated open 1.0T MR-simulation. J Appl Clin Med Phys. 2015;16:5201.
72. Tyagi N, Fontenla S, Zelefsky M, Chong-Ton M, et al. Clinical workflow for MR-only simulation and planning in prostate. Radiat Oncol. 2017;12:119.
73. Tyagi N, Zelefsky MJ, Wibmer A, Zakian K, et al. Clinical experience and workflow challenges with magnetic resonance-only radiation therapy simulation and planning for prostate cancer. Phys Imaging Radiat Oncol. 2020;16:43–9.
74. Cao M, Padgett KR, Rong Y. Are in-house diagnostic MR physicists necessary for clinical implementation of MRI guided radiotherapy? J Appl Clin Med Phys. 2017;18:6–9.
75. Kagawa K, Lee WR, Schultheiss TE, Hunt MA, et al. Initial clinical assessment of CT-MRI image fusion software in localization of the prostate for 3D conformal radiation therapy. Int J Radiat Oncol Biol Phys. 1997;38:319–25.
76. Roach M 3rd, Faillace-Akazawa P, Malfatti C, Holland J, et al. Prostate volumes defined by magnetic resonance imaging and computerized tomographic scans for three-dimensional conformal radiotherapy. Int J Radiat Oncol Biol Phys. 1996;35:1011–8.
77. Guo L, Shen S, Harris E, Wang Z, et al. A tri-modality image fusion method for target delineation of brain tumors in radiotherapy. PLoS One. 2014;9:e112187.
78. Basson L, Jarraya H, Escande A, Cordoba A, et al. Chest magnetic resonance imaging decreases interobserver variability of gross target volume for lung tumors. Front Oncol. 2019;9:690.
79. Batumalai V, Burke S, Roach D, Lim K, et al. Impact of dosimetric differences between CT and MRI derived target volumes for external beam cervical cancer radiotherapy. Br J Radiol. 2020;93:20190564.
80. den Hartogh MD, Philippens ME, van Dam IE, Kleynen CE, et al. MRI and CT imaging for preoperative target volume delineation in breast-conserving therapy. Radiat Oncol. 2014;9:63.
81. Lee E, Park W, Ahn SH, Cho JH, et al. Interobserver variation in target volume for salvage radiotherapy in recurrent prostate cancer patients after radical prostatectomy using CT versus combined CT and MRI: a multicenter study (KROG 13-11). Radiat Oncol J. 2018;36:11–6.

82. White I, Hunt A, Bird T, Settatree S, et al. Interobserver variability in target volume delineation for CT/MRI simulation and MRI-guided adaptive radiotherapy in rectal cancer. Br J Radiol. 2021;94:20210350.
83. Zhang H, Fu C, Fan M, Lu L, et al. Reduction of inter-observer variability using MRI and CT fusion in delineating of primary tumor for radiotherapy in lung cancer with atelectasis. Front Oncol. 2022;12:841771.
84. Weltens C, Menten J, Feron M, Bellon E, et al. Interobserver variations in gross tumor volume delineation of brain tumors on computed tomography and impact of magnetic resonance imaging. Radiother Oncol. 2001;60:49–59.
85. Aoyama H, Shirato H, Nishioka T, Hashimoto S, et al. Magnetic resonance imaging system for three-dimensional conformal radiotherapy and its impact on gross tumor volume delineation of central nervous system tumors. Int J Radiat Oncol Biol Phys. 2001;50:821–7.
86. Aslian H, Sadeghi M, Mahdavi SR, Babapour Mofrad F, et al. Magnetic resonance imaging-based target volume delineation in radiation therapy treatment planning for brain tumors using localized region-based active contour. Int J Radiat Oncol Biol Phys. 2013;87:195–201.
87. Raman S, Chin L, Erler D, Atenafu EG, et al. Impact of magnetic resonance imaging on gross tumor volume delineation in non-spine bony metastasis treated with stereotactic body radiation therapy. Int J Radiat Oncol Biol Phys. 2018;102:735–43.
88. Dalah E, Moraru I, Paulson E, Erickson B, et al. Variability of target and normal structure delineation using multimodality imaging for radiation therapy of pancreatic cancer. Int J Radiat Oncol Biol Phys. 2014;89:633–40.
89. Vorwerk H, Beckmann G, Bremer M, Degen M, et al. The delineation of target volumes for radiotherapy of lung cancer patients. Radiother Oncol. 2009;91:455–60.
90. Cazzaniga L, Marinoni M, Bossi A, Bianchi E, et al. Interphysician variability in defining the planning target volume in the irradiation of prostate and seminal vesicles. Radiother Oncol. 1998;28:293–6.
91. Tsang Y, Hoskin P, Spezi E, Landau D, et al. Assessment of contour variability in target volumes and organs at risk in lung cancer radiotherapy. Tech Innov Patient Support Radiat Oncol. 2019;10:8–12.
92. Unkelbach J, Bortfeld T, Cardenas CE, Gregoire V, et al. The role of computational methods for automating and improving clinical target volume definition. Radiother Oncol. 2020;153:15–25.
93. Das IJ, Compton JJ, Bajaj A, Johnstone PA. Intra- and inter-physician variability in target volume delineation in radiation therapy. J Radiat Res. 2021;62:1083–9.
94. Wong J, Fong A, McVicar N, Smith S, et al. Comparing deep learning-based auto-segmentation of organs at risk and clinical target volumes to expert inter-observer variability in radiotherapy planning. Radiother Oncol. 2020;144:152–8.
95. Wong J, Huang V, Wells D, Giambattista J, et al. Implementation of deep learning-based auto-segmentation for radiotherapy planning structures: a workflow study at two cancer centers. Radiat Oncol. 2021;16:101.
96. Gooding MJ, Smith AJ, Tariq M, Aljabar P, et al. Comparative evaluation of autocontouring in clinical practice: a practical method using the Turing test. Med Phys. 2018;45:5105–15.
97. Loap P, De Marzi L, Kirov K, Servois V, et al. Development of simplified auto-segmentable functional cardiac atlas. Pract Radiat Oncol. 2022;12:533–8.
98. Rhee DJ, Akinfenwa CPA, Rigaud B, Jhingran A, et al. Automatic contouring QA method using a deep learning-based autocontouring system. J Appl Clin Med Phys. 2022;23:e13647.
99. Kawula M, Hadi I, Nierer L, Vagni M, et al. Patient-specific transfer learning for auto-segmentation in adaptive 0.35 T MRgRT of prostate cancer: a bi-centric evaluation. Med Phys. 2023;50:1573–85.
100. Fast MF, Eiben B, Menten MJ, Wetscherek A, et al. Tumour auto-contouring on 2d cine MRI for locally advanced lung cancer: a comparative study. Radiother Oncol. 2017;125:485–91.
101. Liang F, Qian P, Su KH, Baydoun A, et al. Abdominal, multi-organ, auto-contouring method for online adaptive magnetic resonance guided radiotherapy: an intelligent, multi-level fusion approach. Artif Intell Med. 2018;90:34–41.
102. Feng L, Grimm R, Block KT, Chandarana H, et al. Golden-angle radial sparse parallel MRI: combination of compressed sensing, parallel imaging, and golden-angle radial sampling for fast and flexible dynamic volumetric MRI. Magn Reson Med. 2014;72:707–17.
103. Nepal P, Bagga B, Feng L, Chandarana H. Respiratory motion management in abdominal MRI: radiology in training. Radiology. 2023;306:47–53.
104. Curtis AD, Cheng HM. Primer and historical review on rapid cardiac CINE MRI. J Magn Reson Imaging. 2022;55:373–88.
105. Akino Y, Oh RJ, Masai N, Shiomi H, et al. Evaluation of potential internal target volume of liver tumors using cine-MRI. Med Phys. 2014;41:111704.
106. Ho VB, Foo TK. Impact of "Cine MR imaging: potential for the evaluation of cardiovascular function". AJR Am J Roentgenol. 2006;187:605–8.
107. Kim T, Wu Y, Ji Z, Gach HM, et al. Technical note: Institutional solution of clinical cine MRI for tumor motion evaluation in radiotherapy. J Appl Clin Med Phys. 2022;23:e13650.
108. Winkel D, Bol GH, Kroon PS, van Asselen B, et al. Adaptive radiotherapy: the Elekta Unity MR-linac concept. Clin Transl Radiat Oncol. 2019;18:54–9.
109. Wald C, Luchs J, Davila J, Lozano K, et al. Residents' preceptions of MRI training in the United States. J Am Coll Radiol. 2004;1:331–7.

110. McGee KP, Tyagi N, Bayouth JE, Cao M, et al. Findings of the AAPM Ad Hoc committee on magnetic resonance imaging in radiation therapy: Unmet needs, opportunities, and recommendations. Med Phys. 2021;48:4523–31.
111. Singer L, Rosenberg SA. The impact of MRI on radiation oncology graduate medical education. J Am Coll Radiol. 2019;16:859–63.
112. Hasford F, Mumuni AN, Trauernicht C, Ige TA, et al. A review of MRI studies in Africa with special focus on quantitative MRI: historical development, current status and the role of medical physicists. Phys Med. 2022;103:46–58.
113. Hall WA, Paulson ES, van der Heide UA, Fuller CD, et al. The transformation of radiation oncology using real-time magnetic resonance guidance: a review. Eur J Cancer. 2019;122:42–52.
114. Cahoon G, Skehan K, Elwadia D, Rai R. The current and future role of the MRI radiographer in radiation oncology: a collaborative, experiential reflection on the Australian rollout of dedicated MRI simulators. J Med Radiat Sci. 2023;70(Suppl 2):107–13.

# Clinical Necessity and Patient Selection in MR-Linac

3

Poonam Yadav, Filippo Alongi, Bharat B. Mittal, and Indra J. Das

## Introduction

The monumental technological improvements in radiation oncology have ushered in a new paradigm in patient management. In recent times, radiation therapy has benefitted from significant technological advancements that have transformed the field. These advancements, such as intensity-modulated radiation therapy (IMRT), adaptive radiation therapy (ART), and image-guided radiation therapy (IGRT) [1–3], have allowed clinicians to administer precise and highly conformal radiation doses. To achieve the necessary dose conformity and steeper dose gradients used in modern radiation therapy to spare normal tissues, accurate localization and beam targeting techniques are critical. However, the reproducibility of target positions during radiation therapy can be affected by the motion of internal organs during radiation and the displacement of internal organs between radiation sessions. Fortunately, IGRT leverages modern imaging technology to enhance tumor definition, and minimize treatment uncertainties.

Even with modern technology such as onboard imaging, target and normal structures cannot be clearly visualized with cone beam computed tomography (CBCT), which has become a routine process and standard of care. This is due to its lack of soft tissue contrast and real-time imaging. CBCT is used before treatment to ensure that treatment is delivered as planned; however, it cannot confirm if the position is accurate during treatment. Multiple image acquisitions with X-rays also add radiation exposure and are of concern especially for long-term survivors [4]. Magnetic resonance imaging in combination with a linear accelerator (MR-Linac) provides an opportunity to visualize the target and normal structures during treatment, continuously and in real time, for optimal coverage. Additionally, beam gating has been used on an ad hoc basis to keep the target in the field of view, but due to lack of continuous monitoring capability, it has not been widely applied. In contrast, in MR-Linac continuous monitoring is automated and provides seamless visualization without radiation exposure.

## Onboard Imaging

Modern radiation therapy machines are equipped with onboard imaging systems. While these systems have varying designs, the majority of them

P. Yadav (✉) · B. B. Mittal · I. J. Das
Northwestern Memorial Hospital, Northwestern University Feinberg School of Medicine, Chicago, IL, USA
e-mail: poonam.yadav@northwestern.edu; b-mittal@northwestern.edu; indra.das@nm.org

F. Alongi
Università di Brescia, Direttore del Dipartimento di Radioterapia Oncologica Avanzata, Brescia, Italy


I. J. Das et al. (eds.), *A Practical Guide to MR-Linac*, https://doi.org/10.1007/978-3-031-48165-9_3

share one common element: using a kV source and a flat-panel imager. The purpose of these systems is to capture volumetric images with enough detail to visualize soft tissues properly. It is truly remarkable to see how far technology has come to improve imaging. Imaging of soft tissues is crucial for planning of radiation treatment. CT alone or along with other imaging techniques is routinely used for soft tissue delineation for the planning of radiation therapy. There has been continuous improvement in the sensitivity and specificity of CT-based systems. New contrast agents have improved the delineation of boundaries of tumors in CT. CT simulation with three-dimensional (3D) planning is the standard of care in most centers. Dedicated CT simulators have now totally replaced conventional simulators. Jaffrey et al. [5] provided a rationale for IMRT and IGRT and how they will affect the outcome. In a similar manner Ling et al. [6] provided an editorial on these technologies, but lamented that the full potential has not been achieved.

The kV CBCT helps target localization by acquiring 3D images that provide additional details such as size, shape, and organs in proximity. These details are vital for ART procedures, which rely on regular imaging and treatment modification based on imaging results. ART is resource-intensive due to the need for frequent imaging, but the benefits are undeniable. One study reported the feasibility of dose reduction in organs at risk (OAR) and an average dose escalation of >10% to the prostate when the treatment plan was modified based on daily fractional scans [7]. Thus, ART has excellent potential to deliver precise radiation doses as was advocated initially by Yan et al. [3].

Apart from its success, CBCT suffers from various drawbacks such as low image quality, blurred images mainly due to organ motion, and artifacts due to scattering of X-rays. Despite these constraints, CBCT is still used routinely. Van Herk and colleagues reported good accuracy of CBCT as compared with electronic portal imaging and digitally reconstructed radiographs in prostate imaging [8]. Other studies have also reported the geometric reliability of CBCT. However, there is a risk of unwanted radiation being delivered to the tissues adjoining the tumor due to the poor image quality of CBCT imaging. Radiation exposure may also increase the risk of radiation-induced cancer, as reported by some studies. The CBCT imaging dose also depends on several attributes such as field of view, number of projections, and anatomical considerations. Additionally, skin doses within the imaging field are often 2–5 cGy [2, 4], and multiple rounds of imaging will certainly increase this dose.

CT imaging and linear accelerators can be integrated in various ways. Helical tomotherapy is a Linac on a CT ring gantry. Megavoltage (MV) Linac beams can be used for MVCT scans with the help of MV image detectors for reconstructing CT projections. MVCT prevails over planar MV portal radiographs due to improved soft tissue contrast and 3D projections and is comparable to planning CT. From a performance perspective, flat-panel detectors are suitable, and a handful of rotations can scan an entire region of interest. However, increased scatter at the detector surface impairs the image contrast, further requiring imaging dose escalation or an increase in slice thickness [9].

A CT scanner can also be installed on a traditional C-arm Linac gantry, usually using electronic portal imaging systems. In a single rotation, several CT slices can be obtained. However, the patient's daily radiation dose is significant and has limited use. Future advances such as employing multirow detector systems, increasing the number of CT rows on CT scanners, and increasing data processing speed may help. CBCT on conventional accelerators has drawbacks, such as accommodating gantry sag and slower movement of the C-arm gantries [10]. Figure 3.1 shows the inferior image quality (low contrast) where structures are hard to visualize even with optimum techniques and proper filters.

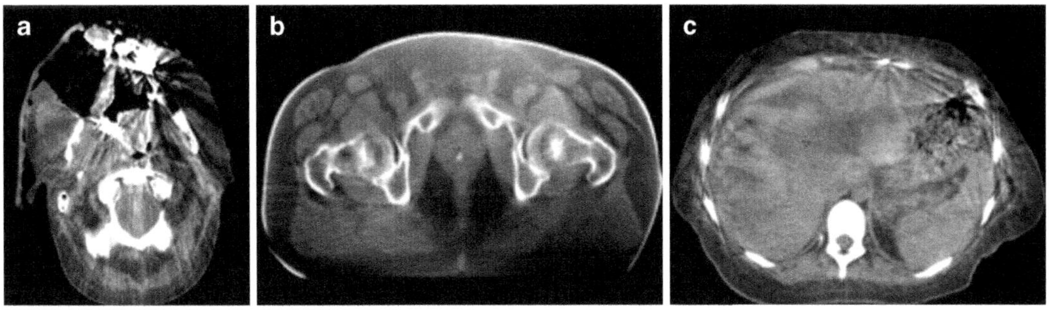


**Fig. 3.1** Pretreatment imaging using CBCT of a head and neck (**a**), prostate (**b**), and liver (**c**) showing poor resolution and contrast of images which make target alignment difficult

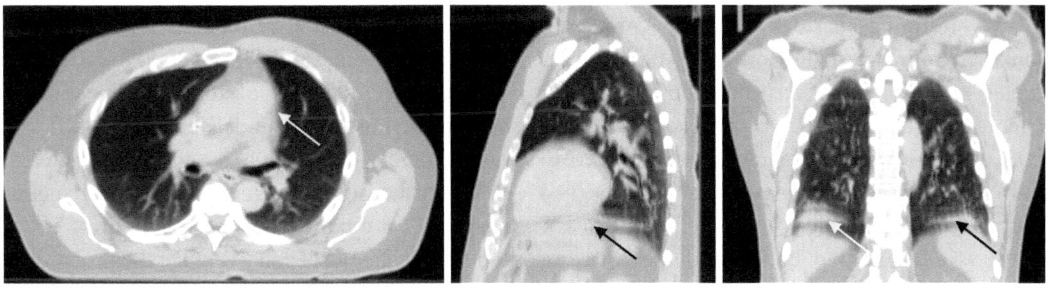

**Fig. 3.2** Motion artifact of a 4DCT lung patient on axial, sagittal, and coronal views. Arrows show the motion artifact

## 4D Computed Tomography

External body movement and the motion of organs may not precisely correlate and may cause inaccuracies in imaging. There might be variations in amplitude proportions between fiducial and tumor displacement. These differences may follow a pattern between cycles, amplitude ratios between fiducial and tumor displacement may vary from cycle to cycle, and thoracic and abdominal points may involve relative phase displacement. Such considerations are essential for treating tumors in the vicinity of the lungs. The breathing cycle can produce significant motion artifacts for large organs such as the liver due to temporal averaging [11].

Additionally, respiratory motion may create artifacts and cause errors during radiation treatment. Breath-hold techniques are mainly used to counteract these respiratory motion artifacts. Different active and passive techniques are employed depending upon the tumor's location, and the radiation dose is administered during the period of breath hold. Examples of breath-hold techniques include active breathing control (ABC), deep inspiration breath hold, and self breath hold. ABC requires shallow breathing and tries to freeze the movement of the tumor for short intervals of time [12, 13].

Respiratory gating methods monitor the natural breathing cycle of patients. The radiation beam can be periodically turned on in a predetermined phase of the patient's breathing cycle. Generally, the beam is turned off at the end of inhalation or exhalation. These respiratory gating methods help offset the motion of the patient. However, they require patient cooperation. The CT images utilized for treatment should represent the actual situation. It has been observed that gating centered on the end of expiration can increase reproducibility, since the post-expiration period may comprise up to 30% of the breathing period, giving an adequate treatment time to leverage delivery in this phase [14]. Figure 3.2

shows an example of motion-related challenges and the adaptive nature of treatment in lung stereotactic body radiation therapy.

## Magnetic Resonance Imaging Systems

The characteristics of magnetic resonance imaging (MRI) for the superior distinction of soft tissue for the abdomen, pelvis, and central nervous system are not serendipitous. They are frequently used for diagnostic and tumor segmentation purposes. Refining various imaging processes makes it intuitive to explore MRI-based treatment planning simulations for radiotherapy procedures. In the presence of a magnetic field, the bulk magnetization of the nuclei in a given voxel through the radiofrequency coils is quantified during the MRI scan. A fusion of MRI and CT scans is evaluated to understand the areas of interest. The diagnostic use and understanding of MRI have gone through a series of technological developments in the past four decades. Until the last decade, most of the research on improving MRI image quality was directed towards increasing the strength of the magnetic field, as is recognized by the installation of 3T and higher field strength scanners in clinics. Organ motion, especially breathing-induced, abdominal, heartbeat-related, and temporal, has always remained a concern for clinicians, and fast cine MRI has evolved as a prevalent solution for such situations.

The success of MRI with soft tissues and nervous systems has also encouraged the exploration of special MRI procedures such as functional MRI, cardiac MRI, MR venography, non-contrast and contrast MRI, MR angiography, MR spectroscopy, and open-field MRI. Functional MRI has gained popularity by detecting small and measurable signal changes in the brain and applying diffusion tensor imaging techniques, which measure diffusion in different directions with the help of fractional anisotropy. A functional map of the brain captures the dynamic neural activity and allows for improved diagnosis of brain tumors [15].

Many radiotherapy procedures are nowadays MRI-assisted, where CT scans and MR images of the tumor site are used in combination for radiotherapy treatment planning. CT scans have remained at the heart of treatment planning, not just for image guidance but also as a prerequisite for electron density calculations, while MR scans help in superior soft tissue isolation.

Uncertainties in gross tumor volume delineation have remained a key constraint in precision treatment delivery [16]. Many factors contribute to variations in tumor delineation, such as target motion, setup errors, and interobserver variability, but tumor visibility remains one of the top concerns. Several studies from 2000 to 2010 have characterized the role of tumor delineation in compromising treatment delivery [8, 17–20].

In contemporary high dose rate procedures, dose escalation to the tumor and simultaneous sparing of normal tissues are required for better locoregional control and to reduce the probability of recurrence. With constrained tumor visibility, adequately delineating the regions of interest becomes impossible, and the associated uncertainties hinder steep dose gradients needed for optimal tumor dose and better OAR sparing. With reduced tumor visibility, the delineation protocols cannot be aggressively set up, and systematic variations are found across treatment planning. The resulting compromise in treatment delivery has been adversely associated with tumor control probability [16, 21].

Inter- and intraobserver variability also has a negative effect on treatment plan quality and is especially noticeable in sites involving soft tissue anatomy and deformable organs. If MR scans are utilized for treatment planning, such observer-related discrepancies can be significantly reduced, mainly due to enhanced soft tissue contrast [18, 22]. Additionally, imaging is crucial in brain tumor management. MRI can recognize and characterize edema [23] and provides accurate contouring of brain tumors compared to CT images [24]. Functional MRI can be used to evaluate the radiotherapy treatment response. MR-CT registration, i.e., mapping of one set of images to the other, can successfully increase the dose to the target and reduce radiation-related severe toxicity due to increased accuracy in the localization of gross tumor volume and neighboring structures [25–27]. For oropharyngeal cancer radiotherapy, MRI scans can be acquired in the treatment setup position with the help of the same immobilization device. Such setup reproducibility is beneficial in reducing MR-CT registration errors and can improve target definition, dose coverage, and dose homogeneity.

Multiparametric MRI scans have been found to be useful in prostate cancer. MR-generated anatomical and functional imaging has increased accuracy in segmenting subvolumes of the prostate and maintained the functionality of sensitive tissues in the apical posterior region, an area of relatively high incidence and hence one for which it is critical to delineate such volumes precisely [28–30]. Haider et al. reported that functional MRI is useful in the treatment planning of recurrent prostate cancer [31]. Quality of life improvements have been reported with MR-based registration, mainly due to conformal coverage [32]. Pötter et al. reported that survival gains of up to 20% could be achieved by dose escalation in cervical cancer cases with a simultaneous reduction in gastrointestinal and urinary morbidity [33]. MRI-based tumor delineation may lead to optimal dosimetry and improvement in therapeutic ratio, as improved delineation can support tumor dose escalation (~7 Gy) and spare the rectal wall [34].

The Radiation Therapy Oncology Group consensus panel on pelvis contouring identified MRI as an appropriate modality for the adnexa, uterocervix, and bladder [35]. Functional MRIs are superior for gynecological malignancies due to inherent metabolic profiling and monitoring of tissue microarchitecture and tumor characteristics and are also suggested by ESTRO guidelines for delineating gynecological tumors [33]. MRI's predictive capabilities have already been proved; treatment response can also be evaluated with MRI [36]. Dynamic contrast-enhanced MRI perfusion studies during different phases of radiotherapy treatment can detect cervical tumor response and help adapt treatment courses in early or mid-radiotherapy phases [37].

Early MR-based IGRT procedures relied on tumor and region of interest definition via the assistance of appropriate image registration algorithms porting the structures from MRI to CT images. MR-CT co-registration can lead to geometric uncertainties in the treatment planning process, and studies have reported geometrical uncertainties of approximately 2 mm for the brain and up to 3 mm for the prostate and gynecological treatment sites [38, 39]. Though the MRI setup taken in the treatment position helps reduce discrepancies, there are still geometric misalignments throughout the process of CT-MR co-registration, which raises concerns regarding geometric miss [8].

Though CT remains the standard for contemporary radiotherapy, parallel MRI-based co-registration has gained significant attention, especially for sites involving soft tissue anatomy. It has paved the way for research on incorporating MRI simulators in radiotherapy treatment planning to address the geometrical mismatch in the co-registration of images from the two modalities. Recent research on MRI-based simulations has shown that MR simulators can be an appealing option for radiotherapy treatment planning [40–42]. This topic is addressed in Chap. 2 of this book. MRI simulators can address various aspects of diagnostic MRI for treatment planning purposes. The simulators can resolve the mismatch in bore sizes, flat tabletop instead of immobilization devices, in-room alignment systems, and imaging algorithms, and a high-precision patient model can be constructed. Therefore, MR-only treatment planning can be accomplished where synthetic CT data is constructed based on MR scans.

So, the radiotherapy workflow can accommodate MRI in several ways, like MRI co-registration with CT scans, MR scans in the treatment position, MR-only treatment planning, and others. In most of these cases, the primary rationale is the superior soft tissue contrast of MRI. Hence, MR-assisted soft tissue delineation is preferred, while CT scans often confirm treatment setup. Thus, MRI can be a crucial part of the pretreatment process, though systematic errors in setup remain a concern, as highlighted in previous sections [43–45].

There are some barriers to MR-based planning, and we can see some considerable limitations. It cannot be overlooked that CT data has a one-to-one relationship with tissue attenuation and, subsequently, dose calculations and setup reproduction purposes. This makes CT often the best choice for geometric accuracy in treatment setup. Any MRI scan process lasts longer and raises the probability of motion and patient distress-related issues, while CBCT-based data collection is relatively faster [8].

The clinically available MR-Linacs utilize a deformable image registration process. This process is integrated with the treatment planning system and facilitates the propagation of structures from the planning CT scan to the MR scan for the treatment session. This reduces the uncertainties, as the deformable registration process is integral to the system. At the same time, a wide range of deformable image registration algorithms has resulted in disparities in results.

MR spectroscopy is another powerful MR-based tool for pinpointing the location of localized prostate cancers during brachytherapy, as noted by Dibase et al. [29]. This technique allows for precise delivery of ionizing radiation and has quickly become a valuable addition to the medical field. The kV CBCT images acquired during this process are also of exceptional quality, offering precise delineation of soft tissue structures. Numerous studies have demonstrated the clinical usability of these images within a short amount of time, with analysis times being reduced with experience. In one study, analysis time was even reduced from 10 min to just 2 min. These advancements are truly remarkable and have the potential to revolutionize the way we treat localized prostate cancers.

## Patient Selection

Despite all the advantages of MR as discussed in previous sections, not every patient can be treated with MR-Linac. This section provides some information on patient selection. The importance of patient selection cannot be overstated and should be started well in advance of initiating the treatment process. Some patients might require support such as relaxation techniques. Preparing patients with knowledge of the workflow and overcoming certain situations has been found to be very effective for successful patient compliance with the instructions of the procedure. The patient should be made aware of the MR screening forms, the noise level that is generated from the MR scanner, the need for wearing headphones or earplugs for hearing protection, the burning effect that maybe caused by the loop generated by connected body parts, and most importantly, that the magnetic field is on all the time.

## MRI Safety

MRIs are routinely utilized for better target definition and are mostly done in diagnostic imaging departments. With the evolution of hybrid MR-Linac, careful planning by the interdisciplinary team, made up of radiation oncologists, medical physicists, nurses, and radiation therapy technologists, is important from various perspectives such as developing standardized treatment procedures, MR safety and training, logistics, and familiarity with treatment and control rooms.

Despite the relatively lower field strength of hybrid MR-Linacs compared to diagnostic scanners, the safety criteria are the same. Therefore, all patients undergoing treatment should be screened. To standardize the MR safety procedures, everyone on the care team is trained by experts in coordination with the vendor. The importance of training cannot be overstated. An American Association of Physicists in Medicine (AAPM) task group, TG-203 [46], has pointed out that the radiotherapy community did not fully utilize measures of education and training with clinical teams on the recommendations of TG-34 [46, 47].

One of the initial MR safety checks involves completion of a checklist or questionnaire, not only for the patients but for everyone entering the treatment room, such as volunteers and visitors.

Existing MR safety checklists from diagnostic imaging departments can serve as a baseline for developing such a document and should include questions about neurostimulators, claustrophobia, pacemakers, intracranial clips, ferromagnetic materials, and other electronic hardware (Fig. 3.3). The installation of ferromagnetic detectors can be helpful in elevating MR safety checks. Recently developed electronic devices such as pacemakers, prostheses, and other implants are designated as MR safe. A thorough check of compliance with MR safety measures should be a routine part of MR-guided radiation therapy (MRgRT) workflows.

## Claustrophobia

Proper patient selection is key to the success of both diagnostic MRI scans and MRgRT treatment procedures. Certain existing conditions such as arrhythmias, claustrophobia, and problems with following breath-hold instructions should be properly addressed. Challenges posed by claustrophobia and anxiety are well documented for MRI procedures, and numerous solutions have been explored [48]. The probability of a patient developing claustrophobia should be evaluated before the start of the treatment, and the clinical team should make due preparations for such accommodations. A light sedation may help, especially for breath-hold compliance and other sudden movements. Other solutions involve installing a mirror in the bore and audio assurances. A well-vetted preselection screening procedure could reduce patient dropout due to MR-related discomfort.

## Metallic Implants

Certain metallic objects that were previously considered contraindicated are now observed as sufficiently safe. Along with reduction in the use of ferromagnetic material in implants, better understanding and identification of implants with clear labeling in distinguishable categories such as MR safe, MR conditional, and MR unsafe helps in the screening process (Fig. 3.3). Figure 3.4 provides additional images of metallic implants where in MR images it produces a blooming effect with loss of anatomical information.

An item with no known hazard in any MR environment and basically nonconducting, nonmetallic, and/or nonmagnetic can be considered

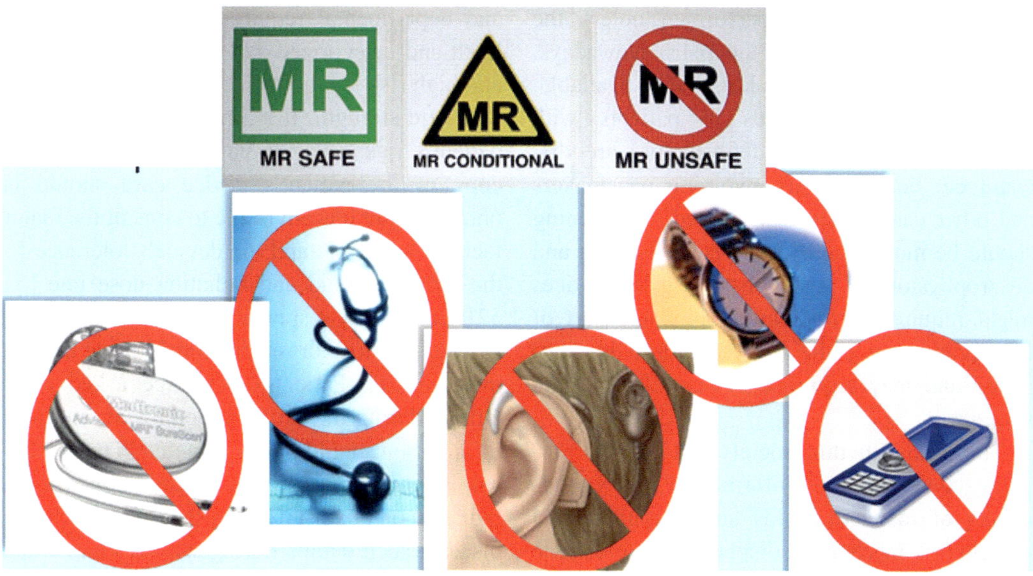


**Fig. 3.3** Figure showing the MR signs and unsafe objects in MRI

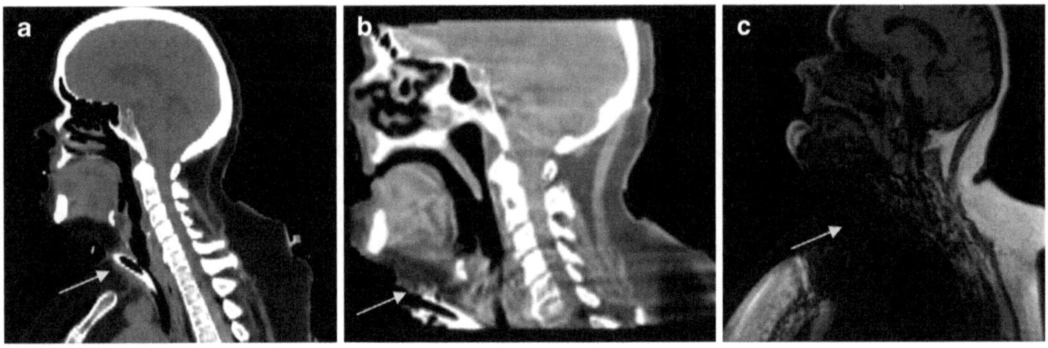


**Fig. 3.4** Sagittal images of a patient with tracheal implant via (a) planning CT, (b) CBCT, and (c) 0.35T MR. As noted, a total loss of image information (blooming artifact) is observed in the MR image. Arrows indicate the position of the trachea

as MR safe. If an item has no known hazards in a specified MRI environment with particular conditions of use like spatial magnetic gradient, radiofrequency fields, related systemic absorption rates, and static magnetic field strength, it is known as MR conditional. An item is considered as MR unsafe if it could be hazardous in MR environments.

Most coronary stents are either MR safe or MR conditional. Prosthetic valves and sternal wires are usually MR safe. Specific exceptions and explicit declaration should be included on the screening forms [49]. Studies have suggested that MR procedures for pacemaker-dependent patients should not be performed unless the potential benefits surpass the risks. Nowadays, MR conditional pacemakers are available. Likewise, MR procedures in patients with implantable cardioverter defibrillators are discouraged. For situations involving pacemakers and other cardiac devices, if needed, screening should be monitored by competent imaging and electrophysiology specialists. Such devices might require reprogramming or some sort of recalibration once the MR procedure is completed and may require a device-related clinic visit.

The Heart Rhythm Society 2017 report provides comprehensive information on various effects of magnetic fields such as unplanned stimulation, mechanical displacement, electrode heating, RF injected disturbances, and imaging artifacts on cardiac implantable electronic devices (CIEDs) [50]. The American Association of Physicists in Medicine TG-203 acknowledges MRI procedures do not significantly increase the probability of CIED-related issues, but still requires that well-documented MR safety protocols should be developed involving such devices with due consideration of multidisciplinary aspects of both devices and procedures [46].

While developing radiotherapy treatment programs for treating cases with CIEDs, radiation oncologists should coordinate their activities with cardiology teams with well-defined responsibilities for the management of such cases. Staff and departmental requirements should be outlined and categorized as low risk, medium risk, and high risk for the radiation dose exposure and MR field strength. It is important to ascertain a patient's risk category early in the process. A copy of the patient's device card should be obtained, and it is advisable to consult the manufacturer to understand the device's tolerance for the magnetic field and radiation dose rate [51, 52]. Emergency response devices should be easily available to address the probability of medical interventions. Patients should be asked about CIEDs during the initial appointment, and such information should be documented in their medical file and verified during consultation with the radiation oncologist. Clinics should preemptively disclose such details during weekly chart round meetings and discuss management plans.

Devices such as implantable loop monitors, cochlear implants, epidural pumps, nerve stimulators, insulin pumps, and programmable intrathecal drug delivery devices should be examined for MR safety concerns.

## Patient Comfort

Traditionally, patient comfort is given high priority in radiation therapy and is usually achieved by providing a comfortable immobilization system. However, patients undergoing treatment in an MR-Linac do not have comfortable immobilization due to eddy currents and the proximity of coils. MR-Linac patients have to spend considerably longer treatment time in an enclosed bore, which is a well-known concern of patients as shown in Fig. 3.5. The narrower bores can further impact such unpleasant experiences. Some patients cannot stay in MR-Linac's bore for an extended period of time. Also, large body habitus can be a constraint in patient selection if coil size or instrument bore size is relatively small.

MR-Linac's longer and narrower bore size and requirements for immobilization along with tolerance for coil placement can increase the patient's discomfort [53]. Klüter et al. have reported such anxiety and distress-related complaints from patients undergoing MRgRT [54]. Adequate consideration and remedies of such concerns is crucial for safe administration of all treatment fractions and avoidance of patient-induced early termination. Patient weight may be a consideration in some cases as some manufacturers have a maximum couch weight limit of 400 lb. (181.4 kg).

Breath-hold imaging and treatment delivery are readily used to overcome challenges posed by respiratory motion. The MRgRT systems from both of the current manufacturers are capable of monitoring two-dimensional target movement during treatment delivery and this real-time tumor imaging is utilized for modulating radiation dose during breathing. Patients should be able to coordinate with the breath-hold instructions so that not only motion artifacts and planning margins are constrained but also treatment can be optimally delivered.

Physiological motion during setup and treatment can adversely impact the overall procedure, as the former results in unrepresentative setup images and the latter forces the team to increase planning margins. Breath-hold techniques are reported to have shorter treatment time [55]. Patient's visual feedback can help facilitate the breath-hold workflows. Also, breath-hold instructions can be provided through audio systems, but visual coaching can be a great help. With the help of equipment such as custom-designed prismatic glasses allowing sight of a computer monitor on the wall behind the MRI bore, patients can actively synchronize the breath-hold and release phases of the respiratory cycle.

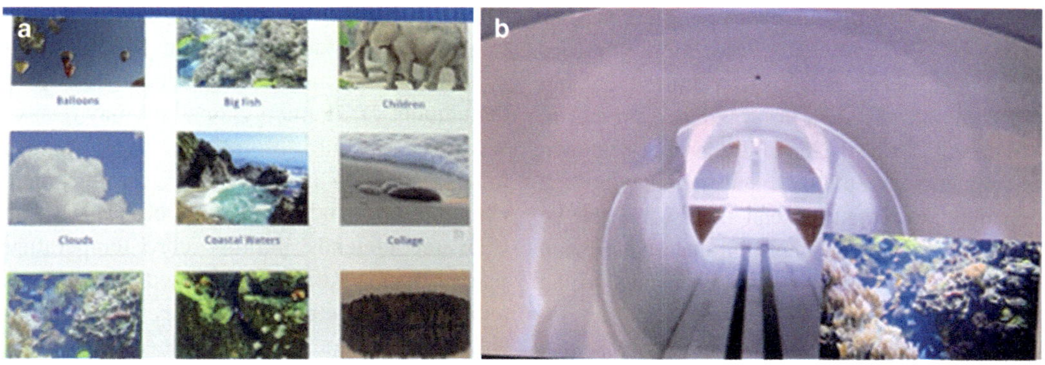


**Fig. 3.5** An example of the resources to provide a soothing environment to the patient during treatment is shown. (**a**) The library of the display includes mountains, oceans, animals, flowers, etc. that a patient can choose. (**b**) Gantry with selected view

## Breathing Fatigue

Modern treatment pathways consider 17 s as the minimum breath-hold duration for adequate imaging, simulation, and reduced treatment duration. This usually involves sites where inter-organ and intra-organ mobility is a challenge such as thoracic and abdominal sites. Also, breath hold can facilitate the delineation of the target with OARs, especially in soft tissue anatomy and sites where organ displacement is a huge concern. In many patients, such a long breath hold is not possible, and hence, it should be tested upfront during patient selection. An optimal breath-hold technique can address the overlap of critical OARs with the target, and with superior soft tissue contrast and real-time gating, not only is treatment duration reduced but dose escalation is also feasible. Coordinated breathing patterns and enhanced visualization and delineation of tumor and healthy tissues facilitates better treatment adaptation, even for cases with complex contours with challenging treatment objectives. These considerations are key to the success of high dose rate single fraction and stereotactic deliveries. So, the benefits of breath hold and motion management should be evaluated during patient selection. This also applies for assessing or screening the cases for online adaptative MRgRT.

## Magnetic Susceptibility

Magnetic fields might temporarily affect various materials. The magnetic susceptibility varies with the material composition and shape, but more importantly with the rotating gantry as in MR-Linac [56–58]. This effect is widely known as magnetic susceptibility, and implants may be impaired due to this; the scan quality may also be affected. Magnetic susceptibility can be reduced or enhanced depending on the sequence selected for the procedure.

Patient safety is of paramount importance and includes many aspects such as awareness of the specific absorption rate, checking hearing protections, ensuring multiple means of communicating with the patient, avoiding loops in coil wires and on patients, and use of medical gowns to avoid any metals in clothing (even underwires in bras can be a hazard). Contrast administration only in the presence of a physician is a good patient safety measure. Patient safety also includes screening the patient, rechecking medical records, and verification of implants for MR safety or compatibility considering the field strength of the scanner. The simple rule here is trust but verify the information. For situations when a patient is unable to communicate or is under anesthesia, protocols should entail extra cautionary measures.

Practicing emergency safety is also equally important. It covers proper signage of emergency response equipment, ascertaining emergency response plans, and training and drilling on removing patients from the couch. There should be an emergency cart with instant access to safety equipment, and monitoring systems should be functional at all times.

## Thermal Effects

When undergoing MR imaging, it is essential to keep in mind that there is a risk of localized heating if patient-created current loops are present. To avoid this, patients should ensure that different parts of their body do not touch during the procedure and that they are not wearing any wet or damp clothing. It is also important to be cautious when using transdermal patches, as they may cause burns to the skin. Modern scanners have safety measures in place to determine the amount of RF energy a patient may receive in terms of specific absorption rate, and clinical limits are set by International Electrotechnical Commission standards (IEC 60601-2-33). The FDA recommends limiting core temperature rise to 1 °C. Patients with medical implants require special attention as they may be at a higher risk for local heating injury. Chances of device failure are higher in MR, and challenges increase as some devices may be compatible with one MR strength and not others. To prevent injury, proper patient preparation guidelines and avoiding noncompliant patient monitoring leads, looping wires, and clothing is important [49].

## Summary

MR-Linac provides a unique opportunity for real-time tumor visualization that can be used for gated therapy. However, not all patients can undergo such treatment. Owing to different anatomical sites with different tissue characteristics in MRI and CT imaging, a proper patient selection protocol should be adopted. Patient selection should be based on magnetic susceptibility, thermal issues, MR safety, claustrophobia, high-Z implants like pacemakers and CIEDs, and most importantly, breathing irregularity and fatigue. Additionally, it is a time-consuming technology with treatment sessions lasting over an hour where patient comfort can be a critical factor. It has been estimated that only 1/3 of the patient population is suitable for this technology [59]. Thus, prudent and proper selection of patients is critical for the success of this technology to provide optimum care to the patients.

## References


1. Supanich M, Siewerdsen J, Fahrig R, Farahani K, Gang GJ, Helm P, et al. AAPM Task Group Report 238: 3D C-arms with volumetric imaging capability. Med Phys. 2023;50(8):e904–945.
2. Ding GX, Alaei P, Curran B, Flynn R, Gossman M, Mackie TR, et al. Image guidance doses delivered during radiotherapy: quantification, management, and reduction: report of the AAPM Therapy Physics Committee Task Group 180. Med Phys. 2018;45(5):e84–99.
3. Yan D, Georg D. Adaptive radiation therapy. Z Med Phys. 2018;28(3):173–4.
4. Murphy MJ, Balter J, Balter S, BenComo JA Jr, Das IJ, Jiang SB, et al. The management of imaging dose during image-guided radiotherapy: report of the AAPM Task Group 75. Med Phys. 2007;34(10):4041–63.
5. Jaffray DA, Das S, Jacobs PM, Jeraj R, Lambin P. How advances in imaging will affect precision radiation oncology. Int J Radiat Oncol Biol Phys. 2018;101(2):292–8.
6. Ling CC, Yorke E, Fuks Z. From IMRT to IGRT: frontierland or neverland? Radiother Oncol. 2006;78(2):119–22.
7. Ghilezan M, Yan D, Liang J, Jaffray D, Wong J, Martinez A. Online image-guided intensity-modulated radiotherapy for prostate cancer: how much improvement can we expect? A theoretical assessment of clinical benefits and potential dose escalation by improving precision and accuracy of radiation delivery. Int J Radiat Oncol Biol Phys. 2004;60(5):1602–10.
8. van Herk M. Errors and margins in radiotherapy. Semin Radiat Oncol. 2004;14(1):52–64.
9. Siewerdsen JH, Jaffray DA. Cone-beam computed tomography with a flat-panel imager: magnitude and effects of x-ray scatter. Med Phys. 2001;28(2):220–31.
10. Mackie TR, Kapatoes J, Ruchala K, Lu W, Wu C, Olivera G, et al. Image guidance for precise conformal radiotherapy. Int J Radiat Oncol Biol Phys. 2003;56(1):89–105.
11. Shirato H, Seppenwoolde Y, Kitamura K, Onimura R, Shimizu S. Intrafractional tumor motion: lung and liver. Semin Radiat Oncol. 2004;14(1):10–8.
12. Dawson LA, Brock KK, Kazanjian S, Fitch D, McGinn CJ, Lawrence TS, et al. The reproducibility of organ position using active breathing control (ABC) during liver radiotherapy. Int J Radiat Oncol Biol Phys. 2001;51(5):1410–21.
13. Wong JW, Sharpe MB, Jaffray DA, Kini VR, Robertson JM, Stromberg JS, et al. The use of active breathing control (ABC) to reduce margin for breathing motion. Int J Radiat Oncol Biol Phys. 1999;44(4):911–9.
14. Xing L, Thorndyke B, Schreibmann E, Yang Y, Li TF, Kim GY, et al. Overview of image-guided radiation therapy. Med Dosim. 2006;31(2):91–112.
15. Hamilton RJ, Sweeney PJ, Pelizzari CA, Yetkin FZ, Holman BL, Garada B, et al. Functional imaging in treatment planning of brain lesions. Int J Radiat Oncol Biol Phys. 1997;37(1):181–8.
16. Njeh CF. Tumor delineation: the weakest link in the search for accuracy in radiotherapy. J Med Phys. 2008;33(4):136–40.
17. Weiss E, Hess CF. The impact of gross tumor volume (GTV) and clinical target volume (CTV) definition on the total accuracy in radiotherapy theoretical aspects and practical experiences. Strahlenther Onkol. 2003;179(1):21–30.
18. Rasch C, Steenbakkers R, van Herk M. Target definition in prostate, head, and neck. Semin Radiat Oncol. 2005;15(3):136–45.
19. Vorwerk H, Beckmann G, Bremer M, Degen M, Dietl B, Fietkau R, et al. The delineation of target volumes for radiotherapy of lung cancer patients. Radiother Oncol. 2009;91(3):455–60.
20. van Mourik AM, Elkhuizen PH, Minkema D, Duppen JC, Dutch Young Boost Study Group, van Vliet-Vroegindeweij C. Multiinstitutional study on target volume delineation variation in breast radiotherapy in the presence of guidelines. Radiother Oncol. 2010;94(3):286–91.

21. Chen AM, Farwell DG, Luu Q, Chen LM, Vijayakumar S, Purdy JA. Marginal misses after postoperative intensity-modulated radiotherapy for head and neck cancer. Int J Radiat Oncol Biol Phys. 2011;80(5):1423–9.
22. Jolicoeur M, Racine ML, Trop I, Hathout L, Nguyen D, Derashodian T, et al. Localization of the surgical bed using supine magnetic resonance and computed tomography scan fusion for planification of breast interstitial brachytherapy. Radiother Oncol. 2011;100(3):480–4.
23. Chang EL, Akyurek S, Avalos T, Rebueno N, Spicer C, Garcia J, et al. Evaluation of peritumoral edema in the delineation of radiotherapy clinical target volumes for glioblastoma. Int J Radiat Oncol Biol Phys. 2007;68(1):144–50.
24. Just M, Rosler HP, Higer HP, Kutzner J, Thelen M. MRI-assisted radiation therapy planning of brain tumors—clinical experiences in 17 patients. Magn Reson Imaging. 1991;9(2):173–7.
25. Khoo VS, Joon DL. New developments in MRI for target volume delineation in radiotherapy. Br J Radiol. 2006;79 Spec No 1:S2–15.
26. Mardor Y, Pfeffer R, Spiegelmann R, Roth Y, Maier SE, Nissim O, et al. Early detection of response to radiation therapy in patients with brain malignancies using conventional and high b-value diffusion-weighted magnetic resonance imaging. J Clin Oncol. 2003;21(6):1094–100.
27. Garcia-Alvarez R, Liney GP, Beavis AW. Repeatability of functional MRI for conformal avoidance radiotherapy planning. J Magn Reson Imaging. 2006;23(2):108–14.
28. Nguyen PL, Aizer A, Assimos DG, D'Amico AV, Frank SJ, Gottschalk AR, et al. ACR appropriateness criteria(R) definitive external-beam irradiation in stage T1 and T2 prostate cancer. Am J Clin Oncol. 2014;37(3):278–88.
29. Debois M, Oyen R, Maes F, Verswijvel G, Gatti G, Bosmans H, et al. The contribution of magnetic resonance imaging to the three-dimensional treatment planning of localized prostate cancer. Int J Radiat Oncol Biol Phys. 1999;45(4):857–65.
30. Chen ME, Troncoso P, Johnston DA, Tang K, Babaian RJ. Optimization of prostate biopsy strategy using computer based analysis. J Urol. 1997;158(6):2168–75.
31. Haider MA, Chung P, Sweet J, Toi A, Jhaveri K, Menard C, et al. Dynamic contrast-enhanced magnetic resonance imaging for localization of recurrent prostate cancer after external beam radiotherapy. Int J Radiat Oncol Biol Phys. 2008;70(2):425–30.
32. McLaughlin PW, Narayana V, Meirovitz A, Troyer S, Roberson PL, Gonda R Jr, et al. Vessel-sparing prostate radiotherapy: dose limitation to critical erectile vascular structures (internal pudendal artery and corpus cavernosum) defined by MRI. Int J Radiat Oncol Biol Phys. 2005;61(1):20–31.
33. Potter R, Dimopoulos J, Georg P, Lang S, Waldhausl C, Wachter-Gerstner N, et al. Clinical impact of MRI assisted dose volume adaptation and dose escalation in brachytherapy of locally advanced cervix cancer. Radiother Oncol. 2007;83(2):148–55.
34. Steenbakkers RJ, Deurloo KE, Nowak PJ, Lebesque JV, van Herk M, Rasch CR. Reduction of dose delivered to the rectum and bulb of the penis using MRI delineation for radiotherapy of the prostate. Int J Radiat Oncol Biol Phys. 2003;57(5):1269–79.
35. Gay HA, Barthold HJ, O'Meara E, Bosch WR, El Naqa I, Al-Lozi R, et al. Pelvic normal tissue contouring guidelines for radiation therapy: a Radiation Therapy Oncology Group consensus panel atlas. Int J Radiat Oncol Biol Phys. 2012;83(3):e353–62.
36. McVeigh PZ, Syed AM, Milosevic M, Fyles A, Haider MA. Diffusion-weighted MRI in cervical cancer. Eur Radiol. 2008;18(5):1058–64.
37. Mayr NA, Wang JZ, Zhang D, Grecula JC, Lo SS, Jaroura D, et al. Longitudinal changes in tumor perfusion pattern during the radiation therapy course and its clinical impact in cervical cancer. Int J Radiat Oncol Biol Phys. 2010;77(2):502–8.
38. Ulin K, Urie MM, Cherlow JM. Results of a multi-institutional benchmark test for cranial CT/MR image registration. Int J Radiat Oncol Biol Phys. 2010;77(5):1584–9.
39. van Herk M, Kooy HM. Automatic three-dimensional correlation of CT-CT, CT-MRI, and CT-SPECT using chamfer matching. Med Phys. 1994;21(7):1163–78.
40. Devic S. MRI simulation for radiotherapy treatment planning. Med Phys. 2012;39(11):6701–11.
41. Kapanen M, Collan J, Beule A, Seppala T, Saarilahti K, Tenhunen M. Commissioning of MRI-only based treatment planning procedure for external beam radiotherapy of prostate. Magn Reson Med. 2013;70(1):127–35.
42. Paulson ES, Erickson B, Schultz C, Allen LX. Comprehensive MRI simulation methodology using a dedicated MRI scanner in radiation oncology for external beam radiation treatment planning. Med Phys. 2015;42(1):28–39.
43. Johnstone E, Wyatt JJ, Henry AM, Short SC, Sebag-Montefiore D, Murray L, et al. Systematic review of synthetic computed tomography generation methodologies for use in magnetic resonance imaging-only radiation therapy. Int J Radiat Oncol Biol Phys. 2018;100(1):199–217.
44. Khoo VS, Padhani AR, Tanner SF, Finnigan DJ, Leach MO, Dearnaley DP. Comparison of MRI with CT for the radiotherapy planning of prostate cancer: a feasibility study. Br J Radiol. 1999;72(858):590–7.
45. Tyagi N, Fontenla S, Zelefsky M, Chong-Ton M, Ostergren K, Shah N, et al. Clinical workflow for MR-only simulation and planning in prostate. Radiat Oncol. 2017;12(1):119.
46. Miften M, Mihailidis D, Kry SF, Reft C, Esquivel C, Farr J, et al. Management of radiotherapy patients with implanted cardiac pacemakers and defibrillators: a report of the AAPM TG-203(dagger). Med Phys. 2019;46(12):e757–e88.

47. Marbach JR, Sontag MR, Van Dyk J, Wolbarst AB. Management of radiation oncology patients with implanted cardiac pacemakers: report of AAPM Task Group No. 34. American Association of Physicists in Medicine. Med Phys. 1994;21(1):85–90.
48. Munn Z, Jordan Z. The effectiveness of interventions to reduce anxiety, claustrophobia, sedation and non-completion rates of patients undergoing high technology medical imaging. JBI Libr Syst Rev. 2012;10(19):1122–85.
49. Levine GN, Gomes AS, Arai AE, Bluemke DA, Flamm SD, Kanal E, et al. Safety of magnetic resonance imaging in patients with cardiovascular devices: an American Heart Association scientific statement from the Committee on Diagnostic and Interventional Cardiac Catheterization, Council on Clinical Cardiology, and the Council on Cardiovascular Radiology and Intervention: endorsed by the American College of Cardiology Foundation, the North American Society for Cardiac Imaging, and the Society for Cardiovascular Magnetic Resonance. Circulation. 2007;116(24):2878–91.
50. Kusumoto FM, Schoenfeld MH, Wilkoff BL, Berul CI, Birgersdotter-Green UM, Carrillo R, et al. 2017 HRS expert consensus statement on cardiovascular implantable electronic device lead management and extraction. Heart Rhythm. 2017;14(12):e503–e51.
51. Bhandiwad AR, Cummings KW, Crowley M, Woodard PK. Cardiovascular magnetic resonance with an MR compatible pacemaker. J Cardiovasc Magn Reson. 2013;15(1):18.
52. Harden SP. MRI conditional pacemakers: the start of a new era. Br J Radiol. 2011;84(1005):773–4.
53. Botman R, Tetar SU, Palacios MA, Slotman BJ, Lagerwaard FJ, Bruynzeel AME. The clinical introduction of MR-guided radiation therapy from a RTT perspective. Clin Transl Radiat Oncol. 2019;18:140–5.
54. Kluter S, Katayama S, Spindeldreier CK, Koerber SA, Major G, Alber M, et al. First prospective clinical evaluation of feasibility and patient acceptance of magnetic resonance-guided radiotherapy in Germany. Strahlenther Onkol. 2020;196(8):691–8.
55. Sahin B, Zoto Mustafayev T, Gungor G, Aydin G, Yapici B, Atalar B, et al. First 500 fractions delivered with a magnetic resonance-guided radiotherapy system: initial experience. Cureus. 2019;11(12):e6457.
56. Wachowicz K, Stanescu T, Thomas SD, Fallone BG. Implications of tissue magnetic susceptibility-related distortion on the rotating magnet in an MR-linac design. Med Phys. 2010;37(4):1714–21.
57. Jackson S, Glitzner M, Tijssen RHN, Raaymakers BW. MRI B (0) homogeneity and geometric distortion with continuous linac gantry rotation on an Elekta Unity MR-linac. Phys Med Biol. 2019;64(12):12NT01.
58. Stanescu T, Wachowicz K, Jaffray DA. Characterization of tissue magnetic susceptibility-induced distortions for MRIgRT. Med Phys. 2012;39(12):7185–93.
59. Moore-Palhares D, Ho L, Lu L, Chugh B, Vesprini D, Karam I, et al. Clinical implementation of magnetic resonance imaging simulation for radiation oncology planning: 5 year experience. Radiat Oncol. 2023;18(1):27.

# Education and Training in MR-Based Radiation Treatment

4

Luca Nicosia, Riccardo Filippo Borgese, Kerryn Brown, and Filippo Alongi

## Introduction

Magnetic resonance image (MR)-guided radiotherapy (MRgRT) is one of the most innovative developments of image-guided radiotherapy (IGRT) [1]. The acquisition of MR images immediately before and after radiotherapy treatments permits daily online planning adaptation with the aim to improve treatment accuracy and tolerability. This new therapeutical approach introduces new specific issues like preparation before device installation, MRI safety program and specific training, development of online adaptive workflow, and treatment delivery. Firstly, a deep knowledge of the available treatment systems and their peculiar differences is necessary to exploit the potential of MRgRT [2].

MRIdian by Viewray (Fig. 4.1) consist of a 0.35T MR scanner with a split magnet design (detail in Chap. 9 of this book) [3]. In this system the force lines of the magnetic field are oriented along the cranio-caudal axis of the patient. The ring gantry is positioned between the two magnets, and it is equipped with a 6 MV flattening filter free linear accelerator (Linac). The Linac is hosted into a dedicated shielded compartment mounted on the gantry. The beam is collimated with a double-focused and stacked multileaf collimator. With this design, the radiation does not pass through the magnet, but it is absorbed by the 5-mm-thick connecting fiberglass placed between the two magnets [4–6].

The original project of the MR-Linac system of Elekta, Unity (Fig. 4.2), was started in 1999 in

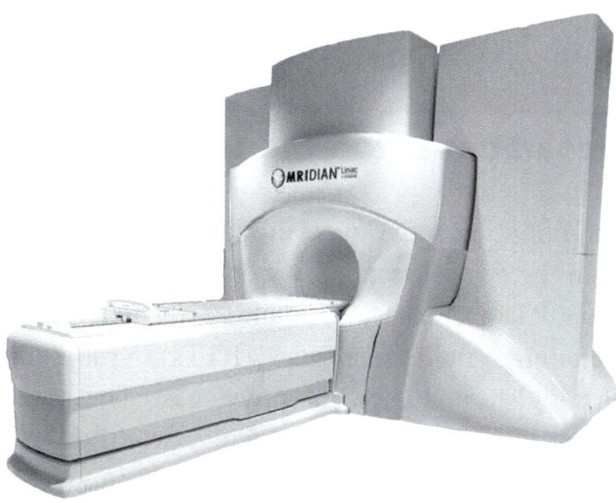


**Fig. 4.1** MRIdian System by Viewray

L. Nicosia (✉) · R. F. Borgese
Department of Advanced Radiation Oncology, IRCCS Sacro Cuore Don Calabria, Cancer Care Center, Negrar di Valpolicella, Italy
e-mail: luca.nicosia@sacrocuore.it; riccardo.borgese@sacrocuore.it

K. Brown
Deparment of Radiation Oncology, ONJ Centre, Austin Hospital, Heidelberg, VIC, Australia
e-mail: Kez.Brown@austin.org.au

F. Alongi
Department of Advanced Radiation Oncology, IRCCS Sacro Cuore Don Calabria, Cancer Care Center, Negrar di Valpolicella, Italy

University of Brescia, Brescia, Italy
e-mail: filippo.alongi@sacrocuore.it


I. J. Das et al. (eds.), *A Practical Guide to MR-Linac*, https://doi.org/10.1007/978-3-031-48165-9_4

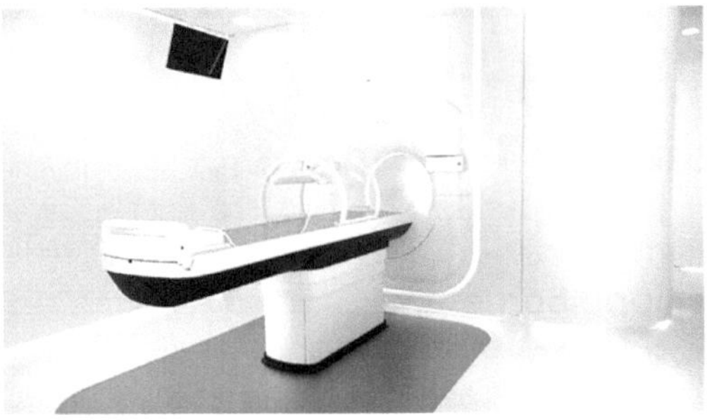

**Fig. 4.2** Unity system by Elekta

UMC Utrecht [7]. The system mounts a 1.5T MRI system with the gantry positioned in the mid-transversal plane around the isocenter. The Linac is a 7 MV flattening filter free [8]. There is an active magnetic shield to magnetically decouple the MRI and the Linac allowing them to operate independently without mutual interaction. Moreover, the lines of magnetic field force are oriented cranio-caudally to the patient [9]. Details of the Elekta Unity system can be found in Chap. 10 of this book.

Both systems have a model within the treatment planning system that considers the impact of magnetic field on the secondary electrons which can be compensated by dedicated optimization strategies [10, 11]. The commissioning and acceptance procedures are similar to those of the traditional RT systems, except for those uniquely linked to the MRgRT aspects. Other than the innovative equipment, one of the most peculiar aspects of this new technology is the possibility of the online adaptive workflow [12].

## Preparation Before Starting

The online adaptive workflow procedure differs from the typical treatment workflow on conventional RT machines and requires dedicated teams, including radiotherapists (RTTs), radiation oncologist, and medical physicists. The three major components of a MR-Linac workflow are (1) MRI safety and imaging acquisition, (2) imaging registration and contouring or recontouring of the organs at risk (OAR), and (3) treatment plan adaptation and verification. The procedures requiring specific training consist of the following:

- MR safety training for all personnel involved in the MR procedures or working in the MR environment with clear guidance, authorization, and responsibilities (i.e., RTTs, radiation oncologist, medical physicists, cleaning staff, visitors).
- Vendor-specific training to learn the different aspects of the online workflow (i.e., alternative workflows in case of system errors) is required to safely administer adaptive plans. Special training might be offered by vendors on dedicated consoles for planning and simulation of adaptive treatment to speed up the learning curve and identify potential challenges and pitfalls. An initial experience with these procedures before clinical implementation is essential.
- Specific training for each role, based on the traditional role and responsibilities including the following: RTTs are a component of the adaptive team and may need dedicated training to improve their skills (i.e., MRI training to ensure patient safety, understand MR sequences, and troubleshoot imaging problems). Radiation oncologists, or dedicated RTTs, must be familiar with the online contouring tools and algorithms and understand the time for contouring edits is limited during online adaptive procedures, since longer treatment sessions might increase treatment uncertainties such as patient fatigue and movement. Guidelines for contouring should be provided to maintain consistency between team members, for example, defining treatment margins or which contours are to be amended during the online adaptive procedures.
- Medical physicists and dosimetrists specialized in treatment planning are responsible for the online treatment planning procedures. Dedicated training should include the knowledge of the different online treatment planning alternatives, pre-planning offline actions, creation of treatment planning templates, and specific plan quality assurance (QA) procedures. The training can be provided in many

ways and can consist of online modules and onsite training sessions. It should include all the different aspects of the respective online workflows such as different solutions for adaptive planning, possible solution in case of errors or system failures, software modules including treatment calendar, MRI console use, tools for data transfer, image registration, contouring and treatment planning, and storage and actionability of data. Dedicated visitation of experienced centers might increase the learning curve on a peer-to-peer basis and offer details of clinical and technical insights.
- Foster close collaboration between medical physicists, MR imaging scientists, and physicians to leverage complementary expertise in order to address those unmet needs related to the clinical use of MR imaging in RT.

Considering the novelty and peculiarity of this new therapeutical approach, it is important that all the team members are familiar with every step of the entire workflow in case of unforeseen events and to rapidly make decisions [13, 14]. As experience develops over time, team roles and responsibilities might be modified according to necessity whereby crossing the traditional boundaries of the respective roles, to enable a prompter and less resource-intensive treatment delivery. Furthermore, due the potential adaptive nature of every treatment session, it is essential that critical online decisions are made swiftly and consistently within the team. To ensure this, close multidisciplinary teamwork is required to define and optimize treatment workflow and thresholds for decision-making and action levels.

The dedicated training must cover all the aspects of the workflow to enable new treatment plans to be created, based on the daily anatomy of target and OARs. Also, dedicated training in MR anatomy should be deemed essential. This new treatment modality has important implication in the typical organization of a radiation oncology department, because of the need of a dedicated team, rapid availability of radiation oncologists, and medical physicists for each treatment fraction, that must be imbued into the usual day activity. In contrast with the typical treatment workflow with conventional radiotherapy, adaptive workflow requires daily recontouring and approval of target and treatment planning, either at the treatment console or remotely.

To reduce potential errors and mitigate risks, a risk analysis for a safe and efficient workflow may be useful [15]. For example, the dedicated multidisciplinary team might implement a process failure mode and effect analysis (P-FMEA) to improve the quality and safety of the treatment and reduce risks [16–18]. This framework comprises a first step in which the clinical team develops a process map for MR-guided adaptive radiotherapy. In a second step, the team identifies possible failure mode and pitfalls and risk probability numbers are assigned based on the probability of occurrence, the severity of the risk, and the detectability. Risk mitigation strategies support the generation of a standardized workflow, clearly defined protocols, and the definition checklist as well as standard operating procedures (SOPs) for a safe and efficient implementation of MRgRT procedures. This adopted model should be tested on volunteers during dry run training sessions. After clinical implementation this model might be improved, as the experience of the team increases [19].

An exhaustive QA guideline for medical physicists was proposed by ESTRO and ASTRO to help commissioning [20].

## MRI-Specific Training and Equipment

Before starting with MRgRT, dedicated SOPs for MRI safety should be written and the staff must be trained according to this, in addition to dedicated training as described above. The following aspects should be addressed:

- Potential hazards and risks (magnetic field risk, radio frequency field risk, cryogen risk, biological effects due to magnetic field, projectile incident risk).
- Acoustic noise (patient compliance, hearing loss).
- Magnet quench (emergency procedure).

- Access restriction (fringe field, safety zones, MR unsafe/conditional/safe, screening of the patient [implants, foreign body metals, pacemaker, or ICD implants], authorized and safety-certified staff). (Additional information can be found in Chap. 3 of this book.)
- Radio frequency (RF) and gradient fields (RF absorption, looped conductors, patient contact with bore or wires).
- Emergencies (guidelines for the staff to ensure a safe evacuation of the patient).

Internally approved guidelines for emergency management must be created. These guidelines must be readily available to all involved staff (i.e., the emergency response team, including physicians and nurses). These guidelines must be consistent with other institutional guidelines on MRI safety as well as national/international guidelines.

Patients must be screened for MRI safety before MR simulation (metal screening, implant screening, pacemaker, ICD screening) and informed about the procedure to familiarize with the equipment (i.e., hearing protection, squeeze bulb alarm) and instructed on how the staff will communicate with them during treatment. All the screening procedures for MR safety are essential to avoid any harm to patient and personnel; therefore, the following indications must be learned by the dedicated team, including also periodic training to refresh them.

All items that are at risk within the MR environment must be removed by the patient before entering the treatment and simulation room. Ferromagnetic detection systems (i.e., entry control systems, handheld metal detectors) can be used additionally to enhance safety, but only after the conventional screening procedure abovementioned and not as a substitute for them. Moreover, for each subsequent fraction, the staff should be informed about any procedure the patient may have had between the last and the current fraction [21].

Regarding positioning equipment, not all the available patient setup devices have been tested or designed for MR use. Therefore, it is critical to ensure the safety of any material that is implemented into the MR-Linac. It is recommended to differentiate between (1) ferromagnetic safety, (2) imaging artefact assessment, (3) proper coil handling and usage, and (4) degree of dose attenuation.

For ferromagnetic safety, vendors should always provide a written confirmation regarding MR-environment compatibility. Nevertheless, as part of risk assessment and QA management strategy, all equipment must be tested on site and any ferromagnetic properties should be excluded. Phantom measurements should be performed to assess both susceptibility and distortion artifacts and to quantify their impact on image quality and spatial accuracy and to evaluate possible dose attenuation. Finally, proper coil placement needs to be tested to ensure optimal distance and location to the area of interest and to avoid imaging deterioration due to insufficient signal generation and detection.

The knowledge of acquisition, processing, and interpretation of MR images and the sequences is a prerequisite for MRgRT [22–24]. Considering the variety in the educational programs for RTTs, radiation oncologists, medical physicists, and dosimetrists across different countries, it is difficult to define a complete core of required skills and training pathways. Nevertheless, some guidance can be derived from the initial experience that is currently being made with the MR-Linac implementation. During the clinical implementation phase, the dedicated team to MRgRT should include RTTs, radiation oncologists, and medical physicists with a specialized training in MR technology. This core team should have received MR training as part of their education course or national certification program or visited specific courses that provided the basic knowledge of MR imaging. A refresher course or an internship in a radiology department can be useful for all members of the team. In addition, courses provided by vendors, universities (i.e., Utrecht), ESTRO, the Institute of Cancer Research, or American College of Radiology (ACR) or by site visits and peer-to-peer training at experienced MR-Linac departments provide additional opportunities to improve skills and experience with MR-Linac technology and adaptive workflow [25–27].

Also, a RT-trained radiographer might be involved in the treatment process and not only in positioning and imaging aspects. With dedicated training is it possible to achieve contouring consistency comparable to that of clinicians. This can help reduce the workload of the team and to acquire mutual skills to speed up the learning curve of the MR team [28].

## Training and Education During the Delivery Phase

The MR-Linac provides the fundamental ability to adapt treatment plans on a daily basis. Is it known that anatomy changes, moment by moment due to breathing motion, peristalsis, or patient movement, as well tumor change over the course of treatment. The use of an MR-Linac may prevent this problem, as OAR motion and morphological changes in the target can be seen in real time. This advantage may indicate the use of real-time biological targeting based on biomarker or differential response, resistance, and progression or even hypoxia and introduces the possibility to translate the radiomics concepts into the clinical practice in radiation oncology [29–31].

The typical treatment workflow on a MR-Linac is constituted by patient positioning, preadaptive imaging acquisition, recontouring of OARs and/or target or superimposition of previous contour, treatment plan adaptation or shifting, plan quality assurance, pre-treatment imaging verification, treatment delivery with the possibility of online delivery verification (i.e., cine-MRI or 3D MRI) or motion management (i.e., gating), and eventual post-treatment imaging.

The goal of daily plan adaptation can be signified by improved target coverage, OAR sparing, or both, depending on whether the tumor coverage or the OAR sparing requires primary consideration during the specific treatment session. The clinical benefit of daily adaptation greatly depends on the tumor site and can be of different importance, for example, head and neck tumors vs pancreatic tumors [32–35]. It is recommended to define internally the minimum requirements for treatment adaptation or which threshold values, or action levels, should be considered. These considerations depend on the tumor site and the clinical relevance of the daily anatomy change, for example, the variation of the tumor and OARs, the type of dose distribution (homogeneous, heterogeneous, SBRT treatments), total dose, and fractionation. As a general rule, the higher the dose per fraction and the proximity to OARs, the smaller the therapeutic window for anatomical variation. These minimum requirements can be implemented in different ways into the clinical practice [34, 36]. One approach is to predefine the type of plan adaptation for each treatment. It is helpful to implement clear dose constraints and target coverage goals to define if a plan is acceptable for delivery (green light), within a range of accepted variability (yellow light), or is an unacceptable parameter (red light) with different levels of action outlined for correction and decision-making. Furthermore, it might be useful to set a time frame for online contouring and a maximum number of optimization cycles to avoid long adaptation times.

Future deformable dose accumulation in combination with daily adaptation can be expected to refine the current knowledge on toxicity parameters, because dosimetry will be based on actual delivered, rather than planned, radiation dose. Also, the possibility to obtain synthetic CTs from MR images might increase the accuracy of parameter calculations. The two available MRgRT systems present differences, but both allow the assessment of the anatomy on MRI prior to shifting or replanning. This assessment can be with respect to reference plan dosimetry (MRIdian) or reference anatomy (Unity). When changes in OARs or targets appear, it is mandatory to replan, especially for ultra hypofractionated regimens, with minimal PTV margins or when steep dose gradients are used.

The type of adaptive workflow depends on the complexity and variation of the daily imaging compared to the reference and can range from a simple shift in MLC position to a full reoptimization, based on patients' anatomy [37, 38].

Similarly, OARs and target contours can be either rigidly translated from the reference plan or fully recontoured, based on daily anatomical differences.

Another important issue is interobserver contouring variation. To account for this, each center should assess the accuracy of its recontouring process against a reference gold standard and implement a QA program for online contouring procedures. In order to make the recontouring process faster, consistent, and compatible with the daily clinical routine, several centers have adopted a workflow with a partial recontouring of OAR only in the proximity of PTV (i.e., the 2–3 cm surrounding area), since the recontouring is performed with the patient waiting in the treatment position [37, 39]. In this case, OAR constraints have to be adapted to this concept (use of absolute volume in cc) for a fast and reliable plan evaluation.

Some pre-treatment procedures can be time consuming, exceeding 15 min. In this case a pre-delivery, or verification, MRI can be acquired to ensure the correct positioning of the planned isocenter. This image can be used to further optimize patient positioning with small couch or plan shifts, or conduct a full reoptimization in the case of major and unacceptable target/OAR changes [40].

Each center should define the level of action and threshold in the case of target movement outside of the high-dose area, which is dependent on, among the other factors, the used PTV margins. The MRIdian system allows for respiratory-gated treatments using breath-hold techniques with automated real-time anatomy structure tracking for lesions in the abdomen and thorax. Several visual feedback modalities for patients have been implemented in the clinical practice (i.e., MRI-compatible monitors in combination with an adjustable mirror in the gantry or prism glasses) [41–44]. These modalities facilitate voluntary breath-hold delivery at the appropriate respiratory phase. Preliminary experience shows that this treatment modality is well tolerated or appreciated by patients and can reduce the duty cycle between 67% and 87% in adrenal and pancreatic tumor-gated MRgRT [41].

Presently, the Unity system is implementing a gating system to account for target movement during the delivery phase. Moreover, this system provides different MRI sequences that can be acquired during beam on phase [45]. These options enable the reconstruction of the delivered dose to be compared with the intended dose delivery for more precise treatment evaluation and/or dose response assessment. This system can be used for offline adaptation, but in a next phase can be used for intra-fraction dose-guided adaptation [46, 47].

## Training and Education Procedures for MR-Specific Patient Selection and Indications for MR-Linac

Patient selection is a crucial point in MRgRT which is provided in detail in Chap. 3. Selection criteria and training are a fundamental part to correctly homogeneously address patients to MR-Linac treatments. Among the most important criteria are patient characteristics, target volume characteristics, and type. Since this is still a novel technology, it is not feasible to address all the possible diseases amenable for treatment on MR-Linac; however, the inclusion of patients within a prospective protocol is suggested. The evaluation of patient's allocation should be based according to the principles of health technology assessment for cost-benefit analysis and the principles of evidence-based medicine. There are several ongoing clinical trials on MRgRT, as well as published data. However, the acquisition of strong evidence through prospective multicenter cohorts (i.e., MOMENTUM study) is of outmost importance.

All patients should be screened for MRI compatibility, according to the adopted guidelines [48]. Patients should be classified as physically incompatible (i.e., non-MRI-conditional pacemaker carriers, severe obesity), clinically incompatible (i.e., severe psychiatric disorder, severe claustrophobia, inability to understand the instructions), borderline compatible (i.e., mild claustrophobia), or fully compatible for MRgRT. Patients unfit for MRgRT should be referred to standard RT, while supportive intervention (i.e., psychological intervention, anesthesia, pharmacological or supportive technique such as music or aromatherapy) could be implemented for borderline compatible patients [48–51].

Other than physical compatibility, physicians should evaluate the general clinical status of the patient, especially regarding the balance between the compliance degree with MRgRT and the expected benefit from it. In fact, treatment slots on a MR-Linac are generally longer than that on a conventional RT in a range between 20 and 60 min [39, 52]. For this reason, elderly or frailty does not represent an automatic exclusion criterion, but a scoring system can be used to support the most suitable treatment modality [53]. Patients with severe obesity or cachexia (body mass index >40 or weight <40 kg) should be evaluated on a case-by-case basis, especially considering the possibility of having artifacts, patient heating, and bore size issue (gantry diameter <70 cm) [6, 54].

The second selection criterion is the target characteristics and their relation with the surrounding normal tissues. The MR-Linac offers higher soft tissue contrast visibility, which may allow safe and accurate dose administration in scenarios with OARs in close proximity to the target. Published clinical data is showing that SBRT might be safely administered to prostate, lymph node, and pancreas sites. In particular, early comparative studies in prostate settings are reporting a lower incidence of side effects, while in the case of pancreas SBRT the rate of grade 3 or higher side effects is very limited using both available technologies [55].

Considering the evidence, the MR-Linac might be best exploited in those situations where CT density is homogeneous and the ability to distinguish the tissue is reduced (i.e., head and neck, upper abdomen) and the higher soft tissue contrast provided by MRI can overcome this problem. A typical case is represented by prostate cancer, where MRI might reduce the daily uncertainties, compared to cone-beam CT, by a more accurate definition of the prostate-rectum interface and the correct identification of the prostate apex and penile bulb [56, 57].

Also, moving targets might benefit from MRgRT, especially when they are close to OARs. In fact with the daily adaptation it is possible to account not only for OAR position, but also follow the tumor using motion management tools and, when available, gating systems [58–60]. This can be of particular interest in lung and pancreatic tumors, liver and kidney lesions, head and neck tumors, oligometastases, lymph nodes, and reirradiation. Moreover, MRgRT is indicated in all the cases where MRI can provide additional information (i.e., tumor shrinkage, radiomics, early toxicity onset).

## Backup Solutions

Considering the elevated cost of an MR-Linac, it is unlikely that a single department might have more than one system. Therefore, since a direct transfer from an MR-Linac to a conventional Linac is not easy to plan, it is important to identify alternative strategies to avoid unnecessary treatment interruptions. The best solution might be represented by a collaboration with nearby centers harboring the same technology with the aim to ensure treatment continuation with the same standards, especially in those situations when long treatment interruption might affect the clinical outcome (i.e., head and neck tumors). Another solution is represented by planning an alternative plan on a conventional Linac, using the same patient setup and the simulation CT used for electron density, or a new simulation CT. This last solution requires new plan summation and dose accumulation modalities that restrict the adoption of this modality as routine.

## References

1. Landry G, Corradini S, Belka C. Magnetic resonance-guided radiation therapy: the beginning of a new era. Radiat Oncol. 2020;15:163.
2. Corradini S, Alongi F, Andratschke N, et al. MR-guidance in clinical reality: current treatment challenges and future perspectives. Radiat Oncol. 2019;14:92.
3. Mutic S, Dempsey JF. The ViewRay system: magnetic resonance-guided and controlled radiotherapy. Semin Radiat Oncol. 2014;24:196–9.

4. Raaymakers BW, Jürgenliemk-Schulz IM, Bol GH, et al. First patients treated with a 1.5 T MRI-Linac: clinical proof of concept of a high-precision, high-field MRI guided radiotherapy treatment. Phys Med Biol. 2017;62(23):L41–50. Published 2017 Nov 14.
5. Acharya S, Fischer-Valuck BW, Kashani R, et al. Online magnetic resonance image guided adaptive radiation therapy: first clinical applications [published correction appears in Int J Radiat Oncol biol Phys. 2016 Sep 1;96(1):243]. Int J Radiat Oncol Biol Phys. 2016;94(2):394–403.
6. Klüter S. Technical design and concept of a 0.35 T MR-Linac. Clin Transl Radiat Oncol. 2019;18:98–101.
7. Lagendijk JJW, Raaymakers BW, Raaijmakers AJE, et al. MRI/linac integration. Radiother Oncol. 2008;86:25–9.
8. Raaymakers BW, Lagendijk JJ, Overweg J, et al. Integrating a 1.5 T MRI scanner with a 6 MV accelerator: proof of concept. Phys Med Biol. 2009;54(12):N229–37.
9. Raaijmakers AJE, Hårdemark B, Raaymakers BW, et al. Dose optimization for the MRI-accelerator: IMRT in the presence of a magnetic field. Phys Med Biol. 2007;52:7045–54.
10. Woodings SJ, Bluemink JJ, de Vries JHW, et al. Beam characterisation of the 1.5 T MRI-linac. Phys Med Biol. 2018;63:85015.
11. Park JM, Park S-Y, Wu H-G, Kim J. Commissioning experience of tri-Cobalt-60 MRI-guided radiation therapy system. Prog Med Phys. 2015;26:193.
12. Winkel D, Bol GH, Kroon PS, van Asselen B, et al. Adaptive radiotherapy: the Elekta Unity MR-Linac concept. Clin Transl Radiat Oncol. 2019;18:54–9.
13. McNair HA, Wiseman T, Joyce E, et al. International survey; current practice in on-line adaptive radiotherapy (ART) delivered using magnetic resonance image (MRI) guidance. Tech Innov Patient Support Radiat Oncol. 2020;16:1–9.
14. McGee KP, Tyagi N, Bayouth JE, et al. Findings of the AAPM Ad Hoc committee on magnetic resonance imaging in radiation therapy: unmet needs, opportunities, and recommendations [published correction appears in Med Phys. 2023 Mar;50(3):1943]. Med Phys. 2021;48(8):4523–31.
15. Klüter S, Schrenk O, Renkamp CK, et al. A practical implementation of risk management for the clinical introduction of online adaptive Magnetic Resonance-guided radiotherapy. Phys Imaging Radiat Oncol. 2021;17:53–7.
16. Younge KC, Wang Y, Thompson J, et al. Practical implementation of failure mode and effects analysis for safety and efficiency in stereotactic radiosurgery. Int J Radiat Oncol Biol Phys. 2015;91:1003–8.
17. Huq MS, Fraass BA, Dunscombe PB, et al. A method for evaluating quality assurance needs in radiation therapy. Int J Radiat Oncol Biol Phys. 2008;71:S170–3.
18. Noel CE, Santanam L, Parikh PJ, et al. Process-based quality management for clinical implementation of adaptive radiotherapy. Med Phys. 2014;41:81717.
19. Botman R, Tetar SU, Palacios MA, et al. The clinical introduction of MR-guided radiation therapy from a RTT perspective. Clin Transl Radiat Oncol. 2019;18:140–5.
20. Tanadini-Lang S, Budgell G, Bohoudi O, et al. An ESTRO-ACROP guideline on quality assurance and medical physics commissioning of online MRI guided radiotherapy systems based on a consensus expert opinion. Radiother Oncol. 2023;181:109504.
21. Green O, Henke LE, Parikh P, et al. Practical implications of ferromagnetic artifacts in low-field MRI-guided radiotherapy. Cureus. 2018;10:e2359.
22. van der Heide UA, Frantzen-Steneker M, Astreinidou E, et al. MRI basics for radiation oncologists. Clin Transl Radiat Oncol. 2019;18:74–9.
23. Jonsson J, Nyholm T, Söderkvist K. The rationale for MR-only treatment planning for external radiotherapy. Clin Transl Radiat Oncol. 2019;18:60–5.
24. Kurz C, Buizza G, Landry G, et al. Medical physics challenges in clinical MR-guided radiotherapy. Radiat Oncol. 2020;15:1–16.
25. ESTRO. In-room MRI-guided RT. n.d. https://www.estro.org/Courses/In-room-MRI-guided-RT.
26. Utrecht U. MRI in radiotherapy course. n.d. https://www.umcutrecht.nl/nl/ziekenhuis/mri-in-radiotherapy-course.
27. Research TI of C. Magnetic resonance image guided radiotherapy (MRIgRT). n.d. http://www.icr.ac.uk/studying-and-training/opportunities-for-clinicians/radiotherapy-and-imaging-training-courses/magnetic-resonance-imageguided-radiotherapy-(mrigrt).
28. Adair Smith G, Dunlop A, Alexander SE, Barnes H, Casey F, Chick J, Gunapala R, Herbert T, Lawes R, Mason SA, Mitchell A, Mohajer J, Murray J, Nill S, Patel P, Pathmanathan A, Sritharan K, Sundahl N, Tree AC, Westley R, Williams B, McNair HA. Evaluation of therapeutic radiographer contouring for magnetic resonance image guided online adaptive prostate radiotherapy. Radiother Oncol. 2023;180:109457.
29. Boldrini L, Cusumano D, Chiloiro G, et al. Delta radiomics for rectal cancer response prediction with hybrid 0.35 T magnetic resonance-guided radiotherapy (MRgRT): a hypothesis-generating study for an innovative personalized medicine approach. Radiol Med. 2019;124:145.
30. Cusumano D, Boldrini L, Yadav P, et al. External validation of early regression index (ERI(TCP)) as predictor of pathologic complete response in rectal cancer using magnetic resonance-guided radiation therapy. Int J Radiat Oncol Biol Phys. 2020;108:1347.
31. Bibault J-E, Xing L, Giraud P, et al. Radiomics: a primer for the radiation oncologist. Cancer Radiother. 2020;24:403–10.
32. Finazzi T, Haasbeek CJA, Spoelstra FOB, et al. Clinical outcomes of stereotactic MR-guided adaptive radiation therapy for high-risk lung tumors. Int J Radiat Oncol Biol Phys. 2020;107:270–8.

33. Finazzi T, Palacios MA, Spoelstra FOB, et al. Role of on-table plan adaptation in MR-guided ablative radiation therapy for central lung tumors. Int J Radiat Oncol Biol Phys. 2019;104:933–41.
34. Finazzi T, Palacios MA, Haasbeek CJA, et al. Stereotactic MR-guided adaptive radiation therapy for peripheral lung tumors. Radiother Oncol. 2020;144:46–52.
35. Boldrini L, Cusumano D, Cellini F, et al. Online adaptive magnetic resonance guided radiotherapy for pancreatic cancer: state of the art, pearls and pitfalls. Radiat Oncol. 2019;14:71.
36. Winkel D, Bol GH, Werensteijn-Honingh AM, et al. Evaluation of plan adaptation strategies for stereotactic radiotherapy of lymph node oligometastases using online magnetic resonance image guidance. Phys Imaging Radiat Oncol. 2019;9:58–64.
37. Bohoudi O, Bruynzeel AME, Senan S, et al. Fast and robust online adaptive planning in stereotactic MR-guided adaptive radiation therapy (SMART) for pancreatic cancer. Radiother Oncol. 2017;125:439–44.
38. Kim J, Park JM, Choi CH, et al. Retrospective study comparing MR-guided radiation therapy (MRgRT) setup strategies for prostate treatment: repositioning vs. replanning. Radiat Oncol. 2019;14:139.
39. Tetar SU, Bruynzeel AME, Lagerwaard FJ, et al. Clinical implementation of magnetic resonance imaging guided adaptive radiotherapy for localized prostate cancer. Phys Imaging Radiat Oncol. 2019;9:69–76.
40. van Timmeren JE, Chamberlain M, Krayenbuehl J, et al. Treatment plan quality during online adaptive re-planning. Radiat Oncol. 2020;15:1–11.
41. van Sörnsen de Koste JR, Palacios MA, Bruynzeel AME, et al. MR-guided gated stereotactic radiation therapy delivery for lung, adrenal, and pancreatic tumors: a geometric analysis. Int J Radiat Oncol Biol Phys. 2018;102:858–66.
42. Green OL, Rankine LJ, Cai B, et al. First clinical implementation of real-time, real anatomy tracking and radiation beam control. Med Phys. 2018;45:3728–40.
43. Tetar S, Bruynzeel A, Bakker R, et al. Patient reported outcome measurements on the tolerance of magnetic resonance imaging-guided radiation therapy. Cureus. 2018;10:e2236.
44. Klüter S, Katayama S, Spindeldreier CK, et al. First prospective clinical evaluation of feasibility and patient acceptance of magnetic resonance-guided radiotherapy in Germany. Strahlenther Onkol. 2020;196:691–8.
45. Jackson S, Glitzner M, Tijssen RHN, Raaymakers BW. MRI B 0 homogeneity and geometric distortion with continuous linac gantry rotation on an Elekta Unity MR-Linac. Phys Med Biol. 2019;64(12):12NT01.
46. Menten MJ, Mohajer JK, Nilawar R, et al. Automatic reconstruction of the delivered dose of the day using MR-Linac treatment log files and online MR imaging. Radiother Oncol. 2020;145:88–94.
47. Kontaxis C, de Muinck Keizer DM, et al. Delivered dose quantification in prostate radiotherapy using online 3D cine imaging and treatment log files on a combined 1.5T magnetic resonance imaging and linear accelerator system. Phys Imaging Radiat Oncol. 2020;15:23–9.
48. Sammet S. Magnetic resonance safety. Abdom Radiol. 2016;41:444–51.
49. Chadderdon AL, Carns DR, Pudalov LR, et al. Underlying mechanisms of psychological interventions in magnetic resonance imaging and image-guided radiology procedures. Top Magn Reson Imaging. 2020;29:157–63.
50. Walworth DD. Effect of live music therapy for patients undergoing magnetic resonance imaging. J Music Ther. 2010;47:335–50.
51. Schellhammer F, Ostermann T, Krüger G, et al. Good scent in MRI: can scent management optimize patient tolerance? Acta Radiol. 2013;54:795–9.
52. Werensteijn-Honingh AM, Kroon PS, Winkel D, et al. Feasibility of stereotactic radiotherapy using a 1.5 T MR-Linac: multifraction treatment of pelvic lymph node oligometastases. Radiother Oncol. 2019;134:50–4.
53. Boldrini L, Colloca GF, Villani E, et al. Magnetic resonance-guided radiotherapy feasibility in elderly cancer patients: proposal of the MASTER scoring system. Tumori. 2021;107(1):26–31.
54. Graves SA, Snyder JE, Boczkowski A, et al. Commissioning and performance evaluation of RadCalc for the Elekta unity MRI-linac. J Appl Clin Med Phys. 2019;20:54–62.
55. Corradini S, Alongi F, Andratschke N, et al. MRguidance in clinical reality: current treatment challenges and future perspectives. Radiat Oncol. 2019;14:1–12.
56. Alongi F, Rigo M, Figlia V, et al. 1.5 T MR guided and daily adapted SBRT for prostate cancer: feasibility, preliminary clinical tolerability, quality of life and patient-reported outcomes during treatment. Radiat Oncol. 2020;15:69.
57. Mazzola R, Figlia V, Rigo M, et al. Feasibility and safety of 1.5 T MR-guided and daily adapted abdominal-pelvic SBRT for elderly cancer patients: geriatric assessment tools and preliminary patientreported outcomes. J Cancer Res Clin Oncol. 2020;146:2379.
58. Park JM, Wu H-G, Kim HJ, et al. Comparison of treatment plans between IMRT with MR-Linac and VMAT for lung SABR. Radiat Oncol. 2019;14:105.
59. Henke L, Kashani R, Robinson C, et al. Phase I trial of stereotactic MR-guided online adaptive radiation therapy (SMART) for the treatment of oligometastatic or unresectable primary malignancies of the abdomen. Radiother Oncol. 2018;126:519–26.
60. Witt JS, Rosenberg SA, Bassetti MF. MRI-guided adaptive radiotherapy for liver tumours: visualising the future. Lancet Oncol. 2020;21:e74–82.

# 5 An Introduction to MR Sequences Used in Oncology with Application to MR-Linac

Hersh Sagreiya, Karthik Sundaram, Jooyoung (James) Sohn, and Poonam Yadav

## Introduction

Magnetic resonance imaging (MRI) is a powerful imaging modality that has several advantages over CT. Unlike CT, MRI does not require the use of ionizing radiation. Radiation exposure is of particular concern for children and young adults. They are more sensitive to the stochastic effects of ionizing radiation due to rapid cell division related to growth and development, their long remaining lifespan allows more time to experience the deleterious effects of radiation, and children could potentially receive a higher effective radiation dose if optimized pediatric protocols are not applied [1]. In addition, cumulative radiation dose also becomes more of a concern if serial imaging is required, particularly in otherwise healthy patients.

MRI also provides the ability to image using various sequences, which highlight different aspects of anatomy. For instance, T2-weighted sequences can highlight areas of fluid or edema, fat saturation images can establish the presence of fat in tissues, and contrast-enhanced sequences can aid in identifying tumors or abscesses. MRI is commonly used as a problem-solving modality, and it is frequently recommended as a follow-up when there are indeterminate findings on CT or ultrasound. In fact, the use of multiple sequences can help to narrow the differential diagnosis to a specific entity. In addition, MRI is able to provide superior soft tissue contrast. In this example, sagittal T1, T2, and T1 postcontrast images of the uterus show excellent soft tissue differentiation between the uterus and endometrium (Fig. 5.1a–c). This highlights structures within the uterus, such as Nabothian cysts, the surrounding fat, and nearby structures (bladder, bowel, spine, and musculature). Moreover, different sequences highlight different aspects of the anatomy, such as the endometrium. This stands in comparison to CT, which only shows one view of the anatomy (Fig. 5.1d). While ultrasound can also show similar anatomy, the relationship between structures is not as clearly demarcated, and there is a potential issue of operator dependence, with image quality related to the level of training of the sonographer (Fig. 5.1e). The improved soft tissue contrast of MRI can potentially be important in certain scenarios. For instance, pancreatic ductal adenocarcinoma can occasionally appear as isodense (similar in density) to the pancreatic parenchyma on CT, with

H. Sagreiya (✉) · K. Sundaram
Hospital of the University of Pennsylvania, Department of Radiology, Division of Abdominal Imaging, Philadelphia, PA, USA
e-mail: Hersh.Sagreiya@Pennmedicine.upenn.edu; Karthik.Sundaram@Pennmedicine.upenn.edu

J. (J) Sohn · P. Yadav
Northwestern University Feinberg School of Medicine, Department of Radiation Oncology, Chicago, IL, USA
e-mail: jooyoung.sohn@nm.org; poonam.yadav@nm.org



I. J. Das et al. (eds.), *A Practical Guide to MR-Linac*, https://doi.org/10.1007/978-3-031-48165-9_5

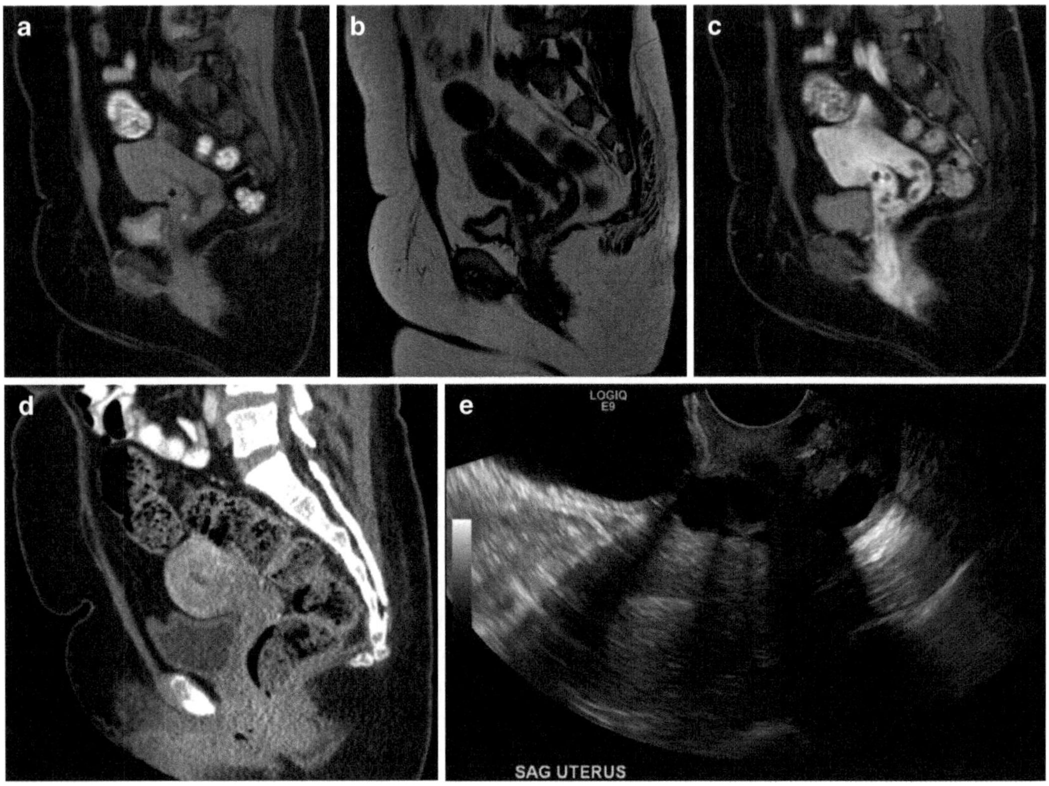


**Fig. 5.1** Differences in soft tissue differentiation between MRI, CT, and ultrasound for the same patient. MRI shows better soft tissue differentiation than CT and can highlight specific soft tissue characteristics through different sequences. Ultrasound highlights different aspects of the image. (**a**) T1 pre-contrast. (**b**) T2. (**c**) T1 postcontrast. (**d**) CT pelvis with contrast. (**e**) Ultrasound pelvis

the only indication being a dilated pancreatic duct if there is resultant obstruction [2]. However, the use of multiple MRI sequences (T1, T2, diffusion, contrast enhanced) will typically readily show the tumor in equivocal CT cases.

MRI does have a few potential disadvantages. First of all, MRI is typically more expensive, time-consuming, and less readily available than CT [3]. However, there has been the development of abbreviated MR protocols, often targeted to a particular indication, which eliminate unnecessary sequences that provide redundant information, and are hence able to decrease imaging time, interpretation time, and ultimately imaging cost [4]. In addition, although certain findings are more obvious on CT, such as calcifications, calcification can be appreciated on MRI [5]. For instance, phase information from susceptibility-weighted imaging can help to discriminate blood products from calcifications [6]. Moreover, CT is typically preferred over MRI for generalized oncologic staging, with MRI typically reserved for further characterization of suspicious findings. However, MRI is particularly useful for local staging for certain cancers, such as prostate, rectal, and gynecologic cancers, due to its superior soft tissue contrast. Whole body MRI is currently used for cancer screening in patients with syndromes that predispose them to developing cancer, including Li-Fraumeni syndrome, constitutional mismatch repair deficiency, and hereditary paraganglioma-pheochromocytoma syndromes [7]. These patients are at high risk for developing tumors, and a major advantage of MRI in this context is the lack of ionizing radiation, which facilitates serial imaging. Along the same lines, while whole body MRI screening of asymptomatic, low-risk individuals has been postulated, there are potential concerns regarding discovering indeterminate or false-positive findings, with potentially deleterious psychological effects for the patient, as well as expensive follow-up [8]. Finally, MRI is also used for screening in specific scenarios, such as for the development of hepatocellular carcinoma in patients with cirrhosis, for which it may have better performance than ultrasound [9].

## Basics of MR Physics

MRI takes advantage of magnetically active nuclei. In order for nuclei to be magnetically active, they must have a net charge. Magnetic resonance imaging typically takes advantage of the hydrogen ion, which is widely available throughout the human body as a fundamental molecular building block. These nuclei have an intrinsic property, conceptually thought of as spin, conceived as the nucleus spinning around its central axis; although the nucleus does not spin in the classical sense, it does have a magnetic moment, creating a local magnetic field with a north and south pole, and the description of this dipolar magnet from quantum mechanics is analogous to the description of spinning objects from classical mechanics [10]. Without a magnetic field, there is a random alignment of the magnetic poles of protons, and there is no net magnetization vector (Fig. 5.2). However, in the presence of a magnetic field ($B_0$), the protons align both parallel and antiparallel to the magnetic field. As more protons align parallel to rather than antiparallel to the magnetic field, there is a net magnetization vector.

In MRI, the strength of the magnetic field is measured in Tesla (T). While 1.5T and 3T are commonly used in clinical scenarios, research indications can use even higher strength magnets (7T and beyond). Greater magnetic field strength increases the signal-to-noise ratio. In the presence of a magnetic field, protons precess or spin along the direction of magnetization (the axis of $B_0$). The angular momentum of the precession of protons in the presence of a magnetic field is related to the strength of the magnetic field by the Larmor equation:

$$\omega = \gamma B_0$$

**Fig. 5.2** (Left) Random alignment of protons without a magnetic field. (Right) Parallel and antiparallel alignment of protons within an external magnetic field ($B_0$). Because more protons are aligned parallel (yellow) rather than antiparallel to the magnetic field (green), there is a net magnetization vector (NMV)

In this equation, $\omega$ is the angular momentum (which is related to the frequency of precession), $\gamma$ is the gyromagnetic ratio, and $B_0$ is the strength of the magnetic field.

A current running through a solenoid or a large coil generates a magnetic field [11]. In order to generate an MR image, a transmit coil must first transmit a radiofrequency (RF) excitation pulse. Protons that are precessing at the frequency of the RF pulse will become excited and go from a low to a high energy state. In doing so, the net magnetization vector tilts away from the axis of the magnetic field and into the transverse plane (Fig. 5.3). As the protons relax and return to their previous state, they will emit an RF pulse. This is what is captured as the MR image. During this process, the net magnetization vector returns to the direction of $B_0$.

In order to localize a particular point in 3D space, magnetic gradients must be applied in the x, y, and z directions. This requires gradient coils in each orthogonal direction that gradually vary the magnetic field. As a result, for any given gradient, the strength of the magnetic field is a function of location along that axis. Since the frequency of precession is directly proportional to the strength of the magnetic field, the set of frequencies of the transmitted RF pulse will excite protons in a particular location in three-dimensional space, and the subsequent relaxation of these protons will emit an RF pulse that eventually leads to the MR image.

The first gradient that is applied along the z-axis, or the axis along $B_0$, is the slice select gradient. This gradient selects a particular axial slice in 3D space. Subsequent gradients localize data

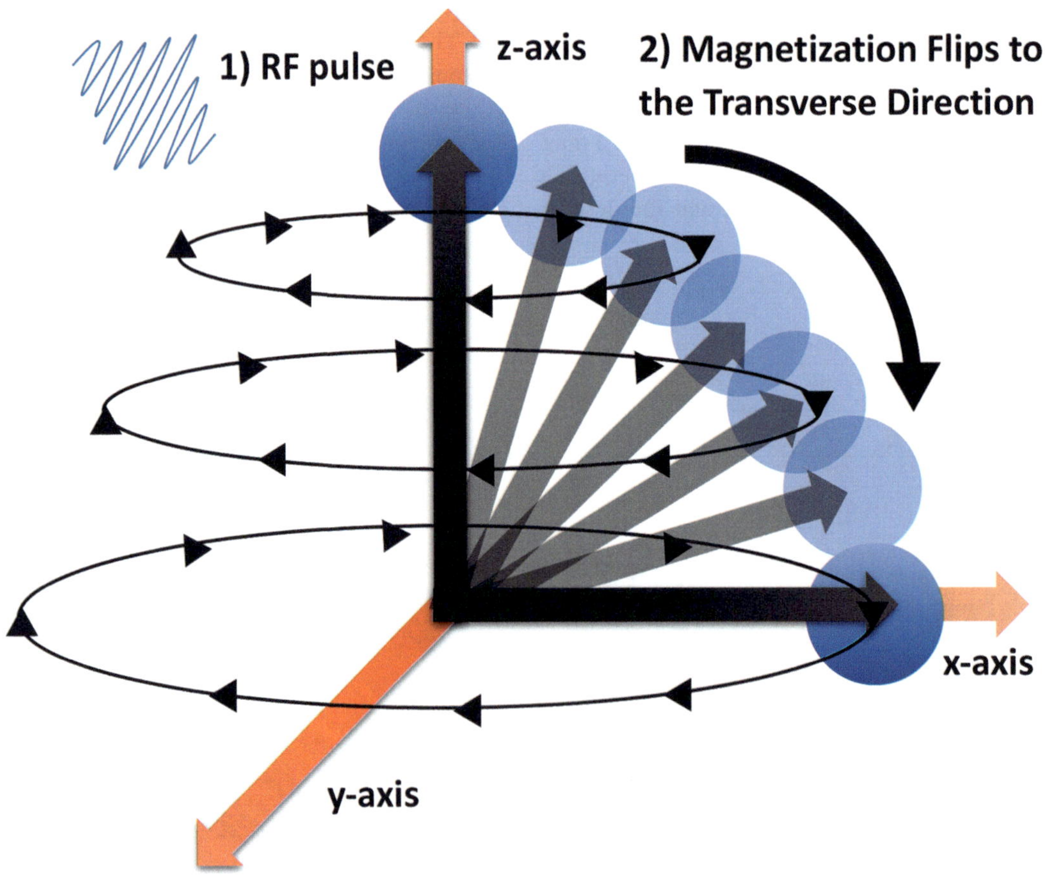


**Fig. 5.3** The net magnetization vector is initially in the direction of $B_0$. After an RF pulse, net magnetization flips into the transverse plane

along the x-axis and the y-axis. First, the gradient along the x-axis, known as the frequency-encoding gradient, is applied, encoding information along the x-axis. This includes two parts, a phasing and a dephasing lobe. Next, the gradient along the y-axis, known as the phase-encoding gradient, is applied, encoding information along the y-axis. The specific pattern in which the RF pulse and various gradients are applied results in different MR sequences, which are discussed later in this chapter (Fig. 5.6). The subsequently emitted RF pulse is detected by a receive coil. Phased-array coils consist of multiple coils that receive the MR signal. Parallel imaging is a technique that can take advantage of phased-array radiofrequency coils by undersampling the data (i.e., reducing the number of sampled k-space lines, further described below) and hence reducing acquisition times [12]. The data from the multiple coils is combined to create the image. Next, the signal is amplified and then converted from analog to digital data. This digital data is subsequently used to create the image.

The Fourier transformation relies on the concept that a complex signal can be rewritten as the sum of a series of waves, and the Fourier transform decomposes this signal into the frequencies and amplitudes of its constituent waves; in MRI, the signal in question is the MR echo containing frequency and phase-encoded spatial information needed to construct the image, and the Fourier transform applied to this complex signal represents it in k-space, a 2D Fourier space organizing information about spatial frequency and amplitude [13]. Going from k-space to the final image is performed using a 2D inverse Fourier transform. K-space contains frequency and phase data, unlike the typical data one sees in images. The center of k-space typically contains low spatial frequency image information about image shape and contrast, consisting of the majority of the signal, whereas the periphery of k-space has high spatial frequency information about fine details (edges, interface between tissues), consisting of a relatively weaker signal [14]. The amount of time that the patient is in the scanner and the type of sequence obtained are both related to the amount of time that it takes to fill k-space. There are newer techniques that allow for faster filling of k-space. For instance, compressed sensing can accelerate MR image acquisition times by undersampling k-space and hence obtaining less data [15]. Even newer techniques have leveraged deep learning for compressed-sensing-based acceleration of MR image acquisition [16]. As previously mentioned, in order to generate an interpretable image, an inverse Fourier transform is performed on the K-space data.

The amount of time that it takes to create an MR image depends on three factors—the repetition time (TR), the number of excitations for phase-encoding (NEX), and the number of signals averaged (NSA) [17]. Since the vast majority of patients are narrower in the anterior-to-poster dimension, it is typically the phase-encoding direction for the sake of minimizing examination time.

## T1-Weighted Imaging, T2-Weighted Imaging, and Pulse Sequences

The ability to differentiate within or between two tissues (contrast) allows a radiological image to be of diagnostic value for clinical, research, or therapeutic purposes. Tissue contrast in modalities such as US and CT can be considered "fixed," as differences between tissues can typically be manipulated after acquiring the raw data or altered by using exogenous contrast. Magnetic resonance imaging (MRI) allows for user alteration of contrast before image acquisition through the choice of pulse sequence and parameters.

All MR images are affected by three constants that affect tissue contrast: a longitudinal relaxation constant (Fig. 5.4), a transverse relaxation constant (Fig. 5.4), and proton density—the density of protons in a voxel. The basic spin-echo sequence best describes the relationship between these parameters and signal intensity (SI) from an image [18]:

$$\mathrm{SI}(\mathrm{TE},\mathrm{TR}) = \rho\left(1 - e^{\frac{-\mathrm{TR}}{\mathrm{T1}}}\right) * e^{\frac{-\mathrm{TE}}{\mathrm{T2}}}$$

where $\rho$ represents a scaling factor proportional to the proton density, T1 represents the longitudinal relaxation, and T2 represents the transverse relaxation. Manipulation of parameters such as echo

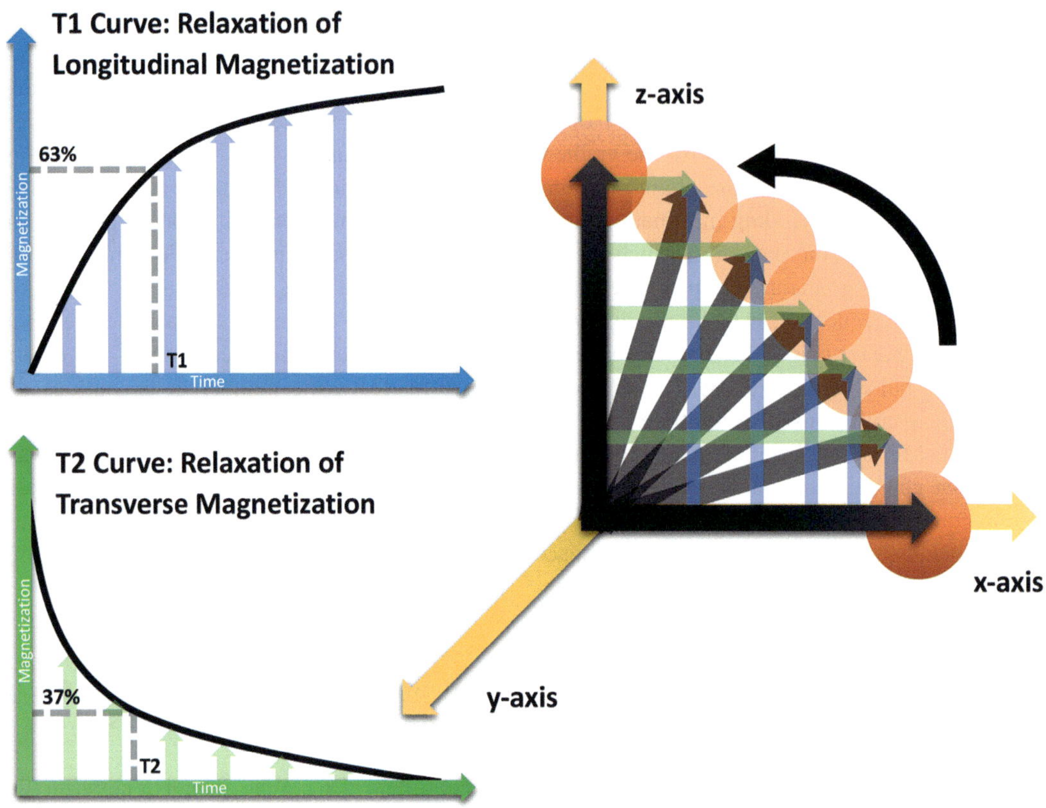


**Fig. 5.4** T1 is related to relaxation of longitudinal magnetization, and T2 is related to relaxation of transverse magnetization

time (TE), the time between excitation and the middle of the acquired echo, and repetition time (TR), the time between each successive excitation, can fundamentally change the type of image. By varying TE and TR, the SI can be made "weighted" to differences in the T1 values of tissues (TR ~ T1, TE $\ll$ T2), T2 values (TR $\gg$ T1, TE ~ T2), or a proton density (TR $\gg$ T1, TE $\ll$ T2).

There are variations from the spin-echo sequence now used in clinical MRI scanners, including gradient-echo sequences, spoiled gradient recalled echo, and inversion recovery sequences, among others [19].

## $T_1$-Weighted Sequences

The magnetization vector realigns with the axis of $B_0$ through the process of T1 recovery, during which the longitudinal magnetization increases in magnitude or recovers (Fig. 5.4). The definition of the T1 time is the time that it takes for approximately 63% of the z-axis magnetization to return to its equilibrium [20].

In T1-weighted sequences, tissues, specifically the hydrogen protons in tissues, with inherently short T1 relaxation time demonstrate high signal intensity compared to tissues with long relaxation times. In the case of spin-echo sequence acquisition, this can be achieved by setting TE and TR to low values.

In general, T1-weighted imaging can provide anatomical details, especially useful for imaging the head and neck, as well as the bones. Tissues that appear bright include fat, blood products (methemoglobin), slow-moving blood, radiation changes, melanin, and paramagnetic contrast agents (i.e., gadolinium) (Fig. 5.5a and b) [21, 22]. Tissues that appear dark include air, bones (noting that marrow in bones will appear bright), and fast-flowing blood. T1 values are affected by magnetic field strength, but T2 values are not.

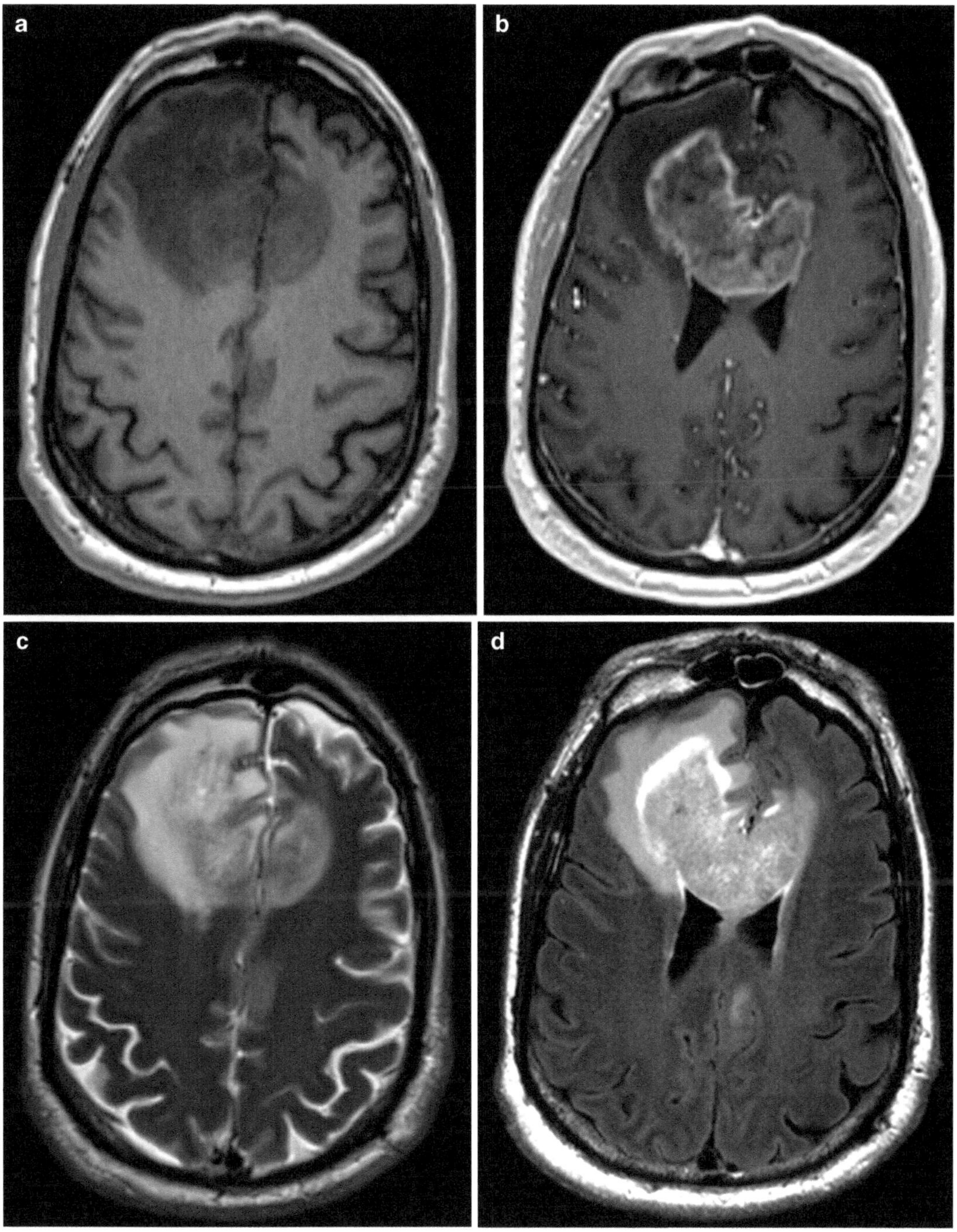


**Fig. 5.5** WHO Grade IV glioblastoma. (**a**) T1. (**b**) T1 postcontrast. (**c**) T2. (**d**) T2 FLAIR

## $T_2$-Weighted Sequences

Transverse relaxation, or spin-spin relaxation, actually represents a decay of the transverse magnetization (loss of signal) through both T2 decay—random interaction of the individual magnetic fields of spinning nuclei—and T2* decay—due to interaction with local magnetic field inhomogeneities. In the spin-echo sequence (often used for T2-weighted images), the 180° pulse (Fig. 5.6) refocuses field inhomogeneities, and hence T2* decay is negligible. This is not the

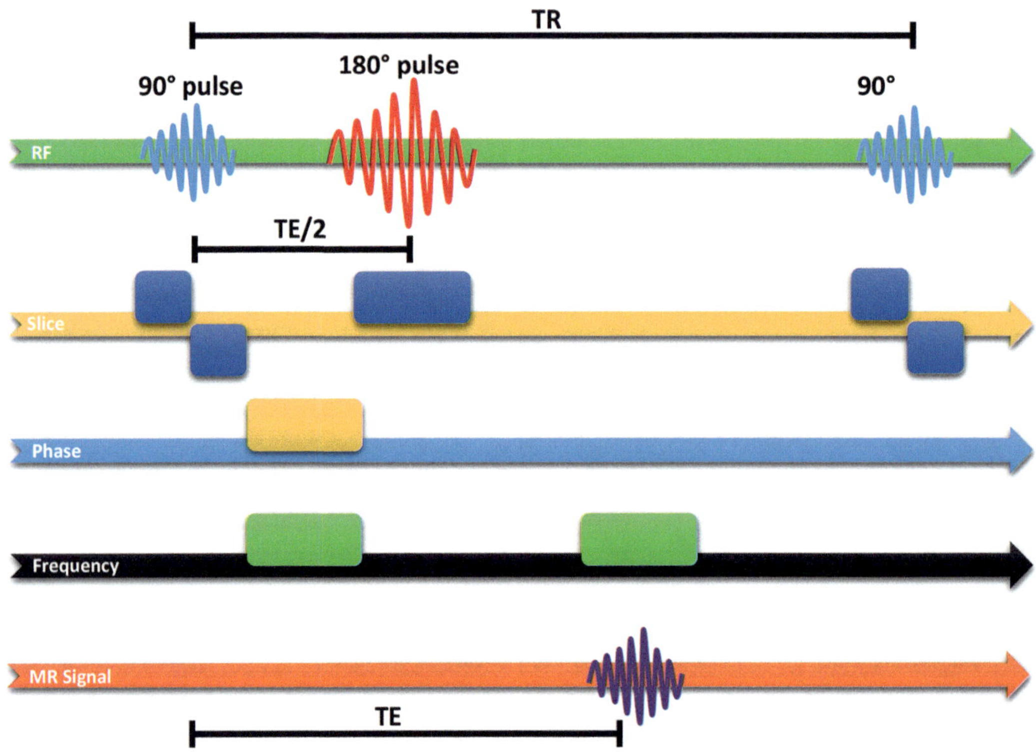


**Fig. 5.6** Spin-echo sequence diagram. This includes the RF pulse, the application of the slice selection, phase encoding, and frequency encoding gradients, and the receipt of the MR signal. The echo time (TE) is the amount of time between the RF pulse and the MR signal. The repetition time (TR) is the amount of time between successive MR pulses

case in gradient-echo recall sequences, which are often used for T1-weighted images. T2 time is defined as the time at which approximately 63% of the nuclei spinning perpendicular to the main field have lost their phase coherence (Fig. 5.4) [19]. Hence, tissues with long T2 relaxation times (fluids) appear brighter than those with short T2 relaxation times. T2-weighted sequences are acquired at long TE and long TR [22].

## Variations in Spin Echo and Gradient Echo

Fast spin echo (FSE) consists of an excitation by the RF pulse, followed by multiple 180-degree refocusing pulses occurring before the next excitation pulse, with resultant echoes more rapidly filling k-space. The echo train length (ETL) represents the number of resultant echoes per RF pulse. Single-shot fast spin echo (SSFSE) consists of an excitation by the RF pulse, followed by a rapid succession of 180-degree refocusing pulses, which results in all the echoes needed to fill k-space for a given slice. The more rapid acquisition of FSE and particularly SSFSE reduces the potential for motion artifact. The rapid succession of refocusing pulses helps to mitigate susceptibility artifact. That is because the refocusing pulses in spin echo minimize T2* effects related to magnetic field inhomogeneities. Like spin echo, gradient-echo sequences used in modern scanners can also leverage multiple echoes to decrease imaging time. For instance, the EPI sequence, used in diffusion-weighted imaging, obtains all necessary k-space data after a single excitation by the RF pulse. Flip angle is an important concept in gradient echo. While spin echo typically uses a 90-degree flip angle, gradient echo can use lower flip angles, which affects the recovery of longitudinal magnetization and ultimately T1-weighting. In gradient echo, increasing the flip angle increases T1-weighting.

In steady-state images, the TR value is short. As a result, there is only partial recovery of longitudinal magnetization and decay of transverse magnetization before the application of the next RF excitation pulse, once again encouraging a higher energy state. The interplay between magnetic relaxation and RF excitation eventually reaches a steady state. In these kinds of rapidly acquired sequences, fluid appears hyperintense, and they are useful for imaging vascular structures.

## Clinical Applications

These sequences are often implemented for anatomic imaging and treatment planning. However, they are also useful in diagnostic imaging to identify pathology, as fluid (cysts, bile, cerebrospinal fluid) and edema appear bright or hyperintense on T2-weighted imaging. Moreover, inflammation (from injury), infections, and radiation changes appear hyperintense on T2-weighted imaging (Fig. 5.5c). For abdomen and pelvis, T2-weighted imaging is the "workhorse" sequence for evaluation of anatomy, pathology, and treatment planning. In a recent study, T2-weighted images were the preferred weighting for radiation treatment in most sites, except the chest wall/breast and the lung/esophagus where T1-weighting was preferred [23].

## Proton Density

In proton density (PD) weighted images, tissues with a higher concentration of protons produce the strongest signal and appear the brightest. This is achieved by minimizing the impact of T1 and T2 differences in tissues by acquiring at long TR and short TE. The signal intensities of tissues are similar to T2-weighted sequences where air and bones are dark, while fluid and fat are bright. PD sequences are useful in the brain to differentiate gray and white matter and are useful in musculoskeletal/extremity imaging [24].

## Diffusion-Weighted Imaging

Diffusion-weighted imaging (DWI) is based on initial observations that spin-echo sequences are sensitive to the diffusion of water molecules [18, 25]. This provides a noninvasive means of detecting changes in the random motion of water molecules due to thermal energy (Brownian motion) by MRI. Molecule movement can be described by a Gaussian function where the mean-squared displacement of water depends on the diffusion coefficient of the molecule (water) and time. The free movement of water becomes more restricted in tissues due to complex factors including cellularity, membrane permeability, and extracellular volume fraction. Hence, the diffusion coefficient of water is lower in tissues that restrict movement.

Acquisition of images and quantification of an apparent diffusion coefficient (ADC) are based on the application of diffusion gradients to pulse sequences to make images sensitive to these molecular changes. An initial T2-weighted image with fat suppression is acquired with low gradients ($b$ = 0–50 s/mm$^2$). Next, one or more additional images are acquired by increasing the diffusion weighting (i.e., $b$ = 100, 500, 1000 s/mm$^2$, etc.), which leads to decreasing signal intensities from tissues that allow more water diffusion, while those that "restrict" water movement demonstrate a higher signal. This allows the quantification of an apparent diffusion coefficient (ADC) and generation of an ADC map image. Hence, tumors often demonstrate high signal intensity on high $b$-value diffusion-weighted images while demonstrating low signal on generated ADC maps. Lesions with intrinsic T2 hyperintensity that demonstrate high signal on both the DWI and ADC maps exhibit T2 shine through and do not represent true restricted diffusion.

In oncological imaging, DWI-MR has been tested in almost all types of tumors in order to differentiate benign from malignant disease, to grade malignancies, to predict and assess treatment responses, and to identify residual or recurrent tumors on follow-up exam [26]. An example case is shown for osteosarcoma of the humerus (Fig. 5.7). Sagittal STIR and sagittal T1 postcontrast images demonstrate an irregular lesion cen-

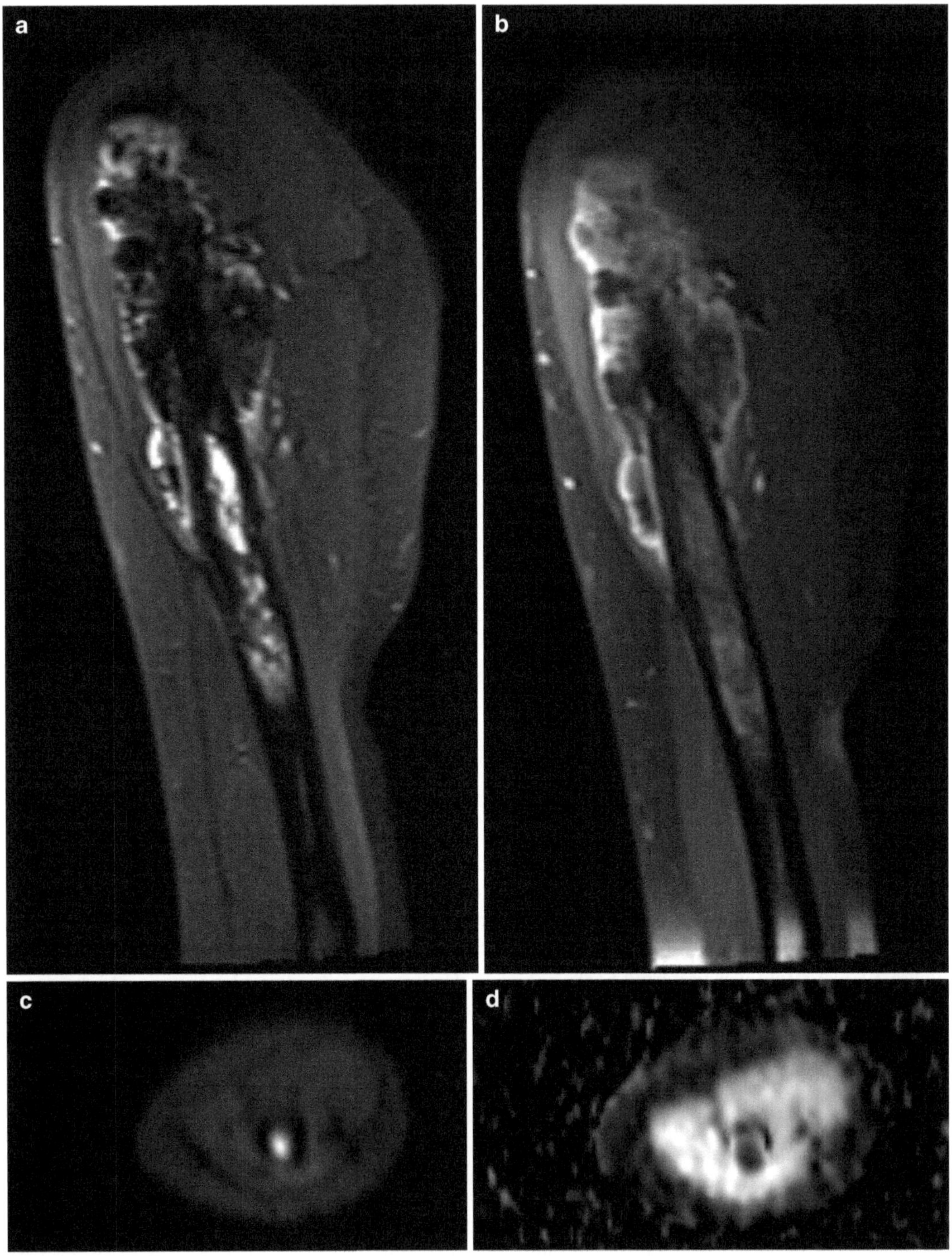

**Fig. 5.7** Osteosarcoma of the humerus with periosteal reaction and extraosseous component. (**a**) Sagittal T1 STIR. (**b**) Sagittal T1 postcontrast. (**c**) Diffusion-weighted imaging. (**d**) Apparent diffusion coefficient

tered on the humerus, with associated enhancement and extension into the soft tissues (a and b). The tumor is hyperintense on diffusion-weighted imaging and hypointense or dark on the apparent diffusion coefficient map (c and d), a pattern that is typical for tumors.

## Fat Reduction Techniques (Chemical Shift, Fat Saturation, STIR)

Fat demonstrates high signal intensity on T1- and T2-weighted images due to the inherent short T1 time and prolonged T2 time, respectively. However, there are situations in which "nulling" the fat signal becomes advantageous to visualize non-fatty tissue to better visualize pathology. Multiple methods exist to "null" fat from sequences, including chemical shift, fat saturation using a spoiler gradient, and inversion recovery techniques [27].

Hydrogen within either water or fat will have different precessional frequencies, related to the Larmor equation previously discussed. As a result, the precessional frequencies for hydrogen within fat and water will sometimes be in-phase and sometimes out-of-phase. In chemical shift imaging, in-phase and out-of-phase imaging consists of paired gradient echoes with different TE values based on whether hydrogen within fat molecules is either in-phase or out-of-phase with hydrogen within water molecules. Fat and water within the same voxel will cancel each other out on out-of-phase imaging, and both add to the signal on in-phase imaging. The timing of the in-phase image (in milliseconds) is always double that of the out-of-phase image. As a result, magnetic susceptibility artifacts are more pronounced on the in-phase images. Of note, the timing of in-phase and out-of-phase varies by the magnetic field strength, with times decreasing with increasing magnetic field strength.

In fat saturation, an RF pulse tuned to the fat resonance frequency is used together with a spoiler gradient to saturate and de-phase fat protons, which eliminates the signal from fat. This causes tissues containing fat to appear darker on the image. Sometimes this process fails, such as in the presence of magnetic field distortions, resulting in a failure of fat saturation. The Dixon technique involves obtaining water-only images and fat-only images in addition to in-phase and out-of-phase images.

Different fat reduction techniques can be useful depending upon the pathology that is being imaged. For instance, in hepatic steatosis, there is a loss of signal on out-of-phase images compared to in-phase images due to the presence of microscopic or intracellular fat (Fig. 5.8a and b). Hepatic iron deposition is the opposite of this, with a loss of signal on in-phase images. This occurs because of the longer TE of the in-phase gradient-echo sequence. In an analogous fashion, metallic clips and coils, in addition to gas as seen in the bowel, will demonstrate a larger region of even more pronounced signal loss on in-phase images, a phenomenon referred to as "blooming." In the example of angiomyolipoma, out-of-phase imaging emphasizes an "India ink" artifact, with an emphasis of the black outline of the lesion due to the interface between tissues (Fig. 5.8c and d). This name comes from an analogous appearance to a staining technique from pathology. However, the loss of signal of the lesion with fat saturation clearly shows that it contains macroscopic or extracellular fat (Fig. 5.8e). Fat saturation can be especially useful for T1 postcontrast sequences, as the suppression of fat can help to emphasize enhancing tissues.

Inversion recovery pulses are techniques for removing signals from tissues such as fat and water. In these sequences, before the initial 90-degree RF excitation pulse, there is a 180-degree inversion pulse that is specifically timed to cancel signals from particular molecules. The time between these pulses is known as the inversion time (TI). For STIR (*short tau inversion recovery*), the TI is timed to occur when protons within fat are in the 90-degree position. By its namesake, the TI for the STIR sequence is relatively short. Since this inversion recovery pulse nulls the signal from materials that demonstrate short T1 relaxation, it can technically suppress the signal from tissues that have similar T1 values to fat, which can include tissues with gadolinium or proteinaceous material. STIR is often used in extremity imaging. For example, T1 STIR sequence of the humerus highlights areas of abnormality associated with osteosarcoma (Fig. 5.7a). On the other hand, the TI can also be timed to remove signal from protons within water, using a much longer TI. This is put to good use in FLAIR (*fluid attenuation inversion recovery*). This is often used in neuroimaging. For

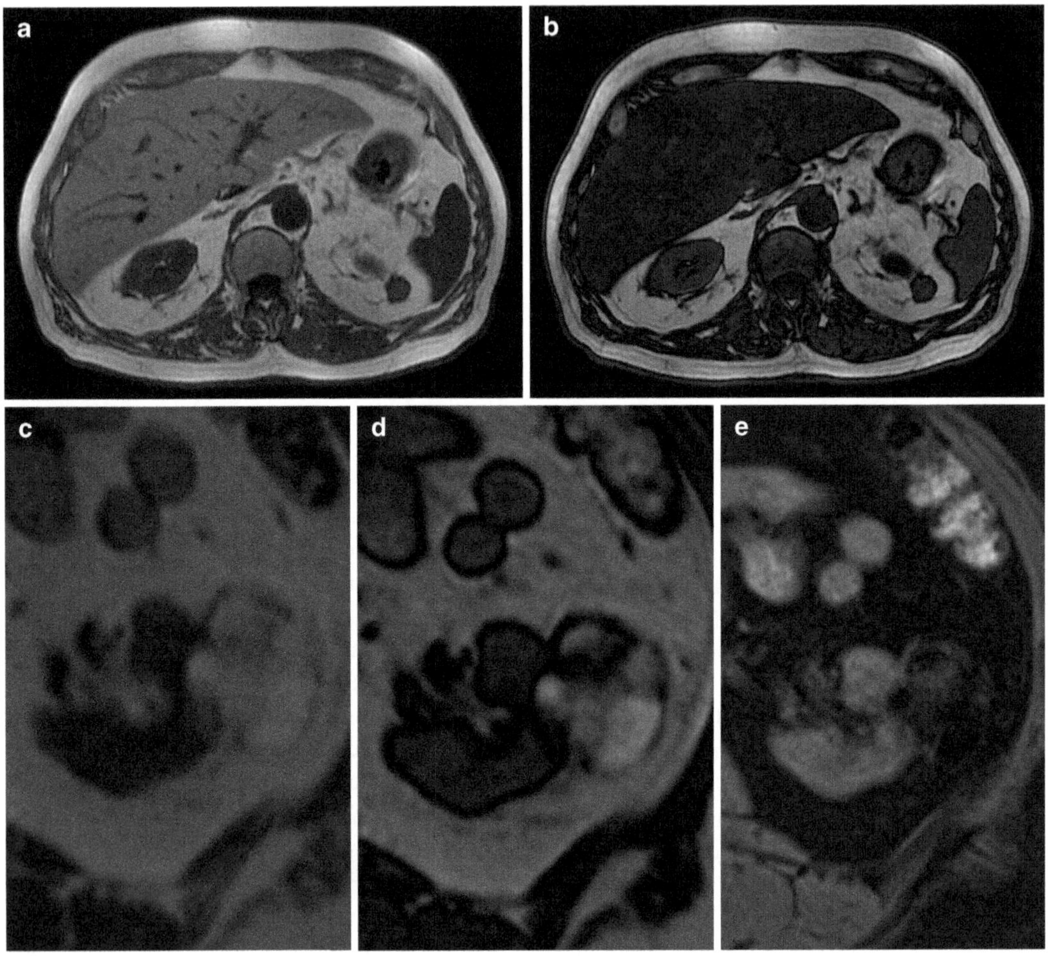


**Fig. 5.8** In-/out-of-phase imaging versus fat saturation. Figure (**a**) and (**b**) demonstrate a patient with fatty liver disease, which shows a loss of signal on out-of-phase imaging. (**a**) In-phase image. (**b**) Out-of-phase image. Figure (**c**–**e**) shows a patient with an angiomyolipoma. Angiomyolipoma shows an "India ink" artifact outlining the lesion on out-of-phase imaging, as well as a loss of signal within the lesion with fat saturation. (**c**) In-phase image. (**d**) Out-of-phase image. (**e**) Fat saturation

instance, a T2 FLAIR image of the brain highlights areas of edema within the brain (Fig. 5.5d).

## Contrast Enhancement

MRI typically takes advantage of gadolinium-based contrast agents. Since gadolinium leads to T1-shortening, causing tissues containing the agent to appear bright, contrast-enhanced sequences are naturally T1-weighted. In addition, fat saturation can also increase lesion conspicuity. After the injection of the contrast agent into the bloodstream, there are different timings that are used for imaging the relevant anatomy that vary by the MR exam protocol. For instance, abdominal imaging protocols can include an arterial phase, a portal venous phase, and a late or delayed phase (Fig. 5.9). In the figure, contrast-enhanced CT shows a lesion in hepatic segment 4A which is technically indeterminate. On T2-weighted imaging and diffusion-weighted imaging, it is hyperintense but still not fully characterized. However, on contrast-enhanced imaging, there is peripheral nodular discontinuous enhancement on arterial phase imaging, with centripetal fill-in on subsequent contrast-enhanced sequences. This firmly establishes the diagnosis of a benign hemangioma. It is important to note that different abdominal lesions have characteristic enhancement patterns, which are critical for diagnosis. For instance, hepatocellular carcinoma will classically demonstrate non-rim arterial phase hyperenhancement with subsequent washout within a cirrhotic liver [28].

Dynamic contrast-enhanced MRI involves obtaining serial MR images before, during, and after the administration of a bolus of gadolinium contrast material [29]. The temporal resolution varies by machine, vendor, and clinical application. This allows for more precise characterization of enhancement patterns, including enhancement curves demonstrating the degree of enhancement as a function of time, allowing for the calculation of specific perfusion parameters. There are several clinical applications of dynamic contrast-enhanced MRI, including neuroimaging, breast, cardiac, and prostate.

Much research has been done on the safety of administering gadolinium-based contrast agents. Nephrogenic systemic fibrosis (NSF) is a rare and life-threatening condition that was linked to

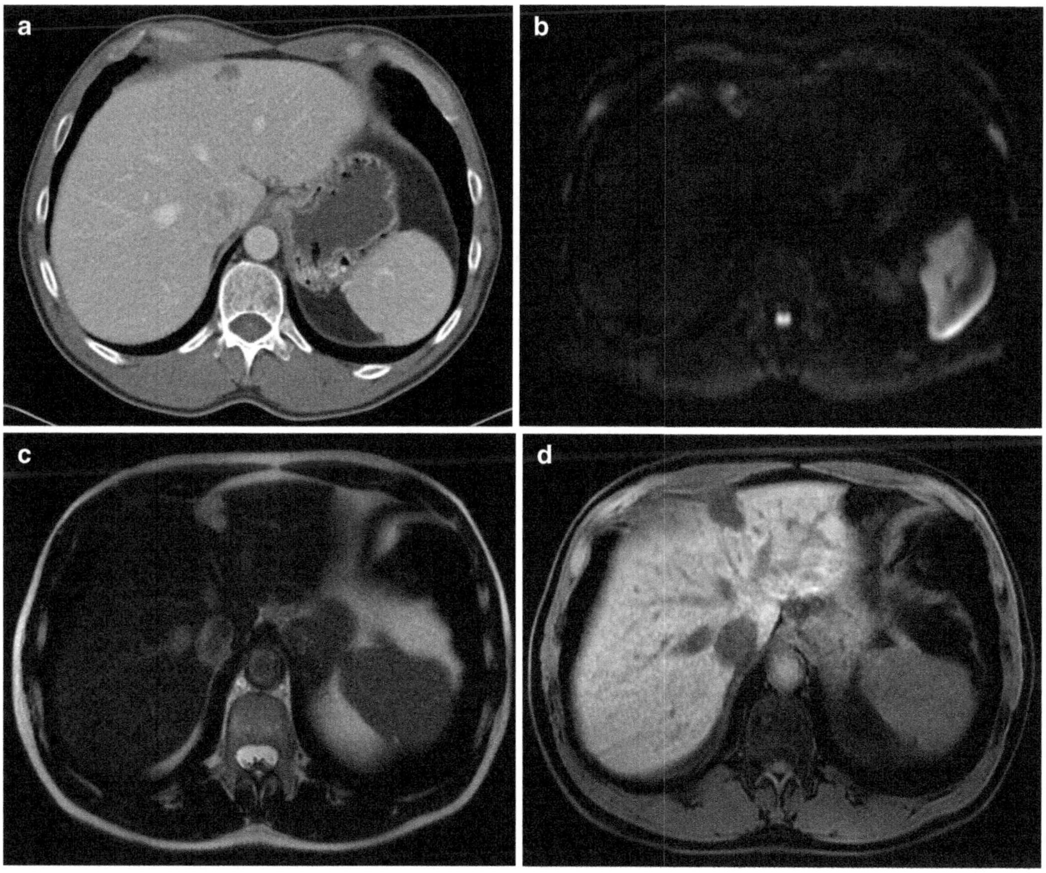


**Fig. 5.9** (**a**) Non-contrast CT shows a lesion with internal enhancement within liver segment 4A. While this could be a benign hemangioma, it is indeterminate on single-phase CT. Diffusion-weighted imaging (**b**) and T2-weighted imaging (**c**) demonstrate a hyperintense lesion in this area, although it is not yet fully characterized. T1-weighted imaging (**d**) and postcontrast series in the arterial phase (**e**), portal venous phase (**f**), late phase (**g**), and delayed phase (**h**) demonstrate peripheral nodular discontinuous enhancement on the arterial phase with progressive centripetal filling on each subsequent postcontrast phase, firmly establishing the diagnosis of hemangioma

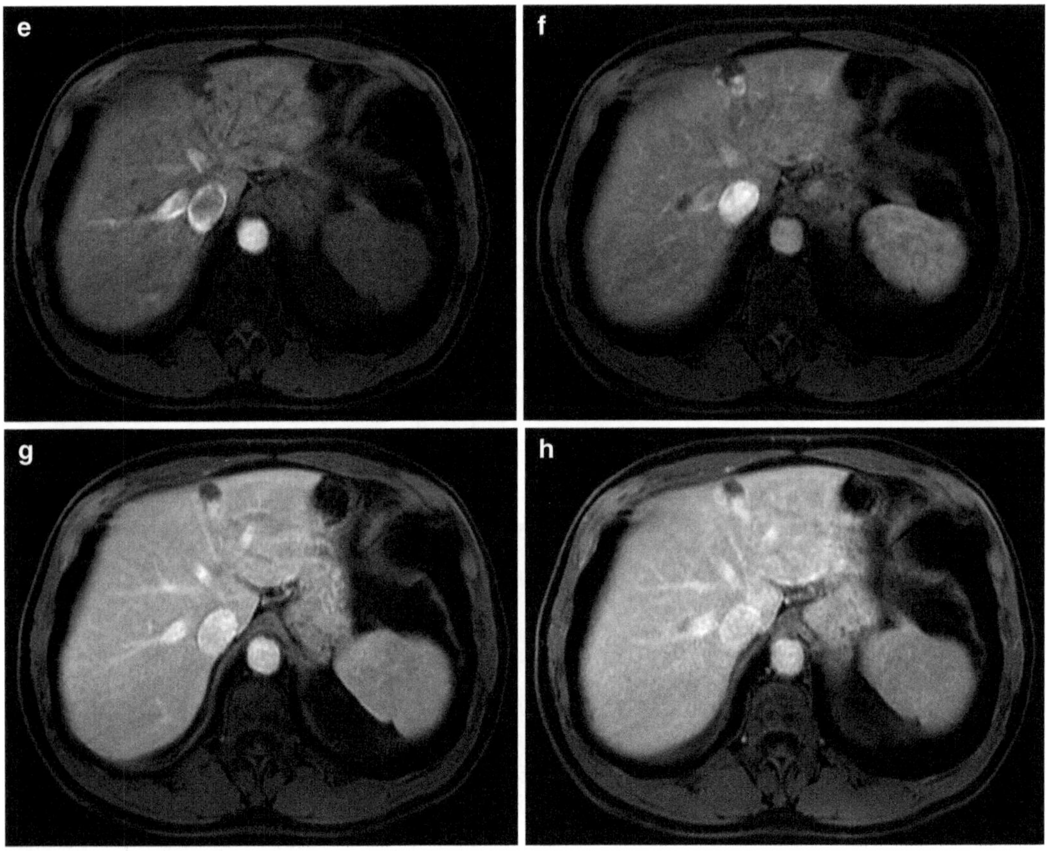


**Fig. 5.9** (continued)

gadolinium administration in patients with acute kidney injury or stage 4–5 chronic kidney disease, which represents an estimated glomerular filtration rate of <30 mL/min/1.73 $m^2$ [30]. However, the risk of NSF has not been found to be the same for all forms of gadolinium-based contrast agents (GBCAs). The American College of Radiology (ACR) Manual on Contrast Media has divided GBCAs into three categories: *group I* (agents associated with the greatest number of NSF cases), *group II* (agents associated with few, if any, unconfounded cases of NSF), and *group III* (agents for which data remains limited regarding NSF risk, but for which few, if any unconfounded cases of NSF have been reported) [31]. In particular, the manual considers the risk of NSF among patients given standard doses of group II contrast agents to be sufficiently low that assessment of renal function with a questionnaire or laboratory testing is optional prior to IV gadolinium administration, as long as there is a true clinical indication. A recent consensus statement from the ACR and the National Kidney Foundation states that the risk of NSF from Group II GBCAs is felt to be very low, and no unconfounded cases of NSF were reported for the only available type of group III GBCAs—gadoxetate disodium; hence, depending upon the clinical scenario, the harms of delaying or withholding GBCAs in patients with acute kidney injury should be balanced against and could very well outweigh the risk of NSF [32].

Transient reactions to contrast such as flushing or nausea are typically self-limited. Allergic reactions are uncommon, and life-threatening reactions such as anaphylaxis are rare. Gadolinium-based contrast agents are typically avoided in pregnancy unless the benefits substantially outweigh the risks, as it is an FDA Class C drug.

## Imaging Protocols

Imaging protocols include several imaging sequences. They can either be general (standard MRI of the abdomen) or tailored to a particular indication (MR elastography, MR pancreatic cyst protocol). Table 5.1 summarizes the different imaging sequences previously discussed, including the concept behind them and their utility/indications.

**Table 5.1** Summary of MRI sequences

| Sequence | Concept | Utility/select indications |
|---|---|---|
| T1 | Relaxation of longitudinal magnetization | Fat, methemoglobin, melanin, and contrast are hyperintense |
| T2 | Decay of transverse magnetization | Fluid, edema, inflammation, infection, and radiation change are hyperintense |
| Proton density | Concentration of protons | Musculoskeletal imaging, gray-white differentiation |
| DWI/ADC | Brownian motion of molecules | Tumor, stroke, inflammation, active demyelination, specific clinical entities (epidermoid cyst) |
| SSFP | Steady-state free precession | Cine images showing movement, useful for vascular structures such as the heart |
| In–/out-of-phase | Paired gradient echo with different TE values | Microscopic fat, iron (hepatic steatosis, hemochromatosis, adrenal adenoma) |
| Fat saturation | Suppress signal from fat | Macroscopic fat (myelolipoma), often used postcontrast to highlight enhancing tissue |
| STIR | Short tau inversion recovery | Suppress signal within fat, often used in extremity imaging |
| FLAIR | Fluid attenuation inversion recovery | Suppress signal within fluid, often used in neuroimaging |
| T1 postcontrast | Gadolinium within tissues | Tumor, abscess. Dynamic technique can highlight the contrast timing of lesions/tissue |

## Sequences in MR-Linac

The clinically used MR scanner strength is typically 3T or 1.5T in the radiology department. However, 0.35T or 1.5T MR scanners are used in radiation oncology in combination with Linac. The basic physics behind all MR scanners is fundamentally the same.

### Sequences Used in Elekta Unity MR-Linac

The Elekta Unity MR-Linac (Elekta, Stockholm, Sweden) includes a 1.5T Philips big-bore MRI (Philips Healthcare, Amsterdam, Netherlands) and a 7 MV linear accelerator [33]. As this system uses standard 1.5T imaging, sequences will appear similar to others presented in this chapter. The sequences used depend upon the clinical application for the precise delineation of the organs at risk and the target; for instance, brain applications may include T2 FLAIR and T1 post-contrast images, as well as DWI and ADC maps for patients receiving MR-guided radiotherapy [34]. ADC values can change with radiotherapy, and a recent study sought to provide Unity-specific recommendations for measuring ADC to increase its consistency, potentially facilitating its use for treatment monitoring as a quantitative imaging biomarker [35]. The role of MR-Linac in different clinical indications is an active area of research [36, 37].

### Sequences Used in ViewRay MRidian

The ViewRay MRIdian MR-Linac (ViewRay Inc., Oakwood, USA) includes a 0.35T split superconducting magnet and a 6 MV linear accelerator [38]. The MRIdian system can be operated in either the clinical or MRI mode. In the clinical mode, the Linac and MRI work simultaneously, and in the MR-only mode, the system behaves as an MR simulator. In the clinical mode, the pulse sequence used for 3D volumetric imaging and 2D cine MRI is a true fast imaging with steady-state precession (TRUFI) sequence, which is a type of balanced steady-state free precession (bSSFP) sequence, yielding a $T_2/T_1$-weighted contrast. Apart from a planning TRUFI sequence, T1-weighted, T2-weighted, and DWI sequences are also available to users.

## Volumetric TRUFI Sequence

A low-resolution volumetric planning image can be acquired in 15 s for verification of patient position. 25-s high-resolution imaging protocols can be used for 3D volumetric images with sufficient contrast for contouring and planning. The system transfers the reconstructed images in the DICOM format to the system database.

*T1*: There are two different T1 sequences that are integrated. The first is a 3D gradient-echo (GRE) sequence for the head and neck, which offers heavy T1 contrast, superior signal-to-noise ratio, and high spatial resolution; this sequence is slightly more sensitive to body motions such as respiratory breathing. For the thorax and abdominopelvic imaging, where patient breathing commonly creates motion artifact, the Turbo FLASH (TFL) sequence, in conjunction with the navigator-triggered prospective acquisition correction (PACE) technique, significantly reduces respiratory motion artifact while achieving a reasonable scan time.

*T2*: The system provides fast T2 imaging for all regions of the body. A rapid HASTE technique with navigator-echo-gating is recommended to reduce motion artifact from the patient's respiratory motion. For imaging in regions without respiratory motion, navigator-echo gating can be turned off to reduce acquisition time.

## Navigator

MRI and Linac integration allows an automatic mode for selecting trigger position and acceptance window, which can be used to improve respiratory motion artifact in the thoracic and abdominal regions. This navigator-triggered imaging technique can be implemented prior to most standard spin-echo and gradient-echo-based acquisitions. The automatic mode of navigator-triggered acquisition is divided into two phases, including a learning scout phase followed by a standard imaging phase. The aim of the learning scout phase is to estimate the patient's respiratory pattern and position based on the first five respiratory cycles by using a series of navigator echoes. The targeted diaphragm position can be determined by averaging the acquired respiratory curve during this period. With an appropriate setting of an acceptance window, a standard MR acquisition starts during the imaging phase when the diaphragm motion is within a targeted position.

*DWI*: Diffusion-weighted imaging can be used in the MR-only mode. This sequence can be used to measure intensity based on the diffusion of water molecules in and out of each voxel and can provide functional imaging to quickly differentiate between tumor and edema.

*2D cine*: The TRUFI sequence is used for real-time tracking and automated beam control or gating.

*Example Images*: Examples of the ViewRay MRIdian MR-Linac are shown for cardiac, pancreatic, and prostatic cases (Figs. 5.10, 5.11, and 5.12).

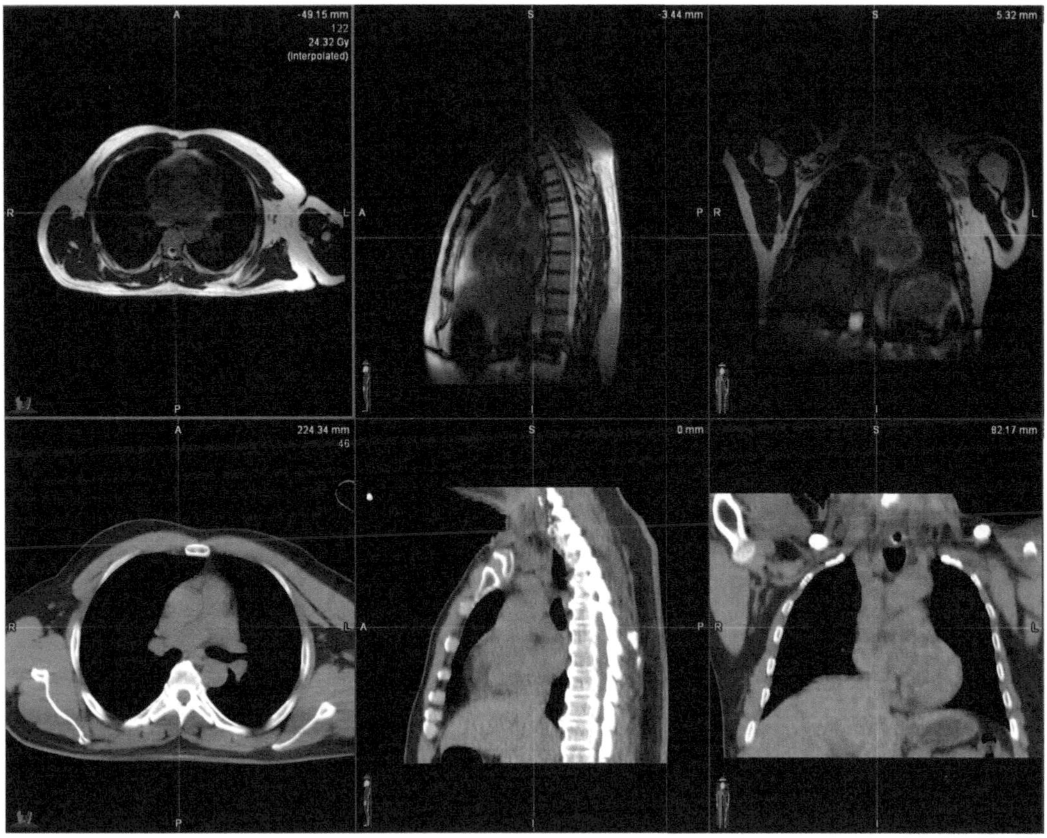


**Fig. 5.10** Cardiac example for MR-Linac. The top row shows ViewRay MRIdian TRUFI imaging with free-breathing (172 s). The bottom row presents the image from CT simulation

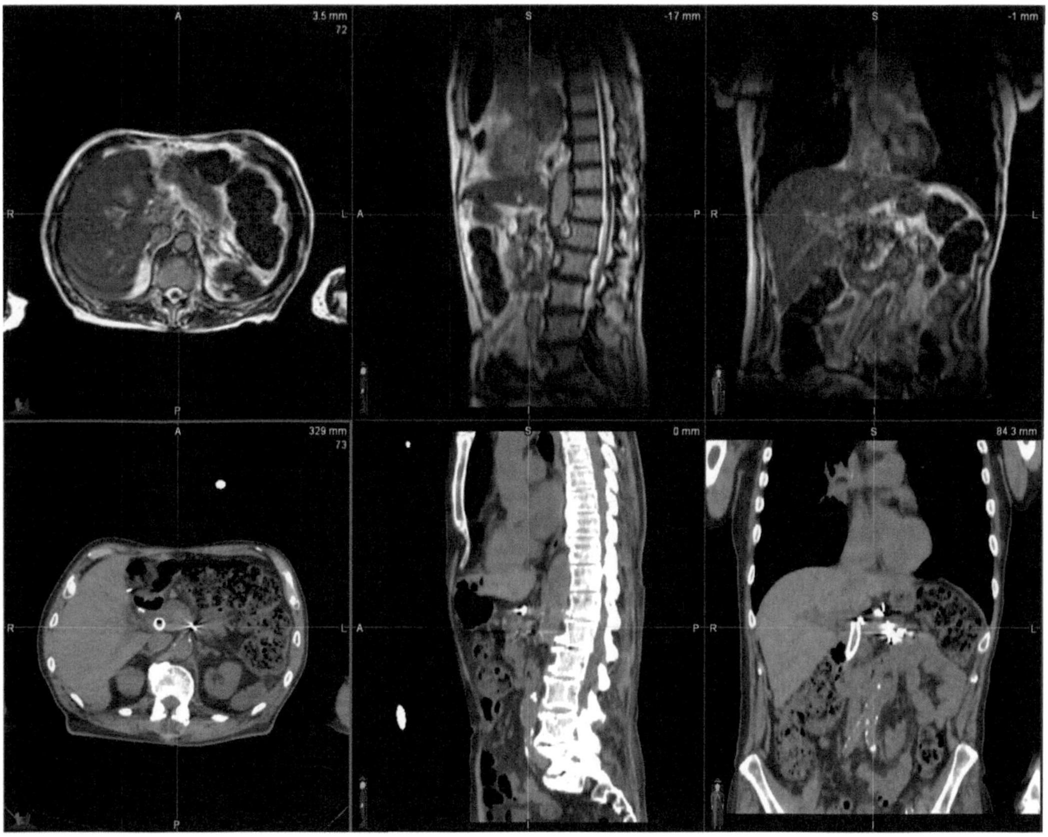

**Fig. 5.11** Pancreatic example for MR-Linac. The top row shows ViewRay MRIdian TRUFI imaging with breath-hold (25 s). The bottom row presents images from CT simulation

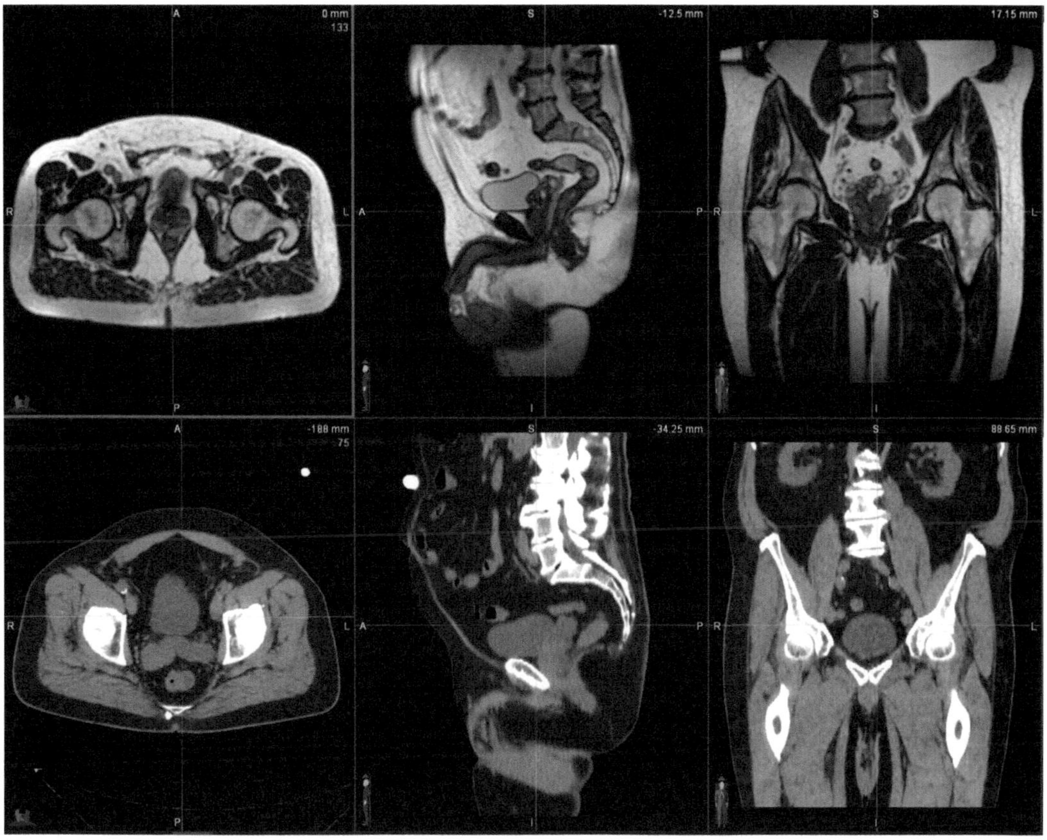

**Fig. 5.12** Prostate example for MR-Linac. The top row shows ViewRay MRIdian TRUFI imaging with free-breathing (128 s). The bottom row presents the image from CT simulation

## MRI Artifacts

This section reviews common MR imaging artifacts. Due to the nature of MR signal acquisition, several artifacts can be created that are worth knowing about. These artifacts can make medical diagnosis more challenging. Aliasing artifact is also known as wrap-around artifact (Fig. 5.13a). This occurs when a body part is outside the field of view, causing it to wrap around the opposite side of the image [39]. This is more of a problem in the phase-encoding direction, and one solution is to increase the field of view in that direction. Moiré fringes, also known as the zebra artifact, are an interference pattern that is related to magnetic field inhomogeneity (Fig. 5.13b); these are more frequently seen with gradient-echo sequences [40]. The dielectric effect demonstrates a decreased MR signal in the center of the image (Fig. 5.13c) and is frequently encountered when there is a large body habitus or a large volume of ascites [41]. However, artifacts can also aid in medical diagnosis. For instance, cerebral cavernous malformations (or cavernomas) demonstrate susceptibility artifact on MRI (Fig. 5.13d), related to the presence of hemosiderin which will cause blooming on gradient-echo sequences [42]. Susceptibility artifacts are more pronounced at higher field strengths. Another example of an artifact being useful for diagnosis is the chemical shift artifact that was discussed earlier (Fig. 5.8).

Gibbs or truncation artifact occurs in areas of significant intensity changes in MR images, such as the boundary between the skull and the brain, and it appears as alternating lines extending from

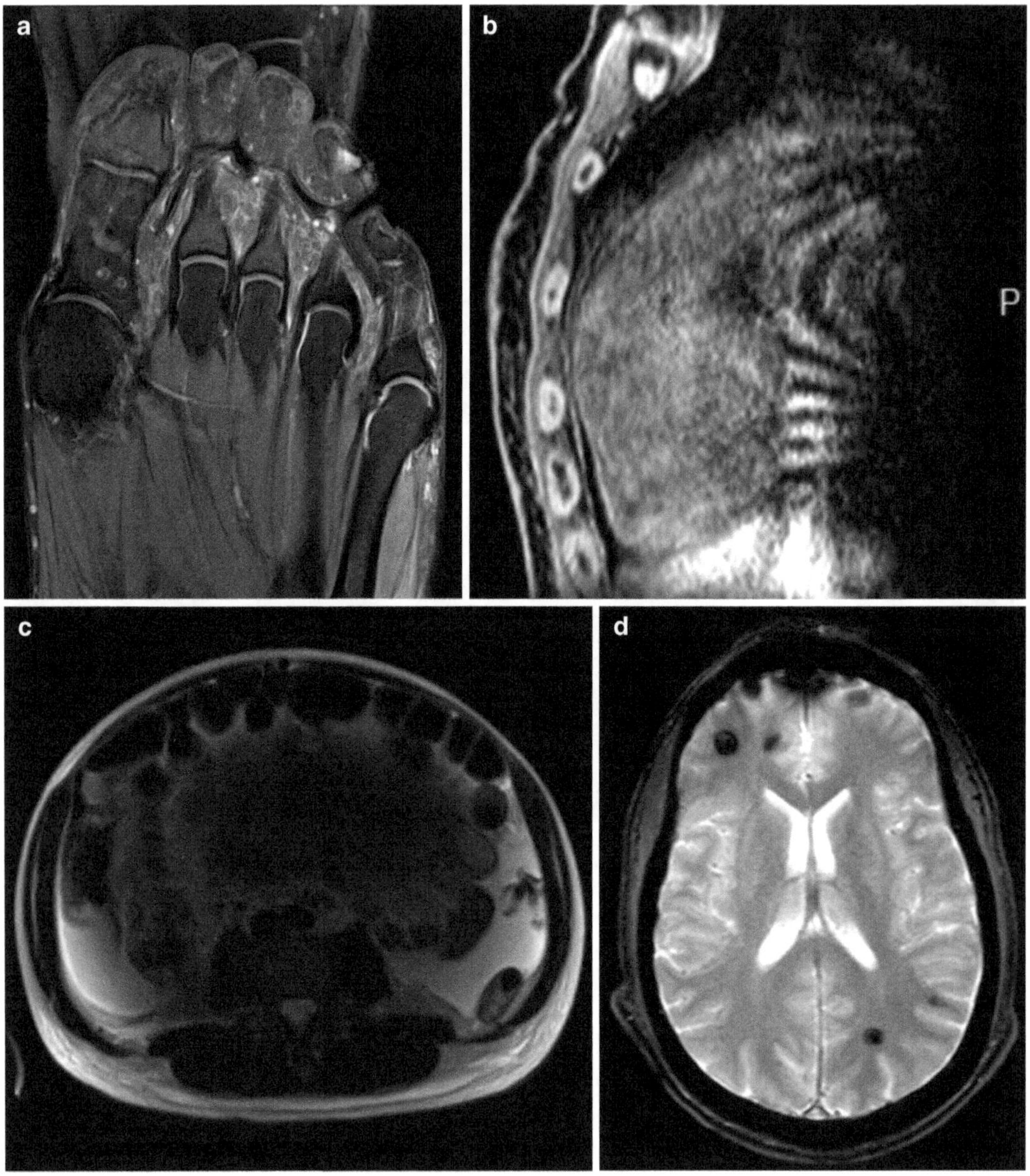


**Fig. 5.13** MR artifacts. (**a**) Proton density imaging of the foot demonstrates aliasing or wrap-around artifact. (**b**) MR breast shows an image with Moiré fringes, related to signal interference. (**c**) T2-weighted imaging in a patient with ascites demonstrates the dielectric effect, with darkening at the center of the image. (**d**) Gradient echo shows foci of susceptibility artifact with areas of blooming in the cerebral hemispheres, reflecting multiple cavernous malformations. This is an example in which an MR artifact can actually help with diagnosis

these tissue interfaces; this artifact is related to the truncation of k-space during MR data acquisition [43]. Gibbs artifact can be reduced by increasing the imaging matrix size. Herringbone artifact (also known as spike or corduroy artifact) is related to abnormal data point(s) in k-space and results in dark bands across the image [44]. It is thusly named because it was originally felt to mimic the appearance of herringbone or corduroy fabric. Zipper artifact is related to unwanted radiofrequency signals corrupting the MR images and can either arise from an external source or internally from the MRI system [45]. This artifact highlights the need for keeping the MR magnet in an RF-shielded enclosure and results in white and black bands across the image.

Imaging artifacts due to metal implants are commonly encountered. Near metal, the magnetic field can vary widely, and the magnetization in a single voxel will precess at fluctuating rates; this leads to dephasing and signal loss, as evidenced by a black area on the image [46]. This is much more of a problem for gradient echo rather than spin echo. Motion artifact is another common problem in MRI, with examples including patient movement, cardiac movement, respiratory motion, bowel peristalsis, vessel pulsation, blood flow, and CSF flow [47]. There are several strategies for mitigating motion artifact. These include fast imaging sequences, reducing patient discomfort within the MRI scanner, sedation in the right clinical scenarios, imaging at specified time points during the cardiac/respiratory cycle, and post-examination motion correction algorithms [48]. For instance, PROPELLER (periodically rotated overlapping parallel lines with enhanced reconstruction) is a technique for filling k-space that is more resistant to motion [49]. K-space is sampled in a rotating fashion analogous to the blades of a propeller, particularly oversampling the central aspect of k-space. The trade-off is that PROPELLER sequences can take slightly longer.

## General MRI Safety Considerations

Robust screening precautions are essential in the vicinity of an MRI machine. Ferromagnetic substances near the vicinity of the scanner can become dangerous projectiles, and this effect increases exponentially as one gets closer to the magnet. Shielding is a strategy that is often employed in the vicinity of the scanner, as its magnetic field can extend through walls and floors. In particular, access to the five Gauss line is strictly monitored. The American College of Radiology has designated four MRI safety zones, with particularly restricted access in Zone III and the magnet itself in Zone IV [50]. Another important consideration is MRI device compatibility. While modern devices are increasingly being made to be MR-compatible, it is important to be mindful of their MRI safety profile [51]. Metallic shrapnel and other foreign bodies should also be carefully considered, particularly if they are near anatomically critical structures.

Magnetic gradients are essential for localization in MRI, but they can create magnetic fields that vary with time. As a result, they can induce a current, particularly in conductive materials, creating the potential for a burn risk. In particular, looping structures should be avoided. RF pulses can also deposit energy, and this is measured by the specific absorption rate (SAR). The FDA maintains a limit of a 1 °C increase of whole body temperature, and this should be rigorously measured by the MR machine. Quenching is a rare, albeit life-threatening scenario, in which heating of cryogenic material converts it from liquid to gas. As a result, helium replaces oxygen gas, increasing the risk of suffocation, also resulting in increased air pressure within the room.

There are also patient-specific factors associated with MRI. Some patients find placement within the MRI scanner to be uncomfortable, and other patients can experience symptoms of claustrophobia. Rapid gradient switching may create a lot of noise, and patients are typically given earplugs to mitigate this. Finally some patient populations, such as infants, small children, and adults with altered mental status, may not be cooperative with the examination, resulting in motion artifact and nondiagnostic images.

## Conclusion

This section provides a brief overview regarding the clinical indications of MRI. First, it covered the basics of MR physics required for the generation of the image. Next, it covered the fundamentals of T1- and T2-weighted imaging. There was a discussion of MR sequences, including spin echo and gradient echo, as well as different fat reduction techniques. This chapter discussed MRI contrast and contrast safety. Sequences were discussed in the context of the Elekta Unity and ViewRay MRIdian MR-Linac systems. There was a discussion of MRI artifacts, which both hinder and aid image interpretation. Finally, general MRI safety concerns were discussed. While it ultimately serves as an introductory blueprint for further study, these fundamental principles are essential for the successful operation of MR-Linac in the clinic.

## References


1. Nagayama Y, Oda S, Nakaura T, Tsuji A, Urata J, Furusawa M, et al. Radiation dose reduction at pediatric CT: use of low tube voltage and iterative reconstruction. Radiographics. 2018;38(5):1421–40.
2. Scialpi M, Reginelli A, D'Andrea A, Gravante S, Falcone G, Baccari P, et al. Pancreatic tumors imaging: an update. Int J Surg. 2016;28(Suppl 1):S142–55.
3. Glover M, Whorms D, Singh R, Almeida RR, Prabhakar AM, Saini S, et al. A radiology-focused analysis of transparency and usability of top U.S. Hospitals' Chargemasters. Acad Radiol. 2020;27(11):1603–7.
4. Canellas R, Rosenkrantz AB, Taouli B, Sala E, Saini S, Pedrosa I, et al. Abbreviated MRI protocols for the abdomen. Radiographics. 2019;39(3):744–58.
5. Wang Y, Osborne MT, Tung B, Li M, Li Y. Imaging cardiovascular calcification. J Am Heart Assoc. 2018;7(13):e008564.
6. Haller S, Haacke EM, Thurnher MM, Barkhof F. Susceptibility-weighted imaging: technical essentials and clinical neurologic applications. Radiology. 2021;299(1):3–26.
7. Petralia G, Zugni F, Summers PE, Colombo A, Pricolo P, Grazioli L, et al. Whole-body magnetic resonance imaging (WB-MRI) for cancer screening: recommendations for use. Radiol Med. 2021;126(11):1434–50.
8. Busacchio D, Mazzocco K, Gandini S, Pricolo P, Masiero M, Summers PE, et al. Preliminary observations regarding the expectations, acceptability and satisfaction of whole-body MRI in self-referring asymptomatic subjects. Br J Radiol. 2021;94(1118):20191031.
9. Kim SY, An J, Lim YS, Han S, Lee JY, Byun JH, et al. MRI with liver-specific contrast for surveillance of patients with cirrhosis at high risk of hepatocellular carcinoma. JAMA Oncol. 2017;3(4):456–63.
10. Grover VP, Tognarelli JM, Crossey MM, Cox IJ, Taylor-Robinson SD, McPhail MJ. Magnetic resonance imaging: principles and techniques: lessons for clinicians. J Clin Exp Hepatol. 2015;5(3):246–55.
11. Roth C, Deshmukh S. Fundamentals of body MRI. 2nd ed. Philadelphia, PA: Elsevier; 2017.
12. Glockner JF, Hu HH, Stanley DW, Angelos L, King K. Parallel MR imaging: a user's guide. Radiographics. 2005;25(5):1279–97.
13. Gallagher TA, Nemeth AJ, Hacein-Bey L. An introduction to the Fourier transform: relationship to MRI. AJR Am J Roentgenol. 2008;190(5):1396–405.
14. Paschal CB, Morris HD. K-space in the clinic. J Magn Reson Imaging. 2004;19(2):145–59.
15. Jaspan ON, Fleysher R, Lipton ML. Compressed sensing MRI: a review of the clinical literature. Br J Radiol. 2015;88(1056):20150487.
16. Chen Y, Schönlieb CB, Liò P, Leiner T, Dragotti PL, Wang G, et al. AI-based reconstruction for fast MRI—a systematic review and meta-analysis. Proc IEEE. 2022;110(2):224–45.
17. Petersein J, Saini S. Fast MR imaging: technical strategies. AJR Am J Roentgenol. 1995;165(5):1105–9.
18. Hahn EL. Spin echoes. Phys Rev. 1950;80(4):580–94.
19. Brown RW, Cheng Y-CN, Haacke M, Thompson MR, Venkatesan R. Magnetic resonance imaging: physical principles and sequence design. 2nd ed. Hoboken, NJ: Wiley-Blackwell; 2014.
20. Bloch F. The principle of nuclear induction. Science. 1953;118(3068):425–30.
21. Lu H, Clingman C, Golay X, van Zijl PC. Determining the longitudinal relaxation time (T1) of blood at 3.0 Tesla. Magn Reson Med. 2004;52(3):679–82.
22. Stanisz GJ, Odrobina EE, Pun J, Escaravage M, Graham SJ, Bronskill MJ, et al. T1, T2 relaxation and magnetization transfer in tissue at 3T. Magn Reson Med. 2005;54(3):507–12.
23. Eccles CL, Adair Smith G, Bower L, Hafeez S, Herbert T, Hunt A, et al. Magnetic resonance imaging sequence evaluation of an MR Linac system; early clinical experience. Tech Innov Patient Support Radiat Oncol. 2019;12:56–63.
24. Westbrook C, Talbot J. MRI in practice. 5th ed. Hoboken, NJ: Wiley-Blackwell; 2018.
25. Carr HY, Purcell EM. Effects of diffusion on free precession in nuclear magnetic resonance experiments. Phys Rev. 1954;94(3):630–8.
26. Yankeelov TE, Pickens DR, Price RR. Quantitative MRI in cancer. 1st ed. Boca Raton, FL: CRC Press; 2012.
27. Bitar R, Leung G, Perng R, Tadros S, Moody AR, Sarrazin J, et al. MR pulse sequences: what every radiologist wants to know but is afraid to ask. Radiographics. 2006;26(2):513–37.
28. Chernyak V, Fowler KJ, Kamaya A, Kielar AZ, Elsayes KM, Bashir MR, et al. Liver imaging reporting and data system (LI-RADS) version 2018: imaging of hepatocellular carcinoma in at-risk patients. Radiology. 2018;289(3):816–30.
29. Verma S, Turkbey B, Muradyan N, Rajesh A, Cornud F, Haider MA, et al. Overview of dynamic contrast-enhanced MRI in prostate cancer diagnosis and management. AJR Am J Roentgenol. 2012;198(6):1277–88.
30. Woolen SA, Shankar PR, Gagnier JJ, MacEachern MP, Singer L, Davenport MS. Risk of nephrogenic systemic fibrosis in patients with stage 4 or 5 chronic kidney disease receiving a group II gadolinium-based

contrast agent: a systematic review and meta-analysis. JAMA Intern Med. 2020;180(2):223–30.
31. ACR. ACR manual on contrast media. Reston, VA: American College of Radiology; 2022.
32. Weinreb JC, Rodby RA, Yee J, Wang CL, Fine D, McDonald RJ, et al. Use of intravenous gadolinium-based contrast media in patients with kidney disease: consensus statements from the American College of Radiology and the National Kidney Foundation. Radiology. 2021;298(1):28–35.
33. Snyder JE, St-Aubin J, Yaddanapudi S, Boczkowski A, Dunkerley DAP, Graves SA, et al. Commissioning of a 1.5T Elekta Unity MR-linac: a single institution experience. J Appl Clin Med Phys. 2020;21(7):160–72.
34. Dunkerley DAP, Hyer DE, Snyder JE, St-Aubin JJ, Anderson CM, Caster JM, et al. Clinical implementational and site-specific workflows for a 1.5T MR-Linac. J Clin Med. 2022;11(6):1662.
35. Kooreman ES, van Houdt PJ, Keesman R, Pos FJ, van Pelt VWJ, Nowee ME, et al. ADC measurements on the Unity MR-linac - a recommendation on behalf of the Elekta Unity MR-linac consortium. Radiother Oncol. 2020;153:106–13.
36. Civil YA, Oei AL, Duvivier KM, Bijker N, Meijnen P, Donkers L, et al. Prediction of pathologic complete response after single-dose MR-guided partial breast irradiation in low-risk breast cancer patients: the ABLATIVE-2 trial-a study protocol. BMC Cancer. 2023;23(1):419.
37. Regnery S, de Colle C, Eze C, Corradini S, Thieke C, Sedlaczek O, et al. Pulmonary magnetic resonance-guided online adaptive radiotherapy of locally advanced: the PUMA trial. Radiat Oncol. 2023;18(1):74.
38. Kluter S. Technical design and concept of a 0.35 T MR-Linac. Clin Transl Radiat Oncol. 2019;18:98–101.
39. Stadler A, Schima W, Ba-Ssalamah A, Kettenbach J, Eisenhuber E. Artifacts in body MR imaging: their appearance and how to eliminate them. Eur Radiol. 2007;17(5):1242–55.
40. Alfudhili K, Masci PG, Delacoste J, Ledoux JB, Berchier G, Dunet V, et al. Current artefacts in cardiac and chest magnetic resonance imaging: tips and tricks. Br J Radiol. 2016;89(1062):20150987.
41. Cornfeld D, Weinreb J. Simple changes to 1.5-T MRI abdomen and pelvis protocols to optimize results at 3 T. AJR Am J Roentgenol. 2008;190(2): W140–50.
42. Campbell PG, Jabbour P, Yadla S, Awad IA. Emerging clinical imaging techniques for cerebral cavernous malformations: a systematic review. Neurosurg Focus. 2010;29(3):E6.
43. Kellner E, Dhital B, Kiselev VG, Reisert M. Gibbs-ringing artifact removal based on local subvoxel-shifts. Magn Reson Med. 2016;76(5):1574–81.
44. Zhuo J, Gullapalli RP. AAPM/RSNA physics tutorial for residents: MR artifacts, safety, and quality control. Radiographics. 2006;26(1):275–97.
45. Graves MJ, Mitchell DG. Body MRI artifacts in clinical practice: a physicist's and radiologist's perspective. J Magn Reson Imaging. 2013;38(2):269–87.
46. Hargreaves BA, Worters PW, Pauly KB, Pauly JM, Koch KM, Gold GE. Metal-induced artifacts in MRI. AJR Am J Roentgenol. 2011;197(3):547–55.
47. Zaitsev M, Maclaren J, Herbst M. Motion artifacts in MRI: a complex problem with many partial solutions. J Magn Reson Imaging. 2015;42(4):887–901.
48. Havsteen I, Ohlhues A, Madsen KH, Nybing JD, Christensen H, Christensen A. Are movement artifacts in magnetic resonance imaging a real problem?-a narrative review. Front Neurol. 2017;8:232.
49. Forbes KP, Pipe JG, Bird CR, Heiserman JE. PROPELLER MRI: clinical testing of a novel technique for quantification and compensation of head motion. J Magn Reson Imaging. 2001;14(3):215–22.
50. ACR. ACR manual on MR safety. Reston, VA: American College of Radiology; 2020.
51. Shellock FG, Crues JV. MRI bioeffects, safety, and patient management. 2nd ed. Los Angeles, CA: Biomedical Research Publishing Group; 2022.

# Motion Management and Tracking

# 6

Elizabeth Huynh, Eenas A. Omari, Sara N. Lim, and Raymond H. Mak

## Introduction

Motion management is a critical component of modern image-guided radiation therapy, and accurate modeling of motion during simulation, planning, and treatment delivery over the past two decades has enabled the advent of high-precision radiation therapy techniques such as stereotactic body radiation therapy (SBRT). Historically, motion management (particularly for respiratory motion) relied on adding large planning margins to account for this motion. The development of multiple motion management approaches with X-ray imaging-based linear accelerators and now more recently MRI-guided linear accelerators (MR-Linacs) has provided clinicians with the ability to reduce uncertainty and target margins. This has expanded the ability to dose escalate with steeper dose gradients and/or hypofractionated treatments while reducing dose exposure to surrounding organs.

Motion of both targets and organs-at-risk (OARs) can occur due to a variety of physiological processes, including breathing, cardiac activity, digestion, swallowing, and bladder filling/voiding, and can lead to significant intra- and inter-fraction variation in the position of the tumor and surrounding normal tissues. Failure to account for this variation during treatment planning and delivery can result in decreased treatment accuracy and increased toxicity to surrounding normal tissues.

E. Huynh (✉)
Department of Physics and Engineering, London Regional Cancer Program, London Health Sciences Centre, London, ON, Canada
e-mail: Elizabeth.Huynh@lhsc.on.ca

E. A. Omari
Department of Radiation Oncology, Medical College of Wisconsin, Milwaukee, WI, USA
e-mail: eomari@mcw.edu

S. N. Lim
Department of Radiation Oncology, Northwestern Memorial Hospital, Northwestern University Feinberg School of Medicine, Chicago, IL, USA
e-mail: Sara.lim@nm.org

R. H. Mak
Department of Radiation Oncology, Brigham and Women's Hospital, Dana-Farber Cancer Institute, Boston, MA, USA
e-mail: rmak@partners.org

## Types of Motion

### Respiratory Motion

Respiratory motion is typically periodic (on average ~ 5 s per breath or 12 breaths per minute) and more predictably modeled, but there can be significant variation in amplitude, period, and regularity with potential significant patient-to-patient, intra-fraction, and inter-fraction variability. Respiratory motion is particularly relevant for thoracic (e.g., breast, lung, intracardiac) and upper abdominal cancers (e.g., gastric, liver, pan-



I. J. Das et al. (eds.), *A Practical Guide to MR-Linac*, https://doi.org/10.1007/978-3-031-48165-9_6

creas, esophageal). There is substantial variation in the amplitude of respiratory motion, and there is a need for motion management techniques for lung tumors, with the greatest motion typically encountered for lower lobe tumors or organs adjacent to the diaphragm [1, 2]. In general, respiratory motion is most substantial in the superior-inferior direction, ranging up to 2–5 cm in reported series, but lateral and anterior-posterior motion can also be substantial in a subset of patients (ranging up to 1–2 cm in selected patients) [2, 3]. There are various respiratory motion management techniques, discussed in section "X-Rray- and CT-Based Techniques for Respiratory Motion Management".

## Cardiac Motion

Cardiac motion has an inherent period (normally 1–2 beats per second) similar to respiratory motion; however, the motion induced by cardiac contraction and relaxation during systole and diastole, respectively, can be volumetrically more complex. While cardiac motion has not been historically accounted for in the treatment of thoracic and upper abdominal tumors due to limitations in the frequency of X-ray imaging technologies relative to the cardiac cycle, the ability for MRI to perform continuous cine imaging allows for real-time imaging of cardiac motion. More recent applications for SBRT to treat intracardiac tumors [4] and for the ablation of cardiac arrhythmias such as ventricular tachycardia [5] provide new indications and impetus to develop novel cardiac motion management techniques.

## Physiological Motion of Gastrointestinal, Urinary, and Head and Neck Organs

Physiological motion of the gastrointestinal organs can be due to distention of the organs and peristalsis of liquids, solid food, and stool content, which can be dependent on the quantity and timing of ingestion. The process of digestion also results in production of gas that can result in significant intra-fraction and inter-fraction movement of the stomach, duodenum, small and large bowel, and rectum. Traditional physiological management techniques for radiation therapy for abdominal and pelvic sites include pre-specified fasting intervals (e.g., nothing by mouth for 4–8 hours before simulation and treatment) to decompress the stomach and duodenum and minimize peristaltic movement, dietary restrictions (e.g., avoiding gas forming foods), and bowel preparation regimens (e.g., daily medications for gas-reduction such as simethicone, laxatives, and/or enemas to decompress the bowels) to optimize reproducibility of gastrointestinal organ location and filling, which are applicable for both X-ray-based and MRI-guided approaches.

Similarly, bladder filling can result in substantial intra-fraction and inter-faction variability in bladder volume and shape which impacts a variety of pelvic tumor sites (e.g., prostate, bladder, gynecological, rectal). In particular, X-ray-based assessments of bladder volume can be particularly susceptible to intra-fraction filling and can result in substantial target shifts as reported in prostate cancer [6]. Conventional methods to mitigate bladder motion issues include bladder regimens with pre-specified timing of bladder voiding and hydration instructions prior to treatment and implantation of fiducials for prostate cancer for intra-fraction monitoring.

Lastly, swallowing and phonation can impact the location of oral, oropharyngeal, pharyngeal, and hypopharyngeal organs and tumor targets for head and neck cancers. The impact of this motion is greatest for laryngeal tumor locations with up to 2 cm of laryngeal movement with swallowing [7], and more recent investigations with MRI-based dynamic imaging approaches have quantified the expected intra-fraction motion [8–10].

These physiological motions can result in significant variability in delivered dose either through displacement of the tumor target or shift in location of the organ itself and can be critical in clinical scenarios with high doses of radiation delivered with steep gradients between tumors and OARs (e.g., pancreatic cancer adjacent to duodenum, stomach, and/or bowel). The continuous imaging and tracking capabilities of MRI-guided linear accelerators provide advantages in real-time monitoring of intra-fraction physiological movement.

## X-Ray- and CT-Based Techniques for Respiratory Motion Management

Historically, motion management with X-ray- and CT-based approaches in radiation therapy planning has focused on respiratory motion. Approaches for respiratory management include motion-encompassing methods (e.g., 4D-CT), respiratory gating, breath-hold, forced shallow breathing with abdominal compression, and real-time tracking (often requiring fiducial placement). These techniques can be further divided into passive and active approaches: motion-encompassing methods and forced shallow breathing are passive approaches, while respiratory gating, breath-hold, and respiratory-synchronization are active. Active approaches account for variations in breathing patterns in real-time, while passive approaches do not.

### Motion Encompassing Methods

4D-CT involves acquiring multiple CT scans of the patient during the breathing cycle and creating a 4D image set that allows the clinician to define an internal target volume (ITV) for the tumor that encompasses the range of tumor location throughout the breathing cycle and also identify normal tissue location during the breathing cycle. Some potential limitations of the 4D-CT and ITV approach is that ITV can still be prohibitively large with significant respiratory motion and expose surrounding organs, such as the lungs for lung tumors, to excess dose. There are also concerns regarding dose blurring and interplay effects [2].

### Respiratory Gating

Respiratory-gated radiation therapy with X-ray approaches typically utilizes an external respiratory surrogate to gate the treatment beam on and off during the patient's breathing cycle, so that the radiation is delivered only when the tumor is in specific phases of the breathing cycle or during a breath-hold. These active motion management approaches in general allow for reduction of the target volume.

While external surrogates can be used for gating, internal surrogates such as implantation of fiducials in and/or around the tumor may be required for internal verification of the gating process. Fiducials (implanted metal markers) are required due to the inherent limitations of cone beam CT or intra-fraction triggered kV imaging in resolving soft tissue or small tumors in the thorax and upper abdomen.

### Breath-Hold

Breath-hold treatments are similar to respiratory gated treatments, but with treatment delivery occurring only when a patient is holding their breath at a pre-specified depth. For X-ray-based approaches, external surrogates and/or implanted fiducials are required as described above. Breath-hold approaches have some advantages to gating approaches including improved duty cycle if the patient can consistently reproduce the breath-hold and the ability to maneuver critical organs away from the target zone (e.g., deep inspiration breath-hold for left-sided breast cancer to position the high-dose region further away from the heart [11]).

### Forced Shallow Breathing with Abdominal Compression

Compression devices can also be utilized to suppress respiratory motion and is most effective for lower lobe lung tumors. Drawbacks of this approach include increased inter-fraction variability and the potentially increased motion in a small proportion of cases [12].

### Tracking

Motion management with fiducials for X-ray-guided radiation therapy involves the implantation of small, visible (i.e., radio-opaque) markers that are placed within or near the target area. These markers are then tracked during treatment to monitor and compensate for any movement of the target.

### Limitations of X-Ray-Based Motion Management

Key limitations of X-ray-based respiratory motion management include higher normal tissue dose exposure with motion encompassing (e.g., 4D-CT) approaches, the dependency on external surrogates, and/or implantation of fiducial markers to enable solutions that can reduce target volume. The technologies are highly dependent on accurate modeling and reconstructions and are not easily applied to non-periodic motion (e.g., physiological motion), and there is a limitation on intra-fraction frequency of imaging due to cumulative imaging X-ray exposure concerns.

## MRI-Guided Linear Accelerator Techniques for Motion Management

The combination of a MRI with a linear accelerator allows for new motion management strategies. MRI-based approaches for motion management in radiation therapy planning include both respiratory-gated MRI and breath-hold MRI, while motion management during treatment delivery can be enhanced with continuous, cine MRI.

### 4D-MRI

Respiratory correlated 4D-MRI can be acquired to generate motion averaged or mid-position images for daily treatment planning and daily motion assessment [13]. This is of importance when treating abdominal and thoracic tumors with a non-gated treatment. The average or mid-position images generated from the respiratory binned 4D-MRI can be used for treatment planning, daily image registration/fusion, and daily re-planning (i.e., contouring and plan reoptimization). The determination between the motion-averaged and mid-position images is based on image quality and the presence of artifacts. For example, Fig. 6.1 shows the reconstructed MR-Linac 4D-MRI for a liver patient where the mid-position image would be the better option due to the blurring artifacts caused by the large respiratory motion as shown on the motion averaged image. The mid-position image takes longer to reconstruct but shows sharper boundaries despite the presence of large respiratory motion. If similar quality is observed between the motion averaged and the mid-position images, the motion averaged image may be the better option due to the faster reconstruction time.

4D-MRI can also be used to determine the maximum extent of the motion with respect to the original planning target volume (PTV) for non-gated treatments [15]. If the extent of the motion lies outside of the planned volume, a new PTV can be generated based on the daily motion to avoid target underdosing.

### Breath-Hold MRI

Similar to X-ray-based breath-hold techniques, MRI-based breath-hold techniques require patient cooperation to reproducibly hold their breath during simulation and treatment delivery. In contrast to X-ray-based breath-hold techniques, MRI-based breath-hold treatments are not dependent on external surrogates or place-

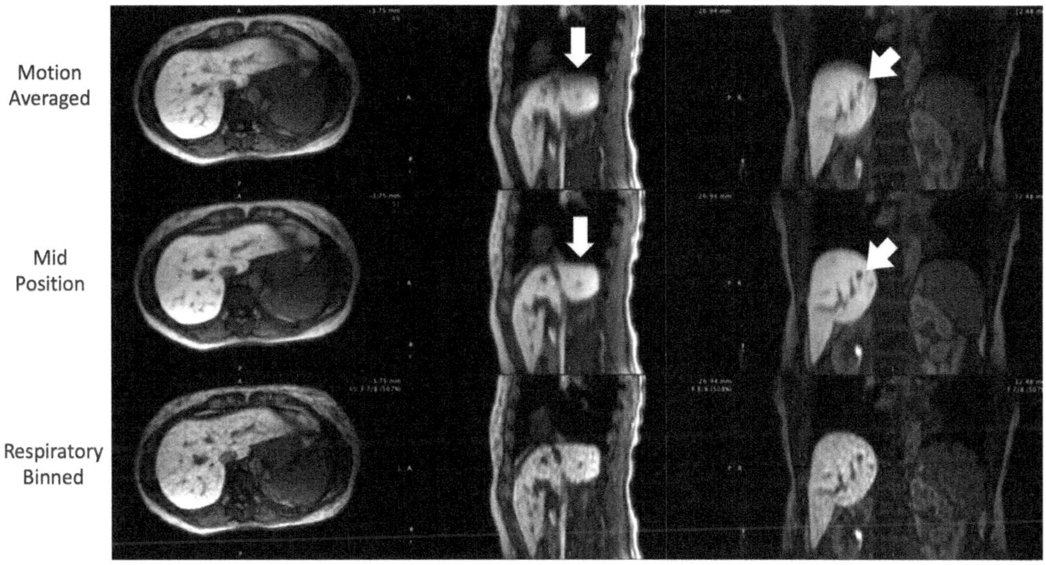


**Fig. 6.1** Elekta Unity MR-Linac 4D-MR images of a liver patient showing the motion averaged (top), mid-position (middle), and respiratory binned (bottom) images [14]

ment of fiducial markers for verification. Furthermore, multiple image acquisitions can be obtained until a favorable and reproducible breath-hold is obtained without additional radiation dose. The ability to select breath-holds that result in favorable geometric relationships is particularly helpful for cases with tumors abutting central organs (e.g., central lung tumor, Fig. 6.2).

Clinically, the patient can be coached to either perform the breath-hold at the end of exhale or end of inhale. Some factors determining the breath-hold phase selection are reproducibility and the duty cycle for which the patient is capable of holding their breath. For MR-Linacs with gating capability, such as the ViewRay MRIdian, the beam is held automatically when the target or surrogate is outside of the breath-hold window. For MR-Linacs without gating capability, such as the early version of the Elekta Unity, the beam is held manually when the target or surrogate lies outside the breath-hold window.

## Cine MRI with Breath-Hold Technique and Target Tracking

Cine MRI involves acquiring multiple MR images of the patient during the breathing cycle and creating a cine image set that shows the tumor and surrounding normal tissues in different positions during the breathing cycle. Since cine MRI does not expose patients to ionizing radiation, the cine MRI can be acquired continuously during radiation delivery for intra-fraction monitoring and gating. While MRI acquisition times were previously a barrier for real-time imaging, selection and optimization of MR sequences for radiation therapy monitoring allow for rapid image acquisition (e.g., 4–8 frames per second) in two-dimensional planes that can be selected to intersect with the tumor target or critical OAR for continuous visualization and tracking.

Furthermore, continuous MRI with tumor tracking enables monitoring for non-periodic motion, such as the detection of bowel gas movement causing a shift in tumor location during treatment delivery (Fig. 6.3).

## Limitations of MRI-Guided Motion Management

Despite the advantages of MRI-guided motion management, there are key limitations associated with the approach. MRI bores are relatively nar-

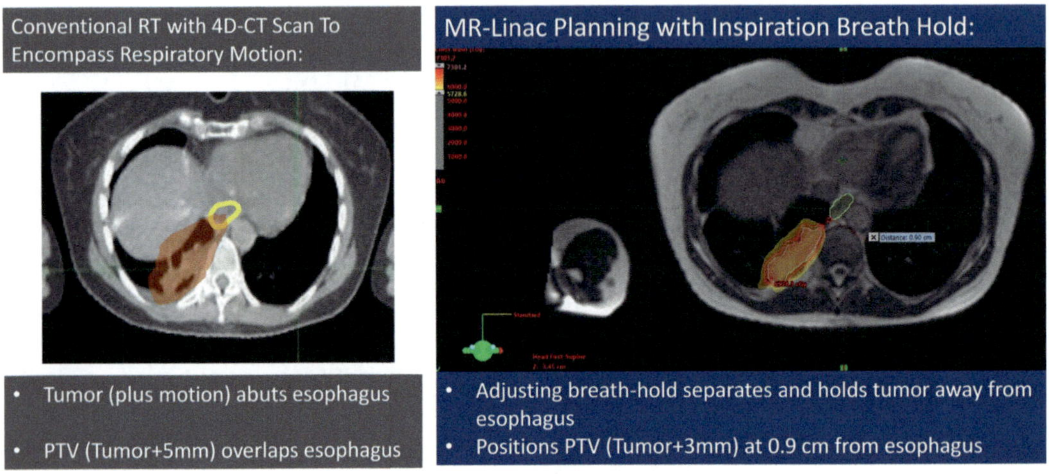


**Fig. 6.2** Comparison of 4D-CT (left) versus MR-guided breath-hold (right) approaches for SBRT for a large right lower lobe lung tumor with significant respiratory motion. Left: 4D-CT-based ITV generation with 5 mm PTV results in high-dose overlap with the esophagus (57 Gy isodose line in color wash, esophagus in yellow). Right: MR-guided breath-hold technique and adjustment of the patient's breath-hold depth allowed for positioning of the tumor away from the esophagus and safe delivery of 12 Gy × 5 fractions (57 Gy isodose line in color wash, esophagus in yellow)

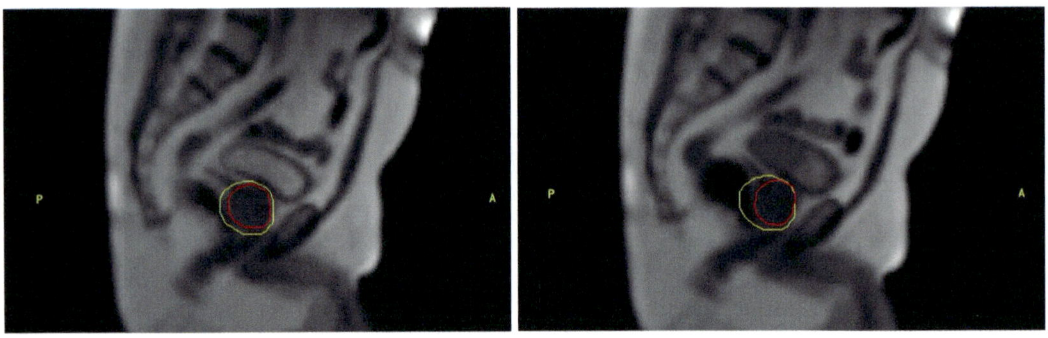


**Fig. 6.3** Intra-fraction physiological motion: Representative frames from continuous sagittal MR cine imaging on the ViewRay MRIdian during beam-on for a prostate cancer received MRI-guided SBRT with the prostate target tracking structure (red) and gating region of interest (yellow). Left: Decompressed rectum that matches rectal contours from initial simulation image and MRI of the day; Right: Intra-fraction shift of gas from bowel into the rectum causing expansion of the rectum and shift of the prostate target during beam delivery

row and more enclosed compared to CT bores, and therefore, the patient must be able to tolerate being in the MRI for an extended period of time. Furthermore, MRI is susceptible to motion artifacts as acquiring 3D volume images require more time than CT scans, and the motion over the MRI acquisition time will cause blurring of the image. Tracking of small lesions is challenging with current MR-Linac systems as the tracking software has difficulty tracking small lesions accurately. Often in these cases, a surrogate will be used for tracking instead. The accuracy of the tracking software can also be compromised by the presence of nearby MR artifacts.

In summary, X-ray-based versus MRI-based motion management approaches both have unique advantages and disadvantages for radiation therapy planning and delivery. The selection of the technology for a given patient and clinical indication needs to be personalized.

## Simulation

Patients treated on an MR-Linac will typically undergo CT simulation, simulation on the MR-Linac, and, at times, an additional simulation on a diagnostic MRI scanner or MR simulator that may have imaging capabilities that the MR-Linac does not possess. The CT scan is used for dose calculation and MR-Linac scan for treatment planning and contouring, and the additional MR scan is usually acquired on a higher field strength MRI scanner, equipped with a flat top table and laser alignment system, with superior image quality and more available sequences compared with the MR-Linac for guiding delineation of the treatment target. While the order in which the simulations are performed may vary across institutions, motion management strategies for inter-fraction and intra-fraction motion are considered at the first simulation and executed in each successive simulation such that the patient's treatment position is reproducible and representative of their internal anatomy and external position at treatment.

## Positioning and Immobilization

For conventional X-ray-based radiation therapy, rigid immobilization devices are heavily relied on for motion management in order to reproduce the patient's position from simulation to treatment and minimize motion during treatment. This is particularly important for conventional Linac treatments because soft tissue is often not well visualized on kV radiographs or cone beam CT, and imaging internal anatomy during treatment is limited to kV images that only provide good visualization of high-density structures such as bone or implanted radio-opaque fiducial markers.

Considerations for immobilization devices for treatments on a MR-Linac are different than for conventional Linac treatments because soft tissue is better visualized on MRI, treatment plans may be adapted each fraction for variations in internal anatomy and external position, and internal anatomy may be visualized and tracked for gating during treatment. Furthermore, patient set-up with immobilization devices on MR-Linacs are limited by the size of the MR bore and also need to include the MR radiofrequency (RF) receive coils. Depending on the MR-Linac configuration, the requirements for incorporation of the receive coils in the immobilization strategy will differ. The Elekta Unity system has posterior RF receive coils integrated in the couch top and an anterior coil mounted on a bridge, whereas the ViewRay MRIdian has flexible surface RF receive coils that need to be positioned anteriorly and posteriorly to the patient [16, 17]. Additional MR-specific equipment such as hearing protection (headphones or earplugs) and emergency squeeze bulb also need to be considered and incorporated into the patient set-up with the immobilization device. Immobilization devices for the MR-Linac must be MR conditional or MR safe. The immobilization devices must also fit through the MR bore with enough clearance to account for variation in patient geometry, primarily patient width, as the diameter of the bore is the limiting factor. For MR-Linacs that allow for lateral couch shifts, such as the ViewRay MRIdian, enough clearance between the immobilization device and the bore must also be considered to allow for lateral shifts at treatment, such as 1–2 cm.

For treatment sites in the thorax, abdomen, or pelvis, rigid immobilization devices such as vacuum cushion-based immobilization may be used (e.g., Vac-Loc Cushions (CIVCO) or VacQfix Vacuum Cushions (Qfix)) to aid in patient positioning reproducibility. For MR-Linacs requiring surface RF receive coils, the receive coil must fit on top or under the vacuum cushion, and the vacuum cushion must be formed to fit through the bore with enough clearance for lateral shifts, if applicable. Some institutions choose not to use any rigid immobilization devices, but instead opt for a non-rigid immobilization such as only the surface RF receive coils. The RF receive coils on the ViewRay MRIdian are flexible coils that are placed on the anterior and posterior of the patient, secured with velcro straps, centered over the treatment area. Many groups find that this is enough immobilization to maintain patient comfort and ensure that the patient remains as still as possible for the duration of the treatment [18]. Having the patient secured in only the RF receive coils also reduces the probability of clearance issues with the bore since there is no additional width added to the patient from a rigid immobilization device. On the Elekta Unity system, there are less issues with RF coil interference since the posterior RF receive coils are integrated in the couch and an anterior coil mounted on a bridge which can slide superiorly-inferiorly and can be adjusted in height to avoid patient and immobilization collisions.

For treatment sites in the brain and head and neck, patients are immobilized with a thermoplastic mask. Headphones are typically used for hearing protection and communication with therapists. When the thermoplastic mask is used for immobilization, hearing protection and communication must also be accommodated through the mask. In this situation, ear plugs can be used in place of headphones that directly connect to the audio system and fit more easily through the mask. When flexible surface RF receive coils are used, the posterior coil may lie flat on the couch top or on top of the headrest beneath the patient's head. The anterior coil is then applied and positioned to avoid touching the patient's head, such as through the use of a holder or grip ring.

Patient positioning should be reproduced at each simulation, CT, MR-Linac, and, if applicable, additional MRI simulation, including the rigid immobilization devices. The MR coils should also be positioned as they would at treatment for each simulation. Some MR-Linacs, such as the ViewRay MRIdian, provide dummy coils which have the same geometry and materials as the RF receive coils, without the electronics in the coil. This ensures that the patient's position at simulation is most similar to treatment to minimize registration error between the images. This reduces the amount of deformation needed when performing a deformable registration with the CT to the MR-Linac MRI and a more accurate rigid registration between the MR-Linac and additional MRI simulation images.

For online adaptive treatments, day-to-day reproducibility of the exact patient position is not absolutely critical as the treatment plan will be adapted for changes in the patient's internal anatomy and external position. Treatment times, especially online adaptive treatments, on an MR-Linac are longer than conventional Linac treatments. Currently, online adaptive treatments on an MR-Linac can take on average 45 min [19], although software and hardware improvements are constantly being made by vendors to decrease MR-Linac treatment times. Given the longer treatment times, patient comfort is very important for safe and effective treatment [20, 21]; if a patient is comfortable, they are less likely to move during treatment. Patients may be treated with their arms above their head as in conventional Linac treatments or treated with their arms down by their sides. For online adaptive treatments, many groups choose to treat patients with their arms down by their sides as this is more comfortable for the long treatment times that are needed for online adaptive treatments. For treatments that use the MR for image guidance and are not online adaptive treatments, many groups choose to treat patients with arms above their head as these treatments are shorter than online adaptive treatments and this position may have treatment planning advantages, allowing accessibility of more gantry angles in the treatment plan.

## Respiratory Motion

Acquisition of MRIs requires tens of seconds to several minutes depending on the sequence parameters, field of view, and resolution. As a result, respiratory motion, especially for treatment areas in the thorax and abdomen, can cause significant motion artifacts when the MRI is acquired while the patient is free breathing. MRIs acquired over these time periods encompass multiple breathing cycles and result in motion artifacts that appear as blurring of the image. Tumor targets in the pelvis may also be subject to the impact of respiratory motion when the patient breathes with their belly as opposed to chest, resulting in blurring of the MRI. An MRI with these motion artifacts for treatment planning creates uncertainty in contouring of the tumor targets and OARs and dosimetry.

On gated MR-Linac systems, such as the ViewRay MRIdian, respiratory motion can be managed with a breath-hold technique, where the patient is instructed to hold their breath at a consistent capacity for tens of seconds during image acquisition and during treatment. At simulation, the patient must first be assessed for breath-hold candidacy. A good candidate for the breath-hold technique is a patient that is able to hold their breath for at least the minimum duration of a 3D image acquisition on the MR-Linac (e.g., 20–30 s) reproducibly at a consistent capacity (i.e., not releasing their breath throughout the breath hold). This can be assessed at CT simulation using systems that are also applied for conventional Linac simulations and treatments for respiratory motion, such as the Varian respiratory gating for scanners device, that characterize and monitor the patient's respiratory pattern. If the patient is a good candidate to be treated with a breath-hold technique, all simulation images (CT and MRIs) should be acquired in the treatment position with breath-hold. If simulation on the MR-Linac precedes CT simulation, breath-hold candidacy would be assessed by acquiring MRIs and evaluating the amount of respiratory motion artifact present. The tumor target and OARs are contoured on the breath-hold MRI, and the treatment plan is subsequently created on this MRI.

During treatment, gated MR-Linac systems are designed to track a target that moves with respiratory motion and deliver radiation only when the target that is being tracked is in the predefined breath-hold position. Thus, patients that use the breath-hold technique will also require an internal target that moves with respiratory motion and can be tracked. Ideally, the target that is tracked is the treatment target itself, that is, the tumor (Fig. 6.4), but may also be a nearby organ as a surrogate for motion. The motion of the selected tracking target can be evaluated at simulation on the MR-Linac, and multiple tracking target candidates for tracking during treatment may be identified. For treatments in the pelvis, identifying a target to track for breath-hold can be difficult as often the treatment target (e.g., diseased lymph nodes) may not move with breathing; however, OARs such as the bowel may be impacted by respiratory motion. However, bowel is not a good candidate for a tracking target as its shape, position, and/or track of movement can vary in a short period of time depending on the presence of gas and/or its content. In these types

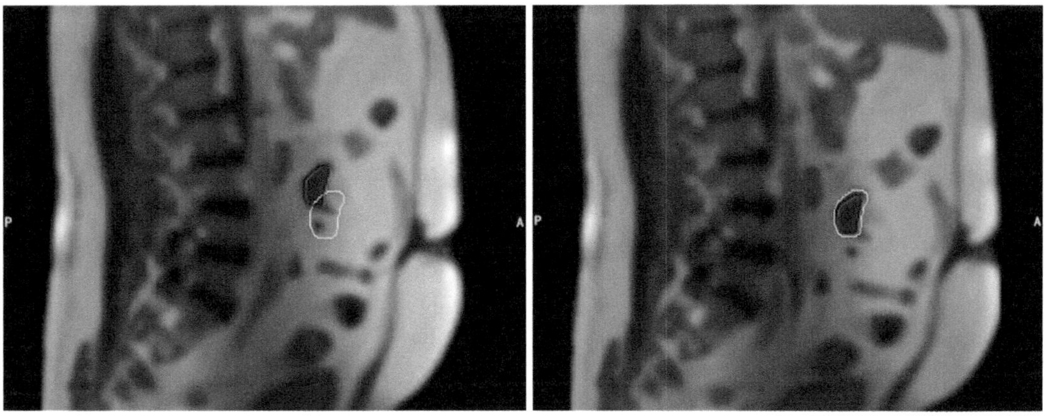


**Fig. 6.4** Representative frames from the sagittal cine during MR-Linac simulation on the ViewRay MRIdian outlining the tumor to be treated (red) when not in the treatment breath-hold position (left) and when the tumor is in the correct breath-hold position for treatment (right)

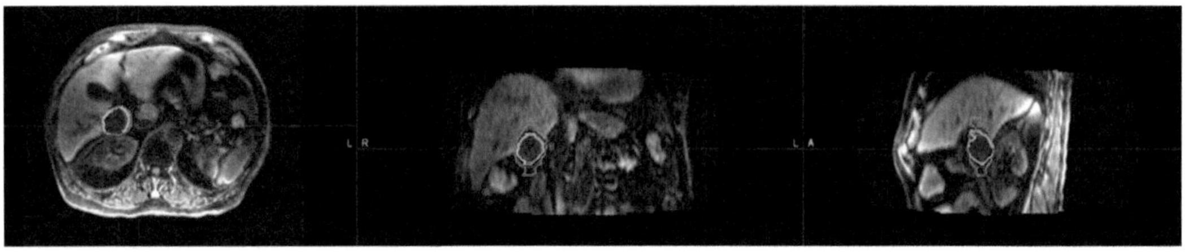

**Fig. 6.5** Motion averaged images on the Elekta Unity for a liver cancer patient showing the planned ITV (blue) vs. daily ITV (pink) drawn from the daily 4D-MRI respiratory binned images. Left, axial view; middle, coronal view; right, sagittal view

of scenarios, the tracking target may not be the tumor itself, but a surrogate for respiratory motion such as a kidney.

If a patient is not a good candidate for breath-hold or there are no sufficient internal targets to track for breath-hold, the patient may be considered for simulation and treatment free breathing. The patient should only be treated free breathing if the target and critical OARs are sufficiently visible despite motion artifacts on the MRI, and additional planning considerations should be incorporated for the uncertainty of tumor and OAR position, such as larger planning target volume (PTV) margins for the tumor or the use of planning organ-at-risk volumes (PRVs) for OARs.

For non-gated treatments on the MR-Linac, for example, the Elekta Unity system with no gating capability, motion can be managed through more conventional techniques. Patient selection should account for the feasibility of treatment on the system depending on the target's motion extent. If the motion assessed during 4D-CT simulation does not exceed the clinical threshold for treatment without gating (e.g., 1.5 cm), then the patient may be a good candidate, given that no other contraindications are present. The extent of the motion to allow for non-gated treatments may vary between clinics. To reduce respiratory motion, a compression belt can be used. The measured motion is then accounted for in the planning process by determining the ITV which is used to generate the PTV.

Prior to treatment, daily motion-average or mid-position images generated from the acquired 4D-MRI is used for image registration and pre-treatment adaptive planning. The respiratory binned images can be used to visualize the extent of the daily motion with respect to the daily PTV. PTV adjustments can be made based on the extent of motion. Figure 6.5 shows an example of an ITV adjustment made based on the daily respiratory binned images for a patient with metastatic melanoma to the liver. In this case, the extent of the motion was larger than the initial planned motion.

Breath-hold techniques are not widely adopted on the Elekta Unity system (non-gating version of the system). However, they can be feasible by coaching the patient and manually holding the beam when the target is outside of the breath-hold window. This technique is more reliant on user operation and will need to be carefully implemented. A recently released upgraded version of the Elekta Unity system allows for breath-hold treatments (deep inspiration breath-hold, inhale, exhale), where the patient can be imaged and treated in a voluntary breath-hold position.

## Tracking and Gating

Cine MR images are single plane images that are taken with very short repetition and echo times, allowing multiple images to be taken per second. This allows close to real-time imaging of a patient's internal anatomy. More importantly, this allows for both visualization of tumors and OARs during treatment and the development of algorithms to track them from cine frame to cine frame [22]. Motion management in MR-guided radiotherapy involves the use of 1 to 3 planes of these cine MR images, a predefined tracking target structure and a boundary that defines the range of motion allowable for the target. The target structure is defined during the initial planning process and often redrawn during the adaptive workflow accounting for the difference in its daily position. During treatment, this structure is deformed first from the adaptive re-planning MRI to the first cine, then from each preceding cine onto the next succeeding cine image. Because of this, care must be taken when defining the tracking structure.

Currently, there are generally two different types of motion management, namely, tracking with manual gating, where users have to actively determine when to turn the beam off manually, and tracking with automatic gating, where the Linac is designed to automatically beam off when certain predefined conditions are met. These conditions are user defined and ideally involve tracking the tumor itself, though this is not always possible depending on the tumor size and relative intensity difference from its environment. In automatic gating, these conditions would generally involve constraints on the motion of the tumor or OAR and the confidence in the deformation of the structure being tracked. Acquisition speed of cine images is on the order of a few hundred milliseconds per image (four or eight frames per second for Viewray MRIdian and five frames per second for Elekta Unity), allowing users to track motion multiple times during a single respiratory cycle.

## Single Plane Vs. Multiplane Tracking

Due to computational limitations, for now, real-time 4D-MRI cannot be generated on MR-Linacs for motion management; instead, single or multiple 2D cine planes are utilized. Early versions of the ViewRay MRIdian enable single plane tracking and gating during treatment in the sagittal plane containing a tracking structure used to gate the radiation beam. The sagittal view is sufficient to visualize the superior-inferior and anterior-posterior motion, thus allowing the user to account for respiratory motion as most evidently seen through the motion of the diaphragm (superior-inferior motion) and chest (anterior-posterior motion). However, lateral movement is generally not visible on the sagittal plane. Consequently, lateral tumor motion out of the treatment region or lateral OAR motion into the treatment region may not also be easily seen and accounted for. Single plane tracking in other planes such as the coronal plane provides information on the lateral motion of the tumor and OARs as well as in the superior-inferior directions, but not in the anterior-posterior direction. Single plane tracking in the axial plane provides lateral motion and anterior-posterior information but not superior-inferior motion. This remains a considerable drawback in machines capable of only single plane cine, as information on motion in at least one direction is lacking.

Tracking in multiple planes provides information on tumor and OAR motion in all three directions, superior-inferior, anterior-posterior, and lateral. A recent upgrade to the ViewRay MRIdian, the A3i system, has multiplane tracking capability. This enables tracking in all three planes, although most often the tumor or OAR is tracked in the sagittal and coronal planes during treatment. Figure 6.6 provides an example of multiplane tracking for a pancreatic tumor on the ViewRay MRIdian A3i system showing the tumor tracked in the sagittal and the coronal planes.

The Elekta Unity system allows for three-plane tracking for non-gated treatments. Figure 6.7 shows a frame with three-plane tracking from continuous MR cine imaging during treatment for a patient with metastatic renal cell carcinoma treating the intra-abdominal lymph nodes.

The upgraded version of the Elekta system allows for automatic gating in two planes, the sagittal and coronal. The two planes cover the motion extents in all three directions: laterally, anterior-posterior, and superior-inferior. Gating strategies on the system are divided into two categories: non-respiratory and respiratory. The non-respiratory “exception gating” allows the beam to be held when the target moves outside of a defined tolerance. This technique can be

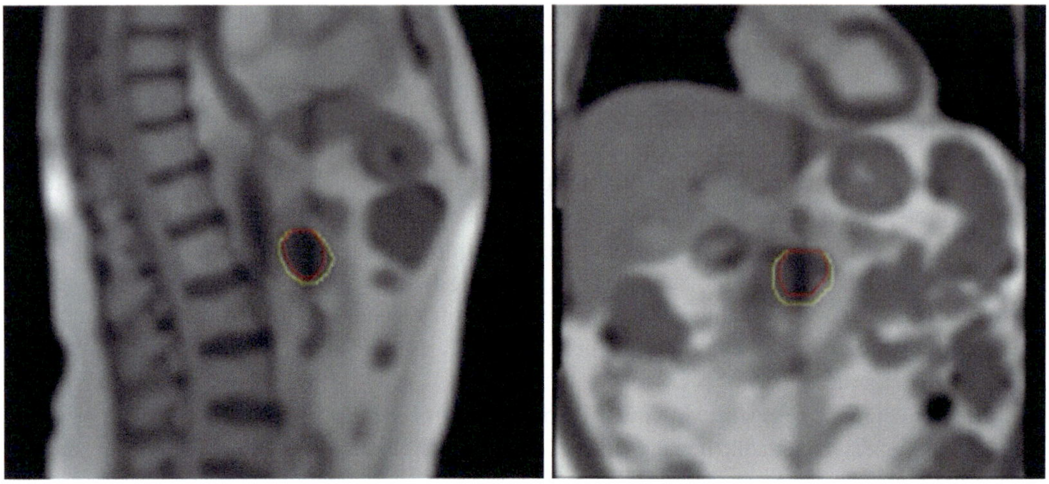

**Fig. 6.6** Representative frames from continuous MR cine imaging during beam-on for a pancreatic tumor receiving MRI-guided SBRT on the ViewRay MRIdian with a pancreatic tumor tracking structure (red) and gating region of interest (yellow). Left, sagittal cine view; right, coronal cine view

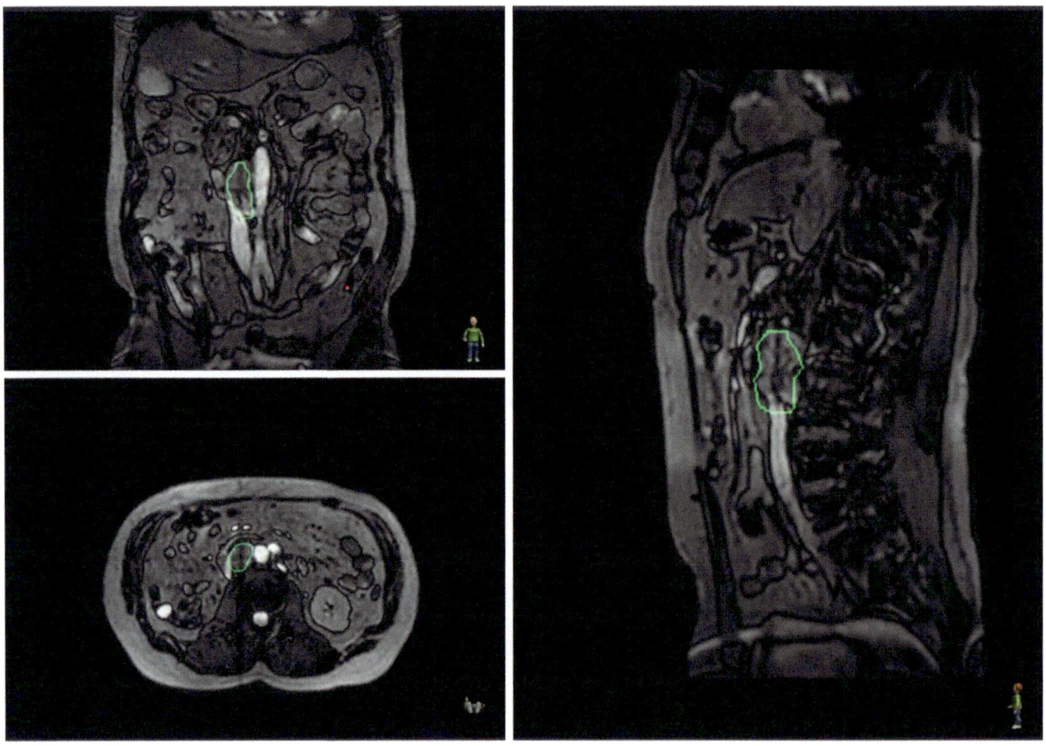

**Fig. 6.7** Representative frames from continuous MR cine imaging during treatment for a patient with metastatic renal cell carcinoma, treating the intra-abdominal lymph nodes, receiving MRI-guided SBRT on the Elekta Unity system. The green contour is the "Mask" used for tracking. Top left, coronal cine view; bottom left, axial cine view; right, sagittal cine view

used for pelvis and head and neck cases. For respiratory gating, three strategies can be used: free breathing (time averaged position), full exhalation, and breath-hold (deep inspiration breath-hold, inhale, exhale). For the free breathing strategy, the patient can breathe freely, and the beam will be held if the target moves outside of the defined tolerance. The full exhalation strategy allows the patient to breathe freely, and the beam will only be enabled at the end of the exhale position. With the breath-hold strategy, the patient can be imaged and treated in a voluntary breath-hold position. These respiratory-based techniques are most suitable for patients with treatment sites in the abdomen and thorax.

## Considerations for Tracking and Gating Structures

There are several important considerations for determining the appropriate structure to track on the real-time cine that will gate the radiation beam on and off.

1. The most important consideration is the relative intensity difference between the region being tracked and its surroundings. There must be a stable, distinct, and well-defined difference in image intensity from surrounding structures in the tracking region. The deformation algorithm is intensity based, so the difference in intensity should be stable regardless of differences in $B_0$ homogeneity due to gantry position. Figure 6.8 shows an example of an ideal, contrast-enhanced liver lesion that can be used for tracking. This lesion was successfully tracked because of the contrast difference between the lesion, abdominal fat, and the surrounding liver. Contrast agents, such as gadoxetic acid, may be used to help increase intensity differences between the tracking region and its surroundings when the difference is not as distinct.

   When tracking contrast-enhanced lesions, the amount of time between injection of contrast and start of treatment should be taken into account during the adaptive workflow to minimize the contrast from washing out at treatment. For gadoxetic acid, since contrast is enhanced for 20 min to 120 min post injection, this translates to an approximately 100 min time limit from start of imaging to the end of treatment before contrast begins to decrease. In addition, while blood vessels/ducts may seem like ideal candidates for tracking on the 3D MR image, the fluid flow in them could cause their brightness to be unstable relative to their surroundings in single plane cine MRI and could result in suboptimal tracking in real time.

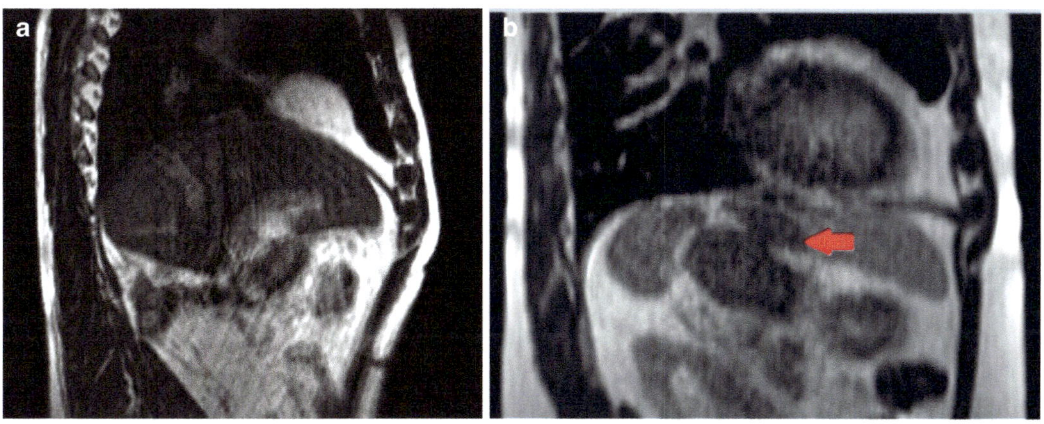


**Fig. 6.8** (**a**) Sagittal view of an MR acquired on the ViewRay MRIdian of a liver lesion that did not require Gd-contrast since the liver lesion had good contrast with the rest of the liver (left) (**b**) and another patient that had Gd-based contrast, Eovist, administered to differentiate the liver lesion (red arrow) from the healthy liver that was successfully used for tracking (right)

2. Imaging speed considerations: of particular interest for this are lung and liver lesions. Because the pixel spacing for cine images are relatively large, it may not be possible to reliably track small lesions. For example, in the lungs, while the intensity difference between lesions and the surrounding lung is large and stable, some lesions are small enough and undergo enough motion that a lower acquisition speed may be required to allow for greater temporal averaging. This would, in turn, allow for better tumor visualization. However, because slower acquisition does result in greater averaging, it can wash out the intensity difference between the tracking region and its surroundings. While usually not an issue for lung lesions due to the large differences in voxel intensity between lung tissue and tumor, this may decrease the intensity difference in other cases (e.g., between liver lesions and healthy liver).
3. In relation to consideration #2, the size of the tracking region cannot be too small (less than ~0.5 cm minimum distance). If the lesion is only a few voxels large, then blurring and motion during cine will make it difficult not only for the machine to reliably identify the lesion but also for users to verify the accuracy of the region being tracked. Figure 6.9 shows an example of a lesion that was not a good candidate for tracking due to both its size and predominance of similar sized dark spots in the surrounding liver. In such instances, it might be necessary to track the whole liver or a surrogate that can be assumed to have correlated motion. In these cases, however, a reconsideration of tolerances such as limits on overall motion of the surrogate might be necessary.
4. Distinctness of the tracking region: ideally, the tracking region would be the only region with differing densities. For instance, in Fig. 6.7, there are multiple dark regions all over the liver that are similar in size and intensity to the gross tumor volume. During beam delivery, the patient's breathing motion can cause deformation of the liver, resulting in the software and users being unable to determine which dark region is the actual lesion. In this instance, the decision was made to track a whole segment of the liver with tighter allowable motion.

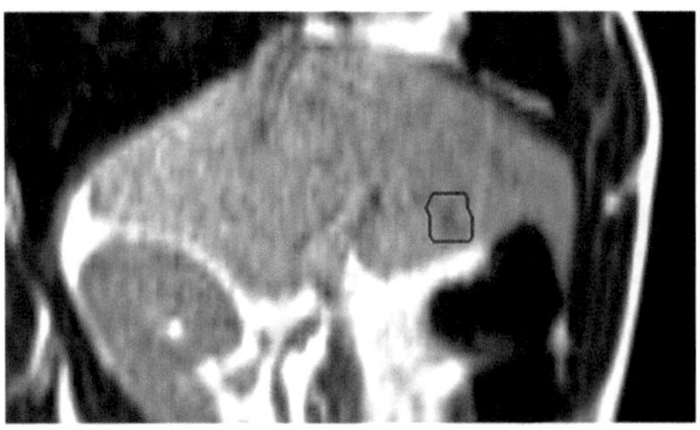

**Fig. 6.9** Sagittal view of a liver lesion treated on the ViewRay MRIdian that was a poor choice for tracking due to the very small area occupied by the hypointense region and the presence of multiple, similarly sized hypointense regions. The gross tumor volume is shown in blue

## Multi-Target Tracking

Current MR-guided linacs can image and track in multiple cine planes. In cases where tracking is limited to a single plane, motion of a target can only reflect motion in two out of three directions (e.g., superior-inferior and anterior-posterior in a sagittal cine MRI). As a consequence of this, tracking multiple targets with only one plane involves either (1) having the targets in the same image plane; (2) tracking only one lesion after determining the correlation in motion between the two targets and adjusting the PTV margin to account for not visualizing the second target during treatment; or (3) having multiple isocenters to treat each target separately.

Some systems such as the ViewRay MRIdian A3i system has the ability to track multiple structures (e.g., target and an adjacent OAR) in different planes and to simultaneously gate on each structure, which allows for enhanced verification of proper positioning for critical structures. For example, diseased abdominal lymph nodes may not move significantly with respiration; however, adjacent critical OARs such as the bowel may move. When these patients are treated with the breath-hold technique, tracking multiple structures ensures that the patient is in the correct breath-hold position during radiation delivery. As seen in Fig. 6.10, the kidney can be tracked as a surrogate for respiratory motion in the sagittal plane, and the abdominal lymph node can be tracked in the coronal plane, ensuring that the treatment target has not moved during treatment. Both structures need to be within the predefined gating region of interest in order for the beam to gate on. If either of the structures are outside of the gating region of interest, the beam will not turn on. A similar concept may be applied to patients where multiple tumors are being treated simultaneously.

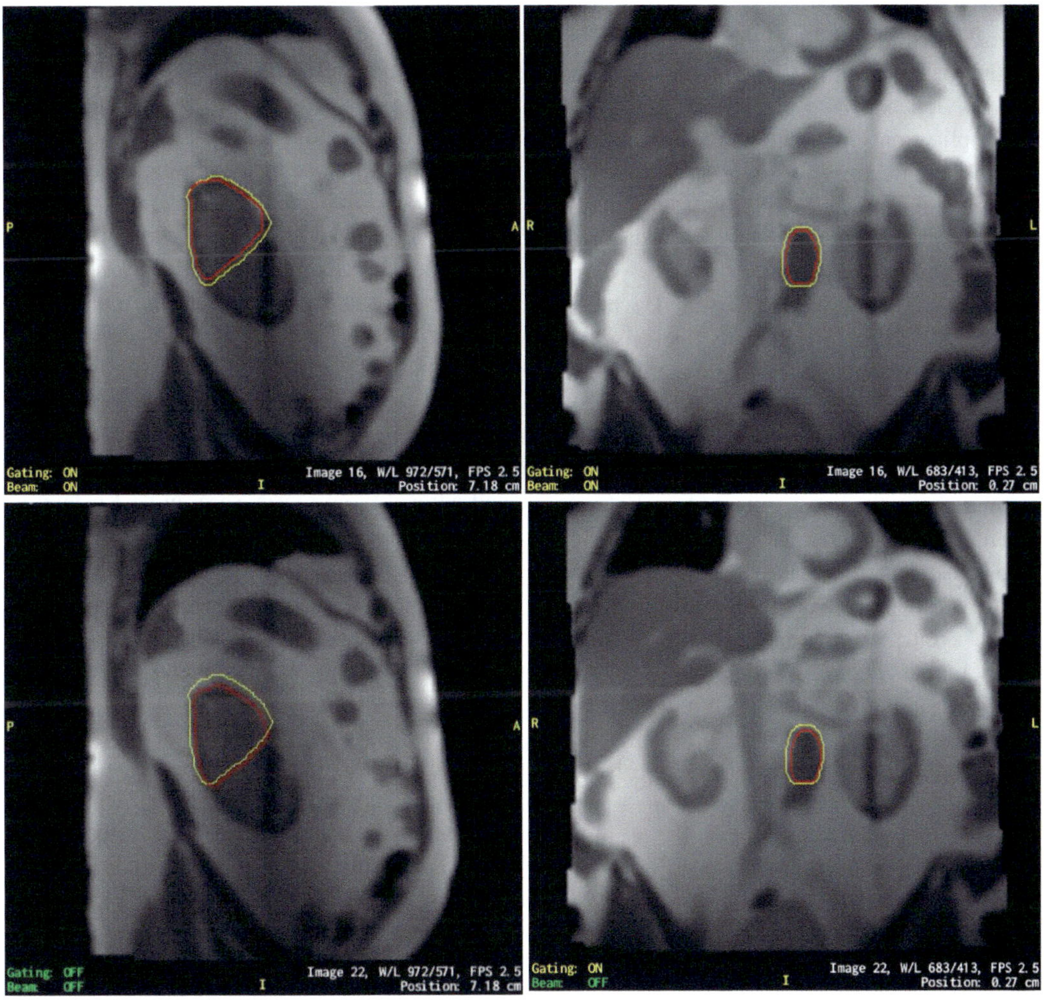


**Fig. 6.10** Representative frames from continuous MR cine imaging during beam-on for a diseased abdominal lymph node receiving MRI-guided SBRT on the ViewRay MRIdian. Multiplane tracking with multiple tracking targets was applied. A portion of the kidney was tracked on the sagittal plane (left) as a surrogate for respiratory motion and consequently OAR motion, and the abdominal node was tracked on the coronal plane (right) to verify that the treatment target was not moving. Top row, both the kidney and abdominal node tracking structures (red) are within the gating region of interest (yellow); the beam is gated on. Bottom row, the kidney tracking structure is outside of the gating region of interest, but the abdominal node remains inside, the beam is gated off

## Planning Considerations

Planning considerations for patient immobilization are similar to conventional Linac treatment plans. As recommended by AAPM TG-176, attenuation through immobilization devices should be accounted for in the dose calculation [23]; this should also include attenuation through the RF receive coils.

For patient positioning, the decision between treating patients with their arms up versus arms down is dependent on patient comfort but also has implications for treatment planning strategies. When treating a patient with their arms positioned down, gantry angles that would allow the beam to traverse through the arms are eliminated due to the uncertainty in arm position. Furthermore, additional gantry angles may be avoided clockwise and counterclockwise from each arm (e.g., within 2 cm in each direction from the arm) since the exact arm position at simulation may not be reproduced at each treatment day, and this reduces the probability that an arm may be in the beam path. Although this further limits the potential gantry angles that can be used in the plan, of note is that some of these gantry angles that are avoided due to the arms are already avoided in an arms up scenario due to the couch corners/rails that some groups already avoid for the high attenuation through the outer edges of the couch. The arms could also purposefully be placed a bit farther from the body to reduce the number of avoidance angles. Arm clearance will need to be accessed at each treatment on MRI. Van den Wollenberg et al. compared plan quality metrics for patients with oligometastatic liver disease treated on a conventional Linac with arms up versus an MR-Linac with a 15 IMRT beam plan with arms down and avoiding the arms. They found that target coverage was comparable between the plans. Dose conformity and dose to organs-at-risk were slightly worse for MR-Linac plans due to the restriction of gantry angles. However, the MR-Linac plans still met all the clinical target objectives and OAR constraints [24].

Given the features available with MR-guided linacs, enhanced soft tissue image-guided set-up, accounting for inter-fraction motion through online adaptive planning, the size of margin required to maintain appropriate target coverage for MR-Linacs may be reduced compared with conventional Linac treatments. For example, a prospective study with rectal cancer patients treated on an MR-Linac initially used conventional Linac PTV margins and then reduced the PTV margins after evaluating motion during treatment. The PTV margins were reduced for both the mesorectum and elective nodal regions from 1 cm to 4–6 mm for the mesorectum and 8 mm to 4–6 mm for the elective nodal regions [25]. Without the use of real-time tumor imaging, the PTV margin must account for some uncertainty when surrogates are used for motion such as fiducials [26]. Under MR guidance with real-time cine imaging and gating during treatment, the tumor itself is tracked in real-time enabling a reduction in the PTV margin. For breath-hold treatments on the MR-Linac for lung, pancreas, and adrenal tumors, van Sornsen de Koste et al. found that a PTV margin of 3 mm was sufficient to account for almost all the uncertainties in the study, however, recommended that a larger PTV margin be considered for free breathing scenarios [27]. PTV margins of 3 mm are typically used on the ViewRay MRIdian with gated breath-hold treatments.

When treating patients on the MR-Linac with a non-gated approach, the total treatment volume will be larger since an ITV must be generated, such as on the Elekta Unity MR-Linac. The selection of PTV margin is site specific, but typically ranges from 3 mm to 1 cm from the ITV. Using a gating approach could reduce the PTV margin used and overall treatment volume since the PTV would be expanded from the GTV or CTV, instead of an ITV.

Patients treated with the breath-hold technique have a significant reduction in tumor motion compared to being treated free breathing. Ehrbar et al. evaluated the residual motion during gating for lung and liver lesions and found that it was small and comparable for all patients (mean 2D root mean square error of 2.4 mm). Furthermore, compared to the static planned dose, with gating, the dose to the CTV was slightly reduced, with a difference of -0.6% for the mean dose and -1.7% for the dose to 95% of the CTV. This study indicates that gating based on internal anatomy minimizes target motion during radiation delivery and safe and robust delivery of high radiation doses to these tumors [28].

## Patient Coaching

One major advantage of motion management with MR-Linacs is that radiation delivery can be triggered by the motion of the lesion itself, instead of using a surrogate such as in surface-guided radiation therapy. Alongside these benefits, however, the importance of proper patient education cannot be overstated. Generally, patients will need to be coached on their breathing patterns and, especially in gating, be trained to take reproducible breaths.

### Feedback Mechanism

In some MR-Linacs, the patient can view the cine images used to track tumor motion. The cine is projected in real time onto a screen located at the wall close to the opposite end of the MRI bore. By using mirrors, patients are able to view the screen throughout the treatment and watch as their tumor moves according to how much air they inhale/exhale. The ViewRay MRIdian A3i system has the capability to display both the cine and tracking structure or only the tracking structure for patient viewing. This feedback mechanism can empower patients to self-coach, deciding when they will breath-hold and when to recover their breath, while the therapist supervises the treatment, as opposed to following instructions dictated by a therapist throughout the duration of treatment. While the feedback mechanism works well for some patients, it does not work for all patients, and some patients will still need coaching from a therapist.

### Patient Education and Coaching

For gated treatments, a patient can perform either deep inspiration, mid-inspiration, or end expiration breath-hold, according to what is clinically determined most suitable for the treatment. For instance, a mid-inspiration breath could allow the patient to adjust the amount of air in the lungs more easily, alternating inhaling or exhaling more air as needed [29]. On the other hand, end expiration or deep inspiration may position the tumor farther away from OARs and/or be easier for patients to maintain [30]. Ultimately, the clinician will need to assess both the breath-hold capabilities of the patient and the resulting location of the tumor relative to OAR before deciding on which breath-hold techniques would be the most ideal before simulation. This assessment can be done at the initial patient consultation where the physician can evaluate the breath-hold capabilities of the candidate patient and educate the patient on the breath-hold expectations at simulation and treatment.

The best opportunity for active patient coaching is during the simulation. First, a patient would be instructed to perform a deep or mid-inspiration or end expiration breath-hold, and a 3D MRI is acquired. A temporary tracking structure is then drawn on the treatment region as verified by the physician, and a cine is performed. While watching the cine, the patient can be coached to attempt to reproduce the breath-hold performed during the acquisition of the 3D MRI. After the simulation, the patient should be reminded to practice breath-hold prior to treatment.

## Conclusion

MR-Linacs offer a new era for motion management strategies with visualization of soft tissue targets and the ability to gate the radiation beam by tracking the actual tumor itself. These improved motion management strategies over X-ray-based image-guided radiation therapy offer the opportunity to reduce target margins, thereby reducing toxicity to normal tissues and yet maintaining accuracy in treating tumors. While there are still several limitations with MR-guided motion management strategies, it is anticipated that the continuous development of MR-Linacs, both hardware and software, will overcome these limitations of current MR-Linacs.

**Acknowledgments** We thank Zhaohui Han and Jennifer Campbell for their assistance in creating ViewRay MRIdian figures.

## References

1. Brandner ED, Wu A, Chen H, et al. Abdominal organ motion measured using 4D CT. Int J Radiat Oncol Biol Phys. 2006;65:554–60.
2. Keall PJ, Mageras GS, Balter JM, et al. The management of respiratory motion in radiation oncology report of AAPM Task Group 76. Med Phys. 2006;33:3874–900.
3. Liu HH, Balter P, Tutt T, et al. Assessing respiration-induced tumor motion and internal target volume using four-dimensional computed tomography for radiotherapy of lung cancer. Int J Radiat Oncol Biol Phys. 2007;68:531–40.
4. Sim AJ, Palm RF, DeLozier KB, et al. MR-guided stereotactic body radiation therapy for intracardiac and pericardial metastases. Clin Transl Radiat Oncol. 2020;25:102–6.
5. Bernstein HM, Leon W, Daly ME, et al. Noninvasive stereotactic radiation for refractory ventricular tachycardia after failure of cardiac sympathetic denervation. JACC Case Rep. 2022;4:1189–94.
6. Dees-Ribbers HM, Betgen A, Pos FJ, et al. Inter- and intra-fractional bladder motion during radiotherapy for bladder cancer: a comparison of full and empty bladders. Radiother Oncol. 2014;113:254–9.
7. Hamlet S, Ezzell G, Aref A. Larynx motion associated with swallowing during radiation therapy. Int J Radiat Oncol Biol Phys. 1994;28:467–70.
8. Li H, Chen H-C, Dolly S, et al. An integrated model-driven method for in-treatment upper airway motion tracking using cine MRI in head and neck radiation therapy. Med Phys. 2016;43:4700.
9. Bruijnen T, Stemkens B, Terhaard CHJ, et al. Intrafraction motion quantification and planning target volume margin determination of head-and-neck tumors using cine magnetic resonance imaging. Radiother Oncol. 2019;130:82–8.
10. Gurney-Champion OJ, McQuaid D, Dunlop A, et al. MRI-based assessment of 3D intrafractional motion of head and neck cancer for radiation therapy. Int J Radiat Oncol Biol Phys. 2018;100:306–16.
11. Sixel KE, Aznar MC, Ung YC. Deep inspiration breath hold to reduce irradiated heart volume in breast cancer patients. Int J Radiat Oncol Biol Phys. 2001;49:199–204.
12. Mampuya WA, Nakamura M, Matsuo Y, et al. Interfraction variation in lung tumor position with abdominal compression during stereotactic body radiotherapy. Med Phys. 2013;40:091718.
13. Stemkens B, Paulson ES, Tijssen RHN. Nuts and bolts of 4D-MRI for radiotherapy. Phys Med Biol. 2018;63:21TR01.
14. Paulson ES, Ahunbay E, Chen X, et al. 4D-MRI driven MR-guided online adaptive radiotherapy for abdominal stereotactic body radiation therapy on a high field MR-Linac: implementation and initial clinical experience. Clin Transl Radiat Oncol. 2020;23:72–9.
15. Omari E, Jassar H, Straza M, et al. Daily 4DMRI assessed motion and its impact on ITV for non-gated MRI-guided adaptive radiation therapy of liver cancer. Hoboken, NJ: Wiley; 2022.
16. Winkel D, Bol GH, Kroon PS, et al. Adaptive radiotherapy: the Elekta Unity MR-linac concept. Clin Transl Radiat Oncol. 2019;18:54–9.
17. Mutic S, Dempsey JF. The ViewRay system: magnetic resonance-guided and controlled radiotherapy. Semin Radiat Oncol. 2014;24:196–9.
18. Cuccia F, Alongi F, Belka C, et al. Patient positioning and immobilization procedures for hybrid MR-Linac systems. Radiat Oncol. 2021;16:183.
19. Güngör G, Serbez İ, Temur B, et al. Time analysis of online adaptive magnetic resonance-guided radiation therapy workflow according to anatomical sites. Pract Radiat Oncol. 2021;11:e11–21.
20. Raaymakers BW, Jürgenliemk-Schulz IM, Bol GH, et al. First patients treated with a 1.5 T MRI-Linac: clinical proof of concept of a high-precision, high-field MRI guided radiotherapy treatment. Phys Med Biol. 2017;62:L41–50.
21. Kurz C, Buizza G, Landry G, et al. Medical physics challenges in clinical MR-guided radiotherapy. Radiat Oncol. 2020;15:93.
22. Menten MJ, Fast MF, Wetscherek A, et al. The impact of 2D cine MR imaging parameters on automated tumor and organ localization for MR-guided real-time adaptive radiotherapy. Phys Med Biol. 2018;63:235005.
23. Olch AJ, Gerig L, Li H, et al. Dosimetric effects caused by couch tops and immobilization devices: report of AAPM Task Group 176. Med Phys. 2014;41:061501.
24. van den Wollenberg W, de Ruiter P, Nowee ME, et al. Investigating the impact of patient arm position in an MR-linac on liver SBRT treatment plans. Med Phys. 2019;46:5144–51.
25. Intven MPW, de Mol van Otterloo SR, Mook S, et al. Online adaptive MR-guided radiotherapy for

rectal cancer; feasibility of the workflow on a 1.5T MR-linac: clinical implementation and initial experience. Radiother Oncol. 2021;154:172–8.
26. Ge J, Santanam L, Yang D, et al. Accuracy and consistency of respiratory gating in abdominal cancer patients. Int J Radiat Oncol Biol Phys. 2013;85:854–61.
27. van Sörnsen de Koste JR, Palacios MA, Bruynzeel AME, et al. MR-guided gated stereotactic radiation therapy delivery for lung, adrenal, and pancreatic tumors: a geometric analysis. Int J Radiat Oncol Biol Phys. 2018;102:858–66.
28. Ehrbar S, Braga Käser S, Chamberlain M, et al. MR-guided beam gating: residual motion, gating efficiency and dose reconstruction for stereotactic treatments of the liver and lung. Radiother Oncol. 2022;174:101–8.
29. Lens E, Gurney-Champion OJ, Tekelenburg DR, et al. Abdominal organ motion during inhalation and exhalation breath-holds: pancreatic motion at different lung volumes compared. Radiother Oncol. 2016;121:268–75.
30. Rodriguez LL, Kotecha R, Tom MC, et al. Impact of breath-hold MR-guided radiotherapy (MRgRTBH) vs. free-breathing CT image-guided radiotherapy (CT-IGRTFB) on gastrointestinal sparing and dose conformality in adrenal SBRT. Int J Radiat Oncol Biol Phys. 2021;111:e527.

# 7 sCT and Dose Calculation

Emilie Alvarez-Andres, Fernanda Villegas, Anaïs Barateau, and Charlotte Robert

## Introduction

### Crucial Role of the CT Scanner in Radiatiotherapy

Conventional 3D radiation therapy (RT) planning, as implemented since the introduction of the CT scanner in the 1990s in RT departments, uses Hounsfield units (HU) to qualify tissue properties in terms of electron densities (ED). This correlation is obtained through a calibration curve for each set of acquisition/reconstruction CT scan parameters, which allows the corresponding HUs to be derived from inserts with known ED. This property is essential in external photon beam RT since it conditions cross sections of photon-matter interactions. In addition to its usefulness for dose calculation, the planning CT also allows the patient to be repositioned at each treatment session using in-room kV X-rays or cone beam CT (CBCT) imaging, which are essential for the implementation of image-guided radiation therapy (IGRT). However, CT suffers from very poor contrast when it comes to identifying structures in volumes containing soft tissue, hence the use of combined CT/MRI imaging for volume delineation, particularly for brain and pelvis [1] (Fig. 7.1, upper part). The recent commercialisation of hybrid machines combining an MR device and a linear accelerator (MR-Linac)

E. Alvarez-Andres
Department of Radiotherapy and Radiation Oncology, Faculty of Medicine and University Hospital Carl Gustav Carus of Technische Universität Dresden, Dresden, Germany

OncoRay – National Center for Radiation Research in Oncology, Faculty of Medicine and University Hospital Carl Gustav Carus, Technische Universität Dresden; Helmholtz-Zentrum Dresden-Rossendorf, Dresden, Germany
e-mail: Emilie.AlvarezAndres@ukdd.de

F. Villegas
Radiotherapy Physics and Engineering, Medical Radiation Physics and Nuclear Medicine, Karolinska University Hospital, Stockholm, Sweden

Department of Oncology-Pathology, Karolinska Institute, Stockholm, Sweden
e-mail:
fernanda.villegas-navarro@regionstockholm.se

A. Barateau
Univ Rennes, CLCC Eugène Marquis, Inserm, LTSI - UMR 1099, F-35000 Rennes, France
e-mail: a.barateau@rennes.unicancer.fr

C. Robert (✉)
UMR 1030 Radiothérapie Moléculaire et Innovation Thérapeutique, ImmunoRadAI, Université Paris-Saclay, Institut Gustave Roussy, Inserm, Villejuif, France

Department of Radiation Oncology, Gustave Roussy, Villejuif, France
e-mail: CH.ROBERT@gustaveroussy.fr



I. J. Das et al. (eds.), *A Practical Guide to MR-Linac*, https://doi.org/10.1007/978-3-031-48165-9_7

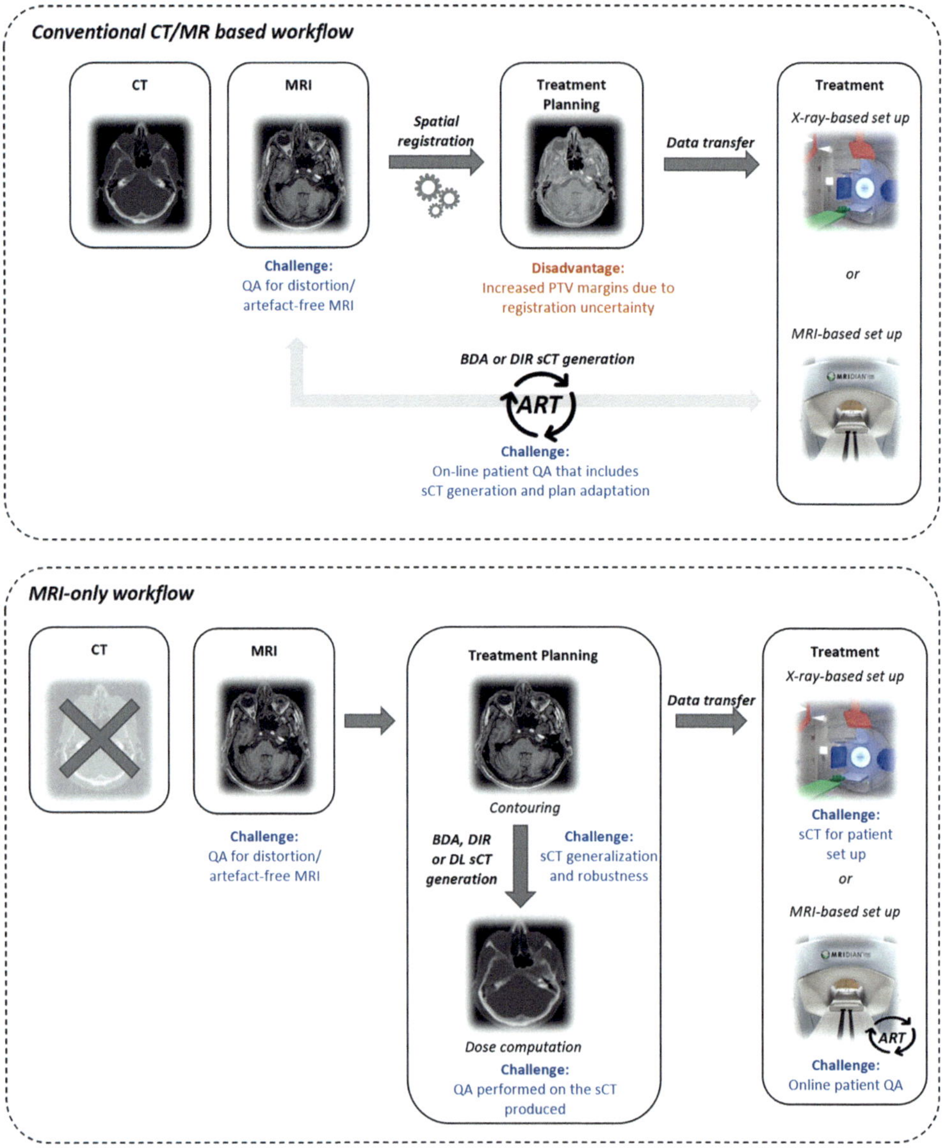


**Fig. 7.1** Comparison of the two possible workflows implemented in RT in 2023. In the first case (top), treatment planning is based on CT imaging to which MRI can be added to facilitate the delineation of target structures, where arising spatial registration uncertainties between the two modalities are addressed by increased margins. In this workflow, the main challenges rely in the production of high-fidelity, artefact-free MR images, and the ability to generate dosimetrically accurate sCT at each treatment session for workflows that include MR-Linac treatments. However, this process can take advantage of the information provided by the CT scan acquired during the planning stage for the implementation of adaptive radiation therapy (ART), i.e. based on the anatomy of the day. Two methods are proposed by vendors for this purpose: methods based on deformable image registration (DIR) or methods based on electronic density assignments in segmented regions (BDA, bulk density assignment). By contrast, the MRI-only workflow (bottom) eliminates the planning CT and therefore uses a sCT as early as at the planning stage, required for dose calculation. In this workflow, geometric fidelity over large fields of view becomes important to correctly restore all volumes and especially the external contour. The implemented method for generating sCT must be generalisable to any new patient anatomy. Patient-specific quality controls must be implemented to ensure that there are no dose-impacting mistakes in the generation of the sCT. As CT imaging is still required to set up the patient in the X-ray-based image-guided RT workflow, it is ultimately necessary to ensure that the generated sCT can perform this task

has brought breakthrough technological solutions to access the patient's anatomy of the day that, together with the subsequent adaptation of treatment plans, and makes real the possibility to "see what you treat". The latter, when applied to an online re-optimisation of treatment plans workflow, has however raised questions, hence bringing the need to generate *synthetic CT scans (sCT)*, also called in the literature "virtual CT", "substitute CT", or "pseudo CT".

## Other Applications of sCT

The concept of generating sCTs is however not new and has already been proposed for the implementation of MRI-only workflows, i.e. without requiring the acquisition of a planning CT, for brain, head and neck, and pelvis applications (Fig. 7.1, lower part). The goals of such MR-only workflows are manifold. First, it allows to limit the imaging departments' burden and patient irradiation by eliminating an X-ray-based examination. Second, it removes anatomy differences deriving from the non-simultaneous CT and MRI examinations, especially for pelvic tumour sites where bladder and rectal fillings may differ. Finally, it reduces the uncertainty underlying multimodal registration – evaluated to be about 2 mm for the brain and the pelvis and up to 5 mm in the abdomen [2–6] – and so the derived margins between the clinical target volume (CTV) and the planning target volume (PTV). MRI-based planning is also widely used in brachytherapy since the introduction of the concept of image-guided adaptive brachytherapy, which has demonstrated benefits in terms of local control in locally advanced cervical cancer [7]. However, in brachytherapy, dose calculations are still currently based on the TG43 formalism and can be performed on MRI without methodological difficulties, as the patient is assumed to be only composed of water [8], thus not requiring sCTs. In nuclear medicine, the need to use sCT already emerged at the time of the introduction of the PET-MRI bi-modality, for which the replacement of the CT modality by MRI raised methodological questions for the implementation of the attenuation correction. In the early 2010s, methods to assign uniform electron densities in previously segmented regions and multi-atlas strategies were already proposed as solutions to this issue [9, 10].

Recently, several groups have evaluated the cost-effectiveness of MR-only workflows, particularly in the management of prostate cancer [11, 12]. Based on ten patients, Persson et al. [11] showed that the elimination of the CT scan and the acceleration of the target delineation stage were the main cost-saving factors, leading to a moderate cost reduction of about 200 EUR/patient. A Finnish team tackled a similar problem, but also considering notions of investment in the purchase of heavy equipment [12]. For a period of 10 years and 300 patients annually managed for prostate cancer, a 2% cost reduction was evaluated between CT + MRI workflow and MRI-only workflow. These studies are, however, merely indicative as the prices of each procedure vary from one country to another and are based on local experiences and workflows. In both studies, costs linked to the step of generating synthetic scans were however not considered.

## Challenges in MRI-Based Workflows

The implementation of MRI-based workflows is not straightforward in many respects, as illustrated in Fig. 7.1, particularly since it requires the development of specific quality assurance (QA) programmes. These challenges are detailed in the following lines which serve as an introduction to all the sections that will follow.

First, when using MR images for treatment planning purposes, their geometric fidelity must be ensured. This is of particular concern for MRI-only workflows, requiring large fields of view (FOV). Indeed, distortions can affect the external contours, which are of major importance for dose calculation, but also the definition of target structures, their accuracy being crucial especially for the implementation of stereotactic treatments. Magnetic susceptibility artefacts or chemical shifts can also affect the images, leading to errors in the generation of sCT images, in addition to difficulties/errors in volumes delineation.

Second, for MRI-only workflows, robust methods for generating sCTs need to be implemented, i.e. that are reliable regardless of the patient anatomy and the quality of the MR images used as input data. In addition to routine QA, patient-specific quality checks must therefore be conducted with the aim of verifying the trustworthiness of the result. Several solutions are possible for this purpose, including visual checks, comparison with a reference CT, the use of a second algorithm for sCT generation, or the generation of uncertainty maps as second output of the tool [13]. For treatments where patient repositioning is based on X-rays, it is also important to verify the ability of the generated sCT to perform this task accurately and most importantly without systematic errors. Finally, the use of MR-Linac requires both rapid generation and quality control of sCTs, with the patient lying on the table. For MRI/CT-based workflows, the availability of the planning CT scanner is a strong advantage.

In this chapter, the first part will present the methods proposed in the literature to generate sCTs, from the basic methods, such as bulk density, to more recent methods based on deep learning. The second part will discuss the metrics used to evaluate the quality of these sCTs in the context of RT. The third part will summarise the key points of MRI quality and recent recommendations for QA programmes for the implementation of safe MRI-based or even MRI-only workflow. A review of implemented academic and industrial solutions in the clinic will conclude the chapter. Throughout the chapter, the focus will be on methods integrated into commercially available MR-Linacs.

## sCT Generation from MRI for Radiotherapy Dose Calculation: State of the art

The first reported way for RT dose calculation in the context of an MRI-only workflow was to perform a bulk density assignment on MRI sub-regions [14]. Then, atlas-based and voxel-based approaches were both first introduced in 2011. Most recent voxel-based methods of sCT generation from MRI are now based on machine learning and particularly deep learning (DL). Table 7.1 presents the global characteristics of these methods and their respective advantages and drawbacks. Some of the sCT generation results produced by research teams will be presented in this section, the performance metric being described in Table 7.2 and in section "Metrics for sCT Evaluation".

### Bulk Density Assignment

The bulk density assignment method requires the segmentation of the MRI into sub-volumes before assigning them with pre-defined ED depending on the tissue categories (such as air, bone, and soft tissue).

According to Johnstone et al. [14], the number of tissue classes considered and the assigned density values differ between studies. Several sCT studies, for prostate and brain RT, have considered one single water class [15–17]. The reported success of this simple approach is, however, limited with dose errors potentially exceeding 2% [18–22]. More complex scenarios were implemented with the aim of identifying the number of tissue categories necessary to achieve acceptable dosimetric errors.

Based on four prostate patients, Lee et al. [23] generated a first water-only sCT ($sCT_W$) by assigning the whole body with the water ED derived from ICRU Report 46 [24]. A second sCT ($sCT_{WB}$), considering both water and bone classes, was computed with the additional bone assignment obtained averaging the bone CT intensities over all the patients. Based on a 64 Gy PTV prescription dose, the maximum root mean square differences (RMSD) inside the 70% CT and sCTs isodoses were equal to 1.33 Gy and 0.88 Gy for $sCT_W$ and $sCT_{WB}$, respectively, proving the crucial need of a minimum of a two class-density assignment approach.

This observation was further confirmed in 2008 by Eilertsen et al. [20], with ten prostate

**Table 7.1** Presentation of the main sCT generation methods with their prerequisites, advantages, and drawbacks

| | Bulk density assignment | Atlas-based approaches | Voxel-based approaches | |
|---|---|---|---|---|
| | | | Statistics | Machine learning |
| Definition | Assignation of homogeneous densities to MRI segmented classes | Performing deformable registration of the given MRI to one or several co-registered MRI-CT atlas(es) | Use of voxel intensity information for sCT prediction | |
| Pre-requisite | Segmentations on MRI + density values per structure | Atlas database (MRI-CT pairs) | Statistic model definition | Training step to establish the MRI to CT translation model |
| Advantages | • Simple<br>• Fast | • Continuous values of HU | • Continuous values of HU if regression | • Continuous values of HU<br>• Fast<br>• Does not require access to matched data for the latest deep-learning methods |
| Inconvenients | • Need of specific MR sequence(s) for bone delineation<br>• Homogeneous areas in the sCT | • Dependent on deformable registration accuracy<br>• Calculation time (depending on the number of patients considered in the database)<br>• Not robust to atypical patient anatomies | • Potential need of specific MR sequence(s) enhancing bone visibility<br>• Model performance dependency | • Potential need of specific MR sequence(s) enhancing bone visibility<br>• Not robust to atypical patient anatomies (if not present in training data)<br>• Learning step computationally costly |

patients. The axial T2-weighted MRI were converted into water-equivalent sCT by assigning 1.02 g/cm$^3$ to the body ($sCT_{WO}$) and two water-bone sCTs with the additional bone segmentation and densities assignment of 1.3 g/cm$^3$ ($sCT_{WB1}$) and 2.1 g/cm$^3$ ($sCT_{WB2}$). After applying a collapsed cone dose algorithm, the dose volume histogram (DVH)-based differences between mean doses to the CTV on CT and sCT were equal to −0.9%, 0.2%, and 4.2% for $sCT_{WO}$, $sCT_{WB1}$, and $sCT_{WB2}$, respectively, highlighting the importance of the adequate bone density assignment.

The same conclusions were obtained in case of head-and-neck tumours for seven patients with a 2D axial T1-weighted MRI [25]. In this study, three sCTs were generated, namely, a water equivalent sCT ($sCT_W$), a water/bone sCT ($sCT_{WB}$), and a water/bone/air sCT ($sCT_{WBA}$), based on bone and air densities equal to 1.53 g/cm$^3$ and 0 g/cm$^3$, respectively. For the PTV receiving 70 Gy, V100% [minimum; maximum] were [6.5%; 65.5%], [−0.2%; 5.7%], and [1.5%; 7.4%], respectively.

The bulk density assignment method is simple to implement but faces one major challenge, being the distinction between bone tissue and air-filled cavities, since both types of tissue show a similar signal on routine anatomical MR sequences. Several sequences have been developed to better visualise tissues with very short T2 relaxation times, in particular bones, tendons, and ligaments. These sequences, called "UTE" for ultrashort echo time or "ZTE" for zero echo time, however, greatly extend the examination times and thus the risks of patients movements. The bulk density assignment method, if manually implemented, can be also tedious and biased by high inter-operator variability. Automatic segmentation solutions were developed very early on, particularly in the context of nuclear medicine, based either on clustering methods or on neural networks [26, 27], and are now commonly used in RT [28, 29].

Several studies have evaluated the bulk density assignment method in the context of MR-Linac treatments, demonstrating its clinical applicability. For breast cancer, Koerkamp et al.

**Table 7.2** Metrics used in the literature for the evaluation of sCT scans generated from MR images. This table is inspired by (Boulanger et al., 2021) [32] and (Vandewinckele et al., 2020) [13]

| | Type of metrics | Metric | Definition | Ideal value | + Advantages/- Disadvantages. |
|---|---|---|---|---|---|
| Image evaluation | Intensity-based metrics | ME: Mean error | $\mathrm{ME} = \frac{1}{N}\sum_{i=1}^{N}(\mathrm{sCT}_i - \mathrm{CT}_i)$ | 0 HU | + Is usually reported<br>+ Easy to be calculated<br>+ Can be calculated within different regions of interest<br>- Does not show the spread in the voxel-wise differences since mean/standard deviation are calculated<br>- Hide patient-wise dispersion as values at the cohort scale are usually reported<br>- No information about the exact location of the error (only organ-wise)<br>- Difficult to compare between studies<br>- Non-trivial association with dosimetric impact<br>- Gives information of absolute error and not of the relative error |
| | | MAE: Mean absolute error | $\mathrm{MAE} = \frac{1}{N}\sum_{i=1}^{N}\lvert \mathrm{sCT}_i - \mathrm{CT}_i\rvert$ | 0 HU | |
| | | MSE: Mean squared error | $\mathrm{MSE} = \frac{1}{N}\sum_{i=1}^{N}(\mathrm{sCT}_i - \mathrm{CT}_i)^2$ | 0 HU | |
| | | RMSE: Root mean squared error | $\mathrm{RMSE} = \sqrt{\frac{1}{N}\sum_{i=1}^{N}(\mathrm{sCT}_i - \mathrm{CT}_i)^2}$ | 0 HU | |
| | | PSNR: Peak signal-to-noise ratio | $\mathrm{PSNR} = \left(\frac{Q^2}{\mathrm{MSE}}\right)$ | Maximum value of dB | + Easy to be calculated<br>+ Gives some information about the relative error<br>- No information about the location of the error<br>- Not representative of the human vision |
| | | SSIM: Structural similarity metric | $\mathrm{SSIM} = \frac{(2\mu_{\mathrm{CT}}\mu_{\mathrm{sCT}} + C_1)(2\sigma_{\mathrm{CTsCT}} + C_2)}{(\mu_{\mathrm{CT}}^2 + \mu_{\mathrm{sCT}}^2 + C_1)(\sigma_{\mathrm{CT}}^2 + \sigma_{\mathrm{sCT}}^2 + C_2)}$ | 1 | + Carries information about inter-dependencies between pixels<br>+ More accurate representation of the human vision compared to PSNR<br>- More difficult to calculate |
| | | NCC: Normalised cross-correlation | $\mathrm{NCC} = \frac{1}{N}\sum_i \frac{(\mathrm{CT}_i - \mu_{\mathrm{CT}})(\mathrm{sCT}_i - \mu_{\mathrm{sCT}})}{\sigma_{\mathrm{CT}}\sigma_{\mathrm{sCT}}}$ | 1 | + Evaluates the level of correlation between two signals<br>+ Less sensitive to linear variations in intensities in the two images compared to conventional cross-validation<br>- Unintuitive interpretation |
| | Geometric fidelity metrics | DSC: Dice similarity coefficient | $\mathrm{DSC} = \frac{2(V_{\mathrm{CT}} \cap V_{\mathrm{sCT}})}{V_{\mathrm{CT}} + V_{\mathrm{pCT}}}$ | 1 | + Relevant for dose calculation/dose reporting<br>+ Relevant for repositioning<br>- DSC is dependent on voxel size<br>- DSC is dependent on the volume of the structures and thresholds/segmentation quality to assess it<br>- Gives no information on actual HU values |
| | | HD: Hausdorff distance | $H(\mathrm{sCT},\mathrm{CT}) = \max\,(h(\mathrm{sCT},\mathrm{CT}), h(\mathrm{CT},\mathrm{sCT}))$ | 0 mm | + Focus on the boundary of the contours = > relevant for repositioning<br>- Sensitive to outliers = > 95th percentile is often preferred<br>- Non-trivial association with dosimetric impact |
| | | MASD: Mean absolute surface distance | $\mathrm{MASD}(A,R) = \frac{d_{\mathrm{ave}}(S_{\mathrm{CT}},S_{\mathrm{sCT}}) + d_{\mathrm{ave}}(S_{\mathrm{CT}},S_{\mathrm{sCT}})}{2}$ | 0 mm | + Focus on the boundary of the contours = > relevant for repositioning<br>- Non-trivial association with dosimetric impact |

| | Type of metrics | Metric | Definition | Ideal value | + Advantages/- Disadvantages. |
|---|---|---|---|---|---|
| Dose evaluation | Dose difference metrics | Voxel-to-voxel dose differences | Difference between the dose distribution computed on the reference CT and on the sCT | 0 Gy or 0% | + Clinically relevant<br>+ Spatial evaluation possible with subtraction of dose distributions<br>- Average can hide potential deviations |
| | | DVH difference | Dose differences on DVH specific points, for a given structure | 0 Gy or 0% | + Easy to calculate if both DVHs are available<br>- No spatial information<br>- Care should be taken in transferring/warping structures sets<br>- A difference in DVH or DVH parameter can be caused by an "error" in the sCT or in the contour or by a difference in the anatomy |
| | Gamma analysis metrics | Mean gamma | Value of the mean gamma | 0 | + Used to give an overall representation of error usability of the sCT |
| | | Gamma pass rate | Percentage of pixels/voxels with a gamma value lower than 1 | 100% | + Analysis can be tweaked to preference (global/local, 2D/3D, dose difference, DTA value and threshold)<br>+ Spatial evaluation possible with gamma maps<br>- Dependency to target volumes, dose algorithm, treatment technique<br>- Difficult to compare between studies |
| Patient repositioning | | Matching accuracy | Compare translational and rotational differences in case of:<br>- CBCT and CT/sCT registration<br>- kV imaging and CT/sCT-derived DRR registration | 0 mm, 0° | + Vital information about ability of the sCT for patient repositioning<br>- True matching value is unknown: No gold standard available<br>- No standard on the procedure, e.g. handling of residual misregistration |

Abbreviations: *N* number of voxels, $sCT_i$ intensity of the ith voxel of the sCT, $CT_i$ intensity of the ith voxel of the CT, *Q* range of voxel values of sCT and reference CT, $\mu_{CT}$ mean value of CT, $\mu_{sCT}$ mean value of sCT, $\sigma_{CT}$ and $\sigma_{sCT}$ standard deviation of the reference CT and sCT, $C_1$ and $C_2$ are expressed as $(k_1Q)^2$ and $(k_2Q)^2$, *V* volume of CT and sCT, $d_{ave}$ absolute Euclidean distance, h(X, Y) is the largest distance from a point in X to the nearest point in Y, $S_{CT}$ surface of the segmentation of a volume of interest in the CT, $S_{sCT}$ surface of the segmentation of a volume of interest in the sCT, *DRR* digitally reconstructed radiograph, *DTA* distance to agreement

[30] collected ten patients from the ABLATIVE trial and assigned mean ED from the CT segmented volumes for the sCT generation. Three scenarios were investigated: lungs-gross tumour volume (GTV)-remaining body ($sCT_{LT}$), lungs-GTV-chest wall-ipsilateral breast-remaining body ($sCT_{LTCB}$), and lungs-GTV-chest wall-ipsilateral breast-ribs-remaining body ($sCT_{LTCBR}$). Intensity modulated radiation therapy (IMRT) treatment plans were directly optimised on the sCT before being re-calculated on CT with a Monte Carlo algorithm simulating a 1.5T MR-Linac. The obtained medians for the body mean absolute error (MAE) were equal to 106HU, 104HU, and 104HU for $sCT_{LT}$, $sCT_{LTCB}$, and $sCT_{LTCBR}$, respectively. For these same images, median gamma pass rates of 95.8%, 96.3%, and 95.9% were obtained for the 3D 2%/0 mm criterion combined with a 10% dose threshold applied to 20 Gy. Thus, it suggested that segmenting the lungs, GTV, chest wall, and ipsilateral breast, and assigning them specific ED, led to the best compromise between image processing complexity and performance. However, this study raises the issue of the need for an alternative workflow in case the scanner is unavailable.

Similarly, Ding et al. [31] investigated the optimal tissue classes in the case of five patients presenting cervix carcinoma treated with the Elekta Unity (Elekta AB, Stockholm, Sweden). In the first scenario, the sacrum, ilium, ischium, pubis, and coccyx structures were set as water-equivalent, whereas the rest of the body, including femoral heads, was assigned with the averages of relative ED of the corresponding CT structures ($sCT_{FB}$). For the second scenario, these CT-derived relative ED averages were assigned to all the volumes ($sCT_{EB}$). Dose calculation on $sCT_{EB}$ provided better results than on $sCT_{FB}$, at least in terms of gamma analysis, which proved the superiority of complete bone segmentation over the conventional approach only including femoral heads. Yet, this higher number of required volumes may result in a time-consuming segmentation task in clinics.

## Atlas-Based Approaches

The atlas (multi-atlas) approach consists of forming one (several) atlas(es), corresponding to a registered pair(s) of CT and MRI, prior to the sCT generation. The MRI of the atlas(es) is (are) deformably registered to the incoming patient MRI, before applying the obtained deformation field(s) to the corresponding CT atlas(es). In case of the multi-atlas approach, the multiple generated warped CT atlases are finally combined to form a final sCT. This method has several advantages [14, 32]. Firstly, it is based on the use of conventional MRI anatomical sequences and, thus, does not involve any increase in acquisition time. Furthermore, this method has been shown to be robust to artefacts that could affect MRI [33]. Finally, the same method can be applied to the deformation of organs-at-risk (OAR), fulfilling two clinical tasks in a single operation. The major difficulties, however, relate to the ability of such methods to accurately recover the external contour, highly impacting the dose computation [34], and its application to atypical anatomies, in particular for brain tumours where a surgical procedure may have led to the removal of part of the skull [35]. The calculations are also very time intensive, with, for instance, 10 min for a 15 patient large atlas computed on a NVIDIA Titan X graphical processing unit (GPU) [36].

According to the review of Johnstone et al. [14], straightforward approaches, involving the use of a single average atlas for pelvis sCT generation, have first been introduced in the literature [34, 37]. In 2015, Sjölund et al. [38] collected CTs and T1-weighted MRIs of ten head tumour patients to generate the multi-atlas database. The Morphon deformable registration resulted in a series of warped CT atlases that were either combined with mean or median operators. The head MAE resulted in differences of 123.2HU ± 18.6HU and 113.4HU ± 17.6HU for the mean and median strategies, respectively, the latter being of superior accuracy to combine multiple warped CT atlases.

A more elaborated approach was proposed by Farjam et al. [39] for prostate patients. It involved 23 pairs of CT and dual fast field echo mDixon MRI. During the fusion of the warped CT atlases, the generalised registration error [40], a metric evaluating the registration inaccuracy, was computed between the deformed MRI atlases and the processed MRI of the incoming patient. The final sCTs were finally obtained through a weighted mean, with weights equal to the metric inverse. For the 13 patients of the testing set, the mean MAE were equal to 47HU, 116HU, 36HU, and 47HU for the body, bone, fat, and muscle regions, respectively. The initial treatment plan recalculation resulted in mean and maximum DVH-extracted doses to the PTV equal to 0.8% ± 0.2% and 0.9% ± 0.3%, respectively.

In this same perspective of similarity-based sCT process optimisation, Uh et al. [41] performed a pioneering study in 2014, involving 26 brain tumour paediatric patients. The fusion of warped CT atlases, following a pattern recognition approach, was performed with the extraction of 9x9 voxels size patches. The final sCT value for a given voxel V was a weighted mean of all corresponding encompassed V-based patches voxels. A similarity evaluation involving the incoming patient MRI and the corresponding multi-atlas MRI patches was used to assess the ponderation factors. To evaluate the role of the multi-atlas set size, either 6 or all 12 patients were used to generate the sCT. The body RMSD resulted in mean performance of 219 HU and 207 HU for the 6- and 12-patient-based sCT approach, respectively ($p < 0.01$). The chi index, a gamma index-inspired metric [42], with a 2%/2 mm criterion, led to rates equal to 98.3% ± 1.5% and 98.5% ± 1.3% for the 6- and 12-patient-based sCT generation, respectively, illustrating in this case the limited added value of the multi-atlas set enlargement.

## Voxel-Based Approaches

The voxel-based methods, first developed in 2011, rely on MRI voxel information to predict sCT. They present the disadvantage of limited clinical translation owing to the potential need of exotic MRI sequences enhancing bone visibility, such as UTE or ZTE [43].

First introduced in 1974 [44], the Gaussian mixture models (GMM) belong to the unsupervised machine learning category and are often used for data clustering. In the context of sCT generation, tissues from MRI and CT are assumed to have Gaussian distributions, enabling the elaboration of a regression between these two intensity types [45]. The main studies relying on this approach were reviewed by Johnstone et al. [14]. In 2011, Johansson et al. collected data from five patients presenting brain CT, UTE, and T2-weighted MRI sequences [46]. The GMM, which was elaborated based on head voxels only, resulted in a MAE of 137HU. The potential of these sCTs for dosimetry and patient setup accuracy was evaluated by Jonsson et al. [47]. The dose calculation on CT and sCT of five glioblastoma patients resulted in maximum mean DVH differences of 0.8% in target volumes and 3%/3 mm gamma pass rates without dose threshold ranging from 99.4% and 100%. Regarding the patient setup accuracy, the volume delineation on sCT-derived digitally reconstructed radiographs (DRR) and then transposition to CT-derived DRR resulted in an absence of visual discrepancies. An improvement of this approach was proposed in 2013 by Johansson et al. [48]. The spatial location details of the voxels were added to the model via the consideration of the distance voxels/external segmentation and voxels 3D coordinates. The database was composed of nine patients, all presenting brain CT and UTE MRI. The model without and with the latter modifications resulted in mean absolute prediction deviations of 140HU ± 22HU and 130HU ± 18HU, respectively, proving the superiority of the updated approach. Based on a non-AI regression method, Pandey et al. [49] designed in 2019 a hybrid model relying on a linear function to predict HU involving MRI intensities and coordinates, as well as a patient-specific random contribution. The database was composed of 12 head tumour patients, with CT and T1-weighted MRI. The head MAE was equal to 71.1HU ± 5.5HU.

The fuzzy c-means (FCM) algorithms have also been widely used in fields such as segmentation of medical images [50], astronomy [51], and economy [52]. In the context of sCT generation, they are used to obtain the probabilities for voxels to belong to given tissue classes before assigning corresponding HU intensities. In 2020, Hsu et al. [53] collected five cases presenting abdominal 3D mDixon sequences that were first pre-processed and then used as FCM algorithm input for classification into low-intensity representing high-density tissue, mid-intensity involving air/bone marrow/lung/solid bone, and high-intensity for fat. Assigned intensities varied between -1000HU for the air and 354HU for the spine. These HU were multiplicated by the probability matrices and then summed to obtain the final sCT. Minimum Dice similarity coefficients (DSC) of 0.24, 0.84, and 0.94 were achieved in air, fat, and lung, respectively. Based on a 4-tissue classification with FCM, namely, low-density, mid-density, bone, and marrow, Bredfeldt et al. [54] enrolled 16 hepatocellular carcinoma patients with CT and 3D Dixon MRI. Median MAE were all below 60HU. After VMAT plans calculation with the anisotropic analytical algorithm (AAA), median PTV DVH differences were between ±0.17Gy. To enhance the accuracy of a FCM approach aiming at classifying six thoracic tissues, Su et al. [55] proved that the acquisition of UTE-mDixon and the association of reconstructed water-only Dixon MRI, fat-only Dixon MRI, and the water-only divided by the sum of water-only and fat-only Dixon images should be used as algorithm input. Body MAE inferior to 50HU were reported. A similar order of magnitude was obtained by Liu et al. [56] with a MAE of 49.1HU ± 17.8HU in pelvic soft tissues and 274.4HU ± 26.9HU for bones. In this study, to prevent the classification into bone of non-bone mask belonging voxels, the five-class-based FCM algorithm was regularised. The dosimetry analysis, involving VMAT plans, resulted in maximum mean DVH differences of 0.5% for PTV.

An additional approach is random forests, i.e. the combination of decision nodes organised into trees that can be both used for classification and regression. However, they are inaccurate when dealing with non-bijective functions such as translation between MRI and CT [57]. Based on five patients with brain CT and 3D T1-weighted MRI, Yang et al. [58] reported a peak signal-to-noise ratio (PSNR) of 27.27 dB ± 1.00 dB based on an approach where features, namely, the local binary pattern and the Gabor wavelet, were first computed in all MRI patches, before disregarding the less suited for sCT generation. In Wang et al. [59], an auto-context approach was adopted, implying that, except for the first sCT training set case computation, the model included sCT-derived features. This model evaluation relied on a cohort of 14 brain lesion patients with CT and T1-weighted MRI. The treatment plans, dynamic conformal arc (DCA) or VMAT, were re-calculated with AAA and resulted in averages inferior or equal to 0.171Gy and 0.019Gy for PTV and OAR DVH differences, respectively. With similar performance, Lei et al. [57] proposed to develop a model for MRI tissue classification into bone, air, and water with a first random forest algorithm, before using the probability map-derived features to build a second random forest model, as additional input to MRI-derived features. Based on 14 brain CT and T1-weighted MRI, body MAE and PSNR were respectively equal to 57.5HU ± 8.5HU and 28.33 dB ± 1.68 dB. Average DVH differences were not superior to 0.123Gy and 0.020Gy for PTV and OAR, respectively. A comparison of multi-atlas, random forests without auto-context and with auto-context sCT approaches was performed by Nie et al. [60] for 22 pelvis cases. It respectively resulted in MAE of 66.1HU ± 6.9HU, 51.2HU ± 3.8HU, and 48.1HU ± 4.6HU, proving the superiority of the voxel-based method over multi-atlas.

DL concept was first introduced in the mid-1980s and is now widespread in many medical and non-medical fields. DL relies on a series of non-linear functions organised into layers, enabling the access to highly deep features [61]. DL networks are trained using an optimisation process that requires a loss function to calculate the model error. The choice of the loss function(s) has an impact on sCT accuracy, as well as the training data and the tuning of network hyperparameters [62]. The DL model can be trained with 2D, 2.5D, or 3D images. 2D images are typically axial slices. A 2.5D model corresponds to three different 2D trainings in axial, sagittal, and coronal planes. For a given MRI, the final sCT is hence generated by averaging (or computing the median value) of the three sCTs obtained from each training. A 3D DL model allows to reduce inter-slice artefact, but it is more computationally costly than 2D or 2.5D.

Two reviews about the sCT generation with DL methods in radiotherapy were published in 2021 [32, 43]. They classified the popular DL architectures into two categories: generator-only and generative adversarial network (GAN) with its variants. Several architectures were proposed in each category. Here, we summarise the most used DL architectures for sCT generation from MRI, with their advantages and drawbacks.

Among the generator-only models, the most famous architecture is the U-Net, displayed in Fig. 7.2 and which was initially designed for microscopy images delineation [63]. It relies on the combination of an encoder and a decoder, respectively, aiming at computing abstract features

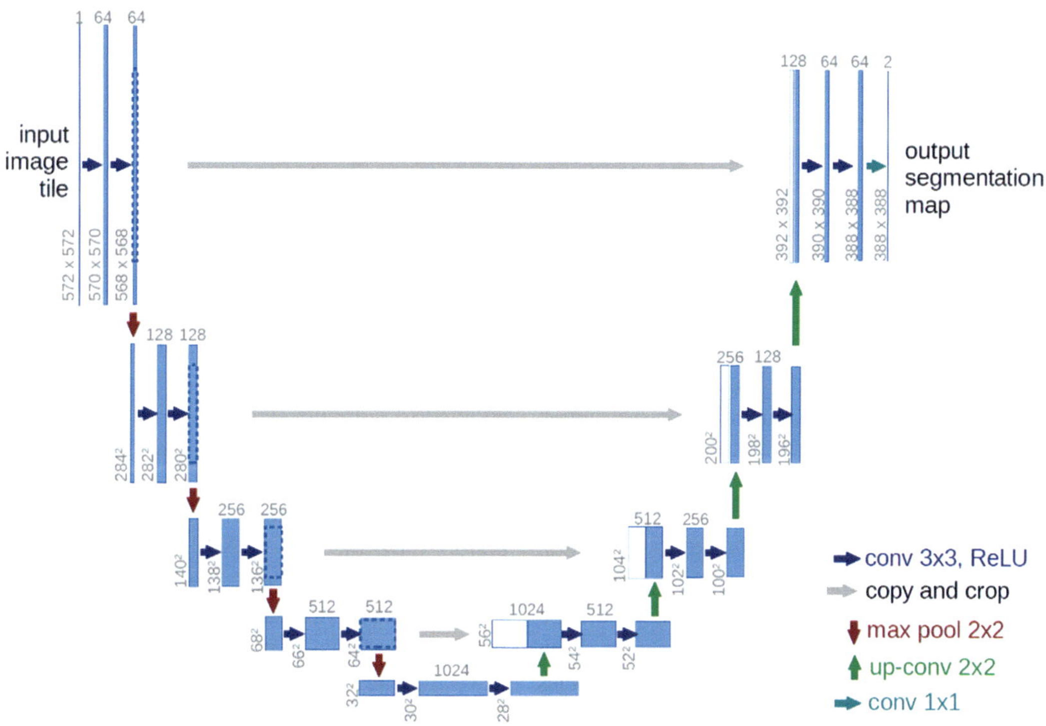


**Fig. 7.2** Original 2D U-Net architecture. The encoder, corresponding to the left branch of the U-shape, is composed of the same block repeated four times. This block is first composed of a convolution with a 3 × 3 kernel size (CONV 3) combined with a rectified linear unit (ReLU) activation function which has the advantage to avoid gradients vanishing. After repeating this step twice, a max pooling with a 2 × 2 kernel size is applied resulting in features map size reduction by a factor 2. Two CONV3, both followed by ReLU, are then applied, before the data reaches the decoder, i.e. the right symmetrical branch of the architecture. It involves four main blocks, which are first composed of a transposed convolution. To further enhance detailed images computation, the equivalent encoder feature maps and the feature maps of the current layer are concatenated. The series of operations CONV 3, ReLU, CONV 3, and ReLU is then applied. At the very end, a convolution with a 1 × 1 kernel size is performed, before generating the output image, corresponding in this context to a sCT. Reproduced from Ronneberger O, Fischer P, Brox T, U-Net: Convolutional Networks for Biomedical Image Segmentation, Medical Image Computing and Computer-Assisted Intervention – MICCAI 2015, MICCAI 2015, Lecture Notes in Computer Science, vol 9351, 234–241, 2015, https://doi.org/10.1007/978-3-319-24574-4_28, with permission from Springer Nature

maps and reconstructing the output images from these features maps. However, this network architecture suffers from certain limitations. First, the convolutions composing the U-Net are not a global operation, hampering the consideration of distant spatial context [64]. Chen et al. [65], therefore, proposed the TransU-Net, a U-shape-based neural network whose encoder is a combination between a convolutional neural network and a transformer, i.e. an attention only-based architecture [66]. Based on the MICCAI 2015 Multi-Atlas Abdomen Labelling Challenge cohort, the segmentation of abdominal CT by Trans-U-Net and the original U-Net resulted in mean DSC equal to 77.48% and 74.68% respectively, proving the potential interest of this new architecture, which however has not yet been tested on the task of sCT generation. It was also proved to lead to brain sCT of 65.6 HU, surpassing the U-Net performance of 80.8 HU [67]. Second, the optimal number of total layers without considering the input, also known as depth, is not straightforward to evaluate since it depends on the network goal [68]. To overcome this issue, Zhou et al. [69] recently introduced the U-Net ++, a deeply supervised segmentation network combining U-Nets of different depths. For the BraTS 2013 challenge focused on brain tumour delineation based on MRI, intersection-over-union score of 91.21% ± 0.68% was reached for U-Net ++, while a rate of 89.21% ± 1.55% was reported for the original U-Net.

The GAN architecture for sCT generation from MRI was first proposed by Goodfellow et al. in 2014 [70]. It consists of the combination of a generator, having the role to compute, from random noise, highly realistic synthetic images to fool the discriminator, and a discriminator aiming at accurately classifying ground truth images and synthetic images as real and false, respectively. Largent et al. compared a machine learning method (patch-based) to a U-Net and a GAN architecture, for prostate sCT generation from T2 3T MRI [71]. The best image quality and dose results were obtained for the GAN DL method. In this study, they also compared performance from different loss functions: L2, single-scale perceptual loss, multiscale perceptual loss, and weighted multiscale perceptual loss. The L2 loss (with U-Net or GAN architecture) offered higher sCT accuracy.

The conditional GAN (cGAN) has also been widely used for sCT generation since its first introduction in 2014 by Mirza et al. [72]. The cGAN particularity is to offer the operator the opportunity to influence the generator output by adding extra data to the generator and discriminator. Despite their higher likelihood to experience mode-collapse in case of small datasets [73], cGAN proved great success in various medical fields. In 2019, Kazemifar et al. [74] proposed a brain sCT study with a 2D cGAN, composed of a U-Net as generator, and a combination of six convolutional layers as well as five dense layers as discriminator. Regarding the cohort, 77 patients presenting a couple of a brain CT and a 2D contrast-enhanced T1-weighted MRI sequence were enrolled. The MAE, air and bone DSC were respectively equal to 47.2HU ± 11.0HU, 0.70 ± 0.07, and 0.80 ± 0.06. After the re-computation of the VMAT treatment plan on the sCT, mean differences between $D_{mean}$, $D_{max}$, $D_{95\%}$, and $D_{5\%}$ metrics were all below 1% for both PTV and OARs.

Still in the GAN category, cycleGANs, introduced by Zhu et al. [75], have been massively adopted to compute sCT since they present the opportunity to use unpaired data. This architecture consists of an image transformation task through the combination of two GANs. The generators have the role to predict Image 1 (CT) from Image 2 (MRI) and Image 2 (MRI) from Image 1 (CT) in case of GAN 1 and 2, respectively. Regarding the discriminators, original and synthetic Images 1 as well as original and synthetic Images 2 are classified by GAN 1 and 2, respectively. One-to-one mappings are assumed by cycleGAN, which may not always be an accurate representation of the clinical reality, in the case, for instance, of the generation of one sCT from multiple MRI [76]. Peng et al. [77] studied sCT generation with cGAN and cycleGAN for 173 nasopharyngeal carcinoma patient data. The generators of the cGAN and the cycleGAN were respectively a modified U-Net and a combination of U-Net with residual neural networks. All discriminators corresponded to a series of six convolutional layers. Matched CT and MRI, after deformable registration, and unmatched data were used as cGAN and cycleGAN input, respectively. A better image accuracy was obtained for cGAN than for cycleGAN, for equivalent dose results in terms of gamma analysis.

The recent studies on sCT for MR-Linacs have mostly focused on thorax, abdomen, and pelvis body regions [78–86]. Indeed, in 2022, Lenkowicz et al. [78] collected 60 patients treated with the MRIdian (ViewRay, California, USA) for lung tumours. All of them underwent a CT without contrast agent and a true fast imaging with steady-state precession (FISP) MRI sequence. To generate the DL model of the sCT ($sCT_{DL}$), axial slices were used as inputs for the 2D cGAN, composed of a U-Net-based network as generator and a 70 × 70 PatchGAN as discriminator [87]. Testing sets 1 and 2 were respectively composed of in-house and outer data. The body MAE were equal to 56.8HU ± 12.4HU and 53HU ± 8.5HU for testing sets 1 and 2, respectively. For the dosimetry analysis, the GTV ED was forced to 1, leading to a second sCT ($sCT_{DL+Bulk}$). When evaluating the $sCT_{DL}$, the global 3D 3%/3 mm with 10% dose threshold gamma index led to 99.1 ± 0.9% and 97.3 ± 5.7% for testing sets 1 and 2, respectively. Similarly, the $sCT_{DL+Bulk}$ analysis resulted in pass rates 99.4% ± 0.7% and 97.6% ± 5.5% for testing sets 1 and 2, respectively.

Kang et al. [79] outperformed the dosimetry performance reported by Lenkowicz et al. [78], potentially due to the implementation of a perceptual loss to ensure highly detailed final sCT [71]. The total cohort was composed of abdominal, breast, and prostate tumour sites, each including 30 patients. The TrueFISP MR sequence was acquired for each patient on the MRIdian. The generator and discriminator of the 2.5D cycleGAN used in this study respectively corresponded to a U-Net and a ResNet. The body MAE achieved for the prostate, breast, and abdomen patients were equal to 55.3HU ± 5.5HU, 63.7HU ± 3.8HU, and 58.8HU ± 4.4HU, respectively. The total mean structural similarity metric (SSIM) was equal to 0.90. After dose computation with Monte Carlo, the 3D 3%/3 mm gamma index with 10% dose threshold were respectively equal to 99.3% ± 0.71%, 99.9% ± 0.10%, and 99.9% ± 0.21% for pelvis, thorax, and abdomen. This dosimetry performance proved the high potential of the perceptual loss-derived sCT.

Similarly, Lapaeva et al. [80] investigated abdominal sCT accuracy from a cycleGAN, and collected 69, 54, 30, 20, 11, and 2 patients, respectively treated for liver, nodes, adrenal glands, pancreas, kidney, and spleen. It led to a total of 144 patients presenting 186 pairs of CT and MRIdian-available true-FISP MRI sequence. The selected network architecture generator and discriminator were, respectively, a HighResNet and a PatchGAN. This architecture offered the opportunity to perform the training based on non-matching images. The external contour-based MAE and SSIM were respectively equal to 70.10HU ± 18.97HU and 0.981 ± 0.009. The Monte Carlo-based dose re-computation on sCT resulted in 3D 2%/2 mm local gamma indices with, respectively, 50% and 90% dose thresholds equal to 99.4% ± 0.6% and 99.5% ± 0.6%. The differences in DVH for the target volume and OAR did not exceed ±2%, proving the high sCT accuracy.

## Metrics for sCT Evaluation

Several metrics have been introduced in the literature to evaluate sCT generation. They are gathered in Table 7.2. These metrics can be classified into four broad categories: (1) intensity-based metrics, which aim at quantifying differences between sCT and CT intensities at the voxel scale or, in a more complex way, their textural differences; (2) metrics to evaluate the geometric fidelity; and metrics of greater clinical interest, either (3) based on the comparison of dose maps or DVH produced by CT and sCT or (4) on the ability of the sCT to provide accurate images to deal with patients setup. The following paragraphs introduce these metrics by illustrating some typical values reported in recent studies, mostly published in 2021, or after.

## Intensity-Based Metrics

First, the mean absolute error (MAE) is by far the most reported metric, mainly due to its computation ease. It corresponds to the average of absolute voxel-wise differences between a reference CT and sCT. Despite its frequent occurrence, MAE should be manipulated with caution, since it is heavily impacted by CT/sCT or CT/MRI registration errors. This metric can be calculated in whole body contour and in volumes of interest (depending on the anatomical localisation). The MAEs in the literature range from 47 HU [74] to 83 HU [88] for the brain and from 24 HU [89] to 59 HU [90] for the pelvis.

The mean error (ME), representing the average of voxel-wise HU differences, has been moderately reported in the literature. It indicates a systematic under/over HU predictions trend of neural networks when used as loss function. It has the major disadvantage to potentially mask intensities differences due to the absence of the absolute value operator, and additionally shares some limitations with the MAE. The reported mean/median ME values in the different studies are between −11 HU and 28 HU [30, 36, 78, 91–94], regardless of the anatomical localisation and the method of sCT generation. DL and multi-atlas-based sCT were also reported to present ME closer to zero compared to bulk density assignment.

A third metric evaluating intensities restoration accuracy is the peak signal-to-noise ratio (PSNR), expressed in dB, which is proportional to the logarithm of the ratio between the maximum intensity of the reference image (here, the CT) squared and the mean squared error. The higher the value of the metric, the better the ability of the algorithm to accurately restore the HU. Although it presents the limitation to not accurately represent the human vision [95], it has been used for brain [88, 96–100], pelvis [85, 94, 101], and abdomen [80, 86] studies with mean/median ranges comprised between 21.422 dB and 39.02 dB.

First introduced in 2004, structural similarity metric (SSIM) aims at evaluating the structural similarity between two images, i.e. the luminance, contrast, and similarity. As reported by Wang 2014 et al. [102], these terms respectively rely on signals mean, signals standard deviation, and standardised signals analyses. The target value is 1, and the range of possible values ranges between −1 and 1. The literature reported mean/median performance varying between 0.70 and 0.98 for brain [88, 96–98, 103], pelvis [85, 94, 101, 104], and head-and-neck [105] sCT, respectively.

## Metrics to Characterise Geometric Fidelity

The Dice similarity coefficient (DSC), computed as twice the cardinality of two volumes intersection divided by the sum of the two volumes cardinalities, quantifies volumes overlap. As specified by Reinke et al. in 2021 [106], this index presents multiple limitations, including underestimated small volume scores or a dependence on voxel size. It also depends on a large set of parameters, including the involved images (sCT against CT/MRI), the investigated sub-volumes (mostly air and bone), and the method used to obtain these sub-volumes (threshold or segmentation). In the literature, air mean indices between 0.65 and 0.92 were reported for brain, pelvis, and abdomen sCT [43, 80, 94, 97, 107]. Mean bone performance were ranging from 0.80 to 0.93 for brain and pelvis-femur studies [43, 74, 90, 94, 97, 107–109].

## Metrics Assessing the Resulting Dosimetric Impact

Regarding dosimetry quality, gamma indices [110], quantifying two dose maps similarity in terms of Euclidean distance and dose discrepancy, have been frequently reported, partly owing to their wide availability in clinics. This metric has the advantage of adding a distance tolerance which may be meaningful due to residual misregistrations. However, this metric suffers from some limitations, such as highly disregarding or penalising errors in small dose regions for the global and local approaches, respectively [111].

Each study uses its own gamma criteria: dose difference value (in %), distance to agreement value (in mm), dose threshold, local or global analysis, 2D or 3D. For pelvis and brain, mean 1%/1 mm and 2%/2 mm with 10% dose threshold were, respectively, reported to range from 97.3% to 99.0% [85, 112] and from 97.95% to 99.9% [85, 96, 112, 113]. For the 3D 2%/2 mm criterion with 10% dose threshold, the dosimetric performance of bulk density assignment, multi-atlas, and cGAN were compared in 2022 by O'Connor et al. [93] in the context of definitive pelvic radiotherapy and resulted in a same mean pass rate of 99.7%.

A second metric assessed for sCT dosimetry quality evaluation is the relative or absolute Dose Volume Histogram-derived criteria differences between CT and sCT, such as the dose received by X% of a given volume denoted DX%. However, no knowledge on the spatial location of the errors is accessible with this metric. Target and OAR maximum means/medians dose errors were equal to −0.8% and 1.6%, respectively, for the brain and pelvis [74, 85, 90, 114]. Values as high as ±30% and ±75% in target and OAR, respectively, were reported for some bulk assignment strategies [115].

As already mentioned, in the aim of dose calculation from MRI, a dose difference in a voxel is not necessarily linked to a HU error in this voxel but can be attributed to HU errors throughout the beam pathway. In this context, a metric reflecting the dosimetric utility of sCT has recently been proposed and consists of assessing the water equivalent depth from the skin surface to the treatment plane isocentre at 36 equidistant gantry angles. This metric demonstrated a better correlation with the dose difference at the isocentre compared to conventional metrics such as MAE or ME in prostate treatments [104].

### Accuracy of Patient Repositioning

The sCT ability to set up the patient as precisely as in the CT-based workflow should be verified [13]. It has been assessed in the literature by registering in-room imaging such as CBCT (or kV) and CT/sCT (or CT/sCT-derived DRR) and evaluating the translational and rotational differences. Maximum means of −0.4 mm and (±)0.1° (when reported) were obtained for pelvis, brain, and head-and-neck sCT when registering CT/sCT-derived DRR and kV imaging [112, 116, 117]. For CT/sCT and CBCT fusion, the maximum means were equal to (±)0.7 mm and 0.5° (when reported) [89, 112, 117–119].

### New Metrics in the Context of MRI-Only Workflow

The metrics presented above require a reference CT to assess the reliability of the sCT scan, and are thus found in the literature as reference metrics. When implementing a MRI-only RT workflow, the CT scan will no longer be acquired, so these metrics will not be available for every sCT scan generated. However, in these configurations, the implementation of a sCT quality assurance process will still be necessary. Several strategies have been proposed for this purpose, including the comparison to reference values obtained from templates to check the plausibility of the predicted HUs [92, 120]. The use of a secondary independent sCT generation algorithm is another path for a more secure implementation of such a workflow [13].

## MRI-QA for Distortion/Artefact-Free Synthetic CT

### Generalities

Geometric fidelity is a key feature in MRI-guided RT as it directly affects the restoration of external contours and the accuracy of OAR and target volume delineation. Inaccurate reconstruction of external contours can have a significant impact on dose evaluation due to errors in X-ray path length and source-to-skin distance estimates. Let's note that coil bridges should be used in CT-free treatment planning to avoid distorting the body contours due to the coils weight, which has been reported to lead to displacements higher than 1 cm [121]. Inaccuracies in the definition of target volumes due to image distortion also affect dose coverage, especially when considering small volumes and stereotactic irradiation conditions [122, 123], with errors up to 32% reported when comparing the minimum dose received by 99% of the target volume in multiple brain metastases radiosurgery [122]. Obviously, the dosimetric impact is reduced when large target volumes are considered, with deviations of less than 1% observed in prostate and head and neck cancers [124, 125].

Several causes affect the geometric fidelity of MR images, including $B_0$ inhomogeneities, residual gradient non-linearities (GNL), eddy currents, and associated trajectory errors in k-space [126, 127]. The following paragraphs briefly review their definitions and the methods proposed to quantify and mitigate them. Several working groups have already defined consensus QA guidelines for MRI, in the context of RT treatment planning incorporating MRI or for MRI-guided online radiotherapy systems [128, 129]. The key points of these reports are described here. The main reconstruction artefacts, e.g. magnetic susceptibility artefacts, which can largely affect as well the quality of the segmentation produced by the radiation oncologist and the performance of the sCT generation tools, are also briefly discussed here.

## $B_0$ Inhomogeneities

The principle of MRI is based on the application of a static polarising magnetic field that is assumed to be spatially invariant. Any deviation from this principle affects, among other things, the uniformity of the image, its spatial integrity, and blurring. The homogeneity of the magnetic field is conventionally expressed in parts per million (ppm) of the static magnetic field value.

$$\text{ppm} = \frac{B}{B_0} = \frac{f}{f_L}$$

where $B_0$ is the magnetic field strength, $f_L$ is the Larmor frequency, $\Delta B$ represents the magnetic field inhomogeneity or offset (e.g. chemical shift), and $\Delta f$ represents the frequency variation or offset. This field strength independent quantity is often provided in a spherical volume of interest described by its diameter (DSV for diameter spherical volume). In 2023, various vendor reports guaranteed values in the order of 1.0–2.0 ppm VRMS (volume root mean square) for 1.5 and 3.0T magnetic fields for a 45 cm DSV for RT dedicated MR devices. A detailed review of the properties of current machines in terms of $B_0$ field homogeneity in the context of radiotherapy, including MR-Linac, has been performed by Gach et al. [127]. The shimming process aims to homogenise the $B_0$ field as much as possible but cannot deal with patient-related $B_0$ inhomogeneities caused by the variations in tissue magnetic susceptibility $\chi$ (in particular, differences in susceptibility between air and biological tissue) which has been shown to be the main cause of residual $B_0$ inhomogeneities today, even with the application of an active shimming [127].

Spin echo-based pulse sequences are less affected than gradient-recalled echo sequences by $B_0$ field inhomogeneities due to their intrinsic principle, which incorporates an additional 180° refocusing pulse. Increasing the readout bandwidth or receiver bandwidth, i.e. the frequency range used when receiving RF pulses, can reduce the impact of these field inhomogeneities on the reconstructed image, however resulting in a higher amount of noise encompassed and thus a lower signal-to-noise ratio. The phase difference mapping technique is recommended to map, at a voxel scale, these field inhomogeneities [130].

## Residual Gradient Non-linearity-Induced Distortions

Spatial encoding in MRI is based on the successive application of magnetic field gradients. Conventional reconstruction methods based on direct fast Fourier transform assume that these gradients are perfectly linear throughout the entire imaging volume. Due to various reasons, including engineering limitations, physiological constraints, or intentional sacrifices to maximise system performance, gradient fields are not perfectly linear and always contain higher-order non-linear components [131], directly impacting geometric fidelity. A priori knowledge of GNL fields can be managed retrospectively by applying post-processing in the image domain or prospectively during image reconstruction by MRI vendors to reduce the impact of these distortions in the final images. However, despite the application of these algorithms, residual distortions remain. In 2019, Gach et al. illustrated effects of GNL for two ViewRay MRI-$^{60}$Co and MRI-Linac systems (Fig. 7.3) for 2D slice excitations [132]. This figure illustrates the fact that distortions increase when the distance from the isocentre of the magnet increases and the major impact of this effect on the restoration of external contours and organs located at the periphery of the FOV. This is even more crucial in radiotherapy where large diameter scanners (≥70 cm) are used in order to manage the clutter of immobilisation devices and to correctly capture the full anatomy of the patient. Geometric distortions resulting from GNL have been shown to be greater than 1 mm at radii >10 cm from the isocentre of the magnet and to reach maximum values of 2–3 mm [129].

The use of the reverse gradient technique, which takes advantage of the application of opposite readout gradients for GNL isolation, is the recommended method for the characterisation of GNL-induced 3D distortion maps [129]. Evaluations should be performed by applying the 3D corrections provided by the vendor and using a large phantom (>80% of the usable FOV) with fully characterised landmarks.

## Susceptibility Artefacts

Magnetic susceptibility, denoted $\chi$, quantifies the magnetisation properties of a material when subjected to a magnetic field. The quantity $\chi$ corresponds to a proportionality constant between the macroscopic magnetisation M of the material and the local magnetic field $B_0$ [133]. The value of $\chi$

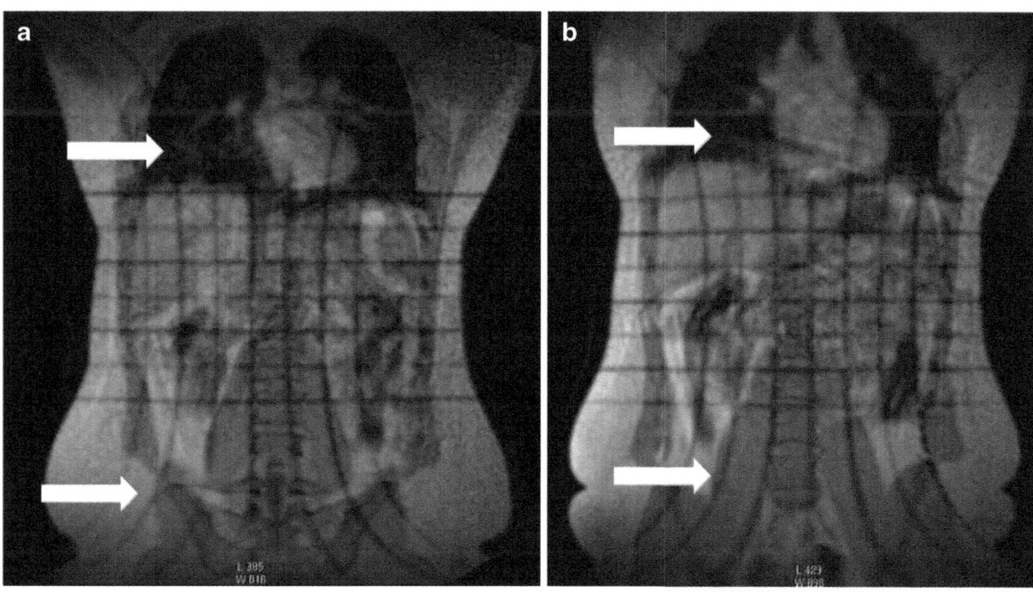


**Fig. 7.3** Illustration of geometric distortions generated by residual gradient non-linearities, highlighting the greater impact at large distances from the isocentre. These distortions are illustrated here for the MRI-$^{60}$Co (**a**) and the MR-Linac (**b**) [132]. Reproduced from Michael Gach H, Curcuru AN, Wittland EJ, Maraghechi B, Cai B, Mutic S, Green OL, MRI quality control for low-field MR-IGRT systems: Lessons learned, Journal of Applied Clinical Medical Physics, 20(10), 53-66, 2019, https://doi.org/10.1002/acm2.12713, CC.BY 4.0

can be positive if the magnetisation is aligned with the field (paramagnetism) or negative, in the case where the two vectors are opposite (diamagnetism). Ferromagnetism characterises materials for which residual magnetism is retained when the external magnetic field is removed. Iron and nickel, for example, are ferromagnetic materials.

As already mentioned, tissue susceptibility can affect both the amplitude and frequency of echo signals, with the total $B$ field becoming a combination of the applied static field $B_0$ and the field offset $\Delta B$ created by the magnetisation $M$ of the object. Frequency shifts in the echo signal can thus be observed as the applied field strength $B_0$ increases, even though variations in $\chi$ are generally small (of the order of 0.1 ppm). Susceptibility-weighted sequences, which consist of the acquisition of one or more gradient echo signals, from which amplitude and frequency are extracted, are used to derive susceptibility maps [134, 135].

Magnetic susceptibility artefacts are thus caused by local gradients created at the interfaces of tissues with very different magnetic susceptibilities (air/tissue, air/bone, metallic implant, calcium deposits). They are responsible for two main effects: image distortions and hyposignal areas. The presence of ferromagnetic materials in or near the FOV also causes distortions of the $B_0$ magnetic field lines, leading to areas of intense hyposignal in MR images. The use of a low magnetic field or the preferential choice of spin echo sequences can reduce the impact of these artefacts. The selection of specific phase and frequency encoding directions, the use of high RF pulse bandwidths, the use of slice encoding for metal artefact correction (SEMAC), or multi-acquisition with variable-resonance image combinations are other ways to achieve this goal [136]. Recent methods to correct susceptibility artefacts have been proposed in the image domain to directly inpaint the artefact regions [136]. Tools to simulate the distortion maps generated by the patient herself/himself by taking the susceptibility values in the different tissues as inputs are also available and could be interesting solutions to generate distortion-free MR images [137].

## Recent Guidelines from Scientific Societies on the Frequency of Quality Controls

Table 7.3 summarises the consensus of the AAPM Working Group 284 established in 2021 regarding monthly quality checks, emphasising the importance of monthly geometric integrity of the system verifications. A daily check is even recommended during the commissioning stage. For this test, the use of a large diameter phantom (>30 cm), which can be a homemade, vendor-supplied, or commercial phantom, is preferred. The acceptability criterion is not strict, with a tolerance equal to 2 mm across a 25 cm FOV. The same group recommends an independent assessment of GNL during commissioning and after any major hardware repair, as well as after hardware and software upgrades.

The ESTRO-ACROP working group focused on the commissioning of online MRI-guided radiotherapy systems and made proposals based on consensus expert opinion [128]. Their recommendations, published in 2023, are in favour of either a weekly or monthly check of $B_0$ field inhomogeneities using vendor-provided 3D phantoms and analysis software. The use of third-party solutions and in-house developed methods is mentioned here as an alternative, used for independent tests. A particularity of the MR-Linac is the presence of a linear accelerator which can interfere with the magnetic field of the MRI. The working group therefore recommends the evaluation of the level of MR distortions as a function of the gantry position on an annual basis.

## Overall Distortion Values Reported on MR-Linacs

Several recent studies have quantified spatial distortions on MR-Linacs [138–142]. For the ViewRay system, Marasini et al. [138] reported geometric distortions of 3.2 mm ± 0.02 mm, 2.9 mm ± 0.02 mm, and 1.8 mm ± 0.01 mm in right-left, anterior-posterior, and superior-inferior directions using the MRID$^{3D}$ phantom

**Table 7.3** Consensus of the AAPM Working Group 284 establishing the monthly quality controls to be implemented [129]. Reproduced from Glide-Hurst CK, Paulson ES, McGee K, Tyagi N, Hu Y, Balter J, Bayouth J, Task group 284 report: magnetic resonance imaging simulation in radiotherapy: considerations for clinical implementation, optimization, and quality assurance, Medical Physics, 48(7), 2021, https://doi.org/10.1002/mp.14695, with permission from John Wiley and Sons

| Test | Suggested equipment | Protocol/parameter/tolerance |
|---|---|---|
| *System* | | |
| Room temperature and humidity | Visual inspection of digital readout | Functional |
| Cold head operation | Audible pump sound | Functional |
| Cryogen level indicator | Digital readout from MRI console | Manufacturer specified (no change from baseline) |
| *Mechanical* | | |
| Table movement smoothness and accuracy | Ruler | ±1.0 mm from set distances |
| Laser alignment with imaging isocenter | Ruler and phantoms with internal landmarks | ±2.0 mm from expected distance offsets |
| Laser movement smoothness and accuracy | Ruler | ±2.0 mm from set distances |
| *Patient marking* | | |
| Laser marking accuracy | Patient marking software | ±2.0 mm |
| *Image quality*[a] | | |
| Central frequency | Digital readout from MRI console after imaging reference phantom (i.e., ACR large MRI phantom or homogeneous phantom) | Manufacturer specified |
| Transmitter gain | | ± 5% from baseline |
| Flexible RF coil testing | Homogeneous phantom | Individual elements, exceeds minimum vendor-provided threshold |
| Geometric accuracy | Manufacturer supplied, in-house, or commercially available phantom >30 cm in diameter/width | Verify ≤2 mm across 25 cm FOV |
| High contrast spatial resolution | ACR large MRI phantom with procedures/tolerances defined in Ref. [83] | ≤1.0 mm |
| Low contrast detectability | | Total number of discernible spokes (for four slices) for fields <3 T should range from 21 (0.3 T) to 36 (1.5 T), and 40 for 3 T. |
| Artifact evaluation | | No obvious image artifacts |
| Percent image uniformity | | Head coil:<br>≥ 87.5% for <3 T<br>≥ 82.0% for 3 T |
| Percent signal ghosting | | Ghosting ratio ≤ 2.5% |

[a] May skip measurement if weekly ACR MRI QA results performed by Diagnostic Radiology Department staff are made available for the QMP to review. Baseline values shall he established following the ACR 2015 QC Manual (collect data for 10 days and determine baseline value). Action limits will be magnet and system-specific as determined by the QMP and manufacturer recommendations

with the vendor-provided distortion correction turned off. The onboard Siemens distortion correction function highly improved the results, with the mean values falling to 0.25 mm ± 0.03 mm, 0.30 mm ± 0.01 mm, and 0.27 mm ± 0.01 mm for the 90° gantry angle. In 2019, based on a home-made phantom, a mean distortion over the whole phantom of

0.60 ± 0.28 mm was reported by Dorsch et al. [141], with 99.8% of the control points presenting distortions below 1.5 mm. In 2022, Damyanovich et al. evaluated the temporal stability of the Elekta Unity system based on 76 weekly measurements [142]. GNL and 3D distorsions were considered in this analysis, which involved a large FOV 3D-phantom containing 1932 markers embedded in seven parallel plates, spaced 25 mm × 25 mm in- and 55 mm through-plane. Time-averaged absolute distortions were reported to be in the range of 0.30–0.49 mm, 0.53–0.80 mm, 1.0–1.4 mm, and 2.28–2.37 mm for DSV respectively, equal to 200, 300, 400, and 500 mm and considering the 98th percentile. The statistical process control they implemented proved that the distortions were stable over time. Significant changes in GNL were however observed following servicing at the 1-year mark.

## sCT Implementation in MR-only Radiotherapy Workflows

### Introduction

Since synthetic CTs are directly used for dose calculation, it is expected that they should perform as good as a CT to be clinically implemented. This section will give an overview on the academic and industrial sCT solutions, both AI and non-AI-based, that have been commissioned and implemented into clinical radiotherapy workflows.

The lack of a consensus QA programme for these images seems to be the main issue that has hindered their wider use in the clinic. Indeed, sCT solutions are not yet generalisable to all situations, limiting their use to one specific patient group or diagnosis. Hospitals that have successfully implemented a sCT in the clinic had to tailor the commissioning process according to their needs. Yet, two aspects are common to all commissioning processes reported in the literature: geometric fidelity and dosimetric performance. Each clinic might have decided to focus more on some aspects than others, choosing different metrics to evaluate.

In this section, a distinction between MR-only workflows with conventional Linac or MR-Linac is made because there are some key differences in the sCT generation process as well as in its role in the workflow (see Fig. 7.1). For example, an essential aspect to evaluate when implementing a sCT product in a conventional Linac workflow is its high patient setup reproducibility. By contrast, in an MR-Linac workflow, this is not an issue as patient positioning is verified by MRI. Rather, one of the biggest challenges the sCT faces when implementing an MR-only MR-Linac workflow will be setting up a fast and robust online QA that will keep up with quick adaptive radiotherapy that these hybrid machines offer.

### MR-Only Workflows with Conventional Linac

In an MR-only RT workflow with conventional Linac, the sCT represents the substitute of a planning CT. Therefore, the straightforward commissioning process should compare the performance of the sCT against the gold standard CT. To assess the ability to correctly recover intensities, metrics such as MAE (Table 7.2) are amongst the most popular, yet the relevance of measurements in the whole body or just in a particular region of interest are still under discussions. Susceptibility artefacts, from bone tissue and air cavities to dental prosthesis or postoperative metal clips, are another point of interest in the commissioning evaluation. The magnitude of the impact on dose calculation from these artefacts may vary depending on the approach used to generate the sCT. Dosimetric evaluation, on the other hand, seems to focus on the analysis of DVH parameters and gamma pass rates (Table 7.2). However, the exact criteria to use are also debatable as parameters such as dose calculation algorithms and dose delivery techniques may play a role.

Patient positioning verification is central for the commissioning of sCT in the case of MR-only RT with conventional Linacs. Correct generation of DRRs from sCT is necessary for 2D kV verification and therefore should be included in the QA procedure. Gold fiducial identification for prostate cancer patients may also become central to commissioning procedures. The available sCT solutions have tackled this issue in different ways forcing quality controls to be accordingly tailored.

Finally, the importance of risk analysis should be highlighted because it allows the clinical team to understand and address the major obstacles in the implementation process. By quantifying the risk involved in each step, appropriate safeguards can be put in place and monitored throughout the implementation process.

## Clinical sCT Generation with Non-DL Solutions

The review paper by Edmund and Nyholm [143] on the status of sCT solutions for RT and PET/MR until November 2015 mentioned two groups that performed an early MR-only radiotherapy workflow implementation for prostate and brain cancers. The Helsinki group used a voxel-based dual regression approach to treat prostate cases, while the Michigan group chose a voxel-based probabilistic approach for their brain cases. Another review [14] that included data until March 2017 reported centres in New York and Turku as the first to use the CE-marked commercial sCT solution developed by PHILIPS (Koniklijke Philips N.V., Eindhoven, The Netherlands) for the treatment of prostate cases. The PHILIPS product, MR for Calculating Attenuation (MRCAT) prostate, is generated from a 3D mDIXON fast field echo sequence and then uses bulk density assignment for five tissues: air, adipose, water, trabecular/spongy bone, and compact/cortical bone. In a review published in 2019 [144], two more centres (Utrecht, Netherlands and Odense, Denmark) were identified to have successfully implemented the MRCAT prostate into clinics. They evaluated different aspects of an MR-only RT workflow, and three common factors were of interest: (1) geometric fidelity of MR images, (2) dosimetric performance of the sCT compared to CT, and (3) patient positioning verification through gold fiducials. The first included an assessment of the individual MR device performance and its stability over time. A maximum distortion of 3 mm in the outer edges of volumes of pelvic examinations was reported [145–147]. In particular, Kemppanien et al. [146] found that such a distortion influenced up to 0.1% the dose to the PTV. On the other hand, each group reported different methodologies for dose distribution comparison between sCT and CT, not facilitating inter-clinic comparison. Among the factors that influenced dose distribution analysis were the image modality selected as base for the dose planning (CT or sCT), the type of registration used to fuse structures and dose plans between image modalities, the treatment technique, and the calculation algorithm of the treatment planning system. Despite these differences in methodology, all groups reported mean dose differences to the target below 1%. Finally, patient position verification was an essential part of the clinical feasibility for an MR-only workflow. Gold fiducials are not visible in the MRCAT prostate sCT; hence they need to be correctly identified on the MRI and thereafter delineated on the sCT. Some groups have introduced automatic localisation of fiducials, but manual observation remains the most reliable method [147].

By 2018, PHILIPS released a second product, MRCAT pelvis, still based on a 3D mDIXON fast field echo sequence but in which the assignment of HU is continuous (PHILIPS brochure). The resulting sCT can be used for other pelvic cancers such as rectal/anal as well as gynaecological. Yu et al. [148] conducted a dosimetric performance evaluation of the MRCAT pelvis generated by three centres with different field strengths and treatment planning systems. They included 77 prostate, 43 rectum, and 27 gynaecological cases. As expected, they reported that MRCAT pelvis better reproduces the CT when compared to the bulk density MRCAT prostate based on HU profile comparisons. The MAE between CT and MRCAT pelvis sCT was equal to 120 ± 15 HU for bone and 33 ± 4 HU for soft tissue. They reported visualisation of gold fiducials and rectal spacer on the MRCAT pelvis sCT, thus being superior to MRCAT prostate. However, they also observed a slight overestimation (but still clinically acceptable) of PTV doses in some rectal cases, which they attributed to the fact that this sCT solution classifies air within the body as soft tissue.

An additional commercially available sCT generator (CE-marked) is the MRIPlanner software by Spectronic Medical AB (Helsingborg, Sweden). This solution requires a T2-weighted MR image, and thereafter a statistical decomposition algorithm is used to assign HU [149]. The MR-OPERA study [150] was designed to validate this product for prostate cancer patients across four Swedish institutions using MR and CT devices from different vendors. A total of 170 patients were recruited with the only exclusion criterion being patients with hip prostheses. No specific MR scan protocol was required other than a T2-weighted with a large FOV. All centres were given access to the MRIPlanner installed on a cloud-based service where the sCT were generated from the uploaded MR images. Dose distributions comparison between the CT and sCT showed a difference below 0.5% for all OARs and targets. The mean gamma pass rate for the 2%/2 mm criterion was found to be 99.12%. These results demonstrated that the sCTs produced with MRIPlanner were dosimetrically accurate and robust across a variety of vendors, field strengths, and treatment techniques.

The MR-OPERA study was followed by the MR-PROTECT study which explored the feasibility of implementing an MR-only radiotherapy protocol for prostate [151]. A total of 40 patients were included between March 2017 and May 2018. All patients had gold markers implanted as part of the setup verification protocol. Patients were planned in an MR-only workflow; however, a CT was acquired directly after the MRI session but only used for QA purposes and retrospective analysis. The retrospective DVH analysis and the gamma pass rates were in line with the results obtained in the MR-OPERA study, after the outliers were corrected for rectal air. They also noted that experience and training of the MR staff mitigated risks associated with MRI acquisition which inevitably impact sCT generation. Their overall results demonstrated that MR-only workflows for prostate cancer can be safely implemented in clinics. Nonetheless, they highlighted the lack of guidelines for QA hinders the widespread implementation of MR-only RT workflows.

Noteworthy is the patient specific QA procedure proposed by Palmer et al. (2018) [152], based on CBCT imaging. To enable this procedure, two criteria need to be evaluated, namely, the CBCT HU stability over time and the comparison of the dose distributions of the CBCT with the reference CT. Periodic phantom measurements over a period of 8 weeks presented a deviation of less than 40 HU for materials relative ED ranging from 0.2 to 1.7. Ten prostate patients from the MR-PROTECT study were analysed in this work. Mean dose deviations were less than 0.9% for either CT-sCT or CBCT-sCT for target DVHs. Finally, the CBCT QA tool was tested by introducing known errors in the sCT. Only a water-equivalent sCT failed to be detected by the QA. In conclusion, the proposed patient-specific QA proved to be clinically feasible enabling a CT free QA. However, this QA cannot be performed before the first CBCT acquisition, preventing this solution to be applicable in all dose regimes such as hypofractionation.

In the review by Bird et al. [144], the Helsinki and Michigan groups were mentioned for their early clinical results after implementing their in-house sCT solutions. The Helsinki group [153], who treated prostate cancer with MR-only workflow since 2012, was the first to report similar clinical outcomes between patients treated either with MR-only or CT-based workflows by monitoring the early prostate-specific antigen and the radiation therapy oncology group grades. They pointed out that the main limitations of using sCT in a clinical environment were the identification of the gold markers, large artefacts due to hip implants and obese patients. The Michigan group, on the other hand, developed a voxel-based weighted summation method for their sCT generation of the pelvic region. Kim et al. [154] evaluated the MR-only workflow implementation through a failure mode and effects analysis that helps to develop a robust quality management programme. Particularly for sCT, bone segmentation and air pocket identification led to improvements in QA and quality control checklists, as well as limited risks for gross anatomic errors leading to incorrect HU assignment and impacting dose calculation.

The HIPSTER (high precision prostate substitute CT-based external beam radiotherapy) trial is a sCT prospective clinical study consisting of 30 prostate patients recruited between 2017 and 2019 from two Australian institutions. The authors used an in-house developed sCT using a hybrid atlas-voxel method [155]. The acceptance criteria for the dose calculation were isocentre dose within 2% and gamma pass rate larger than 90% for the 2%/2 mm criterion. The study demonstrated that the specially tailored QA programme was robust for MR-only RT workflow implementation as 90% of the cases showed a mean isocentre dose difference below 1.5% and mean gamma pass rate of 99.7% [156].

## Clinical sCT Generation with DL Solutions

DL approaches to generate sCT are an expanding area of research. Their popularity stems from their ability to bypass some of the intrinsic limitations of non-AI sCT methods. For example, the fast sCT generation, potentially from a single MR sequence, reduces scan time or the superiority it has in reproducing uncommon anatomies compared to atlas-based algorithms.

Palmer et al. [157] were the first to validate the commercial AI-based sCT product released by Spectronic Medical AB for head and neck radiotherapy. A convolutional neural network (CNN)-based transfer function estimation algorithm [158] was used to automatically generate sCT from a T1-weighted Dixon Vibe MR sequence. They reported a MAE of 67 ± 14 HU for the overall body and 195 ± 27 HU for bone particularly. As for dose distributions comparison, the difference in absorbed dose mean ranged from −0.3% to 0.02% for the tested DVH parameters. An additional validation study performed by the Lund group reported similar promising results for the sCT generated with MRIPlanner (Spectronic Medical AB) but applied to brain lesions and focusing on gliomas and metastases [159]. Of special interest was data from 14 patients who presented missing bone skull due to surgery performed before RT. They found a MAE of 177 ± 19 HU for bone in a volume of interest (VOI) around the resected bone which was of the same order as the MAE for the entire cohort of patients. The average DSC for the whole skull was 0.92 compared to 0.90 in the VOI. Unfortunately, no specific data on the impact on dose distribution was reported. The overall positive results from the two validation studies encouraged the Lund group to conduct the first prospective clinical study for the implementation of an MR-only workflow for brain radiotherapy [119]. A total of 21 patients with grades III and IV gliomas were treated between March 2021 and March 2022. A QA programme was specifically developed to ensure the safe use of the sCT. The prospective part of the QA consisted of four control steps. The first was performed by a script that checked that the MRI acquisition parameters were those required for sCT generation. Thereafter, the sCT was visually inspected for artefacts and misalignments against MRI. Then, a HU comparison to the CT was performed. The areas where bone resection was observed in the sCT were also controlled based on the CT. Lastly, calculated plans were accepted if they fulfilled either one of these two criteria: (1) dose distribution differences between sCT and CT should be below 1% for the target and OAR, and (2) OAR absolute dose should be 10% below clinical tolerance. The retrospective part of the QA consisted of verifying the rigid registration equivalence between CBCT to sCT and CBCT to CT. Overall, the success of this study relied on the establishment of acceptance criteria tailored to local clinical routines and this solution limitations evaluation.

An additional commercially available sCT is provided by Siemens Healthineers (Erlangen, Germany) as part of syngo.via RT Image Suite. This sCT is generated by a multi-layer neural network (CNN for segmentation and a cGAN for image reconstruction) with a T1-weighted Dixon Vibe MR sequence as input [160]. In 2022, Masitho et al. [114] performed the validation of this product for brain MR-only workflow. Apart from evaluation of geometry fidelity and dose calculation performance, they also focused on the ability to produce DRR from sCTs that can be used for 2D/2D non-coplanar kV X-ray patient setup verification. Results showed a MAE of 136 ± 13 HU for the brain. After adding extra fixation to the mask to prevent variation in the rotational lateral direction when acquiring the MR sequence, they concluded that patient setup with sCT DRR as reference was equivalent to CT DRR-based setup. The clinical implementation evaluation for MR-only RT brain workflow is planned for the near future.

PHILIPS (Koniklijke Philips N.V.) has also launched two AI-based products, MRCAT brain and MRCAT head and neck. Both use as input a 3D T1-weighted mDIXON sequence and generate a sCT with continuous HU directly on the MR console. As part of a clinical implementation of the MRCAT brain, Emin et al. [161] conducted a two-phase process consisting of a retrospective study (CT-based RT workflow with sCT on the background) and a prospective study (MR-only RT workflow with CT in the background). They reported a mean dose difference below 1% for PTV D2%, PTV D98%, and PTV Dmean, based on 60 glioblastoma patients (30 patients per phase). As the MRCAT brain has a fixed voxel size (1.1 mm × 1.1 mm × 1 mm), the importance of re-gridding the images to matching voxel sizes before performing dose comparisons was emphasised due to smaller structures (e.g. cochleae and chiasma) being particularly sensitive to the rigid registration process between CT and sCT. Indeed, a clear reduction in the spread of the DVH difference distributions was reported when comparing pre- and post re-gridded data for the OAR studied.

Academic AI-based sCT solutions for image reconstruction of other more difficult sites such as breast, liver and abdomen can be found in the literature [32]. MRI-only clinical implementation using such solutions is still in its infancy, even though the geometric and dosimetric evaluations seem to be favourable. The main drawbacks for these sCT solutions can be reduced to two reasons: the lack of knowledge about the failure modes and the lack of a general consensus in QA criteria [162]. An increase in prospective studies with larger cohorts is expected to happen in the next few years as more hospitals access these solutions.

## MR-Only Workflows with MR-Linac

Unlike the MR-only workflow with conventional Linac, the MR-only workflow with an MR-Linac hybrid system opens the possibility for fast treatment adaptation, both intrafraction and interfraction, as anatomy changes can be observed in real time. In this context, the role of the sCT is to provide fast and accurate 3D ED maps for interfraction dose calculation based on the MRI of the day. On the other hand, the sCT impacts neither the target and OAR delineation nor the patient position verification, as this is entirely handled by the MRI. Thus, the major challenge of the implementation process of sCT is to develop a fast (on-the-fly) QA for evaluation of image fidelity.

In the 2020 review paper by Kurtz et al. [163], six MR-Linac approaches were mentioned, and only two were commercialised at that time, namely, the MRIdian by ViewRay Technologies, Inc. (California, USA) and the Unity by Elekta AB (Stockholm, Sweden). The workflows of these machines are different because of the differences in their designs. Yet, none of them operate in a true MR-only workflow as there is always a CT image acquired before the start of the treatment. The dose plan optimised on this CT image is referred to as the base plan. With the MRIdian system, one may choose between modifying the table position and delivering the base plan or performing a full re-optimisation [163]. On the other hand, the Unity system does not allow for table correction. So the base plan must be adapted in one of the two ways: (1) adapt-to-position (ATP) where a rigid translational registration and limited optimiser is used to reproduce DVH from the base plan or (2) adapt-to-shape (ATS) where recontouring and full re-optimisation are possible [164].

Plans re-optimisation is currently performed on a bulk density sCT generated by different methods depending on the system. The MRIdian system method consists of using DIR to fuse the MRI and the planning CT thereby creating a deformed version of the CT. Discrepancies between image modalities such as gas or body shape are compensated by dedicated structures where ED is overridden. The second method, used in the ATS workflow of the Unity system, starts by extracting the contours along with their average ED from the planning CT. These values are then propagated to the daily MR image by performing a DIR. ATS workflow allows for manual recontouring before the average ED is assigned to each contour [165]. Tang et al. [165] added a step to the workflow where the reference plan (on CT) and the re-calculated plan (on sCT) were compared on-the-fly. If major discrepancies were found, then the sCT was modified, and the plan re-calculated in an iterative fashion until the dose difference would be within the acceptable clinical tolerance. For 54 out of 61 patients with various treatment sites, the gamma pass rate was 99.7% with the criteria 3%/3 mm and a DVH difference of less than 1.5% when comparing against the first generated sCT. Major differences were found in two lung cases, where the ED gradient within the target and the adjoining tissue could not be reproduced by bulk density method, and thus, these patients were treated with ATP workflow instead. Their study demonstrated the importance of extra dosimetric checks on the sCT to enable a quick identification and correction of the sources of errors that otherwise would remain undetected in pre-treatment check or plan adaptation phase. A prospective risk study was conducted by the Memorial Sloan Kettering Cancer Center department for the ATP workflow, highlighting the need to train planners/therapists to review ED in all three planes of the entire image [166].

In 2023, more sophisticated sCT generation methods are not yet clinically implemented in any commercially available MR-Linac. As pointed out by Ng et al. [167], the real challenge is the development of a generalisable and reliable sCT solution that would be applicable to various anatomical and clinical scenarios. Thus, sCT is still a future development for a use in MR-Linac workflows. The advent of AI has increased this area of research, and of interest are the DL sCT solutions generated from MRI acquired in low magnetic fields such as the MRIdian system. Apart from the advantage of offering a lower electron return effect, the images produced by this system are particularly convenient for lung cases [168]. Lenkowicz et al. [78], previously introduced, were the first to produce an in-house sCT based on cGAN architecture trained with 32 patients presenting lung lesions. They concluded that their approach was suitable for palliative cases; however improvements such as using a larger training cohort and more complex neural networks needed to be performed to obtain a better dosimetry performance which is required for curative purposes.

In 2022, Therapanacea (Paris, France) introduced the MR-Box option in their ART-Plan™ toolbox. The MR-Box is an AI-powered, CE-marked, and FDA-cleared software providing sCT images and automated OAR delineation from MRI. GAN architecture is used to generate high-fidelity sCT images that match the underlying distribution of the training CT scans while maintaining a voxel level anatomical correspondence with the input MRI. Particularly, a 3D cGAN architecture, similar to the Pix2Pix architecture, is used for brain imaging, while for pelvis, the underlying DL model is a cycleGAN. The solution can be used in MRI-only workflows or workflows based on an MR-Linac. In 2022, a multi-centric evaluation was conducted considering pelvic anatomies and 20 test patients from 8 different institutions using a 0.35 T MR [169]. Reconstruction performance was assessed using the OARs delineated for the treatment planning. The results showed that the model was able to reliably restore the intensities of the femoral heads, while those of the rectum and the sigmoid were more difficult to reconstruct. In 2022, a clinical trial was launched at Gustave Roussy Cancer Campus (Villejuif, France) to evaluate the performance of the sCT in patients being treated for glial tumours. Results are expected by the end of 2023.

DL-generated sCT promise to shorten the re-planning phase in an MR-Linac workflow while increasing accuracy compared to atlas-based or bulk density methods. To be compatible with online adaptive workflows, the time between generation and quality control of these images cannot exceed a few minutes [170]. Indeed, some studies have reported that generation of sCT can be achieved in less than 10 s if GPU-based networks are used [171]. However, time for MRI acquisition must also be accounted for, and currently the dedicated MR sequences that are used for sCT generation are time-consuming and, thus, not yet applicable for online adaptive MRgRT [172].

Once these hurdles are overcome, the last remaining challenge for the adoption of AI-based sCT in MR-linac workflows is the development of a robust image QA that can be carried out online within the span of a few minutes. A good starting point for relevant QA criteria and tolerance levels can be drawn from the QA programmes already developed for conventional Linac MR-only workflows. Adjustments should be done considering the hospitals needs and assess the limitations of the MR-Linac devices. As already mentioned, several strategies can also be implemented for this purpose, such as comparison with reference values obtained from models to check the plausibility of the predicted HUs or the use of a second independent algorithm similar to the second monitor unit calculation algorithms. As long as a consensus in QA is not achieved, a planning CT will still be needed for patient-specific QA purposes for the foreseeable future. A novel idea, however, has been put forward by Goodburn et al. [173] where a potential solution to save time would consist of removing the sCT visualisation step and directly use the k-space data as treatment planning software input.

## Conclusion

In conclusion, the implementation of sCT in clinics, which was rather experimental and anecdotal a few years ago, is in the process of being democratised today, at least for workflows based on MRI-only. Like in numerous fields, the arrival of DL, and its associated robustness, has largely contributed to this change in practices. However, workflows based on MR-Linac still have reduced benefits from these advances, since no DL solution was yet commercially integrated. A major shortcoming widely identified by several authors is the lack of recommendations concerning the commissioning stage and QA procedures. At least the absence of high magnitude distortions and artefacts should be checked in this context. The literature review also highlighted the need for new profiles of physicists with mixed knowledge in RT physics, MRI physics, and AI for the implementation of safe workflows.

## References


1. Owrangi AM, Greer PB, Glide-Hurst CK. MRI-only treatment planning: benefits and challenges. Phys Med Biol. 2018;63(5):05TR01.
2. van Herk M, Kooy HM. Automatic three-dimensional correlation of CT-CT, CT-MRI, and CT-SPECT using chamfer matching. Med Phys. 1994;21(7):1163–78.
3. Ulin K, Urie MM, Cherlow JM. Results of a multi-institutional benchmark test for cranial CT/MR image registration. Int J Radiat Oncol Biol Phys. 2010;77(5):1584–9.
4. Dean CJ, Sykes JR, Cooper RA, Hatfield P, Carey B, Swift S, u. a. An evaluation of four CT-MRI co-registration techniques for radiotherapy treatment planning of prone rectal cancer patients. Br J Radiol 2012;85(1009):61–8.
5. Korsager AS, Carl J, Østergaard LR. Comparison of manual and automatic MR-CT registration for radiotherapy of prostate cancer. J Appl Clin Med Phys. 2016;17(3):294–303.
6. Brock KK. Deformable registration accuracy consortium. Results of a multi-institution deformable registration accuracy study (MIDRAS). Int J Radiat Oncol Biol Phys. 2010;76(2):583–96.
7. Pötter R, Tanderup K, Kirisits C, de Leeuw A, Kirchheiner K, Nout R, u. a. The EMBRACE II study: the outcome and prospect of two decades of evolution within the GEC-ESTRO GYN working group and the EMBRACE studies. Clin Transl Radiat Oncol. 2018;9:48–60.
8. Rivard MJ, Coursey BM, DeWerd LA, Hanson WF, Huq MS, Ibbott GS, u. a. Update of AAPM Task Group No. 43 report: a revised AAPM protocol for brachytherapy dose calculations. Med Phys. 2004;31(3):633–74.

9. Burgos N, Cardoso MJ, Modat M, Pedemonte S, Dickson J, Barnes A, u. a. Attenuation correction synthesis for hybrid PET-MR scanners. Med Image Comput Comput Assist Interv. 2013;16(Pt 1):147–54.
10. Martinez-Möller A, Souvatzoglou M, Delso G, Bundschuh RA, Chefd'hotel C, Ziegler SI, u. a. Tissue classification as a potential approach for attenuation correction in whole-body PET/MRI: evaluation with PET/CT data. J Nucl Med. 2009;50(4):520–6.
11. Persson E, Svanberg N, Scherman J, Jamtheim Gustafsson C, Fridhammar A, Hjalte F, u. a. MRI-only radiotherapy from an economic perspective: can new techniques in prostate cancer treatment be cost saving? Clin Transl Radiat Oncol. 2023;38:183–7.
12. Keyriläinen J, Sjöblom O, Turnbull-Smith S, Hovirinta T, Minn H. Clinical experience and cost evaluation of magnetic resonance imaging -only workflow in radiation therapy planning of prostate cancer. Phys Imaging Radiat Oncol. 2021;19:66–71.
13. Vandewinckele L, Claessens M, Dinkla A, Brouwer C, Crijns W, Verellen D, u. a. Overview of artificial intelligence-based applications in radiotherapy: recommendations for implementation and quality assurance. Radiother Oncol. 2020;153:55–66.
14. Johnstone E, Wyatt JJ, Henry AM, Short SC, Sebag-Montefiore D, Murray L, u. a. Systematic review of synthetic computed tomography generation methodologies for use in magnetic resonance imaging-only radiation therapy. Int J Radiat Oncol Biol Phys. 2018;100(1):199–217.
15. Prabhakar R, Julka PK, Ganesh T, Munshi A, Joshi RC, Rath GK. Feasibility of using MRI alone for 3D radiation treatment planning in brain tumors. Jpn J Clin Oncol. 2007;37(6):405–11.
16. Chen L, Price RA, Wang L, Li J, Qin L, McNeeley S, u. a. MRI-based treatment planning for radiotherapy: dosimetric verification for prostate IMRT. Int J Radiat Oncol Biol Phys. 2004;60(2):636–47.
17. Weber DC, Wang H, Albrecht S, Ozsahin M, Tkachuk E, Rouzaud M, u. a. Open low-field magnetic resonance imaging for target definition, dose calculations and set-up verification during three-dimensional CRT for glioblastoma multiforme. Clin Oncol. 2008;20(2):157–67.
18. Ramsey CR, Oliver AL. Magnetic resonance imaging based digitally reconstructed radiographs, virtual simulation, and three-dimensional treatment planning for brain neoplasms. Med Phys. 1998;25(10):1928–34.
19. Wang C, Chao M, Lee L, Xing L. MRI-based treatment planning with electron density information mapped from CT images: a preliminary study. Technol Cancer Res Treat. 2008;7(5):341–8.
20. Eilertsen K, Vestad LNTA, Geier O, Skretting A. A simulation of MRI based dose calculations on the basis of radiotherapy planning CT images. Acta Oncol. 2008;47(7):1294–302.
21. Karotki A, Mah K, Meijer G, Meltsner M. Comparison of bulk electron density and voxel-based electron density treatment planning. J Appl Clin Med Phys. 2011;12(4):3522.
22. Lambert J, Greer PB, Menk F, Patterson J, Parker J, Dahl K, u. a. MRI-guided prostate radiation therapy planning: investigation of dosimetric accuracy of MRI-based dose planning. Radiother Oncol. 2011;98(3):330–4.
23. Lee YK, Bollet M, Charles-Edwards G, Flower MA, Leach MO, McNair H, u. a. Radiotherapy treatment planning of prostate cancer using magnetic resonance imaging alone. Radiother Oncol. 2003;66(2):203–16.
24. ICRU Report 46. Photon, electron, proton and neutron interaction data for body tissues. 1992.
25. Chin AL, Lin A, Anamalayil S, Teo BKK. Feasibility and limitations of bulk density assignment in MRI for head and neck IMRT treatment planning. J Appl Clin Med Phys. 2014;15(5):4851.
26. Zaidi H, Montandon ML, Slosman DO. Magnetic resonance imaging-guided attenuation and scatter corrections in three-dimensional brain positron emission tomography. Med Phys. 2003;30(5):937–48.
27. Wagenknecht G, Kops ER, Tellmann L, Herzog H. Knowledge-based segmentation of attenuation-relevant regions of the head in T1-weighted MR images for attenuation correction in MR/PET systems. In: 2009 IEEE nuclear science symposium conference record (NSS/MIC). Orlando, FL: IEEE; 2009. [zitiert 27. März 2023]. S. 3338–43. Verfügbar unter: http://ieeexplore.ieee.org/document/5401751/.
28. Harrison K, Pullen H, Welsh C, Oktay O, Alvarez-Valle J, Jena R. Machine learning for auto-segmentation in radiotherapy planning. Clin Oncol. 2022;34(2):74–88.
29. Robert C, Munoz A, Moreau D, Mazurier J, Sidorski G, Gasnier A, u. a. Clinical implementation of deep-learning based auto-contouring tools-experience of three French radiotherapy centers. Cancer Radiother. 2021;25(6–7):607–16.
30. Groot Koerkamp ML, de Hond YJM, Maspero M, Kontaxis C, Mandija S, Vasmel JE, u. a. Synthetic CT for single-fraction neoadjuvant partial breast irradiation on an MRI-linac. Phys Med Biol. 2021;66(8).
31. Ding S, Liu H, Li Y, Wang B, Li R, Liu B, u. a. Assessment of dose accuracy for online MR-guided radiotherapy for cervical carcinoma. J Radiat Res Appl Sci. 2021;14(1):159–70.
32. Boulanger M, Nunes JC, Chourak H, Largent A, Tahri S, Acosta O, u. a. Deep learning methods to generate synthetic CT from MRI in radiotherapy: a literature review. Phys Med. 2021;89:265–81.
33. Wyatt J, McCallum H. Applying a commercial atlas-based synthetic computed tomography algorithm to patients with hip prostheses for prostate magnetic resonance-only radiotherapy. Radiother Oncol. 2019;133:100–5.
34. Dowling JA, Lambert J, Parker J, Salvado O, Fripp J, Capp A, u. a. An atlas-based electron density mapping method for magnetic resonance imaging (MRI)-alone treatment planning and adaptive MRI-

based prostate radiation therapy. Int J Radiat Oncol Biol Phys. 2012;83(1):e5-11.
35. Demol B, Boydev C, Korhonen J, Reynaert N. Dosimetric characterization of MRI-only treatment planning for brain tumors in atlas-based pseudo-CT images generated from standard T1-weighted MR images. Med Phys. 2016;43(12):6557.
36. Han X. MR-based synthetic CT generation using a deep convolutional neural network method. Med Phys. 2017;44(4):1408–19.
37. Greer PB, Dowling JA, Lambert JA, Fripp J, Parker J, Denham JW, u. a. A magnetic resonance imaging-based workflow for planning radiation therapy for prostate cancer. Med J Aust. 2011;194(4):S24-27.
38. Sjölund J, Forsberg D, Andersson M, Knutsson H. Generating patient specific pseudo-CT of the head from MR using atlas-based regression. Phys Med Biol. 2015;60(2):825–39.
39. Farjam R, Tyagi N, Deasy JO, Hunt MA. Dosimetric evaluation of an atlas-based synthetic CT generation approach for MR-only radiotherapy of pelvis anatomy. J Appl Clin Med Phys. 2019;20(1):101–9.
40. Farjam R, Tyagi N, Veeraraghavan H, Apte A, Zakian K, Hunt MA, u. a. Multiatlas approach with local registration goodness weighting for MRI-based electron density mapping of head and neck anatomy. Med Phys. 2017;44(7):3706–17.
41. Uh J, Merchant TE, Li Y, Li X, Hua C. MRI-based treatment planning with pseudo CT generated through atlas registration. Med Phys. 2014;41(5):051711.
42. Bakai A, Alber M, Nüsslin F. A revision of the gamma-evaluation concept for the comparison of dose distributions. Phys Med Biol. 2003;48(21):3543–53.
43. Spadea MF, Pileggi G, Zaffino P, Salome P, Catana C, Izquierdo-Garcia D, u. a. Deep convolution neural network (DCNN) multiplane approach to synthetic CT generation from MR images-application in brain proton therapy. Int J Radiat Oncol Biol Phys. 2019;105(3):495–503.
44. Thompson M, Duda RO, Hart PE. Pattern classification and scene analysis. Leonardo. 1974;7(4):370.
45. Edmund JM, Kjer HM, Van Leemput K, Hansen RH, Andersen JAL, Andreasen D. A voxel-based investigation for MRI-only radiotherapy of the brain using ultra short echo times. Phys Med Biol. 2014;59(23):7501–19.
46. Johansson A, Karlsson M, Nyholm T. CT substitute derived from MRI sequences with ultrashort echo time. Med Phys. 2011;38(5):2708–14.
47. Jonsson JH, Akhtari MM, Karlsson MG, Johansson A, Asklund T, Nyholm T. Accuracy of inverse treatment planning on substitute CT images derived from MR data for brain lesions. Radiat Oncol. 2015;10:13.
48. Johansson A, Garpebring A, Karlsson M, Asklund T, Nyholm T. Improved quality of computed tomography substitute derived from magnetic resonance (MR) data by incorporation of spatial information—potential application for MR-only radiotherapy and attenuation correction in positron emission tomography. Acta Oncol. 2013;52(7):1369–73.
49. Pandey A, Sa Y, Guo B, Zhang R. Feasibility of generating synthetic CT from T1-weighted MRI using a linear mixed-effects regression model. Biomed Phys Eng Express. 2019;5(4):047004.
50. Ranjbarzadeh R, Saadi SB. Automated liver and tumor segmentation based on concave and convex points using fuzzy c-means and mean shift clustering. Measurement. 2020;150:107086.
51. Bandyopadhyay S, Das S, Datta A. Detection of coronal holes in solar disk image using fast fuzzy C-means algorithm. In: Das NR, Sarkar S, Herausgeber. Computers and devices for communication [Internet]. Singapore: Springer Singapore; 2021 [zitiert 27. März 2023]. S. 29–35. (Lecture Notes in Networks and Systems; Bd. 147). Verfügbar unter: http://link.springer.com/10.1007/978-981-15-8366-7_5.
52. Jahangoshai Rezaee M, Jozmaleki M, Valipour M. Integrating dynamic fuzzy C-means, data envelopment analysis and artificial neural network to online prediction performance of companies in stock exchange. Phys A. 2018;489:78–93.
53. Hsu SH, DuPre P, Peng Q, Tomé WA. A technique to generate synthetic CT from MRI for abdominal radiotherapy. J Appl Clin Med Phys. 2020;21(2):136–43.
54. Bredfeldt JS, Liu L, Feng M, Cao Y, Balter JM. Synthetic CT for MRI-based liver stereotactic body radiotherapy treatment planning. Phys Med Biol. 2017;62(8):2922–34.
55. Su KH, Friel HT, Kuo JW, Al Helo R, Baydoun A, Stehning C, u. a. UTE-mDixon-based thorax synthetic CT generation. Med Phys. 2019;46(8):3520–31.
56. Liu C, Zhang X, Si W, Ni X. Multiview self-supervised segmentation for OARs delineation in radiotherapy. Evid Based Complement Alternat Med. 2021;2021:8894222.
57. Lei Y, Harms J, Wang T, Tian S, Zhou J, Shu HK, u. a. MRI-based synthetic CT generation using semantic random forest with iterative refinement. Phys Med Biol. 2019;64(8):085001.
58. Yang X, Lei Y, Shu HK, Rossi P, Mao H, Shim H, u. a. Pseudo CT estimation from MRI using patch-based random forest. Proc SPIE Int Soc Opt Eng. 2017;10133:101332Q.
59. Wang T, Manohar N, Lei Y, Dhabaan A, Shu HK, Liu T, u. a. MRI-based treatment planning for brain stereotactic radiosurgery: dosimetric validation of a learning-based pseudo-CT generation method. Med Dosim. 2019;44(3):199–204.
60. Nie D, Cao X, Gao Y, Wang L, Shen D. Estimating CT image from MRI data using 3D fully convolutional networks. Deep Learn Data Label Med Appl. 2016;2016:170–8.

61. LeCun Y, Bengio Y, Hinton G. Deep learning. Nature. 2015;521(7553):436–44.
62. Tahri S, Barateau A, Cadin C, Chourak H, Ribault S, Nozahic F, u. a. A high-performance method of deep learning for prostate MR-only radiotherapy planning using an optimized Pix2Pix architecture. Phys Med. 2022;103:108–18.
63. Ronneberger O, Fischer P, Brox T. U-Net: convolutional networks for biomedical image segmentation. 2015 [zitiert 28. April 2023]; Verfügbar unter. https://arxiv.org/abs/1505.04597.
64. Liang J, Yang C, Zhong J, Ye X. BTSwin-Unet: 3D U-shaped symmetrical swin transformer-based network for brain tumor segmentation with self-supervised pre-training. Neural Process Lett [Internet]. 2022 [zitiert 28. April 2023]; Verfügbar unter. https://link.springer.com/10.1007/s11063-022-10919-1.
65. Chen J, Lu Y, Yu Q, Luo X, Adeli E, Wang Y, u. a. TransUNet: transformers make strong encoders for medical image segmentation. 2021 [zitiert 28. April 2023]; Verfügbar unter: https://arxiv.org/abs/2102.04306.
66. Vaswani A, Shazeer N, Parmar N, Uszkoreit J, Jones L, Gomez AN, u. a. Attention is all you need [Internet]. arXiv; 2017 [zitiert 28. April 2023]. Verfügbar unter: http://arxiv.org/abs/1706.03762.
67. Weijie Chen, Seyed Iman Zare Estakhraji, Alan B McMillan. Exploration of vision transformer models in medical images synthesis.
68. Iqbal A, Sharif M, Khan MA, Nisar W, Alhaisoni M. FF-UNet: a U-shaped deep convolutional neural network for multimodal biomedical image segmentation. Cognit Comput. 2022;14(4):1287–302.
69. Zhou Z, Siddiquee MMR, Tajbakhsh N, Liang J. UNet++: redesigning skip connections to exploit multiscale features in image segmentation. IEEE Trans Med Imaging. 2020;39(6):1856–67.
70. Goodfellow IJ, Pouget-Abadie J, Mirza M, Xu B, Warde-Farley D, Ozair S, u. a. Generative adversarial networks. 2014 [zitiert 1. Mai 2023]; Verfügbar unter: https://arxiv.org/abs/1406.2661.
71. Largent A, Barateau A, Nunes JC, Mylona E, Castelli J, Lafond C, u. a. Comparison of deep learning-based and patch-based methods for pseudo-CT generation in MRI-based prostate dose planning. Int J Radiat Oncol Biol Phys. 2019;105(5):1137–50.
72. Mirza M, Osindero S. Conditional generative adversarial nets. 2014 [zitiert 28. April 2023]; Verfügbar unter: https://arxiv.org/abs/1411.1784.
73. Shahbazi M, Danelljan M, Paudel DP, Van Gool L. Collapse by conditioning: training class-conditional GANs with limited data. 2022 [zitiert 28. April 2023]; Verfügbar unter: https://arxiv.org/abs/2201.06578.
74. Kazemifar S, McGuire S, Timmerman R, Wardak Z, Nguyen D, Park Y, u. a. MRI-only brain radiotherapy: assessing the dosimetric accuracy of synthetic CT images generated using a deep learning approach. Radiother Oncol. 2019;136:56–63.
75. Zhu JY, Park T, Isola P, Efros AA. Unpaired image-to-image translation using cycle-consistent adversarial networks. In Proceedings of the IEEE international conference on computer vision. 2017;2223–32.
76. Brou Boni KND, Klein J, Gulyban A, Reynaert N, Pasquier D. Improving generalization in MR-to-CT synthesis in radiotherapy by using an augmented cycle generative adversarial network with unpaired data. Med Phys. 2021;48(6):3003–10.
77. Peng Y, Chen S, Qin A, Chen M, Gao X, Liu Y, u. a. Magnetic resonance-based synthetic computed tomography images generated using generative adversarial networks for nasopharyngeal carcinoma radiotherapy treatment planning. Radiother Oncol. 2020;150:217–24.
78. Lenkowicz J, Votta C, Nardini M, Quaranta F, Catucci F, Boldrini L, u. a. A deep learning approach to generate synthetic CT in low field MR-guided radiotherapy for lung cases. Radiother Oncol. 2022;176:31–8.
79. Kang SK, An HJ, Jin H, Jin K, Chie EK, Park JM, u. a. Synthetic CT generation from weakly paired MR images using cycle-consistent GAN for MR-guided radiotherapy. s. 2021;11(3):263–71.
80. Lapaeva M, La Greca Saint-Esteven A, Wallimann P, Günther M, Konukoglu E, Andratschke N, u. a. Synthetic computed tomography for low-field magnetic resonance-guided radiotherapy in the abdomen. Phys Imaging Radiat Oncol. 2022;24:173–9.
81. Garcia Hernandez A, Fau P, Wojak J, Mailleux H, Benkreira M, Rapacchi S, u. a. Synthetic computed tomography generation for abdominal adaptive radiotherapy using low-field magnetic resonance imaging. Phys Imaging Radiat Oncol. 2023;25:100425.
82. Olberg S, Chun J, Su Choi B, Park I, Kim H, Kim T, u. a. Abdominal synthetic CT reconstruction with intensity projection prior for MRI-only adaptive radiotherapy. Phys Med Biol. 2021;66(20).
83. Nousiainen K, Santurio GV, Lundahl N, Cronholm R, Siversson C, Edmund JM. Evaluation of MRI-only based online adaptive radiotherapy of abdominal region on MR-linac. J Appl Clin Med Phys. 2022;24:e13838.
84. Farjam R, Nagar H, Kathy Zhou X, Ouellette D, Chiara Formenti S, DeWyngaert JK. Deep learning-based synthetic CT generation for MR-only radiotherapy of prostate cancer patients with 0.35T MRI linear accelerator. J Appl Clin Med Phys. 2021;22(8):93–104.
85. Hsu SH, Han Z, Leeman JE, Hu YH, Mak RH, Sudhyadhom A. Synthetic CT generation for MRI-guided adaptive radiotherapy in prostate cancer. Front Oncol. 2022;12:969463.
86. Fu J, Singhrao K, Cao M, Yu V, Santhanam AP, Yang Y, u. a. Generation of abdominal synthetic CTs from 0.35T MR images using generative adversarial networks for MR-only liver radiotherapy. Biomed Phys Eng Express. 2020;6(1):015033.

87. Isola P, Zhu JY, Zhou T, Efros AA. Image-to-image translation with conditional adversarial networks. ArXiv161107004 Cs [Internet]. 2018 [zitiert 24. Juni 2021]; Verfügbar unter: http://arxiv.org/abs/1611.07004.
88. Sreeja S, Mubarak DMN. Pseudo computed tomography estimation from brain MRI using anatomic signature and joint dictionary learning. Proc Comput Sci. 2023;218:1946–57.
89. Ahangari S, Hansen NL, Olin AB, Nøttrup TJ, Ryssel H, Berthelsen AK, u. a. Toward PET/MRI as one-stop shop for radiotherapy planning in cervical cancer patients. Acta Oncol. s2021;60(8):1045–53.
90. O'Connor LM, Dowling JA, Choi JH, Martin J, Warren-Forward H, Richardson H, u. a. Validation of an MRI-only planning workflow for definitive pelvic radiotherapy. Radiat Oncol. 2022;17(1):55.
91. Zhao S, Geng C, Guo C, Tian F, Tang X. SARU: a self-attention ResUNet to generate synthetic CT images for MR-only BNCT treatment planning. Med Phys. 2023;50(1):117–27.
92. Chourak H, Barateau A, Tahri S, Cadin C, Lafond C, Nunes JC, u. a. Quality assurance for MRI-only radiation therapy: a voxel-wise population-based methodology for image and dose assessment of synthetic CT generation methods. Front Oncol. 2022;12:968689.
93. O'Connor LM, Choi JH, Dowling JA, Warren-Forward H, Martin J, Greer PB. Comparison of synthetic computed tomography generation methods, incorporating male and female anatomical differences, for magnetic resonance imaging-only definitive pelvic radiotherapy. Front Oncol. 2022;12:822687.
94. Bahrami A, Karimian A, Arabi H. Comparison of different deep learning architectures for synthetic CT generation from MR images. Phys Med. 2021;90:99–107.
95. Mori S. A real-time, single-exposure, dual-energy subtraction mask for markerless tumor tracking in radiotherapy: proof of concept. Phys Med. 2019;63:63–9.
96. Wang J, Yan B, Wu X, Jiang X, Zuo Y, Yang Y. Development of an unsupervised cycle contrastive unpaired translation network for MRI-to-CT synthesis. J Appl Clin Med Phys. 2022;23(11):e13775.
97. Wang C, Uh J, Merchant TE, Hua CH, Acharya S. Facilitating MR-guided adaptive proton therapy in children using deep learning-based synthetic CT. Int J Part Ther. 2022;8(3):11–20.
98. Ranjan A, Lalwani D, Misra R. GAN for synthesizing CT from T2-weighted MRI data towards MR-guided radiation treatment. MAGMA. 2022;35(3):449–57.
99. Wang CC, Wu PH, Lin G, Huang YL, Lin YC, Chang YPE, u. a. Magnetic resonance-based synthetic computed tomography using generative adversarial networks for intracranial tumor radiotherapy treatment planning. J Pers Med. 2022;12(3):361.
100. Jabbarpour A, Mahdavi SR, Vafaei Sadr A, Esmaili G, Shiri I, Zaidi H. Unsupervised pseudo CT generation using heterogenous multicentric CT/MR images and CycleGAN: dosimetric assessment for 3D conformal radiotherapy. Comput Biol Med. 2022;143:105277.
101. Reaungamornrat S, Sari H, Catana C, Kamen A. Multimodal image synthesis based on disentanglement representations of anatomical and modality specific features, learned using uncooperative relativistic GAN. Med Image Anal. 2022;80:102514.
102. Wang Z, Bovik AC, Sheikh HR, Simoncelli EP. Image quality assessment: from error visibility to structural similarity. IEEE Trans Image Process. 2004;13(4):600–12.
103. Zimmermann L, Knäusl B, Stock M, Lütgendorf-Caucig C, Georg D, Kuess P. An MRI sequence independent convolutional neural network for synthetic head CT generation in proton therapy. Z Med Phys. 2022;32(2):218–27.
104. Hyuk Choi J, Asadi B, Simpson J, Dowling JA, Chalup S, Welsh J, u. a. Investigation of a water equivalent depth method for dosimetric accuracy evaluation of synthetic CT. Phys Med. 2023;105:102507.
105. Zhao Y, Wang H, Yu C, Court LE, Wang X, Wang Q, u. a. Compensation cycle consistent generative adversarial networks (Comp-GAN) for synthetic CT generation from MR scans with truncated anatomy. Med Phys. 2023;50(7):4399.
106. Reinke A, Tizabi MD, Sudre CH, Eisenmann M, Rädsch T, Baumgartner M, Acion L, Antonelli M, Bakas S, Bankhead P, Benis A, Cardoso MJ, Cheplygina V, Christodoulou E, Cimini B, Collins GS, Farahani K, Ginneken BV, Glocker B, Maier-Hein L. Common limitations of image processing metrics: A Picture Story. ArXiv; 2021. https://doi.org/10.48550/arXiv.2104.05642
107. Kaushik S, Bylund M, Cozzini C, Shanbhag D, Petit SF, Wyatt JJ, u. a. Region of interest focused MRI to Synthetic CT Translation using Regression and Classification Multi-task Network. arXiv; 2022.
108. Lena B, Florkow MC, Ferrer CJ, van Stralen M, Seevinck PR, Vonken EJPA, u. a. Synthetic CT for the planning of MR-HIFU treatment of bone metastases in pelvic and femoral bones: a feasibility study. Eur Radiol. 2022;32(7):4537–46.
109. Eshraghi Boroojeni P, Chen Y, Commean PK, Eldeniz C, Skolnick GB, Merrill C, u. a. Deep-learning synthesized pseudo-CT for MR high-resolution pediatric cranial bone imaging (MR-HiPCB). Magn Reson Med. 2022;88(5):2285–97.
110. Low DA, Harms WB, Mutic S, Purdy JA. A technique for the quantitative evaluation of dose distributions. Med Phys. 1998;25(5):656–61.
111. Yu L, Baker A, Kairn T, Livingstone A, Trapp J, Crowe S. A structure-based gamma evaluation method for identifying clinically relevant dose differences in organs at risk. Phys Eng Sci Med. 2023;46(3):1033.

112. Liu X, Emami H, Nejad-Davarani SP, Morris E, Schultz L, Dong M, u. a. Performance of deep learning synthetic CTs for MR-only brain radiation therapy. J Appl Clin Med Phys. 2021;22(1):308–17.
113. Maspero M, Bentvelzen LG, Savenije MHF, Guerreiro F, Seravalli E, Janssens GO, u. a. Deep learning-based synthetic CT generation for paediatric brain MR-only photon and proton radiotherapy. Radiother Oncol. 2020;153:197–204.
114. Masitho S, Szkitsak J, Grigo J, Fietkau R, Putz F, Bert C. Feasibility of artificial-intelligence-based synthetic computed tomography in a magnetic resonance-only radiotherapy workflow for brain radiotherapy: two-way dose validation and 2D/2D kV-image-based positioning. Phys Imaging Radiat Oncol. 2022;24:111–7.
115. Ding S, Liu H, Li Y, Wang B, Li R, Huang X. Dosimetric accuracy of MR-guided online adaptive planning for nasopharyngeal carcinoma radiotherapy on 1.5 T MR-Linac. Front Oncol. 2022;12:858076.
116. Kemppainen R, Suilamo S, Ranta I, Pesola M, Halkola A, Eufemio A, u. a. Assessment of dosimetric and positioning accuracy of a magnetic resonance imaging-only solution for external beam radiotherapy of pelvic anatomy. Phys Imaging Radiat Oncol. 2019;11:1–8.
117. Palmér E, Nordström F, Karlsson A, Petruson K, Ljungberg M, Sohlin M. Head and neck cancer patient positioning using synthetic CT data in MRI-only radiation therapy. J Appl Clin Med Phys. 2022;23(4):e13525.
118. Maspero M, Tyyger MD, Tijssen RHN, Seevinck PR, Intven MPW, van den Berg CAT. Feasibility of magnetic resonance imaging-only rectum radiotherapy with a commercial synthetic computed tomography generation solution. Phys Imaging Radiat Oncol. 2018;7:58–64.
119. Lerner M, Medin J, Jamtheim Gustafsson C, Alkner S, Olsson LE. Prospective clinical feasibility study for MRI-only brain radiotherapy. Front Oncol. 2022;11:812643.
120. Wang C, Uh J, Patni T, Merchant T, Li Y, Hua CH, u. a. Toward MR-only proton therapy planning for pediatric brain tumors: synthesis of relative proton stopping power images with multiple sequence MRI and development of an online quality assurance tool. Med Phys. 2022;49(3):1559–70.
121. Buatti JS, Gallagher KJ, Bailey I, Griglock T, Heard M. An evaluation of quality assurance guidelines comparing the American College of Radiology and American Association of Physicists in Medicine task group 284 for magnetic resonance simulation. J Appl Clin Med Phys August 2022;23(8):e13730.
122. Karaiskos P, Moutsatsos A, Pappas E, Georgiou E, Roussakis A, Torrens M, u. a. A simple and efficient methodology to improve geometric accuracy in gamma knife radiation surgery: implementation in multiple brain metastases. Int J Radiat Oncol Biol Phys. 2014;90(5):1234–41.
123. Pappas EP, Alshanqity M, Moutsatsos A, Lababidi H, Alsafi K, Georgiou K, u. a. MRI-related geometric distortions in stereotactic radiotherapy treatment planning: evaluation and dosimetric impact. Technol Cancer Res Treat. 2017;16(6):1120–9.
124. Adjeiwaah M, Bylund M, Lundman JA, Karlsson CT, Jonsson JH, Nyholm T. Quantifying the effect of 3T magnetic resonance imaging residual system distortions and patient-induced susceptibility distortions on radiation therapy treatment planning for prostate cancer. Int J Radiat Oncol Biol Phys. 2018;100(2):317–24.
125. Adjeiwaah M, Bylund M, Lundman JA, Söderström K, Zackrisson B, Jonsson JH, u. a. Dosimetric impact of MRI distortions: a study on head and neck cancers. Int J Radiat Oncol Biol Phys. 2019;103(4):994–1003.
126. Curcuru AN, Lewis BC, Kim T, Yang D, Michael GH. Effects of $B_0$ eddy currents on imaging isocenter shifts in 0.35-T MRI-guided radiotherapy (MR-IGRT) system. Med Phys. 2021;48(6):2929–38.
127. Gach HM, Curcuru AN, Mutic S, Kim T. $B_0$ field homogeneity recommendations, specifications, and measurement units for MRI in radiation therapy. Med Phys 2020;47(9):4101–4114.
128. Tanadini-Lang S, Budgell G, Bohoudi O, Corradini S, Cusumano D, Güngör G, u. a. An ESTRO-ACROP guideline on quality assurance and medical physics commissioning of online MRI guided radiotherapy systems based on a consensus expert opinion. Radiother Oncol. 2023;181:109504.
129. Glide-Hurst CK, Paulson ES, McGee K, Tyagi N, Hu Y, Balter J, u. a. Task group 284 report: magnetic resonance imaging simulation in radiotherapy: considerations for clinical implementation, optimization, and quality assurance. Med Phys. 2021;48(7):e636–70.
130. Jackson E, Bronskill M, Drost D, Och J, Pooley R, Sobol W, u. a. Acceptance testing and quality assurance procedures for magnetic resonance imaging facilities [internet]. AAPM; 2010 Dez [zitiert 2. Mai 2023]. Verfügbar unter: https://www.aapm.org/pubs/reports/detail.asp?docid=101.
131. Tao S, Trzasko JD, Gunter JL, Weavers PT, Shu Y, Huston J, u. a. Gradient nonlinearity calibration and correction for a compact, asymmetric magnetic resonance imaging gradient system. Phys Med Biol. 2017;62(2):N18–31.
132. Michael Gach H, Curcuru AN, Wittland EJ, Maraghechi B, Cai B, Mutic S, u. a. MRI quality control for low-field MR-IGRT systems: lessons learned. J Appl Clin Med Phys. 2019;20(10):53–66.
133. Duyn JH, Schenck J. Contributions to magnetic susceptibility of brain tissue. NMR Biomed. 2017;30(4).
134. Kames C, Doucette J, Rauscher A. Multi-echo dipole inversion for magnetic susceptibility mapping. Magn Reson Med. 2023;89(6):2391–401.
135. Tyagi N, Fontenla S, Zhang J, Cloutier M, Kadbi M, Mechalakos J, u. a. Dosimetric and workflow evaluation of first commercial synthetic CT soft-

ware for clinical use in pelvis. Phys Med Biol. 2017;62(8):2961–75.
136. Xie K, Gao L, Lu Z, Li C, Xi Q, Zhang F, u. a. Inpainting the metal artifact region in MRI images by using generative adversarial networks with gated convolution. Med Phys. 2022;49(10):6424–38.
137. Lundman JA, Bylund M, Garpebring A, Thellenberg Karlsson C, Nyholm T. Phys Imaging Radiat Oncol. 2017;1:41–5.
138. Marasini S, Quinn B, Cole M, Flores R, Kim T. System-dependent image distortion related to gantry positions of a 0.35 T MRgRT: characterization and the corresponding correction. J Appl Clin Med Phys. 2023;24(1):e13826.
139. Lewis BC, Shin J, Quinn B, Barberi E, Sievert D, Kim JS, u. a. First clinical experience of correcting phantom-based image distortion related to gantry position on a 0.35T MR-Linac. J Appl Clin Med Phys. 2021;22(11):21–8.
140. Lewis B, Guta A, Mackey S, Gach HM, Mutic S, Green O, u. a. Evaluation of diffusion-weighted MRI and geometric distortion on a 0.35T MR-LINAC at multiple gantry angles. J Appl Clin Med Phys. 2021;22(2):118–25.
141. Dorsch S, Mann P, Elter A, Runz A, Spindeldreier CK, Klüter S, u. a. Measurement of isocenter alignment accuracy and image distortion of an 0.35 T MR-Linac system. Phys Med Biol. 2019;64(20):205011.
142. Damyanovich AZ, Tadic T, Foltz WD, Jelveh S, Bissonnette JP. Time-course assessment of 3D-image distortion on the 1.5 T Marlin/Elekta Unity MR-LINAC. Phys Med. 2022;100:90–8.
143. Edmund JM, Nyholm T. A review of substitute CT generation for MRI-only radiation therapy. Radiat Oncol. 2017;12(1):28.
144. Bird D, Henry AM, Sebag-Montefiore D, Buckley DL, Al-Qaisieh B, Speight R. A systematic review of the clinical implementation of pelvic magnetic resonance imaging-only planning for external beam radiation therapy1. Int J Radiat Oncol Biol Phys. 2019;105(3):479–92.
145. Tyagi N, Fontenla S, Zelefsky M, Chong-Ton M, Ostergren K, Shah N, u. a. Clinical workflow for MR-only simulation and planning in prostate. Radiat Oncol. 2017;12(1):119.
146. Kemppainen R, Suilamo S, Tuokkola T, Lindholm P, Deppe MH, Keyriläinen J. Magnetic resonance-only simulation and dose calculation in external beam radiation therapy: a feasibility study for pelvic cancers. Acta Oncol. 2017;56(6):792–8.
147. Kerkmeijer LGW, Maspero M, Meijer GJ, van der Voort van Zyp JRN, de Boer HCJ, van den Berg CAT. Magnetic resonance imaging only workflow for radiotherapy simulation and planning in prostate cancer. Clin Oncol. 2018;30(11):692–701.
148. Yu V, Keyrilainen J, Suilamo S, Beslimane I, Dresner A, Halkola A, u. a. A multi-institutional analysis of a general pelvis continuous Hounsfield unit synthetic CT software for radiotherapy. J Appl Clin Med Phys. 2021;22:207.
149. Siversson C, Nordström F, Nilsson T, Nyholm T, Jonsson J, Gunnlaugsson A, u. a. Technical note: MRI only prostate radiotherapy planning using the statistical decomposition algorithm. Med Phys. 2015;42(10):6090–7.
150. Persson E, Gustafsson C, Nordström F, Sohlin M, Gunnlaugsson A, Petruson K, u. a. MR-OPERA: a multicenter/multivendor validation of magnetic resonance imaging-only prostate treatment planning using synthetic computed tomography images. Int J Radiat Oncol Biol Phys. 2017;99(3):692–700.
151. Persson E, Jamtheim Gustafsson C, Ambolt P, Engelholm S, Ceberg S, Bäck S, u. a. MR-PROTECT: clinical feasibility of a prostate MRI-only radiotherapy treatment workflow and investigation of acceptance criteria. Radiat Oncol. 2020;15(1):77.
152. Palmér E, Persson E, Ambolt P, Gustafsson C, Gunnlaugsson A, Olsson LE. Cone beam CT for QA of synthetic CT in MRI only for prostate patients. J Appl Clin Med Phys. 2018;19(6):44–52.
153. Tenhunen M, Korhonen J, Kapanen M, Seppälä T, Koivula L, Collan J, u. a. MRI-only based radiation therapy of prostate cancer: workflow and early clinical experience. Acta Oncol. 2018;57(7):902–7.
154. Kim J, Miller B, Siddiqui MS, Movsas B, Glide-Hurst C. FMEA of MR-only treatment planning in the pelvis. Adv Radiat Oncol. 2019;4(1):168–76.
155. Dowling JA, Sun J, Pichler P, Rivest-Hénault D, Ghose S, Richardson H, u. a. Automatic substitute computed tomography generation and contouring for magnetic resonance imaging (MRI)-alone external beam radiation therapy from standard MRI sequences. Int J Radiat Oncol Biol Phys. 2015;93(5):1144–53.
156. Greer P, Martin J, Sidhom M, Hunter P, Pichler P, Choi JH, u. a. A multi-center prospective study for implementation of an MRI-only prostate treatment planning workflow. Front Oncol. 2019;9:826.
157. Palmér E, Karlsson A, Nordström F, Petruson K, Siversson C, Ljungberg M, u. a. Synthetic computed tomography data allows for accurate absorbed dose calculations in a magnetic resonance imaging only workflow for head and neck radiotherapy. Phys Imaging Radiat Oncol. 2021;17:36–42.
158. Cronholm RO, Karlsson A, Siversson C. Whitepaper: MRI only radiotherapy planning using the transfer function estimation algorithm. 2020.
159. Lerner M, Medin J, Jamtheim Gustafsson C, Alkner S, Siversson C, Olsson LE. Clinical validation of a commercially available deep learning software for synthetic CT generation for brain. Radiat Oncol. 2021;16(1):66.
160. Hoesl M, Escobar Corral, Mistry N. Whitepaper: MR-based synthetic CT reimagined, an AI-based algorithm for continuous Hounsfield units in the pelvis and brain – with syngo.via RT Image Suite (VB60). Siemens-Healthineers; 2022.

161. Emin S, Rossi E, Gagliardi G, Hedman M, Villegas F. Uncertainty-aware MR-based CT synthesis for robust proton planning of skull-based tumour. ESTRO. 2023.
162. Jonsson J, Nyholm T, Söderkvist K. The rationale for MR-only treatment planning for external radiotherapy. Clin Transl Radiat Oncol. 2019;18:60–5.
163. Kurz C, Buizza G, Landry G, Kamp F, Rabe M, Paganelli C, u. a. Medical physics challenges in clinical MR-guided radiotherapy. Radiat Oncol. 2020;15(1):93.
164. Dunkerley DAP, Hyer DE, Snyder JE, St-Aubin JJ, Anderson CM, Caster JM, u. a. Clinical implementational and site-specific workflows for a 1.5T MR-Linac. J Clin Med. 2022;11(6):1662.
165. Tang B, Liu M, Wang B, Diao P, Li J, Feng X, u. a. Improving the clinical workflow of a MR-Linac by dosimetric evaluation of synthetic CT. Front Oncol. 2022;12:920443.
166. Liang J, Scripes PG, Tyagi N, Subashi E, Wunner T, Cote N, u. a. Risk analysis of the Unity 1.5 T MR-Linac adapt-to-position workflow. J Appl Clin Med Phys [Internet]. 2023 [zitiert 3. Mai 2023];24(3). Verfügbar unter: https://onlinelibrary.wiley.com/doi/10.1002/acm2.13850.
167. Ng J, Gregucci F, Pennell RT, Nagar H, Golden EB, Knisely JPS, u. a. MRI-LINAC: a transformative technology in radiation oncology. Front Oncol. 2023;13:1117874.
168. Das IJ, Yadav P, Mittal BB. Emergence of MR-Linac in radiation oncology: successes and challenges of riding on the MRgRT bandwagon. J Clin Med. 2022;11(17):5136.
169. Güngör G, Azria D, Balermpas P, Boldrini L, Chuong MD, De Ridder M, u. a. A multi-centric evaluation of AI-driven synthetic CT generation form low field magnetic resonance imaging. Int J Radiat Oncol. 2022;114(3):S163.
170. Cusumano D, Boldrini L, Dhont J, Fiorino C, Green O, Güngör G, u. a. Artificial intelligence in magnetic resonance guided radiotherapy: medical and physical considerations on state of art and future perspectives. Phys Med. 2021;85:175–91.
171. Maspero M, Savenije MHF, Dinkla AM, Seevinck PR, Intven MPW, Jurgenliemk-Schulz IM, u. a. Dose evaluation of fast synthetic-CT generation using a generative adversarial network for general pelvis MR-only radiotherapy. Phys Med Biol. 2018;63(18):185001.
172. Thorwarth D, Low DA. Technical challenges of real-time adaptive MR-guided radiotherapy. Front Oncol. 2021;11:634507.
173. Goodburn RJ, Philippens MEP, Lefebvre TL, Khalifa A, Bruijnen T, Freedman JN, u. a. The future of MRI in radiation therapy: challenges and opportunities for the MR community. Magn Reson Med. 2022;88(6):2592–608.

# Treatment Planning Considerations for an MR-Linac

# 8

Justine M. Cunningham, Jennifer L. Dolan, Kate Aldridge, and Ergys Subashi

## Treatment Planning System

MR-guided radiation therapy (MRgRT) delivered with MR-Linacs is motivated by advanced simulation and on-board imaging. MRI-primary contouring, treatment planning, localization, and motion management enhance the targeting accuracy of this treatment modality. The gain in targeting accuracy facilitates the shrinking of target margins, sparing of organs at risk (OAR), online adaptive treatment planning and delivery, and furthermore escalation or de-escalation of prescribed dose. Thus, MR-Linacs have largely become specialized treatment machines with utilization focused on hypo-fractionation and stereotactic style treatments [1]. Particularly, MR-Linacs have proven impactful in abdominal, pelvic, and cranial treatment sites [2–4].

Complex treatments, such as stereotactic body radiation therapy (SBRT), demand advanced treatment planning systems (TPS) to meet the intricate and diverse treatment planning needs. Such needs largely rely on inverse optimization algorithms that result in intensity modulated radiation therapy (IMRT) deliveries paired with accurate dose calculation algorithms (further discussed in sections "Planning Geometry and Optimization and Dose Calculation"). Additional imaging (e.g., CT, PET) is often required for electron density (section "Electron Density") or delineation purposes (section "Contouring"). Furthermore, the effective clinical application of MRgRT relies on ultra-fast optimization and dose calculation to the point where such operations can be performed while the patient waits on the table for treatment.

## Hardware and Software Requirements

TPS hardware and software system requirements align with the safety and security necessitated by any commercial planning system. High-performance computational abilities are required to enable Monte Carlo optimizations and dose calculations with computation times at less than 1 min. This requires powerful computers and code which are optimized for CPUs and/or GPUs. An example of MR-Linac TPS has dual CPUs with a minimum of 14 processor cores each, 128 GB RAM, 512 GB hard disk space, and 4 GB of video memory.

Additional emphasis should be placed on data transfer and interconnectivity quality assurance (QA), as the MR-Linac TPS and treatment deliv-

J. M. Cunningham (✉) · J. L. Dolan · K. Aldridge
Henry Ford Health, Department of Radiation Oncology, Detroit, MI, USA
e-mail: JCunnin5@hfhs.org; jdolan1@hfhs.org; kaldrid1@hfhs.org

E. Subashi
Memorial Sloan Kettering Cancer Center, Department of Medical Physics, New York, NY, USA
e-mail: subashie@mskcc.org



I. J. Das et al. (eds.), *A Practical Guide to MR-Linac*, https://doi.org/10.1007/978-3-031-48165-9_8

ery system (TDS) are likely to reside separately from a clinic's primary record and verify system and other necessary subsystems [5, 6].

## Integration with Treatment Delivery Software

From a system architecture perspective, the novelty for MR-Linacs is the presence of a TPS in parallel to the TDS. Since MRgRT TDSs are capable of on-table adaptive therapy (section "Adaptive Planning Considerations"), they are responsible for more than just treatment delivery. In a common adaptive work flow, the original CT is deformed to the daily MRI for electron density, and then the original plan is calculated onto the current anatomy. Dose evaluation statistics are presented to the clinician to review and decide on whether to adapt or not (section "Plan Evaluation"). In the background, the system can begin optimizing, generating a new plan, and present a dose distribution by Monte Carlo with excellent accuracy, all while the patient waits inside the machine. The integration of the treatment planning activities with the treatment delivery software presents unique considerations in the commissioning and QA of the planning systems [5, 7].

## Patient Database

The TPS and TDS will require secure access to a patient and machine delivery database. Each subsystem in the integrated treatment planning and treatment delivery system must accurately access beam and patient data. Commissioning and routine QA should aim at the evaluation for the accuracy and security of such connectivity [5, 6].

# Commissioning and Quality Assurance

Commissioning and QA of the treatment planning systems (TPS) associated with MR-Linacs can follow established guidelines for conventional TPSs to assess the non-dosimetric and dosimetric TPS commissioning and for ongoing QA of the TPS [8]. Additional TPS testing can include data acquisition and processing, beam model iteration, and validation following the American Association of Physicists in Medicine (AAPM) Medical Physics Practice Guidelines [9, 10]. Data acquisition and processing includes CT calibration data and treatment beam data following AAPM report of Task Group 106 [11]. Validation tests can include basic photon dose calculation, as well as dose calculations in heterogeneous media and for intensity modulated radiation therapy (IMRT). IMRT commissioning tests can include the tests outlined in AAPM Report of Task Group 119 [12]. Dose calculation for MR-Linacs is achieved utilizing Monte Carlo-based dose calculation algorithms; therefore beam modeling and beam model commissioning can follow standards outlined in the AAPM Report of Task Group 157 and 105 [13, 14].

More sophisticated treatment techniques result in increased complexity of plans and more advanced software. Therefore, there are testing requirements in addition to MRI-specific imaging QA requirements, to consider when implementing MR-guided radiation therapy. Specifically, the magnetic field may cause non-negligible dosimetric effects, and the beam model within the TPS can account for these effects, necessitating thorough validation. Current practices have included this testing within end-to-end verification tests [15] or specifically designed phantoms [16].

Online adaptive treatment techniques offer additional TPS commissioning challenges. If the treatment delivery system planning platform is unique from the TPS, the same processes and verification of treatment planning must be performed utilizing the online adaptive platform. End-to-end verification tests isolating couch attenuation, electron density conversion, contour generation, and optimization robustness are some of the many tests that can be considered during adaptive treatment planning validation [17–19].

## Contouring

The integration of an MRI scanner with a linear accelerator offers unique advantages compared to conventional CT-based on-board imaging. The soft tissue contrast achievable on MR-Linacs has been shown to provide better visualization of radiotherapy targets and OARs compared to on-board CT-based imaging techniques [20]. Functional imaging during simulation aids in improving the precision of target delineation and OARs [21]. Additionally, the incorporation of multi-planar MR-cine image acquisition introduces new motion management techniques such as real-time motion monitoring and gating which may facilitate the use of reduced margins [22–24].

Standard contouring tools available across various treatment planning platforms can be utilized for MR-Linac treatment planning. MRI-based target volume and OAR contouring recommendations exist for various locations throughout the body [25–28]. Additionally, several auto-segmentation techniques are being studied not only for their use in initial segmentation for planning but also to be utilized during online adaptive radiotherapy [29–32]. Auto-segmentation methods can be categorized into methods that do and do not use prior knowledge. Methods that do not use prior knowledge include intensity-based segmentation where voxels are classified into different tissue types based on intensity levels. Methods that require prior knowledge include atlas based, statistical models, and models based on machine learning and artificial intelligence. Atlas-based segmentation uses deformable or rigid registration to transform a structure of interest from a reference image to test image. Segmentation using machine learning is achieved through using a training dataset to establish features for tissue segmentation and applying it to a test image [33]. Auto-segmentation techniques are an invaluable tool for increased planning efficiency, specifically in relation to online adaptive treatment planning.

## Treatment Planning Imaging Datasets

A combination of treatment planning CTs and MRIs are used for OAR and target contouring. MR-Linac platforms offer both T1-weighted and T2-weighted images. High-field MR-Linacs have been evaluated for the preference of MR-derived sequences including mDIXON and T1- and T2-weighted images for different anatomical regions [34, 35]. Low-field MR-Linacs require a true fast imaging with steady-state precession (TRUFI) sequence yielding a T2–/T1-weighted contrast for planning [36]. The possibility of short acquisition times for TRUFI image sequences allows for images to be acquired at breath-hold, facilitating the ability to treat at breath-hold using cine image real-time motion monitoring and gating during treatment delivery. MRI sequences with longer acquisition times can use 4D-MRI to facilitate treatment planning on motion averaged images [28]. These may be used with 4DCT (four-dimensional CT) to account for breathing motion using an internal target volume. Additionally, motion encompassing methods with abdominal compression have been used to reduce the amplitude of breathing motion [37]. Recommended slice thicknesses of one to three millimeters are achievable for conventional and stereotactic treatment planning. Although MRI is the gold standard for soft tissue visualization, contrast-enhanced CT-based imaging or F-fluorodeoxyglucose PET imaging may prove useful in the contouring of targets in various regions throughout the body and can be registered to the treatment planning MRI to aid in tissue delineation.

### Contrast

To aid in the visualization and delineation of the disease being treated, gadolinium-based contrast agents can be used. Spinal cord tumors, cranial tumors, and liver metastasis are common anatomical sites where contrast agents are used to better visualize the disease. These contrast agents cannot only be used for target delineation but can also enhance online gating capabilities. For example, when treating liver metastasis, contrast can be distributed, if kidney function allows, to the patient on the day of treatment, and the lesion itself can be used for real time motion monitoring and gating during treatment delivery. An example of this can be seen in Fig. 8.1.

## Artifacts and Geometric Distortions

Two primary categories of geometric distortions exist in MR imaging including system level, resulting from gradient nonlinearity in the spatial encoding gradients and magnetic field inhomogeneities, as well as patient-level distortions due to chemical shift artifacts and susceptibility artifacts [38]. Due to the high dependence of geometric accuracy in radiation delivery, the reduction of geometric distortions and artifacts through quantification and mitigation is vital for the delineation of target volumes and OARs.

## Margins

Planning target volume margins are applied to account for uncertainties in target volume delineation and positioning, organ motion, variations in patient positioning, and uncertainties in dose delivery and target coverage. The required margin size is impacted by image-guided technique and frequency of use. In MRgRT direct visualiza-

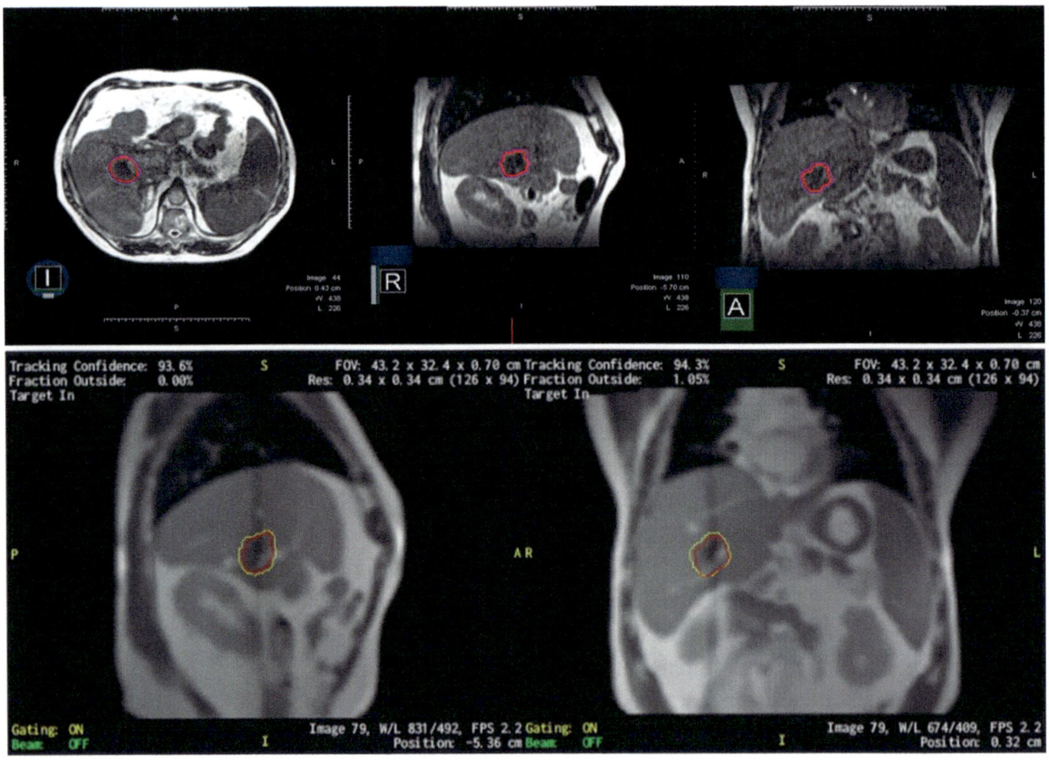


**Fig. 8.1** Top images, liver lesion with the CTV outlined in red and PTV outlined in pink in the axial sagittal and coronal view of a 14 s TRUFI image taken at end exhale with an image resolution of 1.5 mm × 1.5 mm × 3 mm. The bottom image includes a sagittal and coronal view of the cine MRI with 3.5 mm × 3.5 mm resolution and four frames per second used for gating. The smoothed CTV volume, shown in red falls within the gating boundary (yellow), which was generated using a 3 mm isotropic margin from the smoothed CTV volume

tion of target volumes and OAR in conjunction with functional imaging during simulation and treatment delivery may facilitate the reduction in margins due to the reduced uncertainties in the delineation of target volume and patient localization at time of treatment. Additionally, planar cine MR images are acquired continuously during treatment for delivery verification. These cine images can be used for motion monitoring with manual gating or real-time motion management with continuous feedback. Frequently, real-time motion management is performed utilizing a contour of interest that can be outlined as the target volume in the MR cine image; however one may also choose to monitor an organ at risk to ensure it does not enter the radiation field or a surrogate contour if the target volume is not visible. For mobile targets, a large PTV margin was previously used to provide sufficient coverage of the moving lesion. However, continual monitoring of the precise location of the target makes it feasible to reduce PTV treatment margins, therefore reducing the dose to OARs. These motion management techniques may further facilitate the reduction in margins due to reducing the uncertainties in dose delivery [22, 24, 39, 40]. Margins will be specific to the anatomical area being treated, immobilization devices used, and the motion management technique selected for treatment delivery.

## Motion Management with Cine Imaging

During motion monitoring, two-dimensional cine MRIs are acquired continuously in up to three planes. During delivery, the contour of interest is displayed on the cine MRI allowing the treatment delivery team to assess any changes in anatomy and intervene when necessary. This method relies on manual monitoring of the changing anatomy, without automatic beam gating. During real-time motion management with continuous feedback, the system can gate the beam automatically. An anatomical structure is outlined on each plane with a desired margin generating a gating boundary. A defined percentage of the structure of interest allowed outside of the gating boundary is used to gate the beam off. For each frame of cine MRI's taken, the gating structure is deformably registered to a selected key frame outlining the desired treatment position. If the defined parameters are met, the beam is turned on. This technique can be used for gated breath-hold, gated free-breathing treatments, and gated treatments that are unaffected by respiratory motion but may be affected by peristalsis in the gastrointestinal track, changes in the genitourinary track, or patient movement [36]. The delivery technique is crucial to determining the size of the margins required for various clinical sites. Depending on the contour of interest chosen for motion monitoring, the gating boundary may correspond to the planning target volume margins.

## Online Adaptive Considerations

During the online adaptive treatment planning workflow, efficiency in re-contouring and replanning is vital to reduce treatment times. Limiting the number of organs that require recontouring to the immediate regions adjacent to the target volume can facilitate this. Additionally, defining those structures that can be strictly deformed and reviewed from the planning image to the daily image will aid in reducing the number of contours that will be manually recontoured online.

Gross tumor volumes and clinical tumor volumes are frequently redrawn daily. Planning rules imbedded into the treatment plan can be used to automatically generate the resulting planning target volume based on pre-established margin expansions. These rules can include the generation of target volumes using pre-established margins, optimization structures that are frequently used in planning, and planning at risk volumes. The predefinition of margin expan-

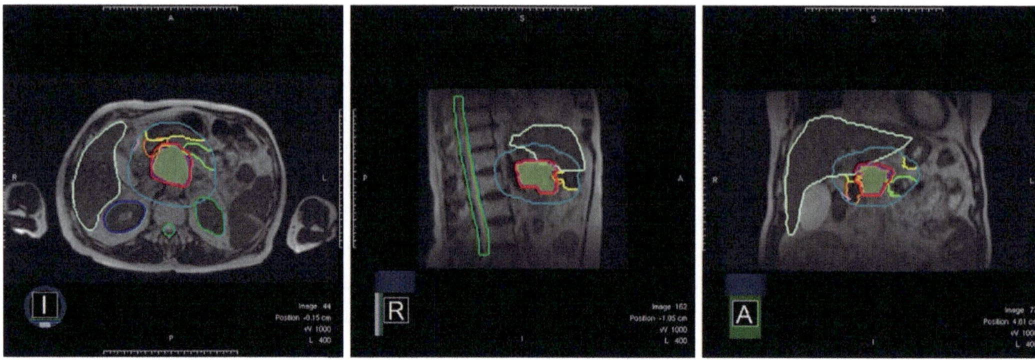

**Fig. 8.2** Adaptive pancreas TRUFI MRI shown in the axial, sagittal, and coronal orientations. Structures outlined for the adaptive procedure include the liver (pale green), right kidney (dark blue), left kidney (dark green), and spinal canal (dark green) which will be deformed onto the MRI of the day and reviewed and edited, as necessary. The gastrointestinal OAR include the colon (pink), stomach (yellow), duodenum (orange), and small bowel (green) which will be redrawn within the normal tissue contouring ring (light blue) daily during the adaptive procedure. Finally, the CTV OPT (color wash green) and PTV OPT (color wash purple) are automatically regenerated for daily adaptive treatments by cropping the rigidly copied PTV (red) and CTV (orange) 3 mm away from the gastrointestinal structures

sions within the plan not only improves efficiency in the generation of these structures but also reduces the opportunity for error associated with the manual generation of these structures. Due to their impact on the adaptive workflow, it is imperative that these structures and rules are outlined during the initial planning process. An example of a normal tissue contouring ring and contours generated through treatment planning rules is shown in Fig. 8.2.

## Electron Density

MR imaging is not sufficient for radiotherapy treatment planning due to its intrinsic lack of electron density (ED) information. Conventionally, radiation dose is calculated using CT images where voxel Hounsfield units are converted to ED using a CT to ED conversion curve. In MRgRT, there are distinct options available to overcome this limitation. Plans can be developed on a patient CT as opposed to an MRI, allowing for the dose calculation process to be performed on the gold standard dataset for dose calculation. For MR only options, ED information can be propagated onto the MRI via bulk density overrides, deformable registration of a CT simulation to the MR, and synthetic CT (sCT) generation.

### CT Primary

At the initiation of the treatment planning process, the primary imaging dataset is chosen. In some MRgRT treatment planning software versions, the user can choose the patient's CT dataset to be used as the primary image dataset. Supplemental datasets, such as the MRI simulation, can be fused to the CT dataset for contouring purposes. However, this workflow may lead to image registration errors as significant as 1-2 mm, where the magnitude of these errors depends on the co-registration technique used as well as the anatomical site [41]. These image registration errors are systematic and will be propagated throughout the treatment planning and delivery process, which could result in a geometric miss. This method allows the user to calculate dose in the conventional fashion, but this will also result in MRI to CT co-registration errors during daily image guidance.

## MR Primary

The MR image can also be chosen as the primary imaging dataset. In this scenario, electron density must be propagated to the treatment planning image for dose calculation. This can be done through bulk density overrides or a deformable registration of the patient's CT simulation or sCT to the treatment planning MRI. In each scenario, the electron density map should be reviewed by a physicist prior to the final dose calculation.

MR-only treatment planning can be performed using sCT datasets generated from MRI and used for dose calculation. A significant review of the production of sCT has been done by Owrangi et al. [38] and Chap. 7. Voxel, atlas-based, and deep learning methods have been under development in recent years. Voxel-based methods use classification of tissues on a voxel-by-voxel basis. Categories of tissue include water, air, adipose, trabecular bone, and cortical bone. For each classification of tissue, a bulk HU value is assigned. The simplest scenario involves assigning a water-equivalent HU value to the entirety of the patient volume; however, in the presence of heterogeneous media including bone and air, the uncertainty in dose calculation will increase. Therefore, various MRI sequences can be used and combined to segment air, bone, and soft tissue in different anatomical regions of the body and assign the corresponding HU value. Both single and multi-atlas-based techniques have been developed. In single atlas-based techniques, the average patient anatomy in the form of a CT or CT-MRI pair is deformably registered to the treatment planning MRI to produce the sCT. Multi-atlas methods can take into consideration the variation in size and shape of individual patient anatomy, where the most similar CT or CT-MRI based on registration metrics, to the current anatomy, will be deformably registered to the treatment planning MRI and used to generate the sCT. The generation of sCT through machine learning is an ongoing research initiative. Various neural networks have been developed utilizing both paired MRI/CT patient images and unpaired images, where co-registration errors are not propagated to the synthetic CT generation with promising results. An example of a sCT generated using MRCAT (MR for calculation attenuation), an FDA-approved commercial sCT software, is shown in Fig. 8.3.

The ED map can be generated from the CT simulation image dataset for dose calculation using a deformable image registration. The CT simulation and MR simulation are typically taken sequentially with consistent immobilization devices to ensure similar anatomical properties between the two imaging datasets. In abdominal and pelvic anatomy, there are frequently differences in the air present in each of the imaging datasets. One method that has been introduced to overcome this issue has been to implement contours that can be used for bulk density overrides of the tissues present in each of the images to air or water. Specifically, the air in the CT image can be contoured and overridden to water, while the air in the MRI can be contoured and overridden to air. An example of this is shown in Fig. 8.4. This method can also be used to account for any differences in air-tissue interfaces between the synthetic CT and treatment planning MRI.

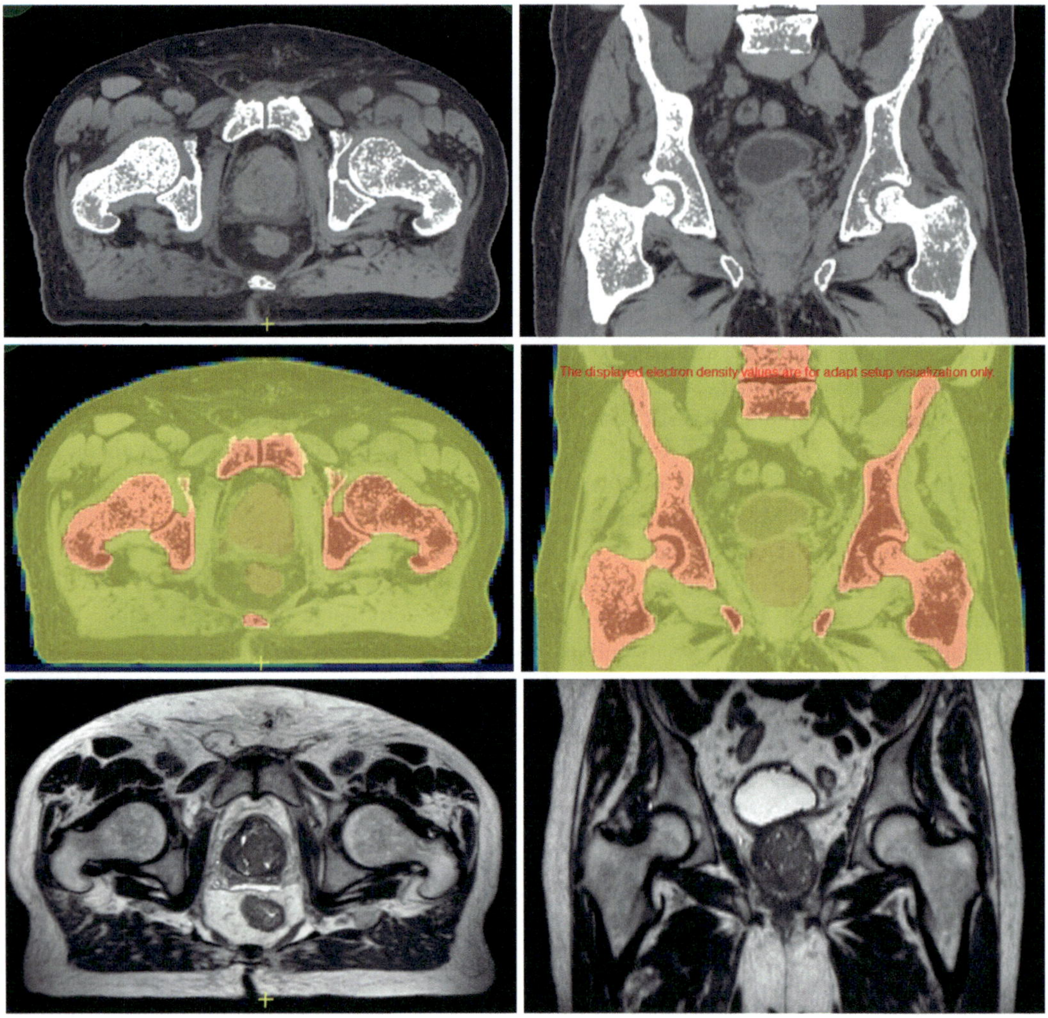


**Fig. 8.3** MRCAT sCT shown on the top, with the associated electron density heat map (middle) and treatment planning MRI (bottom)

**Fig. 8.4** Axial and coronal views of the deformed CT (top), electron density map (middle), and treatment planning MRI (bottom). The override to water structure in dark blue is used to bulk density override the visible air in CT to an electron density of water. The override to air structure in light blue is used to bulk density override the visible air in MRI to an electron density of air. The override to air structure is set to a higher priority than that of the override to water structure

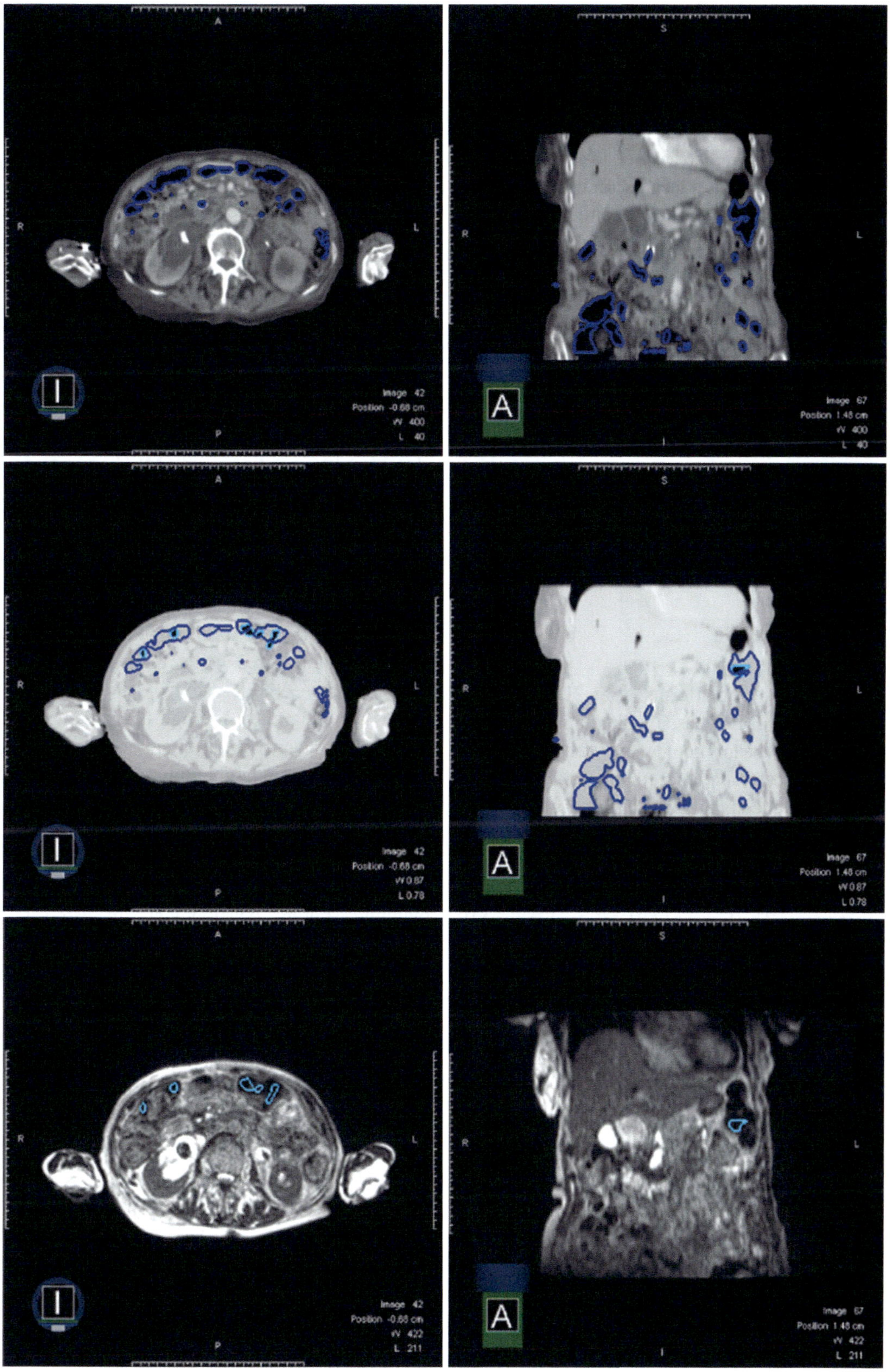
A
R
L
P
I
Image 42
Position -0.68 cm
W 400
L 40
S
A
Image 67
Position 1.48 cm
W 400
L 40
W 0.87
L 0.78
W 422
L 211

## Planning Geometry and Optimization

### Isocenter Placement

In conventional radiation therapy, the isocenter location is optimally placed within the center of the target volume or the radiation field. However, MR-Linacs have a limited bore size resulting in either limited or prohibited table movement in the anterior/posterior and lateral directions. As a result, the isocenter location is typically placed in the center of the target or radiation field in the superior-inferior direction but is centrally located in the axial plane. Depending on the laterality of the tumor location, this may result in large off-axis segment openings. To reduce the distance from the isocenter to the target volume center, the patient is positioned with the treatment site as close to midline as possible.

### Beam Geometry

Current delivery platforms offer 3D conformal and static IMRT-based planning methods. Conventional beams geometries used in CT-based planning methods are acceptable for MRI-based planning while keeping in mind additional limitations due a decreased size in the field-of-view (FOV), limited beam angles, and lack of collimator rotation, off-axis treatments, and longer treatment times. The largest FOV achievable for MR simulation images is smaller than the FOV achievable on CT simulation machines. This may result in edges of the patient body contour being cut off in simulation images, causing limited choices of beam angles due to missing electron density information. Additionally, due to manufacturer designs, some MR-Linacs have an associated dead zone where a range of beam angles cannot be used for treatment planning. It is also recommended to avoid treatment through the edges of the couch due to uncertainties associated with patient position and variation in attenuation through the edges of the couch. MLC motion is limited to a single plane (e.g., left-right or superior-inferior) as collimator rotation is disabled on MR-Linac treatment platforms. As mentioned above, the isocenter location is sometimes placed outside of the treatment volume, which results in frequent off-axis treatments to lateral targets where beam angles must be chosen carefully, taking into consideration the MLC position limitations and ensuring the target volume is adequately encompassed in the beams eye view (BEV). An example of beam geometry is shown in Fig. 8.5, including the off-axis beam eye view showing the limitations of the MLC movement and beam angle choice.

Conventional treatment delivery times range from 15 to 30 min, depending on the number of beams angles and complexity (e.g., number of segments) of the treatment delivery. This is due to slow gantry rotation and MLC movement. However, the length of the adaptive treatment process can be much longer and range from 45 to 90 min due to the need to generate and review new contours and treatment plans. Patient positioning for adaptive treatments is often modified to increase patient comfort, including simulating patients immobilized with their arms down, resulting in the requirement of choosing beam angles avoiding entry through the patient arms. Additional adaptive treatment considerations include maintaining suboptimal beam angles in the base plan, as they may be required to maintain plan quality due to inter-fraction motion. A pseudo arc technique, with equally spaced static IMRT beams, is frequently adopted for treatment planning. Future developments in MR-Linac treatment delivery platforms may lead to arc-based treatment options, where these limitations will still need to be considered during the beam selection process.

### Treatment Delivery Techniques

#### 3D Conformal

Three-dimensional conformal treatment planning on an MR-Linac follows a similar concept to that of CT-based planning techniques. A set beam geometry is chosen following the limitations outlined in section "Beam Geometry", MLCs can either be manually positioned or automatically fit

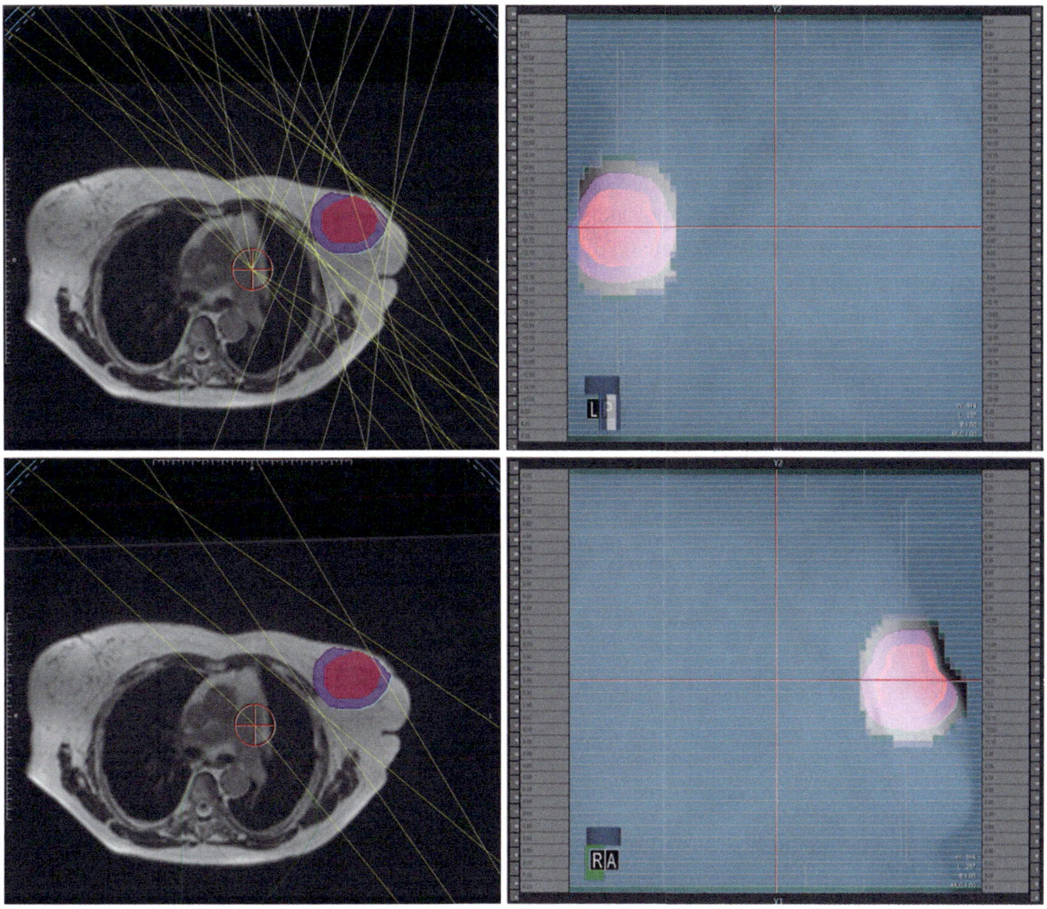

**Fig. 8.5** Left breast APBI off-axis beam geometry with the isocenter shown as a red cross, the CTVsb color washed red, and PTVsb color washed purple. Two BEVs of the beams at gantry angle 140° (top) and 312° (bottom) indicate the off-axis nature of the beams required due to the limited couch movement. These beams are also shown in the bottom left of the figure

to a target structure with a specified margin. The dose is calculated to a normalization or weighting point, and fixed beams can be weighted to create the desired dose distribution.

### IMRT

Current delivery platforms are limited to step-and-shoot IMRT delivery techniques, allowing for highly conformal dose distributions improving target coverage while reducing dose to OAR. IMRT plans are generated by dividing the beams into bixels, then optimizing and calculating the bixel fluence based on user defined objectives and constraints as shown by Das et al. [42]. Fluences are then converted into deliverable segments, re-optimized and re-calculated to consider the physical limitations of the MLC. During the optimization, various parameters can be chosen that will impact the dose calculation including the bixel size, turning the magnetic field on/off for dose calculation, editing the grid resolution and selection of MLC sequencing options. Bixel size is limited in one dimension by the width of the MLC. A smaller bixel size will result in a finer resolution of the fluence; however it may result in increased modulation. A larger bixel size will reduce the fluence resolution, decreasing the modulation of the plan; however, it may result in sub-optimal plans. Once the desired bixel fluence is achieved, MLC sequencing options can be chosen to put an emphasis on plan quality by creating a best matched deliverable fluence to the desired bixel fluence, degree of modulation through discretization of the fluence's in the bixel map, or simply limiting the number of segments. However, step-and-shoot when compared to sliding window and arc-based treatment delivery techniques is less efficient and often results in longer treatment times. Due to the inherent longer treatment times, the planning process often includes treatment delivery time reduction techniques, such as limiting the number of MLC segments while maintaining satisfactory plan quality. However, choosing to use too few segments will result in a suboptimal dose distribution. Optimization strategies, such as allowing the user to generate a series of plans for a single cost function with varying number of segments, may be useful in balancing plan complexity with plan quality.

### Optimization

The goal of the optimization process is to reduce the cost function to near zero (minimum). However, this is typically not achievable, and an acceptable solution is found by reducing the cost function until negligible improvements can be made and the optimizer converges. The cost function is used to define the difference between the desired dose distribution and the obtained one. Each treatment planning platform may define their cost function differently; however, in each scenario the cost function parameters are defined by the user and may depend on dose-based criteria or biological criteria. A physical cost function is mainly used and defined using objectives and constraints based off dose volume histogram criteria. Once the problem is defined within the cost function, the optimizer then begins to search for a minimum in the cost function to generate the desired dose distribution. Inherently, static IMRT results in low-dose streaking in the beam path. This streaking can be minimized through avoidance and ring optimization structures.

Adaptive treatment planning often includes an isotoxic approach. Instead of adaptation based on avoiding under dosing the target or overdosing OARs, an isotoxic approach involves prescribing to a predefined dose-volume constraint to adjacent OARs while escalating the dose to treatment targets as much as technically possible to obtain a favorable therapeutic ratio. This is frequently adopted in the treatment of pancreatic cancers; however, it is being adopted to other anatomical areas including in stereotactic liver radiation therapy and tumor regression in the treatment of lung cancer [43–45].

## Dose Calculation

Integrated MR-Linac TPSs have the unique requirement and challenge of calculating radiation dose in magnetic environments. Magnetic fields influence the trajectory of electrons liberated by megavoltage-level photons and thus impact the spatial distribution of their energy deposition and furthermore radiation therapy dose. This influence, known as the Lorentz force, can manifest as alterations in beam characteristics (e.g., depth versus dose, penumbra) and clinically is of concern at interfaces of significant electron density change (e.g., air/skin or lung/tissue) resulting in the electron return effect (ERE). The magnitude of this force is related to the magnetic field strength. An example of the dose deposition changes for a 0.35T MR-Linac with and without the consideration of the magnetic field in the dose calculation for a left breast treatment is shown in Fig. 8.6, where prominent changes in the surface dose are observed. The effect of the magnetic field on dose deposition is dependent on field strength, patient anatomy, and beam geometry, necessitating a review of common approaches used in conventional treatment planning [46]. The application of magnetic field effects on the mechanics of radiation transport and dose deposition is best handled in Monte Carlo dose calculation algorithms where each particle history can be appropriately altered by the field's presence. Broadly used Monte Carlo

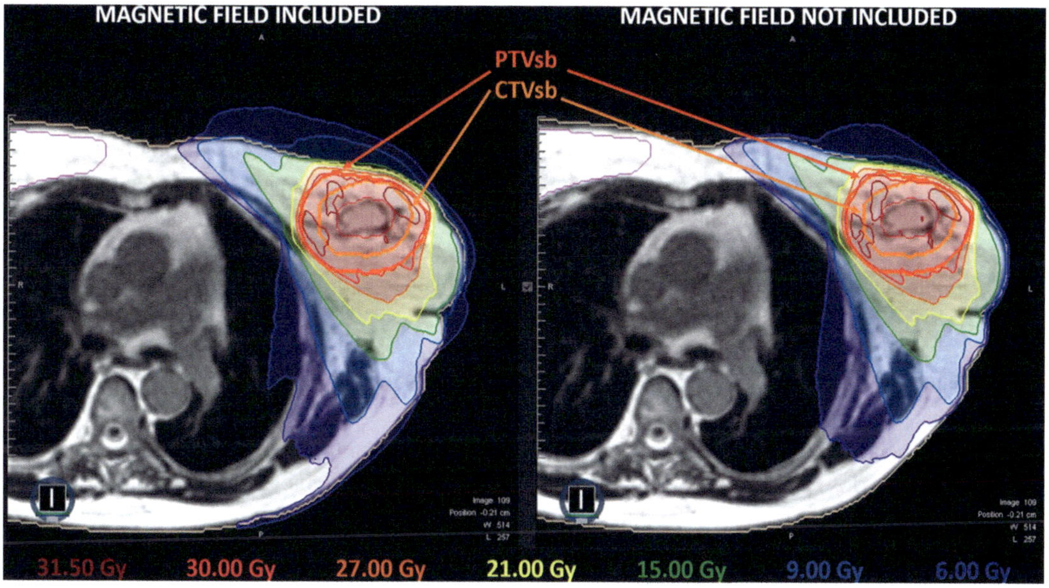


**Fig. 8.6** An example of the dose deposition changes for a 0.35T MR-Linac with and without the consideration of the magnetic field in the dose calculation for a left breast treatment. Notable changes in 6Gy isodose line illustrating the surface dose can be seen anteriorly and posteriorly

codes including GEANT4, PENELOPE, MCNP, and EGS have demonstrated the utility of Monte Carlo dose calculation techniques in the magnetic field environment. Therefore, MR-Linacs rely on such algorithms [14, 47].

In addition, MR-Linac planning systems require an enhanced calculation speed due to the time-limited nature of on-table adaptive radiation therapy treatment planning. To transition from the general-purpose Monte Carlo codes to algorithms that are suitable for this application, variance reduction techniques and efficiency enhancing methods must be implemented [14]. These well-established Monte Carlo engines have since been used to validate commercial packages and their chosen balance between accuracy and efficiency [48–51].

To improve efficiency, the Monte Carlo dose calculation algorithms typically use a machine-specific phase space that acts as the source or starting point to each history in the calculation. The output of the forward or inverse optimization process described in 8.5.3 and 8.5.4 is a series of MLC segments at user-defined gantry angles. These segments are then used as the first patient-specific step in the final dose calculation. The patient-specific collimated beam is then transported to and through the patient anatomy. The electron density map described in section "Electron Density" provides the data needed to query radiation interaction cross sections that dictate the radiation transport and dose deposition.

However, a balance must be obtained between efficiency and accuracy. Dose grid size and the number of particle histories become important parameters to statistically converge the dose calculation to a resolution level that is meaningful for treatment planning and evaluation.

## Electron Return Effect

As mentioned above, a strong motivator behind the use of Monte Carlo dose calculation algorithms in the presence of a magnetic field is the influence of the B-field on the trajectory of liberated electrons. The magnitude of this effect increases with increasing magnetic field strength, along with its clinical impact, and the need to take these effects into consideration during dose calculation [52]. The impact of the magnetic field includes a reduction in the depth of maximum dose, an asymmetric dose kernel (due to a deflected electron path), and changes in the beam penumbra. It also introduces the electron return effect at in-field tissue-air interfaces, which is due to electrons being forced into a circular path in air and re-entering the tissue interface leading to an increase in local dose [53–55]. The ERE results in an increase dose at the proximal end of an air cavity and a decreased dose at the distal end; this can directly impact patient treatments in the nasal and oral cavities and thorax and abdominal regions with tissue-lung interfaces or gas pockets. These effects are considered through the simulation of the dose perturbation from the electron return effect within the TPS dose calculation algorithms. However, there are additional mitigation strategies that can be considered during planning, including opposing beams or increasing the number of beams to balance or reduce hot and cold spots, introducing additional OAR contours such as a skin contour to evaluate and limit dose to these areas, as well incorporating bolus to reduce exit skin dose [56–58].

## Heterogeneity Corrections

Most modern treatment planning systems utilize deterministic model-based dose calculation algorithms that rely on the scaling of dose kernels to account for the impact of tissue heterogeneities on secondary electron dose clouds. Monte Carlo dose calculation algorithms are based on probabilistic models and are inherently more accurate in heterogeneous environments, as they directly account for the impact of heterogeneous mediums on electron transport [14]. Theoretically, their superior handling of inhomogeneity is clear; additionally, significant experimental verification efforts have been undertaken to prove their validity in clinical environments [49, 50, 59].

A low field MR-Linac currently operating clinically has the option to take a static magnetic field into account utilizing fast Monte Carlo dose calculation algorithm [36]. A clinical high field MR-Linac similarly employs a Monte Carlo dose calculation algorithm that models the effects of the magnetic field, cryostat, and modified scattering on the dose distribution that have been validated in treatment planning studies [60–62]. This algorithm utilizes graphical processing units to increase calculation speeds through parallelization of the process and has shown within 1% agreement in a homogenous phantom along the beam central axis and within 2% agreement in a heterogeneous phantom [50, 63].

## Adaptive Planning Considerations

Treatment planning relies on a detailed model of patient anatomy, including all clinical assessments needed to characterize the disease and all physical measurements needed to characterize the relevant properties affecting the distribution of radiation dose. An accurate patient model must account for spatiotemporal changes in any clinical findings or physical properties, and the process of designing a radiotherapy treatment plan must allow for revisions that adapt to such changes. In this context, adaptive planning is the process of adjusting the radiation treatment plan to account for changes in clinical findings and patient physical properties. Currently, the selection of a set of clinical criteria (or triggers) for determining the need for adaptation is still work in progress, with initial research providing promising results [64–66].

### Reference Plans

While it is recognized that a robust treatment plan must account for spatiotemporal changes in the patient model, current methods and measurements for fully characterizing this model are site-specific and will improve with standardization. Therefore, an initial treatment plan is typically devised based on a snapshot of the patient's anatomy from the day of simulation. Depending on treatment site, the measurements may include an estimate of the magnitude of changes due to physiological motion (e.g., cardiac, respiratory, or gastrointestinal motility) or processes (e.g., bladder, rectum, and stomach filling/emptying).

The inclusion of functional imaging during simulation aids in improving the precision of target delineation and OARs [21]. The reference plan is the treatment plan created based on data available at the time of simulation. Generally, the planning procedures for creating a reference plan are the same as, or similar, the institutional procedures established for that treatment site. This also applies to safety checks and QA procedures.

## Adaptation Timescales

Adaptive planning strategies have traditionally focused on spatiotemporal changes categorized over three timescales and assessed by longitudinal anatomic and functional images of the target and OARs. These changes are often observed (1) after the delivery of several fractions, (2) at each fraction, and (3) during the time of irradiation. These timescales correspond, respectively, to changes over several days, daily, and over a few minutes or seconds. The planning methods corresponding to each category are commonly known as offline (or inter-fraction), online (or on-table), and real-time (or dynamic) adaptive radiotherapy [7, 67, 68]. Figure 8.7 shows a schematic diagram presenting the relevant steps for each adaptation method.

### Offline Adaptive Planning

Offline (or inter-fraction) adaptive planning monitors change after the delivery of one to several fractions and accordingly revise the treatment

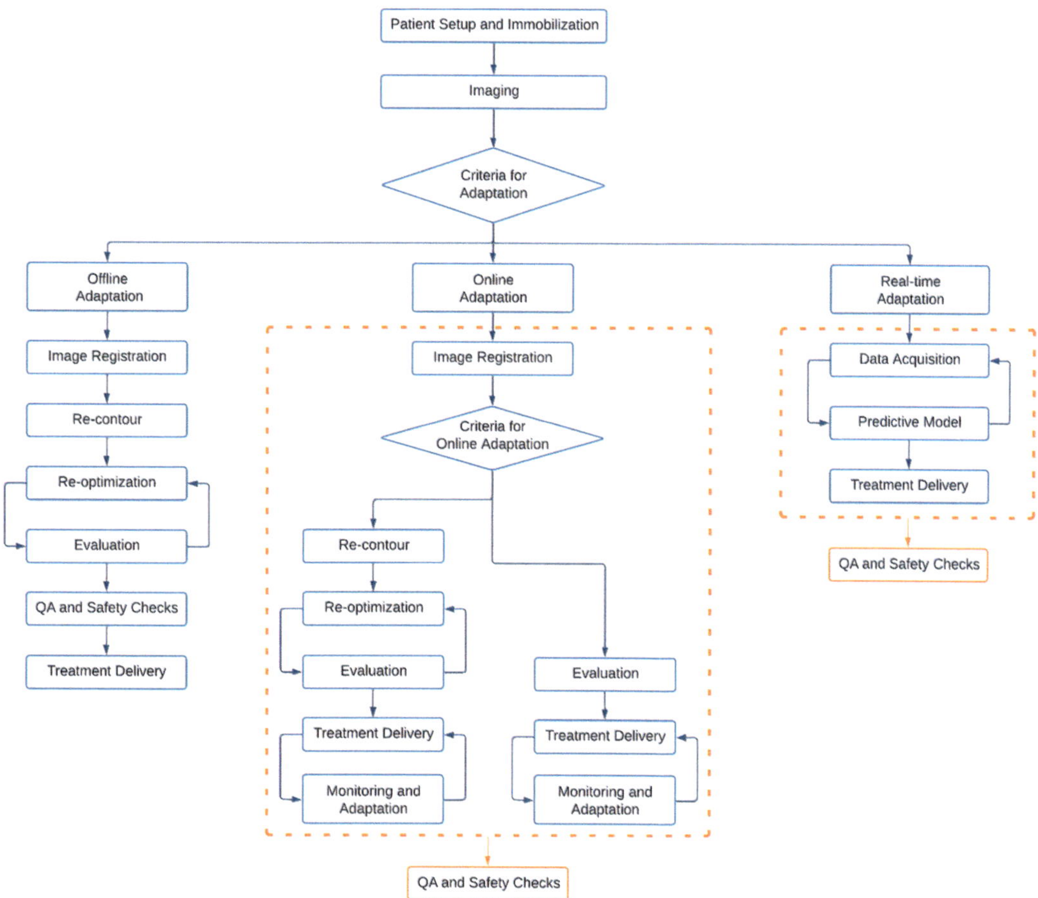


**Fig. 8.7** Schematic diagram of workflow steps for offline, online, and real-time adaptive planning for each treatment session

plan to optimize radiation dose. While safety margins are often employed when preparing the reference plan, additional modifications may be needed to account for the magnitude of variability in target and OAR position, shape, and volume [69–74]. The requirements and constraints for offline plan adaptation and QA are typically the same as for reference plans. Offline adaptation targets slow, systematic, or abrupt changes in the patient model, can use the imaging modality of choice for the given treatment site and is easy to implement.

### Online Adaptive Planning

The most widely implemented adaptation strategy in MR-Linacs is online (or on-table) adaptive planning. Online adaptive planning accounts for daily changes in the patient model and accordingly revises the treatment plan to optimize the delivered radiation dose. This adaptation method is employed in treatment sites where daily variability is expected to often be beyond the safety margins used in the reference plan. The online adapted plan uses an image acquired immediately after the patient is positioned on the treatment couch and is commonly generated with semi-automatic methods that rely on the reference plan as the starting point. In this approach, the images used for daily planning are acquired in the MR-Linac and do not necessarily match the image contrast from the reference plan. Adaptation is achieved in two ways, broadly based on user workflows that allow for a physical or virtual couch shift and no contour editing or full re-planning with daily re-contouring and optimization [23, 75, 76]. A common, semi-automatic technique for online adaptation is based on the generation of a library of reference plans that are optimized for anticipated changes in patient anatomy (e.g., bladder, rectum filling/emptying) [77–80]. Planning templates that attempt to provide robustness against changes in patient anatomy, known as class solution templates, have also shown initial feasibility [81–83]. The generation of safety margins typically follows the same rules as in reference plans [84, 85]. The expectation that anatomical changes may occur during and after online plan adaptation necessitates verification and motion monitoring images in a workflow that includes closed-loop feedback [86, 87]. This in turn prolongs patient on-table time and requires automation at each step of adaptation. Online re-planning remains a resource-intensive method as it routinely requires the presence of the entire treatment team (dosimetrists, physicists, physicians) at the machine during the process of adaptation.

### Real-Time Adaptive Planning

Real-time (or dynamic) adaptive planning monitors changes during the delivery of radiotherapy and revises the delivered plan accordingly to optimize radiation dose. This adaptation strategy monitors changes in patient anatomy that happen in short timescales (seconds to minutes) and for which a predictive model can be generated [88–90]. Common predictive models for real-time adaptation include those for baseline drifts and physiological motion due to cardiac, respiratory, and GI tract motility [91, 92]. Recent work has demonstrated that clinical implementation of real-time adaptive radiotherapy can improve delivery [93]. Real-time adaptation requires the system latencies, i.e., the time from detection to prediction and execution, to be smaller than the time for the anatomical change to occur [94]. Therefore, the temporal requirements for imaging and data acquisition are dependent on the treatment site and the magnitude of changes in patient anatomy. Adaptation is typically implemented through dynamic modifications of MLC patterns to account for predicted changes in patient model. Currently, the predictive models are mainly based on changes in target position and shape, instead of OAR or functional changes. While still primarily a research method, the feasibility of real-time adaptation in MR-Linacs has been successfully demonstrated in multiple sites [90, 91, 95, 96].

## Adaptation to Anatomical Changes

The adaptive planning methods described above are predominantly driven by target and OAR anatomical variability. The variability is due to changes in position, size, and/or shape and can happen at various timescales. Therefore, based on the timescales for these changes, all methods of plan adaptation (offline, online, real-time) can be clinically used. The most common triggers for adaptation to anatomical variability involve non-rigid deformations of the target/OARs at a magnitude beyond the safety margins used in the reference plan. The availability of MR images with various contrast mechanisms allows for improved delineations and can enable methods for adaptation triggers that combine contouring, optimization, and dosimetric evaluation [64, 65, 84, 85]. However, robust and fully automated criteria for triggering adaptation are still needed, with initial work providing promising results [64–66].

## Adaptation to Functional Changes

Multiparametric (mp) MRI is increasingly being used for diagnosis, staging, and treatment response assessment in several disease sites [97–105]. Imaging sequences within mpMRI protocols commonly include T1/T2-weighted, diffusion-weighted (DW), perfusion, dynamic contrast-enhanced (DCE), and spectroscopic MRI. While the definition of functional MRI is dependent on treatment site and will improve with standardization, the most often used sequences for functional MRI in radiation therapy include DWI-MRI, perfusion, and DCE-MRI. These methods, respectively, estimate the local rate of water diffusion, blood flow and oxygenation, and vessel permeability into the extravascular/extracellular space. Recent work has demonstrated the feasibility of functional imaging in MR-Linacs and suggested consensus protocols for future clinical use [106–112]. Functional imaging provides complementary information to anatomical imaging and may help improve the precision of target/OAR delineation and offline or online adaptive planning [21, 84, 85, 113–115]. Figure 8.8 presents an example of combined anatomical and functional imaging used in online adaptive radiotherapy to the prostate and dominant intraprostatic lesion [85, 116]. Additionally, targeting based on functional imaging is feasible in a multi-institutional setting, and treatments can provide improvements in outcomes [117]. It is worth noting that other non-MRI-derived functional biomarkers can be used for offline adaptation and could be combined to increase the precision of online or real-time adaptation.

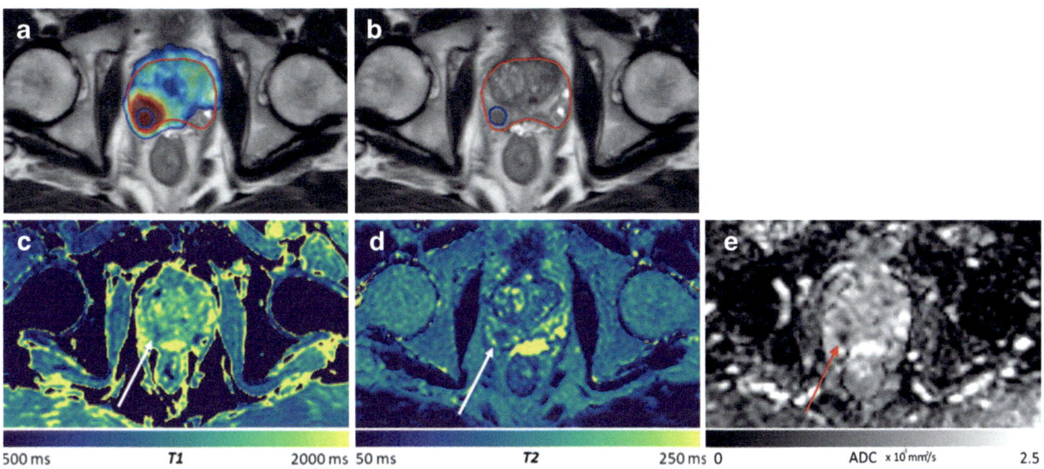


**Fig. 8.8** Use of functional biomarkers for treatment planning (**a**). Dose overlay on anatomical T2-weighted image (**b**). Contours for CTV (red, prostate gland) and GTV (blue, dominant intraprostatic lesion) overlaid on anatomical T2-weighted MRI (**c**). T1 relaxometry map (**d**). T2 relaxometry map (**e**). Apparent diffusion coefficient (ADC) map. Arrows point to the location of the dominant intraprostatic lesion

# Plan Evaluation

## Dose Calculation Accuracy

The point-spread kernel for secondary electrons involved in dose deposition is significantly affected by the presence of the magnetic field. The impact on dose distribution is a decreased depth for build-up, a shifted and asymmetric beam profile and the ERE [53, 55, 118]. The magnitude of these effects increases with magnetic field strength and requires accurate modeling for dose calculations [52]. Currently, clinical treatment planning systems for any MR-Linac must employ a dose calculation engine based on Monte Carlo (MC) computation methods [63, 119–121]. Analytical or model-based dose calculation engines with sufficient accuracy have yet to be implemented for routine use [122]. Guidelines and recommendations for the clinical evaluation of the accuracy in MC-based algorithms, including acceptance testing and commissioning, are described by AAPM TG 105 and TG 157 [13, 14]. Particular attention is devoted to the differences with conventional TPS algorithms, especially as it relates to statistical uncertainty, as well as the conversion from dose-to-water and dose-to-medium [123]. Additional validation may be necessary at tissue interfaces with high/low density and at specific treatment sites where the ERE may be of concern [16, 46, 49, 50, 58, 59, 61, 124–126]. Of note, prescription dose in MC-based TPS systems should be volume-based (e.g., to the PTV) as opposed to point-based (e.g., $D_{max}$) to avoid a systematic bias that can arise from statistical uncertainty in point doses [13, 14].

## Evaluation Metrics

The clinical evaluation metrics used in assessing the quality of conventional plans remain some of the most used metrics for also assessing the quality of MR-Linac plans. Given the statistical uncertainty associated with MC-based calculation engines, point-based evaluation metrics must be used with caution, particularly for serial organs in which max dose is a clinically important measure for tissue response (e.g., spinal cord). DVH (dose volume histogram) metrics and dose indices that model treatment response (e.g., TCP/NTCP) have been found to be less sensitive to statistical noise, and calculations with uncertainty less than 2% are sufficiently accurate for reporting and plan evaluation [14, 127]. The concept of the uncertainty volume histogram (UVH) is an equally valuable tool in the evaluation of plan quality as it allows physicist and physician to assess whether the dose uncertainty in targets/OARs is clinically acceptable [127]. While it is recognized that dose accumulation for time-varying anatomy will provide critical information for treatment response assessment, the implementation of image registration algorithms for routine clinical use remains challenging [128, 129].

## Plan Summation

Accumulated dose has been shown to be a better predictor of treatment response in stereotactic body radiotherapy [130]. Dose accumulation requires summation of the dose from the adapted plans delivered at each fraction, which in turn necessitates implementation of rigid or deformable registration [128]. The accuracy and utility of plan summations are critically dependent on the accuracy of the deformation vector fields estimated with image registration, on the assumptions of outcome modeling for time-varying dose delivery, and on the specific changes in patient anatomy (e.g., tumor response) [131–135]. Note that each step in this process can be achieved using multiple methods with varying degrees of precision [136]. Despite the variability in method implementation, dose summation in MR-Linac treatments has been successfully reported in multiple sites [87, 91, 137–142]. These findings highlight the potential for substantial improvements in response/toxicity assessment and adaptation strategies. Consensus protocols for dose accumulation in MR-guided adaptive radiotherapy will provide valuable help to the community [143].

## Quality Assurance

The implementation of adaptive radiotherapy, particularly for online and real-time adaptive planning, requires efficient automatic and semi-automatic tools at each step of the process. Given the time constraints with the patient on the table, QA for online and real-time planning presents challenges not encountered in conventional treatment planning methods [7]. As shown in Fig. 8.7, while QA for offline adaptation is identical to conventional planning, QA and safety checks for online and real-time adaptation are integrated and performed in parallel with the planning process [144, 145]. Specific to real-time adaptation, QA and safety checks need to be incorporated in the predictive model for robust testing of tracking accuracy, target/OAR detection, and delivery. Furthermore, IMRT plan verification is traditionally based on measurements prior to treatment delivery which is practically infeasible for online and real-time planning. For adaptive planning, calculation-based tools that combine integrity and independent dose checks become the primary verification methods, followed by ad hoc measurement-based methods [122, 146–150]. Additional QA and safety checks for the semi-automated steps during plan adaptation may be feasible [151, 152]. Failure mode and effects analysis (FMEA) will enable the clinical team to identify all critical and impactful steps in the treatment process and has been shown to provide improvements in safety and efficiency [144, 145, 153, 154].

## Summary

A general overview of the requirements of an MR-Linac treatment planning system, as well as the commissioning and QA practices, was reviewed. Followed by current contouring initiatives and practices regarding offline and online adaptive planning. A practical guide to methods for estimating electron density was included for MR-based planning with an overview of treatment plan optimization, dose calculation, and algorithm requirements in MR-Linac treatment planning. This was followed by a detailed discussion of adaptive planning considerations to account for functional and anatomical changes and concluded with practical considerations for plan evaluation techniques. The techniques described in this chapter are applicable to low- and high-field MR-Linacs, but there are certain benefits and limitations associated with each. For example, high-field MR-Linacs provide a better signal-to-noise ratio compared to low-field MR-Linacs which may be beneficial in some anatomical locations for contouring and adaptive treatment planning. However, low-field MR-Linacs have compensated through image reconstruction software to enhance image quality. Similar optimization and delivery techniques are implemented on both available units, allowing for comparable plan quality. However, there is an increased electron return effect in high-field MRI which requires more considerations when developing dose calculation algorithms and planning techniques. Both platforms offer adaptive planning capabilities with different workflows, allowing for each individual clinic to choose which system is more suitable to their needs. The benefits of MRgRT are numerous including enhanced soft tissue visualization, incorporating functional imaging, adaptive planning capabilities, and real-time motion monitoring utilizing 2D cine MRI.

## References

1. Corradini S, Alongi F, Andratschke N, Azria D, Bohoudi O, Boldrini L, et al. ESTRO-ACROP recommendations on the clinical implementation of hybrid MR-linac systems in radiation oncology. Radiother Oncol. 2021;159:146–54.
2. Kishan AU, Ma TM, Lamb JM, Casado M, Wilhalme H, Low DA, et al. Magnetic resonance imaging-guided vs computed tomography-guided stereotactic body radiotherapy for prostate cancer: the MIRAGE randomized clinical trial. JAMA Oncol. 2023;9(3):365.
3. Rudra S, Jiang N, Rosenberg SA, Olsen JR, Roach MC, Wan L, et al. Using adaptive magnetic resonance image-guided radiation therapy for treatment of inoperable pancreatic cancer. Cancer Med. 2019;8(5):2123–32.

4. Tan H, Stewart J, Ruschin M, Wang MH, Myrehaug S, Tseng CL, et al. Inter-fraction dynamics during post-operative 5 fraction cavity hypofractionated stereotactic radiotherapy with a MR LINAC: a prospective serial imaging study. J Neuro-Oncol. 2022;156(3):569–77.
5. Fraass B, Doppke K, Hunt M, Kutcher G, Starkschall G, Stern R, et al. American Association of Physicists in Medicine Radiation Therapy Committee Task Group 53: quality assurance for clinical radiotherapy treatment planning. Med Phys. 1998;25(10):1773–829.
6. Siochi RA, Balter P, Bloch CD, Santanam L, Blodgett K, Curran BH, et al. Report of Task Group 201 of the American Association of Physicists in Medicine: quality management of external beam therapy data transfer. Med Phys. 2021;48(6):e86–e114.
7. Glide-Hurst CK, Lee P, Yock AD, Olsen JR, Cao M, Siddiqui F, et al. Adaptive radiation therapy (ART) strategies and technical considerations: a state of the ART review from NRG oncology. Int J Radiat Oncol Biol Phys. 2021;109(4):1054–75.
8. IAEA TRS-430, Commissioning and quality assurance of computerized planning systems for radiation treatment of cancer. Vienna: International Atomic Energy Agency; 2004.
9. Smilowitz JB, Das IJ, Feygelman V, Fraass BA, Kry SF, Marshall IR, et al. AAPM medical physics practice guideline 5.a.: commissioning and QA of treatment planning dose calculations - megavoltage photon and electron beams. J Appl Clin Med Phys. 2015;16(5):14–34.
10. Geurts MW, Jacqmin DJ, Jones LE, Kry SF, Mihailidis DN, Ohrt JD, et al. AAPM medical physics practice guideline 5.b: commissioning and QA of treatment planning dose calculations—megavoltage photon and electron beams. J Appl Clin Med Phys. 2022;23(9):e13641.
11. Das IJ, Cheng C-W, Watts RJ, Ahnesjö A, Gibbons J, Li XA, et al. Accelerator beam data commissioning equipment and procedures: report of the TG-106 of the Therapy Physics Committee of the AAPM. Med Phys. 2008;35(9):4186–215.
12. Ezzell GA, Burmeister JW, Dogan N, LoSasso TJ, Mechalakos JG, Mihailidis D, et al. IMRT commissioning: multiple institution planning and dosimetry comparisons, a report from AAPM Task Group 119. Med Phys. 2009;36(11):5359–73.
13. Ma CMC, Chetty IJ, Deng J, Faddegon B, Jiang SB, Li J, et al. Beam modeling and beam model commissioning for Monte Carlo dose calculation-based radiation therapy treatment planning: report of AAPM Task Group 157. Med Phys. 2020;47(1):e1–e18.
14. Chetty IJ, Curran B, Cygler JE, DeMarco JJ, Ezzell G, Faddegon BA, et al. Report of the AAPM Task Group No. 105: issues associated with clinical implementation of Monte Carlo-based photon and electron external beam treatment planning. Med Phys. 2007;34(12):4818–53.
15. Chen X, Paulson ES, Ahunbay E, Sanli A, Klawikowski S, Li XA. Measurement validation of treatment planning for a MR-Linac. J Appl Clin Med Phys. 2019;20(7):28–38.
16. Shortall J, Vasquez Osorio E, Aitkenhead A, Berresford J, Agnew J, Budgell G, et al. Experimental verification the electron return effect around spherical air cavities for the MR-Linac using Monte Carlo calculation. Med Phys. 2020;47(6):2506–15.
17. Elter A, Rippke C, Johnen W, Mann P, Hellwich E, Schwahofer A, et al. End-to-end test for fractionated online adaptive MR-guided radiotherapy using a deformable anthropomorphic pelvis phantom. Phys Med Biol. 2021;66(24).
18. Hoffmans D, Niebuhr N, Bohoudi O, Pfaffenberger A, Palacios M. An end-to-end test for MR-guided online adaptive radiotherapy. Phys Med Biol. 2020;65(12):125012.
19. Kisling K, Keiper TD, Branco D, Kim GG, Moore KL, Ray X. Clinical commissioning of an adaptive radiotherapy platform: results and recommendations. J Appl Clin Med Phys. 2022;23(12):e13801.
20. Noel CE, Parikh PJ, Spencer CR, Green OL, Hu Y, Mutic S, et al. Comparison of onboard low-field magnetic resonance imaging versus onboard computed tomography for anatomy visualization in radiotherapy. Acta Oncol. 2015;54(9):1474–82.
21. Bauman G, Haider M, Van der Heide UA, Ménard C. Boosting imaging defined dominant prostatic tumors: a systematic review. Radiother Oncol. 2013;107(3):274–81.
22. Dunlop A, Mitchell A, Tree A, Barnes H, Bower L, Chick J, et al. Daily adaptive radiotherapy for patients with prostate cancer using a high field MR-linac: initial clinical experiences and assessment of delivered doses compared to a C-arm linac. Clin Transl Radiat Oncol. 2020;23:35–42.
23. Winkel D, Bol GH, Kroon PS, van Asselen B, Hackett SS, Werensteijn-Honingh AM, et al. Adaptive radiotherapy: the Elekta Unity MR-linac concept. Clin Transl Radiat Oncol. 2019;18:54–9.
24. Winkel D, Bol GH, Werensteijn-Honingh AM, Intven MPW, Eppinga WSC, Hes J, et al. Target coverage and dose criteria based evaluation of the first clinical 1.5T MR-linac SBRT treatments of lymph node oligometastases compared with conventional CBCT-linac treatment. Radiother Oncol. 2020;146:118–25.
25. Hall WA, Heerkens HD, Paulson ES, Meijer GJ, Kotte AN, Knechtges P, et al. Pancreatic gross tumor volume contouring on computed tomography (CT) compared with magnetic resonance imaging (MRI): results of an international contouring conference. Pract Radiat Oncol. 2018;8(2):107–15.
26. Heerkens HD, Hall WA, Li XA, Knechtges P, Dalah E, Paulson ES, et al. Recommendations for MRI-based contouring of gross tumor volume and organs at risk for radiation therapy of pancreatic cancer. Pract Radiat Oncol. 2017;7(2):126–36.

27. Lukovic J, Henke L, Gani C, Kim TK, Stanescu T, Hosni A, et al. MRI-based upper abdominal organs-at-risk atlas for radiation oncology. Int J Radiat Oncol Biol Phys. 2020;106(4):743–53.
28. Vasmel JE, Groot Koerkamp ML, Kirby AM, Russell NS, Shaitelman SF, Vesprini D, et al. Consensus on contouring primary breast tumors on MRI in the setting of neoadjuvant partial breast irradiation in trials. Pract Radiat Oncol. 2020;10(6):e466–e74.
29. Chen X, Ma X, Yan X, Luo F, Yang S, Wang Z, et al. Personalized auto-segmentation for magnetic resonance imaging-guided adaptive radiotherapy of prostate cancer. Med Phys. 2022;49(8):4971–9.
30. Ding J, Zhang Y, Amjad A, Xu J, Thill D, Li XA. Automatic contour refinement for deep learning auto-segmentation of complex organs in MRI-guided adaptive radiation therapy. Adv Radiat Oncol. 2022;7(5):100968.
31. Kawula M, Hadi I, Nierer L, Vagni M, Cusumano D, Boldrini L, et al. Patient-specific transfer learning for auto-segmentation in adaptive 0.35 T MRgRT of prostate cancer: a bi-centric evaluation. Med Phys. 2023;50:1573.
32. Liang Y, Schott D, Zhang Y, Wang Z, Nasief H, Paulson E, et al. Auto-segmentation of pancreatic tumor in multi-parametric MRI using deep convolutional neural networks. Radiother Oncol. 2020;145:193–200.
33. Sharp G, Fritscher KD, Pekar V, Peroni M, Shusharina N, Veeraraghavan H, et al. Vision 20/20: perspectives on automated image segmentation for radiotherapy. Med Phys. 2014;41(5):050902.
34. Eccles CL, Adair Smith G, Bower L, Hafeez S, Herbert T, Hunt A, et al. Magnetic resonance imaging sequence evaluation of an MR Linac system; early clinical experience. Tech Innov Patient Support Radiat Oncol. 2019;12:56–63.
35. Whiteside L, McDaid L, Hales RB, Rodgers J, Dubec M, Huddart RA, et al. To see or not to see: evaluation of magnetic resonance imaging sequences for use in MR Linac-based radiotherapy treatment. J Med Imaging Radiat Sci. 2022;53(3):362–73.
36. Klüter S. Technical design and concept of a 0.35 T MR-Linac. Clin Transl Radiat Oncol. 2019;18:98–101.
37. Heerkens HD, Reerink O, Intven MPW, Hiensch RR, van den Berg CAT, Crijns SPM, et al. Pancreatic tumor motion reduction by use of a custom abdominal corset. Phys Imaging Radiat Oncol. 2017;2:7–10.
38. Owrangi AM, Greer PB, Glide-Hurst CK. MRI-only treatment planning: benefits and challenges. Phys Med Biol. 2018;63(5):05tr1.
39. Yang B, Yuan J, Poon DMC, Geng H, Lam WW, Cheung KY, et al. Assessment of planning target volume margins in 1.5 T magnetic resonance-guided stereotactic body radiation therapy for localized prostate cancer. Precision. Radiat Oncol. 2022;6(2):127–35.
40. Corradini S, Alongi F, Andratschke N, Belka C, Boldrini L, Cellini F, et al. MR-guidance in clinical reality: current treatment challenges and future perspectives. Radiat Oncol. 2019;14(1):92.
41. Devic S. MRI simulation for radiotherapy treatment planning. Med Phys. 2012;39(11):6701–11.
42. Das IJ, Sanfilippo NJ, Fogliata A, Cozzi L, Physics Io. Intensity modulated radiation therapy: a clinical overview. IOP Publishing; 2020.
43. Hawranko R, Sohn JJ, Neiderer K, Bump E, Harris T, Fields EC, et al. Investigation of isotoxic dose escalation and plan quality with TDABC analysis on a 0.35 T MR-Linac (MRL) system in ablative 5-fraction stereotactic magnetic resonance-guided radiation therapy (MRgRT) for primary pancreatic cancer. J Clin Med. 2022;11(9):2584.
44. Hansen AT, Poulsen PR, Høyer M, Worm ES. Isotoxic dose prescription level strategies for stereotactic liver radiotherapy: the price of dose uniformity. Acta Oncol. 2020;59(5):558–64.
45. Warren M, Webster G, Ryder D, Rowbottom C, Faivre-Finn C. An isotoxic planning comparison study for stage II-III non-small cell lung cancer: is intensity-modulated radiotherapy the answer? Clin Oncol (R Coll Radiol). 2014;26(8):461–7.
46. Kim A, Lim-Reinders S, McCann C, Ahmad SB, Sahgal A, Lee J, et al. Magnetic field dose effects on different radiation beam geometries for hypofractionated partial breast irradiation. J Appl Clin Med Phys. 2017;18(6):62–70.
47. Kirkby C, Stanescu T, Fallone BG. Magnetic field effects on the energy deposition spectra of MV photon radiation. Phys Med Biol. 2009;54(2):243–57.
48. Wang Y, Mazur TR, Green O, Hu Y, Li H, Rodriguez V, et al. A GPU-accelerated Monte Carlo dose calculation platform and its application toward validating an MRI-guided radiation therapy beam model. Med Phys. 2016;43(7):4040.
49. Khan AU, Simiele EA, Lotey R, DeWerd LA, Yadav P. Development and evaluation of a GEANT4-based Monte Carlo Model of a 0.35 T MR-guided radiation therapy (MRgRT) linear accelerator. Med Phys. 2021;48(4):1967–82.
50. Ahmad SB, Sarfehnia A, Paudel MR, Kim A, Hissoiny S, Sahgal A, et al. Evaluation of a commercial MRI Linac based Monte Carlo dose calculation algorithm with GEANT4. Med Phys. 2016;43(2):894–907.
51. Li Y, Ding S, Wang B, Liu H, Huang X, Song T. Extension and validation of a GPU-Monte Carlo dose engine gDPM for 1.5 T MR-LINAC online independent dose verification. Med Phys. 2021;48(10):6174–83.
52. Raaijmakers AJ, Raaymakers BW, Lagendijk JJ. Magnetic-field-induced dose effects in MR-guided radiotherapy systems: dependence on the magnetic field strength. Phys Med Biol. 2008;53(4):909.
53. Raaijmakers AJ, Raaymakers BW, van der Meer S, Lagendijk JJ. Integrating a MRI scanner with a 6 MV radiotherapy accelerator: impact of the surface orientation on the entrance and exit dose due

to the transverse magnetic field. Phys Med Biol. 2007;52(4):929–39.
54. Kurz C, Buizza G, Landry G, Kamp F, Rabe M, Paganelli C, et al. Medical physics challenges in clinical MR-guided radiotherapy. Radiat Oncol. 2020;15(1):93.
55. Raaijmakers AJ, Raaymakers BW, Lagendijk JJ. Integrating a MRI scanner with a 6 MV radiotherapy accelerator: dose increase at tissue-air interfaces in a lateral magnetic field due to returning electrons. Phys Med Biol. 2005;50(7):1363–76.
56. Huang C-Y, Yang B, Lam WW, Geng H, Cheung KY, Yu SK. Magnetic field induced dose effects in radiation therapy using MR-linacs. Med Phys. 2023;50:3623.
57. Oborn BM, Metcalfe PE, Butson MJ, Rosenfeld AB. Monte Carlo characterization of skin doses in 6 MV transverse field MRI-linac systems: effect of field size, surface orientation, magnetic field strength, and exit bolus. Med Phys. 2010;37(10):5208–17.
58. Godoy Scripes P, Subashi E, Burleson S, Liang J, Romesser P, Crane C, et al. Impact of varying air cavity on planning dosimetry for rectum patients treated on a 1.5 T hybrid MR-linac system. J Appl Clin Med Phys. 2020;21(7):144–52.
59. Paudel MR, Kim A, Sarfehnia A, Ahmad SB, Beachey DJ, Sahgal A, et al. Experimental evaluation of a GPU-based Monte Carlo dose calculation algorithm in the Monaco treatment planning system. J Appl Clin Med Phys. 2016;17(6):230–41.
60. Uilkema S, van der Heide U, Sonke JJ, Moreau M, van Triest B, Nijkamp J. A 1.5 T transverse magnetic field in radiotherapy of rectal cancer: impact on the dose distribution. Med Phys. 2015;42(12):7182–9.
61. Menten MJ, Fast MF, Nill S, Kamerling CP, McDonald F, Oelfke U. Lung stereotactic body radiotherapy with an MR-linac–quantifying the impact of the magnetic field and real-time tumor tracking. Radiother Oncol. 2016;119(3):461–6.
62. van Heijst TC, den Hartogh MD, Lagendijk JJ, van den Bongard HJ, van Asselen B. MR-guided breast radiotherapy: feasibility and magnetic-field impact on skin dose. Phys Med Biol. 2013;58(17):5917–30.
63. Bol G, Hissoiny S, Lagendijk J, Raaymakers B. Fast online Monte Carlo-based IMRT planning for the MRI linear accelerator. Phys Med Biol. 2012;57(5):1375.
64. McDonald BA, Vedam S, Yang J, Wang J, Castillo P, Lee B, et al. Initial feasibility and clinical implementation of daily MR-guided adaptive head and neck cancer radiation therapy on a 1.5 T MR-linac system: prospective R-ideal 2a/2b systematic clinical evaluation of technical innovation. Int J Radiat Oncol Biol Phys. 2021;109(5):1606–18.
65. McCulloch MM, Lee C, Rosen BS, Kamp JD, Lockhart CM, Lee JY, et al. Predictive models to determine clinically relevant deviations in delivered dose for head and neck cancer. Pract Radiat Oncol. 2019;9(4):422–31.
66. Tyran M, Jiang N, Cao M, Raldow A, Lamb JM, Low D, et al. Retrospective evaluation of decision-making for pancreatic stereotactic MR-guided adaptive radiotherapy. Radiother Oncol. 2018;129(2):319–25.
67. Green OL, Henke LE, Hugo GD. Practical clinical workflows for online and offline adaptive radiation therapy. Semin Radiat Oncol. 2019;29(3):219–27.
68. Hunt A, Hansen VN, Oelfke U, Nill S, Hafeez S. Adaptive radiotherapy enabled by MRI guidance. Clin Oncol. 2018;30(11):711–9.
69. Langen KM, Jones DT. Organ motion and its management. Int J Radiat Oncol Biol Phys. 2001;50(1):265–78.
70. Yan D, Lockman D. Organ/patient geometric variation in external beam radiotherapy and its effects. Med Phys. 2001;28(4):593–602.
71. Van Herk M, Witte M, Van Der Geer J, Schneider C, Lebesque JV. Biologic and physical fractionation effects of random geometric errors. Int J Radiat Oncol Biol Phys. 2003;57(5):1460–71.
72. Bortfeld T, Jiang SB, Rietzel E. Effects of motion on the total dose distribution. Semin Radiat Oncol. 2004;14(1):41–51.
73. Purdy JA. Current ICRU definitions of volumes: limitations and future directions. Semin Radiat Oncol. 2004;14(1):27–40.
74. Schaly B, Kempe JA, Bauman GS, Battista JJ, Van Dyk J. Tracking the dose distribution in radiation therapy by accounting for variable anatomy. Phys Med Biol. 2004;49(5):791–805.
75. Mutic S, Dempsey JF. The ViewRay system: magnetic resonance–guided and controlled radiotherapy. Semin Radiat Oncol. 2014;24(3):196–9.
76. Gupta A, Dunlop A, Mitchell A, McQuaid D, Nill S, Barnes H, et al. Online adaptive radiotherapy for head and neck cancers on the MR linear accelerator: introducing a novel modified adapt-to-shape approach. Clin Transl Radiat Oncol. 2022;32:48–51.
77. Heijkoop ST, Langerak TR, Quint S, Bondar L, Mens JWM, Heijmen BJM, et al. Clinical implementation of an online adaptive plan-of-the-day protocol for nonrigid motion management in locally advanced cervical cancer IMRT. Int J Radiat Oncol Biol Phys. 2014;90(3):673–9.
78. Meijer GJ, van der Toorn P-P, Bal M, Schuring D, Weterings J, de Wildt M. High precision bladder cancer irradiation by integrating a library planning procedure of 6 prospectively generated SIB IMRT plans with image guidance using lipiodol markers. Radiother Oncol. 2012;105(2):174–9.
79. Vestergaard A, Muren LP, Søndergaard J, Elstrøm UV, Høyer M, Petersen JB. Adaptive plan selection vs. re-optimisation in radiotherapy for bladder cancer: a dose accumulation comparison. Radiother Oncol. 2013;109(3):457–62.
80. Beekman C, van Triest B, van Beek S, Sonke JJ, Remeijer P. Margin and PTV volume reduction using a population based library of plans strategy for rectal cancer radiotherapy. Med Phys. 2018;45(10):4345–54.

81. Korsholm ME, Waring LW, Edmund JM. A criterion for the reliable use of MRI-only radiotherapy. Radiat Oncol. 2014;9(1):16.
82. Ruggieri R, Rigo M, Naccarato S, Gurrera D, Figlia V, Mazzola R, et al. Adaptive SBRT by 1.5 T MR-linac for prostate cancer: on the accuracy of dose delivery in view of the prolonged session time. Phys Med. 2020;80:34–41.
83. Naccarato S, Rigo M, Pellegrini R, Voet P, Akhiat H, Gurrera D, et al. Automated planning for prostate stereotactic body radiation therapy on the 1.5 T MR-Linac. Adv Radiat Oncol. 2022;7(3):100865.
84. Tyagi N, Liang J, Burleson S, Subashi E, Godoy Scripes P, Tringale KR, et al. Feasibility of ablative stereotactic body radiation therapy of pancreas cancer patients on a 1.5 Tesla magnetic resonance-linac system using abdominal compression. Phys Imaging Radiat Oncol. 2021;19:53–9.
85. Brennan VS, Burleson S, Kostrzewa C, Godoy Scripes P, Subashi E, Zhang Z, et al. SBRT focal dose intensification using an MR-Linac adaptive planning for intermediate-risk prostate cancer: an analysis of the dosimetric impact of intra-fractional organ changes. Radiother Oncol. 2023;179:109441.
86. Paulson ES, Ahunbay E, Chen X, Mickevicius NJ, Chen G-P, Schultz C, et al. 4D-MRI driven MR-guided online adaptive radiotherapy for abdominal stereotactic body radiation therapy on a high field MR-Linac: implementation and initial clinical experience. Clin Transl Radiat Oncol. 2020;23:72–9.
87. Alam S, Veeraraghavan H, Tringale K, Amoateng E, Subashi E, Wu AJ, et al. Inter- and intrafraction motion assessment and accumulated dose quantification of upper gastrointestinal organs during magnetic resonance-guided ablative radiation therapy of pancreas patients. Phys Imaging Radiat Oncol. 2022;21:54–61.
88. Fast MF, Nill S, Bedford JL, Oelfke U. Dynamic tumor tracking using the Elekta Agility MLC. Med Phys. 2014;41(11):111719.
89. Glitzner M, Woodhead P, Borman P, Lagendijk J, Raaymakers B. MLC-tracking performance on the Elekta unity MRI-linac. Phys Med Biol. 2019;64(15):15NT02.
90. Feldman AM, Modh A, Glide-Hurst C, Chetty IJ, Movsas B. Real-time magnetic resonance-guided liver stereotactic body radiation therapy: an institutional report using a magnetic resonance-linac system. Cureus. 2019;11(9):e5774.
91. Akdag O, Borman PTS, Woodhead P, Uijtewaal P, Mandija S, Van Asselen B, et al. First experimental exploration of real-time cardiorespiratory motion management for future stereotactic arrhythmia radioablation treatments on the MR-linac. Phys Med Biol. 2022;67(6).
92. Zhang Y, Cao Y, Kashani R, Lawrence TS, Balter JM. Real-time prediction of stomach motions based upon gastric contraction and breathing models. Phys Med Biol. 2022;68(1):015001.
93. Keall P, Nguyen DT, O'Brien R, Hewson E, Ball H, Poulsen P, et al. Real-time image guided ablative prostate cancer radiation therapy: results from the TROG 15.01 SPARK trial. Int J Radiat Oncol Biol Phys. 2020;107(3):530–8.
94. Keall PJ, Sawant A, Berbeco RI, Booth JT, Cho B, Cerviño LI, et al. AAPM Task Group 264: the safe clinical implementation of MLC tracking in radiotherapy. Med Phys. 2021;48(5):e44–64.
95. Uijtewaal P, Borman PT, Woodhead PL, Kontaxis C, Hackett SL, Verhoeff J, et al. First experimental demonstration of VMAT combined with MLC tracking for single and multi fraction lung SBRT on an MR-linac. Radiother Oncol. 2022;174:149–57.
96. NCT03621644. Stereotactic MRI-guided On-table Adaptive Radiation Therapy (SMART) for locally advanced pancreatic cancer. Stereotactic MRI-guided On-table Adaptive Radiation Therapy (SMART) for Locally Advanced Pancreatic Cancer.
97. Marino MA, Helbich T, Baltzer P, Pinker-Domenig K. Multiparametric MRI of the breast: a review. J Magn Reson Imaging. 2018;47(2):301–15.
98. Hoetker AM, Garcia-Aguilar J, Gollub MJ. Multiparametric MRI of rectal cancer in the assessment of response to therapy: a systematic review. Dis Colon Rectum. 2014;57(6):790–9.
99. Gürses B, Böge M, Altınmakas E, Balık E. Multiparametric MRI in rectal cancer. Diagn Interv Radiol. 2019;25(3):175.
100. Mahajan A, Engineer R, Chopra S, Mahanshetty U, Juvekar S, Shrivastava S, et al. Role of 3T multiparametric-MRI with BOLD hypoxia imaging for diagnosis and post therapy response evaluation of postoperative recurrent cervical cancers. Eur J Radiol Open. 2016;3:22–30.
101. Clemente EJI, Navarro OM, Navallas M, Ladera E, Torner F, Sunol M, et al. Multiparametric MRI evaluation of bone sarcomas in children. Insights Imaging. 2022;13(1):1–18.
102. Necchi A, Bandini M, Calareso G, Raggi D, Pederzoli F, Farè E, et al. Multiparametric magnetic resonance imaging as a noninvasive assessment of tumor response to neoadjuvant pembrolizumab in muscle-invasive bladder cancer: preliminary findings from the PURE-01 study. Eur Urol. 2020;77(5):636–43.
103. Rossi SH, Prezzi D, Kelly-Morland C, Goh V. Imaging for the diagnosis and response assessment of renal tumours. World J Urol. 2018;36:1927–42.
104. Haider M, Yao X, Loblaw A, Finelli A. Multiparametric magnetic resonance imaging in the diagnosis of prostate cancer: a systematic review. Clin Oncol. 2016;28(9):550–67.
105. Padhani AR, Miles KA. Multiparametric imaging of tumor response to therapy. Radiology. 2010;256(2):348–64.
106. Kooreman ES, van Houdt PJ, Nowee ME, van Pelt VWJ, Tijssen RHN, Paulson ES, et al. Feasibility and accuracy of quantitative imaging on a 1.5 T MR-linear accelerator. Radiother Oncol. 2019;133:156–62.

107. Kooreman ES, van Houdt PJ, Keesman R, Pos FJ, van Pelt VWJ, Nowee ME, et al. ADC measurements on the Unity MR-linac - a recommendation on behalf of the Elekta Unity MR-linac consortium. Radiother Oncol. 2020;153:106–13.
108. van Houdt PJ, Saeed H, Thorwarth D, Fuller CD, Hall WA, McDonald BA, et al. Integration of quantitative imaging biomarkers in clinical trials for MR-guided radiotherapy: conceptual guidance for multicentre studies from the MR-Linac Consortium Imaging Biomarker Working Group. Eur J Cancer. 1990;2021(153):64–71.
109. Kooreman ES, van Houdt PJ, Keesman R, van Pelt VWJ, Nowee ME, Pos F, et al. Daily intravoxel incoherent motion (IVIM) in prostate cancer patients during MR-guided radiotherapy-a multicenter study. Front Oncol. 2021;11:705964.
110. Kooreman ES, Tanaka M, Ter Beek LC, Peters FP, Marijnen CAM, van der Heide UA, et al. T(1ρ) for radiotherapy treatment response monitoring in rectal cancer patients: a pilot study. J Clin Med. 2022;11(7):1998.
111. Kooreman ES, van Pelt V, Nowee ME, Pos F, van der Heide UA, van Houdt PJ. Longitudinal correlations between intravoxel incoherent motion (IVIM) and dynamic contrast-enhanced (DCE) MRI during radiotherapy in prostate cancer patients. Front Oncol. 2022;12:897130.
112. Mickevicius NJ, Kim JP, Zhao J, Morris ZS, Hurst NJ Jr, Glide-Hurst CK. Toward magnetic resonance fingerprinting for low-field MR-guided radiation therapy. Med Phys. 2021;48(11):6930–40.
113. Steenbergen P, Haustermans K, Lerut E, Oyen R, De Wever L, Van den Bergh L, et al. Prostate tumor delineation using multiparametric magnetic resonance imaging: inter-observer variability and pathology validation. Radiother Oncol. 2015;115(2):186–90.
114. Dinh CV, Steenbergen P, Ghobadi G, van der Poel H, Heijmink SW, de Jong J, et al. Multicenter validation of prostate tumor localization using multiparametric MRI and prior knowledge. Med Phys. 2017;44(3):949–61.
115. van Schie MA, Steenbergen P, Dinh CV, Ghobadi G, van Houdt PJ, Pos FJ, et al. Repeatability of dose painting by numbers treatment planning in prostate cancer radiotherapy based on multiparametric magnetic resonance imaging. Phys Med Biol. 2017;62(14):5575–88.
116. Subashi E, LoCastro E, Apte A, Zelefsky MJ, Tyagi N. Quantitative relaxometry for target localization and response assessment in ultra-hypofractionated MR-guided radiotherapy to the prostate and DIL. Int J Radiat Oncol Biol Phys. 2022;114(3, Supplement):S33.
117. Kerkmeijer LGW, Groen VH, Pos FJ, Haustermans K, Monninkhof EM, Smeenk RJ, et al. Focal boost to the intraprostatic tumor in external beam radiotherapy for patients with localized prostate cancer: results from the FLAME randomized phase III trial. J Clin Oncol. 2021;39(7):787–96.
118. Raaymakers BW, Raaijmakers AJ, Kotte AN, Jette D, Lagendijk JJ. Integrating a MRI scanner with a 6 MV radiotherapy accelerator: dose deposition in a transverse magnetic field. Phys Med Biol. 2004;49(17):4109–18.
119. Hissoiny S, Ozell B, Bouchard H, Després P. GPUMCD: a new GPU-oriented Monte Carlo dose calculation platform. Med Phys. 2011;38(2):754–64.
120. Hissoiny S, Raaijmakers AJ, Ozell B, Després P, Raaymakers BW. Fast dose calculation in magnetic fields with GPUMCD. Phys Med Biol. 2011;56(16):5119–29.
121. Wooten HO, Green O, Yang M, DeWees T, Kashani R, Olsen J, et al. Quality of intensity modulated radiation therapy treatment plans using a $^{60}$Co magnetic resonance image guidance radiation therapy system. Int J Radiat Oncol Biol Phys. 2015;92(4):771–8.
122. Li Y, Wang B, Ding S, Liu H, Liu B, Xia Y, et al. Feasibility of using a commercial collapsed cone dose engine for 1.5T MR-LINAC online independent dose verification. Phys Med. 2020;80:288–96.
123. Kry SF, Feygelman V, Balter P, Knöös T, Charlie Ma C-M, Snyder M, et al. AAPM Task Group 329: reference dose specification for dose calculations: dose-to-water or dose-to-muscle? Med Phys. 2020;47(3):e52–64.
124. Shortall J, Vasquez Osorio E, Chuter R, McWilliam A, Choudhury A, Kirkby K, et al. Assessing localized dosimetric effects due to unplanned gas cavities during pelvic MR-guided radiotherapy using Monte Carlo simulations. Med Phys. 2019;46(12):5807–15.
125. Xia W, Zhang K, Li M, Tian Y, Men K, Wang J, et al. Impact of magnetic field on dose distribution in MR-guided radiotherapy of head and neck cancer. Front Oncol. 2020;10:1739.
126. Chen X, Prior P, Chen GP, Schultz CJ, Li XA. Dose effects of 1.5 T transverse magnetic field on tissue interfaces in MRI-guided radiotherapy. Med Phys. 2016;43(8 Part 1):4797–802.
127. Chetty IJ, Rosu M, Kessler ML, Fraass BA, Ten Haken RK, Kong FM, et al. Reporting and analyzing statistical uncertainties in Monte Carlo-based treatment planning. Int J Radiat Oncol Biol Phys. 2006;65(4):1249–59.
128. Brock KK, Mutic S, McNutt TR, Li H, Kessler ML. Use of image registration and fusion algorithms and techniques in radiotherapy: report of the AAPM Radiation Therapy Committee Task Group No. 132. Med Phys. 2017;44(7):e43–76.
129. Rong Y, Rosu-Bubulac M, Benedict SH, Cui Y, Ruo R, Connell T, et al. Rigid and deformable image registration for radiation therapy: a self-study evaluation guide for NRG oncology clinical trial participation. Pract Radiat Oncol. 2021;11(4):282–98.
130. Swaminath A, Massey C, Brierley JD, Dinniwell R, Wong R, Kim JJ, et al. Accumulated delivered dose response of stereotactic body radiation therapy for liver metastases. Int J Radiat Oncol Biol Phys. 2015;93(3):639–48.

131. Jaffray DA, Lindsay PE, Brock KK, Deasy JO, Tomé WA. Accurate accumulation of dose for improved understanding of radiation effects in normal tissue. Int J Radiat Oncol Biol Phys. 2010;76(3 Suppl):S135–9.
132. Sonke JJ, Aznar M, Rasch C. Adaptive radiotherapy for anatomical changes. Semin Radiat Oncol. 2019;29(3):245–57.
133. Chetty IJ, Rosu-Bubulac M. Deformable registration for dose accumulation. Semin Radiat Oncol. 2019;29(3):198–208.
134. Veiga C, Lourenço AM, Mouinuddin S, van Herk M, Modat M, Ourselin S, et al. Toward adaptive radiotherapy for head and neck patients: uncertainties in dose warping due to the choice of deformable registration algorithm. Med Phys. 2015;42(2):760–9.
135. Heukelom J, Fuller CD. Head and neck cancer adaptive radiation therapy (ART): conceptual considerations for the informed clinician. Semin Radiat Oncol. 2019;29(3):258–73.
136. Kadoya N, Nakajima Y, Saito M, Miyabe Y, Kurooka M, Kito S, et al. Multi-institutional validation study of commercially available deformable image registration software for thoracic images. Int J Radiat Oncol Biol Phys. 2016;96(2):422–31.
137. Zhang Y, Balter JM, Dow JS, Cao Y, Lawrence TS, Kashani R. Development of an abdominal dose accumulation tool and assessments of accumulated dose in gastrointestinal organs. Phys Med Biol. 2023;68.
138. Kontaxis C, Bol G, Lagendijk J, Raaymakers B. A new methodology for inter-and intrafraction plan adaptation for the MR-linac. Phys Med Biol. 2015;60(19):7485.
139. Menten MJ, Mohajer JK, Nilawar R, Bertholet J, Dunlop A, Pathmanathan AU, et al. Automatic reconstruction of the delivered dose of the day using MR-linac treatment log files and online MR imaging. Radiother Oncol. 2020;145:88–94.
140. Fast M, van de Schoot A, van de Lindt T, Carbaat C, van der Heide U, Sonke J-J. Tumor trailing for liver SBRT on the MR-Linac. Int J Radiat Oncol Biol Phys. 2019;103(2):468–78.
141. Lim SY, Tran A, Tran ANK, Sobremonte A, Fuller CD, Simmons L, et al. Dose accumulation of daily adaptive plans to decide optimal plan adaptation strategy for head-and-neck patients treated with MR-Linac. Med Dosim. 2022;47(1):103–9.
142. Willigenburg T, van der Velden JM, Zachiu C, Teunissen FR, Lagendijk JJ, Raaymakers BW, et al. Accumulated bladder wall dose is correlated with patient-reported acute urinary toxicity in prostate cancer patients treated with stereotactic, daily adaptive MR-guided radiotherapy. Radiother Oncol. 2022;171:182–8.
143. McDonald BA, Zachiu C, Christodouleas J, Naser MA, Ruschin M, Sonke JJ, et al. Dose accumulation for MR-guided adaptive radiotherapy: from practical considerations to state-of-the-art clinical implementation. Front Oncol. 2022;12:1086258.
144. Noel CE, Santanam L, Parikh PJ, Mutic S. Process-based quality management for clinical implementation of adaptive radiotherapy. Med Phys. 2014;41(8Part1):081717.
145. Cai B, Green OL, Kashani R, Rodriguez VL, Mutic S, Yang D. A practical implementation of physics quality assurance for photon adaptive radiotherapy. Z Med Phys. 2018;28(3):211–23.
146. Yang J, Zhang P, Tyagi N, Scripes PG, Subashi E, Liang J, et al. Integration of an independent monitor unit check for high-magnetic-field MR-guided radiation therapy system. Front Oncol. 2022;12:747825.
147. Chen X, Ahunbay E, Paulson ES, Chen G, Li XA. A daily end-to-end quality assurance workflow for MR-guided online adaptive radiation therapy on MR-Linac. J Appl Clin Med Phys. 2020;21(1):205–12.
148. Hackett S, van Asselen B, Feist G, Pencea S, Akhiat H, Wolthaus J, et al. Collapsed cone algorithm can be used for quality assurance for Monaco treatment plans for the MR-Linac. Med Phys. 2016;43(6 Part 11):3441.
149. Lim SB, Godoy Scripes P, Napolitano M, Subashi E, Tyagi N, Cervino Arriba L, et al. An investigation of using log-file analysis for automated patient-specific quality assurance in MRgRT. J Appl Clin Med Phys. 2021;22(9):183–8.
150. Peng C, Chen G, Ahunbay E, Wang D, Lawton C, Li XA. Validation of an online replanning technique for prostate adaptive radiotherapy. Phys Med Biol. 2011;56(12):3659.
151. Altman MB, Kavanaugh JA, Wooten HO, Green OL, DeWees TA, Gay H, et al. A framework for automated contour quality assurance in radiation therapy including adaptive techniques. Phys Med Biol. 2015;60(13):5199–209.
152. Chen HC, Tan J, Dolly S, Kavanaugh J, Anastasio MA, Low DA, et al. Automated contouring error detection based on supervised geometric attribute distribution models for radiation therapy: a general strategy. Med Phys. 2015;42(2):1048–59.
153. Huq MS, Fraass BA, Dunscombe PB, Gibbons JP Jr, Ibbott GS, Mundt AJ, et al. The report of Task Group 100 of the AAPM: application of risk analysis methods to radiation therapy quality management. Med Phys. 2016;43(7):4209.
154. Liang J, Scripes PG, Tyagi N, Subashi E, Wunner T, Cote N, et al. Risk analysis of the Unity 1.5 T MR-Linac adapt-to-position workflow. J Appl Clin Med Phys. 2022;24:e13850.

# 9 The First Low-Field MRI-Guided Radiation Therapy Hybrid Integrated System: MRIdian

Poonam Yadav, Jooyoung (James) Sohn, and Ahtesham U. Khan

## Introduction

A significant difficulty in radiation therapy is to provide high radiation doses to malignant tissue while reducing the damage to nearby healthy tissue. Real-time awareness of the tumor's location is necessary to accomplish this, but current X-ray-based imaging systems cannot consistently provide this information. Although magnetic resonance imaging (MRI) has the ability to produce precise and thorough images of malignancies, combining MRI with radiation therapy has not proven practical until recently. Several developments have resulted in the creation of MR-Linac systems, which allow for MRI-guided radiation therapy in real time. Originally developed and put into practice by Lagendijk et al., this groundbreaking technology has now been successfully used for Co-60 beams and linac beams [1, 2]. In this chapter, we will focus on the ViewRay MRIdian system, which is a state-of-the-art MR-Linac system that provides enhanced precision and accuracy in radiation therapy delivery.

P. Yadav (✉) · J. Sohn · A. U. Khan
Department of Radiation Oncology, Northwestern University Feinberg School of Medicine, Chicago, IL, USA
e-mail: Poonam.yadav@nm.org; jooyoung.sohn@nm.org; ahteshamullah.khan@nm.org

## What Is an MR-Linac?

MR-Linacs are hybrid machines that combine an MRI scanner with a radiation delivery device such as a linear accelerator or a radioisotope irradiator. Including an imaging modality with the radiation delivery machine enables image-guided radiation therapy (IGRT) and adaptive radiotherapy (ART). IGRT leads to routine visualization of the patient's anatomy and allows accurate patient positioning based on daily images. ART aims to modify the treatment plan based on the changes in the patient's anatomy or tumor biology. Imaging modalities such as computed tomography (CT), positron emission tomography (PET), or MRI can be employed to implement IGRT and ART. MRI has several advantages over other imaging modalities that have been described in Chap. 2 of this book, but mainly including exquisite soft-tissue contrast, the ability to acquire both functional and anatomical information, and not employing ionizing radiation. Cine MRI images can also be acquired over a long period without any adverse health risks. The radiation isocenter coincides with the MRI isocenter for MR-Linacs and can be used to align the patient on the couch using pretreatment volumetric images. During the treatment delivery, cine images are usually acquired to monitor the target's motion especially due to respiratory motion. Beam gating can be employed to accurately deliver the radiation to the desired location.



I. J. Das et al. (eds.), *A Practical Guide to MR-Linac*, https://doi.org/10.1007/978-3-031-48165-9_9

While MR-Linacs provide many advantages over conventional treatment delivery systems, the presence of a strong magnetic field in the vicinity of the beam delivery system creates unique challenges related to treatment planning, delivery, dosimetry, training, and safety. There are many MR-Linac systems that have been described in Chapters 10 and 11. This chapter focuses on details related to the MRIdian 0.35T low-field MR-Linac.

## Electron Return Effect (ERE)

High magnetic field strengths are desired in MRI to enhance the signal-to-noise ratio (SNR) of the images. However, any charged particles experiencing magnetic fields are also subjected to the Lorentz force ($\vec{F}$):

$$\vec{F} = q\left(\vec{v} \times \vec{B}\right) \tag{9.1}$$

where $q$ is the charge, $v$ is the velocity, and $B$ is the magnetic field. The Lorentz force impacts the trajectory of the charged particles (mainly electrons that are responsible for radiation dose) and causes them to move along a curved path with a radius of

$$r = \frac{E_{\text{total}}\beta}{qBc} \tag{9.2}$$

where $E_{\text{total}}$ is the total energy of the particle, $\beta$ is the relativistic velocity, $c$ is the speed of light, and $r$ is the radius of curvature. The radius of the curvature is inversely proportional to the magnetic field strength; therefore, higher magnetic fields impact the incident particle trajectory more drastically than lower fields. This chapter focuses on MR-Linacs that treat with a photon beam and deposit energy via the generation of secondary electrons. It is of note that the path of the secondary electrons is subjected to the Lorentz force. The trajectory of the generated electrons is impacted greatly at high physical density gradient interfaces such as tissue-air interfaces. When crossing an interface with different material compositions or densities, the circular motion of the charged particles induces the electron return effect (ERE) illustrated in Fig. 9.1. An enhance-

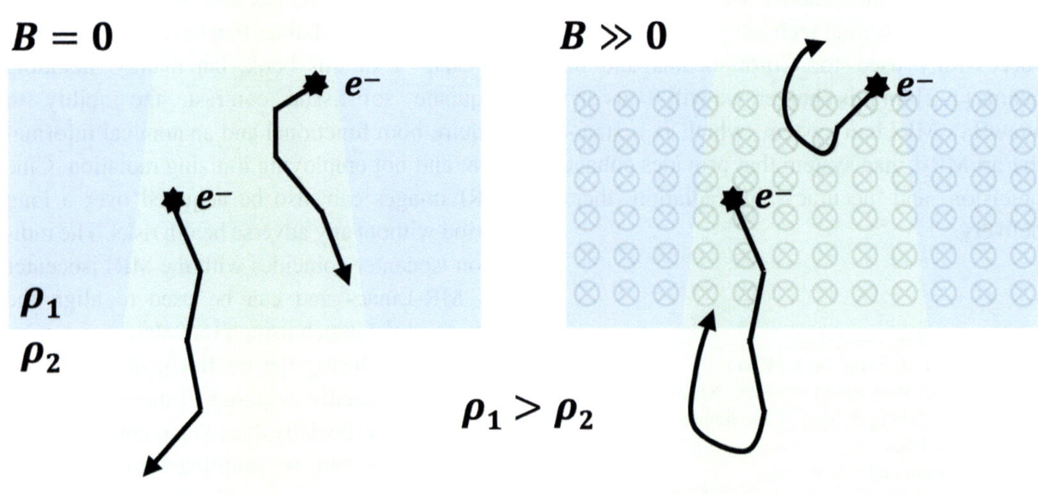


**Fig. 9.1** The impact of a magnetic field on the charged particle trajectories is illustrated

ment or reduction of the radiation dose can occur, relative to a zero magnetic field, depending on the magnetic field strength and direction, density gradient, electron energy, and angle of incidence.

Based on Eq. (9.2), the computed radius of an electron beam is shown in Fig. 9.2 in terms of the Lorentz force as a function of electron kinetic energy. For media with lower stopping power, where the range of the secondary electrons is large relative to the radius of curvature, a large impact on the dose distribution can be expected due to the external magnetic field. Further, the ERE can be significant for thoracic and abdominal treatment sites, where high-density gradients are often encountered by the radiation beam [3, 4]. The presence of the ERE may increase the skin dose, especially for the skin near the region where the beam exits the patient. Therefore, a bolus is typically placed on the skin of the patient during delivery to reduce the skin dose due to the ERE [5]. The impact of the magnetic field on dose distribution can also be noted in dosimetric data such as percent depth dose (PDD) and lateral beam profiles. For MR-Linacs with the magnetic field orthogonal to the radiation beam, a smaller buildup region was observed [6]. The ERE curves the secondary electrons toward the proximal region of the medium creating a transient charged particle equilibrium (TCPE) at shallower depths. An asymmetry was found in the lateral beam profile orthogonal to the magnetic field direction leading to an increase in penumbra toward the direction of the Lorentz force [6]. The impact of the magnetic field on output factors has been reported to be negligible for depths greater than the depth of the maximum dose [7].

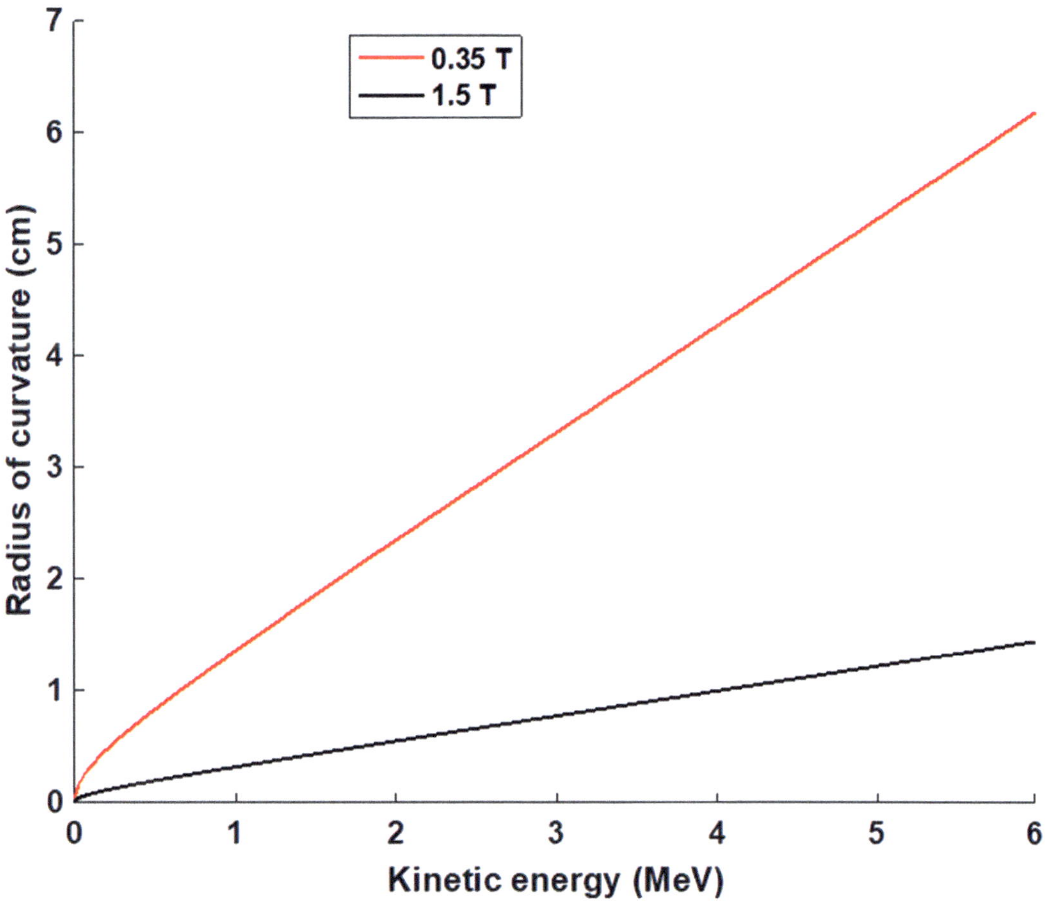


**Fig. 9.2** The radius of curvature of electrons, due to the Lorentz force, as a function of kinetic energy

## Planning and Dosimetry Challenges for MR-Linacs

Several challenges arise when utilizing MR-Linacs for radiation therapy treatment. Besides considering the conventional problems or workflow requirements that are encountered in operating an MRI scanner or a linac separately, the presence of both machines in the same vault puts forth additional considerations and challenges for clinicians. All equipment must be MR-safe, and the vault must be constructed considering both the impact of external radiofrequency (RF) noise on image quality and the shielding requirements for high-energy photon beams. The personnel working with the machine must be trained on relevant aspects of MRI safety and radiation safety. Appropriate signs and warnings must be placed outside the vault to indicate the use of a magnet and a high-energy radiation beam. The linac components generate RF noise that can induce artifacts in the MRI images. Although MRI scanners look similar to CT scanners, the scan acquisition procedure for these modalities differs in various aspects such as couch motion, gantry rotation, scanning time, and noisiness from the scanner. To achieve high SNR for the acquired images, site-specific receiver coils might be employed, which may intervene with patient comfort and attenuate the incident radiation beam. Additionally, the total treatment time on an MR-Linac can differ significantly from conventional linacs, especially if ART is used.

Pretreatment and peri-treatment imaging is performed using MRI, which typically has poor spatial integrity due to B-field inhomogeneities and nonlinear gradients when compared to X-ray-based modalities. Therefore, monitoring and maintaining high spatial fidelity for MR images is of paramount importance when utilizing MR-Linacs. Since MRI is based on the nuclear spin properties of the imaged media, information related to electron density is not obtainable from MRI images, which is required for dose calculation. However, the initial treatment plan can be made using the CT image set from the CT simulation data. If ART is implemented, the daily pretreatment MRI images can be co-registered with the planning MRI images, and dose calculation can be accomplished using the Hounsfield unit (HU) information from the corresponding CT images after a deformation map is applied based on the daily image set. This method circumvents the usage of daily CT imaging when an adapted plan is created for the patient.

The presence of a magnetic field also creates dosimetry challenges. The accuracy of the conventional dose calculation algorithms, such as convolution superposition or pencil beam, is inadequate under magnetic fields for clinical use. Therefore, Monte Carlo (MC) methods are employed to calculate absorbed dose in the presence of an external magnetic field. It is also challenging to acquire 3D scans with a dosimeter since the water tank and the associated motion stages must be MR-compatible and fit inside the bore of the machine. Air-filled ionization chambers (ICs) are the preferred detectors when performing reference dosimetry in radiotherapy. Due to the low density of the air cavity inside the IC, the response of the chamber usually has a dependence on the magnetic field strength and direction [8]. This dependence calls for additional correction factors for ICs when measuring absorbed dose in MR-Linacs. Solid-state dosimeters typically have a much smaller dependence on the magnetic field but are usually not employed as reference detectors for absolute absorbed dose measurement [9].

## System Design of MRIdian

The 2014 version of the MRIdian machine from ViewRay housed three $^{60}Co$ sources spaced 120° apart generating a dose rate of 550 cGy/min at the isocenter [10]. Utilizing a radioactive material that can be exploited by any malevolent parties poses safety concerns and requires repeated replenishment of the source over time. The smaller buildup region of this machine due to the lower energy spectrum of the $^{60}Co$ source, relative to megavoltage photon beams, also leads to lower skin sparring. The collimation of the beam requires three independent sets of multileaf collimators (MLCs), which adds complexity to beam delivery.

ViewRay transitioned all of the $^{60}$Co-based machines into linacs over the course of the past few years. The latest MR-Linac leverages the acceleration of electron beams using S-band microwaves in a standing-wave waveguide. The linac is capable of producing a 6 MV flattening filter-free (FFF) photon beam with a propagation direction orthogonal to the direction of both the MLC motion and the main magnetic field [11]. The beam is collimated by dual-stacked and dual-focused MLC banks that move in an arc to follow the photon beam divergence. The upper and lower MLC stacks are capable of full overtravel and comprise of leaves with a width of 0.83 cm projected at the isocenter. All leaves have a height of 5.5 cm and are composed of a 17.7 g/cc non-ferromagnetic tungsten alloy. The leaves are tilted to match the divergence of the photon beam in the IEC-Y direction. The upper and lower stacks house 68 and 70 dynamic leaves, respectively, and the stacks are offset by half a leaf width to increase the collimation resolution to 0.415 cm in the IEC-Y direction while also mitigating the tongue and groove effect. The minimum and maximum field sizes are $0.2 \times 0.415$ cm$^2$ and $27.4 \times 24.1$ cm$^2$, respectively, projected at the isocenter with a source-to-axial (SAD) distance of 90 cm. The MR-Linac has a nominal dose rate of 600 MU/min. The couch for this linac has a uniform thickness and is only capable of translational motion limiting the delivery to coplanar radiation beams. Schematic of linac and its subcomponents is shown in Fig. 9.3.

The MRIdian system allows 3D conformal radiation therapy (3DCRT), also referred to as fixed conformal, and step-and-shoot intensity-modulated radiation therapy (IMRT), referred to as optimized conformal. A combination of IMRT and 3DCRT can also be used in some cases. Since the beam attenuation through the couch can be significant of up to 20%, gantry angles that shoot the beam through the edges of the couch should be avoided.

The MRIdian MR-Linac uses split-bore superconducting magnets that generate a uniform 0.345T magnetic field along the superior-inferior direction (IEC-Y). The linac components such as the accelerator, magnetron, pulse transformer, port circulator, and gun driver are mounted on a gantry ring placed between the two magnet halves. The beam-generating components operate under a magnetic field strength of <100 Gauss, while the linac itself operates under a magnetic field strength of <2 Gauss. Magnetic fields stronger than 2 Gauss can impact the trajectory of the electron beam inside the waveguides and can disrupt the acceleration process. To mitigate the impact of the strong magnetic field on the linac operation, the beam-generating components are placed inside shielding "buckets" comprised of five concentric cylindrical ferromagnetic steel in order to create a magnetic free zone. The impact of the RF noise, generated by the linac components, on MR image quality is reduced by utilizing RF-absorbing carbon fiber and RF-reflecting copper shields.

The bore of the MR-Linac is 70 cm in diameter, while the maximum imaging field of view (FOV) is 54 cm. To facilitate patient positioning using MRI images, the scanner and the linac share a common isocenter. The RF excitation $B_1$ transmit pulse is generated using a 16-rung quadrature birdcage coil with a 75 cm diameter. The receiver signal can be collected using the body coil or site-specific coils if a higher SNR or parallel image acquisition is needed. The receiver coils use radiolucent phased arrays placed inside low-density foam. The torso and the head and neck (H&N) receiver coils contain $2 \times 6$ and $2 \times 5$ channels, respectively, one for anterior and posterior each. The gradient coils can generate strengths of 18 mT/m with a maximum slew rate of 200T/m/s.

The $B_0$ field homogeneity is improved using five higher-order superconducting shims as well as passive and gradient shimming. The "tune-up" mode for image acquisition uses a constant or zero gradient offset across all gantry angles leading to varying $B_0$ homogeneity depending on the gantry angle. The "standard shim" mode reduces inhomogeneities by calculating gradient offsets using a pre-scan acquisition. This method maximizes $B_0$ homogeneity at the expense of increased scan time. The "MRI-GO" mode

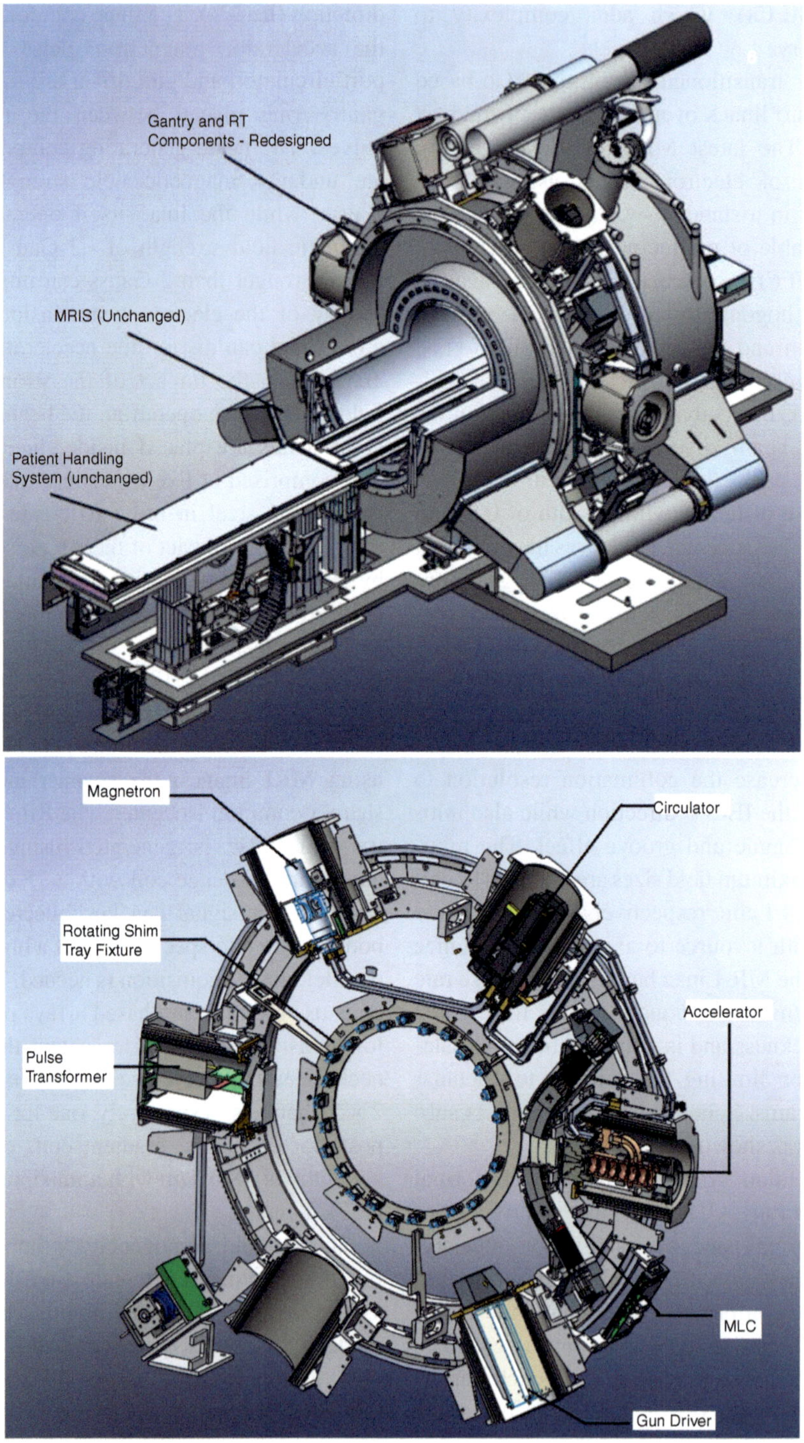


**Fig. 9.3** The schematic of the MRIdian MR-Linac (top). The linac subcomponents inside the shielding buckets (bottom). Figure courtesy of ViewRay, Incorporated

tweaks the gradient current as a function of gantry angles based on homogeneity measurements acquired with a spherical phantom. This mode reduces $B_0$ inhomogeneity, albeit not as well as the standard shim mode, without significantly increasing the scan time.

A localization scan can be performed within 15 s with an SNR of >30 to register the patient or a phantom. True Fast Imaging with Steady State Precession (TRUFI) pulse sequence is the default acquisition sequence for both volumetric and cine imaging utilizing a mixed $T_1$ and $T_2$ weighted contrast. Scans with $T_1$ contrast can be acquired using either a 3D gradient-echo pulse sequence or a turbo flash (TFL) sequence. The latter was developed for thorax and abdomen scans and significantly reduces motion artifacts. Other pulse sequences, such as turbo spin echo (TSE) and diffusion-weighted imaging (DWI), are also available for image acquisition. The volumetric imaging can be acquired with a slice thickness of 1.5 mm or 3 mm with an in-plane resolution of 1.5x1.5 $mm^2$. Cine imaging has a lower spatial resolution of $3.5 \times 3.5$ $mm^2$ or $2.4 \times 2.4$ $mm^2$ with slice thickness options of 5 mm, 7 mm, or 10 mm. Real-time 2D multi-planar imaging can be performed with FOV options of $27 \times 27$ $cm^2$, $35 \times 35$ $cm^2$, and $40 \times 40$ $cm^2$, with acquisition rates of 4 FPS or 8 FPS.

## MR Safety for MRIdian

It is pertinent that MR safety guidelines from relevant organizations, such as the American College of Radiology (ACR), are followed to ensure safe operation of the MR-Linac [12]. The patients can be exposed to several possible hazards from the operation of the MRI scanner in the treatment room. The presence of a magnetic field can induce translational and rotational forces on ferromagnetic objects leading to hazards related to projectile damage. The rapid switching of the gradient fields generates Eddy currents in tissue depending on the rate of change of the field, the electrical conductivity of the media, and the cross-sectional area of the tissue. Heating or burning effects might be induced by Eddy currents depending on the magnitude. Peripheral nerve and muscle stimulation is also possible due to Eddy currents requiring a threshold ceiling on the rate of change of the gradient fields. The rapidly changing gradients also produce strong acoustic noise, which can disrupt patient comfort and may pose a concern to patients with auditorial complications. The RF energy transmitted to the patient tissue to excite the magnetization can act as a localized heating source. The intensity of the heating is proportional to the electrical conductivity of the tissue, patient size, and square of the electric field induced by the RF pulse.

Several measures can be taken to prevent or reduce the hazardous effects produced by the MRI scanner. Since the magnetic field strength drops rapidly as a function of distance from the magnets, multiple zones can be created based on the field strength to isolate regions of high magnetic fields. The zones can then be categorized, and zone-specific protocols can be used to restrict or control different regions depending on the posed concerns. Figure 9.4 describes different MRI zones recommended for the MRIdian system. For diagnostic scanners, it is important to visually indicate the 5 Gauss line inside the treatment vault. The MRIdian system specifies the 50 Gauss line to be near the foot of the couch. A visual indicator of the high magnetic field region also aids in instruction to personnel regarding the storage of equipment related to radiotherapy, such as immobilization or quality assurance (QA) devices, that can generate RF noise or can be impacted by the magnetic field. Prior to the treatment, the patient's medical data must be reviewed to check for any implanted medical devices or other ferromagnetic material inside the patient's body. Any questionable devices must be researched thoroughly by the personnel, and strict screening protocols must be placed to only allow equipment that is considered "MR safe." Besides reviewing medical records, it is good practice to employ a ferromagnetic detection system before each treatment fraction to ensure safe usage of the MRI.

The heating effects, especially hotspots, due to Eddy currents and RF pulses can be reduced by avoiding contact of conductors with the

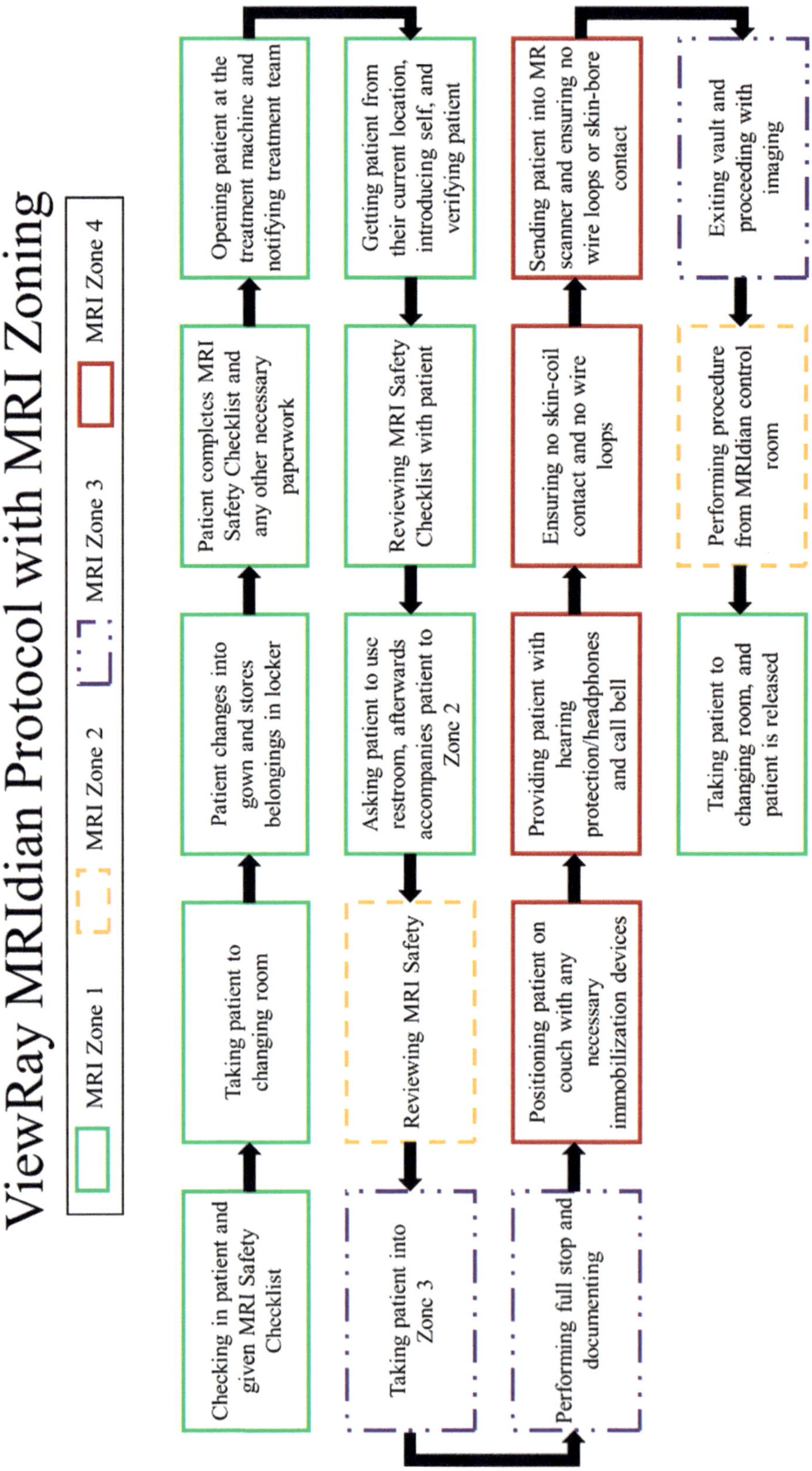


**Fig. 9.4** A safety protocol with a process map for ViewRay MRIdian. The colors of each box indicate the MRI zoning for that step

patient's skin. Such a scenario can occur when using immobilization or other accessory devices for treatment or scanning. Compliance with the US Food and Drug Administration (FDA) or other relevant authorities must be maintained to ensure that the rate of change of the gradient fields is below the threshold of palpable peripheral nerve stimulation and that the acoustic noise levels are safe for each patient. The specific absorption rate (SAR), defined as the energy dissipated in tissue per unit mass, due to the RF pulse must be well below the established thresholds and should be monitored for the utilized pulse sequence. The International Electrotechnical Commission (IEC) limits the average SAR to 4 W/kg for the whole body and 3.2 W/kg for the head. For the MRIdian system, the SAR value is typically <1.2 W/kg and therefore is well below the specified tolerances. Besides health concerns, it is important to ensure that the equipment inside the room has minimal impact on MRI image quality, noise, and the presence of artifacts.

## Logistics and Economics of Acquiring an MR-Linac

Frequently, the purchase of an MR-Linac is considered when an existing linac needs replacing. Otherwise, the decision to instantiate an MR-Linac program can be based on gaining strategic advantage by including advanced techniques such as image-guided radiation therapy (IGRT) and adaptive radiotherapy (ART). Some clinics decide to acquire MR-Linacs to expand on the disease sites or the number of patients that can be treated at their clinic. Evaluation might be performed by the clinics to consider their market share of the potential patients that require treatment using MR-Linacs. Increase in patient load, financial reimbursement for these treatments, and return on investment (ROI) data play an important role when an acquisition of MR-Linacs is considered. Often, physicians or physicists desire to have an MR-Linac at their clinic to pursue ART. Academic centers might consider MR-Linacs to increase their research output, gain access to research grants related to this technology, or boost their academic status. An MR-Linac vendor might reach out to a clinic with a proposal for implementing this technology at the clinic.

There are multiple MR-Linac vendors with distinct linacs and MRI scanning systems. Once the decision of acquiring an MR-Linac is made, presentations and demos are provided by the vendors to summarize the benefits and costs associated with each system. Site visits are scheduled by the vendors to evaluate vault requirements for their MR-Linac, and any additional construction costs are added to the budget proposal. It is of note that the MRIdian MR-Linac can fit inside a conventional vault. The vendor might facilitate site visits to existing users of their MR-Linac for a Q&A session. The selection criteria to choose a vendor can be dependent on the (1) disease sites that will be treated with the machine, (2) clinical trial data showing outcomes for patients treated with the MR-Linac, (3) initial cost related to the MR-Linac acquisition, (4) operational costs and staffing requirements to maintain the MR-Linac program, (5) cost and maintenance of the auxiliary equipment required, (6) potential research grants and projects that can be acquired, and (7) treatment time or physician time required at the treatment console. Following the purchasing order, it can take 1–2 years until a patient is treated with the acquired MR-Linac.

## Teams and Training for MRIdian

### ViewRay Training

ViewRay offers on-site and remote training programs for MRIdian. The on-site training program involves hands-on training sessions conducted by trained staff members, covering MR simulation, data transferring, treatment planning, and treatment procedures. Virtual training includes online courses and webinars on various aspects of MRIdian, including system operation, clinical applications, maintenance, and QA. The training programs are designed to ensure that the staff members possess the knowledge and skills necessary to efficiently operate and maintain the MRIdian systems.

## In-House Training for Physicians, Physicists, Therapists, and Dosimetrists

The following topics can be covered in in-house courses and hands-on training sessions for physicians, physicists, therapists, and dosimetrists for ViewRay MRIdian:

- MRI physics.
- CT vs. MRI anatomy.
- MRI safety.
- Dose calculation under magnetic fields.
- Appropriate beam arrangement for different treatment sites.
- Beam angles to avoid.
- Optimization techniques.
- Adaptive treatment workflow.
- ViewRay MRIdian hardware and software.
- Quality assurance (patient-specific QA, periodic QA, etc.)
- Troubleshooting and maintenance of the device (support package, etc.)

## Practicing Contouring and Workflow

The contouring training session can be set up using a single box (simulation box provided by ViewRay) or MIM. The simulation box is an integrated system that provides MR simulation and on-table adaptation. MIM is a comprehensive image analysis software that provides advanced tools for contouring. Contouring practice should involve a hands-on approach, where each program member should be given the opportunity to contour organs at risk for different clinical sites.

The staff members should receive feedback from their peers and the team leaders. For competency evaluation, five cases should be selected, including the case that will be used for treatment. Members of each team, including radiation oncologists, medical physicists, radiation therapists, and dosimetrists, need to be involved in the training program to ensure that everyone understands their role in the workflow process.

To start the program, various team members should work together to generate checklists that cover each stage of the workflow process. The following checklist may be helpful:

1. Simulation checklist: scanning with a low resolution for the setup verification and field of view, choosing a proper slice for tracking structure.
2. Planning checklist: correct site of treatment, appropriate contours, and optimization.
3. Adaptive checklist: checking the rules, electron density, density override, plan review, and tracking algorithm.

A flowchart describing (example Fig. 9.5 shows the workflow of ViewRay MRIdian in terms of process map including an adaptive treatment workflow) the clinical workflow for the program should be developed. The flowchart should include information about each stage of the process, including imaging, contouring, planning, and treatment delivery. It should also identify the individuals responsible for each task and the timeline for completing each task. The workflow practice should involve different group members to ensure that everyone understands their role in the process. Note some tasks involve different group members working together. During the workflow practice, each group member should be given the opportunity to practice their respective tasks and receive feedback from their peers.

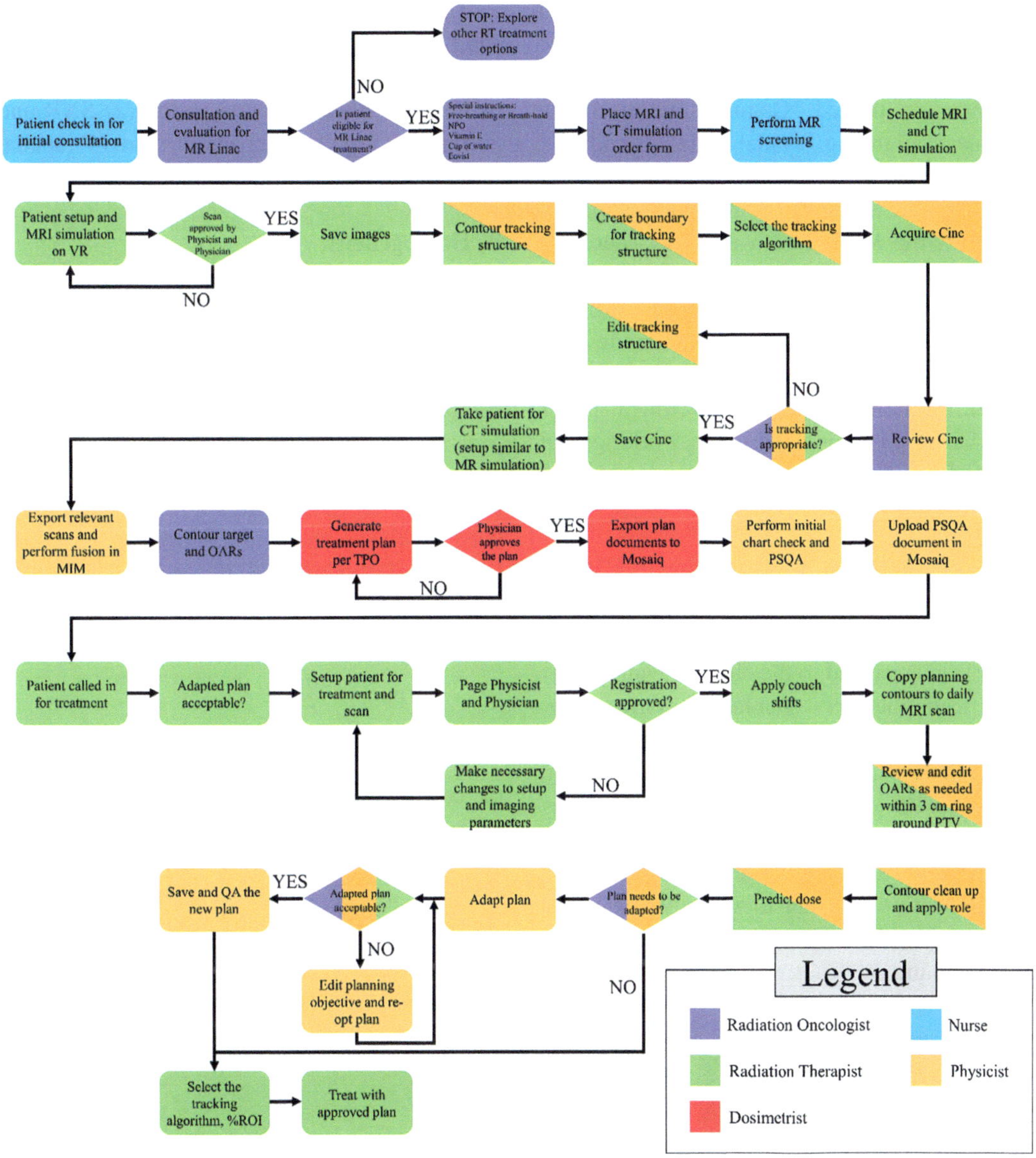


**Fig. 9.5** Process map of ViewRay MRIdian workflow. The colors of each flowchart box indicate the responsible staff member for that step. Multiple colors indicate that multiple staff may be responsible for completing that step

## MRIdian Acceptance, Commissioning, and QAs

The MRIdian MR-Linac is a complex machine with multiple components that work in synergy to treat patients. It is crucial to ensure that all individual subsystems are working as intended and the communication between the subsystems remains intact. Shielding calculations are performed prior to the installation to ensure that the radiation exposures outside the vault are below the National Council on Radiation Protection and Measurements (NCRP) specified tolerances. RF penetration through the walls is also evaluated to ensure adequate RF shielding inside the vault. Following the construction of the treatment vault and installation of the machine, functional and acceptance testing must be performed by the team before any patients are treated. Commissioning of the treatment planning system (TPS) and QA devices is also performed at this stage. Baseline data are also acquired to monitor the performance of the MR-Linac over time. There is a paucity of protocols and guidelines related to the commissioning and QA of MR-Linacs. Therefore, the following sections describe either the vendor-recommended testing protocols or guidelines provided by clinical physicists based on their experience working with the MRIdian MR-Linac.

## Acceptance Testing for MRI

Once the installation is complete and the MR-Linac is handed to the clinical team, the relevant team members should familiarize themselves with the hardware components and any connections between them as well as software interfaces. It's also best to be aware of the accessory components and their place of storage. Before the acceptance testing protocol is launched, it's important to visually inspect the machine and the accessory components for no physical damage.

The acceptance testing for the MRI scanner can be performed using the equipment provided by the vendor and is divided into four distinct phases: (1) $B_0$ field homogeneity, (2) SNR and uniformity, (3) ACR phantom, and (4) spatial integrity.

### $B_0$ Homogeneity

The $B_0$ field homogeneity can be evaluated using a 24 cm diameter of spherical volume (DSV) phantom, shown in Fig. 9.6a), aligned with the isocenter. A localizer scan is used to align any equipment placed on the couch. The spectral peak method is used to test field homogeneity, and a free induction decay (FID) signal is obtained. The signal in the spectral domain is utilized to measure the field homogeneity by calculating the full width at half maximum (FWHM) of the signal profile. The $B_0$ field homogeneity is calculated by

$$\text{ppm} = \frac{\Delta f}{f_o}$$

where $\Delta f$ is the FWHM of the profile and $f_o$ is the central frequency. The tune-up mode is typically utilized for these measurements, and the homogeneity is evaluated over gantry angles from 30° to 330° in 30° intervals. Since each of the six buckets mounted on the gantry contains a different linac subcomponent, the $B_0$ homogene-

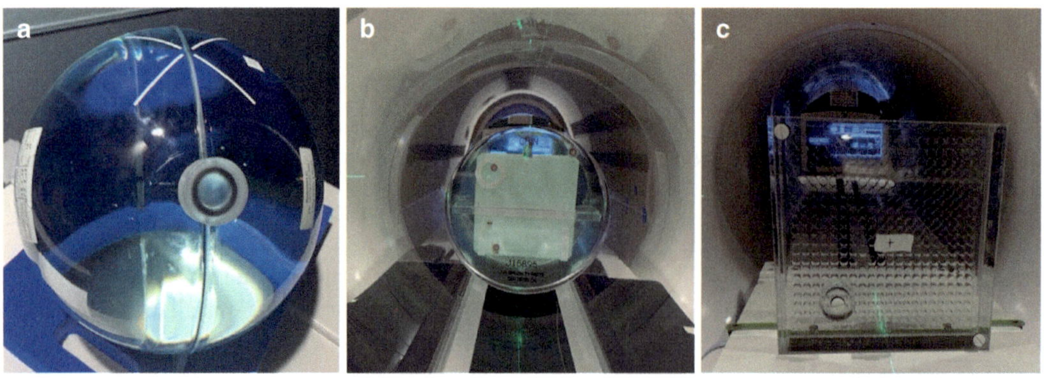


**Fig. 9.6** (**a**) The spherical phantom used for $B_0$ field homogeneity and SNR/uniformity testing. (**b**) The ACR phantom placed inside the 0.35sT MR-Linac is shown. (**c**) The spatial integrity phantom utilized in this work placed in the axial orientation

ity is distinct at each gantry angle. The gantry angle with the greatest homogeneity is usually selected for all pretreatment scans. Using the tune-up mode, the homogeneity is typically <5 parts per million (PPM) over all gantry angles as shown in Fig. 9.7. However, better homogeneity can be achieved using the standard shim mode or the MRI-GO mode.

**Receiver Coil SNR and Uniformity**

For all of the receiver coils, SNR and uniformity are evaluated using the same spherical phantom that was used for field homogeneity testing. The acquisition parameters are provided by the manufacturer for this test and are listed in Table 9.1. For the site-specific receiver coils such as the torso and H&N coils, the SNR is measured for each coil element in addition to the cumulative signal. The SNR and uniformity are evaluated in all three anatomical planes. Region of interests (ROIs) are placed on the scanned images, and the SNR/uniformity metrics are calculated by

$$\mathrm{SNR} = \frac{0.66 \times S_{\mathrm{mean}}}{\sigma}$$

$$\mathrm{Uniformity} = 100 \times 1 - \left[\frac{S_{\max} - S_{\min}}{S_{\max} + S_{\min}}\right]$$

where $\sigma$ is the standard deviation of the ROI on the noise image and $S_{\mathrm{mean}}$, $S_{\min}$, and $S_{\max}$ are the mean, minimum, and maximum on the signal ROI, respectively. The manufacturer-specified SNR thresholds are 12 for the body coil and 30 for the torso and H&N coils. These thresholds are even greater for the SNRs measured with the individual coil elements. SNRs for diagnostic scanners with higher $B_0$ field strengths are typically greater than these values. The vendor specifies the uniformity thresholds to be 50% for the image acquired with the body coil and 60% for the images acquired with the torso and H&N coils. The SNR and uniformity for the MRIdian machines are typically well above the specified thresholds [13].

**Table 9.1** Acquisition parameters for the SNR and uniformity measurements

| Parameter | Value |
|---|---|
| Repetition time (TR) | 1500 ms |
| Echo time (TE) | 15 ms |
| Resolution | 1.2 × 1.2 mm² in-plane and 10 mm slice thickness |
| Flip angle | 90° for signal image and 0° for noise images |

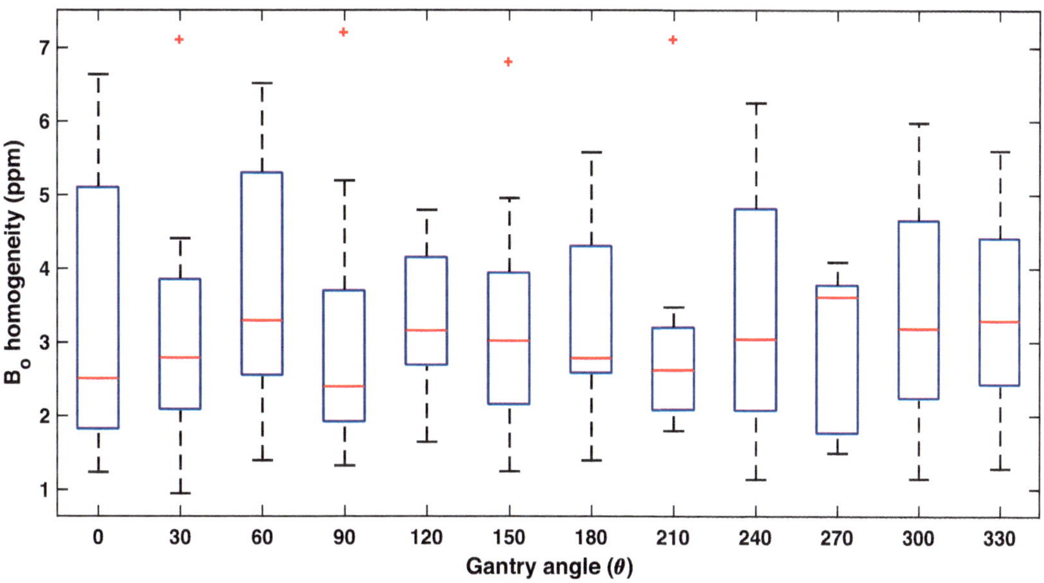


**Fig. 9.7** Multi-institutional data on $B_0$ field homogeneity as a function of gantry angle (adapted from Khan et al. [13])

**ACR Phantom Tests**

The ACR phantom comprehensively evaluates the image quality and positional accuracy of the MRI scanner using a single phantom. The phantom itself has several fixtures and inserts that are used to measure geometrical accuracy, slice positional and thickness accuracy, high contrast spatial resolution, low-contrast object detectability, image uniformity, and percent ghosting. For this test, the phantom is aligned using a localizer scan and imaged with the torso coil. Both $T_1$ and $T_2$ images are acquired with the parameters listed in Table 9.2. The ACR specifies thresholds for each of the tests. The geometrical accuracy, determined by the difference in the known dimensions of the phantom and the measured dimensions on the images, must be greater than ±3 mm. The slice positional and thickness accuracy is measured using two 45° wedges and must be within 5.0 mm and 0.7 mm, respectively, of the known values. The high contrast spatial resolution threshold is 1 mm and is measured using hole arrays of different sizes ranging from 0.9 to 1.1 mm in 0.1 mm increments. Several objects of varying contrast and sizes are embedded inside the ACR phantom that can be used to test the low-contrast object detectability of the scanner. The total number of visible spokes on the two slices must be greater than 9 to meet the ACR specifications for this test. A large ROI on a slice consisting of uniform media is used to measure the uniformity of the images, which has a specification of 87.5%. Four background ROIs are placed on the peripheral regions of the uniform ROI to measure the percent ghosting effect. The intensity of the ghosting must be <2.50% to meet the ACR criteria. ViewRay offers software to analyze and calculate the ACR metrics following the scan to improve the speed of the evaluation [14]. Table 9.3 compiles ACR phantom results for a multi-institutional study [13].

**Table 9.2** Acquisition parameters for the ACR phantom tests

| Parameter | $T_1$ value | $T_2$ value |
|---|---|---|
| Repetition time (TR) | 500 ms | 2000 ms |
| Echo time (TE) | 20 ms | 20 ms |
| Number of averages | 25 | 9 |
| Resolution | 0.98 × 0.98 mm² | 0.98 × 0.98 mm² |
| Flip angle | 90° | 90° |

**Spatial Integrity**

Spatial integrity is crucial in MR-guided radiotherapy (MRgRT) to accurately localize the target using the MR images. ViewRay offers a spatial integrity acrylic phantom, shown in Fig. 9.6c, consisting of cylinders filled with signal-generating media, such as $CuSO_4$ or mineral oil, forming a 20 × 20 grid with a spacing of 1.45 cm. The location of the individual cylinders is known and verified using a CT scan. The Magphan® RT phantom can be utilized as an alternative to the ViewRay phantom to perform this test. The spatial fidelity of each of the clinical pulse sequences is verified by imaging the phantom in all three anatomical planes and calculating the offsets between the actual and imaged locations of the cylinders. Spatial accuracy is also evaluated in the sagittal planes with offsets of ±7 cm and ±12.5 cm from the isocenter since this anatomical plane is used for real-time target tracking. The vendor-provided software is used to calculate the percentage of cylinders located accurately by the MR images in a 10 cm and a 17.5 cm radius ROI. Figure 9.8 displays spatial

**Table 9.3** The mean values of the multi-institutional ACR measurements along with the ACR specifications

| ACR test | Mean | ACR specification |
|---|---|---|
| Slice positional accuracy (mm) | | |
| Slice 1 | 0.97 (T1), 1.03 (T2) | <5.00 |
| Slice 11 | 1.08 (T1), 1.14 (T2) | <5.00 |
| Slice thickness accuracy (mm) | 5.29 (T1), 5.26 (T2) | 5.0 ± 0.7 |
| High contrast spatial resolution (mm) | 1.00 (T1), 1.00 (T2) | ≤1.00 |
| Low-contrast object detectability | 31 (T1), 22 (T2) | >9 |
| Image uniformity (%) | 97.55 (T1), 97.79 (T2) | >87.5 |
| Percentage ghosting (%) | 1.47 (T1), 0.06 (T2) | <2.50 |

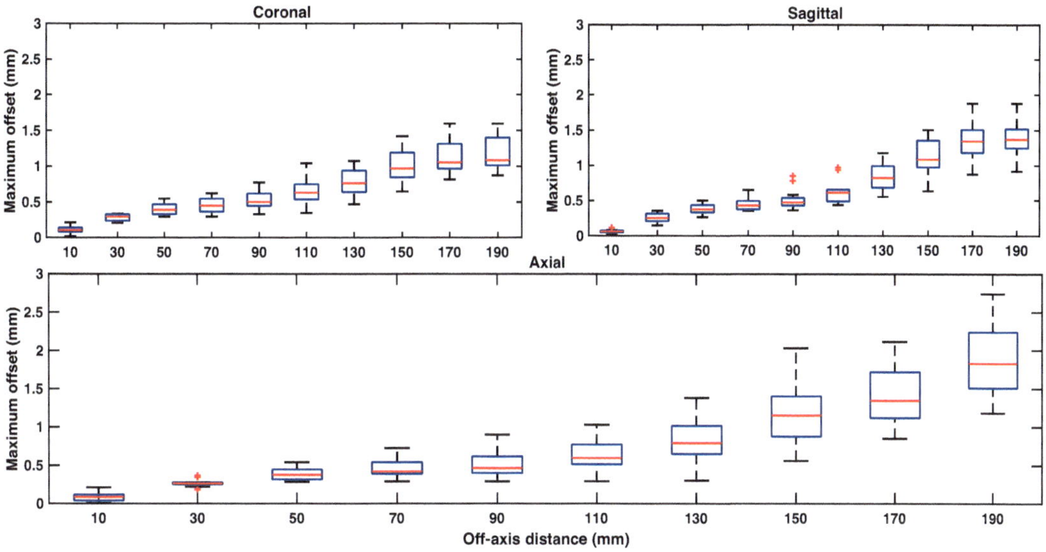


**Fig. 9.8** The spatial integrity results, in the form of maximum offset between the actual and measured locations of the cylinders, as a function of off-axis distance from the isocenter (adapted from Khan et al. [13])

fidelity results for a multi-institutional study. The spatial accuracy typically decreases as the distance from the isocenter increases due to the nonlinearity of the gradients. Therefore, the passing criteria for accurately locating a cylinder are 1 mm and 2 mm for the 10 cm and the 17.5 cm radii ROIs, respectively.

## Acceptance Testing for Linac

Before initiating the acceptance testing for the linac, it is well worth it for the clinicians to familiarize themselves with the software and the provided accessories. The vendor makes a recommendation for all of the equipment required for this testing protocol. However, it is the responsibility of clinical physicists to understand the reasoning behind the recommendation and consequences of using alternative equipment. It must also be ensured that all equipment is MR-safe and compatible. The vendor also provides specifications for each of the tests performed at this stage. Most of these criteria are based on the AAPM TG-142 [15]. Unlike conventional linacs where TPS beam modeling is performed by the clinical physicists using the acquired experimental data, the MRIdian vendor tunes the photon beam to match the TPS beam model. Therefore, the acceptance testing protocol requires a comparison between the measured and the TPS-planned absorbed dose distributions to ensure a good match between the two systems. The beam characteristics of the MRIdian MR-Linacs were previously found to be similar in a multi-institutional study [16].

### Termination of Radiation

The radiation dose to deliver in linacs is quantified using charge accumulated inside the sensitive volume of the monitor chamber defined as monitor units (MUs). Prior to reference and relative dosimetry, the reliability of the monitor chambers inside the MR-Linac head must be evaluated. This test verifies the ability of the secondary dose monitoring system to terminate the beam based on the planned MUs. Various planned MUs are considered, and the delivered MUs are recorded to calculate the deviations between the planned and delivered MUs. According to the vendor-provided specifications, the radiation beam must terminate when the planned MUs exceed by

$$\begin{cases} 1.10MU_{\text{planned}} + 1MU, & MU_{\text{planned}} < 10 \\ \min\left\{1.10MU_{\text{planned}}, 25MU\right\}, & MU_{\text{planned}} \geq 10 \end{cases}$$

### Laser Alignment

It is important to ensure that the lasers are correctly aligned and isocenter coincidence is maintained. For this test, a laser alignment device, such as the ViewRay daily QA phantom or a detector array, must be used to align the lasers accurately without any translational or rotational shifts along all three axes. It is of note that the laser isocenter is offset −155 cm along the IEC-Y axis due to the presence of the bore. The lateral lasers must be checked to ensure a tilt or shift of <1 mm across the couch width. Once the lasers are aligned, the laser coincidence can be visually evaluated using a thin semitransparent sheet of paper and must be within 1 mm.

### Gantry Isocenter Tolerance

Due to the mechanical tolerances on the gantry rotation, the isocenter is typically an ellipsoid or a sphere rather than a point. AAPM TG-142 recommends that the isocenter for a linac is within 1.0 mm considering radiation beams from all gantry angles. The size of the isocenter can be evaluated using a star shot plan. The vendor recommends using radiochromic films placed inside the ViewRay QA phantom provided by the manufacturer. The phantom is made of plastic and houses a cylindrical volume filled with distilled water. The cylindrical volume contains a central insert for an ion chamber as well as several peripheral inserts for additional chambers. The isocenter size in the IEC-XZ plane can be determined by irradiating a piece of circular film from various gantry angles embedded inside the cylindrical volume and calculating the radius of the smallest tangent circle. The vendor suggests utilizing a treatment plan consisting of five radiation beams with a field size of 1.0 × 6.3 $cm^2$ (IEC-X by IEC-Y) with gantry intervals of 72°. Since the photon beam generated by the MRIdian linac is FFF, the center of the beam in the IEC-Y direction can be determined by the peak of the beam profile. Therefore, the variation in the beam's center as a function of gantry angle can be measured by wrapping a sheet of film around the cylindrical volume. Throughout this measurement, the film pieces are marked based on the crosshairs and scribes on the QA phantom, which is aligned using the lasers. The vendor recommends the use of the RIT® film analysis software to calculate the size of the isocenter. The radius of the smallest tangent circle must be within 1.0 mm for the film embedded inside the phantom, and the deviation of the beam center along the IEC-Y direction for the film wrapped around the phantom must be within 1.0 mm of the center value.

### Coincidence of Radiation, Imaging, and Laser Isocenters

Although the isocenters for the imaging system and the linac coincide nominally, an offset between the two isocenters is possible but must be within 1 mm according to the AAPM TG-142 and TG-179 reports [15, 17]. Since patients or phantoms can be aligned on the couch using the lasers or the MRI images, it is crucial that the laser, the MRI scanner, and the linac isocenters are coincident within 1 mm of each other. For the radiation isocenter measurement described above using radiochromic films, the ViewRay QA phantom can be aligned on the couch using the lasers, and the irradiated film can be marked to indicate the phantom crosshairs. Therefore, the offset between the lasers and the radiation isocenters can be measured using the same film analysis software. The embedded film inside the water phantom can be used to determine the offsets in the IEC-X and IEC-Z directions, while the wrapped film can be used to calculate the offset in the IEC-Y direction. The offsets between the MRI and the laser isocenters can be measured by aligning the ViewRay QA phantom using the lasers and acquiring MRI images of the phantom. The acquired images can be co-registered with the planning images, and the offset between the lasers and the MRI isocenters can be calculated in all three directions. The radiation, laser, and MRI isocenters are considered coincident if the offset between the three is within 1 mm.

### Dose and Timer Linearity

The delivered absorbed dose must be proportional to the planned MUs. Any deviations from this linearity must be within 2% for prescriptions over 5 MUs and within 5% for prescriptions under 5 MUs. This relationship can be evaluated by irradiating a dosimeter with various MUs and ensuring a constant response when normalized by the delivered MUs. It is crucial to employ a dosimeter with a linear response over a large range of absorbed dose. The vendor recommends using an Exradin A28 MR safe ionization chamber inside the water-filled cylindrical volume of the ViewRay QA phantom. The constancy of the charge accumulated by the ion chamber normalized by MUs must be verified over a range of MUs.

## Reference Dosimetry

No official guidelines currently exist for reference dosimetry under the influence of strong magnetic fields. Reference class ion chambers remain absolute dosimeters for absorbed dose measurements in radiotherapy. The absolute absorbed dose to water, $D_w$, for photon beams can be measured by

$$D_w = M_Q N_{D,w} k_{Q,Q_o}^{B_o}$$

where $M_Q$ is the measured charge for a set number of MUs, $N_{D,w}$ is the chamber-specific calibration coefficient typically measured in $^{60}$Co beams, and $k_{Q,Q_o}^{B_o}$ corrects for the difference in the chamber response between the calibration beam quality and the MR-Linac beam quality in the presence of a magnetic field [8]. Before the absolute absorbed dose is measured, the clinical beam quality must be determined to calculate the $k_{Q,Q_o}^{B_o}$ correction factor for the utilized ion chamber. The reference dosimetry for the MR-Linac must be performed in a water phantom. MR-safe 1D water tanks (PTW, CNMC, etc.) and 3D tanks (PTW, LAP, etc.) are available for dosimetry. The vendor recommends using an Exradin A28 ion chamber for reference dosimetry. After aligning the chamber on the surface of the water tank visually by partially submerging it until a perfect circle is formed, the beam quality is determined by measuring the percent depth dose (PDD) at 10 cm depth with a source-to-surface distance (SSD) of 80 cm and a field size of 10.04 × 9.96 cm$^2$. A Mayneord F factor is applied to the beam quality measurement to determine the PDD at 10 cm depth with an SSD of 100 cm due to a smaller SAD of the MRIdian machine compared to conventional linacs. Using the corrected beam quality, the $k_{Q,Q_o}^{B_o}$ correction factor can be extracted. It is currently assumed that the impact of a 0.35T magnetic field on the response of the A28 chamber is minimal and can be considered negligible when the chamber is oriented along the direction of the magnetic field.

## Percent Depth Dose

According to the vendor specifications, the PDD at 10 cm and the depth of the maximum dose must be within 1.0% and 1.5 mm of the planned dose. These metrics are already measured when performing reference dosimetry to determine the beam quality and can be compared with the planning system. PDD for three field size is shown in Fig. 9.9. The vendor also recommends evaluating the stability of the monitor chamber. This can be achieved by comparing the output of the radiation beam in a water tank over a long period. The output must be within 1% of the initial value.

## Couch Transmission

For gantry angles between 140° and 220°, the photon beam traverses the couch before reaching the patient. Therefore, the attenuation through the couch must be quantified and compared with the planned value. The relative attenuation can be measured by irradiating an ion chamber, placed inside the ViewRay QA phantom, from various gantry angles using the same MUs. The charge readings as a function of gantry angle can then be normalized by the charge reading acquired with the unattenuated beam. Couch attenuation for the MRIdian MR-Linac can be up to 15% and must be considered during treatment planning. The vendor requires the measured and the planned couch attenuation values to be within 3% of each other. Using the same irradiation setup, the output constancy as a function of gantry angle can be measured. For gantry angles not obstructed by

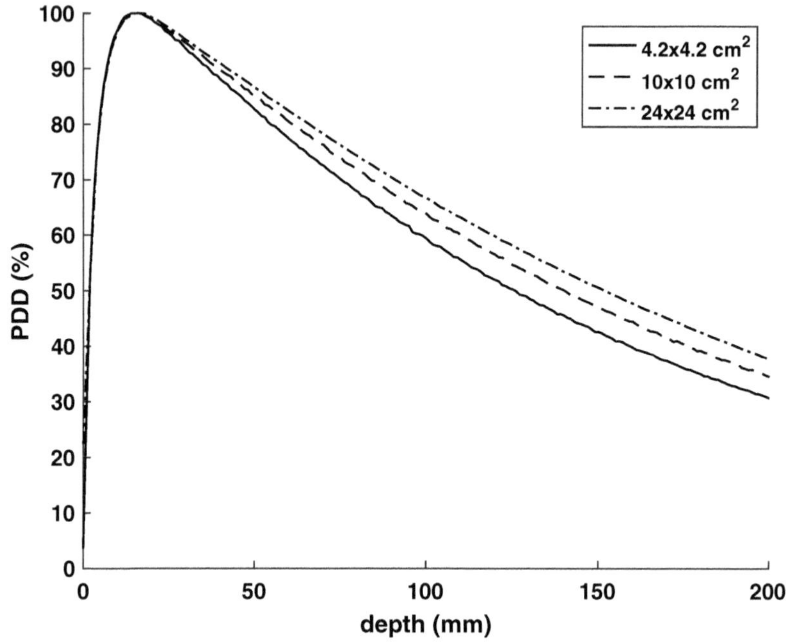


**Fig. 9.9** The percent depth dose (PDD) measured for the MRIdian MR-Linac for three field sizes

the couch attenuation, the output must be within 1% of each other in accordance with AAPM TG-142 [15].

## MLC Transmission and Leakage

The radiation leakage through and between the MLC leaves must be determined to ensure patient safety and minimal out-of-field dose. This task is best achieved when using a 2D dosimeter such as radiochromic films. The vendor recommends irradiating an EBT3/EDR2 film sheet with the upper and lower MLC stacks at −0.2 cm and 0.2 cm, respectively. A large number of MUs, such as 6000–10000 MUs, are delivered to the film with at least 2 cm buildup and 5 cm backscatter provided by solid water slabs. This irradiation is repeated at each cardinal gantry angle. Using the same geometrical setup, a known absorbed dose is delivered to another piece of film using a 10 × 10 $cm^2$ field size for reference. Following the leakage irradiations, the average leakage is calculated based on a large 30 × 30 $cm^2$ ROI on the film sheet, and maximum leakage is calculated based on a small 1x1 $cm^2$ ROI around the hottest spot on the film sheet. The International Electrotechnical Commission (IEC) requires the average and maximum leakages to be <0.375% and 1.0%, respectively. The dual stack design of the MRIdian system significantly reduces the radiation leakage compared to other linear accelerators [16].

## MLC Linearity and Positional Accuracy

The positional accuracy of the MLC leaves must also be evaluated and compared to the TPS. The maximum deviation between the TPS and the measured MLC positions must be within 1 mm, as specified in AAPM TG-142 [15]. The vendor recommends using radiochromic films to measure the positional accuracy since the MRIdian MR-Linac lacks an electronic portal imaging device (EPID) for megavoltage (MV) imaging. The positional accuracy of the MLC banks can be assessed by irradiating 1 × 24.1 $cm^2$ (IEC-X by IEC-Y) rectangular fields centered at −10 cm, −5 cm, 0 cm, 5 cm, and 10 cm IEC-X positions. A treatment plan is created to irradiate a sheet of film placed at 90 cm SAD with 5 cm depth and at least 5 cm backscatter using 40x40 $cm^2$ solid water slabs. This setup is repeated for 90° and 270° gantry angles. The film is exposed to the planned dose distribution consisting of five 1 × 24.1 $cm^2$ rectangular dose regions. The irradiation is repeated for the 0° gantry angle. A reference sheet of film is also typically irradiated using an identical setup but with a single 10 × 10 $cm^2$ field. Following the irradiations, the absorbed dose on the film sheets is scanned and analyzed using film analysis software such as RIT. Several line profiles, parallel to the IEC-X axis, are placed on the scanned film with adequate IEC-Y spacing. The deviation between the measured beam centers and the planned beam centers is calculated for each line profile, and maximum deviations are noted. For all the measurements, the deviations must be within 1 mm.

## Radiation Field Size, Shape, Symmetry, and Penumbra Accuracy

The dosimetric metrics related to lateral beam profiles must also be measured and compared to the TPS. The measured beam shape, symmetry, width, and penumbra must match the TPS data within the specified tolerances. A scanning ion chamber and a 3D water tank are desired for this task for fast acquisition of beam profiles. However, the vendor recommends the use of an ion chamber array, such as the IC profiler by SunNuclear, if a 3D tank is unavailable. For field sizes smaller than 4.16 × 4.16 $cm^2$, small volume detectors such as the EDGE diode, PTW microDiamond, or Exradin W2 can be used in a 1D water tank, and the profile can be measured by moving the couch laterally in small increments. Ideally, the beam profiles would be measured at multiple depths; however, the vendor recommends a single depth for each of the measurements to reduce the total acquisition time. The field size is recommended to be measured at 90 cm SAD with at least 1.4 cm of buildup material. Following the irradiations, the planned and measured field sizes are compared by calculating the FWHM of the profiles which must be within 2 mm of each other. The geometrical setup for the measured and planned dose distributions must be identical. For the penumbra evaluation, the profiles for the larger field sizes are normalized to the absorbed dose at the inflection point. The distance between the 20% and the 80% isodose line is defined as the penumbra. The measured and planned penumbras must be within 1 mm of each other for all of the beam profiles. Since the profile has a forward peaked shape, the conventional definition of flatness cannot be applied. Therefore, the vendor specifies that the shape of the measured and planned profiles must be compared. This can be achieved by comparing measured off-axis ratios at several off-axis distances with the TPS-calculated off-axis ratios. Any deviations between the two values must be within 2%. The beam symmetry can be determined by calculating the difference between the absorbed dose values at off-axis points equidistant from the central axis (CAX). The symmetry must be within 2% according to the vendor recommendations.

## IMRT Functionality

The IMRT functionality must be evaluated to ensure that the MR-Linac is able to deliver the planned absorbed dose within acceptable accuracy. This task requires measurement of the absorbed dose using a 2D or a 3D dosimeter. The vendor recommends the use of a diode array embedded inside a cylindrical phantom such as the ArcCHECK® or the Delta4® phantom. Therefore, this task can be performed following the installation and calibration of the diode-array phantom. The vendor recommends the use of two treatment plans for this task, but this number can be increased at the discretion of the physicists. The planned dose must be recalculated using the CT images of the diode-array phantom. The absolute point absorbed dose can be measured and compared with the planned point dose by inserting an ion chamber inside the phantom. It is of note that the volume averaging effect can lead to large deviations between the planned and measured absorbed dose. The phantom can be aligned on the couch using a conformal plan consisting of beams at 0° and 90°/270° gantry angles with a 10 × 10 $cm^2$ field size. The couch translational shifts that maximize the gamma pass rate with a 1%/1 mm criterion can then be applied. Following the setup, the QA plans must be delivered, and the measured dose is obtained. During the treatment delivery, the MRIdian MR-Linac is capable of recording the delivery parameters in a log file. These parameters can be imported into the TPS, and the delivered absorbed dose can be predicted using the TPS beam model dose calculation. Using a gamma criterion of 3%/3 mm, the gamma pass rate must be ≥95% when comparing the measured and planned absorbed dose and when comparing the measured and log file predicted absorbed dose. The planned and measured absorbed dose using the ion chamber must be within 3%.

## MR Tracking Latency

Tracking moving tumors and accurate gating requires a sufficiently low latency so that the radiation beam is turned off instantly following the target volume moving outside the delineated boundaries. The MRIdian system tracks the tumor motion using cine MRI in the sagittal plane and triggers the radiation beam off when the tracked tumor margin is outside the reference margin. The vendor requires that the gating latency of the MR-Linac is within 0.35 s over 20 gating cycles. The gating latency is affected by the MRI acquisition speed, image reconstruction and processing time, tracking algorithm processing time, and radiation beam signal latency. In case a dosimeter is employed to measure the latency, the delay time between the radiation beam turning off and the dosimeter readout signal must be considered as well. Alternatively, the radiation beam trigger time can be extracted from the beam control system directly. The vendor recommends the use of a CIRS motion phantom to measure the system latency. Other phantoms, such as the MRI$^{4D}$ QUASAR motion phantom, are also commercially available to perform this test. The CIRS motion phantom consists of a cylindrical phantom with ion chamber inserts and small volumes embedded inside the large volume. The smaller volume inside the large cylindrical volume appears as a circle in the sagittal plane and is a surrogate for tumors. The tumor surrogate is driven by a long rod connected to a motion stage, which is kept far away from the isocenter to minimize the impact of the external magnetic field on the functionality of the stage. Tumor motion can be replicated by using an oscillating waveform for the motion stage. The amplitude, phase shift, and starting position of the surrogate can be altered. The CIRS system utilizes optical sensors for triggering when the surrogate volume is outside of the delineated margins. This is achieved by using two optical sensors placed in the superior-inferior direction with a set spacing between them. At the home position, the distal edge of the rod is equidistant from the superior and inferior sensors. When the edge of the rod and the tumor surrogate volume moves outside of the bounds in either direction, the sensors are triggered, and an electrical signal is transmitted to the oscilloscope. The CIRS phantom is imaged using volumetric MRI, and a contour is drawn around the surrogate volume in the sagittal plane. Before the latency measurements are acquired, a preview cine scan is typically acquired to evaluate the accuracy of the tracking algorithm. Following the preview scan, real-time imaging is performed, and the tumor surrogate is driven using the desired waveform. A multichannel oscilloscope is used to input the trigger signal from the gating phantom as well as the beam-off signal from the treatment control cabinet or the dosimeter. The time difference between the gating signal indicating the distal edge of the rod moving out of margins and the beam-off trigger is noted as the latency value.

## Dosimetry Consistency with Gating

In addition to measuring gating latency, it is important that the dose delivered to a moving target using gating remains similar to a stationary target. This can be evaluated by irradiating an ion chamber embedded inside the gating phantom such as the CIRS phantom. A conformal plan is created to deliver a set absorbed dose to the cavity of the ion chamber using a contour drawn around the cavity. Two irradiations are performed. For the first irradiation, the ion chamber is driven resembling tumor motion using an oscillation waveform. The charge accumulated over the irradiation is measured, and gating is enabled by turning off the beam when the ion chamber cavity moves outside the margins. The second irradiation is performed with the motion turned off. The difference between the two charge readings must be within 2% of each other to satisfy the TG-142 criteria.

## Commissioning of the MRIdian MR-Linac

The commissioning of the MRIdian MR-Linac is performed following the acceptance testing and evaluates the linac using tests not performed during acceptance testing. At this stage, the baseline data for ongoing QA are also acquired using the equipment employed by the clinic. For the MRI component of the system, the photon beam attenuation through the receiver coils must be measured. Output factors (OFs) must be measured and compared with the TPS during commissioning. The vendor recommends using either the EDGE diode, the PTW microDiamond, or the Exradin W2 detectors to measure output factors for small fields. The vendor also provides the field- and detector-specific correction factors calculated using MC simulations. Although the gating accuracy and latency are evaluated during the acceptance testing stage, a more extensive evaluation of the gating tracking algorithm can be performed during commissioning using a range of motion traces. Any tilts in the MLC carriages must also be corrected using the beam profiles acquired with the IC profiler. The profile shape and symmetry metrics can be used to assess the alignment of the MLC banks. The mechanical checks must also be performed during commissioning. The digital indicators must be verified using independent measurements. Translational and rotational shifts in the couch can be evaluated using handheld tools.

Extensive validation of the TPS must be performed during commissioning. The guidelines outlined in the AAPM MPPG 5a and TG-53 reports can be used to evaluate the TPS beam model [18, 19]. Dose calculation in heterogenous media should especially be evaluated due to the presence of the ERE. A variety of complex MLC shapes can be created and assessed using 2D profiles and point dose measurements. In addition to the two IMRT QA plans recommended by the vendor for measurement-based QA evaluation, several additional IMRT QA plans can be created and measured using the diode-array phantom depending on the commonly treated sites. Further, independent end-to-end testing using phantoms can be performed to ensure the accuracy of the TPS beam model.

### Patient-Specific QA

An inverse treatment planning technique is used for IMRT deliveries that optimizes beam parameters based on user-defined dose objectives. The complexity of these treatment plans warrants pretreatment QA to assess the plan deliverability. A comparison is performed between the patient-specific planned and machine-delivered dose distributions. The delivered dose distribution can either be measured or calculated using the machine log files. Gamma analysis is used to compare the dose distributions considering both dose differences and distance to agreement. Similar to conventional IMRT QA, several MR-safe diode-array phantoms are commercially available, such as ArcCHECK and Delta4, to acquire the absorbed dose measurements. These phantoms must be installed and commissioned according to vendor guidelines before use. For measurement-based QA, a QA plan is created for each treatment plan before the delivery by recalculating the planned dose distribution using the CT images of the homogeneous phantom. Absorbed dose is sampled using an array of diodes, and the planned and measured dose distribution is compared. A gamma pass rate is calculated to quantify the treatment plan quality, and a threshold value is used to approve the plan. The gamma criteria and the pass rate threshold vary between clinics and treatment techniques. For example, most clinics consider a 95% pass rate with a 3%/3 mm criterion to be adequate. The gamma criterion for SBRT is typically reduced to 3%/2 mm.

MR-Linacs are often used for adaptive radiotherapy (ART) during which the treatment plan gets modified daily based on the changes in the treatment parameters. During the adaptation process, the patient is typically on the table and measurement-based QA is not feasible. In such cases, calculation-based QA can be performed by extracting the delivery parameters from the log file and using an independent dose calculation system. The MRIdian system offers an independent dose calculation software called Zeus that can perform adaptive QA (AQA) [20]. Zeus uses a phase space file from the TPS beam model above the MLC plane and uses an independent fast Monte Carlo-based platform to calculate absorbed dose in patient geometry. The effect of the external magnetic field is considered in the dose calculation. Since the MRIdian system lacks daily CT data for dose calculation, the daily volumetric MRI is co-registered with the CT simulation data, and a deformation matrix is applied to the CT data. The calculated delivered dose distribution is compared with the planned dose using a 3D gamma analysis. A QA report is created indicating the gamma pass rates and other calculation metrics.

## Daily, Monthly, and Annual QA

Following the initial characterization of the MR-Linac, it is important to evaluate the machine at regular intervals to ensure acceptable performance. Table 9.4 shows the recommended daily QA tasks for the MRIdian MR-Linac based on the AAPM TG-142 report [15]. All safety features must be evaluated and functional to ensure the utmost patient safety. The output and dose rate of the photon beam can be evaluated using an ion chamber inserted into the ViewRay DQA phantom or by using other commercially available solutions such as the Sun Nuclear Daily QA 3 phantom. The ViewRay DQA phantom was described earlier in this chapter. The Exradin

**Table 9.4** Suggested daily QA tasks for the MRIdian MR-Linac

| Procedure | Tolerance |
|---|---|
| Dosimetry | |
| X-ray output constancy | 3% |
| Dose-rate constancy | 2% |
| Mechanical and imaging | |
| Laser and imaging coincidence | 1 mm |
| Radiation beam and imaging coincidence | 1 mm |
| Patient positional accuracy | 1 mm |
| Receiver coil functionality | Functional |
| Safety | Functional |
| Door interlocks | Functional |
| Radiation area monitor | Functional |
| Beam-off functionality | Functional |
| In-room camera | Functional |
| In-room audio | Functional |
| Panic bulb | Functional |

A1SL MR-safe ion chamber fits well inside the chamber insert of the phantom with its centroid aligned with the radiation isocenter. A 10.04 × 9.96 cm$^2$ field size is typically used to measure the output, and the MU to time ratio is used to determine the dose rate. The DQA phantom is aligned using the lasers and the scribes on the phantom. By acquiring an MRI image of the phantom and co-registering the acquired volume with the reference volume, the positional shifts between the lasers and the MRI can be calculated. Following the localization of the ion chamber using the MRI images, the radiation isocenter and MRI isocenter coincidence can be evaluated. This is achieved by leveraging the cylindrical shape of the ion chamber cavity with a diameter of 4 mm and a length of 4.4 mm. By irradiating the ion chamber with a small field size of 0.40 × 0.83 cm$^2$ at each cardinal gantry angle, a partial volume irradiation effect is induced inside the cavity, which is highly sensitive to any positional offsets in the radiation beam isocenter [21]. By introducing known shifts to the ion chamber, the relationship between the charge readings and the positional offset can be determined a priori. Daily monitoring of the charge reading with the 0.40 × 0.83 cm$^2$ field size can yield an estimate of the positional offset between the radiation isocenter and the MRI isocenter. The patient's positional and repositional accuracy can be determined by imaging the phantom after introducing a known shift and aligning the phantom based on the peripheral physical landmarks inside the phantom. The shifts between the acquired images and the planned images determine the patient positional accuracy as well as the spatial fidelity of the MR images.

Table 9.5 shows the recommended monthly QA tasks for the MRIdian MR-Linac. In lieu of using the ViewRay DQA phantom for dosimetry tasks, a 1D water tank or solid water slabs are recommended for these measurements. Using the baseline measurements acquired following the acceptance testing and commissioning of the MR-Linac, the constancy of the photon beam must be verified. Output and dose rate can be verified by acquiring absorbed dose measurements at a single depth. The beam energy can be quantified using either the PDD or the TPR metric by acquiring the ratio of dose at two different depths. If a scanning water tank is unavailable, the IC profiler or other detector arrays can be employed to acquire lateral beam profiles. The CIRS or QUASAR gating phantoms can be utilized to measure the gating latency. TPS QA can be performed by creating and comparing absorbed dose distributions for mock IMRT plans from AAPM TG-119 [22].

**Table 9.5** Suggested monthly QA tasks for the MRIdian MR-Linac

| Procedure | Tolerance |
|---|---|
| Dosimetry | |
| X-ray output constancy | 2% |
| Dose-rate constancy | 2% |
| Beam profile constancy | 1% |
| Beam energy constancy | 1% |
| Gating latency | 100 ms of expected |
| TPS QA | Constancy |
| Mechanical and safety | |
| Emergency stop | Functional |
| Treatment couch position indicators | 2 mm/1° |
| Localizing lasers | 1 mm |
| MLC positional accuracy | 1 mm |
| MRI | |
| $B_0$ field homogeneity | Constancy |
| ACR image quality | ACR specifications (see Table 9.3) |
| SNR and uniformity | NEMA specifications |
| Spatial integrity | 1 mm (10 cm radius), 2 mm (17.5 cm radius) |

It is essential that the lasers are accurate and coincident with the radiation beam and the imaging system. The offsets between these systems can be measured using the procedure described in the previous sections. Phantoms with ruler markings can be used to evaluate the couch position indicators. The MLC positional accuracy is typically assessed using a picket fence test. However, due to a lack of an imaging panel, this test can be performed using radiochromic films, which is an arduous process with a passive readout. Therefore, the high sensitivity of the ion chamber response to positional offsets when using a partial volume irradiation technique can be leveraged to estimate the positional accuracy of the MLC leaves. This test involves irradiating half of each of the ion chambers in the array using a half beam block (HBB) technique and establishing a relationship between the detector response and known induced MLC positional shifts for all banks [23]. During monthly QA, the measured detector response can then be used to calculate the expected MLC positional error. This test should be repeated at all four cardinal gantry angles to evaluate the effect of differences in the gravitational force on the MLC banks. The MRI tasks mimic the acceptance testing protocol described in the earlier sections.

In addition to the monthly and daily QA tasks, the annual QA procedures include reference dosimetry in a water tank where an absorbed dose to water calibration must be maintained within 1%. For the TG-51 output calibration, a water tank is recommended instead of solid water slabs. The beam profiles and output factors can be measured using the procedures described earlier. The beam profile size, flatness, and symmetry constancy must be within 1% as well. Field output factors should be measured and a constancy of within 1% should be ensured. The average and maximum MLC leakage must be measured for the cardinal gantry angles to be within 0.375% and 1%, respectively. The laser, MRI, and radiation isocenter coincidence must be within 1 mm.

## IT and Other Considerations

The hospital and/or department IT team is required to install relevant QA software for ViewRay MRIdian as well as to set up the IP address for the system. The QA software that needs to be installed can be used to monitor the system performance and ensure main components are functioning properly. Setting the IP address refers to the process of assigning a unique network address to the ViewRay MRIdian subsystems so that it can communicate with other devices on the network. For the QA software to be installed and the IP address to be set, the hospital's IT support team will need to be familiar with the ViewRay MRIdian system, as well as with the installation and configuration of medical imaging equipment. In order to ensure that the software is installed and configured correctly, they may need to work with the vendor of the QA software and the ViewRay MRIdian system.

A database management system is used to store patient data. The data is periodically backed up to a Synology Drive with a capacity of 10 terabytes, specifically Racket. Data is then transferred to the hospital's server for further storage and management. A multiple backup system is a best practice for data protection and disaster recovery. Furthermore, it is imperative that the data is stored and transmitted securely in order to prevent unauthorized access or theft. By utilizing encryption, secure networks, and access controls, this can be accomplished. The hospital should also have policies and procedures in place to ensure compliance with regulations such as HIPAA regarding the management and protection of patient data. Data backup includes patient data (Images, Dose, Reoptimized, predicated, and Cine, etc.), machine data: configuration file, software version, Core and Database computer.

Using the ViewRay treatment delivery system (TDS), patient data can be accessed and restored using PAR (plan analysis record) files only. In addition to the fact that the PAR files are proprietary and can only be accessed by authorized personnel using the ViewRay TPS, this format is specific to the ViewRay MRIdian system. In order to prevent data loss or corruption, it is imperative that PAR files are stored and managed properly. The PAR files should be regularly backed up and validated, as well as protected by strict access controls to prevent unauthorized access. Performing of backups that usually take about 1 hour and a full backup in Synology Drive at the end of the day (fixed time) can help ensure that data is protected and can be quickly restored in the event of a data loss or corruption. Cine data as well as other types of patient data need to be backed up. Log files provide a detailed record of all treatment and imaging activities, including timestamps and dose information. A software team can identify areas for improvement by analyzing log files and creating patches to resolve any issues that may arise. Performing this type of analysis will help ensure that the healthcare system is operating optimally and that patients are receiving the best possible care.

## Summary

Real-time MRI-guided radiation therapy is made possible by the groundbreaking ViewRay MRIdian system, which combines magnetic resonance imaging (MRI) and linear accelerators (linac). The history and underlying physics of MR-Linac technology, as well as the practicalities and financial aspects of purchasing and putting into use a ViewRay MRIdian system, have all been covered in this chapter. The impact of the magnetic field on the dose distributions and dosimetric data was discussed in the context of the electron return effect (ERE). The system design of the 0.35 T MRIdian MR-Linac was described in detail including the ancillary components used for treatment delivery and imaging. Concerns related to MRI safety were accentuated including outlining a zoning system specific to the MRIdian system. Relevant training topics for all personnel involved in the MR-Linac treatments were discussed including workflow maps and teams involved at each step. Vendor-provided acceptance testing protocols for both the linac and the MRI portions of the machine were compiled. Discussion of additional commissioning tasks and ongoing QA recommendations based on existing protocols were included. The advantages of the ViewRay MRIdian system are obvious, despite the difficulties involved in putting such a cutting-edge technology into practice: increased precision and accuracy in the delivery of radiation therapy, improved imaging capabilities for tumor detection and localization, and reduced radiation exposure to healthy tissue.

## References


1. Lagendijk JJW, Raaymakers BW, van Vulpen M. The magnetic resonance imaging–linac system. Semin Radiat Oncol. 2014;24:207–9.
2. Raaymakers BW, Jürgenliemk-Schulz IM, Bol GH, et al. First patients treated with a 1.5 T MRI-Linac: clinical proof of concept of a high-precision, high-field MRI guided radiotherapy treatment. Phys Med Biol. 2017;62:L41–50.
3. Uilkema S, van der Heide U, Sonke J-J, et al. A 1.5 T transverse magnetic field in radiotherapy of rectal cancer: impact on the dose distribution. Med Phys. 2015;42:7182–9.
4. Menten MJ, Fast MF, Nill S, et al. Lung stereotactic body radiotherapy with an MR-Linac – quantifying the impact of the magnetic field and real-time tumor tracking. Radiother Oncol. 2016;119:461–6.
5. Huang C, Yang B, Lam WW, et al. Magnetic field induced dose effects in radiation therapy using MR-linacs. Med Phys. 2023;50(6):3623.
6. Woodings SJ, Bluemink JJ, de Vries JHW, et al. Beam characterisation of the 1.5 T MRI-linac. Phys Med Biol. 2018;63:085015.
7. Khan AU, Simiele EA, Lotey R, et al. Development and evaluation of a GEANT4-based Monte Carlo Model of a 0.35 T MR-guided radiation therapy (MRgRT) linear accelerator. Med Phys. 2021;48:1967–82.
8. O'Brien DJ, Roberts DA, Ibbott GS, et al. Reference dosimetry in magnetic fields: formalism and ionization chamber correction factors. Med Phys. 2016;43:4915–27.
9. O'Brien DJ, Dolan J, Pencea S, et al. Relative dosimetry with an MR-Linac: response of ion chambers, diamond, and diode detectors for off-axis, depth dose, and output factor measurements. Med Phys. 2018;45:884–97.
10. Mutic S, Dempsey JF. The ViewRay system: magnetic resonance–guided and controlled radiotherapy. Semin Radiat Oncol. 2014;24:196–9.
11. Klüter S. Technical design and concept of a 0.35 T MR-Linac. Clin Transl Radiat Oncol. 2019;18:98–101.
12. Hu Q, Yu VY, Yang Y, et al. Practical safety considerations for integration of magnetic resonance imaging in radiation therapy. Pract Radiat Oncol. 2020;10:443–53.
13. Khan AU, DeWerd LA, Das IJ, et al. A multi-institutional comparison of acceptance testing data for a 0.35 T MRI scanner. Phys Med Biol. 2022;67:235005.
14. Gao Y, Lotey R, Low DA, et al. Technical note: validation of an automatic ACR phantom quality assurance tool for an MR-guided radiotherapy system. Med Phys. 2021;48:1540–5.
15. Klein EE, Hanley J, Bayouth J, et al. Task Group 142 report: quality assurance of medical acceleratorsa. Med Phys. 2009;36:4197–212.
16. Khan AU, Lotey R, DeWerd LA, et al. A multi-institutional comparison of dosimetric data for a 0.35 T MR-Linac. Phys Med Biol. 2022;67:05NT01.
17. Bissonnette J-P, Balter PA, Dong L, et al. Quality assurance for image-guided radiation therapy utilizing CT-based technologies: a report of the AAPM TG-179. Med Phys. 2012;39:1946–63.
18. Fraass B, Doppke K, Hunt M, et al. American Association of Physicists in Medicine Radiation Therapy Committee Task Group 53: quality assurance for clinical radiotherapy treatment planning. Med Phys. 1998;25:1773–829.
19. Jacqmin DJ, Bredfeldt JS, Frigo SP, et al. Implementation of the validation testing in MPPG 5.a "Commissioning and QA of treatment planning dose calculations—megavoltage photon and electron beams". J Appl Clin Med Phys. 2017;18:115–27.
20. Khan AU, Simiele EA, Lotey R, et al. An independent Monte Carlo–based IMRT QA tool for a 0.35 T MRI-guided linear accelerator. J Appl Clin Med Phys. 2023;24:e13820.
21. Mittauer KE, Dunkerley DAPP, Yadav P, et al. Characterization and longitudinal assessment of daily quality assurance for an MR-guided radiotherapy (MRgRT) linac. J Appl Clin Med Phys. 2019;20:27–36.
22. Ezzell GA, Burmeister JW, Dogan N, et al. IMRT commissioning: multiple institution planning and dosimetry comparisons, a report from AAPM Task Group 119. Med Phys. 2009;36:5359–73.
23. Mittauer KE, Yadav P, Paliwal B, et al. Characterization of positional accuracy of a double-focused and double-stack multileaf collimator on an MR-guided radiotherapy (MRgRT) Linac using an IC-profiler array. Med Phys. 2020;47:317–30.

# Elekta Unity System

# 10

Jochem W. H. Wolthaus, Eenas A. Omari, Xinfeng Chen, and Bram van Asselen

## Introduction

Accurate target delineation which provides both *actual* anatomical information (superb soft-tissue contrasts) and functional imaging information (biological function and physiology) just prior to and during delivery is essential for radiation treatment. Including the MRI during radiation delivery allows for reduction of the treatment margins and volumes and opens new possibilities for treatment delivery techniques such as hypofractionation. Furthermore, with fast image acquisition, intrafraction motion can be monitored allowing for real-time adaptive radiotherapy by correcting for short-term geometrical changes, e.g., due to breathing or drift.

The MR-Linac, combining a high voltage linear accelerator with onboard 1.5T MR imaging system, was originally invented at the University Medical Center Utrecht [1, 2] and was developed in collaboration with Elekta (linear accelerator) and Philips (MR unit) [3]. The first patient treatment with MR-Linac was in 2017 [4], and it proved the geometrical and dosimetric accuracy. The Elekta Unity MR-Linac (Fig. 10.1) was released in 2018 worldwide for commercial use [5].

For the linear accelerator (linac) of Unity MR-Linac, a standing wave guide design is used to produce a 7 MV FFF beam. The linac is mounted on a ring rotating around a closed bore MRI scanner with a maximum continuous rotation speed of 6 rpm. The sensitive accelerator parts, such as the waveguide and the magnetron, are placed in a shielded torus with negligible magnetic field strength. This area is created using active shielding. The magnetic field is perpendicular to the photon beam and points out of the bore along the -y direction (IEC61217). The source-to-isocenter distance in this configuration is 143.5 cm. Gantry bore diameter is 70 cm. Since a closed bore design is used, the beam is attenu-

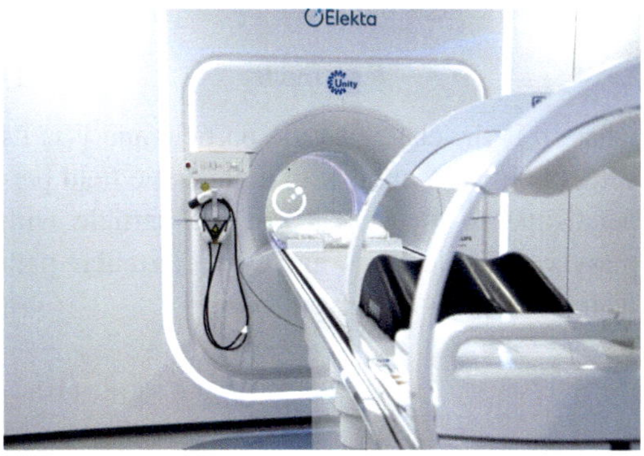


**Fig. 10.1** The Elekta Unity 1.5T MR-Linac system

J. W. H. Wolthaus (✉) · B. van Asselen
Department of Radiotherapy, University Medical Center Utrecht, Utrecht, The Netherlands
e-mail: J.Wolthaus@umcutrecht.nl; B.vanAsselen@umcutrecht.nl

E. A. Omari · X. Chen
Department of Radiation Oncology, Medical College of Wisconsin, Milwaukee, WI, USA
e-mail: Eomari@mcw.edu; xinchen@mcw.edu



I. J. Das et al. (eds.), *A Practical Guide to MR-Linac*, https://doi.org/10.1007/978-3-031-48165-9_10

ated by the cryostat. Dose rate at isocenter is approximately 425 MU per minute, calibrated at the dose maximum (1.3 cm depth). The MRI components have been optimized to create a window with a uniform high transmission but reduce the maximum field size to 22 cm in longitudinal direction. The MLC is based on the design of the Agility MLC but with a fixed collimator angle of 90 degrees allowing for leaf travel to be only in the cranio-caudal direction. Field sizes ranging from 0.7 × 0.7 $cm^2$ to 57 × 22 $cm^2$ can be created by the multi leaf collimator consisting of 80 leaf pairs with a projection of 7.2 mm at the isocenter. Currently, only step-and-shoot IMRT is available on the MR-Linac system, but VMAT delivery is being developed and will be available in future system control software releases.

## Basic Physics in a Transverse Magnetic Field

In MRI guided radiotherapy, patients are treated in the presence of a magnetic field which affects the dose deposition in the patient. The most common photon-electron interactions in radiotherapy are photoelectric effect, Compton scatter, and pair production. Meanwhile, the primary high-energy photons produced by the accelerator are not influenced by the magnetic field. On the flip side, the high-energy electrons produced inside the patient will experience the Lorenz force which changes their trajectories affecting the dose deposition. The Lorenz force ($\vec{F}_L$) for a present magnetic field $\vec{B}$ is given by

$$\vec{F}_L = q\vec{v}x\vec{B} \tag{10.1}$$

where $q$ is the charge of the particle and $\vec{v}$ is its velocity. In a vacuum, with a magnetic field perpendicular to the velocity of the particle with mass $m$, the electrons move in a circular path with gyro-radius ($r_g$) given by

$$r_g = \frac{mv}{qB} \tag{10.2}$$

Positrons are also produced in pair production interactions; however, we will only consider electrons in this chapter. In a medium, the electrons lose energy while interacting with the medium as can be described by the stopping power. The slowing down process of the electrons results in a lower Lorenz force interacting on the electron. As a result, the electron moves over a spiraling trajectory with a gyro-radius which decreases along its path [6]. Therefore, the change of local dose deposit results in a difference in absorbed dose distribution relative to the absence of a magnetic field.

## Beam Characteristics

Several aspects of the beam change in the presence of a magnetic field [7]. The surface dose changes due to the change in contamination electrons and different trajectories of the electrons. For the entrance of beam, the electrons created above the surface are captured by the magnetic field moving out of the beam in a spiraling path reducing the surface dose [8, 9]. However, the secondary electrons at the surface are swept sideways and backward due to the Lorenz force increasing the local dose [10]. The net effect gives a similar surface dose as without magnetic field.

The curved electron paths cause the dose to be deposited closer to the photon-electron interaction point and result in a shallower dose buildup [11, 12] and a shallow depth of dose maximum. Furthermore, this implies that the dose is closer to the kinetic energy released in matter (KERMA) and therefore slightly lower than when the magnetic field is absent [13]. The penumbra region of beam profiles is changed since the electrons are swept out of one side of the field, and they move toward the field center [12, 14, 15]. This results in asymmetrical lateral profiles as shown in Fig. 10.2.

On the exit side of the beam, electrons leaving the patient are either returned to the patient or start spiraling around the magnetic field lines resulting in a stream of electrons. The phenomenon where the electrons return to the surface causing an increase in the exit dose is known as

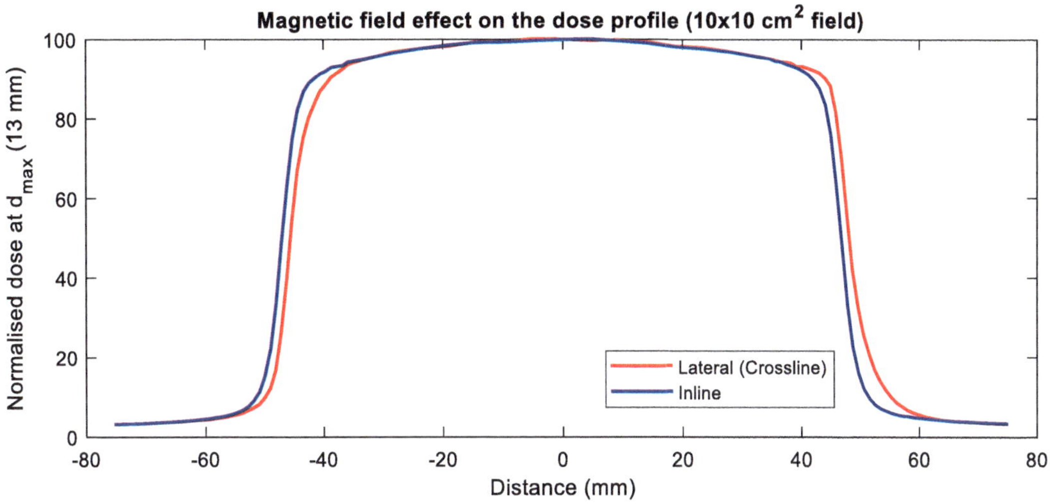


**Fig. 10.2** Magnetic field effect on the inline (blue) and crossline/lateral (red) dose profiles of a 10 × 10 $cm^2$ beam

the electron return effect (ERE) [10, 11, 16]. However, some electrons "escape" and stream away from the beam and deposit dose elsewhere outside the patient [17, 18]; this effect is known as the electron streaming effect (ESE). These differences in dose deposition are taken into account in the treatment planning system with proper modeling by the Monte Carlo algorithm including the magnetic field.

## Logistics and Economics of Acquiring an MR-Linac

### Economic Analysis and Cost

It is difficult to make a realistic estimate of the cost and effort for implementing a real-time MRgRT program in a department as it depends on various circumstances, for example, the available facilities, staff, hospital size, and commitment to research [19–22]. Furthermore, revenue depends on the patient treatment reimbursement that might change as cost per fraction increases with the number of treatment fractions. A 5 fraction MR-Linac treatment is equivalent to a 20 to 30 fraction scheme [19]. Patient selection and fractionation schemes focus on patients that may benefit from high doses with high geometrical accuracy and deformable and/or moving targets [5]. Depending on the chosen workflow, the average time of treatment can be variable between 20 and 60 min per patient [23, 24].

The rapid technological advances give access to real-time MRgRT to all kinds of radiotherapy centers. However, note that introducing MRgRT into the department also involves adaptation to building constructions, new treatment planning software, quality assurance (QA) systems, staffing, and training and may include an introduction of an MRI simulator. This may lead to additional cost depending on existing resources within the department and the degree of desired efficiency and reliability of the clinical workflow. For example, a dedicated MRI simulator for treatment workflow preparation increases the patient throughput on the treatment machine. Furthermore, the MRI simulator can be used for tumor staging and localization regardless whether the patients will be treated on the MR-Linac or a conventional system [25]. In addition, the MRI scans can be converted to pseudo-CT images allowing for MRI only treatment planning [26].

Reliability of the clinical workflow is increased if there are backup MRI systems at the radiology department that can be used in case of system failures. However, these "fallback" MR systems have to be prepared to allow for radiotherapy treatment simulation, e.g., flat tabletop, coils, and position indexing. Another fallback scenario in the case of MR simulator downtime is to use the MR-Linac as a backup simulator.

When an MRgRT system enters a radiotherapy department, staff should be trained or recruited to support the usage of MR imaging. Training involves safety aspects, physics, system operation, workflows, treatment planning system (TPS), and connectivity. Well-trained staff is also important for proper patient referral for groups that may benefit the most from the advanced online MRgRT program. Training and staff responsibilities will be discussed in section "Staffing and Training". The reader can also refer to Chaps. 2 and 4 of this book on this topic.

Depending on the level of vendor support and service agreements, physics and engineering support is crucial. MRgRT requires additional software or modules for existing software to support MRgRT, for example, to account for the magnetic field, additional upgrades are required to the treatment planning software. In addition, licensing will be required to support extra workstations for the TPS or record-and-verify (R&V) systems to improve workflow efficiency. Although operating the MR-Linac in an isolated environment is possible, integration of the MRgRT system into the department's infrastructure is advised.

Architecture and construction should also be considered when installing an MR-Linac system. This is discussed in more detail in the next subsection. Finally, existing hardware and phantoms for QA purposes are generally not compatible with MR or MRgRT systems due to the magnetic field and limited bore diameter, which requires the purchase of new equipment. Examples of new MRI compatible equipment are (1) water scanning phantom, waterproof ion chambers, and phantoms for reference dosimetry; (2) a 2D/3D patient delivery QA phantom for acceptance testing, commissioning, calibration, and verification of the dose delivery; (3) secondary dose calculation software for daily treatment plan verification; and (4) MRI imaging QA phantoms and analysis software.

The initial cost of establishing an MRgRT program is significantly higher than any other new introduced technique or innovation in X-ray radiotherapy. However, expanding the number of MR scanners or MRgRT systems makes use of knowledge from existing staff at the department. Synergy and efficiency can be achieved by implementing better patient workflows, treatment scheduling, and QA and maintenance scheduling. It might also be advantageous to reconsider the vendor support agreements for maintenance.

## Architectural and Structural Planning

A generic structural outline for a Unity MRgRT program consists of the treatment room, machine room, maze, control room, technical room, and patient dressing and waiting rooms, as shown in Fig. 10.3. The treatment room contains the actual MR-Linac, which is isolated from external RF noise by a Faraday cage. The Faraday cage is custom-built close to the wall, floor, and ceiling of the vault to optimize the functional inner space. The machine room is located within the treatment room within the space of the concrete vault and is accessed from via a separate RF door.

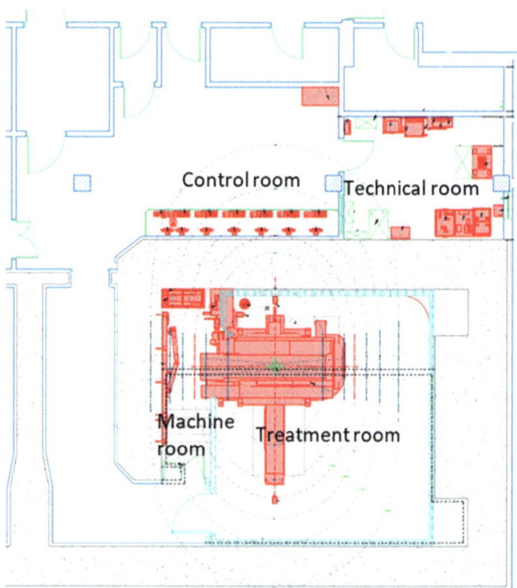


**Fig. 10.3** The layout of a Unity MR-Linac vault at the UMC Utrecht, which includes the treatment room, machine room, maze, control room, technical room, and patient dressing/waiting rooms

This room is utilized for machine maintenance. The maze is the corridor between treatment room and control room. The maze may have reduced number of corners by taking into consideration the maximum energy, shielding, and leakage. All user interface workstations for imaging, delivery and workflow controls, and patient CCTV monitors are located in the control room. The control room should be treated as MR-Zone 3. The technical room contains all computers, servers, switches, routers, and UPS systems used for the MR-Linac system. Finally, waiting rooms and patient dressing rooms are not explicitly part of the official layout, but it is recommended to consider them for workflow efficiency.

The floor plan of a Unity MR-Linac, as defined by Elekta, is such that it can be retrofitted within an existing conventional linac vault. Costs for the rebuilding and construction of an existing vault, prepared to install a Unity MR-Linac, are slightly higher. This is due to reconstruction since the foundation for a conventional linac has to be removed and an additional pit (hole) has to be drilled to fit the large gantry ring and add the Faraday cage. Furthermore, it might be required to reconstruct the maze to allow the delivery of the large components like MRI and gantry ring of the system. In preparation of the installation, environmental conditions of the facility regarding interference from electromagnetic radiation, vibrations, magnetic fringe fields, or radiation leakage should be considered. A survey by the vendor must be performed prior to installation.

The dimensions of the existing vault might be less than preferable. If the available space to store equipment such as MRI crash cart, patient masks, phantoms, or MRI coils is limited, practical procedures and workflows should be adapted. In case of a newly constructed facility, it is advised to consider extra space for these items.

A helium safety vent pipe, quench pipe, is required to exhaust helium gas. The pipe and venting system must be capable of exhausting the large amount of helium gas that may result from a quench of the MRI when the magnetic field is terminated quickly. Although these events are very rare, a quench can occur spontaneously or by manual activation in case of an emergency.

The use of medical gases (e.g., to support anesthesia) is optional for the Elekta Unity system, and if there is any intent to use these gases in the future, supply pipes should be installed at the end of the machine room. The outlets for medical gases are placed on the left side from the bore entrance of the treatment room. Locating the outlets in this location allows for free access behind the bore in case of an emergency in order to avoid tripping over cables and tubes.

Even with proper (active) shielding, magnetic fringe fields could still be present outside the treatment room, at a different level below or above the vault, or outside the building [27]. All adjacent areas have to be identified and magnetic fringe fields measured. Additional attention must be taken for the linear accelerators in adjacent spaces to make sure that they lie outside of the 0.05 mT field. It is important to note that electrons travel through tubes and can distort the target due to the magnetic fringe fields and can vary for different gantry angles; however, the remaining magnetic fringe fields can be considered static and therefore be solved by gantry angular dependent calibration of the beam. In case of a quench, adjacent linacs should immediately be considered out of service and may require recalibration of the beam. It is important to include this information in the emergency procedures.

## Construction and Installation Timeline

Regardless of using an existing or a new bunker, the vendor will perform a site inspection following the completion of construction. If the bunker is considered to be ready, the first stage is the delivery and assembly of all MR-Linac basic components: MRI, the gantry, and linac. After approximately 10 weeks, the linac is mounted on the gantry ring and aligned; this will allow for radiation beam production and to perform radiation survey. Subsequently, in about 6 weeks, the RF Faraday cage is built, the quench pipe and the medical gases are connected, and some remaining electrical works are completed. Finally, the couch, also known as 'patient positioning system', is installed and the MR-Linac is ready to become "on field." After the ramp-up of the magnetic field, all existing (conventional) linacs in the adjacent areas must be recalibrated for gantry angle dependent beam steering and dosimetry to correct for the new emerged magnetic fringe fields. Following the installation of the accompanied software (TPS, Mosaiq patient managements and R&V software, imaging), the device acceptance tests are performed as a joint activity between the vendor and customer. The MR-acceptance testing schedule includes $B_0$ field strength, direction, stability, and homogeneity, as well as an assessment of the $B_0$ inhomogeneity induced by the gantry. The gantry may introduce field distortion, which is induced by the ferromagnetic material on the gantry ring but can be removed with a gantry shimming procedure. Furthermore, cryogen helium level, RF cage sealing, image orientation to ensure that the correct patient positioning is propagated from the MRI control system, and testing for ghosting artifacts to validate the fidelity of the gradient system are performed. At this stage, testing the MR imaging quality may be considered by imaging healthy volunteers or patients. This imaging data may be valuable to use in workflow development and testing in the subsequent stages of the installation.

The linac-acceptance testing schedule includes safety, dosimetry (beam alignment), and mechanical (MLCs, treatment table alignment, MR-to-MV) checks. Acceptance of the separate parts of the MR-Linac, in this phase, will be according to guidelines and procedures of Elekta (or provided by Philips regarding the MR system).

The second stage of the MR-Linac installation will be the 'clinical readiness phase' in which training to operate the system (MR imaging, delivery, TPS, workflow, physics) will be given, beam data is acquired to build a system specific beam model, and the physics of the system is validated. In this phase, physics commissioning is completed where the customer's own QA procedures and tests can be performed. It is recommended that the customer's QA activities are planned at an early stage in order to save time. For example, the selection and purchase of QA equipment (e.g., ion chambers, electrometers, cables, water tank and/or 2D detectors, and patient QA phantoms) should be considered at an early stage as it might take a considerable amount of time to have all required equipment delivered and validated for clinical use. Depending on the customer activities, after about 15 weeks, the clinical readiness phase is completed with an end-to-end workflow test validating the complete patient treatment workflow from start to end.

The installation of the Unity is mainly coordinated and performed by Elekta engineers and technical support. However, a local staff group of the various involved disciplines and expertise is advised to support the installation at any stage or phase of the project, including the acceptance and commissioning according to local regulations.

## Staffing and Training

### Team Member Roles and Responsibilities

It is important to properly staff the MR-Linac program in order to guarantee seamless and efficient workflows. Depending on how many patients are intended to be treated on the machine, the minimum recommended staffing requirements for daily use are two radiation therapists, two dosimetrists, two medical physicists, and two radiation oncologists.

The dosimetrist generates the offline reference plans to be used for daily adaptation: adapt-to-position or adapt-to-shape. Once the reference plan during the initial planning process is reviewed and approved by the radiation oncologist, the dosimetrist's role, depending on the clinic, can be extended into the online adaptive workflow process by being involved with online contouring and optimization. The radiation oncologist's role within the online adaptive planning process is crucial; they focus on editing the planning target volumes and recontouring some major organs at risk, for example, gastrointestinal structures near the targets. The radiation oncologist also reviews the online adaptive plans and makes the final approvals.

The medical physicist plays an important role in various aspects of the MR-Linac program. The medical physicist works closely with the vendor team during acceptance testing and ensures the quality and the safety of the system through commissioning and performing routine QA. They also develop various workflows and perform staff training. Their role also extends into both the offline and online adaptive process. The medical physicist works closely with the dosimetrist to ensure the quality of the reference plan; they also perform patient-specific QA offline prior to patient treatment. During the online adaptive process, the medical physicist assists with contouring by focusing on contours which could affect the dose calculation, such as high-density materials, air cavities, calcifications, and external contours. They also perform the optimization and online QA. The radiation therapist's roles include the performance of the daily quality assurance, patient safety screening, patient setup, image acquisition and registration, and machine operation. In more advanced programs, the radiation therapist can play a role in the online adaptive workflow by performing daily image segmentation and contouring.

The availability of linac engineers is institutional specific; however, direct access to technical support from the vendor to address any machine needs is of importance especially in the early phase of starting the MRgRT program. With regard to IT support, depending on the complexity of the program, IT support specialists should work closely with the MR-Linac physics team to reduce any communication issues that may occur and to support with any software upgrades. IT should ensure that the MR-Linac is fully integrated within the hospital system and patient image data can be routinely archived to avoid storage issues which may cause software speed slowness. The medical physicist works closely with the technical teams to ensure any software and hardware upgrades are safely implemented within the clinic and to provide any additional staff training.

## Staff Training and Education

Training and education for implementing any new technology comes at phases. First is manufacturer or vendor-based training and the second is institutional training. Vendor training is geared toward medical physicists, dosimetrists, and radiation therapists. The medical physicist is expected to play a major role for ongoing institutional training, so it is highly recommended that they would be involved in all various manufacturer course training for all treatment team members.

One major difference between a conventional linear accelerator and the MR-Linac is the existence of the magnetic field, adding another layer for safety training which most therapy physicists are not familiar with [28, 29]. It is important to learn from the radiology department on the clinical safety practices pertaining to the existence of a magnetic field within the environment. These procedures can be transferred to the radiation oncology and implemented within the MR-Linac program. If an existent MRI simulation program is present at the radiation oncology department, similar practices are to be implemented. An institutional annual MRI safety refresher training is recommended for all MR-Linac team members including the radiation oncologists, radiation therapists, and medical physicists. This annual training is generally conducted by a specialized therapist or a medical physicist.

It is also important for the radiation therapy physicist to work closely with an MRI physicist to implement a quality assurance program pertaining to the MRI imaging system and to learn to identify artifacts which may occur, for example, metal artifacts. It is also important to identify whether distortion corrections have been applied to the images, among other image quality factors.

The medical physicist's role is to train the radiation therapists on daily QA procedures and to set daily thresholds. The medical physicist is expected to also work closely with the dosimetrists and train them on various aspects of planning which deviate from conventional planning due to construction differences and the presence of a magnetic field.

## Workflow Setup and Considerations

### Pretreatment Preparations and Online Workflow

The pretreatment process (Fig. 10.4) is similar to the conventional pretreatment process in radiotherapy. It starts with the imaging of a patient, which can either be a CT or MRI as primary image. Other image data sets such as CT, MRI, and PET can be registered to help with tumor and organ delineation. This can all be performed in imaging or planning software. It is prudent to invest in image registration software that can help streamline transfer, fusion, and segmentation to provide seamless operation.

After structure delineation (e.g., GTV/CTV, organ-at-risks, body), the CT or MRI images, including all structures, are imported into the Unity treatment planning system for offline planning. Next, couch and coil structures are added to the data set. Electron density information is assigned to each structure to be used for online MRI planning using a bulk density override technique. Electron density information can be extracted from the simulation CT. An IMRT step-and-shoot plan is then created and evaluated to be used as the daily reference plan (Fig. 10.4).

Prior to treatment, it is important to screen patients daily for MRI compatibility and safety prior to entering the treatment vault. The patient is then positioned on the couch using a couch

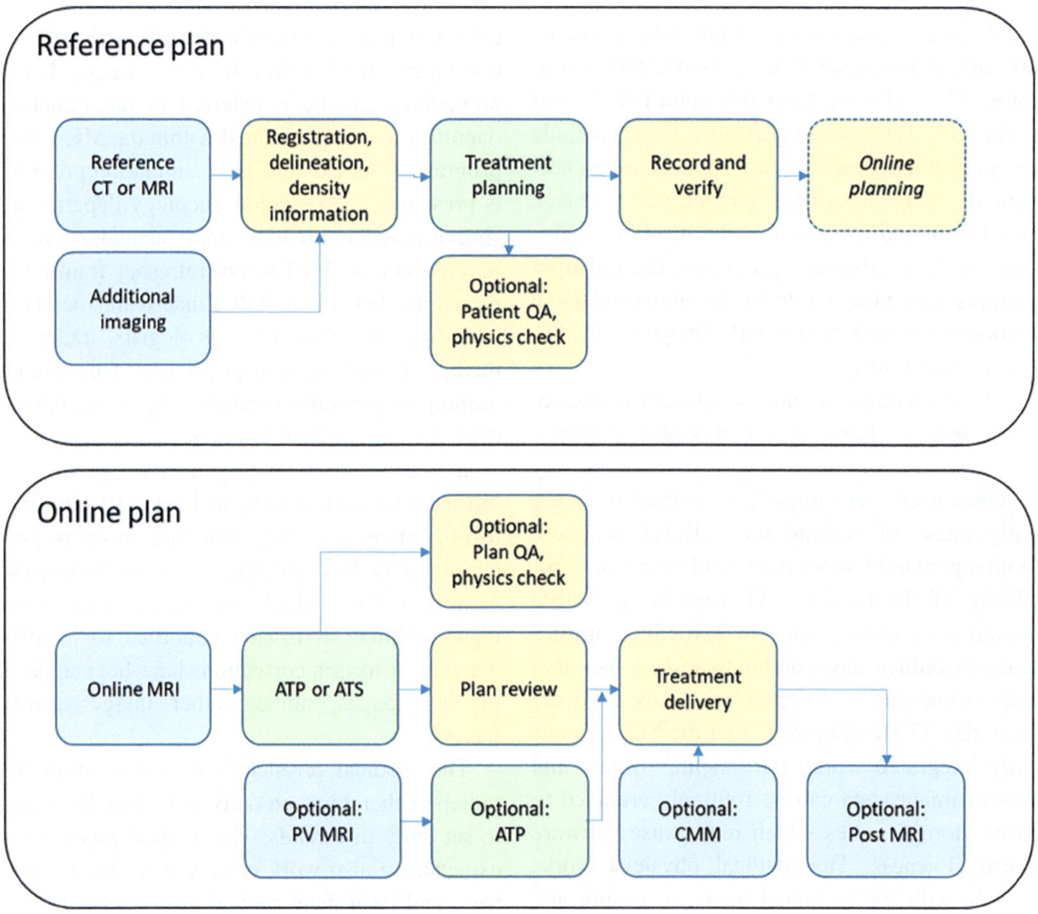


**Fig. 10.4** MR-Linac clinical treatment workflows. Above, the preparation phase to create the patient-specific reference plan. Below, the online workflow to correct for the daily patient translations (adapt-to-position) or to incorporate the daily anatomy deformations (adapt-to-shape)

index system. Next, the coil is placed over the target area for optimal MR imaging. After positioning the patient, an MRI image is taken, with a preselected patient-specific 3D sequence or an average image reconstructed from a 4D MRI, and imported to the online treatment planning system. The MRI is registered to the planning image (CT or MRI), and the plan is adapted to the current anatomical situation. This can be done using two different workflows: adapt-to-position (ATP) or adapt-to-shape (ATS) [30].

The ATP workflow is analogous to image guided radiotherapy on a conventional linac as no contouring and replanning is performed. However, since the Unity couch is static, the isocenter on the pretreatment planning image is shifted using the online registration results. The plan is then adapted to the shifted isocenter by using either the original segments, adaptation of the segments using the changing projection of the structures, or optimizing the weight and/or shape of the segments. The dose calculation is carried out on reference plan image set.

In the ATS workflow, the structures are propagated to the online MR image using deformable registration, and the structures can be evaluated and edited if necessary. An unconventional alternative to the vendor's standard workflow is to use external software for parallel contouring, where contour assignments can be divided among team members to generate contours simultaneously. Once the contours are generated, they are merged and transferred back to the online treatment planning system [31]. The adapted plan is then created on the MRI starting either from fluence or from segments of the reference plan. For the latter, the options are to use the original segments, adaptation of the segments using the changing projection of the structures, or optimizing the weight and/or shape of the segments.

After the plan (Fig. 10.5) is evaluated and approved by the radiation oncologist, the plan is ready to be delivered. At this point, there is an option to evaluate the target position using a secondary verification scan taken during optimization and, if necessary, adapt the plan just before the actual radiation for any additional shifts. Patient-specific QA is performed following plan acceptance [32]. During delivery, three-plane 2D real-time imaging is used for tracking where it is possible to use gating or to correct for target drifts; motion management has been described in detail in Chap. 6. However, in an unconventional workflow, tracking can be disabled if it's not needed (depending on treatment site) to allow for additional volumetric MRI sequence acquisition on the imaging system. Following treatment, posttreatment scans can be acquired for evaluation.

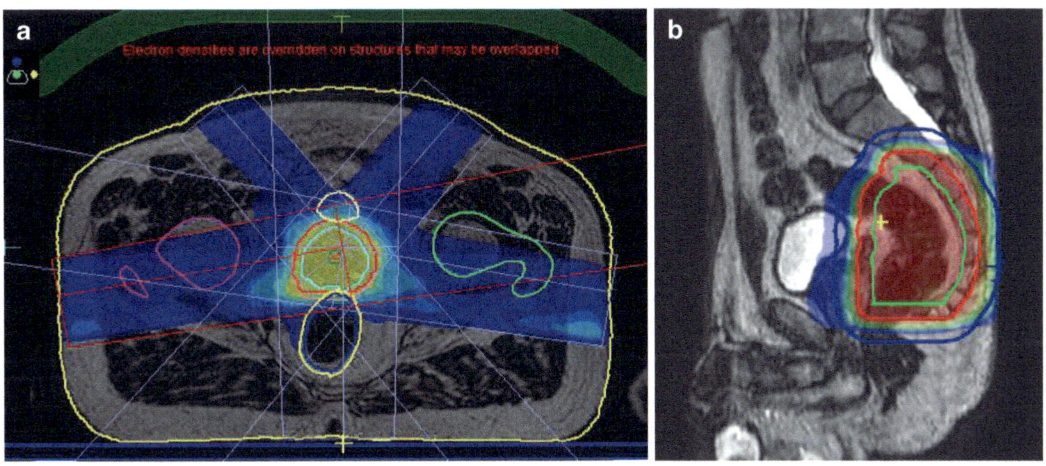


**Fig. 10.5** Examples of MR-Linac patient treatment planning for (**a**) prostate and (**b**) rectum patient

## Implementation of New Sites

Implementation of new workflows depends on the availability of software, hardware, staff, and other resources. However, some general items can be considered before treating patients on the MR-Linac.

First, the patient positioning and immobilization devices being used should be considered. Physical limitations such as the bore size should be taken into consideration when selecting patients and creating/purchasing immobilization devices. All devices must be MRI safe and compatible. MRI sequences can be chosen based on the particular site where image quality, acquisition time, and geometrical fidelity are optimized. Consensus for target and organs at risk delineation should be established in advance to save time during the online workflow. Personnel training for contour delineation should take place following establishing guidelines especially for those who lack experience. In the online process, contours can be propagated to the online MRI in the ATS workflow, but if adaptation of the contours is necessary, the acceptable time needed and who will adapt the contours should be decided up front. PTV margins are based on the total uncertainties in the online planning process such as the remaining intrafraction motion and might differ for different patient groups.

During optimization, MLC field shapes are optimized for the predefined gantry angles, taking the magnetic field into account in the dose calculation. For online MRI planning, an adaptive workflow has to be developed focusing on plan quality and optimization time, thus considering ATP versus ATS but also the way to adapt the plan [24, 33].

Various methods can be used to perform online patient QA. Most clinics use an independent calculation software or perform a plan parameter check during the plan evaluation process. Also, plans could be measured after irradiation. Some backup procedures should be implemented. The systems provide a procedure to finish a plan after the workflow is stopped during radiation (described below). Finally, the workflow should be trained with the personnel and task should be clear to each group. Also several tasks might be passed on to other professions to smoothen the workflow, such as contour adaptation or plan evaluation.

## Fallback Scenarios

In some cases, the radiation delivery can be interrupted during treatment due to machine issues or a patient emergency. Depending on the severity of the issues, the radiation treatment may be resumed if the issues can be resolved within a reasonable time such as a quick system reset or restarting the treatment control software while maintaining the treatment plan and patient position. However, if the issues can't be resolved within a reasonable time and the treatment must be aborted, a completion plan will need to be generated by following vendor's instruction. The completion plan only contains the remaining gantry angles and MUs and uses the ATP workflow assuming no anatomical changes. It is not possible to generate an adaptive plan using the completion plan. In such scenario, it is difficult to estimate the dosimetric quality of the completion plan in the Monaco TPS, but it is possible to export the delivered plan dose and daily completion adaptive plan dose to a third-party image analysis software package to estimate the summed 3D dose distribution.

## Acceptance, Commissioning, and Routine QA

Acceptance testing, commissioning, and routine QA for MR-Linac include tests for the MRI imaging system [34] and Linac radiation delivery system [35, 36]. Most of these tests and routine QA for the radiation delivery system follow published clinical practice guideline such as that from the American Association of Physicists in Medicine (AAPM) for medical accelerators used for radiation therapy. For example, task group reports TG-45 [37] and TG-142 [38] can be used for major guidance. However, several adaptions to the existing protocols have to be applied to account for the influence of the magnetic field from the onboard MRI, the nonstandard source axis distance (SAD = 143.5 cm), a beamline passing through a cryostat wall, and an MR anterior coil before reaching patient. Furthermore, the magnetic field will affect the properties of the QA devices [6, 39]. Acceptance testing, commissioning, and routine QA items and procedures for radiation delivery system will need to be tailored or changed to account for these differences. QA devices should be MR safe and MR compatible. Setup of the dosimeters during testing and QA may be different due to the magnetic field effect on the effective position of measurement. A customized water tank and nonstandard beam data should be collected for the treatment planning system due to geometry limitations of the MR-Linac and magnetic field effect. Additional test such as cryostat characteristics must be performed due to the beamline passing through the cryostat wall.

## Verifying the Onboard MRI Imaging System

Contrary to diagnostic MRI, which is used for detection, characterization, and staging of disease, the onboard MRI is used during daily patient treatment for contouring in adaptive replanning, motion monitoring, and acquiring functional images for offline evaluation. It requires high geometric fidelity and high image quality for patient registration and contour delineation. However, no formal guidelines on commissioning and routine QA for the onboard MRI system have been yet published. TG-284 [40] provides recommendations for an MR simulator for acceptance and commissioning testing, as well as periodic QA. Some tests can be tailored for QA of the onboard MR system of the MR-Linac. However, additional tests are needed to check interference and isocenter coincidence and to verify the communication between the imaging system and the linac.

A multi-institutional study on MRI commissioning of the 1.5T MR-Linac system was published by Tijssen et al. [34] based on the MRI performances of four MR-Linacs from members of the Elekta MR-Linac Consortium. A set of tests for the onboard MRI system were suggested for the commissioning, quality control (QC), and QA measurements as well as system configuration and connectivity (SCC). The SCC includes software configuration checks, receiving coils and other peripherals devices of the MR scanner, DICOM image storing and transferring, and connectivity to hospital picture archiving and communication systems (PACS). QC tests include assessment of the main static $B_0$ field inhomogeneities, gradient field nonlinearity, and RF field and hybrid tests to check the interference between the linac and MR scanner (e.g., direction of the $B_0$ field, gantry rotation, RF interference). QA measurements periodically check the general image quality, such as the signal to noise ratio, low contrast detectability, and gradient fidelity. Most MR tests can be done following the guidance of TG-284. Current phantoms used on the MR-Linac for MRI image quality QA include Philips uniformity disc phantom, GNL phantom, periodical image quality test (PIQT) phantom, ACR phantom, and Modus 3D geometry distortion phantom.

## Verifying the Linac Delivery System

Machine QA of the MR-Linac includes geometrical and dosimetric verification, but the effort is limited as there is just a single energy (7 MV), a beamline mounted on a stiff gantry ring, and a fixed collimator angle (90 degree). The delivery system is aligned to the MRI coordinate system using a dedicated MR-to-MV alignment phantom (Fig. 10.6b).

PTW BEAMSCAN MR water phantom (Fig. 10.6a) is currently the only commercially available MR compatible water tank for radiation beam data collection on the Unity MR-Linac. Due to the bore size limitation, the percentage depth dose (PDD) and beam profiles are collected not only at beam angle 0° (depth can only reach at about 10 cm with SSD =133.5 cm at this direction) but also at 270°.

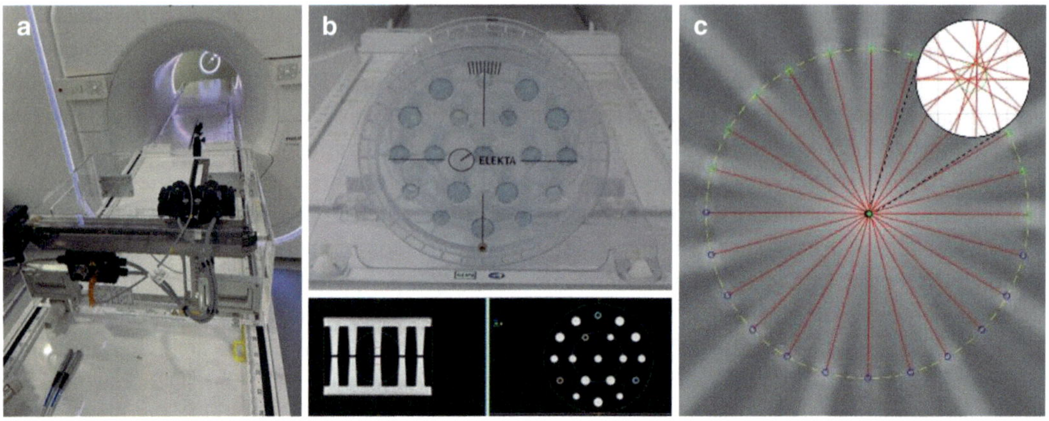


**Fig. 10.6** (**a**) The PTW BEAMSCAN MR water phantom. (**b**) MR-to-MV alignment phantom, provided by Elekta, used to align the delivery and the imaging system. (**c**) Spoke film showing a locus of radius of gantry isocenter of 0.4 mm. The film was acquired on a Unity MR-Linac using a phantom with copper rings to avoid the influence of the Lorentz force on each spoke

MR-to-MV alignment test is a unique measurement for the Unity MR-Linac, which is used to determine the origin-shift of MR imaging to MV radiation. The origin-shift values on three-dimensional orthogonal coordinate axis x, y, and z are measured during commissioning and applied to the TPS for daily adaptive planning. This test needs to be repeated periodically to make sure the drifting of the values is within tolerance and is stable over time. The phantom used for MR-to-MV alignment consists of several ceramic balls surrounded with MR visible solution (Fig. 10.6b) and is provided with analysis software by the vendor.

Verification of the beam direction to the isocenter (isocenter accuracy) is performed by an adapted starshot film measurement [41]. The adapted setup uses a film, sandwiched between two copper rings to prevent a dose print by the secondary electrons affected by the magnetic field (Fig. 10.6c). The spokes, drawn between the two opposing dose prints at the copper ring, should intersect the isocenter. The distance over all spokes to the isocenter can be calculated and is about half the distance compared to a conventional linac.

MLC leaf positions are verified using a standard picket-fence test of adjacent small field slits with a 1 mm gap in between (Fig. 10.7). Measurements can be performed with film or the onboard EPID, but due to the large maximum field size (57 cm × 22 cm), only a portion of the full extent can be captured on the EPID (22 cm × 8.5 cm). Analysis of the measurement quantifies deviation from this 1 mm for each leaf pair separately over the leaf trajectory.

Magnetic field must be taken into account during output calibration of Unity MR-Linac with an ion chamber since the magnetic field will change the trajectory path of the secondary electrons in the surrounding medium and inside the gas filled volume of the detector, thus the response of the detector [13, 42]. Magnetic field correction factors are introduced to account for dose changes (fixed value for the 1.5T Unity setup) and detector response. Correction values for detector response depend on the ion chamber type (available for the commonly used detectors) and beam quality $k_Q$ [43–45]. Both O'Brien [42] and Malkov [46] pointed out that TPR20/10 (IAEA TRS-398 [47] suggestion for $k_Q$ determination) is more stable than $PDD_{10}$ (TG-51 [48] suggestion

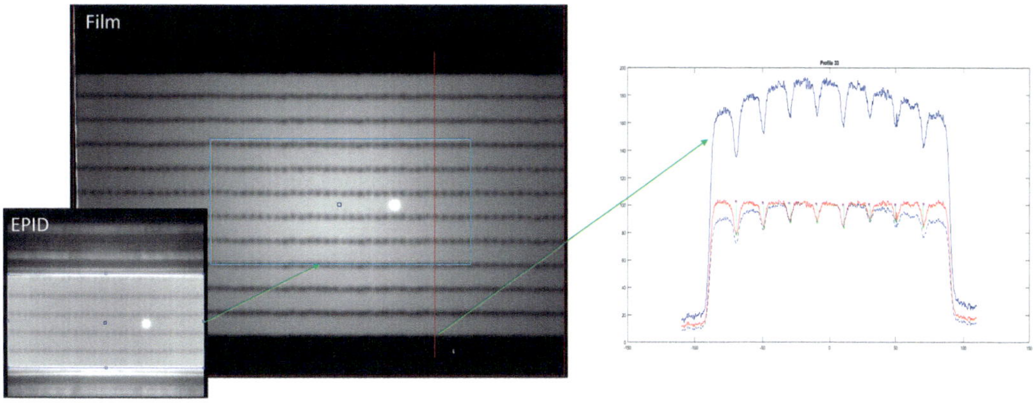


**Fig. 10.7** Film and EPID image (covering the area within the blue rectangle) of the picket-fence test to verify the MLC positions on the MR-Linac. The right plot shows the profile for a single leaf throughout its trajectory

for $k_Q$); thus, it is recommended to use TPR20/10 to obtain the $k_Q$. No consensus of the location of the reference point for the dose calibration has been made; therefore, one can use a reference point at the depth of the maximum dose, or 5 cm and 10 cm at beam angle 0° or 90°.

A waterproof Farmer-type ion chamber should be used for reference dosimetry as measurements have to be performed in water to prevent air gaps around the detector [46, 49–51].

## Routine QA

The full list of checks and measurements, including their frequency and tolerances, was suggested in the report of machine QA for the Elekta Unity system from Elekta MR-Linac Consortium [35] (see Table 10.1). For daily QA, photon beam output constancy (±3%) can be measured with an ion chamber, daily QA device (e.g., Sun Nuclear DailyQA3 MR), onboard MV imaging detector (center pixel value is representative and stable), or the linac secondary (backup) monitor unit chamber. Safety checks include functionality of the audio-video equipment, couch panel emergency button, interlocks, and beam on indicator. MR checks include devices (coil and accessories, call bulb) and image quality checks (SNR, scaling). Additional daily QA can be performed including functionality checks for metal detectors, helium level and MRI cold head, bore inspection for any metal parts, and linac running parameters (magnetron power, waveguide water temperatures, SF6 pressure). Due to the complexity of the treatment workflow, a daily workflow test is favorable in order to verify communication between every component of the prior to daily treatments.

A weekly image quality test with three scans is recommended using the PIQT phantom to evaluate the key properties of the MR system, i.e., the flood-field uniformity, spatial linearity, slice profile, spatial resolution, and noise (acceptance levels are given [35]). The MR-to-MV alignment test can either be performed weekly or monthly, where the maximum deviation from baseline not to exceed ±0.5 mm and ±0.3°. Linac output and backup monitor consistency should be verified weekly with an ion chamber or detector array (acceptance level is ±2%). MLC and diaphragm positions should be verified either by the onboard EPID or film (tolerance ±1 mm at isocenter plane).

Monthly linac QA includes photon beam profile and beam quality constancy verification. Mechanical QA includes checks on gantry angle, table position, MLC field size, leaf position and radiation isocenter size, QA platform alignment, laser alignment (if any), and EPID image quality. Safety checks include radiation door interlock checks, pause and terminate key function on the keypad, and stop motors. Monthly MR checks include MR geometric accuracy tests in volumes of 200, 300, and 400 mm diameter of spherical volume (DSV). It may also be feasible to separately measure the $B_0$ field uniformity. The acceptance levels for monthly QA checks are given by Roberts et al. [35].

Annual dosimetry QA is more comprehensive and includes verification of the beam using the 3D water tank and ionization chambers and/

**Table 10.1** An overview of all checks and measurements, including their frequency and tolerances, as suggested in the report of machine QA for the Elekta Unity system from Elekta MR-Linac Consortium [35]

| Frequency | Category | Test | Device | Acceptance level | Note |
|---|---|---|---|---|---|
| Daily | Dosimetry | Beam output | EPID<br>IC with water<br>IC with solid water | ±3% | Suggested by David Robert et al. [35] |
| | | Backup dose monitor | Linac console display | ±3% | |
| | Safety | Audio-visual monitor | | Functional | |
| | | CCTV | | Functional | |
| | | Intercome system | | Functional | |
| | | Equipment check (coil, PPD) | | Functional | |
| | | Table emergency stop button | | Functional | |
| | | Radiation door interlock | | Functional | |
| | | Radiation area monitor (if used) | | Functional | |
| | | Beam on indicator | | Functional | |
| | MR | Signal to noise ratio | PIQT phantom | Vendor provided | |
| | | Scaling test (TRA and COR) | | Vendor provided | |
| | | Call bulb | | Functional | |
| | Additional | Metal detector | | Functional | More tests performed in some clinics |
| | | Cold head | | Functional | |
| | | Helium level | | Check level | |
| | | End-to-end workflow test | | Functional | |
| Weekly | Dosimetry | Beam output | IC with 1D water tank<br>IC with water equivalent plastic<br>Multi-axis linear or 2D detector array | ±2% | Suggested by David Robert et al. [35] |
| | Mechanicl | MLC and diaphragm position | EPID<br>Gafchromic film<br>Multi-axis linear or 2D detector array | ±1 mm | |
| | MR | PIQT | PIQT phantom | Vendor provided | |
| | MR-to-MV | MR-to-MV aligment test | MR-to-MV alignment phantom | ±1 mm | |

**Table 10.1** (continued)

| Frequency | Category | Test | Device | Acceptance level | Note |
|---|---|---|---|---|---|
| Monthly | Dosimetry | Beam output | IC with water equivalent plastic | ±2% | Suggested by David Robert et al. [35] |
| | | Beam profile | EPID or multi-axis linear detetctor array | ±2% | |
| | | Beam quality | IC with water equivalent plastic | ±1% | |
| | Mechanical | Gantry angle | Level | 0.3 degree | |
| | | Table position | MV alignment phantom | ±1 mm | |
| | | BLD field size check | EPID, Gafchromic film, 2D detector array | ±2 mm | |
| | | Leaf position accuracy | | ±1 mm | |
| | | Radiation isocenter size | MV alignment phantom | ±0.5 mm | |
| | Safety | Safety interlock | | Functional | |
| | MR | GNL test | Philips GNL distortion phantom | 200 mm DSV $\leq$ 1 mm<br>300 mm DSV $\leq$ 2 mm<br>400 mm DSV $\leq$ 4 mm | |
| | Additional | $B_0$ uniformity | Philips 40 cm disc phantom | Ccompared to baseline | More tests performed in some clinics |
| Annual | Dosimetry | Beam output | IC with 1D water tank | ±1% | Suggested by David Robert et al. [35] |
| | | Beam profile comparison from baseline | IC with 3D water tank | ±2% | |
| | | Output factor | IC with 3D water tank | ±2% for field size$\leq 4 \times 4$ cm$^2$<br>±1% for field size$>4 \times 4$ cm$^2$ | |
| | | Beam quality | IC with water equivalent plastic | ±1% | |
| | | MU linearity | IC with water equivalent plastic | 5% for 2–4 MU<br>2% for $\geq$5 MU | |
| | | Output vs dose rate | IC with water equivalent plastic | ±2% from baseline | |
| | | Output vs gantry angle | IC with water equivalent plastic | ±2% from baseline | |
| | Mechanical | Similar to monthly QA leaf position repeatability | | ±0.5 mm | |
| | MR | $B_0$ uniformity vs gantry angle | Philips 40 cm disc phantom | Compared to baseline | More tests performed in some clinics |

or microdiamond detector. Dosimetry related checks are machine dose output, output factors and profiles with a combination of field sizes and gantry angles, beam quality, MU linearity, X-ray output constancy vs. dose rate, and X-ray output constancy vs. gantry angle. Mechanical checks are similar to the monthly QA. MR QA is covered in the daily, weekly, and monthly QA items also in addition to the verification of the MR main frequency variation vs. gantry rotation.

## End-to-End Verification and Patient-Specific Plan QA

Due to the complexity of the daily MR-Linac system and daily treatment workflow, any machine or communication issues could significantly increase patient-on-couch time or result in interruption/aborting of the patient treatment. Issues might occur at and between the various parts of the linac and MR imaging system, the treatment session manager, treatment planning system (Monaco), and record-and-verify system (Mosaiq). Workflow tests are recommended to be performed daily to verify the functionality and communication between all parts of the MR-Linac used in the clinical workflow [52].

After any major hardware or software updates, modifications of the system, or implementation of a new workflow, an end-to-end test should be performed using a phantom to mimic patient treatment which includes testing the various components of the workflow. This includes patient (or phantom) setup, MR pretreatment imaging, adaptive RT plan generation, MR imaging during treatment, delivery, and posttreatment MR imaging. Although not specified and analyzed separately, the "overall" end-to-end test includes all uncertainties related to imaging, beam model (e.g., beam profiles, MLC transmissions, couch), dose delivery, and system communication. Examples on phantoms which can be utilized for end-to-end testing are SNC StereoPHAN, Standard Imaging Lucy Phantom, IROC Houston phantom, MR-to-MV phantom, CIRS STEEV, and Anthropomorphic PRIME phantom. Results should be presented as absolute deviations in position and dose with respect to the intended values.

Patient-specific QA is closely related to end-to-end testing but only verifies specific parts of the workflow, namely, treatment planning and delivery. The generated plan (derived from pretreatment preparations or online workflow) is recalculated on the phantom geometry and compared to the measurement of the delivered dose on the phantom. Positional errors of the phantom setup are generally disregarded. Phantoms for patient-specific plan QA include PTW Octavius 1500MR, SNC ARcCHeckMR, ScandiDos Delta4 Phantom, Gafchromic film, and radiochromic gel.

Patient-specific plan QA is particularly challenging regarding online adaptive radiotherapy. The patient cannot leave the treatment position on the couch for any verification. Quality assurance is often performed on the pretreatment plan, being a representative plan for the online workflow. Online QA might include visual inspection of new contours and electron density map, independent dose calculation of the online treatment plan using third-party software [53], and verification of integrity of transferred data during the workflow [32, 52]. Postdelivery verification can be performed using the treatment delivery log files [54–57] or physical measurements of the treatment plans [39, 58–61].

## Summary

The Elekta MR-Linac, combining a linear accelerator with onboard 1.5T MR imaging system, allows for accurate target definition with accurate radiation delivery. Daily treatment adaptation, based on the online acquired MRI, is performed using an online workflow. The system is in its early phases of adoption of real-time gating and tracking. Although dose delivery is affected by the magnetic field, it is considered in the dose calculation engine and plan optimization. The Unity system has become mature over the last few years and is becoming more widely available worldwide. Installation and quality assurance procedures (described in literature and reports) have been developed and implemented for both the imaging and the dose delivery components of the system. Appropriate training of the staff is necessary for MR-Linac operation, treatment planning, safety due to the presence of the magnetic field, support and maintenance, online replanning, and workflow efficiency.

## References


1. Lagendijk JJW, Bakker CJG. MRI-guided radiotherapy: a MRI based linear accelerator. Radiother Oncol. 2000;56(S1):S60.
2. Lagendijk JJW, et al. MRI/linac integration. Radiother Oncol. 2008;86(1):25–9.
3. Raaymakers BW, et al. Integrating a 1.5 T MRI scanner with a 6 MV accelerator: proof of concept. Phys Med Biol. 2009;54(12):N229–37.
4. Raaymakers BW, et al. First patients treated with a 1.5 T MRI-Linac: clinical proof of concept of a high-precision, high-field MRI guided radiotherapy treatment. Phys Med Biol. 2017;62(23):L41–50.
5. Werensteijn-Honingh AM, et al. Feasibility of stereotactic radiotherapy using a 1.5T MR-linac: multi-fraction treatment of pelvic lymph node oligometastases. Radiother Oncol. 2019;134:50–4.
6. de Pooter J, et al. Reference dosimetry in MRI-linacs: evaluation of available protocols and data to establish a Code of Practice. Phys Med Biol. 2021;66(5):05tr02.
7. Raaijmakers AJ, Raaymakers BW, Lagendijk JJ. Integrating a MRI scanner with a 6 MV radiotherapy accelerator: dose increase at tissue-air interfaces in a lateral magnetic field due to returning electrons. Phys Med Biol. 2005;50(7):1363–76.
8. Hackett SL, et al. Spiraling contaminant electrons increase doses to surfaces outside the photon beam of an MRI-linac with a perpendicular magnetic field. Phys Med Biol. 2018;63(9):095001.
9. Malkov VN, et al. Monte Carlo simulations of out-of-field skin dose due to spiralling contaminant electrons in a perpendicular magnetic field. Med Phys. 2019;46(3):1467–77.
10. Oborn BM, et al. High resolution entry and exit Monte Carlo dose calculations from a linear accelerator 6 MV beam under the influence of transverse magnetic fields. Med Phys. 2009;36(8):3549–59.
11. Raaijmakers AJ, Raaymakers BW, Lagendijk JJ. Magnetic-field-induced dose effects in MR-guided radiotherapy systems: dependence on the magnetic field strength. Phys Med Biol. 2008;53(4):909–23.
12. Woodings SJ, et al. Beam characterisation of the 1.5 T MRI-linac. Phys Med Biol. 2018;63(8):085015.
13. van Asselen B, et al. A formalism for reference dosimetry in photon beams in the presence of a magnetic field. Phys Med Biol. 2018;63(12):125008.
14. Raaymakers BW, et al. Integrating a MRI scanner with a 6 MV radiotherapy accelerator: dose deposition in a transverse magnetic field. Phys Med Biol. 2004;49(17):4109–18.
15. Ahmad SB, et al. Evaluation of a commercial MRI Linac based Monte Carlo dose calculation algorithm with GEANT4. Med Phys. 2016;43(2):894–907.
16. Ahmad SB, et al. Backscatter dose effects for high atomic number materials being irradiated in the presence of a magnetic field: a Monte Carlo study for the MRI linac. Med Phys. 2016;43(8):4665.
17. Malkov VN, et al. Monte Carlo simulations of out-of-field surface doses due to the electron streaming effect in orthogonal magnetic fields. Phys Med Biol. 2019;64(11):115029.
18. Park JM, et al. Air-electron stream interactions during magnetic resonance IGRT: skin irradiation outside the treatment field during accelerated partial breast irradiation. Strahlenther Onkol. 2018;194(1):50–9.
19. Hehakaya C, et al. Early health economic analysis of 1.5 T MRI-guided radiotherapy for localized prostate cancer: decision analytic modelling. Radiother Oncol. 2021;161:74–82.
20. Palm RF, et al. Assessment of MRI-Linac Economics under the RO-APM. J Clin Med. 2021;10(20):4706.
21. Yu JB, et al. National patterns of early adoption of magnetic resonance imaging-guided linear accelerators in 2018 to 2019. Adv Radiat Oncol. 2023;8(3):101167.
22. Hehakaya C, et al. Implementation of magnetic resonance imaging-guided radiation therapy in routine care: opportunities and challenges in the United States. Adv Radiat Oncol. 2022;7(5):100953.
23. Jagt TZ, et al. Benchmarking daily adaptation using fully automated radiotherapy treatment plan optimization for rectal cancer. Phys Imaging Radiat Oncol. 2022;24:7–13.
24. Dunkerley DAP, et al. Clinical implementational and site-specific workflows for a 1.5T MR-Linac. J Clin Med. 2022;11(6):1662.
25. Speight R, et al. IPEM topical report: a 2018 IPEM survey of MRI use for external beam radiotherapy treatment planning in the UK. Phys Med Biol. 2019;64(17):175021.
26. Kim J, et al. Implementation of a novel algorithm for generating synthetic CT images from magnetic resonance imaging data sets for prostate cancer radiation therapy. Int J Radiat Oncol Biol Phys. 2015;91(1):39–47.
27. Kok JG, et al. Installation of the 1.5 T MRI accelerator next to clinical accelerators: impact of the fringe field. Phys Med Biol. 2009;54(18):N409–15.
28. Cao M, Padgett KR, Rong Y. Are in-house diagnostic MR physicists necessary for clinical implementation of MRI guided radiotherapy? J Appl Clin Med Phys. 2017;18(5):6–9.
29. McGee KP, et al. Findings of the AAPM Ad Hoc committee on magnetic resonance imaging in radiation therapy: Unmet needs, opportunities, and recommendations. Med Phys. 2021;48(8):4523–31.

30. Winkel D, et al. Adaptive radiotherapy: The Elekta Unity MR-linac concept. Clin Transl Radiat Oncol. 2019;18:54–9.
31. Paulson ES, et al. 4D-MRI driven MR-guided online adaptive radiotherapy for abdominal stereotactic body radiation therapy on a high field MR-Linac: implementation and initial clinical experience. Clin Transl Radiat Oncol. 2020;23:72–9.
32. Chen X, et al. A daily end-to-end quality assurance workflow for MR-guided online adaptive radiation therapy on MR-Linac. J Appl Clin Med Phys. 2020;21(1):205–12.
33. Gupta A, et al. Online adaptive radiotherapy for head and neck cancers on the MR linear accelerator: introducing a novel modified Adapt-to-Shape approach. Clin Transl Radiat Oncol. 2022;32:48–51.
34. Tijssen RHN, et al. MRI commissioning of 1.5T MR-linac systems - a multi-institutional study. Radiother Oncol. 2019;132:114–20.
35. Roberts DA, et al. Machine QA for the Elekta Unity system: a report from the Elekta MR-linac consortium. Med Phys. 2021;48(5):e67–85.
36. Woodings SJ, et al. Acceptance procedure for the linear accelerator component of the 1.5 T MRI-linac. J Appl Clin Med Phys. 2021;22(8):45–59.
37. Nath R, et al. AAPM code of practice for radiotherapy accelerators: report of AAPM Radiation Therapy Task Group No. 45. Med Phys. 1994;21(7):1093–121.
38. Klein EE, et al. Task Group 142 report: quality assurance of medical accelerators. Med Phys. 2009;36(9):4197–212.
39. Houweling AC, et al. Performance of a cylindrical diode array for use in a 1.5 T MR-linac. Phys Med Biol. 2016;61(3):N80–9.
40. Glide-Hurst CK, et al. Task group 284 report: magnetic resonance imaging simulation in radiotherapy: considerations for clinical implementation, optimization, and quality assurance. Med Phys. 2021;48(7):e636–70.
41. van Zijp HM, et al. Minimizing the magnetic field effect in MR-linac specific QA-tests: the use of electron dense materials. Phys Med Biol. 2016;61(3):N50–9.
42. O'Brien DJ, et al. Reference dosimetry in magnetic fields: formalism and ionization chamber correction factors. Med Phys. 2016;43(8):4915.
43. de Prez L, et al. First water calorimetric Dw measurement and direct measurement of magnetic field correction factors, KQ, B, in a 1.5 T B-field of an MRI Linac. Med Phys. 2016;43(6):3874.
44. de Prez LA, et al. Direct measurement of ion chamber correction factors, $k_Q$ and $k_B$, in a 7 MV MRI-linac. Phys Med Biol. 2019;64(10):105025.
45. Malkov VN, Rogers DWO. Monte Carlo study of ionization chamber magnetic field correction factors as a function of angle and beam quality. Med Phys. 2018;45(2):908–25.
46. Malkov VN, Rogers DWO. Sensitive volume effects on Monte Carlo calculated ion chamber response in magnetic fields. Med Phys. 2017;44(9):4854–8.
47. Andreo P, et al. Absorbed dose determination in external beam radiotherapy - an international code of practice for dosimetry based on standards of absorbed dose to water, in technical report series. Vienna: International Atomic Energy Agency; 2000.
48. Almond PR, et al. AAPM's TG-51 protocol for clinical reference dosimetry of high-energy photon and electron beams. Med Phys. 1999;26(9):1847–70.
49. Hackett SL, et al. Consequences of air around an ionization chamber: are existing solid phantoms suitable for reference dosimetry on an MR-linac? Med Phys. 2016;43(7):3961.
50. O'Brien DJ, et al. Small air-gaps affect the response of ionization chambers in the presence of a 1.5 T magnetic field. Med Phys. 2015;42(6):3724.
51. O'Brien DJ, Sawakuchi GO. Monte Carlo study of the chamber-phantom air gap effect in a magnetic field. Med Phys. 2017;44(7):3830–8.
52. Malkov VN, et al. MR-linac daily semi-automated end-to-end quality control verification. J Appl Clin Med Phys. 2023;24(5):e13916.
53. Wang Y, et al. Development of a fast Monte Carlo dose calculation system for online adaptive radiation therapy quality assurance. Phys Med Biol. 2017;62(12):4970–90.
54. Agnew CE, Irvine DM, McGarry CK. Correlation of phantom-based and log file patient-specific QA with complexity scores for VMAT. J Appl Clin Med Phys. 2014;15(6):4994.
55. Menten MJ, et al. Automatic reconstruction of the delivered dose of the day using MR-linac treatment log files and online MR imaging. Radiother Oncol. 2020;145:88–94.
56. Lim SB, et al. An investigation of using log-file analysis for automated patient-specific quality assurance in MRgRT. J Appl Clin Med Phys. 2021;22(9):183–8.
57. Kontaxis C, et al. Delivered dose quantification in prostate radiotherapy using online 3D cine imaging and treatment log files on a combined 1.5T magnetic resonance imaging and linear accelerator system. Phys Imaging Radiat Oncol. 2020;15:23–9.
58. Yang, B., et al. Initial clinical experience of patient-specific QA of treatment delivery in online adaptive radiotherapy using a 1.5 T MR-Linac. Biomed Phys Eng Express. 2021;7(3).
59. Xhaferllari I, et al. Clinical utility of Gafchromic film in an MRI-guided linear accelerator. Radiat Oncol. 2021;16(1):117.
60. Mönnich D, et al. Quality assurance of IMRT treatment plans for a 1.5 T MR-linac using a 2D ionization chamber array and a static solid phantom. Phys Med Biol. 2020;65(16):16nt01.
61. de Vries JHW, et al. Characterization of a prototype MR-compatible Delta4 QA system in a 1.5 tesla MR-linac. Phys Med Biol. 2018;63(2):02NT02.

# The Alberta Rotating Biplanar Linac-MR, a.k.a., Aurora-RT™

11

B. Gino Fallone, Satyapal Rathee, Nicola de Zanche, Eugene Yip, Keith Wachowicz, and Jihyun Yun

## Introduction

MagnetTx Oncology Solutions (www.MagnetTx.com, Edmonton, Alberta, Canada) which is a start-up from the Alberta Health Services (AHS) and the University of Alberta (UofA) has exclusive license to commercialize the rotating open biplanar linac-MR system developed and installed at the Cross Cancer Institute (CCI), as the Aurora-RT™. The open biplanar MR, with a patient bore size of 110 cm by 60 cm, improves patient experience by reducing claustrophobia, provides space for significant lateral and vertical shifts, and allows perpendicular as well as parallel $B_0$-field to radiation beam configuration. The parallel configuration not only avoids electron-return effect (ERE) [1] but with significant horizontal and vertical shifts allows the positioning of any tumor (including those offset) to be positioned at the 2D planned position or at isocenter. The parallel configuration also allows, with couch movement and gantry rotation to position in 3D, any offset tumor to the planned location or to the isocenter.

B. G. Fallone (✉)
Department of Medical Physics, Cross Cancer Institute, Edmonton, AB, Canada

Department of Oncology, Medical Physics Division, University of Alberta, Edmonton, AB, Canada

MagnetTx Oncology Solutions, Edmonton, AB, Canada
e-mail: Gino.Fallone@albertahealthservices.ca

S. Rathee · N. de Zanche · E. Yip · K. Wachowicz · J. Yun
Department of Medical Physics, Cross Cancer Institute, Edmonton, AB, Canada

Department of Oncology, Medical Physics Division, University of Alberta, Edmonton, AB, Canada
e-mail: srathee@ualberta.ca; dezanche@ualberta.ca; eyip@ualberta.ca; kwachowi@ualberta.ca; jyun@ualberta.ca

## Historical Milestone and Development

Significant scientific and engineering milestones were achieved in developing the Aurora-RT. The Alberta group reported the first proof of principle of a successful linac-MR hybrid system in 2008 which consisted of an open biplanar 0.2T MRI with about a 30 cm gap between the poles connected in perpendicular configuration to a Varian 6 MV medical linac and a Varian 52 leaf MLC, both decommissioned by the CCI [2]. The system allowed extensive (1) experimental studies in basic linac RF noise [3, 4], RF shielding [5], magnetic shielding [2] and MLC operation within a magnetic field, and (2) theoretical investigations that side coupling of a 6 MV linac provided proof of the clinical applicability and the ability to study linac performance within $B_0$ [6]. Linac performance was not affected by electron beam



I. J. Das et al. (eds.), *A Practical Guide to MR-Linac*, https://doi.org/10.1007/978-3-031-48165-9_11

loss within the accelerating waveguide due to magnetic fields that persist in the perpendicular configuration even after shielding the linac [7]. The active shielding design provides magnetic decoupling of the linac from the imager, while the use of magnetically insensitive MLC motors would avoid the need of passive shielding [8, 9].

The parallel configuration was proposed in 2010 [1] with the unobstructed linac beam going through a circular opening in one of the poles [10] to avoid ERE, all together. A maximum magnetic field that sufficiently avoided ERE was 0.5 T, with >0.5 T $B_0$ still resulting in larger ERE. We then evaluated the performance of the linac due to the parallel fringe magnetic field of a 0.5 T open imager. The longitudinal magnetic field on the electron beam only affected in the electron gun optics, which can be eliminated by a redesign of the electron gun optics or through magnetic shielding [9, 11].

The preferred design was parallel configuration with a $B_0$ of 0.5 T which provided acceptable avoidance of ERE and maximum contrast-to-noise for fast MR imaging [12]. A first whole-body rotating biplanar linac-MR (LMR 1000) was built at the CCI/UofA in 2013–2014 using a high-temperature superconducting custom open biplanar 0.5 MRI system with a decommissioned Varian 600C FFF and a Varian Mark 80 MLC for the LMR 1000. The linac-MR gantry is supported by two A-frames allowing +180 and −180 ° rotation. The MRI iron-yoke structure is designed to shape the magnetic field optimally and to minimize the focusing of the contaminate. A beam stop on the gantry also provides counterbalance for the gantry. An unobstructed radiation beam path goes through a 25-cm-diameter opening at the center of MRI planar poles. The LMR 1000 bore size is only 58 cm diameter.

A much larger bore size (110 × 60 cm) was provided by the LMR 2000 assembled in around 2020 as shown in Fig. 11.1. It received Class II Certification by the Canadian Nuclear Safety Commission in March 2021. Aurora-RT received US FDA clearance in May 2022. The LMR 2000 uses a custom 120 double-focused MLC projecting 0.475 mm leaf width at 120 cm SAD, with two independent Y rotating jaws. The dose rate at 120 cm SAD from the custom FFF 6 MV linac is 600 MU/min.

The linac-MR gantry has a rotational range of ±185° from the 0° at 0.1° resolution and position accuracy with maximum nominal speed of 6°/s. The two A-frames holding the gantry have a rear 60-cm-diameter patient-size bore and a front 110-cm-diameter patient-size bore. This provides a useable rectangular-elliptical rotating bore of 110 cm by 60 cm. The Kevlar couch has three translational degrees of freedom with maximum vertical and horizontal speeds of 5 cm/s and 15 cm/s, respectively. It provides a longitudinal range of 260 cm, a vertical range of 65–140 cm above the floor, and, more importantly, lateral ranges of ±23 cm. The latter allows significant

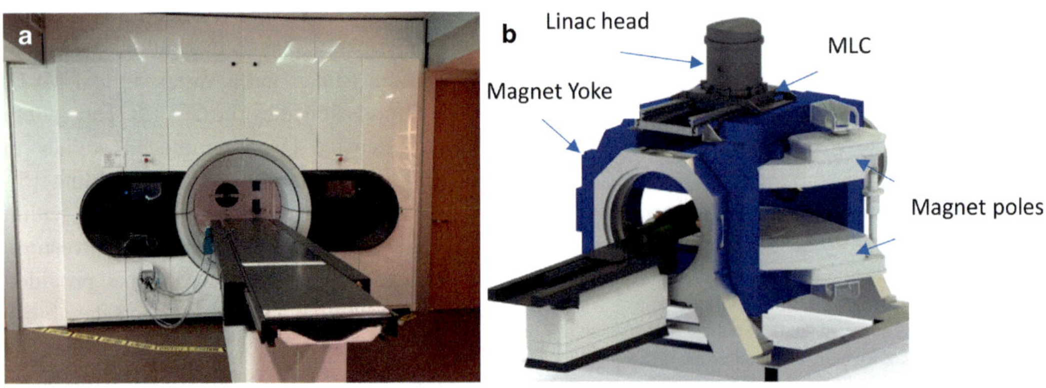


**Fig. 11.1** (**a**) Photo of LMR 2000 (a.k.a. Aurora-RT) installed at the Cross Cancer Institute, Edmonton, Alberta, Canada. (**b**) Schematic design, patient bore opening of 110 cm by 60 cm

2D positioning, while when used in tandem with gantry rotation, 3D positioning of offset targets to any planned position (including isocenter). Because of an open MR, while the gantry is at 90°, during pre-MR imaging used for positioning patient prior to treatment. A patient lying on the couch inside the bore has a direct view of the vault ceiling reducing possible claustrophobia. In addition, when the gantry is at 0°, the left and right side of the patient lying on the couch inside the bore is completely opened allowing staff the direct access to the patient or to a phantom for QA.

Real-time imaging during treatment is possible at 4 fps. Being a high-temperature superconductive MRI, cryogens and exhaust vents are not required and the magnetic field $B_0$ can be safely turned on/off in minutes.

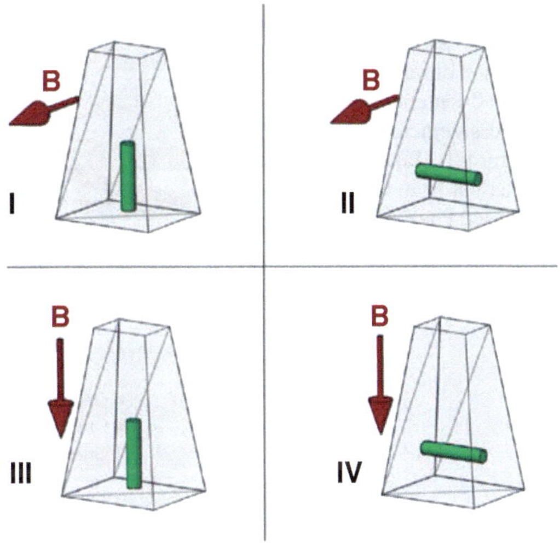


**Fig. 11.2** Various geometrical orientations for radiation detector's long axis (central cylinder) magnetic field (arrow) and radiation beam (transparent rectangular prism) as used in Monte Carlo simulation and experimental measurements. Taken with permission from Ref. [13], Med Phys. 2013;40(4):042102

## Change in Radiation Response of Detectors in Magnetic Field

### Ion Chambers

Ion chambers are generally used for photon beam reference dosimetry to calibrate radiation beams. Electrons traversing the sensitive volume ionize the air; these ions are collected as signal proportional to the energy fluence. Magnetic fields exert Lorentz force on electrons and change their path within the sensitive volume. Therefore, the relative response of ion chambers to photon beam irradiation will change in the presence of magnetic field. Our group has investigated the relative response of cylindrical ion chambers, NE2571 (NE Technology Ltd) and PR06C (Capintec), to magnetic field in 6 MV photon beams [13]. Two magnetic field orientations, either perpendicular (Fig. 11.2 top quadrants) or parallel (Fig. 11.2 bottom quadrants) to the radiation beam direction, with magnetic field strengths up to 2.0T were used in Monte Carlo simulations using PENELOPE. The long axis of chambers was either parallel (Fig. 11.2 left quadrants) or perpendicular (Fig. 11.2 right quadrants) to the radiation beam travelling in both orientations of magnetic field. The relative response of PR06C ion chamber was also measured in both magnetic field orientation up to 0.2 T and compared to PENELOPE simulations, to validate the electromagnetic field implementation in PENELOPE and our choice of simulation parameters within the code.

Lorentz force has more of an effect on electrons traversing low density air than those in chamber wall or buildup cap. Orientation I in Fig. 11.2 suggests that the Lorentz force on electrons will likely deflect them away from the long, sensitive air column, thereby decreasing the chamber's relative response, while an opposite effect can be expected in orientation II. In orientations III and IV, Lorentz force will focus the electron path along the magnetic field direction and may bring few electrons deflecting away from the sensitive volume back into the volume. The results of this simulation are presented in Fig. 11.3. The response of PR06C is affected in perpendicular magnetic field, and it can be as high as 1.085 in orientation II and as low as 0.90 in orientation I. In parallel magnetic field, up to 0.5 T, the response of PR06C is nearly identical to no magnetic field case, and its maximum value is 1.02 in orientation III at

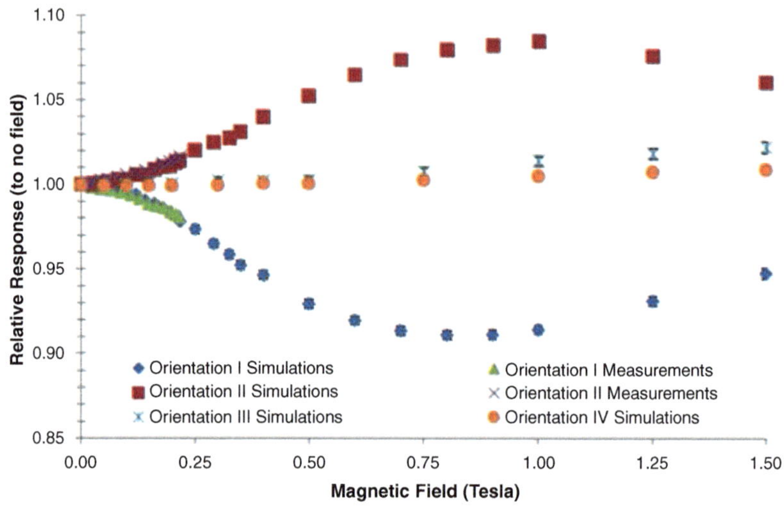


**Fig. 11.3** PR06C relative dose response as a function of magnetic field in orientations I–IV. Measurements are only for I–II. Taken with permission from Ref. [13], Med Phys. 2013;40(4):042102

1.5 T. These results are similar to those obtained by other investigators for other cylindrical ion chambers [14]. As such, no correction for magnetic field induced response change is needed in parallel magnetic field below 0.5 T. Subsequent measurements in parallel oriented magnetic field up to 0.22 T have further corroborated this conclusion [15]. Additionally, if these chambers are used for reference dosimetry, a correction factor is required in perpendicular magnetic field that depends on the magnetic field strength, chamber dimensions, and chamber orientation with respect to magnetic field and photon beam. However, our work further demonstrated that if the chambers' long axis is aligned with the magnetic field direction in the perpendicular magnetic field, the magnetic field response is close to 1.0, and smaller angular deviation of chamber from this orientation does not impact the magnetic field response of chamber. Please note that this orientation is not shown in Fig. 11.3. In other conventional orientations, small 5° misalignments may change the chamber's response up to 2%. In parallel magnetic field, chamber's angular alignment errors in standard orientations III and IV result well below 1% change in magnetic field response [16]. Although one orientation of chamber along the perpendicular magnetic field appears to be suitable in terms of chamber response and its insensitivity to alignment errors, this orientation may be limiting for beam scanning. Inspired by this work, there has been significant amount of work to determine the magnetic field correction factors for perpendicular magnetic field linac-MR's reference dosimetry calibration [17–19].

## Solid State Detectors

Solid state detectors such as PTW 60003 diamond detector and IBA PFD diode detectors are generally used in relative beam dosimetry, i.e., beam scanning. The sensitive volume in these detectors is of significantly higher density than air in ion chambers; thus, the effect of Lorentz force may not be significant. With this hypothesis in mind, we performed Monte Carlo simulation of magnetic field dose response of these detectors by constructing the fine dimensions and materials of the detectors in PENELOPE code system [20]. In addition to the four orientations of long axis of detectors as shown in Fig. 11.2, we considered two additional orientations. Orientation II-2 is the same as II, but detector is rotated by 180 degree such that detectors' stem becomes opposite to orientation II. Another orientation designated as orientation V is obtained by rotating the detector in orientation I or II such that its long axis becomes parallel to magnetic field. The simulated magnetic field response of diamond detector is shown in Fig. 11.4 [20].

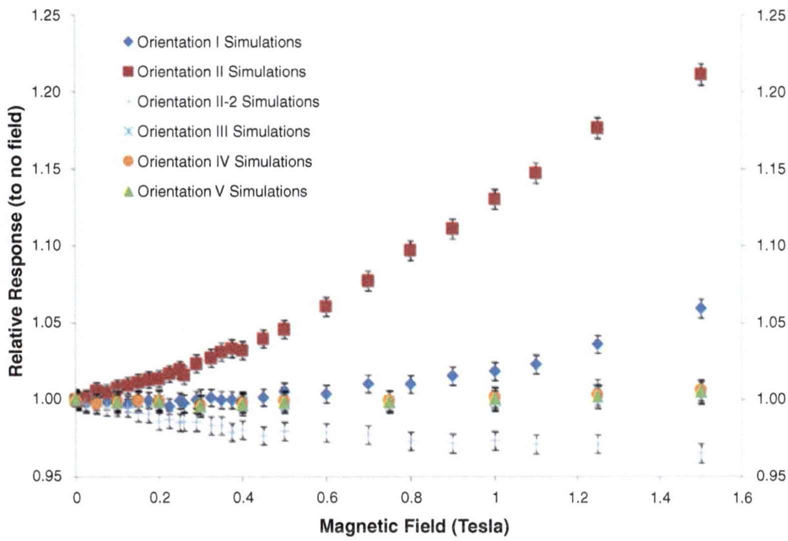


**Fig. 11.4** Relative dose response of PTW60003 diamond detector with magnetic field strength. Taken with permission from Ref. [20], Med Phys. 2014;41(9):09210

Perpendicular field orientations appear to have the largest dose responses, and the parallel magnetic orientations show little response as a function of magnetic field. In the presence of a longitudinal magnetic field, minimal correction is needed at 1.5 T, and no correction appears to be required below a field strength of 1.0 T for both PTW60003 and IBA PFD diode (not shown) detectors. This contrasts with the cases with perpendicular magnetic fields, where complete characterization of detectors on an individual basis is required as a function of relative orientations and field strengths. Our additional simulations mimicking profile measurements at the edges of the photon beam in perpendicular magnetic field further explain the errors, due to detectors' own dose response in magnetic field, in reproducing the relative dose to water under perpendicular magnetic field.

## Summary

Minimal magnetic field responses, in parallel magnetic field orientations, ensure that conventional commissioning and quality assurance detectors and techniques can be used without corrections due to magnetic field with a 0.5 T MRI.

## RF Coils for the Rotating Biplanar Linac-MR

Standard MRI detectors (typically arrays of loop-shaped surface coils) are often placed directly on the patient's body to optimize SNR by placing coils as close to the imaging region as possible. When collisions of the radiation beam with RF coil materials cannot be avoided, their effect on the dose delivered to the underlying tissue must be evaluated. Our measurements [21] have shown that standard coil materials can increase the surface dose to >74% relative to $D_{max}$, compared to 20% with an open beam. This undesired "bolus" effect enhances skin dose, thus increasing the risk of acute reactions, such as erythema, and long-term radiation damage [22].

### Coil Construction

For optimal detector performance, the conductors in an RF coil must have a minimum thickness on the order of one electromagnetic "skin depth" [23] which in copper at 0.48 T (20.6 MHz) is about 15 μm and for aluminum 18 μm. Typical MRI coils, however, also include substantial mechanical supports for the conductors: for example, to maximize durability, allow weight-bearing, etc. A radiation beam may need to travel through several millimeters of these support materials (low-loss dielectrics like polycarbonate or fiberglass composite), which thus contributes substantially to the bolus effect in an RF coil constructed using standard methods.

Coils for linac-MR applications must therefore be constructed with minimal support materials. Potentially, they could be paper thin like those of Corea et al. [24] which are printed on flexible dielectric substrates (50–150 μm thick) using conductive inks. We measured the effects of coils made of thin substrates and conductors with and without a 0.22 T magnetic field (created by a yokeless electromagnet) parallel to the beam [25]. Results show that for thin copper and aluminum foils, in the absence of magnetic field, the surface dose increases linearly with thickness, with slopes of 0.339%$D_{max}$/μm and 0.116%$D_{max}$/μm, respectively, starting from 22.2% of $D_{max}$ for the unobstructed beam. With the magnetic field parallel to the beam, the slopes are lower (0.16%$D_{max}$/μm and 0.06%/μm, respectively) but start from a higher surface dose of 31.9% of $D_{max}$ for the unobstructed beam. The 3.3 times lower density of aluminum compared to copper accounts for the lower slopes both with and without magnetic field, while the higher starting dose in the presence of the magnetic field is due to contaminant electrons trapped by the Lorentz force.

For comparison, the dose beneath a hospital gown was measured to be 36.0% of $D_{max}$ with the 0.22 T field, which corresponds to approximately the same increase as 25-μm-thick copper or 70-μm-thick aluminum. These results suggest that aluminum coil conductors can be chosen to have the same effect on surface dose while being much thicker in terms of skin depth than copper. Because image SNR depends directly on coil resistance, the larger thickness of aluminum could compensate for its slightly higher resistivity and lead to equal or better SNR from coils made of aluminum.

The relationship between coil thickness and its resistance is a complicated function, especially when thickness is near the skin depth. We studied these effects in more detail in Barta (2021) [26] by constructing coils of various thicknesses of copper (17–127 μm) and aluminum (9–127 μm) and measuring their performance (i.e., efficiency, $\eta$, obtained from loaded and unloaded coil quality factors) on the bench and in the 0.48 T scanner (image SNR). For comparison, the thickness of copper in printed circuit boards is 35 μm (standard 1 oz.) and multiples thereof (less common, and rarely >4). To represent manually assembled coil construction (and the maximum SNR achievable in practice), one was also made using 600 μm copper foil. Coil efficiency was in the range of 0.56–0.74 (Cu) and 0.34–0.59 (Al), and it correlated well with image SNR comparisons according to the relationship [27] $\frac{\mathrm{SNR}_A}{\mathrm{SNR}_B} = \sqrt{\frac{\eta_A}{\eta_B}}$. Importantly, there were no differences in SNR achieved by coils made of 127 μm copper and aluminum conductors, and with the much thicker 600 μm copper, the SNR improved only by 5%. Very thin conductors such as 17 μm copper and 13–20 μm aluminum achieved, respectively, 85% and 75–85% of the SNR with 127 μm conductors. If needed, these limited SNR losses can be readily compensated for by corresponding adjustments to the imaging sequences, such as using thicker slices, lower acquisition bandwidth, longer scan time, etc.

## Summary

Thin aluminum conductors can allow surface dose to be controlled (e.g., an enhancement less than that caused by a gown compared to open beam), with very limited loss in SNR (not more than 25% compared to coils made using the thickest common PCB copper thickness). Radiolucent coils made in this way are therefore a practical option for linac-MR applications.

## Accelerated MR Imaging for Real-Time Motion Management

One of the benefits of having a MR/RT hybrid system is the ability to use the online MRI capabilities to provide real-time soft-tissue visualization, to monitor and manage the patient's respiratory motion, particularly in the thorax or abdominal areas during RT. Compared to current state of the art Linacs (e.g., Varian Truebeam, which is capable of real-time orthogonal kV projections), real-time MRI has a couple of key advantages: flexibility of geometry of the imaging slice volume, which is not dependent on the Gantry's position, and the ability to visualize soft tissue, without inserted fiducials. Nevertheless, MRI is inherently a slower imaging method compared to kV imaging. For a typical breathing respiratory motion monitoring, 4 fps is considered to be adequate. For modern MR imagers, the fastest dynamic sequences (i.e., balanced SSFP) should already be capable of providing a single 2D plane with 2–3 mm resolution at that frame speed, without applying any specialized techniques. Nevertheless, the development of special techniques that further increase acquisition speed can improve the spatial or temporal resolution of 2D images, and may even allow real time multi-slice or 3D imaging that more accurately monitor the 3D trajectory of a moving tumor.

## Accelerated MR Acquisition and Compressed Sensing MRI

There has been considerable work in the MR literature to improve acquisition speed. One very well developed and well known is parallel acquisition (i.e., SENSE or GRAPPA), which has specific hardware requirements (i.e., multichannel array coils). Our group decided to pursue more general techniques that can be used in any MR scanners. Compressed sensing (CS) in MRI [28] allows an image to be recovered from significant undersampled k-space data, via optimization, if it (1) has a sparse representation in some known domain and (2) is acquired in an incoherent manner such that any Nyquist artifact appears incoherent and noise like. CS reconstructed image is the solution to an optimization function that maximizes fidelity of its k-space representation to the partially acquired k-space data, via L2 norm minimization, while simultaneously maximizing sparsity in a transform domain (e.g., wavelet, discrete cosine, or total variations) via a L1 norm minimization, weighted by $\lambda_1$. For a total variation transform, a split Bregman solver [29] can be used efficiently to iteratively solve this equation, in approximately 100 ms in a typical CPU (3.4GHz, Intel i7–4770), using MATLAB.

## Prior Data Assisted Compressed Sensing (PDACS)

For 2D Cartesian acquisitions, the ability to undersample k-space is limited to the phase-encode direction only. As a result, the second condition of incoherent sampling is not fully satisfied, usually resulting in poor quality images at high acceleration. An alternative approach is to include dynamic 2D data in the $L_1$ norm minimization term (i.e., perform 3D sparsifying transform on the dynamic data in the optimization function). Unfortunately, these 3D operations in an iterative solver are too slow for real-time applications. The key feature of prior data assisted compressed sensing (PDACS) [30] is to take advantage of *prior data* in a different manner, instead by adding an additional $L_2$ norm minimization term that enforces fidelity of the unacquired k-space to the k-space of some averaged *prior data,* in a 2D operation. PDACS gives considerably better images compared to the original CS method, while the split Bregman solver is still able to provide the final image in fraction of a second. The weights given to $\lambda_1$ and $\lambda_2$ have a significant impact on the quality of the final image – a simple brute force optimization is performed on some training data to give approximate values of $\lambda_1$ and $\lambda_2$ that are adequate.

As these images are ultimately used to monitor a patient's tumor motion, we evaluated the quality of PDACS reconstruction by evaluating its tumor auto-contouring performance of an undersampled image and against a fully sampled image. Relevant metrics include Dice similarity index (DSI), as well as mean centroid displacement (MCD). A high DSI indicates a high quality undersampled image, whereas a low mean centroid displacement indicates a higher quality undersampled image.

### Updating Prior Data with Sliding Window

By using fully acquired images prior to accelerated image acquisitions, the PDACS technique was found to improve image quality of undersampled images. However, we found the benefit decreases gradually over time as the prior data gets progressively more outdated compared to the current data, due to shifts in the baseline MR signal unrelated to patient motion. Over the course of a 3-min acquisition, in terms of tumor autocontouring accuracies for a 5× accelerated acquisition are 0.93, 0.9, and 0.87 for the first, second, and third minutes' data after the acquisition of prior data for the high SNR scenario and 0.91, 0.88, and 0.86 in the low SNR scenario. The mean centroid error increases from 1.1 mm, 1.5 mm, and 2.5 mm for the first, second, third minutes of data after the acquisition of prior data in the high SNR scenario, whereas they are 1.2, 1.9, and 3.2 mm in the low SNR scenario. One should note that when the prior data is 3 min out of date, these results can be even poorer as the 2D-CS technique, without any prior data!

To address this, sliding window PDACS [31] is a technique that requires a different k-space sampling strategy. The CS requires a pseudorandom sampling pattern, but the center of k-space contributes more value to image quality and is sampled more often. In PDACS, the sampling pattern is generated with a probability function such that the central phase encode is sampled at 100% probability, but the function falls off as it moves to the peripheral phase-encode locations at a rate governed by acceleration factor (i.e., the higher the acceleration, the faster the falloff). Once a pattern is determined for a single dynamic, it is used for the entire series, and as such, some peripheral phase-encode locations are never refreshed. For the sliding window technique, the sampling pattern for every dynamic is determined with a dynamic probability function that has three regions (central, intermediate, and peripheral). The central five phase-encode lines are acquired at 100%, while the intermediate region is the region where the probability is falling off from 1 to 0.25, with a falloff rate determined by the acceleration factor and the peripheral region where $p < 0.25$. In the peripheral region, the initial probability for each location is equalized. Once a peripheral phase-encode location is sampled, its probability is uniformly redistributed to other peripheral, non-sampled phase-encode positions for the next dynamic. This forced redistribution of sampling probability ensures that no peripheral dynamic will be acquired consecutively and that sampling pattern eventually cycles through all k-space location, ensuring all k-spaces lines will be periodically refreshed. Prior data now derives from the average of this sliding window, rather than from the very beginning of the acquisition.

In our retrospective study using the same data, our result shows that by refreshing the prior data via a sliding window approach, we are able to maintain a tumor-tracking accuracy of DSI = 0.93 and mean centroid error of 1.0 mm for the entire 3 min of data acquisition in the high SNR scenario and a DSI of 0.91 and centroid displacement of 1.2 mm for a low SNR scenario.

### Prior Data Assisted Compressed Sensing: Summary

Accelerated MRI (reconstructing undersampled MRI with special techniques) has many benefits for real-time MRI in RT. This added speed can be used for increased temporal resolution, spatial resolution, or imaging multiple slices. Accelerated MRI in the context of real-time motion monitoring is quite challenging, in the sense that a long reconstruction time will reduce or negate any of the advantages you gain reducing image acquisition. On the other hand, this problem is made easier by having access to a pool of prior data (due to this being a dynamic scan of a reasonably repetitive motion pattern). The PDACS method is one way in which prior data from a dynamic MR acquisition can be used to improve the image estimate for a undersampled "current" image. Ultimately, this technique can increase acquisition speed by 5×, with the cost of 100 ms reconstruction time, with the 100 ms being an unavoidable aspect of the iterative nature of the compressed sensing technique. Other non-iterative techniques that can take advantage of prior data for accelerated MRI reconstruction with even shorter reconstruction time are highly desirable.

## Compressed Sensing and Principal Component Analysis (CS-PCA)

Principal component analysis (PCA) is a mathematical tool frequently used for reducing the dimensionality of large data sets. In this process, data is mapped onto a new orthogonal basis set defined to correspond with the vectors of maximum variation. In the context of imaging, each pixel or value of acquired k-space can be considered separate variables, and a database of images can be processed through PCA to identify the most common modes of variation between them. Generally, the number of significant modes of variation is vastly smaller than the number of variables, and thus a limited subset of these modes or principal components (PCs) are responsible for the bulk of differences between images. On this assumption, if the database can be considered representative of subsequently acquired images, those new images can be well-represented by a limited number of coefficients representing the magnitudes of significant PCs.

In the context of real-time MRI and reconstruction, if undersampled data can be accurately mapped to a set of PCs from a previously acquired fully sampled database, then the PCs could be used to reconstruct the full image content. Unfortunately, the fact that only a partial amount of data is acquired in the undersampled image means that a simple projection will not be sufficient to determine appropriate PC coefficients for real-time reconstruction. Dietz et al. [32] from our laboratory employed a method to iteratively determine a subset and corresponding amplitudes of the PCs that best represented the undersampled data, analogous to minimizing the $L_1$ norm of a sparsifying transform in common compressed sensing implementations. In this work, although it could have technically been made part of the iterative procedure, the data consistency was forcibly maintained by reinserting the originally undersampled data at every iteration for the purpose of computational efficiency.

This method was tested retrospectively on a set of six patients with non-small cell lung tumors with institutional ethics approval originally acquired at four frames per second on a 3 T MRI scanner using a balanced steady-state free precession (bSSFP) sequence. The first 30 images in the dynamic set were used to create the database that generated representative PCs. A preliminary optimization was performed to select the optimal iteration parameters, including the maximum number of iterations as well as the maximum coefficient threshold below which the PC was dropped from the solution. As in the section above, the visible tumors were auto-contoured on both the original fully sampled images and the retrospectively sampled reconstructions to test for robustness in target delineation. Acceleration rates were tested across a range of 2–10× undersampling. The DSI and MCD of the contours were used to quantify differences between the undersampled reconstructions and those of the fully sampled data. Additional image quality metrics, including normalized mean squared error (NMSE) and structural similarity index (SSIM), were calculated to aid in this comparison.

### CS-PCA Performance

The mean DSI for this technique was reported above 0.9 for both high and low SNR states, even at an acceleration rate of 10×. Specifically at an acceleration of 5×, the DSI was greater than 0.92 for both high and low SNR scenarios. The mean centroid displacement was also reported to be robust up to the tested acceleration of 10×, with the maximum MCD of 1.15 mm across the range.

At 5× acceleration, the MCD was reported at 0.9 mm. The image fidelity and robustness up to high rates of accelerations were quite favorable for this technique. However, as we pointed out in this work, there is a slow quality degradation over time. Since the quality of any given reconstruction depends on how well it can be represented by the extracted features of an initial database, if the database becomes less relevant due to physiologic or anatomic motions, the resulting reconstruction will be less successful. Nevertheless, for the patients examined over a 3-min period in this study, the reported >0.9 DSI was similar or better to that reported for intra-observer contour variability as drawn by expert radiation oncologists when assessing the same data sets at different times (mean DSI = 088) [33]. Reconstruction times were reported to vary between 5 and 20 ms per dynamic image.

### CS-PCA Summary

The CS-PCA strategy as reported had several favorable attributes. Firstly, the algorithm is computationally efficient, with less than 20 ms reconstruction times on a desktop CPU. This is critical for real-time interventional applications which depend on low lag times between acquisition and image availability. Secondly, the method was reported to behave well (based on contouring metrics) in lower SNR regimes as might be expected at lower MR field strengths and sequences for rapid dynamic imaging and maintained its good performance for acceleration rates of up to 10×. A reported limitation was a slow degradation of metrics as the time between the database acquisition and accelerated dynamic frame increases. However, for the patient data sets analyzed in this work, the reported contour metrics were reliable over the observed 3-min dynamic imaging interval.

## Time-Domain PCA-Based Real-Time Reconstruction for Dynamic Undersampled MRI

The previous section discussed a method for using a PCA technique to reconstruct undersampled MRI data based on an initially acquired database. While its performance demonstrated suitability for real-time target tracking applications, anatomic or physiologic changes after the database acquisition could reduce the reconstruction quality. To adapt the technique into one where the database is continually updated, Wright et al. presented a work that uses a PCA characterization in the time domain based on undersampled data alone [34]. In this way, the database can adjust along with potential physiologic or muscle motions to produce a stable output over extended imaging times.

### Implementation Concept

Incoherent undersampling schemes for accelerated MRI generally have high sampling rates in the central low-frequency region of k-space and scattered samples out to the periphery. This work takes advantage of the consistent presence of the low-frequency data to generate a time-based characterization of k-space signal using PCA. With the assumption that time-based evolution in central and peripheral k-space results from the same physical motion sources, the principal characterizations of motion can be used to fit to the less frequently sampled peripheral k-space to fill in missing data with motion-based estimates.

To implement this idea, the authors generated sampling masks that consistently acquired a core of between 8 and 16 central phase encodes, with the periphery undersampled on a regular repeating pattern. Masks for acceleration rates varying between 3× and 8× were generated and used for testing. The time-domain core k-space database was chosen to cover 60 dynamic frames, with the last frame being the one to be reconstructed. As new data is acquired, this 60-frame time-domain window is shifted over, and a new database is formed to reconstruct the new dynamic. Of the PCs generated by the PCA process, only the first few will contain the most valuable motion characterizations with the last ones representing primarily noise.

The optimizations for this technique and subsequent testing were performed on the same six patient data sets as in the previous study with local institutional ethics approval. Both image-based metrics, including NMSE and SSIM, were performed as well as metrics specifically relating to contour accuracy, namely, DSI sand MCD.

### Performance of Time-Domain PCA Reconstruction

The performance of this technique was assessed over multiple minutes to look for trends in terms of degradation of image quality. While the initial dynamic frames generated using this method were slightly poorer in quality metrics than the previously described PCA technique, the performance remained relatively stable over time, as expected due to the continually updated database. All the acceleration rates up to 6× were reported to have DSI over 0.9, with the acceleration rate of 8× yielding a DSI of 0.88. These values are consistent with intra-observer variability when contoured by an expert radiation oncologist and suggest the method is suitable for real-time target tracking. The reconstruction times were reported at approximately 50 ms per dynamic frame on a CPU implementation. This is about twice the reconstruction time required for the previous method, likely due to the extra demands of recalculating the PCs from a new database for each image. However, this reconstruction time is still well within an appropriate range for real-time radiotherapy applications and could likely be improved with higher-spec hardware.

### Summary

The time-domain PCA approach for dynamic undersampled MR reconstruction in real time was shown to perform within acceptable bounds for target tracking. This method does not rely on any fully sampled data being acquired prior to the dynamic acquisition. It functions solely based on the undersampled data itself and uses common modes of variation in the time domain to fully populate k-space for the current frame-of-interest. Importantly, this method provides additional confidence that the geometry has not been misrepresented because of an out-of-date database in the event that there is an unexpected patient or organ motion.

## Patient-Specific CNN Reconstruction for Accelerated MR Imaging in Real-Time Radiotherapy Tracking

### Real-Time CNN Implementation

One of the first publications to apply 3D CNN to dynamic data with a practical implementation for real-time radiotherapy use was published in 2019 from our laboratory by Dietz et al. [35]. This work identified a workflow that would be practical for most radiotherapy patients and worked within that framework to develop a CNN architecture and training paradigm that would allow the radiotherapy process to play out without additional patient visits or requiring them to lay on the couch for prolonged periods during training. In this conceptualization, the patient would be MR imaged with fully sampled dynamic frames on the day of their simulation. A CNN specific to this patient would then be trained based on this data and would be ready to implement on the day of their first treatment to reconstruct accelerated data for target tracking. Provided the training could be easily performed within the span of a day, there would be no alteration of the treatment schedule.

The network itself consisted of a cascade of alternating convolutional networks and data consistency layers. The number of filters and layers in each convolution module and the cascade depth were empirically optimized. Raw MRI data is complex, so the real and imaginary data were fed into the network in separate channels. The 3D nature of the CNN takes advantage of redundancy in much of the slowly varying k-space content to improve the quality of the final 2D reconstruction, rather than having the network process each dynamic frame independently. Data for five dynamic frames were fed into the network at a time (the frame-of-interest preceded by data for the four previous consecutive frames) for processing. An additional data sharing layer was appended to the front of each convolutional module, combining adjacent frames to generate three extra artificial dynamic time points for more reconstruction stability.

Training and testing was performed on six patient dynamic data sets with non-small cell carcinoma, used with local institutional approval. The native frame rate of the source images was ~250 ms/frame, with a matrix of 128 × 128. There were 650 frames for each data set. To simulate undersampling, an incoherent undersampling mask was applied along the phase-encode direction, nulling all the values except for an undersampled fraction. Two different acceleration rates were investigated in this study: 5× and 10×. The first 300 of the available dynamic frames were used for training each patient's individualized CNN. The next 150 were split evenly between validation and testing purposes. The final 200 were used for metric evaluation to determine image and tumor contour accuracy, by comparing the results of the accelerated images to those of the fully sampled. The contour metrics evaluated were DSI and MCD, which are particularly useful parameters for assessing applicability for target tracking. The testing was performed using a single NVIDIA GTX 1080Ti GPU.

### CNN Performance

Training for this optimized network was reported to take roughly 6 h, well within the time window between initial simulation and first treatment day. The reconstruction time for each undersampled frame was reported to be 65 ms. The mean DSI across all CNN reconstructions was 0.94 for 5× acceleration and 0.92 for 10× acceleration. Similar results for 6× SNR-reduced data were reported with 0.92 for 5× and 0.91 for 10× acceleration. One of the patient data sets was reported to have a mean DSI below 0.85, potentially due to an elongated tumor shape to which the autocontouring was particularly sensitive to changes. This was reported as a potential direction for future work.

### Coherent Versus Incoherent Undersampling for CNN Reconstruction

Further work from the same group investigated an idea for improving the stability and accuracy of the CNN from the previous section by fundamentally altering how the data is undersampled [36]. Undersampling strategies such as those for compressed sensing generally require incoherently spaced samples to be distributed throughout k-space into the high-frequency regions. This creates an incoherent artifact pattern superimposed on top of an otherwise high resolution image and allows for an iterative model-based algorithm to remove artifact from the image while maintaining fidelity to the actual sampled data. However, since the CNN above is trained on fully sampled data, the authors proposed that the network can be trained to infer the higher-resolution regions from a strictly low-resolution acquisition. The data in central k-space generally has much higher SNR than does the high-frequency data, potentially offering advantages in terms of reconstruction stability. The same network architecture as in the previous section was employed, although the hyperparameters were reoptimized. The acceleration rates of 5× and 10× were investigated. With this new sampling strategy, the mean DSI was in general much improved across the patient data sets, in addition to the standard deviation and stability across the dynamic frames.

### Summary of Real-Time CNN Reconstruction

This section demonstrates the applicability of CNN reconstruction strategies to accelerated MRI dynamic data. Reconstruction duration was reported to be on the order of 65 ms, with DSI metrics that suggest the images are well-suited for target tracking purposes. While this strategy does require a 6-h training interval, the work points out that this may be feasible to fit into a clinic's regular workflow. The algorithms and hardware available for CNN training and execution are currently evolving, and as they do, the image quality and reconstruction times that such a strategy can produce may continue to improve as well.

## Noninvasive Intra-Fractional Tumor-Tracked Radiotherapy (NifteRT) using the Alberta Linac-MR

### Background

If a tumor is static during irradiation (e.g., brain tumor), modern RT can deliver highly accurate radiation to the tumor (<1 mm geometric error) without much harm to the adjacent normal tissue [37]. Unfortunately, most tumors are mobile and can deform significantly due to the respiratory and cardiac motions (e.g., lung tumor) [38–40], and such accurate RT delivery is challenging due to the uncertainty in tumor localization. RT is typically given over multiple fractions of daily treatment, and the continuous tumor motion and deformation occurring during the therapeutic radiation delivery in a given fraction is known as the intra-fractional tumor motion. In the RT of mobile tumors, intra-fractional tumor motion management and the surrounding critical structure sparing during irradiation are of paramount importance. This is particularly true for the modern stereotactic radiosurgery/stereotactic body radiotherapy (SRS/SBRT) techniques in which much higher dose per fraction is delivered compared to the conventional 3DCRT or IMRT [41].

A traditional solution to manage the intra-fractional tumor motion is to add geometric margins around the tumor ensuring adequate target coverage. This approach, however, can result in (1) unacceptable medical complications due to the excessive normal tissue irradiation adjacent to the tumor [42] or (2) poor disease control caused by limiting the necessary therapeutic dose to avoid those complications. Real-time tumor-tracking RT is an alternative approach that may potentially improve tumor localization. Both "intra-fractional" and "real-time" tumor-tracking RT refer to the methods of continuously tracking the tumor with radiation beam during beam delivery. Several tracking systems are commercially available, estimating the tumor location during irradiation based on invasive internal (X-ray opaque implant surgery) and/or external markers as tumor surrogates [43, 44]. These systems are used in conjunction with additional motion management techniques including abdominal compression, respiratory gating, and breath hold to reduce the geometric margins around the tumor [45].

Despite the variety of techniques, all current tracking methods are based on the indirect tracking using internal/external tumor surrogates. Reliance on surrogates, however, is known to be problematic for accurate tumor tracking because (1) utilizing internal surrogates requires invasive implant procedures that may cause medical complications [46, 47]. Moreover, the implanted surrogates are known to migrate or completely dislodge from their initial positions during the course of RT (typically several weeks) [48, 49]. (2) Tracking with external surrogates must rely on ambiguous correlations between the internal tumor motion and the external surrogate displacement [50]. More importantly, any deformation of tumor shape during irradiation is completely unknown, e.g., >10 mm for lung and liver tumors [40, 51] and up to 6.6 mm for pancreatic tumor [39, 52]. The additional motion management techniques also have disadvantages: (1) Not all patients can tolerate abdominal compression nor breath hold especially for those with impaired pulmonary function. In some cases, abdominal compression is known to work negatively by increasing the tumor motion, and its success rate is largely dependent on the patient and the anatomical location of tumor [53, 54]. (2) Respiratory gating cannot be used for patients with irregular breathing patterns, and it typically requires two to three times longer treatment time than continuous beam delivery, causing patient discomfort and challenges in the clinical workflow.

Most tumors reside in the soft tissue. However, CT, fluoroscopy, and other X-ray based image guidance systems currently available on the RT devices (e.g., linacs) provide good bony but very poor soft-tissue imaging during the therapeutic radiation delivery. This is the fundamental reason why the indirect tumor-tracking methods have long been used with the additional motion management techniques, despite their shortcomings mentioned above. Newly emerging hybrid linac-MRI systems may overcome the limitation by providing simultaneous MR imaging and therapeutic radiation delivery [2, 55, 56]. Our laboratory at the Cross Cancer Institute (CCI) built the world's first hybrid system known as linac-MR in 2008 [2] and a human whole-body system in 2014 [10]. Recently, our latest linac-MR system known as the Alberta linac-MR (LMR 2000) was installed and commissioned for its clinical use at the CCI. This linac-MR can provide continuous MR imaging of the mobile tumor and surrounding anatomy with excellent soft-tissue contrast, concurrent with the therapeutic radiation delivery.

To overcome the abovementioned challenges in RT for mobile tumors, we have been developing a novel RT method known as noninvasive intra-fractional tumor-tracked radiotherapy (nifteRT) using our linac-MR systems. In nifteRT, while a patient is under free breathing, a therapeutic radiation beam will track a moving and deforming tumor by adapting the beam's position and shape in real time.

## Overview of Objectives of the NifteRT Development Program

The objective of nifteRT program is to enhance the geometric accuracy of RT for mobile tumors for the Alberta linac-MR. Specifically, we aim to achieve <3 mm geometric error in radiation delivery treating three mobile tumor sites: lung, liver, and prostate. The 3 mm limit was calculated using the van Herk margin formula [57] by investigating the possible error sources in nifteRT. From the formula, by allowing 3 mm of a planning target volume (PTV) margin for a mobile target in nifteRT, we expect to cover a clinical target volume (CTV) with >95% of prescribed dose without the need of an internal target volume (ITV) margin.

The three tumor sites were chosen with our physician input, to maximize the clinical impact of nifteRT by reducing the PTV margin. The site selection was based on their potential for a large range of intra-fractional motion and the patient population at the CCI (the three tumor sites take 43% of all RT patients). Lung and liver tumors are known to move up to 40 and 50 mm in superior-inferior direction during normal breathing, respectively [58, 59]. Although intra-fractional prostate motion is known to be <7 mm in most cases [60], some patients may exhibit substantially large 3D prostate displacements up to 18 mm [61]. In fact, we observed ~18 mm of abrupt prostate displacement from our patient during 5 min of MR scan.

In general, RT requires a simulation session several days prior to the actual treatment day. NifteRT workflow expects that the simulation occurs in our linac-MR system, and a pretreatment MR scan is performed using the same MR sequence, patient setup, and imaging plane intended to be used during nifteRT. These MR images are used for our algorithm training and risk mitigation. An overall framework of nifteRT is shown in Fig. 11.5.

During nifteRT, our linac-MR acquires continuous MR images of a mobile tumor while the radiation is always on. In Step 1, an intra-fractional MR image of a tumor is acquired. In Step 2, the tumor shape is automatically detected by our auto-contouring algorithm (red boundary), and the centroid of tumor contour is considered as its position (a red dot). In Step 3, a future tumor position is predicted based on the current and previous positions by our motion prediction algorithm. In Step 4, the multi-leaf collimator (MLC) is driven to conform the treatment beam to the auto-contoured tumor shape at its predicted position. Steps 1–4 are continuously run each time a new MR image is acquired at 4 fps.

## NifteRT Step 1: Intra-Fractional MR Imaging

The Alberta linac-MR can provide 2D MR images at 4 fps, i.e., 250 ms acquisition time per frame, using a balanced steady-state free precession (bSSFP) sequence that is a widely used for fast sequence due to its high SNR potential [62]. The 250 ms acquisition time addresses the pro-

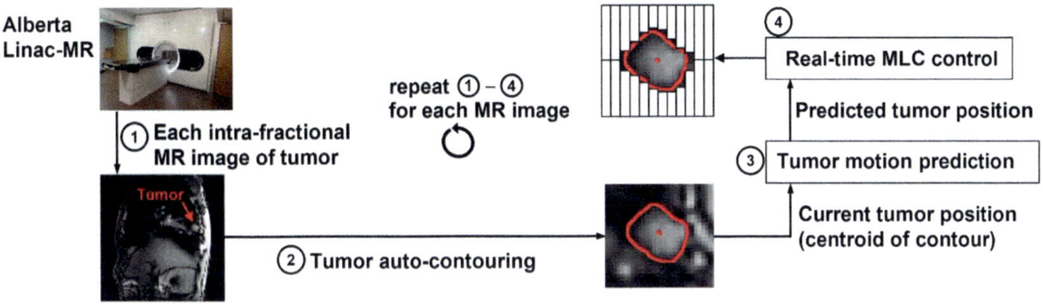


**Fig. 11.5** Overview of nifteRT framework in the linac-MR (radiation is always on during nifteRT)

cessing time in each step of nifteRT, following the American Association of Physicists in Medicine (AAPM) report recommending <500 ms time delay between the tumor detection and RT system response to manage the respiration-related tumor motion [63]. In nifteRT, the time delay is currently ~300 ms including image acquisition time (counted from the center of k-space acquisition, i.e., 125 ms for 250 ms image acquisition), auto-contouring and motion prediction (~60 ms), and MLC motion (~ 100 ms) [64].

Although our system can meet the AAPM recommendation at 250 ms/frame image acquisition time, the AAPM report states that the future tumor positions can be predicted (up to 80% accuracy) only if the time delay is less than 200 ms [63]. One obvious way to decrease the main source of time delay is to decrease the image acquisition time per frame. At the same time, because tumor auto-contouring is unavoidable to cope with the fast frame rate, our imaging techniques must not degrade the auto-contouring accuracy. To satisfy these requirements, we have implemented various partial k-space acquisition schemes to reduce the scan time, combined with novel image reconstruction algorithms to preserve the image quality, e.g., prior data assisted compressed sensing (PDACS) [30], sliding window PDACS [31], compressed sensing and principal component analysis (CS-PCA) [32], fully convolutional neural networks (FCNN) [35, 36], and time-domain PCA [34].

Our most accurate algorithm utilizes FCNN, which is a type of deep-learning neural network (DLNN) applicable to MR image acquisition and reconstruction [35, 36]. We trained FCNN to fill the gaps in partial k-space; thus, the difference in reconstructed images from partial k-space and fully sampled k-space is minimized. Since preserving auto-contouring accuracy at fast frame rate is critical, we used our in-house developed tumor auto-contouring algorithm as a quantitative tool measuring the fidelity of MR images [65]. Ideally, the algorithm generated tumor contours ($ROI_{auto}$) would be identical if the MR images reconstructed from partial k-space acquisition are the same as the ones from fully sampled k-space. Our FCNN algorithm was validated using six non-small cell lung cancer (NSCLC) patients' MR images (bSSFP, 250 ms/frame acquisition). For each patient, $ROI_{auto}$ were generated in 200 images reconstructed from both partial k-space and fully sampled k-space acquisitions. We calculated the Dice coefficient (DC) between each pair of $ROI_{auto}$ as a measure of agreement. Our new FCNN algorithm [35] showed superior performance over our previous CS-PCA algorithm [32]. In case of 5× acceleration in the image acquisition time, FCNN algorithm produced higher mean DC for all patients over CS-PCA algorithm as shown in Fig. 11.6. More importantly, FCNN algorithm showed lower variance in DC, indicating more stable image reconstruction performance.

## NifteRT Step 2: Tumor Auto-Contouring

The accuracy of tumor auto-contouring is crucial in nifteRT, because it dictates the accuracy of

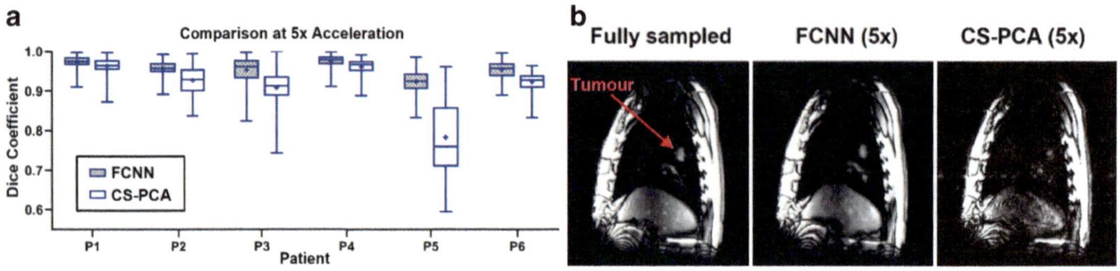


**Fig. 11.6** (**a**) DC comparison: FCNN vs. CS-PCA. (**b**) Image comparison: fully sampled vs. FCNN vs. CS-PCA

tumor localization. We developed and validated several auto-contouring algorithms against human experts' $ROI_{manual}$ in both phantom and in vivo MR images [65–70]. In our studies, we used various metrics to quantify the accuracy of $ROI_{auto}$ against $ROI_{manual}$, e.g., Dice coefficient (DC), Hausdorff distance (HD), and centroid displacement ($\Delta_{centroid}$) between $ROI_{auto}$ and $ROI_{manual}$. Hereinafter, we use DC as a metric indicating the accuracy of $ROI_{auto}$, e.g., 90% auto-contouring accuracy refers to 0.9 DC between $ROI_{auto}$ and $ROI_{manual}$. Also, to find the appropriate level of auto-contouring accuracy required for nifteRT, we investigated the inter- and intra-observer variations in $ROI_{manual}$ drawn by several experts [33]. Our study showed 87% inter- and 88% intra-observer variations in $ROI_{manual}$ drawn on the six NSCLC patients' image data. Because human experts' tumor contour is the gold standard in clinic, the accuracy of $ROI_{auto}$ needs to be comparable to $ROI_{manual}$. Thus, >90% accuracy in $ROI_{auto}$ is considered reasonably acceptable for nifteRT and is set as our general goal in the development of auto-contouring algorithms.

Previously we used pulse-coupled neural network (PCNN), which is a unique 2D neural network applicable to image segmentation tasks [65, 70]. Our PCNN algorithm performed well in lung tumor cases, where the contrast between the tumor and its background is relatively large. However, the algorithm may produce inaccurate contours in more challenging anatomies such as the liver or prostate, where the tumor contrast is low with unclear boundary to its surroundings. To overcome these challenges, our latest algorithm adopted U-net, which is a type of DLNN applicable to image segmentation tasks [66, 69]. We used 12 patients' MR images (11 lung, 10 liver, and 24 prostate cancer patients, 130 images/patient, 250 ms/frame image acquisition) to validate the U-net algorithm. In current RT of lung or liver cancer, a localized tumor lesion is defined, and the therapeutic radiation is focused on the local lesion. For prostate cancer, however, the current standard of care in RT is to irradiate the entire prostate gland. Whether to irradiate the seminal vesicles or not, in addition to the gland, is at each physician's discretion reflected in their $ROI_{manual}$. Hence, we tested our U-net algorithm for both scenarios in prostate auto-contouring.

For each patient, the first 30 images were used for patient-specific U-net algorithm training. In practice, these images will be obtained during the pretreatment RT simulation. The auto-contouring accuracy was evaluated by calculating DC between $ROI_{auto}$ and $ROI_{manual}$ over the remaining 100 images for each patient. From the 45 patients' data, the U-net algorithm showed 91, 90, and 93% mean accuracy in auto-contouring lung, liver, and prostate tumors, respectively. Figure 11.7 emphasizes the strength of U-net algorithm, especially in challenging anatomies: partial omission in tumor shape, diffuse tumor boundary, and low contrast situations. Moreover, our algorithm successfully reproduced each expert's preference whether to include or exclude seminal vesicles for prostate cases.

## NifteRT Step 3: Tumor Motion Prediction

Tumor motion prediction occurs immediately after the auto-contouring that localizes the current tumor position. The prediction is necessary to compensate for the tumor motion during the

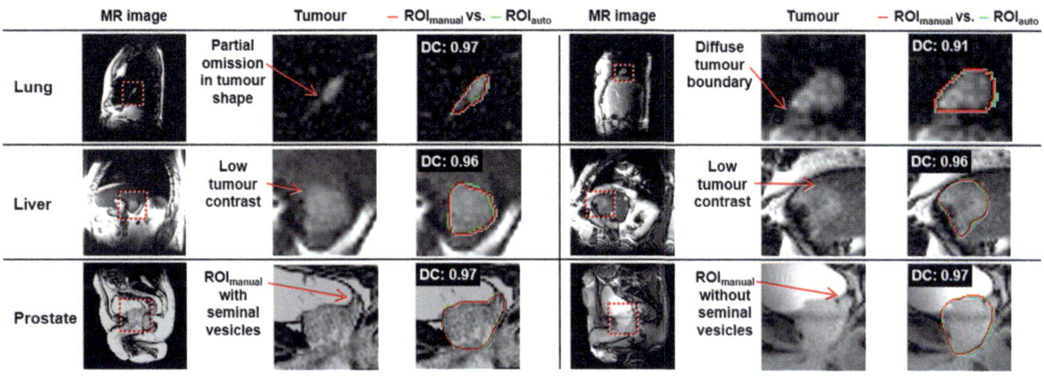


**Fig. 11.7** $ROI_{manual}$ (red contour) vs. $ROI_{auto}$ (green contour) comparisons in lung, liver, and prostate tumors

time delay as mentioned in Sect. "NifteRT step 1: Intra-Fractional MR Imaging." The range for our time delay during nifteRT was approximately 200–500 ms [64]. Without the motion prediction capability, therefore, the therapeutic radiation beam will always be lagging behind the actual tumor motion. Ideally, a future tumor shape should also be predicted if the tumor deforms severely during the time delay. Although this may be feasible by establishing a temporal tumor deformation model during free breathings, it is not addressed in the current nifteRT scheme. This is based on the observation that the deformation between each imaging (250 ms) is small, <0.6 mm in radius from our previous study [65], if each tumor shape is approximated as a circle.

Previously we used a feed forward artificial neural network, specifically multilayer perceptron (MLP), to predict 1D tumor motion [71]. In MLP, however, overfitting is difficult to avoid, and no time factor (previous memory) is involved in the prediction [72]. To address these shortcomings and add 3D motion prediction capability, our latest algorithm adopted long short-term memory (LSTM), which is a type of DLNN applicable to time series predictions [73, 74]. Our algorithm was validated using various 3D tumor motion data from 46 patients (30 lung, 9 pancreas, 2 liver, 2 retroperitoneum, 2 chest wall, 1 internal mammary nodes). In this initial study using LSTM, 46 patients were chosen based on data availability. Each patient's data contains 8 min of 3D tumor motion recorded at 40 ms intervals, which was considered as true 3D tumor positions in the study. The first 0.5 min of data was used for LSTM training, and the remaining 7.5 min of data was used to validate the motion prediction. The prediction error was defined as the distance between true and predicted tumor positions. We calculated the mean error, as well as 99% confidence interval (CI), over the 46 patients to address outliers. We used $= \text{mean} \pm \left(z\_\text{score} \times \text{standard deviation}\right) / \sqrt{\text{sample size}}$, where $z\_score = 2.58$, to calculate 99% CI. Using the LSTM algorithm, the 99% CI for the mean error was from 0.9 ± 0.1 mm to 1.6 ± 0.2 mm for 120–520 ms time delays, respectively. Figure 11.8 presents two patient examples of 3D tumor motion prediction at 280 ms time delay (pancreatic tumor: irregular motion; lung tumor: regular, cyclic motion).

## NifteRT Step 4: Real-Time MLC Control

Immediately after the motion prediction, the MLC is driven to conform the radiation beam to the auto-contoured tumor shape at its predicted position. To control the MLC on-the-fly during nifteRT, we developed the software/hardware components of MLC control system in-house and verified the system using a *Varian 52 leaf MLC* [64]. Also, we investigated the RF noise shielding for proper MR imaging during the MLC operation and the magnetic field effect on the MLC motors [4, 8]. We confirmed the proper MLC motor operation, as well as

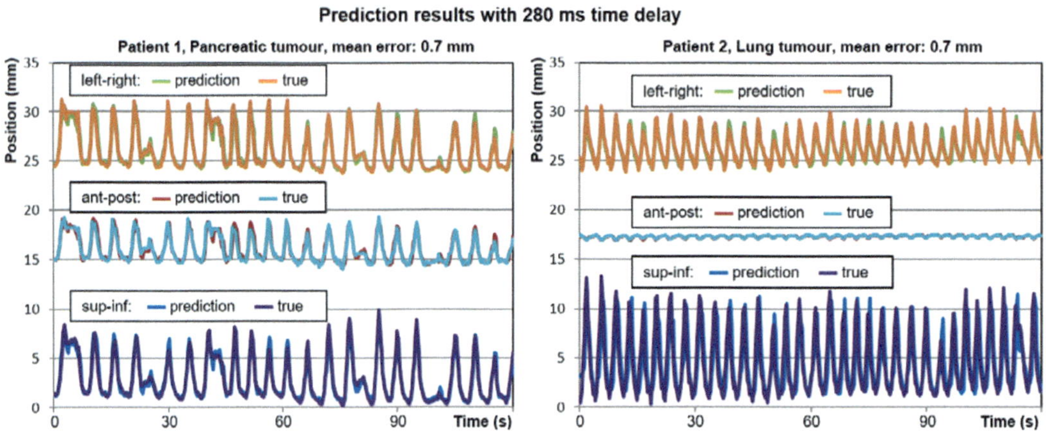


**Fig. 11.8** 3D tumour motion prediction using LSTM algorithm, pancreatic and lung tumour patients. Note: Color overlapping indicates high degree of agreement between predicted and true tumour positions

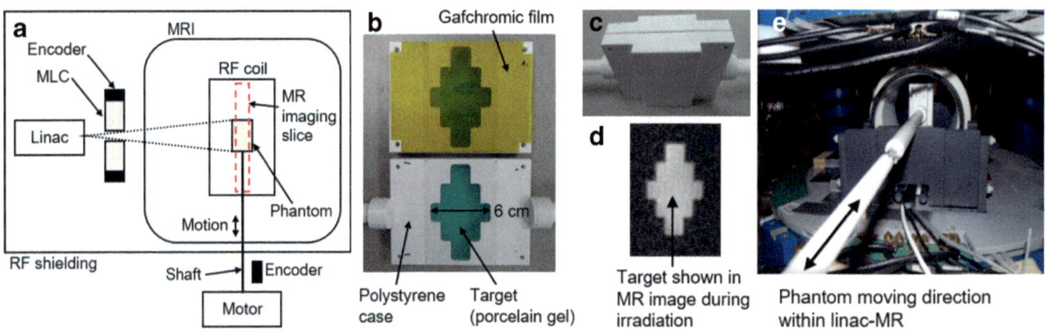


**Fig. 11.9** Experimental setup: (**a**) top-down view diagram of entire setup, (**b**) motion phantom components, (**c**) an assembled phantom, (**d**) MR image of the moving target, (**e**) phantom and RF coil placed in the linac-MR. Taken from Ref. [64]. Med Phys. 2013;40(5):(12pp) 051718

MR imaging during nifteRT in our linac-MR system [64].

## Demonstration of NifteRT in a Prototype Linac-MR

We reported the world's first demonstration of nifteRT utilizing 2D intra-fractional MR imaging and a motion phantom in our prototype linac-MR system [64]. A brief diagram of the experimental setup, motion phantom, and MR image of a moving target acquired during irradiation is shown in Fig. 11.9.

While our phantom was driving a target reproducing 1D lung tumor motion within the linac-MR system (motion range: 4 cm; period: 5 s), we were able to (1) acquire intra-fractional MR images of the moving target at 250 ms/frame image acquisition time, (2) automatically detect the shape and location of the target from each MR image, (3) predict the future target position over the time delay, and (4) control the MLC on-the-fly to conform the radiation beam to the auto-contoured target shape at its predicted position. These four steps were continuously run each time a new MR image was acquired, and the radiation beam was always on during nifteRT. The result is summarized in Fig. 11.10.

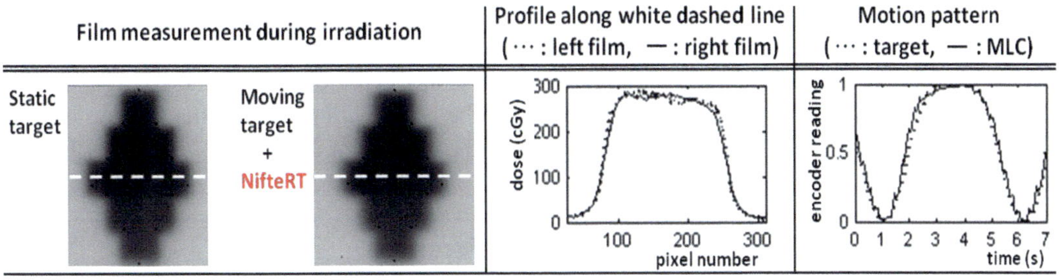


**Fig. 11.10** Film measurements and motion pattern comparisons between target and MLC. From Ref. [64] Med Phys. 2013;40(5):(12pp) 051718T

Our target, with Gafchromic film inserted, was irradiated for 2 min receiving ~2.5 Gy. The radiation delivery to a static or a moving target would be identical if nifteRT tracks the target perfectly. Thus, the film exposed in moving target irradiation was compared to the static target one via dose profiles. Also, two independently recorded motion patterns are compared from the target and the MLC during irradiation. The feasibility of nifteRT was verified by comparing the two dose profiles. We delivered highly accurate radiation to a moving target reproducing 1D lung tumor motion with <1.7 mm error.

## Calculated and Measured Radiation Doses with and without $B_0$

One of the characteristics of the Alberta linac-MR (a.k.a. Aurora-RT) is that the MR's Bo field does not affect the response of the radiation detectors nor the dose distribution perturbations [1, 75, 76]. This simplifies the commissioning process and provides the probability of using treatment planning systems that do not explicitly require code to consider magnetic fields.

Surface $D_{0.2\,cc}$ skin dose [77] is measured and calculated to increase from 46% to 75% when the field is turned on a 20 × 20 cm en face field*s* but increases only by 3–9% for tangents and five-field cases [78]. The increase is caused by contaminant electrons that may be captured and focused in the parallel configuration [79]. BEAMnrc radiation and Opera3D magnetic field were modeled to score the phase space at 120 cm isocenter and the PDDs results validated by measurements. The contaminate electrons originate in air at 60 cm from target near the magnetic pole plate and progressively increase to the isocenter to produce the surface-dose increase [78].

The LMR 2000 has an SAD of 120 cm, and the Bo can safely be on/off in minutes. A conventional TPS (Eclipse v13.6 TPS with doses calculated with the Pencil Beam Convolution algorithm to accommodate the SAD) **without $B_0$** was commissioned for the LMR 2000 **with $B_0$ on** [80]. Calculated tissue maximum ratios (TMR) and the tissue phantom ratios (TPR10) agree with measurements within 0.4% for source-to-point-distance (SPD) of 120 cm and 10 × 10 cm² fields at 1, 3, 5, and 10 cm depth. The calculated TMR and TPR10 agree within 0.48% and 0.26%, respectively, with measurements for the SPD of 127 cm, at depths of 15 and 20 cm for 9.4 × 9.4 cm² field sizes. PDD Gammas (3%|3 mm) are 94.9% and 100% if surface dose is excluded. At all depths to 37 cm, inline and crossline profiles agree within 96.4 and 94.4%, respectively. The Gammas of IMRT plans (3%|3 mm) delivered at 0° are 95.3% (five-field lung), 95.9% (seven-field brain), and 92.5% (five-field prostate). Other more advanced TPSs that can accommodate the 120 cm SAD are being investigated.

Reference dose calibration was performed [81] with EGSnrc simulations to validate TG51 kQ,msr with external validation from IROC (Imaging and Radiation Oncology Core, Houston,

TX). TPR20,10 is converted to equivalent %dd10× to kQ,msr with EGSnrc (MC) validation of the Exradin A12 used with and without the magnetic field. Calibrated dose agrees for the LMR 2000 within 1% for IROCs, OSLDs, and TLDs. Measurements of all TG51 parameters with field are essentially equivalent to that without field. Thus, kQ is 1 which means no magnetic field effects on the dose. Please note that linac-MR with perpendicular configuration can only use TLDs for routine output to IROC as OSLD reports have large variations [82].

The National Research Council (NRC), Canada, performed Co-60 alanine calibration and determined Dread/Dstated of 1.0092 and 0.9986 for water and solid water. The level of agreement between measured and stated is within the $k = 1$ uncertainty (within 1%), and the NRC ruled that no transport correction to $D_{read}$ for the dosimeters irradiated in the linac-MR beam is required.

## Conclusions

The system offers several interesting characteristics:

1. NO ERE with $\mathbf{B_0}$-induced minimal dose perturbation. Conventional commissioning and QA can be performed without the need of $B_0$ corrections, while conventional TPS can be used.
2. The large 110 cm by 60 cm rotating bore with couch motion allows positioning of the target to any planned position including isocenter positioning for offset tumors.
3. Installation is simplified with installation through conventional door or maze of the vault without the need of an exhaust vent. The high-temperature superconducting MR allows safe on/off switching within minutes.
4. Because of item 2 above, there is no need to adapt for every patient because the tumor, even those offset, can be placed at any planned position.
5. Because of item 1 above, any third party TPS and adaptive therapy system can be used. Gating is the current delivery system.
6. 3D MR imaging is about 3 min, while real-time imaging is at 4 fps providing online beam's eye view due to the configuration of the beam with respect to the MR system and the rotation of the gantry.

## References


1. Kirkby C, Murray B, Rathee S, Fallone BG. Lung dosimetry in a linac-MRI radiotherapy unit with a longitudinal magnetic field. Med Phys. 2010;37(9):4722–32.
2. Fallone BG, Murray B, Rathee S, Stanescu T, Steciw S, Vidakovic S, et al. First MR images obtained during megavoltage photon irradiation from a prototype integrated linac-MR system. Med Phys. 2009;36(6):2084–8.
3. Burke B, Lamey M, Rathee S, Murray B, Fallone BG. Radio frequency noise from clinical linear accelerators. Phys Med Biol. 2009;54(8):2483–92.
4. Lamey M, Yun J, Burke B, Rathee S, Fallone BG. Radio frequency noise from an MLC: a feasibility study of the use of an MLC for linac-MR systems. Phys Med Biol. 2010;55(4):981–94.
5. Lamey M, Burke B, Blosser E, Rathee S, De Zanche N, Fallone BG. Radio frequency shielding for a linac-MRI system. Phys Med Biol. 2010;55(4):995–1006.
6. St. Aubin J, Steciw S, Fallone BG. The design of a simulated in-line side-coupled 6 MV linear accelerator waveguide. Med Phys. 2010;37(2):466–76.
7. St Aubin J, Steciw S, Fallone BG. Waveguide detuning caused by transverse magnetic fields on a simulated in-line 6 MV linac. Med Phys. 2010;37(9):4751–4.
8. Yun J, St Aubin J, Rathee S, Fallone BG. Brushed permanent magnet DC MLC motor operation in an external magnetic field. Med Phys. 2010;37(5):2131–4.
9. St Aubin J, Steciw S, Fallone BG. Magnetic decoupling of the linac in a low field biplanar linac-MR system. Med Phys. 2010;37(9):4755–61.
10. Fallone BG. The rotating biplanar linac-magnetic resonance imaging system. Semin Radiat Oncol. 2014;24(3):200–2.
11. St Aubin J, Santos DM, Steciw S, Fallone BG. Effect of longitudinal magnetic fields on a simulated in-line 6 MV linac. Med Phys. 2010;37(9):4916–23.
12. Wachowicz K, De Zanche N, Yip E, Volotovskyy V, Fallone BG. CNR considerations for rapid real-time MRI tumor tracking in radiotherapy hybrid devices: effects of B0 field strength. Med Phys. 2016;43(8):4903.
13. Reynolds M, Fallone BG, Rathee S. Dose response of selected ion chambers in applied homogeneous transverse and longitudinal magnetic fields. Med Phys. 2013;40(4):042102.
14. de Pooter J, Billas I, de Prez L, Duane S, Kapsch RP, Karger CP, et al. Reference dosimetry in MRI-

linacs: evaluation of available protocols and data to establish a code of practice. Phys Med Biol. 2021;66(5):05TR2.

15. Reynolds M, Fallone BG, Rathee S. Technical note: response measurement for select radiation detectors in magnetic fields. Med Phys. 2015;42(6):2837–40.
16. Reynolds M, Rathee S, Fallone BG. Technical note: ion chamber angular dependence in a magnetic field. Med Phys. 2017;44(8):4322–8.
17. Malkov VN, Rogers DWO. Monte Carlo study of ionization chamber magnetic field correction factors as a function of angle and beam quality. Med Phys. 2018;45(2):908–25.
18. Spindeldreier CK, Schrenk O, Bakenecker A, Kawrakow I, Burigo L, Karger CP, et al. Radiation dosimetry in magnetic fields with farmer-type ionization chambers: determination of magnetic field correction factors for different magnetic field strengths and field orientations. Phys Med Biol. 2017;62(16):6708–28.
19. van Asselen B, Woodings SJ, Hackett SL, van Soest TL, Kok JGM, Raaymakers BW, et al. A formalism for reference dosimetry in photon beams in the presence of a magnetic field. Phys Med Biol. 2018;63(12):125008.
20. Reynolds M, Fallone BG, Rathee S. Dose response of selected solid state detectors in applied homogeneous transverse and longitudinal magnetic fields. Med Phys. 2014;41(9):092103.
21. Ghila A, Fallone BG, Rathee S. Influence of standard RF coil materials on surface and buildup dose from a 6 MV photon beam in magnetic field. Med Phys. 2016;43(11):5808.
22. Bray FN, Simmons BJ, Wolfson AH, Nouri K. Acute and chronic cutaneous reactions to ionizing radiation therapy. Dermatol Ther (Heidelb). 2016;6(2):185–206.
23. Jackson JD. Classical electrodynamics, vol. xxi. 3rd ed. New York: Wiley; 1999. p. 808.
24. Corea JR, Lechene PB, Lustig M, Arias AC. Materials and methods for higher performance screen-printed flexible MRI receive coils. Magn Reson Med. 2017;78(2):775–83.
25. Barta R, Ghila A, Rathee S, Fallone BG, De Zanche N. Impact of a parallel magnetic field on radiation dose beneath thin copper and aluminum foils. Biomed Phys Eng Express. 2020;6(3):037002.
26. https://doi.org/10.1002/mrm.28540.
27. Edelstein WA, Glover GH, Hardy CJ, Redington RW. The intrinsic signal-to-noise ratio in NMR imaging. Magn Reson Med. 1986;3(4):604–18.
28. Lustig M, Donoho DL, Pauly JM. Sparse MRI: the application of compressed sensing for rapid MR imaging. Magn Reson Med. 2007;58(6):1182–95.
29. Goldstein T, Osher S. The split Bregman method for L1-regularized problems. Siam J Imaging Sci. 2009;2(2):323–43.
30. Yip E, Yun J, Wachowicz K, Heikal AA, Gabos Z, Rathee S, et al. Prior data assisted compressed sensing: a novel MR imaging strategy for real time tracking of lung tumors. Med Phys. 2014;41(8):082301.
31. Yip E, Yun J, Wachowicz K, Gabos Z, Rathee S, Fallone BG. Sliding window prior data assisted compressed sensing for MRI tracking of lung tumors. Med Phys. 2017;44(1):84–98.
32. Dietz B, Yip E, Yun J, Fallone B, Wachowicz K. Real-time dynamic MR image reconstruction using compressed sensing and principal component analysis (CS-PCA): demonstration in lung tumor tracking. Med Phys. 2017;44(8):3978–89.
33. Yip E, Yun J, Gabos Z, Baker S, Yee D, Wachowicz K, et al. Evaluating performance of a user-trained MR lung tumor autocontouring algorithm in the context of intra-and interobserver variations. Med Phys. 2018;45(1):307–13.
34. Wright M, Dietz B, Yip E, Yun J, Gabos Z, Fallone B, et al. Time domain principal component analysis for rapid, real-time 2D MRI reconstruction from undersampled data. Med Phys. 2021;48(11):6724–39.
35. Dietz B, Yun J, Yip E, Gabos Z, Fallone B, Wachowicz K. Single patient convolutional neural networks for real-time MR reconstruction: a proof of concept application in lung tumor segmentation for adaptive radiotherapy. Phys Med Biol. 2019;64(19):195002.
36. Dietz B, Yun J, Yip E, Gabos Z, Fallone B, Wachowicz K. Single patient convolutional neural networks for real-time MR reconstruction: coherent low-resolution versus incoherent undersampling. Phys Med Biol. 2020;65(8):08NT3.
37. Park J-H, Han JH, Kim C-Y, Oh CW, Lee D-H, Suh T-S, et al. Application of the gamma evaluation method in gamma knife film dosimetry. Med Phys. 2011;38(10):5778–87.
38. Huang F, Ma C, Wang R, Gong G, Shang D, Yin Y. Defining the individual internal gross tumor volume of hepatocellular carcinoma using 4DCT and T2-weighted MRI images by deformable registration. Transl Cancer Res. 2018;7(1):151–7.
39. Feng M, Balter JM, Normolle D, Adusumilli S, Cao Y, Chenevert TL, et al. Characterization of pancreatic tumor motion using cine MRI: surrogates for tumor position should be used with caution. Int J Radiat Oncol Biol Phys. 2009;74(3):884–91.
40. Kyriakou E, McKenzie DR. Changes in lung tumor shape during respiration. Phys Med Biol. 2012;57(4):919–35.
41. Zhang GG, Yu H-HM, Stevens CW, Dilling TJ, Hoffe SE, Moros EG, et al. Motion management in stereotactic body radiotherapy. J Nucl Med Radiat Ther. 2012;S6(012):6.
42. Hendry JH, Jeremic B, Zubizarreta EH. Normal tissue complications after radiation therapy. Rev Panam Salud Publica. 2006;20(2–3):151–60.
43. Vedam SS, Kini VR, Keall PJ, Ramakrishnan V, Mostafavi H, Mohan R. Quantifying the predictability of diaphragm motion during respiration with a noninvasive external marker. Med Phys. 2003;30(4):505–13.

44. Nioutsikou E, Seppenwoolde Y, Symonds-Tayler JRN, Heijmen B, Evans P, Webb S. Dosimetric investigation of lung tumor motion compensation with a robotic respiratory tracking system: an experimental study. Med Phys. 2008;35(4):1232–40.
45. Giraud P, Houle A. Respiratory gating for radiotherapy: main technical aspects and clinical benefits. ISRN Pulmonol. 2013;2013:1–3.
46. Bhagat N, Fidelman N, Durack JC, Collins J, Gordon RL, LaBerge JM, et al. Complications associated with the percutaneous insertion of fiducial markers in the thorax. Cardiovasc Intervent Radiol. 2010;33(6):1186–91.
47. Roberge D, Cabrera T. Liver biopsy in modern medicine. UK: IntechOpen Ltd.; 2011.
48. Kitamura K, Shirato H, Shimizu S, Shinohara N, Harabayashi T, Shimizu T, et al. Registration accuracy and possible migration of internal fiducial gold marker implanted in prostate and liver treated with real-time tumor-tracking radiation therapy (RTRT). Radiother Oncol. 2002;62(3):275–81.
49. Imura M, Yamazaki K, Shirato H, Onimaru R, Fujino M, Shimizu S, et al. Insertion and fixation of fiducial markers for setup and tracking of lung tumors in radiotherapy. Int J Radiat Oncol Biol Phys. 2005;63(5):1442–7.
50. Ozhasoglu C, Murphy MJ. Issues in respiratory motion compensation during external-beam radiotherapy. Int J Radiat Oncol Biol Phys. 2002;52(5):1389–99.
51. Paulsson AK, Yom SS, Anwar M, Pinnaduwage D, Sudhyadhom A, Gottschalk AR, et al. Respiration-induced intraorgan deformation of the liver: implications for treatment planning in patients treated with fiducial tracking. Technol Cancer Res Treat. 2017;16(6):776–82.
52. Chen S, Lin C, Enke C, Zhou S. SU-E-J-122: quantification of respiratory-induced pancreatic head tumor rotation and deformation using 4DCT and fiducial markers. Med Phys. 2012;39(6Part8):3680.
53. Gelder RV, Wong S, Le A, Podreka A, Briggs A, Haddad C, et al. Experience with an abdominal compression band for radiotherapy of upper abdominal tumours. J Med Radiat Sci. 2018;65(1):48–54.
54. Bouilhol G, Ayadi M, Rit S, Thengumpallil S, Schaerer J, Vandemeulebroucke J, et al. Is abdominal compression useful in lung stereotactic body radiation therapy? A 4DCT and dosimetric lobe-dependent study. Phys Med. 2012;29(4):333–40.
55. Dempsey JF, Benoit D, Fitzsimmons JR, Haghighat A, Li JG, Low DA, et al. A device for realtime 3D image-guided IMRT. Int J Radiat Oncol Biol Phys. 2005;63(2):1095.
56. Raaymakers BW, Wlagendijk JJ, Overweg J, Kok JGM, Raaijmakers AJE, Kerkhof EM, et al. Integrating a 1.5 T MRI scanner with a 6 MV accelerator: proof of concept. Phys Med Biol. 2009;54(12):N229–N37.
57. van Herk M, Remeijer P, Rasch C, Lebesque JV. The probability of correct target dosage: dose-population histograms for deriving treatment margins in radiotherapy. Int J Radiat Oncol Biol Phys. 2000;47(4):1121–35.
58. Plathow C, Fink C, Ley S, Puderbach M, Eichinger M, Zuna I, et al. Measurement of tumor diameter-dependent mobility of lung tumors by dynamic MRI. Radiother Oncol. 2004;73(3):349–54.
59. Eccles CL, Patel R, Simeonov AK, Lockwood G, Haider M, Dawson LA. Comparison of liver tumor motion with and without abdominal compression using cine-magnetic resonance imaging. Int J Radiat Oncol Biol Phys. 2011;79(2):602–8.
60. Tong X, Chen X, Li J, Xu Q, Lin M-H, Chen L, et al. Intrafractional prostate motion during external beam radiotherapy monitored by a real-time target localization system. J Appl Clin Med Phys. 2015;16(2):51–61.
61. Sihono DSK, Ehmann M, Heitmann S, Von Swietochowski S, Grimm M, Boda-Heggemann J, et al. Determination of Intrafraction prostate motion during external beam radiation therapy with a transperineal 4-dimensional ultrasound real-time tracking system. Int J Radiat Oncol Biol Phys. 2018;101(1):136–43.
62. Chavhan GB, Babyn PS, Jankharia BG, Cheng H-LM, Shroff MM. Steady-state MR imaging sequences: physics, classification, and clinical applications. Radiographics. 2008;28(4):1147–60.
63. Keall PJ, Mageras GS, Balter JM, Emery RS, Forster KM, Jiang SB, et al. The management of respiratory motion in radiation oncology report of AAPM task group 76. Med Phys. 2006;33(10):3874–900.
64. Yun J, Wachowicz K, Mackenzie M, Rathee S, Robinson D, Fallone BG. First demonstration of intrafractional tumor-tracked irradiation using 2D phantom MR images on a prototype linac-MR. Med Phys. 2013;40(5):051718.
65. Yun J, Yip E, Gabos Z, Wachowicz K, Rathee S, Fallone BG. Improved lung tumor autocontouring algorithm for intrafractional tumor tracking using 0.5 T linac-MR. Biomed Phys Eng Express. 2016;2(6):067004.
66. Yun J, Yip E, Gabos Z, Baker S, Yee D, Wachowicz K, et al. A deep learning-based tumor auto-contouring algorithm for real-time tumor tracking using linac-MR. Med Phys. 2018;45(6):E669.
67. Tahmasebi N, Boulanger P, Yun J, Fallone BG, Punithakumar K. Tracking tumor boundary using point correspondence for adaptive radio therapy. Comput Methods Prog Biomed. 2018;165:187–95.
68. Yun J, Yip E, Wachowicz K, Rathee S, Mackenzie M, Robinson D, et al. Evaluation of a lung tumor autocontouring algorithm for intrafractional tumor tracking using low-field MRI: a phantom study. Med Phys. 2012;39(3):1481–94.
69. Han G, Wachowicz K, Usmani N, Yee J, Wong J, Fallone BG, et al. Deep learning-based autocontouring algorithm for non-invasive intrafractional tumour-tracked radiotherapy (nifteRT) on Linac-MR. Med Phys. 2022;49(8):5634.
70. Yun J, Yip E, Gabos Z, Wachowicz K, Rathee S, Fallone BG. Neural-network based autocontouring algorithm for intrafractional lung-tumor tracking using Linac-MR. Med Phys. 2015;42(5):2296–310.

71. Yun J, Mackenzie M, Rathee S, Robinson D, Fallone BG. An artificial neural network (ANN)-based lung-tumor motion predictor for intrafractional MR tumor tracking. Med Phys. 2012;39(7):4423–33.
72. Lawrence S, Giles CL, editors. Overfitting and neural networks: conjugate gradient and backpropagation. Proceedings of the IEEE-INNS-ENNS International Joint Conference on Neural Networks IJCNN 2000 Neural Computing: New Challenges and Perspectives for the New Millennium; 2000 27-27 July 2000.
73. Yun J, Rathee S, Fallone BG. A deep-learning based 3D tumor motion prediction algorithm for non-invasive intra-fractional tumor-tracked radiotherapy (nifteRT) on Linac-MR. Int J Radiat Oncol Biol Phys. 2019;105(1):S28.
74. Johnson N, Wachowicz K, Rathee S, Fallone BG, Yun J. Accurate, on-demand neural networks for respiratory motion forecasting. Med Phys. 2022;49(8):5640.
75. Yang R, Santos DM, Fallone BG, St-Aubin J. A novel transport sweep architecture for efficient deterministic patient dose calculations in MRI-guided radiotherapy. Phys Med Biol. 2019;64(18):185012.
76. Yang R, Santos DM, Fallone BG, St-Aubin J. Feasibility of energy adaptive angular meshing for perpendicular and parallel magnetic fields in a grid based Boltzmann solver. Biomed Phys Eng Express. 2020;6(2):025006.
77. Hilts M, Halperin H, Morton D, Batchelar D, Bachand F, Chowdhury R, et al. Skin dose in breast brachytherapy: defining a robust metric. Brachytherapy. 2015;14(6):970–8.
78. Oliver P, Yip E, Reynolds M, Burke B, Fallone BG, Murray B, et al. Skin dose investigations on a 0.5 T inline rotating biplanar linac-MR. Med Phys. 2022;49(8):5628.
79. Reynolds M, Oliver PAK, Wood T, Wachowicz K, Burke B, Fallone BG. Contaminant electron origins on a 0.5T inline linac-MR. Med Phys. 2022;49(8):5681.
80. Tari SY, Oliver P, Sinn D, Wood T, Steciw S, Murray B, et al. Initial experience for treatment planning system commissioning of the 0.5 T inline rotating bi-planar Linac-MR system. Med Phys. 2022;49(8):5693.
81. Yip E, Tari SY, Reynolds M, Sinn D, Murray B, Fallone BG, et al. Clinical reference dosimetry on the inline 0.5T rotating biplanar Linac-MR system. Med Phys. 2022;48(8):5630.
82. Alvarez P, IROC, editors. Updates imaging and dosimetry relevant to MRgRT2021. Summer School of the American Association of Physicists in Medicine; 2021.

# 12 Clinical Applications of MR-Linac in Oligometastatic Disease

Katie N. Lee, Martijn Intven, Filippo Alongi, and Mai Anh Huynh

## Disease Overview: Oligometastatic Disease

"Oligometastatic" disease (OMD) has been defined as an intermediate and distinct state of metastatic spread, wherein metastatic lesions are confined to a single or limited number of organs, and metastatic-directed therapy (MDT) can impact the natural history of cancer spread. Originally proposed in 1995 by Hellman and Weichselbaum, they contended that this specific clinical scenario could deserve a radical approach to localized therapy, as in this early state of metastatic disease, the ability to ablate all detectable tumor could translate into prolonged survival and even cure [1]. The oligometastatic state has been increasingly recognized throughout the intervening decades, and its definition continues to be further refined to characterize different disease states within the clinical spectrum of OMD [2]. Currently, advances in methods of oncologic imaging, detection, and treatment have transformed cancer into an increasingly chronic disease with most deaths (greater than 90%) from solid tumors caused by metastasis [3]. The understanding of the metastatic disease continuum and effective therapies for this disease state remains of utmost importance.

K. N. Lee · M. A. Huynh (✉)
Harvard Radiation Oncology Program, Brigham and Women's Hospital/Dana Farber Cancer Center/Massachusetts General Hospital/Boston Children's Hospital, Boston, MA, USA

Department of Radiation Oncology, Dana-Farber/Brigham and Women's Cancer Center and Harvard Medical School, Boston, MA, USA
e-mail: katie_hwang@dfci.harvard.edu; MHUYNH@BWH.HARVARD.EDU

M. Intven
Department of Radiotherapy, Division Imaging and Oncology, University Medical Centre Utrecht, Utrecht, The Netherlands
e-mail: m.intven@umcutrecht.nl

F. Alongi
Advanced Radiation Oncology Department, Cancer Care Center, IRCCS Sacro Cuore Don Calabria Hospital, Negrar, VR, Italy

## Defining Oligometastatic Disease

The definition of OMD continues to evolve since its original conception as a limited number of metastases lying on the spectrum between localized and more widespread tumor dissemination. In clinical trials and meta-analyses, the number of oligometastatic sites has typically been limited to a maximum of three to five lesions [4]. However, some studies have included individual lymph nodes in this number, whereas others have not, and some have limited the number of metastases allowable in a single organ [5–8]. Following a systematic review of the OMD literature, which has predominantly been comprised of retrospective experiences, the European Society for



I. J. Das et al. (eds.), *A Practical Guide to MR-Linac*, https://doi.org/10.1007/978-3-031-48165-9_12

Therapeutic Radiation Oncology and American Society for Therapeutic Radiation Oncology (ESTRO-ASTRO) consensus guidelines defined OMD as one to five lesions, where a controlled primary is optional, but where all metastatic sites must be safely treatable. In the absence of clinical data supporting a maximum number of metastases and organs to define OMD, ESTRO-ASTRO consensus guidelines further specify the ability to deliver safe and clinically meaningful radiotherapy with curative intent to all metastatic sites as a minimum requirement [9].

Beyond a numeric view of OMD, it is important to recognize that OMD could be considered a dynamic state. The magnitude of benefit of MDT for patients may vary depending on the biology underlying the relative timing of metastatic presentation, prior exposure to systemic therapy, or how MDT compares as an alternative to the toxicities associated with systemic therapy. Recognizing the heterogeneity represented within OMD, the ESTRO and European Organization for Research and Treatment of Cancer (EORTC) created a consensus recommendation of nine subgroups [10]. Categories associated with oligometastases include synchronous, in which the primary cancer and oligometastases present at time of initial diagnosis (within 6 months); metachronous, in which OMD develops following treatment of the primary cancer (>6 months); and induced, in which patients with a history of polymetastatic disease are reduced to an oligometastatic number of lesions following systemic treatment (Fig. 12.1). Further defining the OMD state are oligorecurrence, in which OMD develops while systemic therapy-free; oligoprogression, in which oligometastatic sites progress while under treatment with systemic therapy; and oligopersistence, in which there is stable disease or partial response while under treatment with systemic therapy. The prognostic significance of the ESTRO-EORTC classification system remains unknown, although some studies have identified improved PFS and OS among patients with metachronous, synchronous, and repeat presentations vs. induced using this classification system [11].

However, there are likely additional factors characterizing OMD that go beyond this subclassification of the OMD state and may affect patient outcomes. These characteristics, such as lesion number, site, size, and histology, remain controversial, and their exact contribution remains unknown. Furthermore, rapid improvements in diagnostic imaging, specifically in prostate cancer with prostate specific membrane antigen (PSMA)-PET, have led to earlier detection of lower oligometastatic disease burden [12]. With most prior studies defining the OMD state by number of sites based on conventional imaging, now with these improved detection methods, the OMD state and application of these studies will likely need to be further refined [5, 13, 14].

## SBRT Treatment for Oligometastatic Disease

The treatment paradigm for OMD, in general, has been changing with studies demonstrating a survival benefit for definitive treatment of metastases when limited to a maximum of three to five sites, both in the upfront setting and at time of relapse and progression. Hellman and Weichselbaum in 1995 insightfully hypothesized that the application of conformal radiotherapy, involving high doses of radiation delivered precisely to lesions with normal tissue sparing, could be of benefit for those with OMD [1]. Advancements in radiotherapy with image guidance, respiratory motion management, and improved methods of treatment planning and delivery enabled the development of stereotactic body radiotherapy (SBRT), where ablative doses of radiation can be given with rapid dose falloff to surrounding normal tissues. SBRT has increasingly been adopted to ablate common areas of metastatic spread, including the lung, liver, colorectal region, adrenal glands, and bone [13, 15–23]. Draining lymph node regions are also common sites for cancer spread, and SBRT for oligometastatic nodal disease has been associated with greater than 80% local control [24]. Severe (greater than grade 3) toxic-

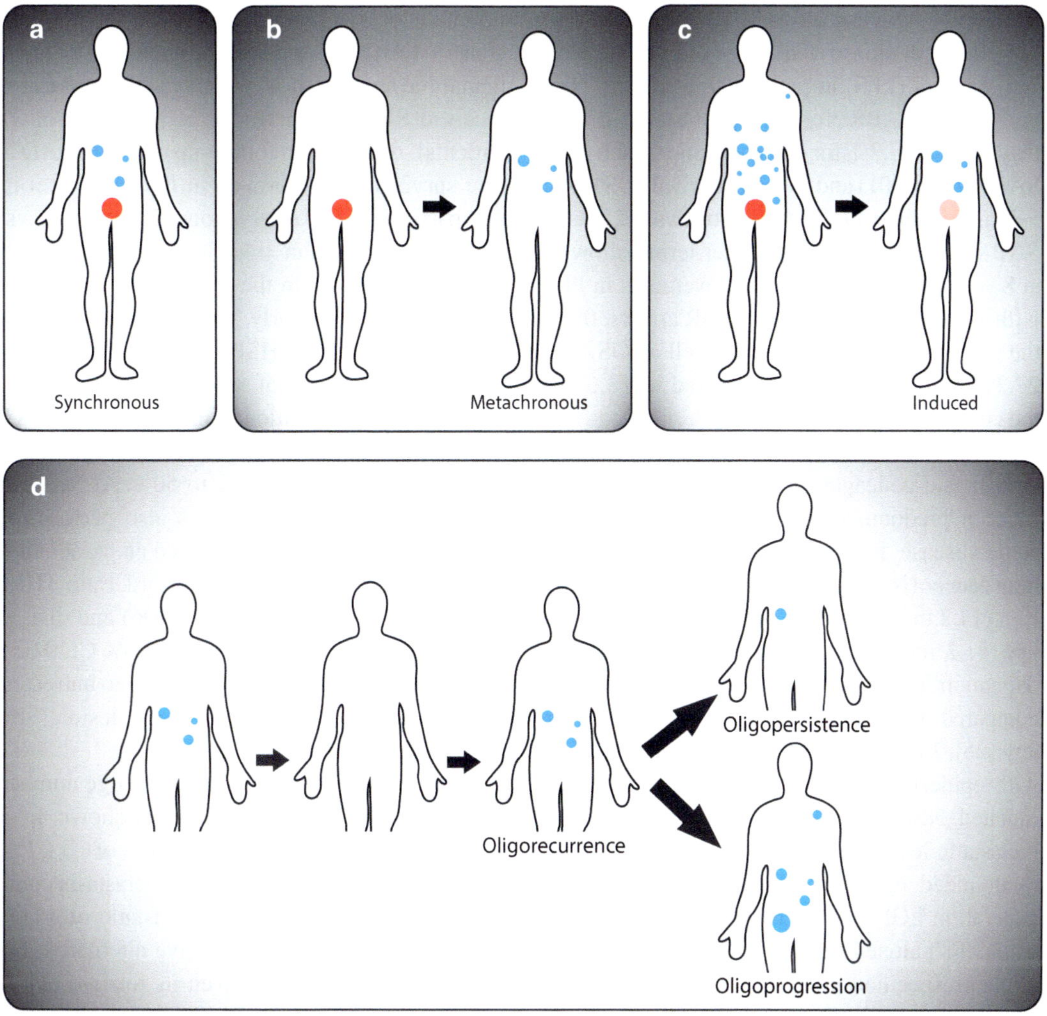


**Fig. 12.1** Oligometastatic disease classification. (**a**) Synchronous OMD where primary (red) and metastatic lesions (blue) present at time of initial diagnosis or within 6 months. (**b**) Metachronous OMD where following treatment of primary (red) and systemic therapy-free interval there is development of OMD (blue). (**c**) Induced OMD where previously diagnosed polymetastatic disease followed by systemic treatment, with or without local treatment of the primary (red), leads to OMD. (**d**) Oligorecurrence where OMD with new or regrowing lesions develops following systemic therapy-free interval; oligoprogression where OMD with new or regrowing lesions develops while undergoing systemic therapy; oligopersistence where OMD lesions are persistent or have partial response while undergoing systemic therapy

ity was mainly GI toxicity and occurred in the setting of re-irradiation. For the purposes of this book chapter, as other chapters address MR-guided treatments of the lung, liver, and bone specifically, we will focus on abdomino-pelvic oligometastatic lesions.

Supporting the role of metastasis-directed therapy (MDT) was the prospective phase II trial SABR-COMET, which randomized patients with a controlled primary solid tumor and up to five distant metastases to either the palliative standard of care (SOC) or standard therapy plus stereotactic ablative body radiation (SABR) in a 1:2 ratio [25]. They demonstrated improved progression-free survival (PFS) in patients treated with comprehensive SABR for oligometastatic disease (5-year PFS 17.3% for control vs. not reached for SABR, $p = 0.001$) and overall survival (OS), with 5-year OS being 17.7% for control vs. 42.3% for SABR ($p = 0.006$) [25]. Longer-term follow-up at 8 years showed durable improvements in PFS, with estimated 21.3% in the SABR arm vs. 0% in the control arm ($p < 0.001$), as well as OS, with 27.2% in the SABR arm compared to 13.6% in the control arm ($p = 0.008$) [7].

In non-small cell lung cancer (NSCLC), Gomez and colleagues looked at definitive treatment of predominantly synchronous oligometastatic disease for patients with non-small cell lung cancer (NSCLC) and showed benefit of both PFS (14.2 months vs. 4.4 months, $p = 0.022$) and OS (41.2 months vs. 17 months, $p = 0.017$) in patients treated with local consolidative therapy compared with observation or maintenance therapy [26]. In the SINDAS phase III trial, patients with epidermal growth factor receptor (EGFR) mutated NSCLC and one to five synchronous metastatic lesions without brain metastases were randomized to either SABR followed by first-generation EGFR tyrosine kinase inhibitor (TKI) versus TKI alone [27]. Due to the efficacy results at the prespecified interim analysis, the trial was stopped at enrollment of 133 patients as the median PFS of 20.2 months in the RT + TKI arm vs. 12.5 months in the TKI only arm ($p < 0.001$). Improvement in median OS was also demonstrated of 25.5 months vs. 17.4 months ($p < 0.001$).

In the phase II randomized STOMP and ORIOLE trials, there was an association of MDT, either surgery or RT, with improved PFS and androgen deprivation therapy (ADT)-free survival in patients with hormone-sensitive oligometastatic prostate cancer (HSPC) [5, 6, 28]. In the STOMP trial, 62 men with asymptomatic oligorecurrent HSPC (1–3 metastases on choline PET) were randomized to either MDT (SBRT or surgical resection) or surveillance [6]. ADT was initiated for symptomatic progression, progression of treated lesions, or progression to more than three lesions. Most patients in the MDT arm received SBRT as the mode of MDT (25 of 31 patients). At a median follow-up of 3 years, ADT-free survival was improved in the MDT arm in comparison with observation (21 months vs. 13 months). Median time to PSA progression was also improved in the MDT arm (10 months vs. 6 months). Similarly, the ORIOLE trial randomized 54 men with HSPC to either observation or SBRT to all sites of metastatic disease (1–3 sites based on conventional imaging) [28]. SBRT improved 6-month PFS from 19% to 61% and biochemical PFS from 11% to 50%. At baseline and day 180, PSMA-PET was also performed, and a *post hoc* analysis suggested an association of significantly better PFS (hazard ratio [HR] 0.26; 95% CI, 0.09–0.76, $p = 0.006$) and distant metastasis-free survival (HR 0.19; 95% CI, 0.07–0.54, $p < 0.001$) among patients with no untreated lesions versus any untreated lesion by PSMA-PET.

In renal cell carcinoma (RCC), a large number of patients present with OMD as shown in a large, population-based analysis of 11,157 patients where 61% had a single metastasis compared to 39% with 2 or more metastatic sites [13, 29]. SBRT has been an attractive alternative for treating metastatic RCC given its relative radioresistance, which may be overcome by delivering a higher dose per fraction. In a prospective phase II single-arm trial of 23 patients with OMD defined as 1–3 extracranial metastatic sites, 1-year PFS was 82.6% (95% CI, 60.1–93.1) and freedom from systemic therapy was 91.3% (95% CI, 69.5–97.8) with 100% LC [30]. Concordantly, in a phase II feasibility trial at MD Anderson Cancer Center with 30 RCC patients with 5 or fewer oligometastatic lesions where all sites were treated with SBRT, including subsequent sites of progression, median PFS was 22.7 months and 1-year systemic therapy-free probability was 82% [31]. A prospective phase II multicenter trial investigated the role of SBRT after TKI therapy for patients with oligopressive RCC with up to five progressive metastases [32]. The median duration of TKI therapy prior to study entry was 18.6 months, and tumors treated with SBRT had 1-year LC of 93%. Following SBRT, the median PFS was 9.3 months (95% CI, 7.5–15.7 months), and median time to change in systemic therapy was 12.6 months. One-year OS was 92%. Hannan and colleagues have also reported a phase II prospective trial of 20 patients with oligoprogressive RCC with up to 3 progressive lesions on first- to fourth-line systemic therapy [33]. At a median follow-up of 10.4 months, SBRT extended the duration of ongoing systemic therapy by more than 6 months in 14 patients (70%), and the median time from SBRT to new systemic therapy or death was 11.1 months. One patient developed possibly treatment-related grade 3 gastrointestinal toxicity.

## Patient Selection Characteristics

As with any treatment choice, appropriate patient selection is critical. In the OMD setting, this is even more imperative given the need to identify those patients that could benefit from a more aggressive approach with MDT and balancing the potential benefit with the risk of toxicity. In the treatment of OMD, there are a number of options including surgery, ablative radiotherapy, interventional radiology (IR)-guided approaches, or systemic therapy. Treatment decisions should be made with a multidisciplinary team to identify the optimal local approach in the context of a controlled primary tumor and a plan to address all sites of disease. SBRT as a treatment option for local ablation of OMD offers a number of attractive advantages in comparison with metastasis-directed surgery, including the ability to target multiple metastatic lesions, limited side effect profile, and the less invasive nature of SBRT, which may expand treatment options for patients with comorbidities and also may reduce time off of systemic therapy. However, it is important to note that SBRT to metastatic lesions is not without risk. Patients with OMD are frequently asymptomatic, and shared decision-making between clinicians and patients regarding an acceptable risk of toxicity in the pursuit of cure is critical. There is limited prospective evidence to justify comprehensive SBRT as superior to the palliative SOC, and the ability to treat all sites safely remains of utmost importance in a population where the intent of treatment frequently remains palliative despite the elusive goal of cure.

Important clinical considerations in determining those who would most benefit from SBRT in OMD are prognosis, performance status (PS), comorbidities, and age. Patients who would benefit from more intensive techniques, such as SBRT, should have good life expectancy, whereas those with poorer prognosis may derive palliative benefit from conventional radiotherapy. Overall PS of the patient is an important measure when deciding on whether to treat oligometastatic disease as most studies have demonstrated a strong correlation between PS and OS [34, 35].

Another important consideration in deciding whether to treat OMD with SBRT is the histology of the primary tumor. While promising results have been seen in the treatment of OMD in a number of primary cancers including prostate, NSCLC, and renal cancer, recent prospective data suggests that MDT does not improve PFS or OS for patients with oligometastatic breast cancer [36, 37]. Recently reported is the NRG-BR002 randomized phase II/III trial (NCT02364557) wherein in the phase II portion, 129 patients with oligometastatic breast cancer with 1–4 extracranial metastases were randomized to SOC systemic therapy (chemotherapy, endocrine therapy, or anti-HER2) with or without SBRT to all metastatic lesions [36]. At median follow-up of 30 months, no difference was seen in PFS (23 months in patients with SOC alone vs. 20 months in patients with SBRT and SOC). Given this result, the trial was terminated and will not proceed to the phase III component. Tsai and colleagues presented the results of their phase II Consolidative Use of Radiotherapy to Block (CURB) study where patients with NSCLC or breast cancer and oligoprogressive disease (one to five oligometastatic sites) were randomized to SBRT to all sites of progressive disease vs. palliative SOC and found an association of SBRT with improved PFS (median PFS 22 weeks for SBRT vs. 10 weeks for palliative SOC, $p = 0.005$) [37]. This benefit was driven by patients with NSCLC [37]. No difference in median PFS was seen in the breast patient cohort (18 weeks for SBRT vs. 17 weeks for palliative SOC, $p = 0.5$). This underscores the possible biological and histological differences in metastatic lesions from different primary cancers. The value of MDT may diminish in the presence of improved systemic therapy or if there is a high competing risk of radiographically occult micrometastatic disease.

In the absence of clinical biomarkers for selecting patients with OMD, a long metastasis-free interval (MFI) has consistently been associated with improved outcomes among patients treated with SBRT. In a large case series of 1033 patients with oligometastatic disease (1–5 lesions) who were treated with SBRT, multivariate analysis identified more than 24 months of MFI since initial diagnosis as a survival factor (HR 0.63; 95% CI, 0.49–0.80, $p < 0.001$) [38]. In consideration of patients that would benefit from SBRT or MDT in the oligometastatic setting, these factors should be considered as part of the multidisciplinary discussion for treatment options.

## Overcoming Challenges with Traditional Approaches

Advances in technology have increasingly allowed for lesions to be targeted with high doses of radiation with a steep dose falloff in all directions to minimize toxicity to surrounding tissues [39]. SBRT requires precise delivery with submillimeter accuracy given these ablative doses. Secure immobilization, accurate repositioning, rigorous accounting for tissue motion, and registration with high-resolution imaging for stereotactic treatment planning are required. However, standard SBRT relies on cone-beam CT (CBCT) image guidance, and while good results can be obtained, there are limits in defining the soft-tissue boundary between target and normal tissues. In cases where imaging quality is insufficient to delineate the target or adjacent organs at risk, reliance on rigid registration, such as in the form of bony landmarks or fiducial markers, may be required. Fiducial marker placement can prolong the patient's timeline to treatment, and there is risk for minor and major complications, such as fiducial marker migration, bleeding, infection, and pneumothorax [40]. Further complicating the accurate delivery of SBRT are variations in the anatomy on a day-to-day or even minute-to-minute basis (e.g., bowel peristalsis, bladder filling, rectal gas, etc.). Given the limits of soft-tissue definition and uncertainties with positioning, the use of CT can result in the necessity of increasing planning target volume (PTV) margins, thereby increasing the dose to adjacent OARs. Conversely, the use of planning organ at risk volume (PRV) margins to buffer for the uncertainty of OAR position and intra- and inter-fraction motion can limit the dose prescribed to target. The close approximation of oligometastatic sites to nearby critical or radiation sensitive organs, such as loops of bowel, esophagus, visceral organs, or nerve structures like the lumbar sacral plexus, frequently limits the potential therapeutic benefit of SBRT by preventing dose escalation to ablative doses due to the risk of dose-limiting toxicity, including risk of perforation or fistula, which can be fatal, or permanent nerve injury. Underscoring the importance of safe delivery and potential toxicities to normal tissues, in the SABR-COMET trial, there were three treatment-related deaths in the SABR treatment arm, which included a subdural hemorrhage following surgery for a perforated gastric ulcer [25].

The advent of MRI-guided radiation, which has only recently been developed with Food and Drug Administration (FDA) approval in 2017, provides a new and improved ability to visualize the lesion and normal tissues with the capability of adapting treatment in real time. MRI provides superior soft-tissue discrimination compared to CT [41]. In oligometastatic disease where metastatic lymph nodes and lesions may be close to organs at risk or vessels, improved imaging with high tissue contrast may be beneficial, especially in the abdomen and pelvis, and allow for highly accurate radiation treatments that can spare surrounding normal tissues and reduce toxicities. Further, diagnostic metabolic imaging, such as PSMA-PET in prostate cancer, is identifying earlier, smaller metastatic lesions that may be better visualized by MRI in comparison with standard CT [12].

Additionally, online-adaptive MRI-guided radiotherapy (MRgRT) enables daily adaptation so that volumes and plan can be changed according to the appearance of the target and OARs on that day compensating for inter-fraction anatomic changes. For example, within the abdomen and pelvis, bowel loops can fall within the treatment field. MRgRT can adapt for these changes and thereby theoretically allow for safer delivery of high doses of radiation therapy. A retrospective dosimetric analysis of 150 sessions for 23 patients who received MRgSBRT on a 1.5 Tesla (T) MR-linac for lymph node oligometastases found bowel Dmax to be significantly reduced with a daily adaptive strategy and single intestinal loop sparing [42].

Further, there is potential for oligometastatic lesions, such as abdominopelvic lymph nodes or soft-tissue deposits, to move with respiratory or patient motion. In traditional CT-based radiotherapy, four-dimensional CT is often utilized to depict the motion of the lesion through the breathing cycle and to generate an internal target volume (ITV). Other strategies to reduce tumor motion include breath hold and abdominal compression. With MR-guided radiotherapy, intra-fraction motion management based on real-time soft-tissue imaging can be further refined using gating and tracking techniques for improved targeting of the lesion and eliminating the need for an internal target volume (ITV) and larger margins around the target. As a result, surrounding normal tissues are able to receive a reduced radiation dose.

Utilizing MR to visualize soft-tissue targets and organs at risk in real time at time of treatment delivery mitigates the issues of needing larger margins around both target and organs at risk to account for this uncertainty. This benefit is especially relevant in the OMD setting where patients often have had prior radiation treatment and increased precision in treatment planning and delivery can further reduce the potential toxicities associated with re-irradiation.

## Contouring, PTV, and OAR

Defining the gross tumor volume (GTV), planning target volume (PTV), and organs at risk (OARs) is an important and critical step for safe and effective delivery of stereotactic MR-guided radiation therapy. During simulation for a 0.35 Tesla (T) MR-linac, patients undergo simulation supine on both MR simulator and CT simulator with motion management via breath hold to reduce motion when necessary, which is typical for abdominal targets. For pelvic targets or on 1.5 T MR-linac, no breath hold is needed. CT simulation may be omitted at centers with an MR pseudo-CT solution [43]. Contouring of the GTV is done with reference to any diagnostic imaging that is available, including PET-CT and diagnostic quality MRI with or without contrast, preferably with fused images, in order to identify targets.

Using an adaptive MR-guided workflow, typically, contouring is done on the MRI sequences which are used during treatment. A breath hold technique with light inspiration or end inspiration is frequently used for abdominal lymph nodes on a 0.35 T MR-linac to reduce respiratory motion and to optimize the capabilities of the MR-linac for real-time monitoring. In most cases, a 3-mm isotropic expansion from the GTV, with or without a clinical target volume (CTV) at the discretion of the treating physician, to the planning target volume (PTV) is used. A CTV may be indicated depending on the relative location of multiple oligometastases, particularly lymph nodes or within the same organ, to encompass areas at increased risk of regional relapse [44]. In cases of spine oligometastases, a CTV should be contoured as per international consensus contouring guidelines [45]. Organs at risk (OARs) that are contoured vary by the location of the targeted lesion but can include the spinal cord, duodenum, small bowel, large bowel, stomach, esophagus, kidneys, heart, liver, femoral heads, bladder, and rectum. OARs are contoured on the breath hold sequence and used for treatment planning. For organs that extend outside of the treatment field, the relevant organ is contoured within 2 cm of PTV in all directions or larger if indicated based on clinical scenario (to encompass relevant lower dose isodose lines in setting of stricter OAR constraints). The spinal cord is expanded with a PRV to 3 mm. Typically, gastrointestinal OARs are grouped together with the small bowel, large bowel, stomach, duodenum, and esophagus and evaluated for overlap with the GTV. If necessary due to overlap, the GTV will then have the GI OARs subtracted from the final GTV volume. In a study of adaptive MR-linac treatment of pancreatic cancer, grouping of GI OARs (stomach, duodenum, small bowel, and large bowel) resulted in outperformance of conventional plans and simplified treatment planning workflow [46]. Similarly, a PTV optimization (PTV_opt) structure can be applied for ultra-hypofractionation cases (one and three fraction treatments) where the spinal cord PRV (PRV_spinal cord) and GI OARs are subtracted if there is overlap with the PTV.

## DVH Constraints

With the treatment of oligometastatic disease with the goal of ablative therapy, the target prescription can be between 20 and 50 Gy typically between one to five fractions dependent on the tolerance of the surrounding OARs (see Dose Escalation section for further discussion of dose). OAR constraints are a critical part of treatment planning, especially in the delivery of SBRT; however, there is wide variability in these constraints in both prior studies and current clinical trials [47]. A key principle in the determination of OAR constraints includes evaluation of the organ type. Serial (e.g., spinal cord) and parallel (e.g., lungs, liver) organs differ in their radiation risks. For serial OARs, constraints are based around a maximal threshold dose that is allowable to a small volume (i.e., point dose). In contrast, OAR constraints for parallel organs typically relate to the maximum organ volume that can receive a specified threshold dose and the minimum critical volume of the OAR that must receive a specified threshold dose or lower [48]. The United Kingdom consensus for SABR constraints was updated from their 2017 publication, in which they took into account the work of the Hypofractionated Treatment Effects in the Clinic (HyTEC) and the systematic review by Gerhard and colleagues (Table 12.1) [47–49]. In the systematic review, 53 trials assessing the role of SBRT in the treatment of oligometastatic disease were examined for OAR constraints and the modal, median, and range reported. Although there was considerable variability, the modal and median constraints reported were often closely aligned to that of other commonly used dose constraint guidelines in SBRT, the American Association of Physicist in Medicine Task Group (AAPM-GT) 101, and the Timmerman constraints [50–52]. As these oligometastatic SBRT trials mature, more information will likely be gleaned as toxicities are reported.

**Table 12.1** DVH constraints (adapted from UK 2022 consensus) [48]

| Organ at risk | Constraint | Number of fractions: 1 Optimal | 1 Mandatory | 3 Optimal | 3 Mandatory | 5 Optimal | 5 Mandatory | Endpoint |
|---|---|---|---|---|---|---|---|---|
| Bowel_Small | $D_{0.1\,cc}$ | | 15.4 Gy | | 25.2 Gy | | 35 Gy | Grade 3+ enteritis/obstruction |
| | $D_{5\,cc}$ | | 11.9 Gy | | 17.7 Gy | | | |
| | $D_{10\,cc}$ | | | | 25 Gy | | | |
| Duodenum | $D_{0.1\,cc}$ | | 12.4 Gy | | 22.2 Gy | 33 Gy | 35 Gy | Grade 3+ ulceration |
| | $D_{10\,cc}$ | | 9 Gy | | 11.4 Gy | 25 Gy | | |
| Kidney_Cortex (individual/combined) | $D_{mean}$ | | | 8.5 Gy | | 10 Gy | | Grade 3+ renal dysfunction |
| Kidney_Cortex (combined) | $D_{\geq 200\,cc}$ | | 8.4 Gy | | 16 Gy | | 17.5 Gy | |
| If solitary kidney or one Kidney_Cortex $D_{mean}$ constraint exceeded | $V_{10\,Gy}$ | | 33% | | 33% | 10% | 45% | |
| Liver (non-liver) lesions and liver – GTV (liver lesions) | $D_{\geq 700\,cc}$ | | 9.1 Gy | 15 Gy | 17 Gy | 15 Gy | | Grade 3+ liver dysfunction, radiation induced liver disease (classic or non-classic) |
| | $V_{10\,Gy}$ | | | | | 70% | | |
| | $D_{mean}$ | | | 13 Gy | 15 Gy | 13 Gy | 15.2 Gy | |
| Stomach | $D_{0.1\,cc}$ | | 12.4 Gy | | 22.2 Gy | 33 Gy | 35 Gy | Grade 3+ ulceration/fistula |
| | $D_{10\,cc}$ | | 11.2 Gy | | 16.5 Gy | 25 Gy | | |
| | $D_{50\,cc}$ | | | | | 12 Gy | | |
| BileDuct | $D_{0.1\,cc}$ | | 30 Gy | 50 Gy | | 50 Gy | | |
| Esophagus | $D_{0.1\,cc}$ | | 15.4 Gy | | 25.2 Gy | | 35 Gy | Grade 3+ stenosis/fistula |
| Bladder | $D_{0.1\,cc}$ | | 18.4 Gy | | 28.2 Gy | | 38 Gy | Grade 3 cystitis/fistula |
| Bowel_Large | $D_{0.1\,cc}$ | | 18.4 Gy | | 28.2 Gy | | 38 Gy | Grade 3+ colitis/fistula |
| Femur_HeadNeck | $D_{10\,cc}$ | 14 Gy | | 21.9 Gy | | 30 Gy | | Grade 3+ necrosis |
| LumbSacPlex | $D_{0.1\,cc}$ | 16 Gy | | 24 Gy | | 32 Gy | | Grade 3+ neuritis |
| | $D_{5\,cc}$ | 14.4 Gy | | 22.5 Gy | | 30 Gy | | |
| Rectum | D | | 18.4 Gy | | 28.2 Gy | | 38 Gy | Grade 3+ proctitis/fistula |
| Ureter | $D_{0.1\,cc}$ | | 35 Gy | | 40 Gy | | | |
| SpinalCord_PRV | $D_{0.035\,cc}$ | 12.4 Gy | 14 Gy | | 20.3 Gy | | 25.3 Gy | Radiation myelopathy (1–5% risk) |
| CaudaEquina and SpinalCanal (below level of cord) | $D_{0.035\,cc}$ | | 16 Gy | | 24 Gy | | 32 Gy | Grade 3+ neuritis |
| | $D_{5\,cc}$ | | 14 Gy | | 21.9 Gy | | 30 Gy | |

## Dose Escalation

Ablative radiotherapy doses to oligometastatic lesions is one of the tenets of localized treatment of OMD, and ablation using SBRT to prolong PFS and overall survival is being increasingly recognized [6, 7, 26, 28]. However, the exact dosing for lesions remains an area of continued investigation as most randomized trials investigating optimal radiotherapy dosing were conducted in polymetastatic disease with nonmobile targets (e.g., bone) with the goal of palliation of symptoms, namely, pain. In determining the dose for OMD lesions, the forefront priority remains minimizing risk of toxicity to surrounding OARs while delivering a therapeutic dose to the target. An additional important factor in consideration of dose and fractionation is histology, as exemplified by RCC. Prior adjuvant RT studies with conventional fractionation in RCC patients demonstrated no benefit and rather increased toxicity; however, with additional evidence of the radioresistant nature of RCC and with the advent of SBRT, meta-analyses of SBRT treatment for oligometastatic RCC have now shown excellent local control rates and safety for select patients [53]. Further, a biologically effective dose (BED) greater than 115 Gy has also been associated with improved outcomes [54]. Another important factor in determination of dose and fractionation is size of the lesion and proximity to OARs. If not able to safely deliver a SBRT fractionation, moderately hypofractionated regimens with or without a simultaneous integrated boost (SIB) can also be considered. In unresectable cholangiocarcinoma, a retrospective series of 79 patients who were mainly treated with moderate hypofractionation and a central tumor boost of 75 Gy in 15 fractions or 100 Gy in 25 fractions to GTV, subtracted from a 5 mm PRV expansion on adjacent OAR, found that a BED greater than 80.5 Gy correlated with improved LC and OS, with LC rates (78%) comparable to resection [55].

Online-adaptive MRgSBRT may allow for an increase in PTV coverage while simultaneously improving OAR sparing compared to non-adaptive SBRT, thereby allowing for dose escalation. In a proof of concept, Henke and colleagues characterized five patients with non-liver abdomen oligometastatic disease and five patients with central thorax disease treated by MR-guided SBRT and compared day-to-day anatomy variations to simulate adaptive planning [56]. Adaptive planning allowed for increased PTV coverage in 70% (21 of 30 treatments), including 14 treatments in which hard OAR constraints were violated by the non-adaptive plan. Dose escalation in treatment of primary pancreatic cancer with use of MRgRT and real-time target localization has been demonstrated, allowing for dose escalation by 36 Gy on average to at least 50 Gy [57]. Initial results from a multicenter, prospective study of ablative adaptive MRgSBRT with a prescribed $BED_{10}$ of 100 Gy (50 Gy in five fractions) for pancreatic cancer demonstrate the safety and feasibility of this approach, with a 2.2% rate of grade 3 toxicity [58]. Extrapolating this to oligometastatic lesions, MRgRT may allow for further increase in dose while sparing normal tissues. In a study of 20 patients with oligometastatic disease or unresectable disease of the abdomen, PTV dose escalation was able to be achieved in 20 of 97 treatments with online daily adaptive plans [59].

Further, there is motivation for utilizing fewer fractions for the delivery of ablative radiation therapy using MRgRT. Some patients travel from far distances specifically for the ability to receive MRgRT and its potential advantages, and fewer fractions may reduce the logistical burdens on patients (*NCT04115254, DF/HCC Protocol #:19-353 SMART-STOP*; *NCT04939246 SMART ONE*). It also reduces the number of sessions that a patient must be on the treatment table and relatively immobilized, which for adaptive planning can be upward of 90 min per session. The MR-linac is also a limited resource for which there is an increasing demand and utility for multiple cancer types. Utilizing fewer fractions may potentially allow for resource optimization and allow a greater number of patients to access this novel technology. Additionally, there is emerging evidence that ultra-high dose fractionation, beginning at a threshold of 12 Gy, may provide distinct advantages compared to lower doses per fraction by inducing microvascular vasoactive dysfunction and inhibition of homologous recombination [60].

A recent randomized phase III trial explored whether a single fraction of 24 Gy would be superior in local control compared to a 3 × 9 Gy fractionation for oligometastatic disease of non-mobile lesions (bone and stationary lymph nodes) [61]. They found that the single fraction increased dose was superior in local control compared to the conventional SBRT regimen (3-year local recurrence 5.8% vs. 22%, $p = 0.0048$). Provocatively, there was also an improvement in distant progression, with the 3-year distant progression being 5.3% vs. 22.5% ($p = 0.010$). However, it is important to recognize the limitations given that most lesions were bone lesions and could not be in close proximity to OARs given the CT-based modality of treatment. It is intriguing that potentially increasing dose could allow for improved oncologic outcomes and advances in MRI-guided technologies could help facilitate this dose escalation safely.

## On Table Adaptation

Adaptive radiotherapy (ART) allows for adjustment of treatment plans based on anatomical and/or biological changes which may occur during treatment, such as respiratory movement, changes in location of bowel loops, and lesion size changes. Online plan adaption allows for changes to be made based on the anatomy of the day while the patient is in the treatment position. As a result, the process for online plan adaptation needs to be performed efficiently and fast with priorities of safety (see Chap. 8 for further technical details). In OMD, there are a number of special cases that highlight the unique challenges and advantages of on-table adaptation.

In OMD treatment, patients may have a history of prior radiation therapy to their primary cancer and/or other metastatic sites necessitating stricter dose cutoffs for OARs. In this setting, adaptive planning can be useful to allow for greater dose delivery to the OMD site with real-time monitoring of OARs. Given the abilities of adaptive planning, OARs can be differentially contoured into regions and set for different constraints to account for prior radiation (Fig. 12.2). In determining the area needed for adaptive recontouring, a ring structure expansion commonly is used, and if OARs fall within this region, these differential constraints can then be applied. By utilizing adaptive recontouring, stricter constraints of OARs can typically be achieved that further mitigate the risk of toxicity.

Depending on anticipated patterns of failure or recurrence based on tumor histology, a CTV may be considered to address the risk of micrometastatic disease in intervening nodal regions. This may be particularly relevant for patients who experience a regional nodal relapse out of field of prior nodal irradiation (Fig. 12.3). The technical capabilities of MRgSBRT to allow for longer areas to be treated while also visualizing nearby OARs have expanded options for patients in whom a protracted course of pelvic nodal irradiation is not clinically or logistically feasible. Given the limited prospective data to compare the benefit of comprehensive nodal radiation to SBRT for oligometastatic nodal recurrence, this should be offered on clinical protocol whenever possible.

OMD treatment is also unique in that multiple lesions tend to be treated contemporaneously for maximal therapeutic benefit. In determining whether more than one lesion can be treated in the same course, the ability to track the lesions, movement of the lesions, proximity to OARs, and placement of the isocenter must be considered. If two or more lesions are adjacent and either are stationary or move together, then creating a single PTV structure incorporating the targets could be employed. However, many lesions move differentially with relation to each other. Until recently, technology for adaptive planning was limited to single planar view for gaiting and tracking. Dependent on the sites of the oligometastases and whether they move differentially with relation to each other, a single plane may not be optimal for tracking of more than one lesion

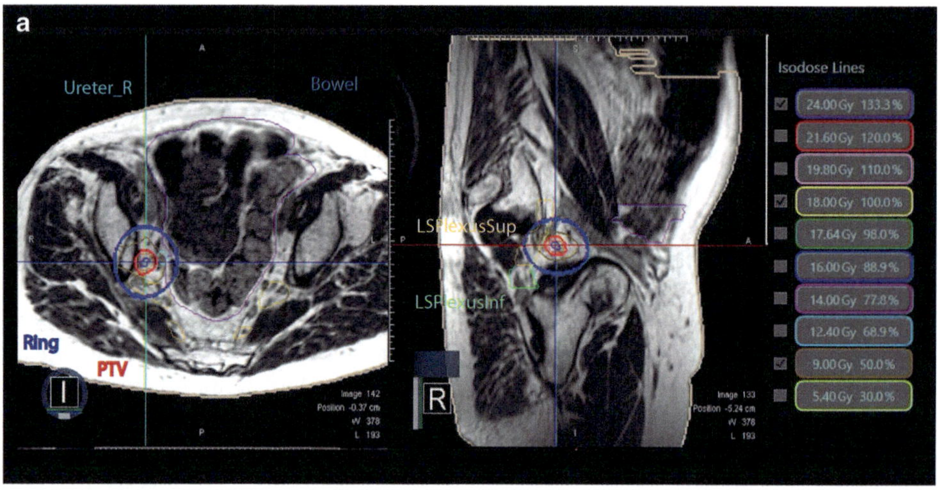


**b**

| Structure | | Min (Gy) | Mean (Gy) | Max (Gy) | Dose to Volume | |
|---|---|---|---|---|---|---|
| PTV_opt | Rx | | | | ≥ 95% | 18 Gy |
| | Plan | 16.60 | 21.55 | 24.52 | 98.77% | 18 Gy |
| PTV | Rx | | | | ≥ 99% | 16 Gy |
| | Plan | 16.60 | 21.55 | 24.52 | 100% | 16 Gy |
| Bowel | Rx | | | | ≤ 0.03 cc | 12.4 Gy |
| | Plan | 0.01 | 0.59 | 11.87 | 0 cc | 12.4 Gy |
| Bowel | Rx | | | | ≤ 5 cc | 11.2 Gy |
| | Plan | 0.01 | 0.59 | 11.87 | 0.03 cc | 11.2 Gy |
| LSPlexusSup | Rx | | | | ≤ 0.03 cc | 14 Gy |
| | Plan | 0.04 | 1.23 | 13.24 | 0 cc | 14 Gy |
| LSPlexusSup | Rx | | | | ≤ 0.35 cc | 10 Gy |
| | Plan | 0.04 | 1.23 | 13.24 | 0.25 cc | 10 Gy |
| LSPlexusInf | Rx | | | | ≤ 0.03 cc | 12 Gy |
| | Plan | 0.03 | 0.26 | 1.76 | 0 cc | 12 Gy |
| LSPlexusInf | Rx | | | | ≤ 5 cc | 8 Gy |
| | Plan | 0.03 | 0.26 | 1.76 | 0 cc | 8 Gy |
| Skin-Organ | Rx | | | | ≤ 0.03 cc | 26 Gy |
| | Plan | 0 | 0.06 | 8.98 | 0 cc | 26 Gy |
| Cauda Equina | Rx | | | | ≤ 1.2 cc | 7 Gy |
| | Plan | 0.06 | 0.20 | 0.90 | 0 cc | 7 Gy |
| Skin-Organ | Rx | | | | ≤ 10 cc | 23 Gy |
| | Plan | 0 | 0.06 | 8.98 | 0 cc | 23 Gy |
| Bladder | Rx | | | | ≤ 0.03 cc | 16 Gy |
| | Plan | 0.03 | 0.15 | 0.64 | 0 cc | 16 Gy |
| Bladder | Rx | | | | ≤ 15 cc | 11.4 Gy |
| | Plan | 0.03 | 0.15 | 0.64 | 0 cc | 11.4 Gy |
| Rectum | Rx | | | | ≤ 0.03 cc | 14 Gy |
| | Plan | 0.04 | 0.12 | 0.30 | 0 cc | 14 Gy |
| Rectum | Rx | | | | ≤ 20 cc | 12 Gy |
| | Plan | 0.04 | 0.12 | 0.30 | 0 cc | 12 Gy |
| FemurHead_L | Rx | | | | ≤ 10 cc | 12 Gy |
| | Plan | 0.01 | 0.08 | 0.25 | 0 cc | 12 Gy |
| FemurHead_R | Rx | | | | ≤ 10 cc | 12 Gy |
| | Plan | 0.06 | 0.18 | 0.40 | 0 cc | 12 Gy |
| UreterR | Rx | | | | ≤ 0.03 cc | 16 Gy |
| | Plan | 0.50 | 5.23 | 15.35 | 0 cc | 16 Gy |
| LSPlexusSup | Rx | | | | ≤ 1.2 cc | 7 Gy |
| | Plan | 0.04 | 1.23 | 13.24 | 2.82 cc | 7 Gy |
| Bowel | Rx | | | | ≤ 10 cc | 9 Gy |
| | Plan | 0.01 | 0.59 | 11.87 | 0.28 cc | 9 Gy |

**Fig. 12.2** Ultra-hypofractionation for oligometastatic right pelvic node. (**a**) Axial and sagittal view of a patient with oligorecurrent prostate cancer with a PSMA-avid right pelvic node treated with a dose of 18 Gy × 1 on SMART-STOP Protocol 19–353. A 3 mm PTV margin was applied to GTV. Organs at risk including the bowel, right ureter, and a lumbar sacral plexus superior (LSPlexSup) and lumbar sacral plexus inferior (LSPlexInf) were contoured to allow for differential constraints, accounting for prior radiation. A 2 cm ring structure was applied to define organs in need of adaptive recontouring on day of treatment. (**b**) Stricter constraints were applied to LSPlexInf due to overlap with prior prostate bed/pelvic EBRT. The adaptive treatment plan met all clinically defined hard constraints. The soft constraint for LSPlexusSup was slightly exceeded but clinically acceptable

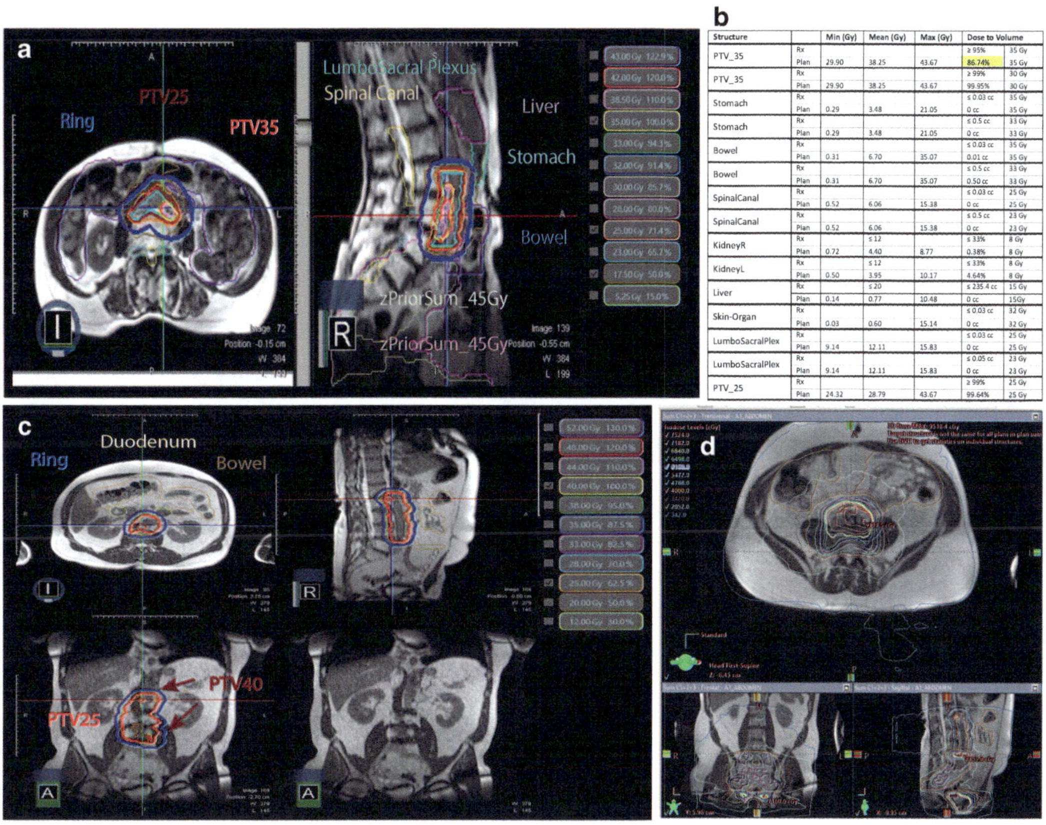


| Structure | | Min (Gy) | Mean (Gy) | Max (Gy) | Dose to Volume | |
|---|---|---|---|---|---|---|
| PTV_35 | Rx | | | | ≥ 95% | 35 Gy |
| | Plan | 29.90 | 38.25 | 43.67 | 86.74% | 35 Gy |
| PTV_35 | Rx | | | | ≥ 99% | 30 Gy |
| | Plan | 29.90 | 38.25 | 43.67 | 99.95% | 30 Gy |
| Stomach | Rx | | | | ≤ 0.03 cc | 35 Gy |
| | Plan | 0.29 | 3.48 | 21.05 | 0 cc | 35 Gy |
| Stomach | Rx | | | | ≤ 0.5 cc | 33 Gy |
| | Plan | 0.29 | 3.48 | 21.05 | 0 cc | 33 Gy |
| Bowel | Rx | | | | ≤ 0.03 cc | 35 Gy |
| | Plan | 0.31 | 6.70 | 35.07 | 0.01 cc | 35 Gy |
| Bowel | Rx | | | | ≤ 0.5 cc | 33 Gy |
| | Plan | 0.31 | 6.70 | 35.07 | 0.50 cc | 33 Gy |
| SpinalCanal | Rx | | | | ≤ 0.03 cc | 25 Gy |
| | Plan | 0.52 | 6.06 | 15.38 | 0 cc | 25 Gy |
| SpinalCanal | Rx | | | | ≤ 0.5 cc | 23 Gy |
| | Plan | 0.52 | 6.06 | 15.38 | 0 cc | 23 Gy |
| KidneyR | Rx | | ≤ 12 | | ≤ 33% | 8 Gy |
| | Plan | 0.72 | 4.40 | 8.77 | 0.38% | 8 Gy |
| KidneyL | Rx | | ≤ 12 | | ≤ 33% | 8 Gy |
| | Plan | 0.50 | 3.95 | 10.17 | 4.64% | 8 Gy |
| Liver | Rx | | ≤ 20 | | ≤ 235.4 cc | 15 Gy |
| | Plan | 0.14 | 0.77 | 10.48 | 0 cc | 15Gy |
| Skin-Organ | Rx | | | | ≤ 0.03 cc | 32 Gy |
| | Plan | 0.03 | 0.60 | 15.14 | 0 cc | 32 Gy |
| LumboSacralPlex | Rx | | | | ≤ 0.03 cc | 25 Gy |
| | Plan | 9.14 | 12.11 | 15.83 | 0 cc | 25 Gy |
| LumboSacralPlex | Rx | | | | ≤ 0.05 cc | 23 Gy |
| | Plan | 9.14 | 12.11 | 15.83 | 0 cc | 23 Gy |
| PTV_25 | Rx | | | | ≥ 99% | 25 Gy |
| | Plan | 24.32 | 28.79 | 43.67 | 99.64% | 25 Gy |

**Fig. 12.3** MRgSBRT to abdominal nodes including a CTV. (**a**) Axial and sagittal view of a patient with vulvar cancer with oligoprogression in the para-aortic lymph nodes abutting prior field of pelvic nodal intensity modulated radiation therapy (IMRT). A simultaneous integrated boost (SIB) was applied to deliver a dose of 35 Gy to the FDG-avid lymph nodes with a 3 mm margin (PTV35) and a dose of 25 Gy to a CTV inclusive of the para-aortic region and involved lymph nodes, cropped from bowel and duodenum anteriorly, with a 3 mm margin (PTV25). A ring using a 2 cm expansion from PTV35 and 1 cm expansion from PTV25 was applied to define the region in which organs at risk (OARs) required adaptive recontouring. zPriorSum_45 Gy and zPriorSum_50 Gy were applied at time of initial contouring to guide target volume delineation and constraints applied to OAR. OARs defined included the lumbosacral plexus, spinal canal, liver, stomach, and bowel. (**b**) The treatment plan was able to meet all predefined hard constraints. PTV35 coverage did not meet the soft constraint of V35 ≥ 95%, but the hard constraint of V30 ≥ 99% was achieved. (**c**) Axial, sagittal, and coronal views of a patient with oligorecurrent prostate cancer involving two PSMA-avid lymph nodes above prior field of pelvic nodal IMRT. An SIB was applied to deliver a dose of 40 Gy to the PSMA-avid lymph nodes with a 3 mm margin (PTV40) and a dose of 25 Gy to a CTV inclusive of the para-aortic region and involved nodes, cropped from duodenum and bowel anteriorly, with a 3 mm margin (PTV25). A 3 cm expansion from GTV and 1 cm expansion from CTV were used to generate a ring structure for adaptive daily recontouring. OARs included bowel and duodenum. (**d**) A sum plan accounting for all prior radiation treatment plans should be generated when possible to account for cumulative dose to OARs. The sum plan graphically displays the dose received to patient treated in (**c**), including prior salvage prostate bed EBRT (68.4 Gy/28 F, 2014), pelvic nodal IMRT (50.4–61.6 Gy/28 F, 2020), and abdominal node MRgSBRT (25–40 Gy/5 F, 2023)

(Fig. 12.4). If determined that sequential treatment of multiple oligometastases would be preferential, then there must be accounting for cumulative OAR dose as though treated in a single course (Fig. 12.5). Further, overlapping hot spots should be avoided, and plan optimization with beam arrangements should be carefully evaluated.

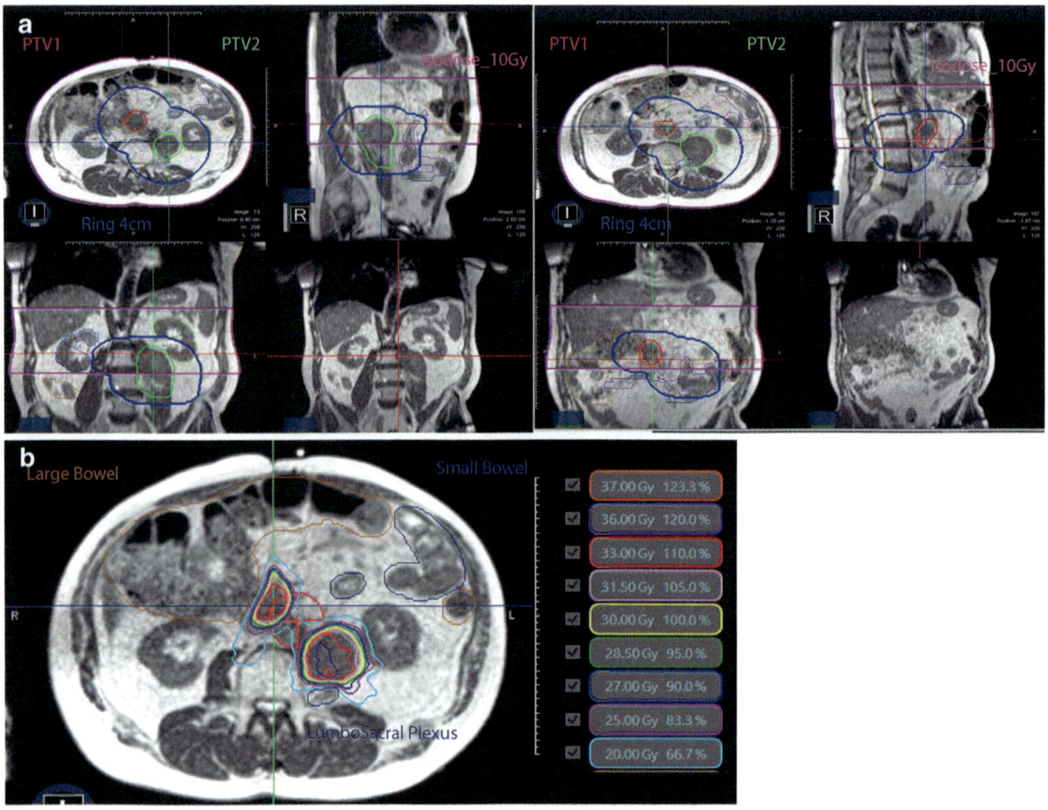


**Fig. 12.4** Single isocenter, multi-metastasis treatment on MR-linac. (**a**) Axial, sagittal, and coronal views of a patient with oligoprogressive renal cell cancer involving a left para-aortic node (PTV1) and left psoas muscle (PTV2) following prior SBRT to a left psoas mass. A 4 cm ring was used to define the organs at risk in need of daily adaptive recontouring due to stricter dose constraints applied in setting of re-irradiation. An "Isodose_10 Gy" was applied each day with alignment checked relative to the vertebral body anatomy to allow for differential dosing to the spinal canal and lumbosacral plexus, based on degree of overlap. (**b**) The adaptive treatment plan to deliver 30 Gy in five fractions for this patient was able to spare dose to the adjacent bowel and lumbosacral plexus and left kidney

**Fig. 12.5** Sequential MRgSBRT for oligometastatic recurrence. (**a**) Axial, sagittal, and coronal view of a treatment plan for a patient with oligorecurrent ovarian cancer s/p prior radical resection and chemotherapy with a vaginal cuff mass and presacral node involvement. The vaginal cuff mass was defined with a 3 mm margin and treated to a dose of 35 Gy in five fractions (PTV). Organs at risk (OARs) including the bowel, bladder, and rectum were defined. An "Isodose_10 Gy" structure was created to allow for differential dose constraints to be generated daily for the bowel and rectum, anticipating overlap from sequential treatment to the presacral nodes. (**b**) Beam arrangement to treat the vaginal cuff mass was intentionally designed to bisect patient to minimize divergence with presacral treatment plan. (**c**) The treatment plan met all hard constraints. PTV coverage was a soft constraint. A bowel and rectum PRV allowed for further sparing of OARs. BowelRectum_Superior, which defined the OAR region that overlapped with Isodose_10 Gy, received minimal dose from the vaginal cuff course. (**d**) Axial view of the sequential treatment plan used to target two oligoprogressive presacral nodes. The OARs included the lumbosacral plexus, cauda equina, and bowel. The presacral nodes were treated to a dose of 40 Gy in five fractions. (**e**) The beam arrangement for the presacral nodes was designed to bisect patient to minimize divergence with prior vaginal cuff treatment plan. (**f**) The adaptive treatment plan was able to optimize coverage to the two oligoprogressive lymph nodes (PTVsup and PTVinf) while meeting the hard constraints for the bowel, lumbosacral plexus, and cauda equina

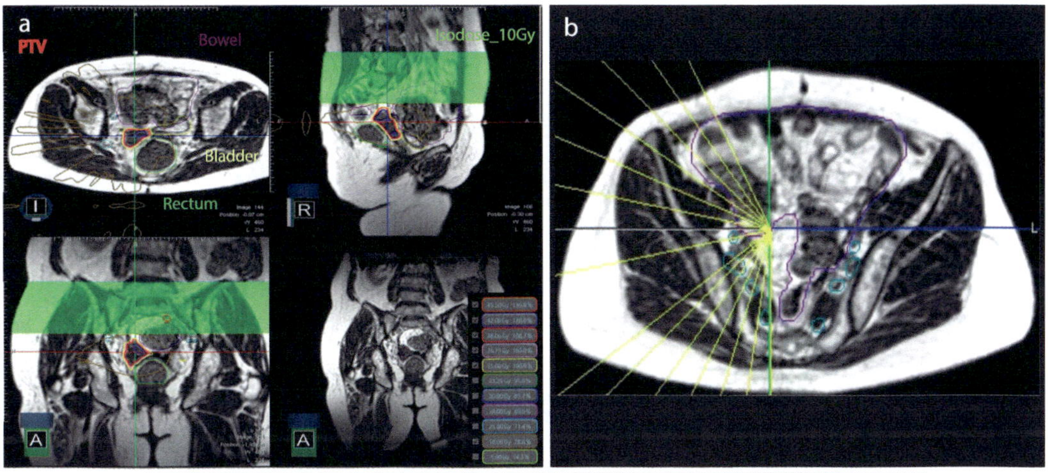


c

| Structure | | Min (Gy) | Mean (Gy) | Max (Gy) | Dose to Volume | |
|---|---|---|---|---|---|---|
| PTV | Rx | | | | ≥ 95% | 35 Gy |
| | Plan | 24.29 | 36.14 | 44.92 | 65.30% | 35 Gy |
| PTV | Rx | | | | ≥ 99% | 25 Gy |
| | Plan | 24.29 | 36.14 | 44.92 | 99.93% | 25 Gy |
| FemurHead_R | Rx | | | | ≤ 1 cc | 19.9 Gy |
| | Plan | 0.50 | 4.83 | 11.83 | 0 cc | 19.9 Gy |
| FemurHead_L | Rx | | | | ≤ 1 cc | 19.9 Gy |
| | Plan | 0.36 | 1.30 | 3.33 | 0 cc | 19.9 Gy |
| CaudaEquina | Rx | | | | ≤ 0.5 cc | 25 Gy |
| | Plan | 0.43 | 0.64 | 0.96 | 0.01 cc | 25 Gy |
| CaudaEquina | Rx | | | | ≤ 0.03 cc | 30 Gy |
| | Plan | 0.43 | 0.64 | 0.96 | 0 cc | 30 Gy |
| Bladder | Rx | | | | ≤ 0.1 cc | 38.06 Gy |
| | Plan | 0.83 | 6.78 | 37.44 | 0 cc | 38.06 Gy |
| Bladder | Rx | | | | ≤ 1 cc | 36.25 Gy |
| | Plan | 0.83 | 6.78 | 37.44 | 0.13 cc | 36.25 Gy |
| SkinOrgan | Rx | | | | ≤ 0.03 cc | 32 Gy |
| | Plan | 0.02 | 0.49 | 13.18 | 0 cc | 32 Gy |
| LSPlexus | Rx | | | | ≤0.5 cc | 25 Gy |
| | Plan | 0.37 | 4.76 | 23.54 | 0 cc | 25 Gy |
| LSPlexus | Rx | | | | ≤ 0.03 cc | 30 Gy |
| | Plan | 0.37 | 4.76 | 23.54 | 0 cc | 30 Gy |
| BowelPRV | Rx | | | | ≤ 0.03 cc | 35 Gy |
| | Plan | 0.25 | 4.28 | 35.87 | 0.03 cc | 35 Gy |
| RectumPRV | Rx | | | | ≤ 0.5 cc | 35 Gy |
| | Plan | 0.97 | 11.63 | 36.21 | 0.01 cc | 35 Gy |
| Bowel | Rx | | | | ≤ 0.5 cc | 33 Gy |
| | Plan | 0.25 | 4.31 | 30.20 | 0 cc | 33 Gy |
| Rectum | Rx | | | | ≤ 2 cc | 30 Gy |
| | Plan | 1.10 | 11.83 | 31.43 | 0.01 | 30 Gy |
| Ureter | Rx | | | | ≤ 0.03% | 45 Gy |
| | Plan | 1.27 | 17.24 | 33.81 | 0% | 45 Gy |
| BowelRectum_SUP | Rx | | | | ≤ 0.03 cc | 10 Gy |
| | Plan | 0.25 | 0.82 | 6.69 | 0 cc | 10 Gy |
| BowelRectum_SUP | Rx | | | | ≤ 0.03 cc | 15 Gy |
| | Plan | 0.25 | 0.82 | 6.69 | 0 cc | 15 Gy |

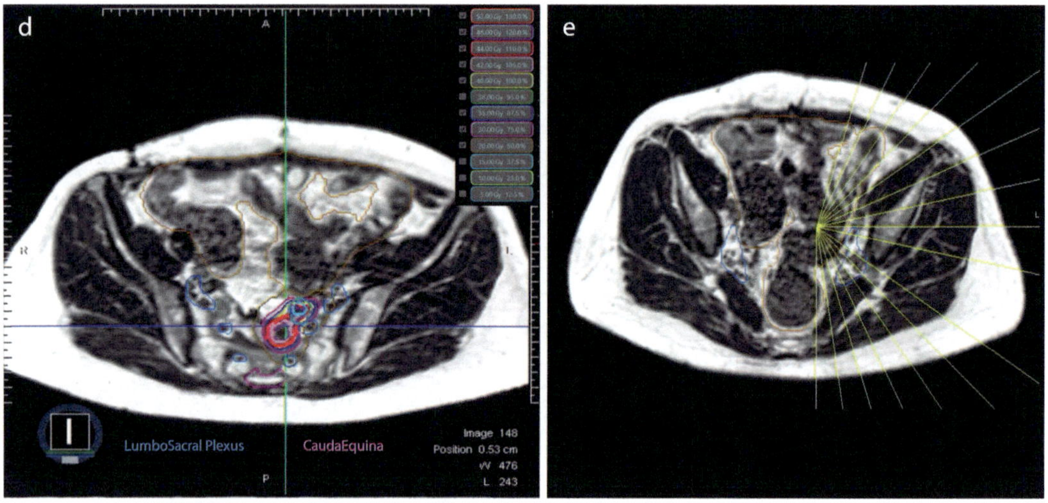


f

| Structure | | Min (Gy) | Mean (Gy) | Max (Gy) | Dose to Volume | |
|---|---|---|---|---|---|---|
| PTV inf | Rx | | | | ≥ 95% | 40 Gy |
| | Plan | 23.53 | 44.31 | 50.70 | 83.48 % | 40 Gy |
| PTV inf | Rx | | | | ≥ 99% | 30 Gy |
| | Plan | 23.53 | 44.31 | 50.70 | 97.14 % | 30 Gy |
| FemurHead_R | Rx | | | | ≤ 1 cc | 19.9 Gy |
| | Plan | 0.23 | 0.32 | 0.44 | 0 cc | 19.9 Gy |
| FemurHead_L | Rx | | | | ≤ 1 cc | 19.9 Gy |
| | Plan | 0.36 | 0.48 | 0.63 | 0 cc | 19.9 Gy |
| Bowel | Rx | | | | ≤ 0. 5 cc | 30 Gy |
| | Plan | 0.32 | 3.13 | 39.04 | 0.22 | 30 Gy |
| Bowel | Rx | | | | ≤ 0. 03 cc | 35 Gy |
| | Plan | 0.32 | 3.13 | 39.04 | 0.03 cc | 35 Gy |
| CaudaEquina | Rx | | | | ≤0.5 cc | 25 Gy |
| | Plan | 0.88 | 5.75 | 18.19 | 0 cc | 25 Gy |
| CaudaEquina | Rx | | | | ≤ 0.03 cc | 30 Gy |
| | Plan | 0.88 | 5.75 | 18.19 | 0 cc | 30 Gy |
| Bladder | Rx | | | | ≤ 0. 1 cc | 38.06 Gy |
| | Plan | 0.53 | 0.71 | 0.92 | 0 cc | 38.06 Gy |
| Bladder | Rx | | | | ≤ 1 cc | 36.25 Gy |
| | Plan | 0.53 | 0.71 | 0.92 | 0 cc | 36.25 Gy |
| SkinOrgan | Rx | | | | ≤ 0.03 cc | 32 Gy |
| | Plan | 0.04 | 0.36 | 13.10 | 0 cc | 32 Gy |
| LSPlexus | Rx | | | | ≤ 0.03 cc | 30 Gy |
| | Plan | 0.50 | 6.00 | 29.41 | 0 cc | 30 Gy |
| LSPlexus | Rx | | | | ≤0.5 cc | 25 Gy |
| | Plan | 0.50 | 6.00 | 29.41 | 0.20 cc | 25 Gy |
| PTVsup | Rx | | | | ≥ 95% | 40 Gy |
| | Plan | 23.58 | 45.73 | 51.76 | 90.29% | 40 Gy |

**Fig. 12.5** (continued)

## Evaluation

Generally, in the treatment of oligometastatic disease, OAR constraints are prioritized over PTV coverage for safety and to minimize harm to the patient. In practice, this can be described with the usage of hard DVH constraints that cannot be breached (where the plan is unacceptable if not met) as well as priority ranking of DVH constraints, with PTV coverage at a lower ranking compared to essential OAR constraints, such as GI and spinal cord OARs. Soft constraints may also be used, where the OAR constraint is a goal but may not be met. In cases where daily treatments are handled by a treating physician and team for the day, a safe handoff of the targeted lesions and OAR priorities must be carefully done to ensure treatment of the proper oligometastatic site and OAR constraints are appropriately met.

In OMD, many patients require multiple courses of MDT throughout the lifetime of their disease. Accounting for prior radiation and calculation of a sum plan for future radiation treatments is necessary for accounting of dose to OARs and to reduce potential toxicities (Figs. 12.2, 12.3, 12.4, and 12.5). However, calculating dose accumulation for MRgRT is challenging given that there are daily adaptive plans. To generate an accurate dose, the daily plans would need to be exported and calculated for total dose accumulation and should be considered in plan evaluation.

## Outcome

The majority of prospective studies that have been published thus far on the role of MRgRT for oligometastatic disease have been phase I trials addressing feasibility; however, long-term impact on oncologic outcomes and toxicity remain under investigation. Since the first clinical applications of MR-radiation therapy systems, there has been expansion into investigations of oligometastatic lesions specifically. In a retrospective review of MRgRT for abdominopelvic oligometastases of 101 patients, online adaptation on a 0.35 T MR-linac resulted in a clinically significant improvement in coverage of the lesion or organ sparing in 86.6% of delivered fractions and showed excellent safety, feasibility, and local control [44]. Henke and colleagues published a phase I trial investigating the use of MRgRT on a 0.35 T MR-linac in oligometastatic or unresectable primary disease of the abdomen and found that daily online-adaptive plans were feasible and created for 81 of 97 patients due to initial plan violation of OAR constraints [59]. All patients in the study had at least one treatment that required adaptive planning. In non-liver lesions, 88% of required plan adaption was due to one or more OAR violations, most commonly of small bowel. Within 6 months of treatment, there were no grade 3 or greater toxicities. Similarly, a phase I study of MRgRT in oligometastatic lesions of ten ovarian cancer patients was conducted where oligometastatic disease was classified as three or less lesions and all radiographic sites were treated, with three lesions in the thorax, six lesions in the abdomen, and eight lesions in the pelvis [62]. This study showed excellent feasibility and significantly improved OAR sparing. Online adaptation was able in 32 of 65 treatments to mitigate unintended OAR constraint violations, most commonly in the pelvis followed by the abdomen, which was generally well tolerated. Dose escalation was allowed, and while most lesions received 35 Gy in five fractions, some received doses upward of 50 Gy in five fractions. Tumor control per lesion at 3 months was 94% and was maintained for 1 year.

Recently the first multicenter experience in the treatment of oligometastatic disease, with lymph node and liver lesions, on the Elekta 1.5 Tesla MR-linac, was published [63]. In total, 168 patients with lymph node metastases were treated mainly in the pelvis (72%), with the primary cancers originating from prostate (60%) or colorectal (12%). Between institutions, there was variation in workflow patterns and differences in dose prescription and coverage criteria, but all within the range of prescriptions set forth in the SABR-COMET trial [7]. Nicosia et al. recently published the largest prospective study of lymph node SBRT on a 1.5 Tesla MR-linac in patients with prostate cancer [64]. A total of 118 lymph node oligometastases were treated in 63 oligometastatic patients. Of the patients, 63.5% were oligorecurrent and 36.5% were oligoprogressive. Lymph node SBRT by 1.5 T MR-linac also provides high local control rates with an excellent toxicity profile. Furthermore, the oligorecurrent state during ADT was correlated with a lower nodal progression-free survival than was the oligoprogressive state. The same group have previously reported preliminary data on feasibility and patient-reported outcomes following PSMA-PET/CT guided SBRT for oligometastatic castration-sensitive prostate cancer by means of 1.5 T MR-Linac [65]. In this series, 25 lesions in 20 castration-sensitive oligorecurrent patients were treated: The most commonly treated anatomic sites were nodal ($n$ = 16) and pelvic bone ($n$ = 9), and encouraging findings in terms of effectiveness and tolerability were confirmed [65].

These studies demonstrate the feasibility and safety of the MR-guided MDT approaches for oligometastases. Nevertheless, more data is needed for understanding longer-term oncologic outcomes, optimization of processes, optimal dose and prescription, long-term toxicities, and patient-reported measures.

## Future Directions

The treatment of OMD is continuing to be refined and phase III clinical trials are ongoing, which will guide recommendations for therapy. These trials will also hopefully evaluate the role of MDT for different histologies, amount of disease, and OMD state (Table 12.2). The role of MRgRT for MDT in OMD is promising with the initial results of the phase I and phase II trials showing feasibility and safety. The advantages of MRgRT with improved soft-tissue visualization and the ability to perform real-time tracking and gating to achieve greater ablative doses while sparing critical OARs are clear benefits for patients receiving MDT in the proper clinical settings.

Improvements in the technology continue to be developed, with most recently the release of MRIdian A3i, which allows for multi-planar tracking. This could potentially allow for ease of treatment of multiple oligometastatic sites in the same radiation course. Further technological advances will also likely improve clinical workflows and time on table with improved efficiency. The role of artificial intelligence to automate current manual operations efficiently is being explored [66].

Utilizing MRgRT for the treatment of OMD is becoming increasingly recognized with the promising results in these initial phase I and phase II trials. Currently, trials for the role of MRgRT in OMD are ongoing, and their results will provide necessary information regarding whether there is a benefit to MRgRT in the OMD setting (Table 12.3).

**Table 12.2** Selected phase II and III ongoing clinical trials investigating metastasis-directed therapy in oligometastatic cancer

| Trial name/ identifier | Histology | Description | Key inclusion criteria | Target enrollment | MDT ablative treatment | Primary endpoint | Estimated study completion date |
|---|---|---|---|---|---|---|---|
| SABR-COMET-3 NCT03862911 [67] | Multiple | Phase III multicenter trial (randomized 1:2) comparing SOC vs. SOC + SBRT | Controlled primary site, one to three metastases with all metastases amenable to SBRT | 330 | SBRT | OS | 2028 |
| SABR-COMET-10 NCT03721341 [68] | Multiple | Phase III multicenter trial (1:2 ratio) comparing SOC vs. SOC + SBRT | Controlled primary site, four to ten metastases with all metastases amenable to SBRT | 204 | SBRT | OS | 2029 |
| EORTC 1945 OligoRARE NCT04498767 | Multiple (exclude lung, breast, colorectal, prostate) | Phase III multicenter trial comparing SOC + palliative RT vs. SOC + SBRT | Controlled primary site, one to five metastases | 200 | SBRT | OS | 2030 |
| CORE NCT02759783 | Breast, prostate, NSCLC | Phase II/III multicenter trial comparing SOC vs. SOC + SBRT | Controlled primary site, one to three extracranial metachronous metastases with all amenable to SBRT and limited to two different organ systems; no prior systemic therapy for metastatic disease | 245 | SBRT | PFS | 2024 |
| PLATON NCT03784755 | Prostate | Phase III multicenter trial comparing SOC ST vs. SOC ST + SBRT | Castration-sensitive with primary site controlled or amenable to definitive or RT or surgery, one to five synchronous or metachronous metastases (detected on conventional imaging or PET) with all amenable to SBRT | 410 | SBRT | FFS | 2025 |
| START-MET NCT05209243 | Prostate | Phase III trial comparing SOC (ADT + RT to primary tumor if not previously treated) + second generation hormonal agent vs. SOC + SBRT | Castration-sensitive, one to three metastases (detected by conventional imaging) or one to five metastases (detected by PET) | 266 | SBRT | rPFS | 2027 |

(continued)

**Table 12.2** (continued)

| Trial name/ identifier | Histology | Description | Key inclusion criteria | Target enrollment | MDT ablative treatment | Primary endpoint | Estimated study completion date |
|---|---|---|---|---|---|---|---|
| PRESTO NCT003678025 | Prostate | Phase III multicenter trial comparing SOC vs. SOC + SBRT | Castration-sensitive, primary site controlled or amenable to definitive RT, one to five synchronous or metachronous metastases (detected on PET) with all amenable to SBRT | 350 | SBRT | CRPC-free survival | 2028 |
| METRO NCT04983095 | Prostate | Phase III multicenter trial (1:2 ratio) comparing SOC vs. SOC + SBRT | Castration-sensitive, primary site controlled or amenable to definitive RT, one to three synchronous or metachronous metastases (detected on PSMA-PET) with all amenable to SBRT | 114 | SBRT | FFS (time to CRPC) | 2029 |
| ARTO NCT03449719 | Prostate | Randomized phase II trial comparing abiraterone vs. abiraterone + SBRT | CRPC; one to three metastases | 174 | SBRT | PSA response | 2022 |
| PILLAR NCT03503344 | Prostate | Randomized phase II trial comparing apalutamide + SBRT vs. SBRT | CRPC with no prior ST for CRPC; one to five metastases (detected by PSMA-PET) | 60 | SBRT | PSA response | 2024 |
| PCS IX NCT02685397 | Prostate | Randomized phase II/III trial comparing LNRH agonist + enzalutamide vs. LNRH agonist + enzalutamide + SBRT | CRPC; one to five metastases (detected by CT, bone scan, and/or MRI) | 130 | SBRT | rPFS | 2025 |
| OMEGA NCT03827577 | NSCLC | Phase III multicenter trial comparing SOC vs. SOC + LAT | Primary site controlled or suitable for definitive treatment, one to three metastases with all amenable to SBRT | 195 | Surgery or SBRT or RFA | OS | 2022 |
| NRG LU-002 NCT03137771 | NSCLC | Phase II/III randomized multicenter trial comparing chemotherapy vs. chemotherapy + SBRT | One to three extracranial synchronous or metachronous metastases with no progression after first line systemic therapy with all amenable to SBRT; if synchronous, primary site amenable to SBRT or hypofractionated RT | 400 | SBRT | Phase II: PFS Phase III: OS | 2031[a] |

| Trial name/ identifier | Histology | Description | Key inclusion criteria | Target enrollment | MDT ablative treatment | Primary endpoint | Estimated study completion date |
|---|---|---|---|---|---|---|---|
| TARGET-02 NCT05278052 [69] | NSCLC with negative oncogene mutation status | Phase III trial comparing SMT vs. SMT + LCRT (SBRT + hypofractionated RT to primary/nodal) | One to five metastases with no progression following four to six cycles of ST | 190 | SBRT | OS | 2028 |
| ECOG-ACRIN EA2183 NCT04248452 [70] | Esophageal and gastric adenocarcinoma (HER2 negative) | Phase III multicenter trial comparing FOLFOX vs. FOLFOX+RT vs. CAPOX vs. CAPOX+RT | One to three synchronous or metachronous metastases all amenable to RT; no evidence of disease progression following induction CT | 314 | RT | OS | 2023 |
| NCT04944914 | Nasopharyngeal carcinoma | Phase III trial comparing camrelizumab vs. camrelizumab + SBRT | Controlled primary site, one to five metastases after first line chemotherapy all amenable to SBRT; exclude patients with immunotherapy failure | 188 | SBRT | PFS | 2026 |
| NCT02956798 | RCC | Phase II single-arm trial investigating SBRT to all metastases | Systemic treatment naïve with one to three extracranial metastases | 23 | SBRT | FFST (time to initiation of systemic therapy) | 2023 |
| GETUG-StORM 01 NCT04299646 | Clear cell renal cancer | Randomized phase II trial (2:1 ratio) comparing ST vs. ST + SBRT | One to three oligoprogressive metastases all amenable to SBRT | 114 | SBRT | PFS | 2023 |

Abbreviations: *CAPOX* capecitabine and oxaliplatin, *CRPC* castration-resistant prostate cancer, *FFS* failure-free survival, *FOLFOX* fluorouracil and leucovorin calcium and oxaliplatin, *LAT* local ablative therapy, *LCRT* local consolidative radiotherapy, *OS* overall survival, *PFS* progression-free survival, *PSA* prostate specific antigen, *RCC* renal cell carcinoma, *rPFS* radiological progression-free survival, *RT* radiation therapy, *SBRT* stereotactic body radiotherapy, *SOC* standard of care, *ST* systemic therapy, *SMT* standard maintenance therapy

[a] Recruitment completed phase II and now paused for scheduled interim analysis to determine if PFS benefit before reopening for phase III recruitment

**Table 12.3** Selected ongoing clinical trials investigating treatment of MR-guided radiation therapy in oligometastatic cancer

| Trial name/ identifier | Included condition(s) | Description | Target enrollment | MDT ablative treatment | Primary endpoint | Estimated study completion date |
|---|---|---|---|---|---|---|
| SMART NCT04115254 | Oligoprogressive nodal metastases<br>Metachronous nodal metastases<br>Synchronous nodal metastases | Phase I/II trial investigating feasibility and efficacy of SMART in multiple cancer types | 1000 (master protocol) | MRIdian | Phase I: Delivery success rate, tumor visualization, plan creation<br>Phase II: Rate of improvement in tumor control | 2024 |
| SMART ONE NCT04939246 | Inoperable primary or metastatic carcinoma | Phase I multisite trial investigating feasibility and safety of one fraction SMART | 30 | MRIdian | Number of successfully delivered single fraction treatments; tolerability (no greater than 4 of 30 patients with grade3+ toxicity within 90 days, no grade 5 toxicity) | 2024 |
| SOFT NCT04407897 | Metastatic cancer with 1-5 synchronous or metachronous metastates or 1-3 oligoprogressive or induced metastases | Phase II multicenter trial investigating safety of MR-guided infra-diaphragmatic oligometastatic disease | 121 | MRIdian | Cumulative toxicity grade 4 or greater | 2023 |
| MOMENTUM NCT04075305 [71] | Oligometastatic cancer | Multi-institutional observational cohort study of multiple cancer types | 6000 | Unity MR-linac | PFS, OS, DFS, patient QoL, toxicity, clinical tumor response, pathologic tumor response | 2024 |
| NCT04172753 [72] | Oligometastatic cancer | Feasibility study investigating imaging and treatment on 1.5 T MR-linac of multiple cancer types | 472 | Unity MR-linac | Feasibility of online MR-guided RT | 2027 |

Abbreviations: *MRgRT* magnetic resonance (MR) guided radiation therapy, *QoL* quality of life, *SMART* stereotactic magnetic resonance guided adaptive radiotherapy

## References


1. Hellman S, Weichselbaum RR. Oligometastases. J Clin Oncol. 1995;13(1):8–10. https://pubmed.ncbi.nlm.nih.gov/7799047/
2. Weichselbaum RR, Hellman S. Oligometastases revisited. Nat Rev Clin Oncol. 2011;8(6):378–82. https://pubmed.ncbi.nlm.nih.gov/21423255/
3. Ganesh K, Massagué J. Targeting metastatic cancer. Nat Med. 2021;27(1):34–44. https://pubmed.ncbi.nlm.nih.gov/33442008/
4. Reyes DK, Pienta KJ. The biology and treatment of oligometastatic cancer. Oncotarget. 2015;6(11):8491–524. https://pubmed.ncbi.nlm.nih.gov/25940699/
5. Liu W, Bahig H, Palma DA. Oligometastases: emerging evidence. J Clin Oncol. 2022;40(36):4250–60. https://pubmed.ncbi.nlm.nih.gov/36306497/
6. Ost P, Reynders D, Decaestecker K, Fonteyne V, Lumen N, DeBruycker A, et al. Surveillance or metastasis-directed therapy for oligometastatic prostate cancer recurrence: a prospective, randomized, multicenter phase II trial. J Clin Oncol. 2018;36(5):446–53. https://pubmed.ncbi.nlm.nih.gov/29240541/
7. Palma DA, Olson R, Harrow S, Gaede S, Louie AV, Haasbeek C, et al. Stereotactic ablative radiotherapy for the comprehensive treatment of oligometastatic cancers: long-term results of the SABR-COMET phase II randomized trial. J Clin Oncol. 2020;38(25):2830–8.
8. Wang XS, Bai YF, Verma V, Yu RL, Tian W, Ao R, et al. Randomized trial of first-line tyrosine kinase inhibitor with or without radiotherapy for synchronous oligometastatic EGFR-mutated NSCLC. J Natl Cancer Inst. 2023;115:742–8. https://pubmed.ncbi.nlm.nih.gov/35094066/
9. Lievens Y, Guckenberger M, Gomez D, Hoyer M, Iyengar P, Kindts I, et al. Defining oligometastatic disease from a radiation oncology perspective: an ESTRO-ASTRO consensus document. Radiother Oncol. 2020;148:157–66. https://pubmed.ncbi.nlm.nih.gov/32388150/
10. Guckenberger M, Lievens Y, Bouma AB, Collette L, Dekker A, deSouza NM, et al. Characterisation and classification of oligometastatic disease: a European Society for Radiotherapy and Oncology and European Organisation for Research and Treatment of Cancer consensus recommendation. Lancet Oncol. 2020;21(1):e18–28. https://pubmed.ncbi.nlm.nih.gov/31908301/
11. Thomas MC, Chen YH, Fite E, Pangilinan A, Bubelo K, Spektor A, et al. Patient and treatment factors associated with improved local control and survival in oligometastatic bone disease: results from a large single-institution experience using stereotactic body radiation therapy. Int J Radiat Oncol Biol Phys. 2022;114(4):747–61. https://pubmed.ncbi.nlm.nih.gov/35840113/
12. Farolfi A, Calderoni L, Mattana F, Mei R, Telo S, Fanti S, et al. Current and emerging clinical applications of PSMA PET diagnostic imaging for prostate cancer. J Nucl Med. 2021;62(5):596–604. https://pubmed.ncbi.nlm.nih.gov/33712536/
13. Huynh MA, Tang C, Siva S, Berlin A, Hannan R, Warner A, et al. Review of prospective trials assessing the role of stereotactic body radiation therapy for metastasis-directed treatment in oligometastatic genitourinary cancers. Eur. Urol Oncol. 2023;6(1):28–38. https://pubmed.ncbi.nlm.nih.gov/36283936/
14. Pacifico P, Colciago RR, de Felice F, Boldrini L, Salvestrini V, Nardone V, et al. A critical review on oligometastatic disease: a radiation oncologist's perspective. Med Oncol. 2022;39(12):181. https://pubmed.ncbi.nlm.nih.gov/36071292/
15. Chetty IJ, Doemer AJ, Dolan JL, Kim JP, Cunningham JM, Dragovic J, et al. MRI-guided radiotherapy (MRgRT) for treatment of oligometastases: review of clinical applications and challenges. Int J Radiat Oncol Biol Phys. 2022;114(5):950–67. https://pubmed.ncbi.nlm.nih.gov/35901978/
16. Fleckenstein J, Petroff A, Schäfers HJ, Wehler T, Schöpe J, Rübe C. Long-term outcomes in radically treated synchronous vs. metachronous oligometastatic non-small-cell lung cancer. BMC Cancer. 2016;16(1):348. https://pubmed.ncbi.nlm.nih.gov/27255302/
17. Milano MT, Katz AW, Zhang H, Okunieff P. Oligometastases treated with stereotactic body radiotherapy: long-term follow-up of prospective study. Int J Radiat Oncol Biol Phys. 2012;83(3):878–86. https://pubmed.ncbi.nlm.nih.gov/22172903/
18. Bezjak A, Paulus R, Gaspar LE, Timmerman RD, Straube WL, Ryan WF, et al. Safety and efficacy of a five-fraction stereotactic body radiotherapy schedule for centrally located non-small-cell lung cancer: NRG oncology/RTOG 0813 trial. J Clin Oncol. 2019;37(15):1316–25. https://pubmed.ncbi.nlm.nih.gov/30943123/
19. Romesser PB, Tyagi N, Crane CH. Magnetic resonance imaging-guided adaptive radiotherapy for colorectal liver metastases. Cancers (Basel). 2021;13(7):1636. https://pubmed.ncbi.nlm.nih.gov/33915810/
20. Rosenberg SA, Henke LE, Shaverdian N, Mittauer K, Wojcieszynski AP, Hullett CR, et al. A multi-institutional experience of MR-guided liver stereotactic body radiation therapy. Adv Radiat Oncol. 2018;4(1):142–9. https://pubmed.ncbi.nlm.nih.gov/30706022/
21. Borghesi S, Casamassima F, Aristei C, Grandinetti A, di Franco R. Stereotactic radiotherapy for adrenal oligometastases. Rep Pract Oncol Radiother. 2022;27(1):52–6. https://pubmed.ncbi.nlm.nih.gov/35402020/
22. de Ruysscher D, Wanders R, Hendriks LE, van Baardwijk A, Reymen B, Houben R, et al. Progression-free survival and overall survival beyond 5 years of NSCLC patients with synchronous oligometastases treated in a prospective phase II trial (NCT 01282450). J Thorac Oncol. 2018;13(12):1958–61. https://pubmed.ncbi.nlm.nih.gov/30253974/

23. Chmura S, Winter KA, Robinson C, Pisansky TM, Borges V, Al-Hallaq H, et al. Evaluation of safety of stereotactic body radiotherapy for the treatment of patients with multiple metastases: findings from the NRG-BR001 phase 1 trial. JAMA Oncol. 2021;7(6):845–52. https://pubmed.ncbi.nlm.nih.gov/33885704/
24. Jereczek-Fossa BA, Ronchi S, Orecchia R. Is stereotactic body radiotherapy (SBRT) in lymph node oligometastatic patients feasible and effective? Rep Pract Oncol Radiother. 2015;20(6):472–83. https://pubmed.ncbi.nlm.nih.gov/26696788/
25. Palma DA, Olson R, Harrow S, Gaede S, Louie AV, Haasbeek C, et al. Stereotactic ablative radiotherapy versus standard of care palliative treatment in patients with oligometastatic cancers (SABR-COMET): a randomised, phase 2, open-label trial. Lancet. 2019;393(10185):2051–8. https://pubmed.ncbi.nlm.nih.gov/30982687/
26. Gomez DR, Tang C, Zhang J, Blumenschein GR, Hernandez M, Jack Lee J, et al. Local consolidative therapy vs. maintenance therapy or observation for patients with oligometastatic non-small-cell lung cancer: long-term results of a multi-institutional, phase II, randomized study. J Clin Oncol. 2019;37(18):1558–65. https://pubmed.ncbi.nlm.nih.gov/31067138/
27. Wang XS, Bai YF, Verma V, Yu RL, Tian W, Ao R, et al. Randomized trial of first-line tyrosine kinase inhibitor with or without radiotherapy for synchronous oligometastatic EGFR-mutated non-small cell lung cancer. J Natl Cancer Inst. 2023;115(6):742–8. https://doi.org/10.1093/jnci/djac015/6517507.
28. Phillips R, Shi WY, Deek M, Radwan N, Lim SJ, Antonarakis ES, et al. Outcomes of observation vs stereotactic ablative radiation for oligometastatic prostate cancer: the ORIOLE phase 2 randomized clinical trial. JAMA Oncologia. 2020;6(5):650. /pmc/articles/PMC7225913/
29. Bianchi M, Sun M, Jeldres C, Shariat SF, Trinh QD, Briganti A, et al. Distribution of metastatic sites in renal cell carcinoma: a population-based analysis. Ann Oncol. 2012;23(4):973–80. https://pubmed.ncbi.nlm.nih.gov/21890909/
30. Hannan R, Christensen M, Christie A, Garant A, Pedrosa I, Robles L, et al. Stereotactic ablative radiation for systemic therapy-naïve oligometastatic kidney cancer. Eur Urol Oncol. 2022;5(6):695–703. https://pubmed.ncbi.nlm.nih.gov/35985982/
31. Tang C, Msaouel P, Hara K, Choi H, Le V, Shah AY, et al. Definitive radiotherapy in lieu of systemic therapy for oligometastatic renal cell carcinoma: a single-arm, single-centre, feasibility, phase 2 trial. Lancet Oncol. 2021;22(12):1732–9. https://pubmed.ncbi.nlm.nih.gov/34717797/
32. Cheung P, Patel S, North SA, Sahgal A, Chu W, Soliman H, et al. Stereotactic radiotherapy for oligoprogression in metastatic renal cell cancer patients receiving tyrosine kinase inhibitor therapy: a phase 2 prospective multicenter study. Eur Urol. 2021;80(6):693–700. https://pubmed.ncbi.nlm.nih.gov/34399998/
33. Hannan R, Christensen M, Hammers H, Christie A, Paulman B, Lin D, et al. Phase II trial of stereotactic ablative radiation for oligoprogressive metastatic kidney cancer. Eur Urol Oncol. 2022;5(2):216–24. https://pubmed.ncbi.nlm.nih.gov/34986993/
34. Yamamoto T, Niibe Y, Aoki M, Shintani T, Yamada K, Kobayashi M, et al. Analyses of the local control of pulmonary oligometastases after stereotactic body radiotherapy and the impact of local control on survival. BMC Cancer. 2020;20(1):997. https://pubmed.ncbi.nlm.nih.gov/33054721/
35. Fode MM, Høyer M. Survival and prognostic factors in 321 patients treated with stereotactic body radiotherapy for oligo-metastases. Radiother Oncol. 2015;114(2):155–60. https://pubmed.ncbi.nlm.nih.gov/25583567/
36. Chmura SJ, Winter KA, Woodward WA, Borges VF, Salama JK, Al-Hallaq HA, et al. NRG-BR002: A phase IIR/III trial of standard of care systemic therapy with or without stereotactic body radiotherapy (SBRT) and/or surgical resection (SR) for newly oligometastatic breast cancer (NCT02364557). J Clin Oncol. 2022;40(16_suppl):1007. https://doi.org/10.1200/JCO20224016_suppl1007.
37. Tsai CJ, Yang JT, Guttmann DM, Shaverdian N, Shepherd AF, Eng J, et al. Consolidative use of radiotherapy to block (CURB) oligoprogression — interim analysis of the first randomized study of stereotactic body radiotherapy in patients with oligoprogressive metastatic cancers of the lung and breast. Int J Radiat Oncol Biol Phys. 2021;111(5):1325–6.
38. Poon I, Erler D, Dagan R, Redmond KJ, Foote M, Badellino S, et al. Evaluation of definitive stereotactic body radiotherapy and outcomes in adults with extracranial oligometastasis. JAMA Netw Open. 2020;3(11):e2026312. https://jamanetwork.com/journals/jamanetworkopen/fullarticle/2772997
39. Timmerman RD, Kavanagh BD, Cho LC, Papiez L, Xing L. Stereotactic body radiation therapy in multiple organ sites. J Clin Oncol. 2007;25(8):947–52. https://pubmed.ncbi.nlm.nih.gov/17350943/
40. Kord M, Kluge A, Kufeld M, Kalinauskaite G, Loebel F, Stromberger C, et al. Risks and benefits of fiducial marker placement in tumor lesions for robotic radiosurgery: technical outcomes of 357 implantations. Cancers (Basel). 2021;13(19):4838. /pmc/articles/PMC8508340/
41. Devic S. MRI simulation for radiotherapy treatment planning. Med Phys. 2012;39(11):6701–11. https://pubmed.ncbi.nlm.nih.gov/23127064/
42. Cuccia F, Rigo M, Gurrera D, Nicosia L, Mazzola R, Figlia V, et al. Mitigation on bowel loops daily variations by 1.5-T MR-guided daily-adaptive SBRT for abdomino-pelvic lymph-nodal oligometastases. J Cancer Res Clin Oncol. 2021;147(11):3269–77. https://pubmed.ncbi.nlm.nih.gov/34268583/
43. Ma X, Chen X, Wang Y, Qin S, Yan X, Cao Y, et al. Personalized modeling to improve pseudo-computed

tomography images for magnetic resonance imaging-guided adaptive radiation therapy. Int J Radiat Oncol Biol Phys. 2022;113(4):885–92. https://pubmed.ncbi.nlm.nih.gov/35462026/
44. Yang DD, Brennan VS, Huynh E, Williams CL, Han Z, Ampofo N, et al. Stereotactic magnetic resonance-guided adaptive radiation therapy (SMART) for abdominopelvic oligometastases. Int J Radiat Oncol Biol Phys. 2022;114(5):941–9. https://pubmed.ncbi.nlm.nih.gov/35598799/
45. Cox BW, Spratt DE, Lovelock M, Bilsky MH, Lis E, Ryu S, et al. International spine radiosurgery consortium consensus guidelines for target volume definition in spinal stereotactic radiosurgery. Int J Radiat Oncol Biol Phys. 2012;83(5):e597–605. https://pubmed.ncbi.nlm.nih.gov/22608954/
46. Olberg S, Green O, Cai B, Yang D, Rodriguez V, Zhang H, et al. Optimization of treatment planning workflow and tumor coverage during daily adaptive magnetic resonance image guided radiation therapy (MR-IGRT) of pancreatic cancer. Radiat Oncol. 2018;13(1):1–8. /pmc/articles/PMC5866525/
47. Gerhard SG, Palma DA, Arifin AJ, Louie AV, Li GJ, Al-Shafa F, et al. Organ at risk dose constraints in SABR: a systematic review of active clinical trials. Pract Radiat Oncol. 2021;11(4):e355–65.
48. Diez P, Hanna GG, Aitken KL, van As N, Carver A, Colaco RJ, et al. UK 2022 consensus on normal tissue dose-volume constraints for oligometastatic, primary lung and hepatocellular carcinoma stereotactic ablative radiotherapy. Clin Oncol. 2022;34(5):288–300. http://www.clinicaloncologyonline.net/article/S0936655522000942/fulltext
49. Hanna GG, Murray L, Patel R, Jain S, Aitken KL, Franks KN, et al. UK consensus on normal tissue dose constraints for stereotactic radiotherapy. Clin Oncol (R Coll Radiol). 2018;30(1):5–14. https://pubmed.ncbi.nlm.nih.gov/29033164/
50. Benedict SH, Yenice KM, Followill D, Galvin JM, Hinson W, Kavanagh B, et al. Stereotactic body radiation therapy: the report of AAPM task group 101. Med Phys. 2010;37(8):4078–101. https://doi.org/10.1118/1.3438081.
51. Timmerman RD. An overview of Hypofractionation and introduction to this issue of seminars in radiation oncology. Semin Radiat Oncol. 2008;18(4):215–22.
52. Kim DWN, Medin PM, Timmerman RD. Emphasis on repair, not just avoidance of injury, facilitates prudent stereotactic ablative radiotherapy. Semin Radiat Oncol. 2017;27(4):378–92.
53. Christensen M, Hannan R. The emerging role of radiation therapy in renal cell carcinoma. Cancers (Basel). 2022;14(19):4693. https://pubmed.ncbi.nlm.nih.gov/36230615/
54. Wang CJ, Christie A, Lin MH, Jung M, Weix D, Huelsmann L, et al. Safety and efficacy of stereotactic ablative radiation therapy for renal cell carcinoma extracranial metastases. Int J Radiat Oncol Biol Phys. 2017;98(1):91. /pmc/articles/PMC5555369/
55. Tao R, Krishnan S, Bhosale PR, Javle MM, Aloia TA, Shroff RT, et al. Ablative radiotherapy doses lead to a substantial prolongation of survival in patients with inoperable intrahepatic cholangiocarcinoma: a retrospective dose response analysis. J Clin Oncol. 2016;34(3):219–26. https://pubmed.ncbi.nlm.nih.gov/26503201/
56. Henke L, Kashani R, Yang D, Zhao T, Green O, Olsen L, et al. Simulated online adaptive magnetic resonance–guided stereotactic body radiation therapy for the treatment of oligometastatic disease of the abdomen and central thorax: characterization of potential advantages. Int J Radiat Oncol Biol Phys. 2016;96(5):1078–86. http://www.redjournal.org/article/S0360301616331510/fulltext
57. Hawranko R, Sohn JJ, Neiderer K, Bump E, Harris T, Fields EC, et al. Investigation of isotoxic dose escalation and plan quality with TDABC analysis on a 0.35 T MR-Linac (MRL) system in ablative 5-fraction stereotactic magnetic resonance-guided radiation therapy (MRgRT) for primary pancreatic cancer. J Clin Med. 2022;11(9):2584. https://pubmed.ncbi.nlm.nih.gov/35566712/
58. Parikh PJ, Lee P, Low D, Kim J, Mittauer KE, Bassetti MF, et al. Stereotactic MR-guided on-table adaptive radiation therapy (SMART) for patients with borderline or locally advanced pancreatic cancer: primary endpoint outcomes of a prospective phase II multicenter international trial. Int J Radiat Oncol Biol Phys. 2022;114(5):1062–3. http://www.redjournal.org/article/S0360301622032680/fulltext
59. Henke L, Kashani R, Robinson C, Curcuru A, DeWees T, Bradley J, et al. Phase I trial of stereotactic MR-guided online adaptive radiation therapy (SMART) for the treatment of oligometastatic or unresectable primary malignancies of the abdomen. Radiother Oncol. 2018;126(3):519–26. https://pubmed.ncbi.nlm.nih.gov/29277446/
60. Bodo S, Campagne C, Thin TH, Higginson DS, Vargas HA, Hua G, et al. Single-dose radiotherapy disables tumor cell homologous recombination via ischemia/reperfusion injury. J Clin Invest. 2019;129(2):786–801. https://pubmed.ncbi.nlm.nih.gov/30480549/
61. Zelefsky MJ, Yamada Y, Greco C, Lis E, Schöder H, Lobaugh S, et al. Phase 3 multi-center, prospective, randomized trial comparing single-dose 24 Gy radiation therapy to a 3-fraction SBRT regimen in the treatment of Oligometastatic cancer. Int J Radiat Oncol Biol Phys. 2021;110(3):672–9. https://pubmed.ncbi.nlm.nih.gov/33422612/
62. Henke LE, Stanley JA, Robinson C, Srivastava A, Contreras JA, Curcuru A, et al. Phase I trial of stereotactic MRI-guided online adaptive radiation therapy (SMART) for the treatment of oligometastatic ovarian cancer. Int J Radiat Oncol Biol Phys. 2022;112(2):379–89. https://pubmed.ncbi.nlm.nih.gov/34474109/
63. Janssen TM, Aitken K, Alongi F, Barry A, Bernchou U, Boeke S, et al. First multicentre experience of SABR for lymph node and liver oligometastatic

disease on the unity MR-Linac. Tech Innov Patient Support Radiat Oncol. 2022;22:50–4. https://pubmed.ncbi.nlm.nih.gov/35586786/
64. Nicosia L, Trapani G, Rigo M, Giaj-Levra N, Mazzola R, Pastorello E, et al. 1.5 T MR-guided daily adapted SBRT on lymph node oligometastases from prostate cancer. J Clin Med. 2022;11(22):6658. https://pubmed.ncbi.nlm.nih.gov/36431135/
65. Mazzola R, Cuccia F, Figlia V, Rigo M, Nicosia L, Giaj-Levra N, et al. Stereotactic body radiotherapy for oligometastatic castration sensitive prostate cancer using 1.5 T MRI-Linac: preliminary data on feasibility and acute patient-reported outcomes. Radiol Med. 2021;126(7):989–97. https://pubmed.ncbi.nlm.nih.gov/33835309/
66. Cusumano D, Boldrini L, Dhont J, Fiorino C, Green O, Güngör G, et al. Artificial intelligence in magnetic resonance guided radiotherapy: medical and physical considerations on state of art and future perspectives. Phys Med. 2021;85:175–91. https://pubmed.ncbi.nlm.nih.gov/34022660/
67. Olson R, Mathews L, Liu M, Schellenberg D, Mou B, Berrang T, et al. Stereotactic ablative radiotherapy for the comprehensive treatment of 1-3 oligometastatic tumors (SABR-COMET-3): study protocol for a randomized phase III trial. BMC Cancer. 2020;20(1):1–12. https://doi.org/10.1186/s12885-020-06876-4.
68. Palma DA, Olson R, Harrow S, Correa RJM, Schneiders F, Haasbeek CJA, et al. Stereotactic ablative radiotherapy for the comprehensive treatment of 4-10 oligometastatic tumors (SABR-COMET-10): study protocol for a randomized phase III trial. BMC Cancer. 2019;19(1):1–15. https://doi.org/10.1186/s12885-019-5977-6.
69. Tibdewal A, Agarwal JP, Srinivasan S, Mummudi N, Noronha V, Prabhash K, et al. Standard maintenance therapy versus local consolidative radiation therapy and standard maintenance therapy in 1-5 sites of oligometastatic non-small cell lung cancer: a study protocol of phase III randomised controlled trial. BMJ Open. 2021;11(3):e043628. https://pubmed.ncbi.nlm.nih.gov/33727268/
70. Uboha NV, Graham N, Rajdev L, Gibson MK, Fisher GA, Lin SH, et al. EA2183: A phase III study of consolidative radiotherapy in patients with oligometastatic HER2-negative esophageal and gastric adenocarcinoma. J Clin Oncol. 2022;40(16_suppl):TPS4162. https://doi.org/10.1200/JCO20224016_supplTPS4162.
71. de Mol van Otterloo SR, Christodouleas JP, Blezer ELA, Akhiat H, Brown K, Choudhury A, et al. The MOMENTUM study: an international registry for the evidence-based introduction of MR-guided adaptive therapy. Front Oncologia. 2020;10:1328.
72. Wegener D, Thome A, Paulsen F, Gani C, Boldt J, Butzer S, et al. First experience and prospective evaluation on feasibility and acute toxicity of online adaptive radiotherapy of the prostate bed as salvage treatment in patients with biochemically recurrent prostate cancer on a 1.5T MR-Linac. J Clin Med. 2022;11(16):4651. https://pubmed.ncbi.nlm.nih.gov/36012885/

# Management of Pancreatic Cancer with MRI-Guided Adaptive Radiotherapy

# 13

Ritchell van Dams, Sylvia S. W. Ng, John Hayes, and Tarita O. Thomas

## Introduction and Disease Overview

The pancreas is a retroperitoneal organ located in the upper, central, posterior abdomen between vertebral levels L1–L2 whose function is complex with both endocrine and exocrine function. It resides in close proximity to multiple adjacent abdominal organs such as the duodenum as well as blood vessels and the celiac nerve plexus. Critical vasculature that extends through this area includes the celiac artery and its main branches (left gastric, splenic, and hepatic arteries), the superior mesenteric artery, and the superior mesenteric and portal veins. Anatomically designated as having four sections—head, neck, body, and tail—the gland is functionally continuous from its head that abuts the second portion of the duodenum to the tail that typically extends toward the splenic hilum (Fig. 13.1).

More than 90% of pancreas cancer derives from ductal epithelium resulting in ductal adenocarcinoma. Globally, pancreatic ductal adenocarcinoma is the 14th most common cancer type but given its high mortality rate represents the seventh most common cause of cancer-related death [1]. Pancreatic cancer has one of the lowest 5-year survival rates of all cancers at 11% and even among patients with localized disease 5-year survival is only 42% [2].

Early diagnosis is uncommon due to the lack of symptoms and signs of earlier stage disease as symptoms are more common when there is local progression of tumor causing obstructive symptoms. Pancreatic ductal carcinomas are also recognized as having a high capacity for early systemic dissemination, so prognosis is guarded even for those patients diagnosed with seemingly localized disease [3]. General consensus is therefore that adenocarcinoma of the pancreas should be treated as a systemic disease even when apparently localized [4]. Yet appreciating the natural history of the disease necessitates the recognition that about one in three patients dies due to progression at the primary site of disease [5]. Optimal treatment strategies for patients are thus dependent on harnessing the benefits of surgery, systemic therapy, and radiation therapy at the appropriate time with a coordinated approach.

R. van Dams (✉)
Department of Radiation Oncology, Dana-Farber Brigham Cancer Center, Harvard Medical School, Boston, MA, USA
e-mail: rvandams@bwh.harvard.edu

S. S. W. Ng
Department of Radiation Oncology, Sunnybrook Odette Cancer Centre, University of Toronto, Toronto, ON, Canada
e-mail: sylvia.ng@sunnybrook.ca

J. Hayes · T. O. Thomas
Department of Radiation Oncology, Northwestern Memorial Hospital, Northwestern University Feinberg School of Medicine, Chicago, IL, USA
e-mail: jphayes@nm.org; tarita.thomas@nm.org



I. J. Das et al. (eds.), *A Practical Guide to MR-Linac*, https://doi.org/10.1007/978-3-031-48165-9_13

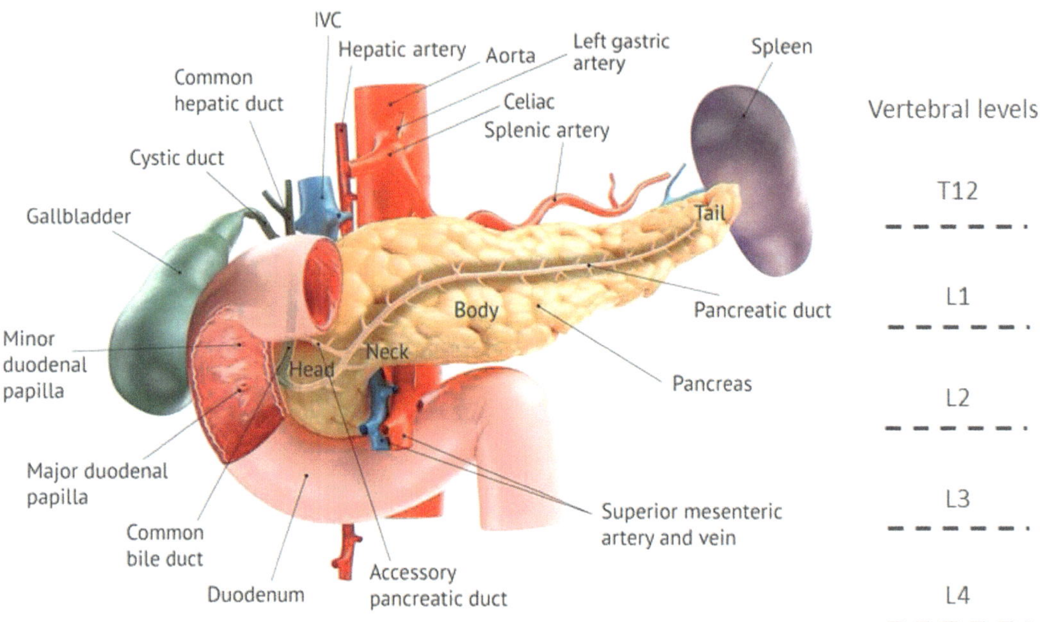


**Fig. 13.1** Anatomy of the pancreas and surrounding vasculature

Although the American Joint Committee on Cancer (AJCC) has developed a traditional tumor, nodal, and metastasis (TNM) staging system for pancreatic cancer, patients are generally stratified and treated according to a surgical classification system that establishes whether a patient has resectable, borderline resectable, or unresectable disease [6]. These differences are critical for appropriate clinical decisions.

Resectable tumors are generally lesions without metastases that do not contact major arteries including the celiac artery (CA), superior mesenteric artery (SMA), or common hepatic artery (CHA). They also have no or minimal contact with the superior mesenteric vein (SMV) or portal vein (PV) and do not deform the normal course of the vein. Pancreatic tail lesions may involve the splenic vasculature in resectable cases as the tumor can be removed with a distal pancreatectomy and splenectomy.

Various definitions of borderline resectable tumors are utilized. Generally, borderline resectable tumors consist of lesions with only abutment (contact ≤180°) of the CA or SMA and/or involvement of the CHA, SMV, or PV that is amenable to vascular reconstruction. Unresectable tumors typically demonstrate either encasement (contact ≥180°) of the CA or SMA or have unreconstructable involvement of the SMV or PV. Involvement of the first jejunal branches of the SMA or SMV, nonregional lymph node involvement, and distant metastases also place a tumor in the unresectable category [6].

## Treatment Background

Patients with resectable pancreatic cancer are treated with upfront resection as surgery offers the only potentially curative treatment strategy for this cancer [7]. Early trials of adjuvant radiation therapy (RT) for resected pancreatic cancer showed mixed findings. These trials, including GITSG 9173, EORTC 40891, and ESPAC-1, were inconclusive regarding whether adjuvant RT provided an overall survival (OS) benefit, no benefit, or decreased OS [8–10]. These historical trials are difficult to interpret as split-course RT was utilized—a technique no longer used in the modern era—did not mandate quality assurance methods to ensure RT was delivered as per protocol, and had crossover of treatment arms. RTOG

9704 was one of the first trials to mandate quality assurance to radiation field design and dosimetry [11]. Although no overall survival difference was detected among the entire cohort for the addition of adjuvant chemoradiotherapy compared to chemotherapy alone, an unplanned secondary analysis demonstrated statistically improved survival if treatment adhered to radiation therapy protocol highlighting the importance of careful target delineation and treatment planning [12].

To date, the single most effective adjuvant therapy for resected pancreatic cancer has been a regimen of modified fluorouracil, leucovorin, irinotecan, and oxaliplatin (mFOLFIRINOX), which increased median survival from 36 to 54 months and 5-year OS from 31 to 43% compared to single-agent gemcitabine [13].

Borderline resectable pancreatic cancer comprises a wide spectrum of disease, from relatively minimal contact with veins that can undergo reconstruction during resection to nearly 180° involvement of major arteries that are at very high risk of a positive margin. Several studies support the role of neoadjuvant therapy for borderline resectable pancreatic cancer; however, the specific benefit of preoperative radiation remains unclear.

Three studies are worth noting in evaluating the overall role of neoadjuvant therapy for patients with borderline resectable disease. First, Katz et al. [14] were the first to provide a formal definition of borderline resectable disease and reported their institutional experience treating these patients with neoadjuvant therapy. Among 160 patients identified with borderline resectable pancreatic cancer, 78% were able to complete neoadjuvant therapy with chemotherapy followed by chemoradiation, and 41% of the total cohort was able to undergo whipple resection with a 94% rate of negative-margin resection. Furthermore, patients who were able to complete all neoadjuvant therapy and undergo surgical resection had a median overall survival of 40 months compared to patients who were unable to complete all planned therapy [14]. The Alliance A021101 was a single-arm prospective trial that treated 22 patients with borderline resectable disease with neoadjuvant mFOLFIRINOX (oxaliplatin 85 mg/m$^2$, irinotecan 180 mg/m$^2$, leucovorin 400 mg/m$^2$, and infusional 5-fluorouracil 2400 mg/m$^2$ over 46 h) followed by chemoradiotherapy to 50.4 Gy in 28 fractions with concurrent capecitabine [15]. A large majority (68%) of patients were able to undergo surgical resection, and 93% had a margin-negative resection. Among patients who underwent surgical resection, the 18-month overall survival rate was 67% compared to 43% for those unable to proceed to resection [15]. Finally, the ESPAC5 trial [16] randomized 90 patients with borderline resectable disease into 4 cohorts: (1) upfront surgery, (2) neoadjuvant gemcitabine/capecitabine, (3) neoadjuvant FOLFIRINOX, or (4) neoadjuvant chemoradiotherapy to 50.4 Gy with concurrent capecitabine. There was no difference in the rate of resection between those randomized to upfront surgery vs. neoadjuvant therapy. Overall survival at 1 year was significantly improved among patients receiving neoadjuvant therapy (39% for upfront surgery vs. 60–84% for neoadjuvant therapy). Although the trial was not powered to compare the neoadjuvant arms to each other, there was a numerical improvement in overall survival for FOLFIRINOX (84%) compared to gemcitabine/capecitabine (78%) and chemoradiotherapy (60%) [16]. These three studies, among others, established the safety, feasibility, and efficacy of neoadjuvant therapy for borderline resectable pancreatic cancer.

Randomized trials investigating the specific role of radiation in borderline resectable pancreatic cancer have shown a more mixed picture. PREOPANC-1 was a randomized study that enrolled both resectable and borderline resectable patients to either upfront surgery or neoadjuvant gemcitabine for 3 cycles with radiation (36 Gy in 15 fractions) given concurrently with the second cycle of chemotherapy [17]. Patients in the upfront surgery arm were planned to receive six cycles of adjuvant gemcitabine, while patients in the preoperative therapy arm were planned for four cycles of adjuvant chemotherapy. Among the cohort with borderline resectable disease, use of neoadjuvant chemoradiotherapy increased the margin-negative resection rate to 79% from 13% and the median overall survival to

18 months from 13 months [17]. One limitation of this study is that single-agent gemcitabine is no longer the standard of care in treatment of pancreatic cancer, raising the question of how chemoradiotherapy compares to mFOLFIRINOX. As a partial answer to this question, the Alliance A021501 trial found no benefit to the addition of neoadjuvant stereotactic body radiotherapy (SBRT) for patients with borderline resectable pancreatic cancer compared to neoadjuvant mFOLFIRINOX [18]. This multicenter trial randomized 126 patients to either 8 cycles of neoadjuvant mFOLFIRINOX followed by surgery or 7 cycles of neoadjuvant mFOLFIRINOX followed by radiation therapy (SBRT to 33–40 Gy in 5 fractions or hypofractionated image-guided RT to 25 Gy in 5 fractions) followed by surgery. All patients were planned to receive four cycles of adjuvant mFOLFOX (oxaliplatin 85 mg/m$^2$, leucovorin 400 mg/m$^2$, and infusional 5-fluorouracil 2400 mg/m$^2$ over 46 h). The SBRT arm was closed to accrual early due to a futility endpoint defined by R0 (margin-negative) resection rate. Patients in the SBRT arm experienced a higher rate of development of distant metastatic disease, which was classified as margin positive on the study. Of note, patients in the SBRT arm had higher carbohydrate antigen (CA) 19–9 scores, a surrogate marker in PDAC used to assess treatment response, than those in the neoadjuvant chemotherapy arm (260 vs. 167, respectively). With these caveats in mind, the Alliance A021501 trial identified a worse overall survival rate at 18 months for SBRT compared to chemotherapy alone (47% vs. 66%, respectively) [18]. Given the disparate results for neoadjuvant radiation for borderline resectable pancreatic cancer, practice patterns vary widely by institution. Of note, the radiation therapy doses employed in this study were non-ablative with a biological effective dose (BED) of 37.5–72 Gy.

Locally advanced pancreatic cancer (LAPC) is defined as nonmetastatic but unresectable disease. It typically presents with either extensive peripancreatic lymph node involvement or substantial vascular involvement of the SMA, celiac axis, IVC, or aorta. Attempts have been made to evaluate the role of conventional chemoradiotherapy for patients with LAPC, but results have been mixed. The ECOG 4201 trial randomized patients to either gemcitabine chemotherapy alone or gemcitabine with conventionally fractionated radiation [19]. This trial identified a mere 2-month survival advantage for chemoradiotherapy at the cost of a fourfold increase in grade 4 toxicity [19]. In contrast, the much larger LAP07 trial randomized 449 patients to gemcitabine vs. chemoradiotherapy with concurrent capecitabine and identified no significant difference in overall survival between the 2 cohorts with a local control benefit seen with chemoradiotherapy compared to chemotherapy alone which was statistically significant [20]. Challenges to identifying a benefit to conventionally fractionated radiation in LAPC have led to development of methods for overcoming those traditional challenges as well as technological improvements allowing for dose escalation to improve clinical outcomes.

## Overcoming Challenges of Radiotherapy in the Treatment of Pancreatic Cancer

Approximately 30% of pancreatic ductal adenocarcinoma (PDAC) patients die with locally destructive disease [5], as progressive locoregional disease can cause pain which may be associated with celiac plexus infiltration, gastric outlet obstruction, and/or duodenal bleeding. These cancer-related complications negatively impact the patient quality of life. As distant metastatic disease is better controlled with current standard-of-care multi-agent systemic therapy, providing durable locoregional control with radiotherapy is integral to the long-term management of PDAC patients. As technical advances in radiation therapy delivery have been made, there are now further studies demonstrating dose-escalated radiation therapy to a BED of 100 Gy has benefit for patients with PADC. Locoregional control can be improved by escalating radiation dose using modern treatment techniques including stereotactic body radiotherapy (SBRT) and magnetic resonance imaging (MRI)-guided adaptive radiotherapy [21]. Early studies evaluating the role of single fraction SBRT in locally advanced PADC demonstrated local control with first progression at distant sites however with significant toxicities [22]. Further studies with fractionated SBRT to lower BED doses reported excellent local control with acceptable toxicity in these [23–28]. Furthermore, cancer-related pain and quality of life were shown to improve in patients with borderline resectable and locally advanced PDAC at 4 months post SBRT [25, 26]. The 2019 American Society for Radiation Oncology (ASTRO) clinical practice guidelines for radiation therapy in PDAC conditionally recommended a definitive treatment regimen consisting of systemic chemotherapy followed by concurrent chemoradiation using conventional fractionation or dose-escalated radiotherapy or multi-fraction SBRT alone in locally advanced disease that is not appropriate for downstaging to eventual surgery [29]. In the 2023 NCCN guidelines, both concurrent chemoradiation and SBRT were included as radiotherapy options to manage locally advanced/unresectable PDAC [30].

It is well recognized that the close proximity of the pancreatic tumor to many of the radiosensitive luminal gastrointestinal structures (duodenum, small bowel, stomach, and large bowel) combined with the significant mobility and highly variable positioning of these structures makes radiation dose escalation extremely challenging. On most conventional linacs, image-guided radiotherapy is delivered using cone beam computed tomography (CBCT) for daily image guidance and radiation treatment verification. The CBCT has suboptimal soft tissue contrast and is not ideal for image-guided SBRT in the treatment of PDAC because the pancreatic tumor cannot be clearly differentiated from the surrounding normal luminal gastrointestinal structures. As such, endoscopic ultrasound guided placement of fiducials within the pancreatic tumor is required for target localization as well as breath hold techniques for improved visualization during safe delivery of SBRT on a conventional linac. Magnetic resonance imaging (MRI) offers superior soft tissue contrast compared to CBCT, thereby enabling better delineation of soft tissue target(s) and normal luminal gastrointestinal organs [31] as shown in Chap. 2 of this book. Using MRI for image-guided radiotherapy, the MR-Linac (MRL) is well suited for dose-escalated SBRT in the treatment of PDAC. Importantly, the MRL allows the implementation of online adaptive treatment planning, whereby the radiation treatment plan can be adapted in real time on the day of treatment based on the patient's luminal gastrointestinal anatomy of the day just prior to radiation delivery. Dose delivery to the intended target is maximized for cancer control while minimizing toxicities to the surrounding luminal gastrointestinal organs. In a single-institution retrospective analysis of inoperable PDAC patients who received induction chemotherapy followed by SBRT to a total dose of 50 Gy in five fractions (biological equivalent dose or BED = 100 Gy) delivered using MRI-guided adaptive radiation therapy. Chuong et al. reported 1-year local control, progression-free survival, and overall survival of 87.8%, 52.4%, and 58.9%, respectively [32]. Updated results by the same authors described median progression-free survival of 20 months, median overall survival of 23 months, and 2-year local control, progression-free survival, and overall survival of 68.8%, 40%, and 45.5%, respectively, in this cohort of PDAC patients [33]. Acute and late grade 3 or higher toxicity rates were both 4.8% [33]. Parikh et al. are the first to report a prospective phase II trial of MRgRT utilizing SBRT for borderline resectable and locally advanced PDAC which demonstrated 1 year LC of 82.9% with 32.4% of patient undergoing surgery following radiotherapy and acute toxicity $\geq$3 GI of 8.8% for the cohort of 136 patients [34]. These data are promising and have led to a phase III randomized controlled trial titled "Locally Advanced Pancreatic Cancer Treated with ABLAtivE Stereotactic MRI-guided Adaptive Radiation Therapy (LAP-ABLATE)" which was launched in 2022 to evaluate overall survival among other endpoints in PDAC patients who receive chemotherapy followed by dose-escalated, MRL-based SBRT compared to those who receive chemotherapy alone [35].

## Contouring, PTV, and OARs

### Contouring, PTV, and OARs for ViewRay MRIdian

Patients treated on the ViewRay MRIdian MRL platform described in Chap. 9 of this book are simulated in the supine position with arms down for both their CT and MRI simulation on the same day. Contours are drawn on the true fast imaging with steady-state free precession (TRUFISP) MR imaging sequence, with reference to the CT simulation sequences with 20-s and 70-s post-contrast administration scans. Prior work has identified sites of local failure along the retroperitoneal margin and nearby vasculature in patients treated to small fields of pancreas SBRT [36]. As a result, the Dana-Farber Brigham Cancer Center typically employs a clinical target volume (CTV) approach informed by the work of Hill et al. who defined a "triangle volume" of at-risk disease near pancreatic head, neck, and uncinate tumors [37]. An example of this triangle volume as contoured on the ViewRay MRIdian can be seen in Fig. 13.2.

The gross tumor volume (GTV) is drawn using simulation TRUFISP and CT sequences in addition to diagnostic triple-phase pancreatic protocol CTs obtained at the time of diagnosis and restaging during initial chemotherapy. Special note is taken of areas of significant vascular contact, as well as areas initially involved by tumor at the time of diagnosis that may have regressed on chemotherapy. Suspicious nodes at risk are also contoured as the GTV. Critical regional vessels are then contoured separately, most often the celiac artery, common hepatic artery, superior mesenteric artery, superior mesenteric vein, and portal vein. The splenic artery and vein are occasionally contoured when there is significant radiographic involvement with the primary tumor.

To create a structure identified as the Drawn_CTV, the GTV and vessels have a margin applied to account for microscopic tumor spread. Most commonly, a 3 mm margin is used for the GTV, common hepatic artery, and portal vein. The celiac artery and superior mesenteric artery and vein are typically expanded with a 7 mm margin. In addition to the expansions above, manual corrections are made to encompass the retroperitoneal margin, especially in the pericaval and portacaval spaces as well as subtraction of the Drawn_CTV where it intersects with the aorta and renal vasculature. OARs are identified including the esophagus, stomach, duodenum, small bowel, and large bowel within 2 cm superior and inferior to all slices of the Drawn_

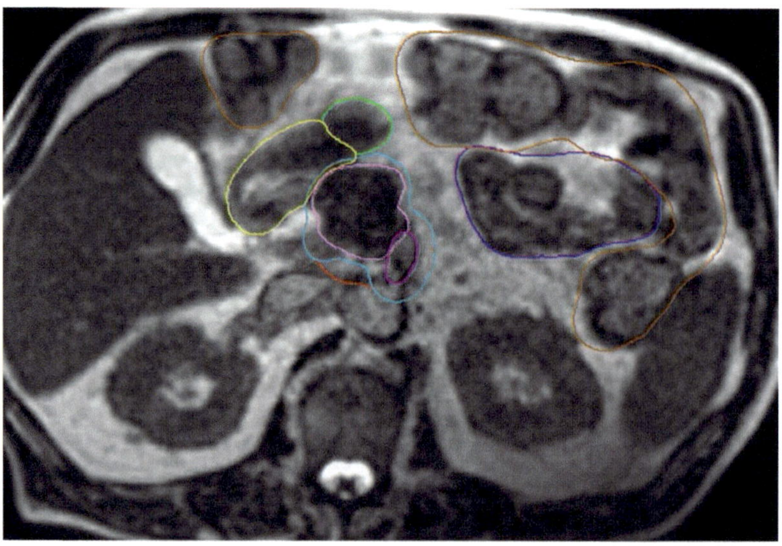

**Fig. 13.2** Various color-coded anatomical structures in the vicinity of the pancreas. "Triangle volume" contour as described by Hill et al. [37] on ViewRay MRIdian. GTV (pink), vasculature (magenta), CTV after expansion (blue), final drawn CTV with manual retroperitoneal margin inclusion (orange), stomach (green), duodenum (yellow), small bowel (purple), and large bowel (brown)

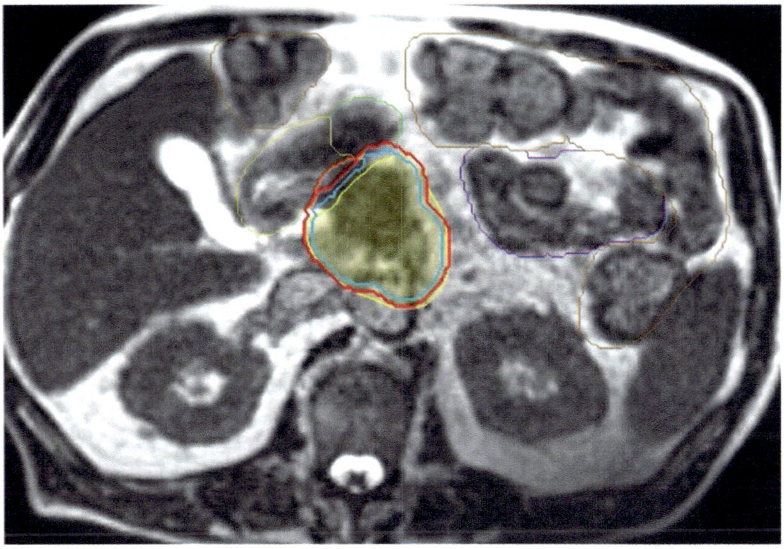

**Fig. 13.3** Locally advanced pancreatic adenocarcinoma plan on ViewRay MRIdian. Drawn_CTV (orange), CTV (blue), PTV (red), stomach (green), duodenum (yellow); small bowel (purple), and large bowel (brown). The 40 Gy isodose line is shown in yellow colorwash

CTV. In addition, the spinal canal, liver, and kidneys are included as OARs for each treatment. No planning risk volume (PRV) expansion is used for OARs.

The Drawn_CTV is permitted to overlap with the adjacent organs-at-risk, including GI OARs. The Drawn_CTV has all GI OARs subtracted to create the final CTV for treatment planning objectives, and a 3 mm margin is added for a final PTV. The use of a Drawn_CTV allows for online adaptive recontouring of the target at risk as well as GI OARs without having to manually subtract structures, thereby saving time. A rule can be implemented in the treatment delivery software to automatically subtract all GI OARs from the Drawn_CTV, thereby recreating the final CTV and PTV through automation rather than through manual implementation. The PTV is treated to a dose of 33 Gy in five fractions for borderline resectable cases planned for surgical exploration and resection or a dose of 40–50 Gy in five fractions for patients with locally advanced disease who are not undergoing a planned surgical exploration or resection. An example of dose coverage achieved on the ViewRay MRIdian can be seen in Fig. 13.3. Other examples of volume delineation are described in the literature as well as ongoing clinical trials as described above.

## Contouring, PTV, and OARs for Elekta Unity

At the Sunnybrook Odette Cancer Centre, PDAC patients who are deemed appropriate for MR-Linac-based SBRT are treated on the Elekta Unity (Elekta AB, Stockholm, Sweden) that has been described in Chap. 10 of this book. In general, most patients have locally advanced/unresectable disease who are not candidates for chemotherapy due to advanced age and/or medical comorbidities, who decline chemotherapy, who tolerate chemotherapy poorly, or who have completed up to 12 cycles of standard-of-care chemotherapy with stable disease or locally progressive disease. Patients with contraindications to MRI, who are unable to lie supine for up to 120 min, or who are >140 kg and/or have a circumference >60 cm are excluded. The radiation treatment workflow and planning and radiation dose-volume constraints described herein represent a high-level overview of the current local institutional practice and should be considered as a guide only and are subject to change as new information and advances may become available in the future.

All patients are positioned supine with both arms down and immobilized in a Vac-Lok cushion. Four-dimensional computed tomography

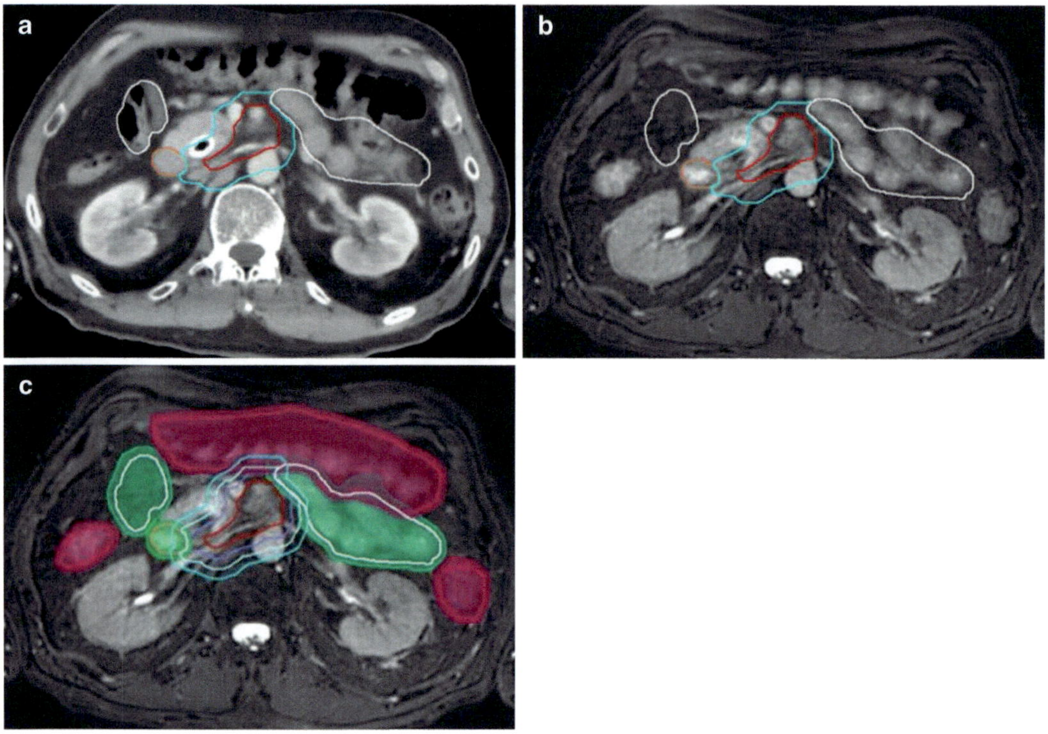


**Fig. 13.4** A representative case of locally advanced/unresectable pancreatic adenocarcinoma being treated on the Elekta Unity. (**a**, **b**) 4DCT simulation with intravenous contrast (left), MR simulation (b3DVANE XD SPAIR sequence) on the Elekta Unity (right). ITV (red), ITV25 (cyan), duodenum (orange), small bowel (white). (**c**) Expansion of target and OAR volumes. ITV (red); PTV = ITV + 5 mm (lavender); ITV25 (aqua); PTV25 = ITV25 + 5 mm (turquoise); duodenum (orange); duodenum PRV = duodenum +3 mm (light green colorwash); small bowel (white); small bowel PRV = small bowel +3 mm (dark green colorwash); large bowel (pink); large bowel PRV = large bowel +3 mm (pink colorwash)

(4DCT) with intravenous contrast and MR simulation is performed on the CT simulator (Philips Healthcare, USA) and the Elekta Unity (Elekta, Stockholm, Sweden), respectively, on the same day. The 4DCT image datasets and the 3DVANE image datasets (b3DVANE XD SPAIR and T1 3DVANE XD) are used for contouring of targets and organs-at-risk. Internal target volume (ITV) is the gross tumor volume (GTV) delineated using the 4DCT and 3DVane image datasets. PTV is a 5 mm expansion of the ITV. PTV is prescribed a dose of 40–50 Gy in five fractions at the discretion of the treating physician. The structure ITV25 is a 1 cm expansion of the ITV, celiac artery, superior mesenteric artery, and portal vein and represents the nodal regions at risk of harboring microscopic disease [38]. PTV25 is a 5 mm expansion of ITV25 and is prescribed a dose of 25 Gy in five fractions. Organs-at-risk (OARs) including the stomach, duodenum, small bowel, large bowel, liver, kidneys, and spinal canal are contoured on the CT datasets. The planning risk volume (PRV) margin for the stomach, duodenum, small bowel, and large bowel is 3 mm. A representative case of locally advanced/unresectable PDAC with target and OAR volumes is illustrated in Fig. 13.4. Reference plans are generated in Monaco 5.4 (Elekta, Stockholm, Sweden) and optimized to maximize PTV coverage by the prescription dose (i.e., 40–50 Gy), with at least 99% of the PTV being covered by

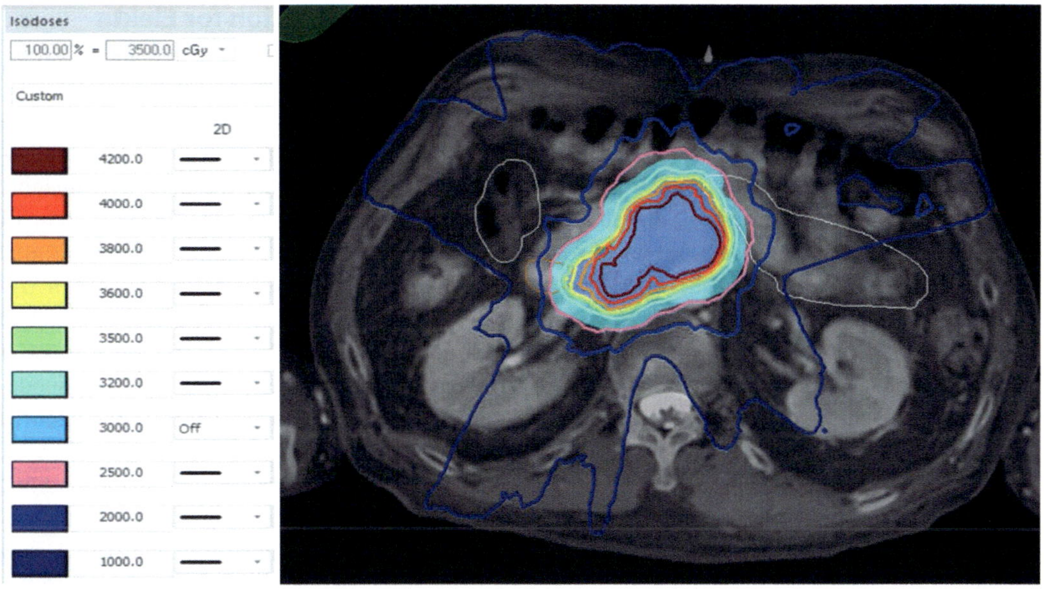


DVH Statistics

Dosimetric Criteria | Statistics | Display

| Structure | Volume (cm$^3$) | Min. Dose (cGy) | Max. Dose (cGy) | Mean Dose (cGy) | Ref. Vol. (cm$^3$) | Ref. Vol. (%) | Ref. Dose (cGy) | Dosimetric Criterion |
|---|---|---|---|---|---|---|---|---|
| OARinPTV | 4.190 | 3408.4 | 3884.8 | 3601.5 | | | | |
| PTV-PRVs | 133.881 | 3406.8 | 4635.9 | 4148.9 | 103.110 | 77.02 | 4000.0 | |
| PTV | 141.561 | 3365.5 | 4635.9 | 4120.6 | 0.030 | 0.02 | 4595.6 | D0.03cm$^3$ < 5500 cGy (+110 cGy) |
| | | | | | 102.860 | 72.66 | 4000.0 | |
| | | | | | 128.171 | 90.54 | 3800.0 | |
| | | | | | 141.509 | 99.96 | 3500.0 | V3500cGy > 99 % (-2 %) |
| PTV25 | 342.276 | 2156.2 | 4635.9 | 3544.2 | 0.030 | 0.01 | 4595.6 | D0.03cm$^3$ < 5500 cGy (+110 cGy) |
| | | | | | 338.878 | 99.01 | 2500.0 | V2500cGy > 99 % (-2 %) |
| External(Unsp.Tiss.) | 13923.829 | 13.8 | 2311.5 | 266.2 | 0.000 | 0.00 | 3625.0 | |
| DUODENUM | 60.429 | 196.5 | 3734.6 | 1690.1 | 0.030 | 0.05 | 3690.3 | D0.03cm$^3$ < 3800 cGy (+76 cGy) |
| | | | | | 1.000 | 1.65 | 3565.7 | D1cm$^3$ < 3600 cGy (+72 cGy) |
| DUODENUM_PRV03 | 110.215 | 183.0 | 3885.1 | 1655.4 | 0.030 | 0.03 | 3853.7 | D0.03cm$^3$ < 4000 cGy (+80 cGy) |
| STOMACH | 158.624 | 267.0 | 3777.5 | 1958.9 | 0.030 | 0.02 | 3697.7 | D0.03cm$^3$ < 3800 cGy (+76 cGy) |
| | | | | | 1.000 | 0.63 | 3594.0 | D1cm$^3$ < 3600 cGy (+72 cGy) |
| STOMACH_PRV03 | 238.885 | 236.6 | 3949.4 | 1897.6 | 0.030 | 0.01 | 3875.0 | D0.03cm$^3$ < 4000 cGy (+80 cGy) |
| Kidney_L | 260.441 | 66.8 | 1805.4 | 628.9 | | | | Dmean < 1100 cGy (+22 cGy) |
| Kidney_R | 258.656 | 67.0 | 2130.8 | 799.0 | | | | Dmean < 1100 cGy (+22 cGy) |
| LARGE BOWEL | 577.380 | 53.4 | 3697.0 | 1137.4 | 0.030 | 0.01 | 3595.0 | D0.03cm$^3$ < 4000 cGy (+80 cGy) |
| | | | | | 1.000 | 0.17 | 3233.1 | D1cm$^3$ < 3600 cGy (+72 cGy) |
| LARGE BOWEL_PRV03 | 843.652 | 51.3 | 4006.8 | 1091.3 | 0.030 | 0.00 | 3949.0 | D0.03cm$^3$ < 4200 cGy (+84 cGy) |
| SMALL BOWEL | 489.360 | 58.4 | 3860.2 | 641.6 | 0.030 | 0.01 | 3775.0 | D0.03cm$^3$ < 3800 cGy (+76 cGy) |
| | | | | | 1.000 | 0.20 | 3345.3 | D1cm$^3$ < 3600 cGy (+72 cGy) |
| SMALL BOWEL_PRV03 | 698.458 | 58.4 | 4032.4 | 654.5 | 0.030 | 0.00 | 3982.0 | D0.03cm$^3$ < 4000 cGy (+80 cGy) |
| SPINAL CANAL | 31.288 | 101.1 | 961.6 | 432.6 | 0.030 | 0.10 | 933.1 | D0.03cm$^3$ < 1500 cGy (+30 cGy) |
| ITV | 65.614 | 3569.0 | 4635.9 | 4252.3 | 60.812 | 92.68 | 4000.0 | |
| LIVER | 1409.497 | 45.6 | 4057.8 | 347.6 | | | | Dmean < 1800 cGy (+36 cGy) |

**Fig. 13.5** An Elekta Unity reference plan is generated and optimized to maximize target coverage while respecting all OAR constraints (top panel). Duodenum (orange), small bowel (white), PTV (lavender colorwash), and PTV25 (turquoise colorwash). Table illustrating dose-volume constraints of the reference plan (bottom panel); these constraints are used in local institutional practice and should be considered as a guide only

35 Gy while respecting all OAR constraints. An arrangement of 12 beams is used for reference plans with 3 mm grid size and 1% statistical uncertainty per plan [39]. Target undercoverage is accepted to respect OAR constraints. The reference plan with a prescription of 40 Gy in five fractions for the representative case is shown in Fig. 13.5, where 72.7% of the PTV is covered by 40 Gy and 100% of the PTV is covered by 35 Gy while meeting all OAR constraints.

## On-Table Adaptation for ViewRay MRIdian

All patients are treated with an online adaptive workflow. A daily TRUFISP sequence is obtained after aligning the patient on the machine. Shifts are identified by aligning the patient to the GTV, taking note of areas of rotation or other deformation of either the GTV or surrounding vasculature. The GTV is rigidly propagated, and most OARs including the GI structures, spinal cord, liver, and kidneys are deformably registered to the daily setup image. On the MRIdian A3i system, the physician, physicist, and therapist role groups can simultaneously interact with the treatment plan, saving time through parallel processing. The physician recontours the Drawn_CTV structure (Fig. 13.2) and all OARs within a 2–3 cm boundary of the PTV. This boundary is chosen due to the high probability that any dose-limiting constraint will be found within 2 cm of the PTV. The physicist ensures fidelity of the dose calculation by ensuring the skin contour is properly drawn and that any new areas of air or soft tissue within the body contour are accounted for. The therapist role group is simultaneously able to modify any tracking structures needed for treatment tracking and gating, as well as being available to the patient for any concerns that may arise while in the MRL bore.

Once all contours are finalized and rules are applied, including automated generation of the CTV and PTV from the Drawn_CTV, the original treatment plan is reoptimized. An assessment is made regarding the dosimetric parameters of the so-called predicted plan, which calculates the dosimetry to adapted target and OAR structures using the original plan MLC positions and beam fluence. In most cases, the predicted plan results either in undercoverage of the PTV or dose to an OAR in excess of a strict constraint. In these cases, the plan is reoptimized by a physicist, often in an iterative fashion in consultation with the physician, until an acceptable plan is obtained.

## On-Table Adaptation for Elekta Unity

All PDAC patients are treated with online adapt-to-shape (ATS) workflow. Clinical ATS workflow is initiated by acquiring the 3DVANE imaging sequences, b3DVANE XD SPAIR, and T1 3DVANE XD. Body contour, vertebrae, spinal canal, liver, and kidneys are deformably or rigidly propagated, visually inspected, and edited. Subsequently, ITV, ITV25, celiac artery, superior mesenteric artery, portal vein, stomach, duodenum, small bowel, and large bowel are contoured online by the treating physician. PTV, PTV25, and PRV margins are recreated from the modified ITV, ITV25, and luminal gastrointestinal structures based on the anatomy of the day. A new ATS plan is generated for each fraction using fluence optimization and bulk electron density assignment derived from the reference plan. Following evaluation and approval of the ATS plan by the treating physician, physics check is performed. Pretreatment verification images are acquired for visual inspection of the plan and target location. Adapt-to-position workflow is to follow only if the target falls outside of PTV on pretreatment verification images. Otherwise, treatment is delivered with CINE motion monitoring. The ATS plan for fraction 5 of the representative case is shown in Fig. 13.6. On the day of fraction 5, improved PTV coverage is achieved with 86.5% of the PTV being covered by 40 Gy (vs. 72.7% on the reference plan in Fig. 13.5) as the small bowel is located farther away from the PTV.

In summary, the ATS workflow on the MRL for pancreas SBRT harnesses the superior soft tissue contrast of MRI and supports adaptive planning in real time based on the patient's anatomy of the day, thereby maximizing target coverage while minimizing toxicities to OARs. The results from the LAP-ABLATE and similar trials are highly anticipated. In addition, it may be interesting to explore whether MRL-based, dose-escalated SBRT rivals surgery in the management of resectable pancreatic cancer.

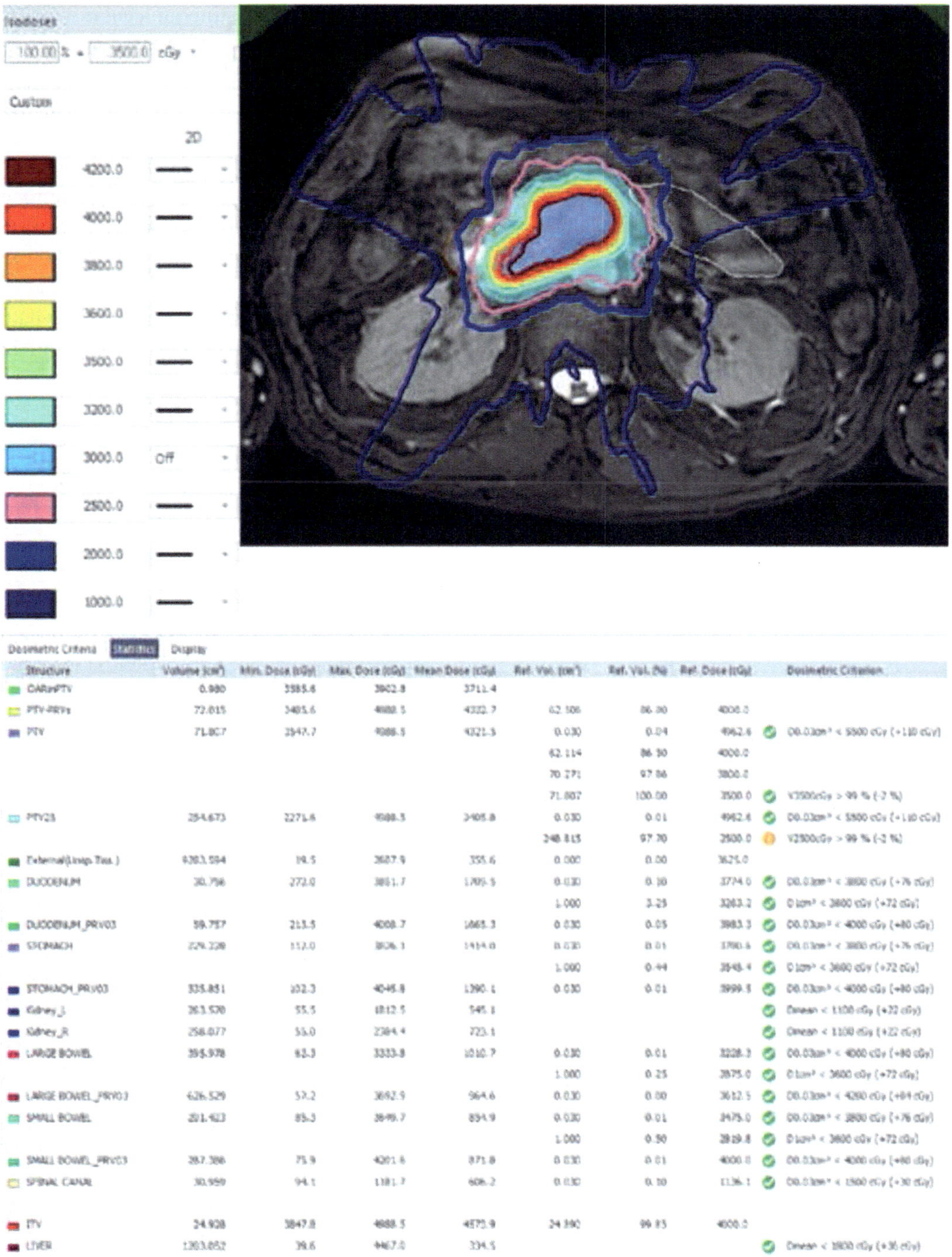


**Fig. 13.6** Adapt-to-shape plan on Elekta Unity for the fifth and last fraction. Duodenum (orange), small bowel (white), PTV (lavender colorwash), and PTV25 (turquoise colorwash). The dose-volume constraints are also shown in the figure

## Plan Evaluation

Plans are evaluated with a goal of maximizing coverage while respecting OAR constraints. When there is a conflict, priority is given to the OAR constraint while sacrificing target coverage. This isotoxic approach ensures that each adaptive fraction of radiation does not exceed tolerances to critical adjacent OARs while attempting to maximize coverage of the target to the maximum safe dose possible each day. This approach is especially prudent in the management of locally advanced, unresectable PDAC as the treatment intent is generally understood to be palliative in nature. Secondary objectives include managing the hotspot of the plan to be reasonably close to the original plan hotspot as well as ensuring the plan can be delivered within a reasonable amount of time and/or monitor units.

## Summary

The limitations for radiation therapy for pancreas cancer have long been recognized as the risk of damage to multiple critical organs intimate with the pancreas. MR-Linac technology allows radiation oncologists to maintain their respect for those adjacent organs-at-risk while significantly increasing the amount of energy administered to the target disease. In the care of patients with adenocarcinoma of the pancreas, there is mounting evidence showing improved local control with the use of dose-escalated radiation therapy given in this way. This has translated in better symptom control and therefore quality of life for patients with locally advanced disease and resectability and negative-margin resections in borderline resectable cases.

Dose-escalated radiotherapy with improved imaging and adaptation capabilities has improved local control and symptoms related to progressive local disease in PDAC but, as yet, has not shown a clear survival benefit. Treatment selection beyond optimized systemic therapy is burdened by a lack of patient and tumor specific data indicating which groups are more likely to have their survival compromised by their primary disease. Incorporating molecular markers such as SMAD4, among others, to stratify patients who would benefit from radiation therapy would be one approach in future clinical trials. Ongoing prospective randomized studies such as LAP-ABLATE will further inform us as to whether dose-escalated MRgRT will improve survival as well as improve quality of life. MRgRT has overcome the technical challenges of tumor location with dose-escalated radiotherapy in PDAC. We eagerly await further data to support the clinical benefit of this tool in PDAC.

Special thanks go to Hyebeen Kim for design of Fig. 13.1.

## References

1. Sung H, Ferlay J, Siegel RL, Laversanne M, Soerjomataram I, Jemal A, et al. Global Cancer Statistics 2020: GLOBOCAN estimates of incidence and mortality worldwide for 36 cancers in 185 countries. CA Cancer J Clin. 2021;71:209–49. https://doi.org/10.3322/caac.21660.
2. Siegel RL, Miller KD, Fuchs HE, Jemal A. Cancer statistics, 2022. CA Cancer J Clin. 2022;72:7–33. https://doi.org/10.3322/caac.21708.
3. Jiang S, Fagman JB, Ma Y, Liu J, Vihav C, Engstrom C, et al. A comprehensive review of pancreatic cancer and its therapeutic challenges. Aging. 2022;14:7635–49. https://doi.org/10.18632/aging.204310.
4. Wolff RA, Crane CH, Li D, Evans DB, Maitra A, Tsai S. Neoplasms of the exocrine pancreas. In: Bast RC, Croce CM, Hait WN, Hong WK, Kufe DW, Piccart-Gebhart M, et al., editors. Holland-Frei cancer medicine. 9th ed. Hoboken: Wiley Blackwell; 2017. p. 1129–50.
5. Iacobuzio-Donahue CA, Fu B, Yachida S, Luo M, Abe H, Henderson CM, et al. DPC4 gene status of the primary carcinoma correlates with patterns of failure in patients with pancreatic cancer. J Clin Oncol. 2009;27:1806–13. https://doi.org/10.1200/JCO.2008.17.7188.
6. Soloff EV, Zaheer A, Meier J, Zins M, Tamm EP. Staging of pancreatic cancer: resectable, borderline resectable, and unresectable disease. Abdom Radiol. 2018;43:301–13. https://doi.org/10.1007/s00261-017-1410-2.
7. Shaib Y, Davila J, Naumann C, El-Serag H. The impact of curative intent surgery on the survival of pancreatic cancer patients: a U.S. population-based study. Off J Am Coll Gastroenterol ACG. 2007;102:1377.
8. Kaiser MH, Ellenberg SS. Pancreatic cancer: adjuvant combined radiation and chemotherapy following cura-

tive resection. Arch Surg. 1985;120:899–903. https://doi.org/10.1001/archsurg.1985.01390320023003.
9. Smeenk HG, van Eijck CHJ, Hop WC, Erdmann J, Tran KCK, Debois M, et al. Long-term survival and metastatic pattern of pancreatic and periampullary cancer after adjuvant chemoradiation or observation: long-term results of EORTC trial 40891. Ann Surg. 2007;246:734–40. https://doi.org/10.1097/SLA.0b013e318156eef3.
10. Neoptolemos JP, Stocken DD, Friess H, Bassi C, Dunn JA, Hickey H, et al. A randomized trial of chemoradiotherapy and chemotherapy after resection of pancreatic cancer. N Engl J Med. 2004;350:1200–10. https://doi.org/10.1056/NEJMoa032295.
11. Regine WF, Winter KA, Abrams R, Safran H, Hoffman JP, Konski A, et al. Fluorouracil based chemoradiation with either gemcitabine or fluorouracil chemotherapy following resection of pancreatic adenocarcinoma: 5-year analysis of the US intergroup/RTOG 9704 phase III trial. Ann Surg Oncol. 2011;18:1319. https://doi.org/10.1245/s10434-011-1630-6.
12. Abrams RA, Winter KA, Regine WF, Safran H, Hoffman JP, Lustig R, et al. Failure to adhere to protocol specified radiation therapy guidelines was associated with decreased survival in RTOG 9704—a phase III trial of adjuvant chemotherapy and chemoradiotherapy for patients with resected adenocarcinoma of the pancreas. Int J Radiat Oncol Biol Phys. 2012;82:809–16. https://doi.org/10.1016/j.ijrobp.2010.11.039.
13. Conroy T, Castan F, Lopez A, Turpin A, Ben Abdelghani M, Wei AC, et al. Five-year outcomes of FOLFIRINOX vs gemcitabine as adjuvant therapy for pancreatic cancer: a randomized clinical trial. JAMA Oncol. 2022;8:1571–8. https://doi.org/10.1001/jamaoncol.2022.3829.
14. Katz MHG, Pisters PWT, Evans DB, Sun CC, Lee JE, Fleming JB, et al. Borderline resectable pancreatic cancer: the importance of this emerging stage of disease. J Am Coll Surg. 2008;206:833–46.; (discussion 846–848). https://doi.org/10.1016/j.jamcollsurg.2007.12.020.
15. Katz MHG, Shi Q, Ahmad SA, Herman JM, de Marsh RW, Collisson E, et al. Preoperative modified FOLFIRINOX treatment followed by capecitabine-based chemoradiation for borderline resectable pancreatic cancer. JAMA Surg. 2016;151:e161137. https://doi.org/10.1001/jamasurg.2016.1137.
16. Ghaneh P, Palmer D, Cicconi S, Jackson R, Halloran CM, Rawcliffe C, et al. Immediate surgery compared with short-course neoadjuvant gemcitabine plus capecitabine, FOLFIRINOX, or chemoradiotherapy in patients with borderline resectable pancreatic cancer (ESPAC5): a four-arm, multicentre, randomised, phase 2 trial. Lancet Gastroenterol Hepatol. 2023;8:157–68. https://doi.org/10.1016/S2468-1253(22)00348-X.
17. Versteijne E, van Dam JL, Suker M, Janssen QP, Groothuis K, Akkermans-Vogelaar JM, et al. Neoadjuvant chemoradiotherapy versus upfront surgery for resectable and borderline resectable pancreatic cancer: long-term results of the Dutch randomized PREOPANC trial. J Clin Oncol. 2022;40:1220–30. https://doi.org/10.1200/JCO.21.02233.
18. Katz MHG, Shi Q, Meyers J, Herman JM, Chuong M, Wolpin BM, et al. Efficacy of preoperative mFOLFIRINOX vs mFOLFIRINOX plus hypofractionated radiotherapy for borderline resectable adenocarcinoma of the pancreas: the A021501 phase 2 randomized clinical trial. JAMA Oncol. 2022;8:1263–70. https://doi.org/10.1001/jamaoncol.2022.2319.
19. Loehrer PJ, Feng Y, Cardenes H, Wagner L, Brell JM, Cella D, et al. Gemcitabine alone versus gemcitabine plus radiotherapy in patients with locally advanced pancreatic cancer: an eastern cooperative oncology group trial. J Clin Oncol. 2011;29:4105–12. https://doi.org/10.1200/JCO.2011.34.8904.
20. Hammel P, Huguet F, van Laethem J-L, Goldstein D, Glimelius B, Artru P, et al. Effect of chemoradiotherapy vs chemotherapy on survival in patients with locally advanced pancreatic cancer controlled after 4 months of gemcitabine with or without Erlotinib: the LAP07 randomized clinical trial. JAMA. 2016;315:1844–53. https://doi.org/10.1001/jama.2016.4324.
21. Koay EJ, Hanania AN, Hall WA, Taniguchi CM, Rebueno N, Myrehaug S, et al. Dose-escalated radiation therapy for pancreatic cancer: a simultaneous integrated boost approach. Pract Radiat Oncol. 2020;10:e495–507. https://doi.org/10.1016/j.prro.2020.01.012.
22. Koong AC, Le QT, Ho A, Fong B, Fisher G, Cho C, et al. Phase I study of stereotactic radiosurgery in patients with locally advanced pancreatic cancer. Int J Radiat Oncol Biol Phys. 2004;58:1017–21. https://doi.org/10.1016/j.ijrobp.2003.11.004.
23. Mahadevan A, Miksad R, Goldstein M, Sullivan R, Bullock A, Buchbinder E, et al. Induction gemcitabine and stereotactic body radiotherapy for locally advanced nonmetastatic pancreas cancer. Int J Radiat Oncol Biol Phys. 2011;81:e615–22. https://doi.org/10.1016/j.ijrobp.2011.04.045.
24. Pollom EL, Alagappan M, von Eyben R, Kunz PL, Fisher GA, Ford JA, et al. Single- versus multifraction stereotactic body radiation therapy for pancreatic adenocarcinoma: outcomes and toxicity. Int J Radiat Oncol Biol Phys. 2014;90:918–25. https://doi.org/10.1016/j.ijrobp.2014.06.066.
25. Rosati LM, Kumar R, Herman JM. Integration of stereotactic body radiation therapy into the multidisciplinary management of pancreatic cancer. Semin Radiat Oncol. 2017;27:256–67. https://doi.org/10.1016/j.semradonc.2017.02.005.
26. Herman JM, Chang DT, Goodman KA, Dholakia AS, Raman SP, Hacker-Prietz A, et al. Phase 2 multi-institutional trial evaluating gemcitabine and stereotactic body radiotherapy for patients with locally advanced unresectable pancreatic adenocarcinoma. Cancer. 2015;121:1128–37. https://doi.org/10.1002/cncr.29161.

27. Moningi S, Dholakia AS, Raman SP, Blackford A, Cameron JL, Le DT, et al. The role of stereotactic body radiation therapy for pancreatic cancer: a single-institution experience. Ann Surg Oncol. 2015;22:2352–8. https://doi.org/10.1245/s10434-014-4274-5.
28. Petrelli F, Comito T, Ghidini A, Torri V, Scorsetti M, Barni S. Stereotactic body radiation therapy for locally advanced pancreatic cancer: a systematic review and pooled analysis of 19 trials. Int J Radiat Oncol Biol Phys. 2017;97:313–22. https://doi.org/10.1016/j.ijrobp.2016.10.030.
29. Palta M, Godfrey D, Goodman KA, Hoffe S, Dawson LA, Dessert D, et al. Radiation therapy for pancreatic cancer: executive summary of an ASTRO clinical practice guideline. Pract Radiat Oncol. 2019;9:322–32. https://doi.org/10.1016/j.prro.2019.06.016.
30. National Comprehensive Cancer Center Network. Pancreatic adenocarcinoma (version 2.2023). 2023. https://www.nccn.org/professionals/physician_gls/pdf/pancreatic.pdf. Accessed 29 June 2023.
31. Caravatta L, Cellini F, Simoni N, Rosa C, Niespolo RM, Lupattelli M, et al. Magnetic resonance imaging (MRI) compared with computed tomography (CT) for interobserver agreement of gross tumor volume delineation in pancreatic cancer: a multi-institutional contouring study on behalf of the AIRO group for gastrointestinal cancers. Acta Oncol Stockh Swed. 2019;58:439–47. https://doi.org/10.1080/0284186X.2018.1546899.
32. Chuong MD, Bryant J, Mittauer KE, Hall M, Kotecha R, Alvarez D, et al. Ablative 5-fraction stereotactic magnetic resonance-guided radiation therapy with on-table adaptive replanning and elective nodal irradiation for inoperable pancreas cancer. Pract Radiat Oncol. 2021;11:134–47. https://doi.org/10.1016/j.prro.2020.09.005.
33. Chuong MD, Herrera R, Kaiser A, Rubens M, Romaguera T, Alvarez D, et al. Induction chemotherapy and ablative stereotactic magnetic resonance image-guided adaptive radiation therapy for inoperable pancreas cancer. Front Oncol. 2022;12:888462. https://doi.org/10.3389/fonc.2022.888462.
34. Parikh PJ, Lee P, Low DA, Kim J, Mittauer KE, Bassetti MF, et al. A multi-institutional phase 2 trial of ablative 5-fraction stereotactic magnetic resonance-guided on-table adaptive radiation therapy for borderline resectable and locally advanced pancreatic cancer. Int J Radiat Oncol Biol Phys. 2023;117(4):799–808. https://doi.org/10.1016/j.ijrobp.2023.05.023.
35. Locally advanced pancreatic cancer treated with ABLAtivE stereotactic MRI-guided adaptive radiation therapy-Full Text. View-ClinicalTrials.gov. https://clinicaltrials.gov/ct2/show/NCT05585554. Accessed 29 June 2023.
36. Zhu X, Ju X, Cao Y, Shen Y, Cao F, Qing S, et al. Patterns of local failure after stereotactic body radiation therapy and sequential chemotherapy as initial treatment for pancreatic cancer: implications of target volume design. Int J Radiat Oncol Biol Phys. 2019;104:101–10. https://doi.org/10.1016/j.ijrobp.2019.01.075.
37. Hill CS, Fu W, Hu C, Sehgal S, Reddy AV, He J, et al. Location, location, location: what should be targeted beyond gross disease for localized pancreatic ductal adenocarcinoma? Proposal of a standardized clinical tumor volume for pancreatic ductal adenocarcinoma of the head: the “triangle volume”. Pract Radiat Oncol. 2022;12:215–25. https://doi.org/10.1016/j.prro.2022.01.005.
38. Kharofa J, Mierzwa M, Olowokure O, Sussman J, Latif T, Gupta A, et al. Pattern of marginal local failure in a phase II trial of neoadjuvant chemotherapy and stereotactic body radiation therapy for resectable and borderline resectable pancreas cancer. Am J Clin Oncol. 2019;42:247–52. https://doi.org/10.1097/COC.0000000000000518.
39. Paulson ES, Ahunbay E, Chen X, Mickevicius NJ, Chen G-P, Schultz C, et al. 4D-MRI driven MR-guided online adaptive radiotherapy for abdominal stereotactic body radiation therapy on a high field MR-Linac: implementation and initial clinical experience. Clin Transl Radiat Oncol. 2020;23:72–9. https://doi.org/10.1016/j.ctro.2020.05.002.

# The Role of MR-Guided Radiation Therapy in the Management of Liver Malignancies

# 14

Nicki Karimi-Mostowfi, Luca Boldrini, Cihan Gani, and Ann Raldow

## Introduction and Disease Overview

Primary liver cancer—most commonly hepatocellular carcinoma (HCC) and intrahepatic cholangiocarcinoma (ICC)—was the sixth most diagnosed cancer and third leading cause of cancer mortality worldwide in 2020 [1]. For decades, liver cancer had the fastest increasing mortality in the United States, but the recent 5 years has shown a plateau in these rates [2]. In the United States in 2022, it was the fifth leading cause of mortality in men and seventh in women, but was not in the top 10 of most commonly occurring cancers [2].

HCC usually arises in the setting of liver cirrhosis and is often associated with alcohol abuse and hepatitis B and C infections [3]. In the West, nonalcoholic steatohepatitis (NASH) associated with metabolic syndrome and diabetes mellitus is also a risk factor [3]. ICC is a malignancy that arises within the cells of the peripheral bile duct and has similar risk factors to HCC [4]. Additional risk factors include primary sclerosing cholangitis (PSC)—characterized by biliary inflammation and fibrosis—congenital malformation of the bile duct, and hepatic parasitic infections, particularly *Opisthorchis viverrini* and *Clonorchis sinensis* in East Asia [4].

Both HCC and ICC are more classically associated with older males [3, 4]. In 2016 in the United States, American Indians/Alaskan Natives had the highest incidence of HCC, followed by Hispanics, Asians/Pacific Islanders, non-Hispanic Blacks, and then non-Hispanic Whites [5]. Worldwide, in 2018, the age-standardized incidence and mortality of HCC was highest in Eastern Asia, followed by Southeast Asia and Africa [1]. The lowest age-standardized incidence and mortality rates are in South Central Asia, followed by Europe and Western Asia [1].

Asians had a higher risk of an ICC diagnosis than Whites, who had a higher risk than Blacks [6]. However, despite this trend, Black patients showed an increased risk of death with less sur-

N. Karimi-Mostowfi
Lewis Katz School of Medicine at Temple University, Philadelphia, PA, USA
e-mail: nicki.karimi-mostowfi@temple.edu

L. Boldrini
Department of Bioimaging, Radiation Oncology and Hematology, Fondazione Policlinico Universitario "Agostino Gemelli" IRCCS, Rome, Italy
e-mail: luca.boldrini@policlinicogemelli.it

C. Gani
Department of Radiation Oncology, University Hospital Tübingen, Tübungen, Germany

German Cancer Research Center (DKFZ) Heidelberg and German Consortium for Translational Cancer Research (DKTK), Partner Site Tübingen, Tübingen, Germany
e-mail: Cihan.Gani@med.uni-tuebingen.de

A. Raldow (✉)
Department of Radiation Oncology, University of California, Los Angeles, Los Angeles, CA, USA
e-mail: ARaldow@mednet.ucla.edu



I. J. Das et al. (eds.), *A Practical Guide to MR-Linac*, https://doi.org/10.1007/978-3-031-48165-9_14

vival time [6]. Worldwide, the incidence of ICC is also increasing—in 2017, higher rates were seen in Eastern Asia when compared to Western countries [4].

The clinical presentation of HCC and ICC can vary. Patients may be asymptomatic, present with nonspecific symptoms—such as abdominal pain, lethargy, nausea/vomiting, and weight loss—or with specific symptoms of jaundice, pruritis, clay-colored stool, and dark urine. Carcinoembryonic antigen (CEA) and carbohydrate antigen (CA) 19–9 and 125 are commonly checked biomarkers in cholangiocarcinoma, and elevations in both correlate with disease [7]. Alpha-fetoprotein (AFP) is useful for HCC diagnosis and screening, although it lacks accuracy, with 32–59% of HCC patients having normal levels [8]. As a result, the American Association for the Study of Liver Diseases (AASLD) has a series of recommendations for surveilling HCC, depending on the risk factors of the individual. Patients with cirrhosis are recommended to undergo ultrasound (US) surveillance, with or without AFP, every 6 months as it improve survival [9]. The diagnosis of HCC can be made via CT/MRI using Liver Reporting and Data System (LIRADS) criteria [10]. Major criteria include (non-rim) arterial phase hyperenhancement, non-peripheral "washout," enhancing capsule/pseudocapsule, and threshold growth. A LIRADS 5 lesion is diagnostic of HCC and does not require a biopsy. ICC is typically diagnosed in later stages when the tumor is unresectable; CT, MRI, magnetic resonance cholangiopancreatography (MRCP), and positron emission tomography (PET) can help delineate the extent of disease [11].

Secondary malignancies are also common in the liver. Cancers that frequently metastasize to the liver include colorectal, pancreatic, lung, breast, and melanoma. In fact, up to 50% of patients with cancer can present with liver metastases at some point during their disease course [12]. More common than primary liver tumors, liver metastases are seen in approximately 5% of patients with cancer at the time of primary diagnosis, although the percentage varies with cancer type and disease progression [12]. Like cancers that originate in the liver, disease manifestation and symptoms can vary; the primary means of diagnosing liver metastases is via imaging including US, CT, MRI, and PET [12]. For patients with limited disease and resectable liver metastases, the preferred treatment is surgical resection [12]. However, radiation therapy with ablative intent is increasingly used with minimal toxicity and high local control [12].

## Overcoming Challenges with Traditional Approaches

Liver cancer is best managed with a multidisciplinary approach. Surgery, liver transplant, and thermal ablation are considered curative treatments for patients with HCC without macrovascular invasion [13]. The most effective therapeutic option for HCC is liver transplantation—however, this is an option available for less than 20% of HCC patients and 15% of patients with ICC [11, 14]. Surgery may not be an option due to advanced local extent of disease, insufficient liver function, or medical inoperability. Liver metastases are only amenable to surgical resection in 10–20% of cases [15]. As a result, external beam radiation therapy (EBRT) has emerged for the treatment of liver metastases.

Historically, liver malignancies were sometimes treated palliatively with low-dose whole liver radiation [16]. HCC, being one of the most radioresistant cancers, is a poor choice for low-dose radiation [13]. An additional challenge with liver-directed EBRT is the risk of radiation-induced liver disease (RILD), a phenomenon associated with a high mortality rate in patients with liver cancer [14]. There are two types of RILD—classic and non-classic. Classic RILD typically occurs between 2 weeks and 3 months of therapy and is characterized by hepatomegaly, anicteric ascites, and more than a twofold increase in alkaline phosphatase (ALP) [14, 17]. Classic RILD usually occurs in patients with fairly normal pretreatment liver function. Non-classic RILD typically occurs between 1 week and 3 months after therapy, develops more often in patients with underlying chronic hepatic disease, and is characterized by elevated liver transaminases or a decline in liver function in the absence of classic RILD [14, 17]. RILD is pathologically similar to sinusoidal obstruction syndrome (SOS), which leads to hepatocellular necrosis, terminal hepatic venule fibrotic scarring, and architectural distortion [18]. The severity of RILD also depends on factors such as total RT dose and fractionation, the volume of liver radiated, and the baseline liver function [19]. In both classic and non-classic RILD, treatment is limited, and both can progress to liver failure and death [17]. To limit RILD, there have been dose constraints controlling the mean radiation dose to the liver [17]. The Child–Pugh (CP) scoring system, a system used to assess hepatic function, has been a helpful parameter in establishing radiation dose constraints. In a 2014 study, an increase in CP score by at least 2 points within 3 months of RT completion was determined to be RILD and shown to affect overall survival, with less survival time and a higher relative risk of mortality [20]. It was determined that in order to minimize such incidence, patients need to be carefully selected and given suitable treatment plans. In liver stereotactic body radiation therapy (SBRT), a classification score of 5 or 6 (class A) is desirable, as it indicates well-compensated disease [21].

EBRT has been deemed to be relatively safe and conditionally recommended with additional treatments in both HCC and ICC for a variety of differing severities, particularly those individuals who are not candidates for curative therapy [13].

SBRT delivers focal and intense doses of RT (fraction size >500 cGy) in 1–5 fractions [13]; however, definitions may vary internationally. Compared to low-dose whole liver radiation, there is less destruction of surrounding normal tissue resulting in better outcomes and less toxicity for the patient [13]. It can be used for both primary and metastatic malignancies and, at high biological doses used with ablative intent (biologically effective doses BED $\geq$100 $Gy_{10}$), is associated with improved local control in the treatment of metastasis [22].

There are several challenges associated with SBRT, and it can be especially difficult in patients with large, invasive tumors, patients with tumors too small to be properly targeted, and/or patients with tumors that are in close proximity to mucosal structures such as bowel [23]. The treatment requires accurate image guidance to properly evaluate the position of the tumor, which can change with breathing motion and the movement of adjacent organs, in order to ensure accurate radiation delivery.

Most facilities deliver SBRT using 4D imaging and treatment under cone-beam CT image guidance, which has limited soft tissue contrast and is susceptible to motion degradation and relies on surrogate tracking with fiducial markers, as detailed below [23]. As a result of respiratory motion, gastric filling, peristalsis, bowel movements, and intrinsic deformation, abdominal organs and tumors are constantly moving and changing position, which can lead to missing of the target and/or higher doses to organs at risk (OARs) [23]. Poor inter- and intra-fraction image guidance can compromise target and OAR visualization, potentially rendering the treatment less effective and more prone to normal tissue damage. Traditionally, an approach used to partially ameliorate this problem was the placement of intra-lesion or peri-lesion fiducial markers to allow for accurate image guidance before treatment and for tracking the target during treatment [22, 24]. Fiducials are small metal markers made of gold or platinum that are inserted in the tumor [25]. However, the problem with fiducials is that they increase the risk of complications such as bleeding, increase the cost of the treatment, and may migrate which renders treatment ineffective [22].

## MRgRT Advantages

There are particular advantages to the magnetic resonance image-guided radiotherapy (MRgRT) system over traditional approaches. MRgRT has improved upon traditional CT-guided SBRT therapy and has been used in the abdomen with success [16]. Broadly speaking, MRI does not subject patients to an increased radiation exposure that would be seen with CT. Given the superior soft tissue contrast of MR images, MRgRT enables providers to directly visualize and localize the tumor and surrounding organs and track the motion of these structures with the patient's breathing, without the need for fiducial markers [16]. In the liver specifically, organ-specific contrast agents such as gadoxetic acid allow for improved differentiation of the liver, liver lesions, and the biliary tree [26]. Therefore, liver tumors are more easily differentiated on MRI, as are some liver OARs, such as the common bile duct and distal biliary tree. It also better allows for adaptive treatment planning and real-time imaging [26]. Additionally, MRgRT alleviates the need for multiphase contrast CT simulation scans in institutions that would otherwise opt to use them [27].

From published data, MRgRT appears to be a safe alternative to traditional therapies. As delineated further in the Evaluation and Outcomes sections below, MRgRT has shown promising clinical outcomes—patients are able to receive higher doses of radiation with a lower incidence of severe toxicity to the liver and surrounding organs of interest. This has been associated with high rates of local control and increased rates of overall survival. However, further prospective clinical trials are necessary to validate all findings.

## MRgRT Contouring, Planning Target Volumes, and Organs at Risk

The two main MRgRT platforms commercially available are the ViewRaw MRIdian system, which uses a 0.35 Tesla MRI scanner, and the Elekta Unity system, which uses a 1.5 Tesla MRI scanner [28]. The main differences fall in construction, imaging, gating, and treatment [28]. The MRIdian is constructed using a split magnet design, uses Trufi Sequence for imaging, has real-time tracking and automatic gating, and treats with a gantry rotation maximum speed of 0.5 rpm [28]. The Elekta Unity system conversely is constructed using a single magnet design, uses range of imaging sequence for imaging and has real-time motion monitoring and introduced gated treatments in 2023. The Elekta Unity treats with a gantry rotation maximum speed of 6.0 rpm [28]. A sample treatment plan using these systems is shown below (Fig. 14.1).

Contouring is the process of outlining the target tumor and OARs slice by slice on a simulation scan and thus is a critical step in the RT treatment plan [29]. A systematic review of contouring guidelines published after 1995 demonstrated that characteristics of consensus contouring guidelines have not been described and that, of the 142 guidelines included in the study, there could be major deviations in tumor target delineation, which can result in increased toxicity and decreased patient survival [29]. An added component is that, as of now, the majority of current guidelines give recommendations for CT only. To ascertain optimal contouring guidelines for MRgRT machines, further studies must be done and a consensus among already available guidelines must be reached. In an effort to standardize OAR contouring, an international collaborative group published a contouring atlas for MR-based segmentation of OAR [26].

OARs with liver-directed therapy include the normal liver, duodenum, small bowel, large bowel, stomach, esophagus, kidneys, spinal cord, biliary tree, large vessels, and heart [30]. Using MRI compared to CT adds superior soft tissue contrast and organ/tumor motion information that can help with positional information as well as the target and delivery of the radiation [30]. As a result, planning target volumes (PTVs), margins encircling the clinical target, can be reliably decreased due to increased confidence in positioning [31]. At UCLA, 3-mm gross target volume (GTV) to PTV margins are routinely employed. Additionally, gadoxetate can be used during MRI-guided therapy as a contrast agent to improve localization, tracking, and treatment of the tumor [26]. MRI delineated target volumes have been shown to be different—often smaller—from those contoured with CT [31].

When available, respiratory gating by MRI is a technique used during radiation that can allow for the adjustment of the radiation beam depending on breathing and the movement of body structures during a respiratory cycle [32]. It could be advantageous for high-dose SBRT because it

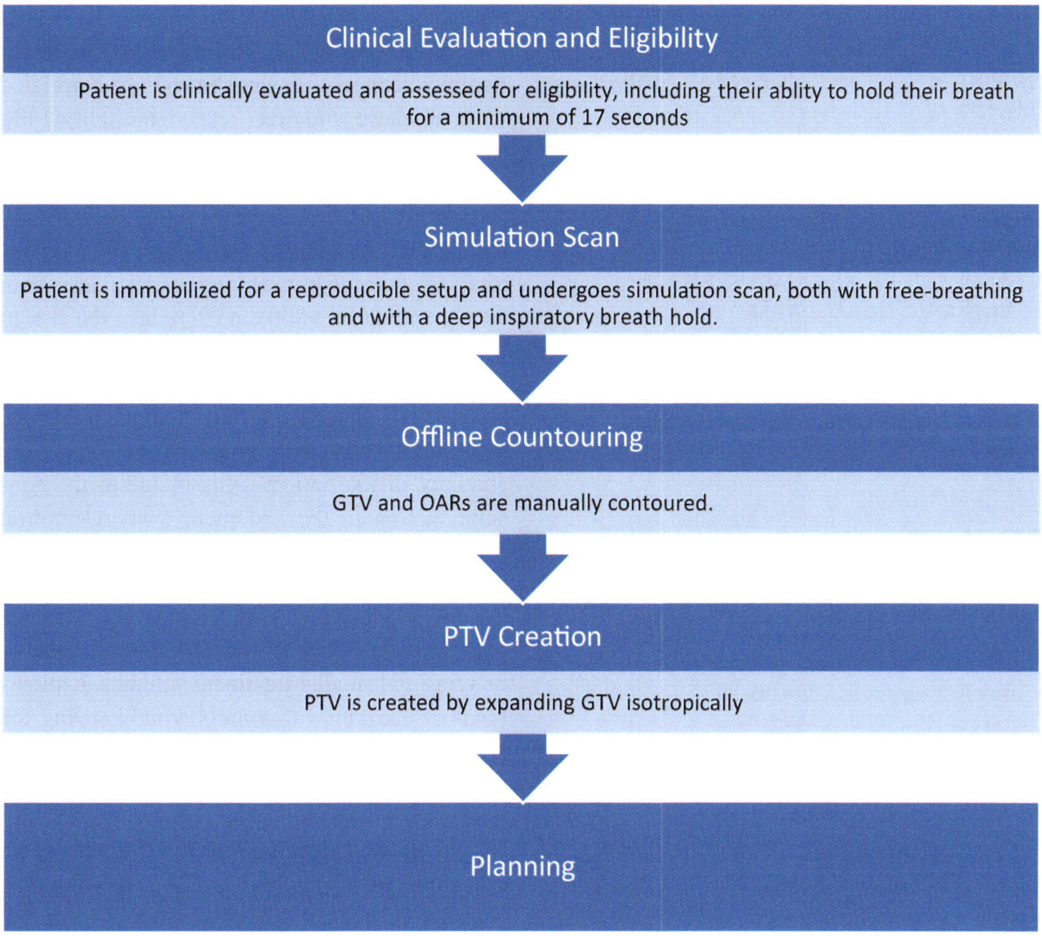


**Fig. 14.1** Sample workflow for the initial setup and creation of a baseline treatment plan

can allow reduction of margins (no internal target volume necessary) and OAR exposure, thereby allowing for a more accurate dose delivery [33].

During treatment with a MR-Linac system, the GTV and PTV are defined, and a gating boundary is defined. When the target structure moves outside the boundary, the beam is automatically stopped. More work is needed to evaluate PTV margins in specific cases, particularly when asymmetric boundaries should be implemented [34]. Visual examples of MRgRT are located in the "Sample Patient Plans" below.

For the 1.5 T MR-Linac systems, ITV or mid-position based systems to ensure sufficient target volume coverage have been used in the past before the implementation of automated gating [35, 36].

## DVH Constraints

Dose-volume histograms (DVHs) summarize the radiation distribution and are important in checking adequate dose levels within a volume of interest in a patient [37]. DVH constraints using MRgRT are more accurate due to the nature of the system allowing for adaptive planning based on the daily anatomy of the patient [38]. Study protocols define parameters that must be adhered to prevent any unwanted effects of radiation [39]. OAR constraints may vary from center to center or depending on the recommendations used. Please see Table 14.1 for an example for 5-fraction liver plan constraints [22].

At UCLA, the ideal normal liver constraint is >1000 cm$^3$ of the normal liver receiving <15 Gy;

**Table 14.1** Structure and OAR dose-volume constraints [22]

| Structure | Goal for three-fraction SBRT |
|---|---|
| Liver | |
| Single-lesion plan | >1000 $cm^3$ of normal liver receive <15 Gy overall |
| Multiple-lesion plan | >800 $cm^3$of normal liver receiving <15 Gy overall |
| OARs | |
| Spinal cord | $D_{max}$ <3 Gy per fraction |
| Esophagus | $D_{.5\ cc}$ <32 Gy |
| Stomach | $V_{20\ Gy}$ <20 $cm^3$ |
| Bowel bag | $V_{20\ Gy}$ <20 $cm^3$ |
| Bowel loop | $V_{35\ Gy}$ <0.35 $cm^3$ |
| Kidney | $D_{max}$ <3 Gy per fraction |
| Skin | $D_{.05\ cc}$ <35 Gy |

**Table 14.2** UCLA SMART constraints

| Structure | Goal |
|---|---|
| Liver | Mean dose <20 Gy<br>Keep 700 cc under 15 Gy |
| Duodenum | V33 ≤0.5 $cm^3$ |
| Stomach | V33 ≤0.5 $cm^3$ |
| Small bowel | V33 ≤0.5 $cm^3$ |
| Large bowel | V33 ≤0.5 $cm^3$ |
| Spinal canal | V33 ≤0.5 $cm^3$ |
| Kidney (each) | Mean <12 Gy<br>2/3 of each kidney <8 Gy |

however, plans where 700 $cm^3$ of the normal liver receive <15 Gy are also accepted. For gastrointestinal mucosal structures (i.e., small bowel, large bowel, esophagus, and stomach), the high-dose constraint is V35 Gy <0.5 $cm^3$. Table 14.2 has the current UCLA liver constraints.

## Dose Escalation

While SBRT is generally well tolerated, there is a risk of toxicity to OARs, which has a historically limited dose escalation, especially in the case of liver tumors in the proximity of luminal organs. There have been many reports of damage to the GI organs in radiation treatments, and while uncommon, potential toxicity is very important to consider in treatment [16].

Higher volumes of radiation have linked with increased toxicity, especially in patients with a more compromised liver. It is important to note that an increase of CP score within 6 months of treatment completion was shown to be correlated with all-cause and liver-related mortality [40]. Son et al. found that when the total liver volume receiving <18 Gy was less than 800 $cm^3$, the probability of CP class progression was abruptly increased. As a result, they suggested that the total liver volume receiving a dose less than 18 Gy should be >800 $cm^3$ to reduce the risk to hepatic function [41].

Other limits should be established. Dose escalation in the abdomen can only be done if the provider can constantly localize and track the anatomy of the patient—this is due to the constant motion in the abdomen, from respiration and peristalsis—as well as due to the large treatment volumes present [42]. MRgRT provides certain advantages over traditional therapy regarding dose escalation. Due to its superior imaging and smaller treatment volumes, it allows for dose escalation to tumors while keeping the dose to normal tissue and surrounding OARs at a minimum [43]. Doses are more accurately tailored and restricted to appropriate locations [16]. It also demonstrates an increased accuracy in ascertaining the anatomical shifts of the abdomen [16]. This can especially be a benefit if the target is close to a sensitive OAR [44].

In patients with pancreatic cancer, MRgRT was shown to be safe for dose escalation, unlike SBRT, which led to dose escalation but with an increase in toxicity [43]. Patients with higher doses of radiation also had a significant survival advantage over lower-dose groups, showing promise for the results of other cancer treatments [43]. Patients with liver cancer could also potentially have an improved local control rate [44]. Hoegen et al. [44] are running a prospective, randomized, three-arm phase II clinical trial—dubbed the MAESTRO trial—which aims to provide evidence for the potential clinical benefit of MRgRT for dose escalation, compared to traditional SBRT, given the costs and lack of long-term data on MRgRT. In the first step of the trial, they are planning on demonstrating the safety and non-inferiority of MRgRT compared

**Table 14.3** Dose constraint values for various thorax and abdominal structures

| | Dose constraints (Gy) | | |
|---|---|---|---|
| Structure | 3 fractions | 6 fractions | 10 fractions |
| Uninvolved liver (=liver CTV) ≥700 cc | <19.2 | <26 | <32 |
| Stomach $D_{.5cc}$ | <22.2 | <37 | <43.5 |
| Duodenum $D_{.5cc}$ | <22.2 | <37 | <43.5 |
| Bowel $D_{.5cc}$ | <25.2 | <37 | <43.5 |
| Sigmoid/rectum $D_{.5cc}$ | <28.2 | <37 | <44 |
| Esophagus $D_{.5cc}$ | <25.2 | <36 | <43.5 |
| Heart $D_{.5cc}$ | <25.0 | <31.5 | <66.0 |
| Kidney mean dose | <8.5 | <10.8 | <12.0 |
| Spinal cord $D_{.1cc}$ | <21.6 | <29.0 | <35 |

to standard SBRT, and in the second step, they are exploring the possibility of dose escalation in critically located hepatic metastasis, where standard SBRT would not be as achievable [44].

Dose constraints for the OARs depend on the number of fractions being delivered. For an uninvolved liver, for example, the dose constraint is <19.2 Gy given in 3 fractions or <32 Gy given in 10 fractions. The values are shown in Table 14.3, adapted from the table published in Hoegen [44].

CTV will be GTV + 5 mm and PTV will be CTV + 3 mm. The dose escalation portion of the study will be in study arm C and has not been delineated yet. Furthermore, at the time of this chapter, this study has yet to be completed and, as a result, has not drawn any conclusions [44].

## On-Table Adaptation

In on-table adaptive radiotherapy (On-ART), an initial radiation plan for a patient is created offline and is then adapted for each fraction depending on the relationship between the target and OARs [45]. A sample flow chart for the process is shown below (Fig. 14.2). In the liver, On-ART with MRgRT improves PTV coverage in 63% of the applied fractions, a value consistent with previous studies [38]. In Mayinger's study, 15 patients assigned to SBRT for oligometastatic liver metastasis underwent daily MR-guided target localization and on-table treatment plan reoptimization, with GTVs and OARs adapted every day [38]. PTV coverage was modified to fit the new constraints and plans were then compared. The reoptimized plan showed improved PTV V100% and V95% coverage in 47 and 45 of the applied fractions, respectively. PTV coverage increased by 4% in patients who had liver metastases close to an OAR (<0.2 cm away), indicating a benefit to such patients. For metastasis distant from an OAR, PTV V100% coverage was not significantly improved, indicating that for such patients, daily adaptive planning is not necessary. This is also supported by a report by Uder et al. who showed a local control of irradiated liver metastasis after 12 months of 89.6% using the "adapt to position" workflow on the 1.5 T MR-Linac [46]. Using the "adapt to position" workflow on the 1.5 T MR-Linac, the target volume and OARs are not resegmented and a virtual shift is performed. In contrast to this, in the "adapt to shape" workflow, a full reoptimization with new segmented structures does take place [47].

The process of on-table adaptation is as follows [22, 38, 47, 48]:

1. Creation of a robust "baseline" plan using the dedicated planning system.
2. Patient arriving for treatment is positioned in the treatment position and MR images are acquired during a breath-hold scan.
3. Newly acquired MR images are "matched" to the original baseline images and any necessary shifts are performed. A decision regarding the necessity for generating a new online adapted plan when inter-fractional anatomical changes are clinically significant is made in real time.
4. If the decision is made to adapt, the target(s) and OARs are recontoured or contours are adjusted based on the pretreatment plan to account for the patient's anatomy of the day.
5. The plan is delivered using respiratory gating with breath hold if necessary.

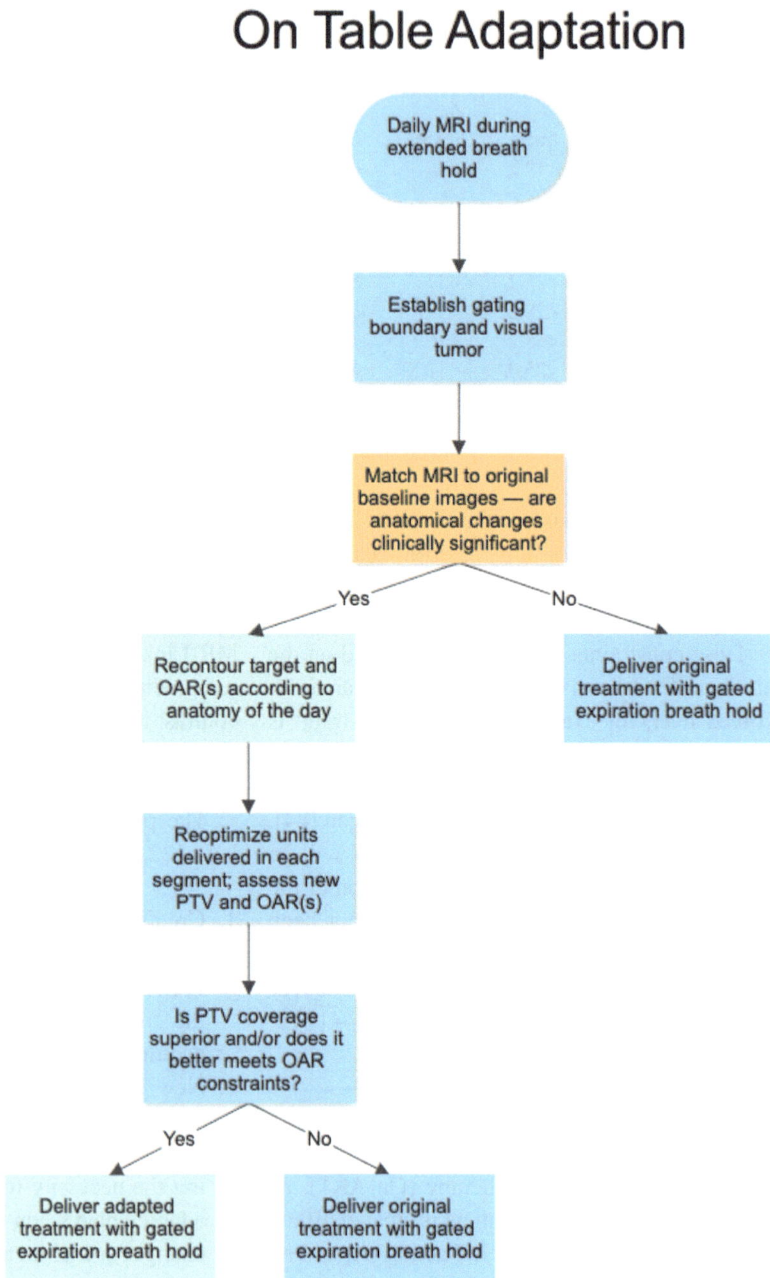


**Fig. 14.2** An example flowchart for on-table adaptation

## Evaluation

MRgRT has shown itself to be a promising addition to RT that can be used for a variety of malignancies, including cancers of the breast, lung, prostate, and liver [49]. There are many advantages of MRgRT over traditional therapies, particularly in the liver. The biggest one is that it allows for superior imaging of liver structures and is well suited to soft tissues and organs, unlike CT [33, 50]. Providers can directly visualize the tumor and anatomy of the patient, allowing for better tracking of the tumor during radiation [33]. Abdominal organs are constantly fluctuating due to breaths, gastric filling, and bowel movements—MRgRT can track these movements and allow better plan adaptation during treatment, although it is still unclear whether that improved dosimetry translates into superior outcomes [38, 49]. There is no doubting, however, that plan adaptation does present the opportunity for the patient to have a more personalized treatment [51].

The increased accuracy, along with gating that automatically stops the radiation beam when a target structure moves outside the boundary, also allows for smaller PTVs and less OAR exposure, thus sparing normal healthy tissues [31, 33, 52]. In a study analyzing 20 consecutive liver SABR patients, it was found that implementing techniques to compensate the movement of tumors and OARs resulted in a reduction in PTV size and ability to escalate dose. The observed increase in BED was up to 21.4 $Gy_{10}$ for a 5-fraction schedule and 37.5 $Gy_{10}$ for a 3-fraction prescription schedule, corresponding to TCP gains of 5.7% and 7.8%, respectively [53].

The disadvantages of MRI treatment start with contraindications to getting an MRI. Non-MR-compatible cardiac implantable electronic devices, metallic intraocular foreign bodies, implantable neurostimulation systems, cochlear implants/ear implants, drug infusion pumps, catheters with metallic components, metallic fragments, cerebral artery aneurysm clips, magnetic dental implants, tissue expanders, artificial limbs, and hearing aids are absolute contraindications [54]. Other limitations include geometric distortion, electron density disruption, and susceptibility artifacts [49]. Geometric distortions can cause inaccurate target volumes and misplaced radiation doses, which would limit the effectiveness of the treatment and even potentially harm the patient [51].

The added cost of MRI treatment has to be considered as well. One study found that the direct costs associated with prostate SBRT were almost $1500 higher with MRgRT than with CTgRT [32]. MRI is also an intrinsically slow imaging modality, taking longer than radiography to acquire images [55]. That, in addition to respiratory gating and on-table adaptation, results in patients spending approximately 10 min longer in the treatment vault per fraction, which can result in other concerns for patient scheduling and waiting list management [38, 51].

MRgRT is also more resource-intensive, requiring increased clinician education to interpret images and more teams trained to operate the systems and be mindful of the safety concerns—MRgRT systems are currently operated by physicians, radiation therapists, dosimetrists, and physicists in multidisciplinary teams operating simultaneously [56].

## Outcome

Table 14.4 details current evidence evaluating clinical outcomes for MRgRT in primary hepatobiliary tumors and/or liver metastases. One of the first studies delineating the MRgRT system was in 2008 at the University Medical Center Utrecht, which aimed to use this prototype system to deliver radiation doses with millimeter precision, based on diagnostic quality 1.5 T MRI images [55]. MRgRT has been clinically implemented from 2014, and since then, it has been used for numerous cancers, from head and neck cancers to abdominal cancers to ovarian cancers [57]. Recently, there has been more clinical studies done for abdominal tumors, and it has been shown that such tumors are responsive to this treatment.

A large retrospective study was done by Henke and colleagues at Washington University School of Medicine in 2018. There were 642 patients recruited between January 2014 and March 2018, 275 of which were in the abdomen with 87 being in the hepatobiliary system [57]. One of the studies that was analyzed was a prospective phase I trial, also done by Henke and colleagues in 2017. Twenty patients with unresectable primary liver and non-liver abdominal malignancies were recruited and received a median radiation dose of 50 Gy in five treatments. Within 1 year, overall survival (OS) was 75%, and 15-month local control (LC) of the cancer was 90%. Through this study, it was concluded that SMART therapy was clinically

**Table 14.4** The prospective and retrospective evidence evaluating clinical outcomes for MRgRT in primary hepatobiliary tumors and/or liver metastases

| Study references | Histology | Year | Design | Number of patients | Median prescribed dose | Median fraction number | Overall survival (OS) | Local control (LC) | Late toxicity (non-hematologic) |
|---|---|---|---|---|---|---|---|---|---|
| [22] | Primary (HCC or ICC) and secondary liver (1–3 metastatic lesions) tumors | 2022 | Prospective phase I | 20 | 54 (range: 11.5–60) | 3 (range: 1–5) | 2 years; 50.7% | 1 year; 94.7%<br>2 years; 79.6% | 0 grade ≥3, 1 late grade 3/4 toxicity |
| [23] | Abdominal and pelvic tumors (primary, locally recurrent, oligometastatic) | 2021 | Retrospective | 106 | 40 Gy | 5 | 2 years; 57% | 2 years; 74%<br>96% for lesions that were treated with $BED_{10}$ ≥100 versus 69% for $BED_{10}$ <100 | 1 grade ≥3<br>0 grade ≥4<br>7 late toxicities |
| [36] | Hepatic metastasis | 2021 | Prospective | 10 | 38.5 Gy | 3 or 5 | N/A | N/A | 0 grade ≥2 |
| [42] | Unresectable malignancies in the upper abdomen (primary and secondary) | 2021 | Prospective | 25 | 35, 40 | 5 | N/A | N/A | 0 grade ≥3 |
| [43] | Hepatic metastasis | 2022 | Phase II | 82 | 19–32 Gy | <10 | 1 year; 84% | 1 year; 88.1% | N/A |
| [57] | Unresectable primary liver and non-liver abdominal malignancies | 2017 | Phase I | 20 | 50 Gy | 5 | 1 year; 75% | 15 months; 90% | 0 grade ≥3 |
| [58] | Unresectable locally advanced cholangiocarcinoma | 2020 | Prospective | 17 | 40 Gy | 5 | 1 year; 76%<br>2 years; 46.1% | 1 year; 85.6%<br>2 years; 73.3% | 1 grade ≥3<br>1 late toxicity |
| [59] | Hepatic metastasis or HCC | 2021 | Prospective | 20 | 50 Gy (range: 45–60 Gy) | 8 (range: 3–12) | 1 year; 84% | 1 year; 88.1% | 0 grade ≥3 |
| [60] | Hepatic primary and metastatic tumors | 2019 | Prospective | 29 | 45–50 Gy in 5 fractions for 27 patients; 27–39 Gy in 2 fractions for 3 patients | 5 or 3 | N/A | N/A | N/A |
| [61] | Primary (HCC or ICC) and secondary liver tumors | 2019 | Retrospective | 26 | 50 Gy (range: 30–60) | 5 | 1 year; 69%<br>2 years; 60% | 21.2 months; 80.4% | 2 grade ≥3<br>0 grade ≥4 |
| [62] | Liver metastatic lesions, HCC | 2020 | Prospective | 10 | 30–60 Gy | 3–5 | N/A | N/A | 0 grade ≥3, 1 late grade 3 toxicity |
| [63] | Hepatic and pancreatic cancers | 2022 | Prospective | 16 (11 liver) | 27.50 Gy and 45 Gy in 5 fractions or 16 Gy and 24 Gy in single fractions | 5 (except 2 liver patients who received 1) | N/A | N/A | N/A |
| [64] | HCC | 2021 | Retrospective | 10 | 50 Gy | 5 | 80% (at time of analysis) | 90% (at time of analysis) | 0 grade ≥3 |

feasible, and ablative radiation doses were able to be delivered with low rates of toxicity [65].

In 2021, a study by van Dams et al. [22] at the University of California, Los Angeles (UCLA) involved 23 patients with primary or metastatic hepatic cancers who underwent liver-directed SBRT from 2015 to 2017. Of these 23, 20 underwent MR-guided SBRT and received a median radiation dose of 54 Gy in three treatments. These patients demonstrated an estimated 2-year LC of 80%, 2-year OS of 50.7%, and no acute grade ≥3 toxicities [22]. A difference in LC between single and multiple lesions or BED ≥100 Gy10 and BED <100 Gy10, respectively, was observed. Luterstein et al. [58] at UCLA reached a similar result, but with 17 patients with unresectable locally advanced cholangiocarcinoma treated between May 2015 and August 2017. These patients received a median radiation dose of 40 Gy in five treatments, with a 2-year LC of 73.3% and a 2-year OS of 46.1%. One patient experienced an acute grade >3 toxicity and one patient experience a late toxicity.

A retrospective study at UCLA by Yoon et al. [23] involved 106 patients with abdominal and pelvic tumors treated from November 2014 to August 2017. The median radiation dose was 40 Gy in five treatments, and the patients demonstrated an overall 2-year LC of 74%, with a breakdown of 96% for lesions that were treated with $BED_{10}$ ≥100 versus 69% for $BED_{10}$ <100, with a 2-year OS of 57% [23]. There was 1 patient who experienced an acute grade ≥3 toxicity and 7 patients who experienced late grade 3 and 4 toxicities. All 3 UCLA studies concluded that MRgRT is viable and safe and demonstrates diminished toxicity with high rates of LC and OS [22, 23, 58].

A trial at Heidelberg University Hospital in Germany by Hoegen et al. [44] treated 20 patients, 18 with liver metastasis and 2 HCC with ablative MR-guided SBRT between January 2019 and February 2020. At 1 year, LC was 88.1% and OS was 84.0%. Grade 1 toxicity occurred in 30.0% and grade 2 in 5.0% of the patients. No patient had grade >3 toxicity. The treatment was overall rated positively, and it was concluded that MRgRT is a well-tolerated and a well-accepted treatment modality [59]. A phase II study at the same hospital recruited 82 patients with hepatic metastasis, delivering a range of doses (between 19 and 32) in less than 10 fractions [44]. This study was simply to demonstrate MRgRT safety, particularly regarding dose escalation. Further studies will analyze the LC, OS, and toxicity. Similar work has been published without delineating LC, OS, and toxicity. Feldman et al. for example, evaluated a treatment plan of 45–50 Gy in five fractions and were able to assess the use of the gated breathing technique on MRI [60]. One-year posttreatment confirmed either stable or decreased size of treated lesions in all but one patient and toxicities were mild [60].

## Sample Patient Plans

Figures 14.3 and 14.4 demonstrate sample patient plans in a patient with metastatic anal cancer and HCC, respectively. As was discussed previously, the MR system allows for smaller targets to be identified, for OARs to be delineated, and for plans to be adapted on-table in order to deliver the maximal dose to the target while keeping the surrounding dose to a minimum.

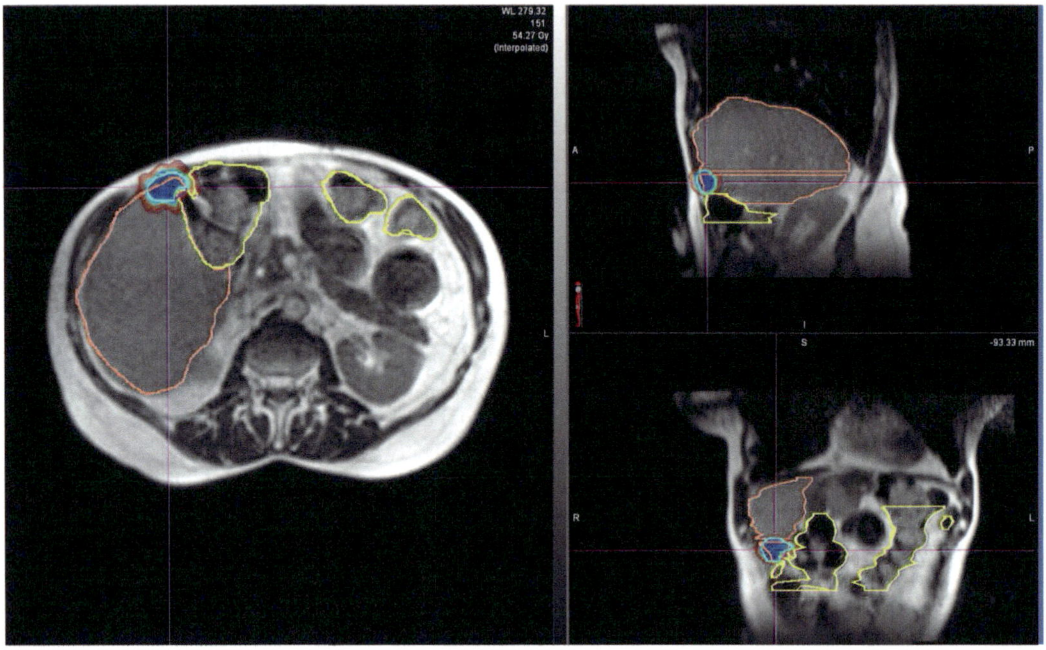


**Fig. 14.3** Patient with metastatic anal cancer to the liver. The target is adjacent to bowel (yellow) and liver (orange)—the OARs in this patient

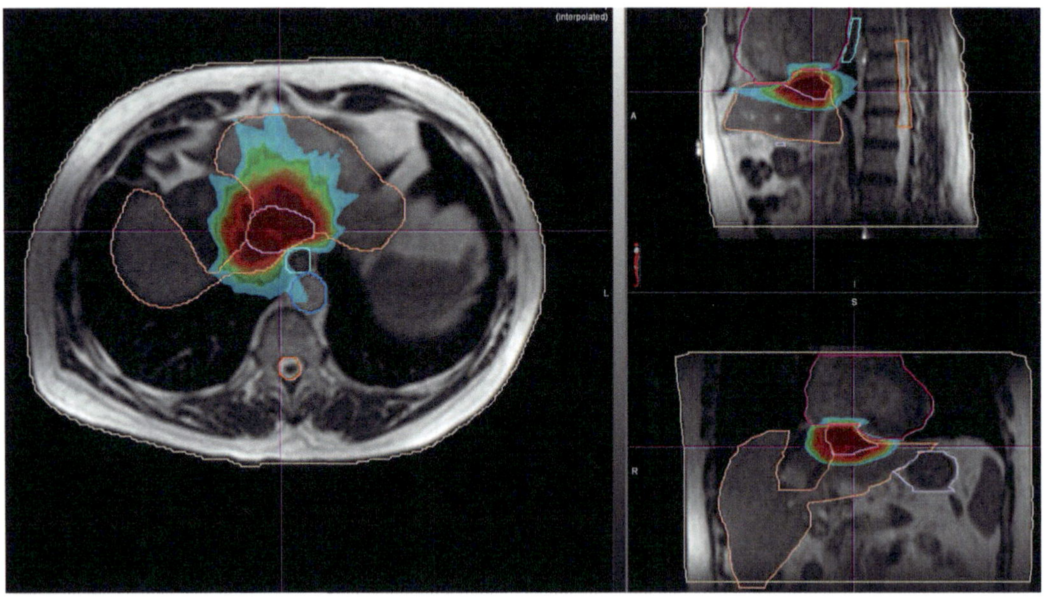


**Fig. 14.4** Patient with HCC status post ablation with recurrence within the ablation zone. Target is adjacent to the esophagus (light blue) and heart (orange)

## Summary

RT is a less invasive alternative to surgery and is possible for patients who are poor candidates for surgery. MRgRT is a promising area of treatment for liver malignancies, which primarily include HCC and ICC. CT-guided RT has traditionally been the mainstay, but it is an especially difficult treatment modality in patients with large tumors, patients with tumors close to other organs, and patients with tumors too small to be properly targeted. There is also the problem with shifting abdominal organs, which can lead to missing of the target and/or inadvertently giving higher doses to nearby organs. MRgRT has improved upon these flaws, allowing physicians to directly visualize the tumor and surrounding organs with higher-quality imaging that can track the movement of the internal structures. On-table adaptation is particularly valuable in the delivery of radiation. With MRgRT, targets can be optimized to the "anatomy of the day," making it possible to deliver higher, tumoricidal doses of radiation to patients, with minimal toxicity to OARs, which has been correlated with improvement in tumor control and an increase in survival for unresectable liver tumors [33, 49]. While all studies mentioned showed varying levels of success, they all agree that future prospective clinical trials to validate their findings are warranted [23]. Future studies additionally have to evaluate MRgRT considering cost, effectiveness, posttreatment quality of life, and long-term outcomes. As of now on the NIH Clinical Trials website, there are 5 actively recruiting studies and 3 completed study to assess MRgRT on hepatic cancers.

## References

1. Singal AG, Lampertico P, Nahon P. Epidemiology and surveillance for hepatocellular carcinoma: new trends. J Hepatol. 2020;72(2):250–61. https://doi.org/10.1016/j.jhep.2019.08.025.
2. Siegel RL, Miller KD, Fuchs HE, Jemal A. Cancer statistics, 2022. CA Cancer J Clin. 2022;72(1):7–33. https://doi.org/10.3322/caac.21708.
3. Llovet JM, Kelley RK, Villanueva A, et al. Hepatocellular carcinoma. Nat Rev Dis Primers. 2021;7(1):1–28. https://doi.org/10.1038/s41572-020-00240-3.
4. Gupta A, Dixon E. Epidemiology and risk factors: intrahepatic cholangiocarcinoma. Hepatobiliary Surg Nutr. 2017;6(2):101–4. https://doi.org/10.21037/hbsn.2017.01.02.
5. McGlynn KA, Petrick JL, El-Serag HB. Epidemiology of hepatocellular carcinoma. Hepatology. 2021;73(S1):4–13. https://doi.org/10.1002/hep.31288.
6. Ali H, Tedder B, Waqar SH, Mohamed R, Cate EL, Ali E. Changing incidence and survival of intrahepatic cholangiocarcinoma based on surveillance, epidemiology, and end results database (2000–2017). Ann Hepato Biliary Pancreat Surg. 2022;26:235. https://doi.org/10.14701/ahbps.21-173.
7. Rodrigues PM, Vogel A, Arrese M, Balderramo DC, Valle JW, Banales JM. Next-generation biomarkers for cholangiocarcinoma. Cancers (Basel). 2021;13(13):3222. https://doi.org/10.3390/cancers13133222.
8. Muscari F, Maulat C. Preoperative alpha-fetoprotein (AFP) in hepatocellular carcinoma (HCC): is this 50-year biomarker still up-to-date? Transl Gastroenterol Hepatol. 2020;5:46. https://doi.org/10.21037/tgh.2019.12.09.
9. Heimbach JK, Kulik LM, Finn RS, et al. AASLD guidelines for the treatment of hepatocellular carcinoma. Hepatology. 2018;67(1):358–80. https://doi.org/10.1002/hep.29086.
10. American College of Radiology Committee on LI-RADS® (Liver). CT/MRI LI-RADS® v2018 CORE. Accessed 28 Mar 2023.
11. Buettner S, van Vugt JL, IJzermans JN, Groot KB. Intrahepatic cholangiocarcinoma: current perspectives. Onco Targets Ther. 2017;10:1131–42. https://doi.org/10.2147/OTT.S93629.
12. Tsilimigras DI, Brodt P, Clavien PA, et al. Liver metastases. Nat Rev Dis Primers. 2021;7(1):1–23. https://doi.org/10.1038/s41572-021-00261-6.
13. Apisarnthanarax S, Barry A, Cao M, et al. External beam radiation therapy for primary liver cancers: an ASTRO clinical practice guideline. Pract Radiat Oncol. 2022;12(1):28–51. https://doi.org/10.1016/j.prro.2021.09.004.
14. Kim J, Jung Y. Radiation-induced liver disease: current understanding and future perspectives. Exp Mol Med. 2017;49(7):e359. https://doi.org/10.1038/emm.2017.85.
15. Goodman KA, Kavanagh BD. Stereotactic body radiotherapy for liver metastases. Semin Radiat Oncol. 2017;27(3):240–6. https://doi.org/10.1016/j.semradonc.2017.02.004.
16. Witt JS, Rosenberg SA, Bassetti MF. MRI-guided adaptive radiotherapy for liver tumours: visualising the future. Lancet Oncol. 2020;21(2):e74–82. https://doi.org/10.1016/S1470-2045(20)30034-6.
17. Pan CC, Kavanagh BD, Dawson LA, et al. Radiation-associated liver injury. Int J Radiat Oncol Biol Phys.

2010;76(3):S94–100. https://doi.org/10.1016/j.ijrobp.2009.06.092.
18. Lo EC, Rucker AN, Federle MP. Hepatocellular carcinoma and intrahepatic cholangiocarcinoma: imaging for diagnosis, tumor response to treatment and liver response to radiation. Semin Radiat Oncol. 2018;28(4):267–76. https://doi.org/10.1016/j.semradonc.2018.06.010.
19. Toesca DAS, Ibragimov B, Koong AJ, Xing L, Koong AC, Chang DT. Strategies for prediction and mitigation of radiation-induced liver toxicity. J Radiat Res. 2018;59(Suppl 1):i40–9. https://doi.org/10.1093/jrr/rrx104.
20. Son SH, Jang HS, Jo IY, et al. Significance of an increase in the Child–Pugh score after radiotherapy in patients with unresectable hepatocellular carcinoma. Radiat Oncol. 2014;9:101. https://doi.org/10.1186/1748-717X-9-101.
21. Tanaka O, Kojima T, Ohbora A, et al. Scores of Child–Pugh classification impact overall survival after stereotactic body radiation therapy for primary and metastatic liver tumors. J Clin Exp Hepatol. 2020;10(2):101–5. https://doi.org/10.1016/j.jceh.2019.05.002.
22. van Dams R, Wu TC, Kishan AU, et al. Ablative radiotherapy for liver tumors using stereotactic MRI-guidance: a prospective phase I trial. Radiother Oncol. 2022;170:14–20. https://doi.org/10.1016/j.radonc.2021.06.005.
23. Yoon SM, Luterstein E, Chu FI, et al. Clinical outcomes of stereotactic magnetic resonance image-guided adaptive radiotherapy for primary and metastatic tumors in the abdomen and pelvis. Cancer Med. 2021;10(17):5897–906. https://doi.org/10.1002/cam4.4139.
24. Parikh NR, Lee PP, Raman SS, et al. Time-driven activity-based costing comparison of CT-guided versus MR-guided SBRT. JCO Oncol Pract. 2020;16(11):e1378–85. https://doi.org/10.1200/JOP.19.00605.
25. Knoll MA, Jagsi R, Rosenzweig K. Focal radiation therapy for cancer. JAMA Oncol. 2019;5(3):442. https://doi.org/10.1001/jamaoncol.2018.5868.
26. Lukovic J, Henke L, Gani C, et al. MRI-based upper abdominal organs-at-risk atlas for radiation oncology. Int J Radiat Oncol Biol Phys. 2020;106(4):743–53. https://doi.org/10.1016/j.ijrobp.2019.12.003.
27. Romesser PB, Tyagi N, Crane CH. Magnetic resonance imaging-guided adaptive radiotherapy for colorectal liver metastases. Cancers (Basel). 2021;13(7):1636. https://doi.org/10.3390/cancers13071636.
28. Ng J, Gregucci F, Pennell RT, et al. MRI-LINAC: a transformative technology in radiation oncology. Front Oncol. 2023;13:1117874. https://doi.org/10.3389/fonc.2023.1117874.
29. Lin D, Lapen K, Sherer MV, et al. A systematic review of contouring guidelines in radiation oncology: analysis of frequency, methodology, and delivery of consensus recommendations. Int J Radiat Oncol Biol Phys. 2020;107(4):827–35. https://doi.org/10.1016/j.ijrobp.2020.04.011.
30. Chandarana H, Wang H, Tijssen RHN, Das IJ. Emerging role of MRI in radiation therapy. J Magn Reson Imaging. 2018;48(6):1468–78. https://doi.org/10.1002/jmri.26271.
31. Hunt A, Hansen VN, Oelfke U, Nill S, Hafeez S. Adaptive radiotherapy enabled by MRI guidance. Clin Oncol. 2018;30(11):711–9. https://doi.org/10.1016/j.clon.2018.08.001.
32. Respiratory Gating | OncoLink. https://www.oncolink.org/cancer-treatment/radiation/support/respiratory-gating. Accessed 10 July 2022.
33. Corradini S, Alongi F, Andratschke N, et al. MR-guidance in clinical reality: current treatment challenges and future perspectives. Radiat Oncol. 2019;14(1):92. https://doi.org/10.1186/s13014-019-1308-y.
34. Green OL, Rankine LJ, Cai B, et al. First clinical implementation of real-time, real anatomy tracking and radiation beam control. Med Phys. 2018;45(8):3728–40. https://doi.org/10.1002/mp.13002.
35. van de Lindt TN, Nowee ME, Janssen T, et al. Technical feasibility and clinical evaluation of 4D-MRI guided liver SBRT on the MR-linac. Radiother Oncol. 2022;167:285–91. https://doi.org/10.1016/j.radonc.2022.01.009.
36. Gani C, Boeke S, McNair H, et al. Marker-less online MR-guided stereotactic body radiotherapy of liver metastases at a 1.5 T MR-Linac—feasibility, workflow data and patient acceptance. ctRO. 2021;26:55–61. https://doi.org/10.1016/j.ctro.2020.11.014.
37. Drzymala RE, Mohan R, Brewster L, et al. Dose-volume histograms. Int J Radiat Oncol Biol Phys. 1991;21(1):71–8. https://doi.org/10.1016/0360-3016(91)90168-4.
38. Mayinger M, Ludwig R, Christ SM, et al. Benefit of replanning in MR-guided online adaptive radiation therapy in the treatment of liver metastasis. Radiat Oncol. 2021;16(1):84. https://doi.org/10.1186/s13014-021-01813-6.
39. Panchal A, Zhang Y, Ruffer J. A dose volume histogram (DVH) constraint analysis system for clinical protocol treatment plan evaluation using DICOM RT. Int J Radiat Oncol Biol Phys. 2010;78(3):S754. https://doi.org/10.1016/j.ijrobp.2010.07.1746.
40. Lee P, Ma Y, Zacharias I, et al. Stereotactic body radiation therapy for hepatocellular carcinoma in patients with Child–Pugh B or C cirrhosis. Adv Radiat Oncol. 2020;5(5):889–96. https://doi.org/10.1016/j.adro.2020.01.009.
41. Son SH, Choi BO, Ryu MR, et al. Stereotactic body radiotherapy for patients with unresectable primary hepatocellular carcinoma: dose-volumetric parameters predicting the hepatic complication. Int J Radiat Oncol Biol Phys. 2010;78(4):1073–80. https://doi.org/10.1016/j.ijrobp.2009.09.009.
42. Daamen LA, de Mol van Otterloo SR, van Goor IWJM, et al. Online adaptive MR-guided stereotac-

tic radiotherapy for unresectable malignancies in the upper abdomen using a 1.5T MR-linac. Acta Oncol. 2022;61(1):111–5. https://doi.org/10.1080/0284186X.2021.2012593.
43. Rudra S, Jiang N, Rosenberg SA, et al. Using adaptive magnetic resonance image-guided radiation therapy for treatment of inoperable pancreatic cancer. Cancer Med. 2019;8(5):2123–32. https://doi.org/10.1002/cam4.2100.
44. Hoegen P, Zhang KS, Tonndorf-Martini E, et al. MR-guided adaptive versus ITV-based stereotactic body radiotherapy for hepatic metastases (MAESTRO): a randomized controlled phase II trial. Radiat Oncol. 2022;17(1):59. https://doi.org/10.1186/s13014-022-02033-2.
45. Finazzi T, Palacios MA, Spoelstra FOB, et al. Role of on-table plan adaptation in MR-guided ablative radiation therapy for central lung tumors. Int J Radiat Oncol Biol Phys. 2019;104(4):933–41. https://doi.org/10.1016/j.ijrobp.2019.03.035.
46. Uder L, Nachbar M, Butzer S, et al. Local control and patient reported outcomes after online MR guided stereotactic body radiotherapy of liver metastases. Front Oncol. 2023;12:1095633. https://doi.org/10.3389/fonc.2022.1095633.
47. Winkel D, Bol GH, Kroon PS, et al. Adaptive radiotherapy: the Elekta Unity MR-linac concept. Clin Transl Radiat Oncol. 2019;18:54–9. https://doi.org/10.1016/j.ctro.2019.04.001.
48. Bohoudi O, Bruynzeel AME, Senan S, et al. Fast and robust online adaptive planning in stereotactic MR-guided adaptive radiation therapy (SMART) for pancreatic cancer. Radiother Oncol. 2017;125(3):439–44. https://doi.org/10.1016/j.radonc.2017.07.028.
49. Tchelebi LT, Zaorsky NG, Rosenberg J, Latifi K, Hoffe S. Integrating MR-guided radiation therapy into clinical practice: clinical advantages and practical limitations. Lung Cancer. 2021;9:1289.
50. Vu LN, Morelli JN, Szklaruk J. Basic MRI for the liver oncologists and surgeons. J Hepatocell Carcinoma. 2018;5:37–50. https://doi.org/10.2147/JHC.S154321.
51. Keall PJ, Brighi C, Glide-Hurst C, et al. Integrated MRI-guided radiotherapy—opportunities and challenges. Nat Rev Clin Oncol. 2022;19(7):458–70. https://doi.org/10.1038/s41571-022-00631-3.
52. Boldrini L, Corradini S, Gani C, et al. MR-guided radiotherapy for liver malignancies. Front Oncol. 2021;11:616027. https://doi.org/10.3389/fonc.2021.616027.
53. Gargett M, Haddad C, Kneebone A, Booth JT, Hardcastle N. Clinical impact of removing respiratory motion during liver SABR. Radiat Oncol. 2019;14(1):93. https://doi.org/10.1186/s13014-019-1300-6.
54. Ghadimi M, Sapra A. Magnetic resonance imaging contraindications. StatPearls Publishing; 2022. http://www.ncbi.nlm.nih.gov/books/NBK551669/
55. Lagendijk JJW, Raaymakers BW, Raaijmakers AJE, et al. MRI/linac integration. Radiother Oncol. 2008;86(1):25–9. https://doi.org/10.1016/j.radonc.2007.10.034.
56. Chin S, Eccles CL, McWilliam A, et al. Magnetic resonance-guided radiation therapy: a review. J Med Imaging Radiat Oncol. 2020;64(1):163–77. https://doi.org/10.1111/1754-9485.12968.
57. Henke LE, Contreras JA, Green OL, et al. Magnetic resonance image-guided radiotherapy (MRIgRT): a 4.5-year clinical experience. Clin Oncol. 2018;30(11):720–7. https://doi.org/10.1016/j.clon.2018.08.010.
58. Luterstein E, Cao M, Lamb JM, et al. Clinical outcomes using magnetic resonance-guided stereotactic body radiation therapy in patients with locally advanced cholangiocarcinoma. Adv Radiat Oncol. 2020;5(2):189–95. https://doi.org/10.1016/j.adro.2019.09.008.
59. Weykamp F, Hoegen P, Klüter S, et al. Magnetic resonance-guided stereotactic body radiotherapy of liver tumors: initial clinical experience and patient-reported outcomes. Front Oncol. 2021;11:610637. https://doi.org/10.3389/fonc.2021.610637.
60. Feldman AM, Modh A, Glide-Hurst C, Chetty IJ, Movsas B. Real-time magnetic resonance-guided liver stereotactic body radiation therapy: an institutional report using a magnetic resonance-linac system. Cureus. 2019;11(9):e5774. https://doi.org/10.7759/cureus.5774.
61. Rosenberg SA, Henke LE, Shaverdian N, et al. A multi-institutional experience of MR-guided liver stereotactic body radiation therapy. Adv Radiat Oncol. 2019;4(1):142–9. https://doi.org/10.1016/j.adro.2018.08.005.
62. Hal WA, Straza MW, Chen X, et al. Initial clinical experience of stereotactic body radiation therapy (SBRT) for liver metastases, primary liver malignancy, and pancreatic cancer with 4D-MRI based online adaptation and real-time MRI monitoring using a 1.5 Tesla MR-Linac. PLoS One. 2020;15(8):e0236570. https://doi.org/10.1371/journal.pone.0236570.
63. Stanescu T, Shessel A, Carpino-Rocca C, et al. MRI-guided online adaptive stereotactic body radiation therapy of liver and pancreas tumors on an MR-Linac system. Cancers. 2022;14(3):716. https://doi.org/10.3390/cancers14030716.
64. Boldrini L, Romano A, Mariani S, et al. MRI-guided stereotactic radiation therapy for hepatocellular carcinoma: a feasible and safe innovative treatment approach. J Cancer Res Clin Oncol. 2021;147(7):2057–68. https://doi.org/10.1007/s00432-020-03480-8.
65. Henke L, Kashani R, Robinson C, et al. Phase I trial of stereotactic MR-guided online adaptive radiation therapy (SMART) for the treatment of oligometastatic or unresectable primary malignancies of the abdomen. Radiother Oncol. 2018;126(3):519–26. https://doi.org/10.1016/j.radonc.2017.11.032.

# 15 Magnetic Resonance Imaging-Guided Radiation Therapy: Advances and Opportunities for Thoracic Malignancies

Julius Weng, Jinzhong Yang, Raymond H. Mak, Mohamed Abazeed, and Percy Lee

## Disease Overview

Non-small cell lung cancer (NSCLC) is the leading cause of cancer-related death in both men and women in the United States [1]. Stereotactic body radiation therapy (SBRT) is a treatment modality that delivers high doses of conformal radiation and is considered standard of care for early-stage NSCLC patients who are medically inoperable or refuse surgery [2]. SBRT is also increasingly utilized for treatment of oligometastatic disease, demonstrating an improvement in overall survival in phase II trials [3–5]. For tumors located away from the proximal bronchi and mediastinum (i.e., in the lung periphery), SBRT has demonstrated excellent local tumor control with limited toxicity [6–8]. However, for centrally located tumors that approximate the proximal bronchial tree, SBRT has been limited by significant toxicity due to complications from adjacent organs at risk (OARs), including the proximal bronchial tree, esophagus, heart/pericardium, and great vessels [9–11]. SBRT for central lung tumors can result in 2-year rates of significant toxicity as high as 46%, including treatment-related deaths [9–11]. As a result, central lung tumors are typically treated with more fractionated regimens (8–15 fx) [12, 13].

RTOG 0813 was a dose escalation study to determine the maximum tolerated dose (MTD) of SBRT for central tumors and included 120 patients [14]. Although the highest dose level of 12 Gy in 5 fractions SBRT schedule was reached, this was associated with a high rate of severe treatment-related toxicity (21% grade ≥3, 7% grade 5). In comparison, RTOG 0236 evaluated SBRT for 59 patients with early-stage peripheral lung tumors and had no grade 5 toxicities [15]. The risk of SBRT may be higher among patients with ultracentral lesions, which have either gross tumor volume (GTV) or planning treatment vol-

J. Weng · J. Yang
Division of Radiation Oncology, The University of Texas MD Anderson Cancer Center, Houston, TX, USA
e-mail: JKWeng@mdanderson.org; JYang4@mdanderson.org

R. H. Mak
Department of Radiation Oncology, Brigham and Women's Hospital, Dana-Farber Cancer Institute, Harvard Medical School, Boston, MA, USA
e-mail: rmak@Partners.org

M. Abazeed
Department of Radiation Oncology, Northwestern University Feinberg School of Medicine, Chicago, IL, USA
e-mail: mohamed.abazeed@nm.org

P. Lee (✉)
Department of Radiation Oncology, City of Hope National Medical Center, Los Angeles, CA, USA
e-mail: PercyLee@coh.org



I. J. Das et al. (eds.), *A Practical Guide to MR-Linac*, https://doi.org/10.1007/978-3-031-48165-9_15

ume (PTV) abutting critical mediastinal OARs [16]. The phase 2 HILUS trial studied SBRT (56 Gy/8 fx) in patients with ultracentral lesions defined as ≤1 cm from the proximal bronchial tree [17]. HILUS similarly demonstrated a high probability of significant treatment-related toxicity (34% grade ≥3, 15% grade 5). Although HILUS utilized a more fractionated SBRT regimen than RTOG 0813, HILUS exceeded standard OAR constraint doses for small/large bronchi and permitted hot spots of 150%. There are also several limitations of CT-based SBRT that may have contributed to the high rates of toxicity observed on these studies. To account for respiratory and cardiac motion, large internal target volumes (ITVs) and planning treatment volumes (PTVs) are often utilized. For central tumors, these large treatment volumes may overlap with OARs and, thus, necessitate the use of very high dose gradients near critical structures. However, inter- and intra-fraction tumor motion can be highly variable, and the baseline planned radiation treatment may not reflect the true dose accumulated by critical OARs over the course of treatment.

Magnetic resonance imaging-guided radiation therapy (MRgRT) is a recent technological innovation integrating a linear accelerator within an MRI machine. The two commercially available MR-linear accelerators are the MRIdian (ViewRay Inc., USA) and Unity (Elekta, Sweden) systems [18]. The first iteration of the MRIdian utilized a low-field 0.35 T MRI with cobalt sources. It was FDA approved in 2012 and was in clinical use by 2014. The second MRIdian system replaced the cobalt sources with a 6MV linear accelerator and was FDA approved in 2017. The Unity system utilizes a high-field 1.5 T MRI with a 7MV linear accelerator. The first patient was treated in 2018 in Europe with CE mark. It received FDA approval in 2018 and the first patient treated in North America was in 2019. Both the MRIdian and Unity are capable of online adaptive therapy and have similar workflows [19, 20]. The MRIdian has received FDA approval for treatment gating functionality and is being used in the clinic. At the time of writing, the Unity motion management has received the European CE mark and is used clinically in Europe, but only recently received FDA approval.

There are several theoretical advantages of MRgRT that can increase the therapeutic ratio of radiation, especially for central tumors adjacent to multiple OARs, high-risk patients undergoing re-irradiation, or patients with pulmonary comorbidities [21, 22]. These advantages address several of the limitations of CT-based SBRT including online adaptation for inter-fraction anatomical changes, intra-fraction motion management, soft tissue visualization, and margin reduction. In this chapter, we discuss the potential benefit, clinical workflow, current limitations, and future directions of MRgRT for thoracic tumors.

## Overcoming Challenges of Traditional Approaches

A significant advantage of MRgRT is the capability for online adaptive radiotherapy, which is the process of optimizing the initial radiation treatment plan based upon "anatomy of the day" obtained from daily onboard MRI. Given the close proximity of central tumors to critical structures, radiation plans must incorporate a sharp dose falloff to ensure that OARs meet dose constraints. Even small changes in the position of tumor or OARs between fractions may result in underdosing of tumor or excessive dose to OARs. Online adaptive MRgRT accounts for inter-fraction changes in patient and tumor/OAR position by applying the original radiation plan to the daily scan and recalculating dose metrics. The plan can then be reoptimized to increase target coverage or further spare normal tissue. An initial in silico study compared adaptive and nonadaptive SBRT plans delivered by MRgRT for 5 patients with centrally located oligometastatic or unresectable early-stage lung cancer [23]. Despite the initial plan meeting all constraints on baseline CT simulation, all 5 patients had at least 1 fraction with ≥1 OAR constraint violation based on daily MR imaging. These constraint violations included the trachea ($n = 6$), esophagus ($n = 5$), and heart/pericardium ($n = 1$). After adaptation, 100% of constraint violations were reversed and, furthermore, PTV dose coverage was improved in 66% of fractions. Based on these promising findings, Henke et al. conducted a subsequent phase I trial of MRgRT for ultracentral thoracic tumors and included 5 patients treated with 50 Gy in 5 fractions. In this trial, 4 out of 5 patients required at least 1 adaptive fraction and 10 out of 25 fractions (40%) were adaptive. Of the adaptive plans, 70% were for constraint violations and 30% were to increase PTV coverage. The OARs that exceeded dose constraints included the heart, trachea, esophagus, stomach, and spinal cord. All the constraint violations were resolved with the adaptive plans and overall PTV coverage was improved. Importantly, there were no acute grade ≥3 toxicities in the first 6 months and only one radiation-related grade 3 toxicity at 15 months. These studies highlighted the safety and dosimetric benefits of online adaptive MRgRT for high-risk thoracic tumors.

MRgRT allows for continuous, real-time imaging during radiation delivery, which permits tracking of intra-fraction tumor motion due to respiration. Whereas CT-based motion management strategies often require fiducial markers or external surrogates, MR cine imaging allows for direct tumor tracking without the need for an invasive procedure. Management of respiratory motion is a significant challenge in treating thoracic malignancies. Accounting for changes in tumor position is particularly important for centrally located tumors in which small differences in adjacent OAR position may greatly impact toxicity. Current strategies, such as contouring an ITV on 4DCT, generating an ITV from the maximum intensity projection (MIP), or utilizing deep inspiratory breath-hold (DIBH) technique, assume that tumor position(s) are relatively stable from baseline simulation. However, there is likely deviation of tumor position both across and within treatment fractions [24]. In a pilot study, 5 lung tumor patients being treated on MRgRT were simulated with 4DCT and MRI (real-time cine imaging for 172 s with −0.35 T MRI). During each fraction, a 20-min cine-mode MR free-breathing scan was obtained. These images were then used to contour and track GTV motion using offline optical flow-based deformable registration on a graphics processing unit cluster. The MR ITV demonstrated differences in intra- and inter-fractional motion that may not otherwise be accounted for by standard 4D CT imaging, which suggests that real-time MR imaging for thoracic cancers may be necessary to ensure accurate delivery of SBRT. In addition, there is ongoing research into the development of 4D MRI, which may further improve resolution of tumor and OAR motion [25].

An alternative to an ITV motion management approach is DIBH with gating [26]. Instead of an ITV to cover the entire path of the tumor during respiration, DIBH with gating allows for smaller margins by limiting the treatment volume to a designated phase in the respiratory cycle (Fig. 15.1). The reduction in treated normal tissue is especially relevant for patients with preexisting medical comorbidities as decreasing the treatment volume may preserve cardiopulmonary function. Patients can be coached for optimal breath holds through audio-video methods or even directly view their tumor motion using a monitor displaying real-time MR-cine imaging [20]. A treatment boundary margin can be created, so that treatments are automatically gated if a predefined percentage of the GTV is outside of the boundary. Although breath hold may not be feasible for some patients due to poor lung function, if tolerated, it can also reduce the frequency of transitions between beam delivery being turned on and off.

An additional potential advantage of MRgRT is the use of single-modality, simulation-free, MR-only treatment planning. In general, a CT simulation is necessary for radiation therapy treatment planning as an MRI does not include electron density data needed for dose calculations. However, there is increasing interest in the generation of synthetic CT scans from MRI using atlas-based, voxel-based, or deep-learning methods [27]. Eliminating the need for a CT simulation has several benefits including reduced resource utilization, financial costs, patient radiation exposure, and errors in image registration. The MR-linear accelerator is well-suited for a MR-only approach given the onboard MR imaging and online adaptive therapy capabilities.

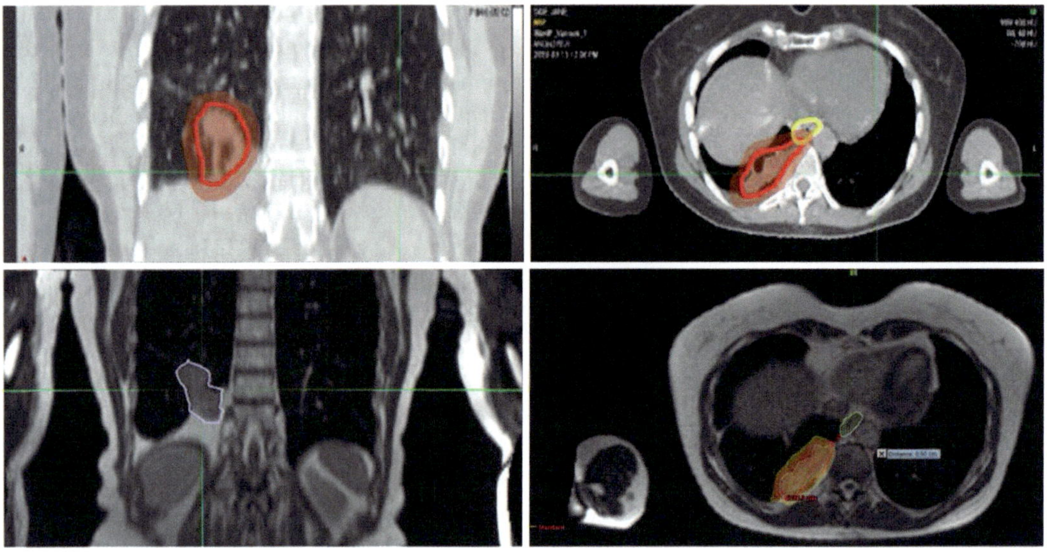

**Fig. 15.1** Comparison of an ITV vs DIBH motion management approach. For this patient with a 5-cm, mobile, central tumor, an ITV approach using 4DCT (upper panels) resulted in a large target volume abutting the esophagus. By utilizing an MR-Linac with DIBH (lower panels), target volumes could be significantly reduced. The decreased proximity of the target to the esophagus allowed for treatment with 5-fraction SBRT

## Contouring, PTV, and OAR

Accurate tumor and OAR delineation is an essential aspect of thoracic radiation treatment planning but can be challenging with conventional CT-based imaging alone. The advent of positron emission therapy (PET)/CT scans has improved the ability to identify malignant metabolically active tumor; however, PET/CT requires injection of a radiotracer, has limited spatial resolution, and is susceptible to false-positive findings [28]. In comparison, MRI substantially improves soft tissue visualization, can highlight functional tumor characteristics, and is obtained without the use of ionizing radiation (Fig. 15.2) [29, 30]. Preliminary studies have demonstrated improvements in the consistency of GTV contours using thoracic MRI. Basson et al. conducted a study including 5 physicians who independently contoured GTV for 14 challenging lung cases. GTV was contoured in three phases: using CT only, adding PET/CT, and MRI with PET/CT [31]. The study demonstrated decreased inter-observer variability in GTV contour with the addition of MRI compared to CT or PET/CT alone, which suggests MRI may improve the consistency of target delineation for difficult cases. Similarly, Zhang et al. compared the contours of 5 radiation oncologists for 15 patients with lung tumors adjacent to atelectasis [32]. They found that MRI improved the delineation of GTV, resulting in reduced inter-observer variability and decreased GTVs. Thoracic MRI may be particularly helpful for challenging clinical scenarios such as tumors located in the superior sulcus, adjacent to atelectatic lung, abutting central mediastinal structures, invading the chest wall, or if the patient has a contraindication to contrast. The improved soft tissue resolution of MRgRT is further highlighted when compared to onboard cone-beam CT (CBCT). An initial study compared OAR and target visualization using low-field onboard MRI and CBCT for 221 OARs and 10 targets [29]. Three radiation oncologists reviewed the images and ultimately rated the onboard MRI images as superior for 63% of the OARs of interest, inferior for the ribs, and equivalent for the remaining structures. The improved image quality of MRgRT may provide greater assurance of accurate targeting and delivery of SBRT.

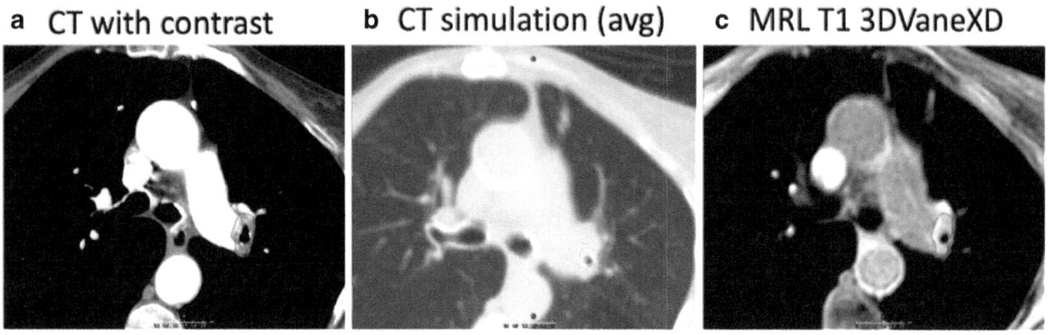


**Fig. 15.2** Comparison of diagnostic CT with IV contrast (**a**), CT simulation (avg) (**b**), and Unity onboard MR imaging (**c**) for a patient with ultracentral lung cancer

Given the ability to obtain continuous, real-time imaging, MRgRT enables tumor tracking with treatment gating and may allow for reduction of PTV margins due to decreased treatment uncertainties. Finazzi et al. conducted a study comparing PTVs generated using 4DCT and MRgRT for 14 patients with peripheral lung tumors [33]. The 4DCT PTVs were created using a motion-encompassing internal target volume with a 5-mm expansion, while the MRgRT PTVs were an isotropic 5-mm expansion from GTV that functioned as a gating boundary. They found that MRgRT PTVs were approximately 50% reduction compared to the volume of 4DCT PTVs. A similar study compared 41 patients treated with online adaptive MRgRT with treatment gating and 40 patients treated with conventional SBRT [34]. They found that the PTVs in the online adaptive MRgRT cohort were 40% smaller and did not compromise local control at 1-year follow-up. The reduction in treated normal tissue may be most beneficial to patients who have multiple lung lesions or preexisting pulmonary comorbidities, such as interstitial lung disease [35]. There is evidence in other disease sites that reducing PTV margins may also decrease treatment-related toxicity. A prospective randomized trial in prostate cancer patients of MRgRT vs conventional CT-based SBRT demonstrated that using MRgRT to reduce PTV margins significantly decreased acute genitourinary toxicity [36]. A similar approach of decreasing PTV margins for central lung tumors may reduce severe toxicity for these high-risk patients.

The high precision of MRgRT may also be beneficial for single-fraction treatment by adapting for geographic misses or excessive OAR dose [37]. Finazzi et al. reported on their initial clinical experience evaluating 17 patients for single-fraction SBRT utilizing MRgRT [38]. Seven of the patients were determined to be unsuitable due to difficulty tracking their small tumors. Out of the remaining 10 patients, nine were able to successfully complete their treatment in a single session and one patient required two sessions due to back pain. Although breath-hold patterns were variable and mean duty cycle efficiency was 51%, GTV coverage was 99.6%. A study conducted by Palacios et al. investigated MR-guided single-fraction SBRT, but with consultation, simulation, and treatment conducted in a single day [39]. All 10 patients in their cohort successfully completed their treatment with excellent coverage of the GTV (94.4%) using treatment gating. Of note, patients were surveyed and reported overall high levels of satisfaction with their experience. By leveraging the advantages of MRgRT, it may be possible to more safely and precisely deliver single-fraction lung SBRT.

## Dose Escalation

There have been several significant imaging advancements that have increased the ability for radiation oncologists to safely dose-escalate radiation treatments for lung tumors. With the advent of PET/CT, 4DCT, and volumetric onboard image guidance, it became feasible to deliver highly conformal SBRT regimens that achieve high biologically effective dose (BED). These innovations in imaging technology and SBRT delivery have translated to improved clinical outcomes for patients. The initial studies utilizing CT-based SBRT demonstrated very promising local control rates of over 90% [15, 40]. In addition, the CHISEL study randomized 101 patients with early-stage peripheral NSCLC to standard RT (66 Gy/33 fx or 50 Gy/20 fx) or SBRT (54 Gy/3 fx or 48 Gy/4 fx). At 2 years, SBRT was found to improve local control (89% vs 65%) and OS (88% vs 59%) [41]. Although SBRT for peripheral tumors has excellent local control with minimal toxicity, central and ultracentral tumors have been associated with severe treatment-related toxicity [9–11]. Given these concerning rates of toxicity, patients with central tumors often are treated with more prolonged radiation regimens (10–15 fractions), which may not achieve equivalent BED. This is concerning as studies have demonstrated that local control of NSCLC has a dose-dependent relationship with BED and that a BED of at least 100 Gy is necessary to achieve 90% local control [42–45]. Thus, the reduction of OAR doses while maintaining high BED would represent a meaningful advance in the treatment of central lung tumors.

The advanced imaging capabilities of MRgRT such as improved soft tissue visualization, online adaptive therapy, and real-time tumor tracking may allow for more precise and safer delivery of SBRT to central lesions. Potentially, OAR violations that occur during traditional treatment delivery could be prevented using online adaptive therapy. A study by Finazzi et al. included 25 patients with central lung tumors who underwent online adaptive MRgRT in 60 Gy/8 fx or 55 Gy/5 fx. They found that physicians selected a reoptimized plan in 92% of fractions and that online adaptive therapy significantly reduced OAR constraint violations [46]. A larger prospective study of online adaptive MRgRT in 50 Gy/5 fx or 60 Gy/8 fx for central and ultracentral lesions is currently ongoing (NCT04917224). There is also an ongoing phase I dose escalation trial using online adaptive MRgRT for ultracentral tumors, which will deliver 55 Gy/10 fx and up to 65 Gy/10 fx (NCT04925583). An additional potential advantage of MRgRT for dose escalation is the incorporation of functional imaging in treatment adaptation. MR sequences can provide not only anatomical, but also functional information that may be utilized as guidance for treatment adaptation and to evaluate treatment response. For example, diffusion weighted imaging (DWI) and apparent diffusion coefficients (ADCs) have been studied in improving tumor delineation, assessing treatment response, and identifying residual tumor [47]. As an increase ADC values during therapy can reflect treatment-related cell death, it may be a useful biomarker to guide dose escalation for poorly responding tumors [48]. Dynamic contrast-enhanced (DCE) MRI is used to evaluate tissue perfusion, blood flow, and tumor vasculature. Early studies in other disease sites have suggested that increased enhancement during treatment may be associated with improved outcomes [49]. The use of MRgRT is novel and dose escalation using MR anatomical and functional imaging in the context of historically high-risk central lung tumors warrants prospective study.

## On-Table Adaptation (ViewRay and Unity)

The following outlines the general workflow for on-table adaptation, but this may vary depending on the system and individual institutional practice. On the MRIdian system, adaptation begins by rigid registration of the online MRI to the simulation scans. If needed, the treatment table is then shifted to account for translational positional differences. The reference plan contours can be rigidly or deformably propagated to the online MRI for modification. On the Unity system, adaptation also starts with a rigid registration of online MRI to the simulation scan. Based on the rigid registration, clinicians can determine to use either an adapt to position (ATP) or adapt to shape (ATS) approach [19]. The ATP method is suitable for cases in which contours do not need to be edited and the relative positions of target/OARs are consistent with the reference treatment plan. The rigid shift from the initial rigid registration is used to move the plan isocenter accordingly. The reference plan is then reoptimized with the new plan isocenter based on the original simulation CT and unedited contours. In contrast, the ATS method is utilized for cases in which the "anatomy of the day" differs significantly from the initial simulation. The initial contours are transferred to the online MRI using deformable registration and edited as necessary within a 2–3 cm region adjacent to the tumor based on a clinical decision. An adaptive plan of the day is then reoptimized based upon the new contours using the online MRI scan. Dose calculation is based on a synthetic CT created using the bulk density override method. For both the MRIdian and Unity systems, after the adaptive plan is generated, dose metrics are compared with the initial baseline plan and a treatment plan is selected. Prior to dose delivery, on-table quality assurance is performed and tumor respiratory motion is reverified using cine imaging.

## DVH Metrics/Evaluation

It is well-established that a high BED of ≥100 Gy is required to ensure optimal local control for lung tumors [42–44]. Although it is preferred that PTV is fully covered by tumoricidal prescription dose, it may be necessary to compromise coverage of PTV to meet adjacent critical OAR constraints. Reassuringly, retrospective data suggests local control rates are maintained despite reduced PTV coverage so long as ITV dose is adequate [50]. Prioritizing critical OAR doses and secondarily maximizing tumor dose is particularly relevant for patients with central lung tumors undergoing online adaptive therapy given the potential for multiple plans. For these high-risk patients, it is critical to thoroughly evaluate that OAR constraints are being met with each fraction and target doses are being optimized. There is increasing interest in developing methods to accurately assess dose accumulation over the course of treatment [51]. This would allow for monitoring of dose per fraction and cumulative dose to OARs and tumor. Daily dose accumulation can be used to inform optimization parameters of subsequent adaptive plans and adjust prioritization of OAR sparing vs target coverage [52].

## Outcomes

Although there are several completed and ongoing clinical trials of CT-based SBRT of central lung tumors, few studies have evaluated the role of MRgRT with daily adaptive plans for these patients [14, 53, 54]. Prior simulation trials and retrospective studies of MRgRT have demonstrated dosimetric improvements including increased PTV coverage and OAR sparing [23, 46]. A phase I trial including 5 ultracentral patients treated with online adaptive MRgRT in 50 Gy/5 fx met 100% of OAR constraints using online adaptive therapy and demonstrated no acute grade ≥3 toxicities in the first 6 months of follow-up [55]. A retrospective study of SBRT using 0.35 T MRgRT reviewed outcomes for 50 consecutive patients with high-risk lung tumors defined as central location, prior thoracic radiation, or interstitial lung disease [22]. A BED ≥100 Gy to 95% PTV was achieved in 93% of patients and local control was 95.6% at 1 year. Importantly, at a median follow-up of 21.7 months, there were only 8% grade 3 and no grade ≥4 toxicities. Lastly, a prospective study of online adaptive MRgRT evaluated 21 patients with peripheral ($n$ = 10), central ($n$ = 2), and ultracentral ($n$ = 11) tumors [56]. Prescriptions were risk-adapted and were 50 Gy/5 fx for peripheral, 60 Gy/8 fx for central, and 50–60 Gy in 10 fx for ultracentral tumors. Overall, 93% of fractions were adapted (154/165). The most common reason for plan adaptation was to increase PTV coverage (86%), whereas 17% of adaptations for OAR constraint violation. Of note, all OAR constraint violations were reversed except for one. These promising results have prompted the authors to initiate a prospective, phase I dose escalation trial of MRgRT for central lung tumors (NCT04925583). There are also other prospective treatment registries and ongoing clinical trials utilizing MRgRT approaches for thoracic tumors (NCT04917224, NCT05609331, NCT04939246, NCT04115254, NCT04789486). While early outcomes are promising, there remains a need for additional prospective studies to better characterize treatment outcomes, ensure that dosimetric optimization results in meaningful improvement in clinical outcomes, and guide patient selection for thoracic MRgRT.

## Challenges

While there are significant advantages of MRgRT for lung tumors, there remain several limitations unique to thoracic MRI [57]. Unlike many other disease sites, the lungs inherently have a low MR signal to noise ratio (SNR) due to low tissue density and high air content [58]. The visualization of small parenchymal lesions and ground-glass opacities can be suboptimal, particularly for smaller tumors <1 cm. Low field strength (0.35 T) on a MRIdian system also limits the SNR of MRI. In addition, MRI body coil design is limited by the immobilization devices used during treatment, which results in a body coil that is suboptimal for MRI image quality. Furthermore, respiratory, cardiac, and great vessel pulsatile motion degrades MR image quality due to motion artifacts and image blurring. These field distortion artifacts can be particularly problematic for tumors located at soft tissue and air interfaces (e.g., the pericardium or diaphragm) and impact tumor/OAR delineation for initial contouring as well as treatment gating. There are ongoing efforts to develop and optimize MR sequences for low-field and standard-field MRI to address these challenges for MRgRT [59, 60]. Specifically, the turbo field-echo (TFE) or fast field-echo (FFE) sequences with a radial sampling in k-space, also known as 3DVaneXD or multi-VaneXD, has been found to greatly reduce the impact from respiratory motion in free-breathing acquisition [61, 62]. Breath-hold scans using a balanced FFE sequence can acquire a 3D MRI scan within 20 s with reasonable image quality, thus reducing the impact from respiratory motion. Also, the compressed SENSE technique has been used in clinic for fast MRI acquisition to further reduce the impact from respiratory motion [63]. Dosimetric drawbacks of MRgRT include the lack of volumetric modulated arc therapy (VMAT) capabilities, which limits dose conformality and extends treatment duration. In addition, MR-gated treatments introduce beam-off latency dosimetric errors [64]. These errors occur due to the time required for the MR cine image to be acquired and the beam to be turned off appropriately when the tumor moves beyond the gating boundary. Beam-off latency may reduce dose to the PTV and inadvertently increase dose to nearby OARs; however, this can be minimized through development of faster MR imaging and limiting the transition between gated and ungated states. An additional dosimetric drawback of MRgRT is that the Lorentz force produced by magnetic fields leads to an electron returning effect. The electron returning effect results in higher doses at air-tissue interfaces, such as the lung, and must be corrected for by the treatment planning algorithm. Lastly, MRgRT with online plan adaptation workflow usually requires an extended treatment time with the patient on the treatment table [65]. This extended treatment time is not only the result of complicated workflow with online plan adaptation and patient-specific QA, but also due to a low dose rate and low duty cycle with treatment gating during beam delivery. Patients are more likely to move when laying on the couch for a long time, resulting in unexpected movement and making motion management more difficult. Real-time motion monitoring and treatment gating can alleviate the impact from this unexpected motion. Additional fast plan adaptation may be necessary to account for large changes in position during treatment delivery.

## Future Directions

MRgRT is a promising technologic innovation with several key advantages for thoracic radiation therapy. This includes clinical scenarios that require precise OAR dose monitoring such as patients with high-risk tumor locations (central/ultracentral) or recurrent tumors requiring reirradiation. MRgRT may also facilitate the simultaneous treatment of primary NSLCL or oligometastatic disease with multiple thoracic lesions with a single isocenter [66]. In addition to SBRT, patients with locally advanced NSCLC undergoing longer courses of RT may benefit from weekly treatment adaptation as tumor regresses on therapy or due to anatomical distortions. The precision of MRgRT may also facilitate delivery of hypofractionated regimens for patients with locally advanced NSCLC who have poor prognostic factors [67]. As many candidates for radiation therapy are not surgical candidates due to poor lung function or pulmonary comorbidities such as interstitial lung disease, MRgRT with real-time tracking and gated treatment delivery may be beneficial to reduce the overall integral dose to the lungs. Future research directions also include MRI sequence optimization and development for lung imaging. Recent advances in MRI have greatly improved the image SNR for lung MRI; however, there is still room for further improvement (e.g., improving the MR body coil design for better image SNR and radiation treatment guidance). Another potential improvement is the latency of cine MR imaging. Although there are currently strategies to reduce the impact of latency, the duty cycle for treatment gating or breath-hold treatment is low and further extends treatment time. In the future, development of faster cine MR imaging is desirable. Lastly, in the future, integration of functional MRI sequences may aid in tumor delineation, monitoring treatment response, and guiding dose escalation or de-escalation studies [68]. Overall, MRgRT for thoracic tumors has demonstrated promising technical results with several active investigations actively seeking to improve the therapeutic ratio.

## References

1. Siegel RL, Miller KD, Wagle NS, Jemal A. Cancer statistics, 2023. CA Cancer J Clin. 2023;73(1):17–48.
2. NCCN Clinical Practice Guidelines. Non-small cell lung cancer. (Version 1.2023). https://www.nccn.org/professionals/physician_gls/pdf/nscl.pdf.
3. Bergsma DP, Salama JK, Singh DP, Chmura SJ, Milano MT. Radiotherapy for oligometastatic lung cancer. Front Oncol. 2017;7:210.
4. Gomez DR, Tang C, Zhang J, Blumenschein GR, Hernandez M, Lee JJ, et al. Local consolidative therapy vs. maintenance therapy or observation for patients with oligometastatic non-small-cell lung cancer: long-term results of a multi-institutional, phase II, randomized study. J Clin Oncol. 2019;37(18):1558–65.
5. Palma DA, Olson R, Harrow S, Gaede S, Louie AV, Haasbeek C, et al. Stereotactic ablative radiotherapy for the comprehensive treatment of oligometastatic cancers: long-term results of the SABR-COMET phase II randomized trial. JCO. 2020;38(25):2830–8.
6. Timmerman R, Paulus R, Galvin J, Michalski J, Straube W, Bradley J, et al. Stereotactic body radiation therapy for inoperable early stage lung cancer. JAMA. 2010;303(11):1070–6.
7. Baumann P, Nyman J, Høyer M, Wennberg B, Gagliardi G, Lax I, et al. Outcome in a prospective phase II trial of medically inoperable stage I non-small-cell lung cancer patients treated with stereotactic body radiotherapy. J Clin Oncol. 2009;27:3290–6.
8. Ricardi U, Filippi AR, Guarneri A, Giglioli FR, Ciammella P, Franco P, et al. Stereotactic body radiation therapy for early stage non-small cell lung cancer: results of a prospective trial. Lung Cancer. 2010;68(1):72–7.
9. Timmerman R, Mcgarry R, Yiannoutsos C, Papiez L, Tudor K, DeLuca J, et al. Excessive toxicity when treating central tumors in a phase II study of stereotactic body radiation therapy for medically inoperable early-stage lung cancer. J Clin Oncol. 2006;24:4833–9.
10. Bral S, Gevaert T, Linthout N, Versmessen H, Collen C, Engels B, et al. Prospective, risk-adapted strategy of stereotactic body radiotherapy for early-stage non-small-cell lung cancer: results of a phase II trial. Int J Radiat Oncol Biol Phys. 2011;80(5):1343–9.
11. Song SY, Choi W, Shin SS, Lee SW, Ahn SD, Kim JH, et al. Fractionated stereotactic body radiation therapy for medically inoperable stage I lung cancer adjacent to central large bronchus. Lung Cancer. 2009;66(1):89–93.
12. Chang JY, Li QQ, Xu QY, Allen PK, Rebueno N, Gomez DR, et al. Stereotactic ablative radiation therapy for centrally located early stage or isolated parenchymal recurrences of non-small cell lung cancer: how to fly in a "no fly zone". Int J Radiat Oncol Biol Phys. 2014;88(5):1120–8.
13. Haasbeek CJA, Lagerwaard FJ, Slotman BJ, Senan S. Outcomes of stereotactic ablative radiotherapy for

centrally located early-stage lung cancer. J Thorac Oncol. 2011;6(12):2036–43.
14. Bezjak A, Paulus R, Gaspar LE, Timmerman RD, Straube WL, Ryan WF, et al. Safety and efficacy of a five-fraction stereotactic body radiotherapy schedule for centrally located non-small-cell lung cancer: NRG Oncology/RTOG 0813 trial. J Clin Oncol. 2019;37(15):1316–25.
15. Timmerman RD, Hu C, Michalski JM, Bradley JC, Galvin J, Johnstone DW, et al. Long-term results of stereotactic body radiation therapy in medically inoperable stage I non-small cell lung cancer. JAMA Oncol. 2018;4(9):1287–8.
16. Chen H, Laba JM, Zayed S, Boldt RG, Palma DA, Louie AV. Safety and effectiveness of stereotactic ablative radiotherapy for ultra-central lung lesions: a systematic review. J Thorac Oncol. 2019;14(8):1332–42.
17. Lindberg K, Grozman V, Karlsson K, Lindberg S, Lax I, Wersäll P, et al. The HILUS-trial—a prospective Nordic multicenter phase 2 study of ultracentral lung tumors treated with stereotactic body radiotherapy. J Thorac Oncol. 2021;16(7):1200–10.
18. Liney GP, Whelan B, Oborn B, Barton M, Keall P. MRI-linear accelerator radiotherapy systems. Clin Oncol. 2018;30(11):686–91.
19. Winkel D, Bol GH, Kroon PS, van Asselen B, Hackett SS, Werensteijn-Honingh AM, et al. Adaptive radiotherapy: the Elekta Unity MR-linac concept. Clin Transl Radiat Oncol. 2019;18:54–9.
20. Bryant JM, Sim AJ, Feygelman V, Latifi K, Rosenberg SA. Adaptive hypofractionted and stereotactic body radiotherapy for lung tumors with real-time MRI guidance. Front Oncol. 2023;13:1061854. https://doi.org/10.3389/fonc.2023.1061854.
21. Padgett KR, Simpson GN, Llorente R, Samuels MA, Dogan N. Feasibility of adaptive MR-guided stereotactic body radiotherapy (SBRT) of lung tumors. Cureus. 2018;10(4):e2423.
22. Finazzi T, Haasbeek CJA, Spoelstra FOB, Palacios MA, Admiraal MA, Bruynzeel AME, et al. Clinical outcomes of stereotactic MR-guided adaptive radiation therapy for high-risk lung tumors. Int J Radiat Oncol Biol Phys. 2020;107(2):270–8.
23. Henke L, Kashani R, Yang D, Zhao T, Green O, Olsen L, et al. Simulated online adaptive magnetic resonance-guided stereotactic body radiation therapy for the treatment of oligometastatic disease of the abdomen and central thorax: characterization of potential advantages. Int J Radiat Oncol Biol Phys. 2016;96(5):1078–86.
24. Thomas DH, Santhanam A, Kishan AU, Cao M, Lamb J, Min Y, et al. Initial clinical observations of intra- and interfractional motion variation in MR-guided lung SBRT. Br J Radiol. 2018;91(1083):20170522.
25. Paulson ES, Ahunbay E, Chen X, Mickevicius NJ, Chen GP, Schultz C, et al. 4D-MRI driven MR-guided online adaptive radiotherapy for abdominal stereotactic body radiation therapy on a high field MR-Linac: implementation and initial clinical experience. Clin Transl Radiat Oncol. 2020;23:72–9.
26. Dhont J, Harden SV, Chee LYS, Aitken K, Hanna GG, Bertholet J. Image-guided radiotherapy to manage respiratory motion: lung and liver. Clin Oncol. 2020;32(12):792–804.
27. Olberg S, Zhang H, Kennedy WR, Chun J, Rodriguez V, Zoberi I, et al. Synthetic CT reconstruction using a deep spatial pyramid convolutional framework for MR-only breast radiotherapy. Med Phys. 2019;46(9):4135–47.
28. Cobben DCP, de Boer HCJ, Tijssen RH, Rutten EGGM, van Vulpen M, Peerlings J, et al. Emerging role of MRI for radiation treatment planning in lung cancer. Technol Cancer Res Treat. 2016;15(6):NP47–60.
29. Noel CE, Parikh PJ, Spencer CR, Green OL, Hu Y, Mutic S, et al. Comparison of onboard low-field magnetic resonance imaging versus onboard computed tomography for anatomy visualization in radiotherapy. Acta Oncol. 2015;54(9):1474–82.
30. Raaymakers BW, Lagendijk JJW, Overweg J, Kok JGM, Raaijmakers AJE, Kerkhof EM, et al. Integrating a 1.5 T MRI scanner with a 6 MV accelerator: proof of concept. Phys Med Biol. 2009;54(12):N229.
31. Basson L, Jarraya H, Escande A, Cordoba A, Daghistani R, Pasquier D, et al. Chest magnetic resonance imaging decreases inter-observer variability of gross target volume for lung tumors. Front Oncol. 2019;9:690. https://doi.org/10.3389/fonc.2019.00690.
32. Zhang H, Fu C, Fan M, Lu L, Chen Y, Liu C, et al. Reduction of inter-observer variability using MRI and CT fusion in delineating of primary tumor for radiotherapy in lung cancer with atelectasis. Front Oncol. 2022;12:841771.
33. Finazzi T, Palacios MA, Haasbeek CJA, Admiraal MA, Spoelstra FOB, Bruynzeel AME, et al. Stereotactic MR-guided adaptive radiation therapy for peripheral lung tumors. Radiother Oncol. 2020;144:46–52.
34. Kang HJ, Kwak YK, Kim M, Lee SJ. Application of real-time MRI-guided linear accelerator in stereotactic ablative body radiotherapy for non-small cell lung cancer: one step forward to precise targeting. J Cancer Res Clin Oncol. 2022;148(11):3215–23.
35. Chen H, Senan S, Nossent EJ, Boldt RG, Warner A, Palma DA, et al. Treatment-related toxicity in patients with early-stage non-small cell lung cancer and coexisting interstitial lung disease: a systematic review. Int J Radiat Oncol Biol Phys. 2017;98(3):622–31.
36. Kishan AU, Ma TM, Lamb JM, Casado M, Wilhalme H, Low DA, et al. Magnetic resonance imaging-guided vs computed tomography-guided stereotactic body radiotherapy for prostate cancer: the MIRAGE randomized clinical trial. JAMA Oncol. 2023;9:365. https://doi.org/10.1001/jamaoncol.2022.6558.
37. Ng SSW, Ning MS, Lee P, McMahon RA, Siva S, Chuong MD. Single-fraction stereotactic body radiation therapy: a paradigm during the coronavirus disease 2019 (COVID-19) pandemic and beyond? Adv Radiat Oncol. 2020;5(4):761–73.
38. Finazzi T, van de Koste JRS, Palacios MA, Spoelstra FOB, Slotman BJ, Haasbeek CJA, et al. Delivery of magnetic resonance-guided single-fraction stereotac-

tic lung radiotherapy. Phys Imaging Radiat Oncol. 2020;14:17–23.
39. Palacios MA, Verheijen S, Schneiders FL, Bohoudi O, Slotman BJ, Lagerwaard FJ, et al. Same-day consultation, simulation and lung stereotactic ablative radiotherapy delivery on a magnetic resonance-linac. Phys Imaging Radiat Oncol. 2022;24:76–81.
40. Timmerman RD, Paulus R, Pass HI, Gore EM, Edelman MJ, Galvin J, et al. Stereotactic body radiation therapy for operable early-stage lung cancer. JAMA Oncol. 2018;4(9):1263–6.
41. Ball D, Mai GT, Vinod S, Babington S, Ruben J, Kron T, et al. Stereotactic ablative radiotherapy versus standard radiotherapy in stage 1 non-small-cell lung cancer (TROG 09.02 CHISEL): a phase 3, open-label, randomised controlled trial. Lancet Oncol. 2019;20(4):494–503.
42. Onishi H, Araki T, Shirato H, Nagata Y, Hiraoka M, Gomi K, et al. Stereotactic hypofractionated high-dose irradiation for stage I nonsmall cell lung carcinoma: clinical outcomes in 245 subjects in a Japanese multi-institutional study. Cancer. 2004;101(7):1623–31.
43. Guckenberger M, Wulf J, Mueller G, Krieger T, Baier K, Gabor M, et al. Dose–response relationship for image-guided stereotactic body radiotherapy of pulmonary tumors: relevance of 4D dose calculation. Int J Radiat Oncol Biol Phys. 2009;74(1):47–54.
44. Onimaru R, Fujino M, Yamazaki K, Onodera Y, Taguchi H, Katoh N, et al. Steep dose–response relationship for stage I non-small-cell lung cancer using hypofractionated high-dose irradiation by real-time tumor-tracking radiotherapy. Int J Radiat Oncol Biol Phys. 2008;70(2):374–81.
45. Zhang J, Yang F, Li B, Li H, Liu J, Huang W, et al. Which is the optimal biologically effective dose of stereotactic body radiotherapy for stage I non-small-cell lung cancer? A meta-analysis. Int J Radiat Oncol Biol Phys. 2011;81(4):e305–16.
46. Finazzi T, Palacios MA, Spoelstra FOB, Haasbeek CJA, Bruynzeel AME, Slotman BJ, et al. Role of on-table plan adaptation in MR-guided ablative radiation therapy for central lung tumors. Int J Radiat Oncol Biol Phys. 2019;104(4):933–41.
47. Datta A, Aznar MC, Dubec M, Parker GJM, O'Connor JPB. Delivering functional imaging on the MRI-Linac: current challenges and potential solutions. Clin Oncol. 2018;30(11):702–10.
48. Chin S, Eccles CL, McWilliam A, Chuter R, Walker E, Whitehurst P, et al. Magnetic resonance-guided radiation therapy: a review. J Med Imaging Radiat Oncol. 2020;64(1):163–77.
49. van Houdt PJ, Yang Y, van der Heide UA. Quantitative magnetic resonance imaging for biological image-guided adaptive radiotherapy. Front Oncol. 2021;10:615643. https://doi.org/10.3389/fonc.2020.615643.
50. Shaverdian N, Tenn S, Veruttipong D, Wang J, Hegde J, Lee C, et al. The significance of PTV dose coverage on cancer control outcomes in early stage non-small cell lung cancer patients treated with highly ablative stereotactic body radiation therapy. Br J Radiol. 2016;89:20150963.
51. McDonald BA, Zachiu C, Christodouleas J, Naser MA, Ruschin M, Sonke JJ, et al. Dose accumulation for MR-guided adaptive radiotherapy: from practical considerations to state-of-the-art clinical implementation. Front Oncol. 2022;12:1086258.
52. Lim SY, Tran A, Tran ANK, Sobremonte A, Fuller CD, Simmons L, et al. Dose accumulation of daily adaptive plans to decide optimal plan adaptation strategy for head-and-neck patients treated with MR-Linac. Med Dosim. 2022;47(1):103–9.
53. Adebahr S, Collette S, Shash E, Lambrecht M, Le Pechoux C, Faivre-Finn C, et al. LungTech, an EORTC phase II trial of stereotactic body radiotherapy for centrally located lung tumours: a clinical perspective. Br J Radiol. 2015;88(1051):20150036.
54. Giuliani M, Mathew AS, Bahig H, Bratman SV, Filion E, Glick D, et al. SUNSET: stereotactic radiation for ultracentral non-small-cell lung cancer—a safety and efficacy trial. Clin Lung Cancer. 2018;19(4):e529–32.
55. Henke LE, Olsen JR, Contreras JA, Curcuru A, DeWees TA, Green OL, et al. Stereotactic MR-guided online adaptive radiation therapy (SMART) for ultracentral thorax malignancies: results of a phase 1 trial. Adv Radiat Oncol. 2019;4(1):201–9.
56. Regnery S, Buchele C, Weykamp F, Pohl M, Hoegen P, Eichkorn T, et al. Adaptive MR-guided stereotactic radiotherapy is beneficial for ablative treatment of lung tumors in high-risk locations. Front Oncol. 2022;11:757031.
57. Sim AJ, Kaza E, Singer L, Rosenberg SA. A review of the role of MRI in diagnosis and treatment of early stage lung cancer. Clin Transl Radiat Oncol. 2020;24:16–22.
58. Wild JM, Marshall H, Bock M, Schad LR, Jakob PM, Puderbach M, et al. MRI of the lung (1/3): methods. Insights Imaging. 2012;3(4):345–53.
59. Groot Koerkamp ML, Vasmel JE, Russell NS, Shaitelman SF, Anandadas CN, Currey A, et al. Optimizing MR-guided radiotherapy for breast cancer patients. Front Oncol. 2020;10:1107.
60. Kurz C, Buizza G, Landry G, Kamp F, Rabe M, Paganelli C, et al. Medical physics challenges in clinical MR-guided radiotherapy. Radiat Oncol. 2020;15(1):93.
61. Eccles CL, Adair Smith G, Bower L, Hafeez S, Herbert T, Hunt A, et al. Magnetic resonance imaging sequence evaluation of an MR Linac system; early clinical experience. Tech Innov Patient Support Radiat Oncol. 2019;12:56–63.
62. Serai SD, Hu HH, Ahmad R, White S, Pednekar A, Anupindi SA, et al. Newly developed methods for reducing motion artifacts in pediatric abdominal MRI: tips and pearls. AJR Am J Roentgenol. 2020;214(5):1042–53.
63. Kocaoglu M, Pednekar AS, Wang H, Alsaied T, Taylor MD, Rattan MS. Breath-hold and free-breathing quantitative assessment of biventricular volume and function using compressed SENSE: a clinical valida-

tion in children and young adults. J Cardiovasc Magn Reson. 2020;22(1):54.
64. Hu P, Li X, Liu W, Yan B, Xue X, Yang F, et al. Dosimetry impact of gating latency in cine magnetic resonance image guided breath-hold pancreatic cancer radiotherapy. Phys Med Biol. 2022;67(5):055008.
65. Botman R, Tetar SU, Palacios MA, Slotman BJ, Lagerwaard FJ, Bruynzeel AME. The clinical introduction of MR-guided radiation therapy from a RTT perspective. Clin Transl Radiat Oncol. 2019;18:140–5.
66. Chen F, Cheng YK, Chiang CH, Lu TY, Huang CJ. Simultaneous treatment of two lung cancer lesions with stereotactic MR-guided adaptive radiation therapy: a case report. Medicine. 2023;102(2):e32626.
67. Eze C, Lombardo E, Nierer L, Xiong Y, Niyazi M, Belka C, et al. MR-guided radiotherapy in node-positive non-small cell lung cancer and severely limited pulmonary reserve: a report proposing a new clinical pathway for the management of high-risk patients. Radiat Oncol. 2022;17(1):43.
68. Keall PJ, Brighi C, Glide-Hurst C, Liney G, Liu PZY, Lydiard S, et al. Integrated MRI-guided radiotherapy—opportunities and challenges. Nat Rev Clin Oncol. 2022;19(7):458–70.

# MRI-Guided Radiation Therapy for Prostate Cancer: Less Is More When Technology and Evidence Intersect

16

Luca Valle, Rosalyne Westley, Filippo Alongi, Sean Sachdev, and Amar Kishan

## Disease Overview

Prostate cancer is the most common non-cutaneous malignancy affecting men, with an estimated 268,490 cases diagnosed in the United States in 2022 [1]. The definitive treatment options for localized prostate cancer include radical prostatectomy (RP) or radiation therapy (RT), both of which have been found to yield equivalent long-term survival outcomes [2]. The natural history of prostate cancer is protracted, with very high overall survival rates and a number of salvage treatment options available for men who undergo management with either definitive RP [3] or definitive RT [4].

RT for the definitive treatment of prostate cancer can be delivered to the prostate and, if warranted, to the surrounding pelvic lymph nodes as external beam radiotherapy (EBRT) or to the prostate alone with brachytherapy and is often delivered in conjunction with hormonal therapy, depending on the prognostic risk stratification group of the patient [5].

For men who elect to receive definitive RT, the treatment of prostate cancer has improved dramatically due to advances in imaging and technology as well as improvements in our understanding of the biology of prostate cancer, particularly regarding the proposed low $\alpha/\beta$ ratio of prostate cancer cells [6] (per the linear quadratic model). As a consequence, there has been a trend toward both radiation dose escalation of the prostate gland and simultaneously shortening the delivery of definitive prostate cancer RT by delivering larger doses of radiation per fraction through hypofractionation.

Seminal trials have demonstrated the merits of dose escalation in prostate cancer. For example, in MRC RT01, 74 Gy in 37 fractions delivered alongside androgen deprivation therapy (ADT) showed superior biochemical control rates com-

L. Valle
Department of Radiation Oncology, David Geffen School of Medicine, University of California Los Angeles, Los Angeles, CA, USA

Veterans Affairs Greater Los Angeles Healthcare System, Los Angeles, CA, USA
e-mail: LFValle@mednet.ucla.edu

A. Kishan (✉)
Department of Radiation Oncology, David Geffen School of Medicine, University of California Los Angeles, Los Angeles, CA, USA
e-mail: AUKishan@mednet.ucla.edu

R. Westley
Clinical Oncology Academic Urology Unit, Institute of Cancer Research, The Royal Marsden Hospital, London, UK
e-mail: Rosalyne.Westley@rmh.nhs.uk

F. Alongi
Department of Advanced Radiation Oncology, IRCCS Ospedale Sacro Cuore Don Calabria, Negrar, Verona, Italy

University of Brescia, Brescia, Italy

S. Sachdev
Department of Radiation Oncology, Northwestern University Feinberg School of Medicine, Chicago, IL, USA
e-mail: Sean.Sachdev@nm.org



I. J. Das et al. (eds.), *A Practical Guide to MR-Linac*, https://doi.org/10.1007/978-3-031-48165-9_16

pared to the standard dose fractionation of 64 Gy in 32 fractions [5]. Parallel data from the RTOG 0126 study in the United States established the 7-year biochemical control benefit to dose escalation to 79.2 Gy over 70.2 Gy (70% vs 55%, $p < 0.001$) [7], thereby establishing the former as the standard of care dosing regimen for men with all stages of prostate cancer.

In parallel to these dose escalation studies, there has been similar interest in harnessing the differential response of prostate cancer cells to larger radiation doses per fraction through hypofractionation. In the UK, the CHHiP trial showed that hypofractioned radiotherapy using 60 Gy in 20 fractions was non-inferior to 74 Gy in 37 fractions [8]. Additionally, RTOG 0415 demonstrated that 70 Gy in 28 fractions was non-inferior to 73.8 Gy in 41 fractions in low-risk prostate cancer, although a late increase in gastrointestinal and genitourinary adverse events were observed in the hypofractionated group [9]. PROFIT then demonstrated that the treatment of low- to intermediate-risk prostate cancer with 60 Gy in 20 fractions was non-inferior to conventional radiotherapy with 78 Gy in 39 fractions over 8 weeks in the absence of androgen deprivation therapy (ADT) [10].

Given the aforementioned level 1a evidence, moderately hypofractionated radiotherapy to the prostate is acknowledged to be a standard treatment option, yet the boundaries continue to be pushed. HYPO-RT-PC demonstrated that the treatment of prostate cancer with ultrahypofractionated radiotherapy (also known as stereotactic body radiation therapy [SBRT]) is non-inferior to conventionally fractionated prostate cancer in the context of failure-free survival [11] and patient-reported outcomes also supported the long-term tolerability of SBRT for these patients [12]. PACE B has also tested the efficacy and toxicity profile of ultrahypofractionated radiotherapy to the prostate with 36.35 Gy in 5 fractions over 2 weeks, demonstrating the safety and efficacy of ultrahypofractionated radiotherapy as well as similar physician-reported GU and GI toxicity between SBRT and conventional fractionation [13, 14].

The advent of hypo- and ultrahypofractionated courses have added considerable convenience [15] and cost savings [16] for healthcare systems. However, the move to delivering ultrahypofractionated regimens has only become possible due to ongoing advances in radiation technology. SBRT (Sterotactic Body Radiation Therapy) is technically more challenging to deliver, requiring precise geometric targeting with in-room image guidance. And while the toxicity results so far have been promising, given the favorable disease-specific outcomes observed in men following prostate radiotherapy, patients tend to live many years after completing their treatment, and thus, minimization of the long-term sequelae of treatment remains a significant priority in the delivery of modern definitive prostate RT.

In addition to deploying RT in the definitive treatment setting, it is also indicated in the salvage setting in the event of a biochemical recurrence (BCR) following RP. In these scenarios, the prostate bed serves as the primary target volume, with the option to electively include draining pelvic lymph nodes in the radiation field depending on patient- and tumor-specific features. While RT delivery with conventional fractionation remains the mainstay of treatment for these patients, there has been parallel interest [17, 18] in shortening RT courses in the postoperative setting as well.

Regardless of the treatment setting, the ultimate aim of RT delivery is to maximize the therapeutic ratio by ensuring robust tumor control by maintaining or increasing dose to the target while simultaneously minimizing treatment toxicity by decreasing the radiation dose to surrounding normal tissues. Further advances in radiotherapy technology is necessary to achieve this objective.

The hybridization of a magnetic resonance imaging (MRI) scanner and the modern linear accelerator has resulted in a new form of prostate cancer treatment, termed MRI-guided RT (MRIgRT). MRIgRT may offer several advantages in the treatment of prostate cancer compared to the more standard modalities used to deliver image-guided EBRT, such as with a CT-guided linear accelerator, Calypso™ tracking markers, and CyberKnife™. There are currently four MRIgRT systems, two of which are available commercially. The Unity™ MR-Linac (Elekta AB, Stockholm, Sweden) combines a

1.5-Tesla Philips Magnetic Resonance (MR) scanner (Philips Medical Systems, Best, the Netherlands) and a 7-MV FFF Elekta linear accelerator [19], while the MRIdian™ system (ViewRay Inc., Mountain View, CA) combines a 6-MV linear accelerator with a 0.35 T MR imaging (Siemens, Munich, Germany) system [20]. Both of these systems have been studied and implemented for the treatment of prostate cancer in both the definitive and the post-prostatectomy setting and will be the focus of this chapter.

## Overcoming Challenges with Traditional Approaches

There are many challenges associated with traditional CT-guided delivery of radiation for prostate cancer, particularly in the era of ultrahypofractionated treatments. Given the close proximity of the prostate to surrounding organs at risk (OARs), particularly the bladder and rectum, dosimetric tradeoffs between prostate target coverage and sparing of OARs are frequent. The urethra is another critical OAR that courses within the prostate gland itself and can be challenging to visualize and spare appropriately with CT-guided EBRT. Additional challenges that limit the therapeutic benefits of the more traditional approaches to delivering prostate RT include concerns with intrafraction motion management, the inability to measure dose accumulation, exposure to ionizing radiation during image-guided RT (IGRT), and the inability to functionally evaluate and capitalize on tumor response. Each of these considerations will be discussed in turn, as well as the limited scope of ultrahypofractionated RT in recurrent prostate cancer and how these hurdles may be overcome using MRIgRT.

### Motion Management

It is well documented that the prostate, seminal vesicles, and OARs are subject to both interfraction and intrafraction motion. There are continuous changes in shape and position of the bladder, bowel, and rectum which in turn influence the position of the prostate and the seminal vesicles [21, 22]. Moreover, a recent systemic review reported that although rectal and bladder filling status was the main cause for seminal vesicle motion, no apparent effect on magnitude of motion was seen when different bladder and rectal preparation protocols were used [23], which can render motion challenging to predict a priori. In addition to OAR motion, the prostate itself has also been shown to undergo changes in volume and dimensions during a course of radiotherapy [24, 25], with the addition of androgen deprivation therapy further altering prostate gland size and tumor characteristics [26]. Previous image-guided techniques such as CyberKnife, cone beam imaging, and fiducial matching have allowed for the treatment to be matched to the position of the target, but never to the position and shape of both the target and adjacent OARs online. Even the 4D detection of implanted electromagnetic transponders [27] (the Calypso™ system) cannot provide true "real-time" motion evaluation that is actionable in the event that surrounding OARs encroach upon the prostate target volume. A further limitation of the Calypso system is that the implanted transponders may preclude the use of MRI in posttreatment assessment [28].

One approach for mitigating target motion due to anatomic changes between fractions is adaptive radiotherapy. The MRIgRT system allows for the original offline-generated treatment plan to be adapted through recontouring or a daily shift to account for any changes in anatomy as seen on the MRI taken on the day of treatment (Fig. 16.1). The new plan which is generated online in real time is then delivered to the patient, negating any interfraction deformations that may have occurred since the planning scan and between each day of radiotherapy. The ability to simultaneously visualize and customize treatments based on deformation of both tumor targets and OARs represents a major theoretical advantage of adaptive planning with MRIgRT.

Depending on the treatment workflow adopted, there is also the potential to account for intrafraction organ motion which occurs during the delivery of a single fraction of radiotherapy

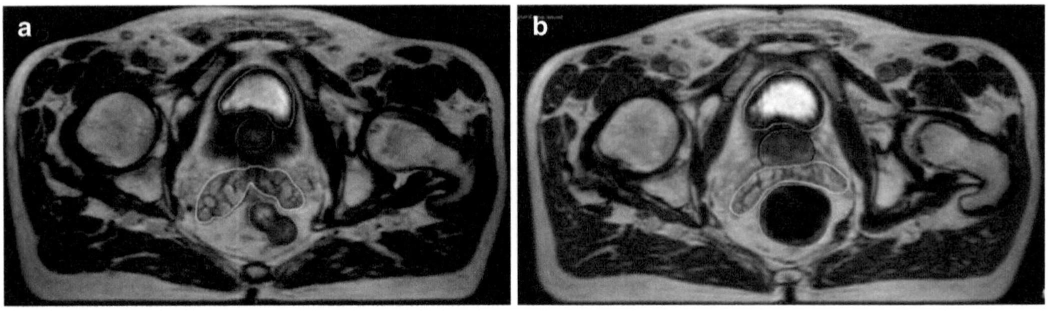


**Fig. 16.1** Changes in anatomy as visualized on the daily MRI as the rectum fills with air and squashes the seminal vesicles anteriorly and laterally. (**a**) axial MR image without rectal distention. (**b**) axial MR image with rectal distention

[29]. This is achieved by creating a gating margin around the prostate CTV and assigning a tolerance for the acceptable proportion of the CTV that is allowable outside of the gating boundary. The intrafraction motion of the prostate and surrounding OARs is then monitored by sagittal images taken during treatment at four frames per second, which might be more aptly termed "real-time" monitoring. Moreover, as part of the automatic beam hold feature of the MRIdian device, treatment is paused if intrafraction motion leads to unacceptable displacement of the prostate gland from the original setup position and resumes when the prostate returns to the original setup position. This provides additional reassurance regarding the accuracy of treatment delivery when treating with ultrahypofractionated regimens [30]. These gating boundaries and tolerance thresholds can be established and modified at the discretion of the treating physician. These features of MRIgRT thus provide a possible advantage over platforms such as Cyberknife™, on which "real-time" imaging and beam compensation occurs via kv planar images taken every 30–60 s depending on the threshold set.

While gating is primarily a feature of the ViewRay device, as of early 2023, the Unity device received FDA 510(k) clearance for its "Comprehensive Motion Management with True 3D Tracking and Automatic Gating" technology, which will allow tracking, gating, and baseline shift correction. At the time of writing, this is not widely clinically implemented, and thus, the use of gating to account for larger degrees of intrafraction motion on the Unity device is more limited. However, Unity does allow the radiotherapy plan to be shifted after the initial recontouring phase to match to the most up-to-date position of the prostate as seen on MRI taken immediately prior to treatment. This accounts for any motion that may have occurred during the contouring and plan adaption phase of the workflow [31]. This is known as adapt-to-position and is outlined in greater detail in "On Table Adaption for Unity ("Unity") and MRIdian ("ViewRay") below.

## Inability to Measure Dose Accumulation

Conventional radiotherapy delivery approaches assume that the organ motion is static during treatment, leading to potential inaccuracies in the assessment of true delivered dose to targets and OARs. Volumetric imaging acquired during the treatment of a patient with a fraction of radiotherapy also offers the opportunity for more accurate "real-time" calculation of the dose accumulation that occurs throughout a treatment course [32]. This enables the opportunity to consider prostate dose escalation for future fractions if the dose received by the surrounding OARs over the preceding fractions is below the allowable dose constraints. It similarly enables dose de-escalation in patients with high-risk anatomy, for whom daily dosimetry to critical OARs is higher than expected. A patient with small bowel adjacent to the prostate, for example, may be more safely treated with MRIgRT as the accumulated dose to the bowel can be calculated throughout the treatment and the prescription dose or PTV coverage modified, if required [31].

## Exposure to Ionizing Radiation During Image-Guided RT

MRI as a tool for image-guided radiotherapy removes the additional ionizing radiation that patients are exposed to using other forms of image-guided radiotherapy. Although the dose for daily X-ray-based image guidance may be considered negligible (8–18 mGy daily) in comparison to the prescription doses for therapeutic RT, this increases if X-ray tracking is used during each fraction over a whole treatment course [33]. Repeated CT-based imaging has thus sometimes been avoided due to the risks associated with increased integral radiation dosing. In comparison, MRIgRT allows for constant motion monitoring throughout treatment and offers the ability to visualize the position of the prostate in relation to the CTV without the concomitant delivery of any additional ionizing radiation.

## Inability to Measure Functional Response

Functional MRI of the prostate can detect changes in the microenvironment of the tumor and surrounding stroma, but with conventional RT delivery and image guidance, the opportunity to assess functional changes in prostate tumor biology throughout treatment becomes forfeit. The 1.5 T MRI of the Elekta Unity MR-Linac offers the potential to guide personalized RT due its biomarker capabilities, and there exists the potential to establish the cumulative effect of RT on cancer cells and surrounding normal tissue [34]. Intuitively, the stronger the magnet, the greater the signal-to-noise ratio and the greater the separation between water, fat, and other metabolites meaning that biomarker techniques become more effective (fat suppression and spectroscopy) [35]. Although yet to be clinically implemented, it is hoped that quantitative and qualitative intrafraction visualization of the real-time biological impacts of prostate radiotherapy will help to tailor the consecutive dose per fraction [34].

## Limited Scope of Ultrahypofractionation in Recurrent Prostate Cancer

As described above, ultrahypofractionation is a well-established treatment technique for the delivery of definitive RT to localized disease. However, RT for prostate cancer is delivered in a variety of clinical settings across the entire continuum of prostate cancer, from early-stage disease to oligometastatic disease to widely metastatic castrate-resistant disease. Thus, there has naturally been interest and enthusiasm in bringing the benefits of ultrahypofractionation to these treatment settings as well and MRIgRT may offer the ideal platform to usher in the expanding scope of ultrahypofractionation in prostate cancer.

Post-prostatectomy RT is one such setting where ultrahypofractionated radiation regimens are being explored. Post-prostatectomy RT is delivered in either the adjuvant setting (when a constellation of high-risk features including node-positive disease, pT3/4 disease, positive margins, or high genomic classifier scores are present at the time of surgery) or as early salvage RT for a rising PSA following RT. In the event of biochemical recurrence, postoperative RT provides the sole curative option after RP. Challenges of hypofractionation RT in this context include the highly deformable "negative space" of the postoperative bed, raising concerns related to elevated toxicity with hypofractionation. As a result, postoperative RT (most commonly delivered as early salvage RT [36–38]) is standardly delivered over 9 weeks of treatment, and thus, the convenience of RT in the definitive setting has yet to be fully realized in the postoperative setting. As will be discussed below, MRIgRT may offer a solution [39].

Analogous to recurrence following RP, patients who underwent definitive RT can also develop recurrent disease in the prostate gland, regionally, or distantly. While less than 15% of men treated with EBRT will develop a biochemical recurrence at 5 years, cure is still attainable for the patients who recur locally [4]. When imaged with MRI or positron emission tomography/computed tomography (PET/CT), local recurrences can be detected in up to 33% of these men before the PSA has risen above 2.0 ng/mL [40]. Moreover, PSMA PET in combination with mpMRI have considerably improved the detection of local-only recurrent prostate cancer [41–44], which invites the therapeutic dilemma of how best to salvage these local failures, if at all. A recent meta-analysis of local salvage therapies identified reirradiation with either ultrahypofractionated EBRT or HDR brachytherapy as the two local salvage options that carried the lowest aggregate risks of GU and GI toxicity [4], thus increasing enthusiasm for offering curative local reirradiation in appropriately selected patients with a low risk of distant metastases and a favorable toxicity course with their initial course of radiation. One of the larger SBRT series reporting long-term outcomes appreciated the importance of sparing normal tissue with reirradiation and therefore used no PTV margins out of concern for inciting toxicity with local reirradiation [45]. Intuitively, the risk of toxicity could be lowered even further when PTV margin reduction *and* MRIgRT are combined. Of note, intrafractional planar imaging which is often used for triangulation of fiducial markers as a surrogate of prostate location during ultrahypofractioned treatments can prove to be challenging with numerous permanently implanted LDR seeds, highlighting the possibility of more benefit with MRIgRT techniques in the population of patients previously treated with LDR brachytherapy who are undergoing local salvage reirradiation.

As described above, CT-based treatment delivery has been employed for safe reirradiation of localized disease, either with a whole-gland approach or a focal-gland approach [46]. Given the promise of MRIgRT in the upfront treatment setting, there has been interest in evaluating its role in the setting of local salvage reirradiation. One-year clinical data from 37 men who had intraprostatic relapsed disease re-treated with MRIgRT found the treatment to be well tolerated with no >2 acute toxicities and only 1 grade 3 late toxicity. In this series, the GTV was treated with a 3-mm isotropic margin to create the PTV [47]. Applying the same concept, gross local relapses following prostatectomy and conventionally fractionated salvage RT might also be visualized [48] and treated with SBRT on the MR-Linac while sparing the dose to the surrounding organs.

## Contouring, PTV, and OAR

### Target Volumes

Standardized guideline-concordant target volume delineation is followed for the prostate [49] and pelvic lymph nodes [50] in the definitive localized setting [51] as well as for the prostate bed and pelvic lymph nodes in the postoperative setting [52]. However, because of superior image resolution, the contouring of the prostate on MR imaging allows for more accurate and confident contouring which in turn lends to reduced interobserver variability and greater speed of contouring compared to CT imaging [53]. This improved accuracy results in the contouring of smaller prostate volumes [54], consequently leading to more attainable target coverage and OAR avoidance, as has been observed for other pelvic malignancies [55].

The soft tissue differentiation of MRI also enables the dominant intraprostatic lesion(s) (DIL[s]) to be visualized and accurately targeted during planning and treatment (Fig. 16.2). This process has been adopted on the MRIdian system using rigid propagation while the lesion can be visualized and recontoured directly on the Elekta Unity MR-Linac [56], given higher magnet strength on this device.

The benefit of being able to contour and target the dominant lesion was demonstrated in the FLAME trial, where it was shown that a simultaneous integrated boost to the dominant intraprostatic lesion(s) improves biochemical disease-free survival (bDFS) without impacting toxicity and quality of life in localized intermediate- and high-risk patients [57]. While FLAME employed conventional fractionation, Kerkmeijer et al. are also exploring focal lesion boosting with SBRT in the hypo-FLAME trial, which recently demonstrated favorable rates of acute grade 2 GU and GI toxicity of 34% and 5%, respectively, when treating the entire prostate to 35 Gy and boosting MRI-defined gross disease to 50 Gy in 5 fractions [58]. Long-

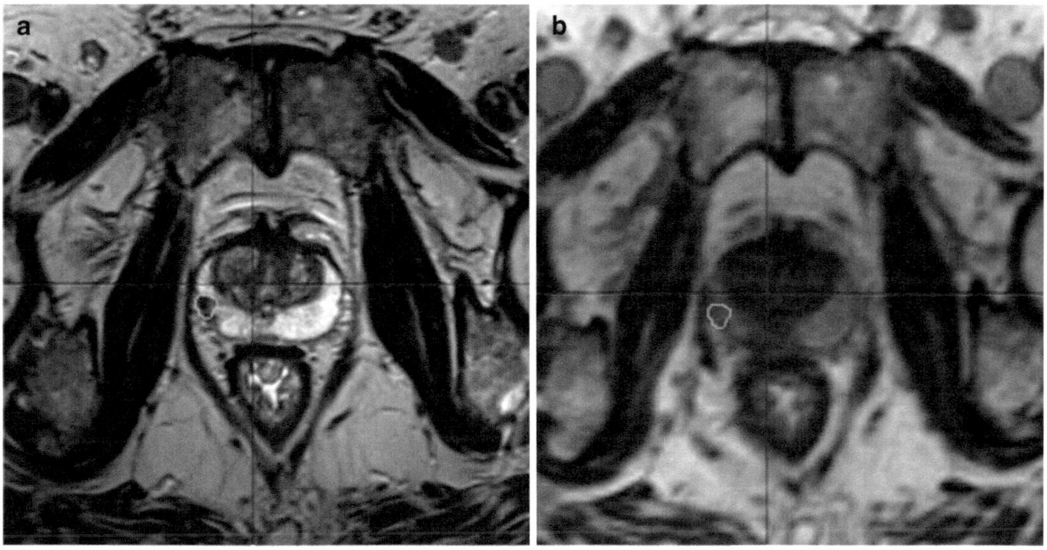


**Fig. 16.2** Visualization of the dominant intraprostatic lesions seen on (**a**) fused diagnostic MRI and (**b**) 0.35 T planning image

term oncologic outcomes associated with these dose-escalated treatments are eagerly anticipated. Further extending the reach of DIL boosting in the treatment of definitive localized prostate cancer, the recently published 5-year outcomes from the DELINEATE [59] trial establish the safety and efficacy of these treatments in the moderately hypofractionated setting as well, where simultaneous integrated boosting to 67 Gy in 20 fractions was similarly found to be safe and effective.

## PTV

In the treatment of prostate cancer, the whole prostate is standardly treated as the clinical target volume (CTV) with a surrounding PTV margin added to account for setup error, delineation error, and inter- and intrafraction motion. This PTV margin has varied across prospective clinical trials, but usually ranges from 5 to 7 mm, with allowances for reduction in PTV margin to no less than 3 mm posteriorly, where overlap with the rectal OAR is all but assured. While MRIgRT does not alter CTV delineation, it does enable considerable reductions in PTV margin. For example, in the MIRAGE trial which will be discussed extensively below, the safety and efficacy of PTV margin reduction to 2 mm was demonstrated in a prospective setting, with commensurate decreases in both GU and GI toxicity [60].

## OAR

The advantages of MRIgRT in the pelvis are further appreciated by the superior visualization of OARs. Specifically, the boundaries of the bladder and the rectum are difficult to visualize on CT but can be more easily visualized on MRI (Fig. 16.3), further enhancing OAR avoidance during treatment planning.

Urethral avoidance also remains paramount to ensure acceptable rates of late toxicity following radiotherapy [61]. However, a limitation of external beam radiotherapy compared to HDR brachytherapy is suboptimal visualization of the urethra. The urethra can however be visualized on the Unity device (Fig. 16.4a) and its delineation has been adopted to reduce the dose it receives and hence improve GU toxicity. Furthermore, on the MRIdian device, investigational 0.35 T MRI sequences are under development that would also enable superior visualization and delineation of the urethra in an effort to minimize long-term genitourinary toxicity (Fig. 16.4b) [62].

Vascular structures can also be better visualized on MRI, such that treatment with MRIgRT

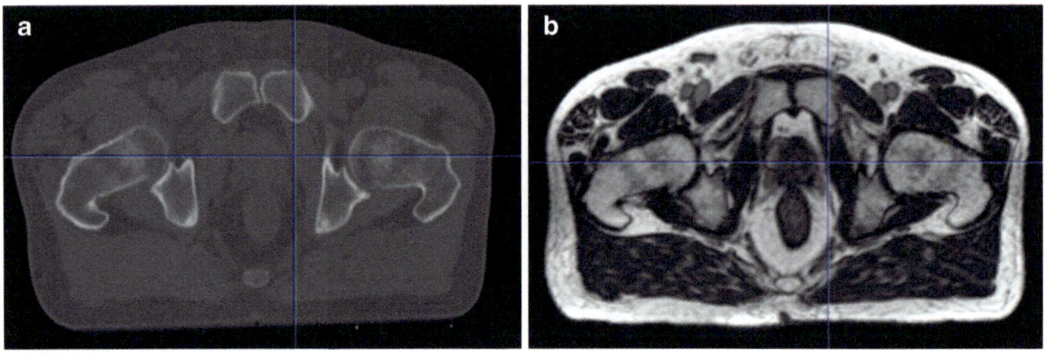

**Fig. 16.3** Prostate and the surrounding normal tissue as visualized with (**a**) CT-based techniques vs (**b**) MR-based techniques

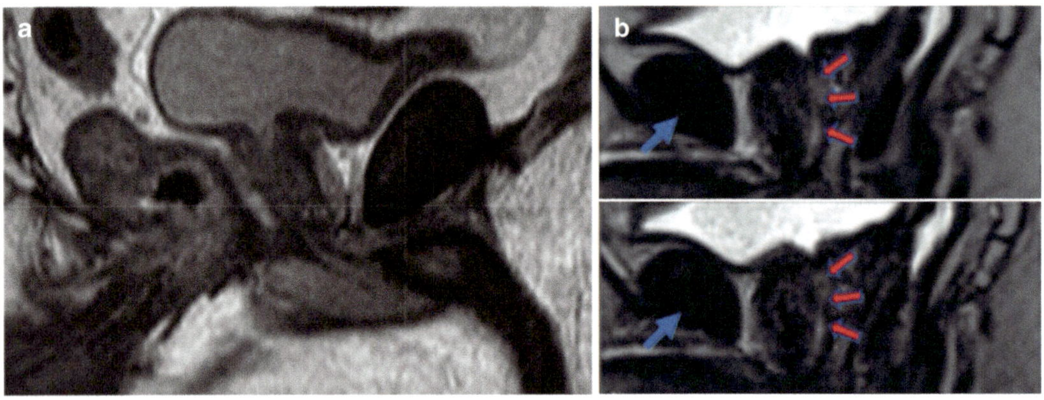

**Fig. 16.4** Visualization of the urethra on (**a**) the Unity device as observed on the 1.5 T T2 3D Tra and (**b**) the MRIdian MRIgRT platforms. MRIdian sequences with urethral visualization include MRIgRT 3D HASTE (top) and MRIgRT 3D TSE (bottom). (MRIdian images borrowed from Pham et al. [63])

offers the potential to outline and limit dose to such structures [33]. For example, sparing of the internal pudendal artery has been shown to improve sexual dysfunction [63], which is now the most prevalent long-term side effect of prostate radiotherapy [64]. The success of the sparing of critical vascular structures has been previously demonstrated and thus highlights a need for MRI-based radiation planning [65], where such structures can be reliably visualized. MRIgRT not only allows for offline MR-guided planning but also the ability to visualize these vital structures on the day of treatment and therefore enable maximal sparing of dose to important vascular structures, particularly when adaptive planning is pursued.

## DVH Constraints

As with other anatomic treatment sites in radiation oncology, there exist no universally accepted or standardized dose constraints for ultrahypofractionated prostate RT. However,

**Table 16.1** Dosimetric objectives for target volumes and OARs for definitive ultrahypofractionated treatment of intact prostate cancer on the MIRAGE clinical trial

| Structure | Dosimetric objective |
|---|---|
| PTV | V40 Gy ≥95% |
| PTV | V42 Gy <30% |
| Rectum | V20 Gy <50% |
| Rectum | V32 Gy <20% |
| Rectum | V36 Gy <10% |
| Rectum | V40 Gy <5% |
| Rectum | V42 Gy <0.03 cc |
| Rectum | V38 Gy <2 cc |
| Rectal wall | V24 Gy <50% |
| Posterior rectum | Max <16 Gy |
| Anal canal | V30 Gy <0.03 cc |
| Anal canal | V20 Gy <5 cc |
| Small bowel | V25 Gy <0.03 cc |
| Small bowel | V20 Gy <30 cc |
| Sigmoid colon | Max <38 Gy |
| Sigmoid colon | V30 Gy <1 cc |
| Sigmoid colon | V25 Gy <30 cc |
| Bladder | V42 Gy <0.03 cc |
| Bladder | V39 Gy <2 cc |
| Bladder | V20 Gy <40% |
| Bladder | V40 Gy <5% |
| Right femoral head | V20 Gy <5 cc |
| Left femoral head | V20 Gy <5 cc |
| Skin | V15 Gy <0.03 cc |
| Penile bulb | V24.8 Gy <5% |
| Penile bulb | Mean <16 Gy |
| Urethra | V42 Gy <0.03 cc |

dose constraints employed in prospective studies where MRIgRT was employed are recapitulated in Tables 16.1, 16.2 and 16.3 below. The dose constraints in Table 16.1 reference those used to treat intact localized prostate cancer with ultrahypofractionated RT on the MIRAGE trial. The dose constraints in Table 16.2 reference those used to guide treatment of intact localized prostate cancer with moderately hypofractionated RT on the GARUDA trial. And the dose constraints in Table 16.3 reference those used to guide ultrahypofractionated salvage RT to the prostate bed on the EXCALIBUR trial. Corresponding DVHs for approved plans are shown in Figs. 16.5, 16.6 and 16.7.

**Table 16.2** Dosimetric objectives for target volumes and OARs for definitive moderately hypofractionated treatment of intact prostate cancer on the GARUDA clinical trial

| Structure | Dosimetric objective |
|---|---|
| PTV | V60 Gy ≥95% |
| Rectum | V24 Gy <80% |
| Rectum | V32 Gy <65% |
| Rectum | V40 Gy <50% |
| Rectum | V48 Gy <35% |
| Rectum | V52 Gy <30% |
| Rectum | V56 Gy <15% |
| Rectum | V60 Gy <3% |
| Rectal wall | V30 Gy <50% |
| Posterior rectum | Max <24 Gy |
| Small bowel | V50 Gy <17 cc |
| Bladder | V40 Gy <50% |
| Bladder | V48 Gy <25% |
| Bladder | V59 Gy <5% |
| Right femoral head | V50 Gy <5% |
| Left femoral head | V50 Gy <5% |
| Penile bulb | V50 Gy <50% |
| Penile bulb | V60 Gy <10% |
| Urethra | Max <63 Gy |

**Table 16.3** Dosimetric objectives for target volumes and OARs for postoperative ultrahypofractionated salvage treatment of recurrent prostate cancer on the EXCALIBUR clinical trial

| Structure | Dosimetric objective |
|---|---|
| PTV | V100% ≥95% |
| Rectum | V27.5 Gy <45% |
| Rectum | V32.5 Gy <30% |
| Rectum | V33.75 Gy <25% |
| Rectal wall | V35.7 Gy <0.03 cc |
| Rectal wall | V24 Gy <50% |
| Anal canal | V30 Gy <0.03 cc |
| Anal canal | V20 Gy <5 cc |
| Small bowel | V25 Gy <0.03 cc |
| Small bowel | V20 Gy <30 cc |
| Sigmoid colon | V30 Gy <1 cc |
| Bladder | V35.7 Gy <0.03 cc |
| Bladder | V32.5 Gy <35% |
| Bladder | V30 Gy <45% |
| Bladder wall | V30 Gy <16 cc |
| Bladder wall | V25 Gy <18 cc |
| Bladder wall | V20 Gy <20 cc |
| Bladder wall | V20 Gy <34% |
| Right femoral head | V20 Gy <5 cc |
| Left femoral head | V20 Gy <5 cc |
| Penile bulb | V24.8 Gy <5% |

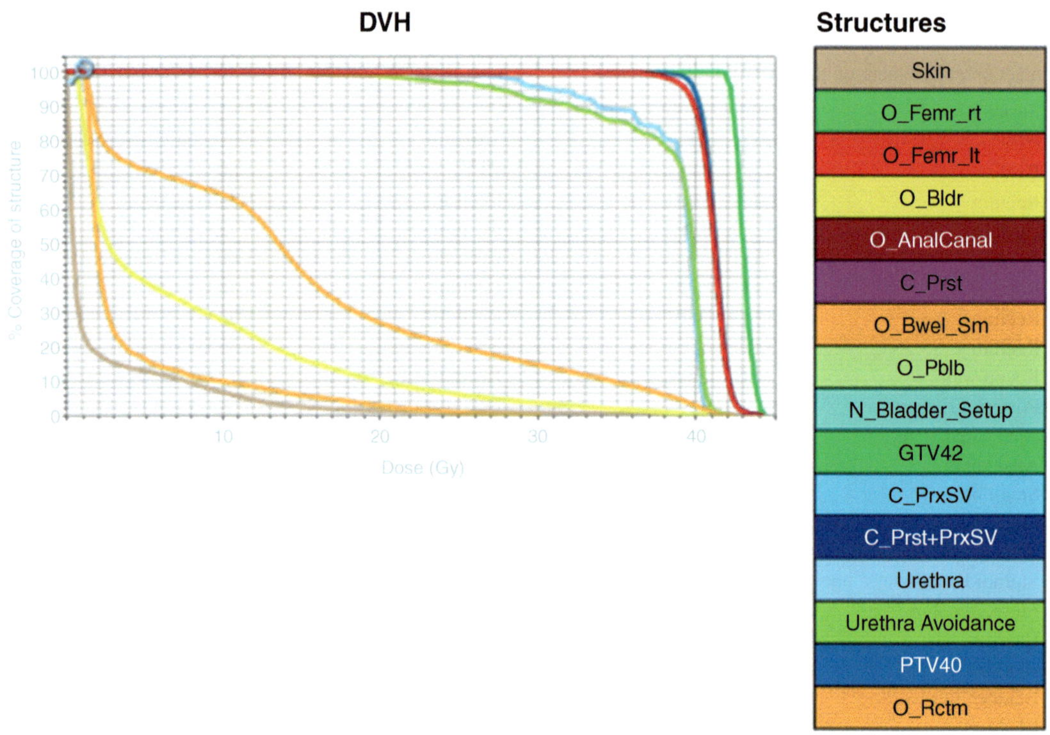


**Fig. 16.5** Dose volume histogram (DVH) from a patient with intact prostate cancer treated with ultrahypofractionated RT on the MIRAGE trial

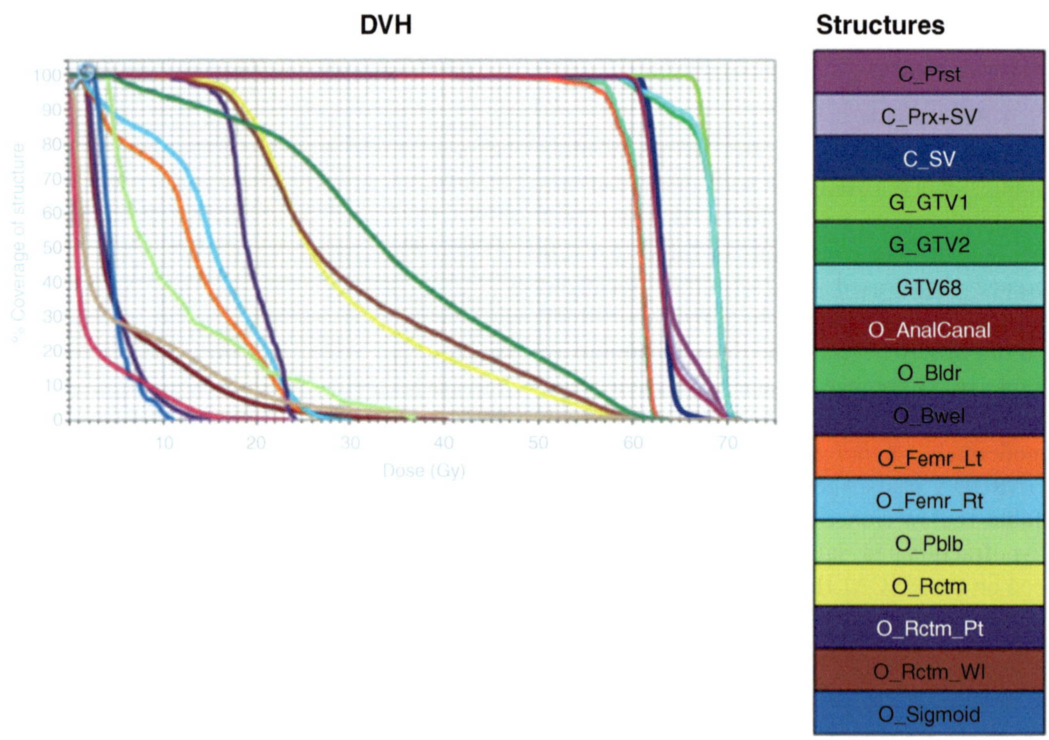


**Fig. 16.6** Dose volume histogram (DVH) from a patient with intact prostate cancer treated with moderately hypofractionated RT on the GARUDA trial

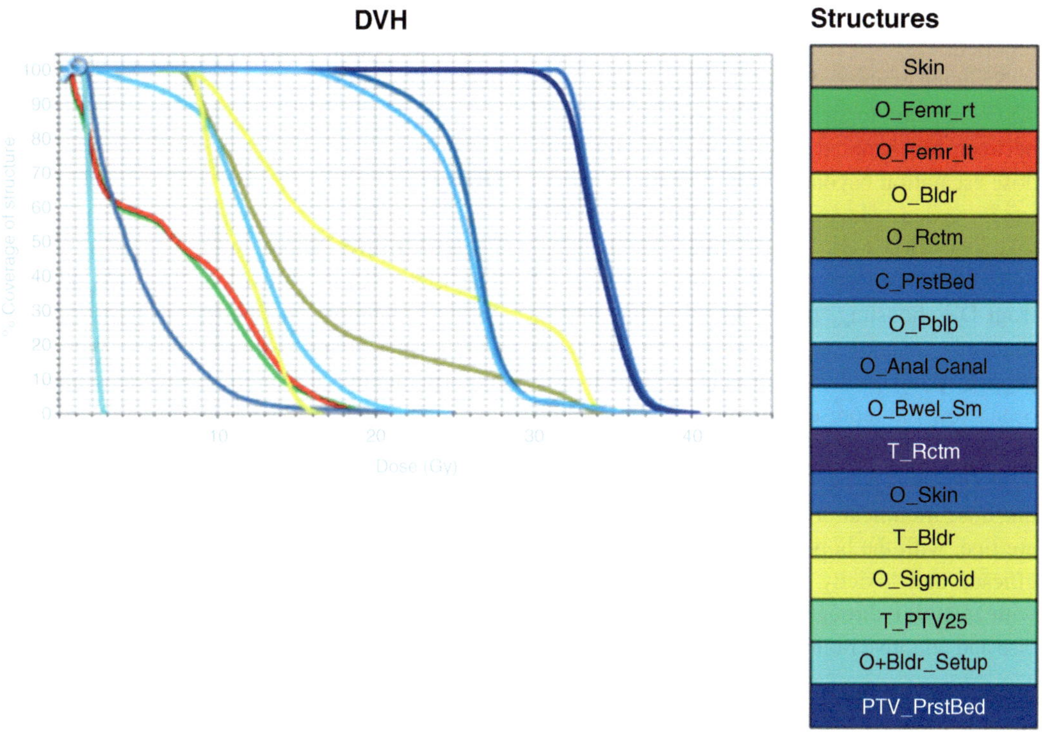


**Fig. 16.7** Dose volume histogram (DVH) from a patient with intact prostate cancer treated with postoperative ultrahypofractionated RT on the EXCALIBUR trial

## Dose Escalation

SBRT dose escalation trials have been carried out using CT-guided delivery with mixed success. A phase I/II dose escalation study from UT Southwestern famously concluded that "the sky is not the limit" after observing an increase in acute and late severe (grade 3 or greater) toxicity in patients receiving 50 Gy in 5 fx. In their study, dose was incrementally increased from a total prescription dose of 45 Gy to 47.5 Gy–50 Gy [66]. Of note, the investigators on this study took particular care to minimize toxicity by requiring Fleet enema and rectal balloon placement for each treatment, 4 mg dexamethasone for each treatment, as well as alpha-adrenergic antagonist usage. Yet, in the 47 patients who received 50 Gy in 5 fractions, within 500 days of treatment, late grade 4 GI and GU toxicities were reported at 3.3% and 1.6%, respectively, whereas late grade 3 GI and GU toxicities were reported in 6.6% and 4.9% of men enrolled, respectively [67]. This has prompted caution for whole-gland dose escalation above 40 Gy, at least with CT-based delivery approaches.

At the same time, data from Memorial Sloan Kettering demonstrates that incremental dose escalation from 32.5 to 35–37.5 Gy to 40 Gy improves biochemical failure rates, with 5-year cumulative incidence rates of PSA failure for each of these four dose levels corresponding to 15%, 6%, 0%, and 0%, respectively. The trial also mandated posttreatment biopsies at 2 years and found that the incidence of a positive posttreatment biopsy was 47.6%, 19.2%, 16.7%, and 7.7%, respectively, for the 4 dose arms ($p = 0.013$). Moreover, dose escalation to 40 Gy was well-tolerated, with no grade 3 or 4 acute GU or GI toxicities observed, only 1 late grade 3 urinary toxicity (urethral stricture) in the 40-Gy dose arm (the stricture was corrected with transurethral resection), no grade 4 late GU toxicity and no late grade 3 or 4 GI toxicity reported.

So in the context of CTgRT, dose escalation to 40 Gy appears to strike the appropriate balance between biochemical control and toxicity, with dose escalation beyond 40 Gy potentially carrying increased risks of acute and late GI and GU toxicity. However, given the biochemical recurrence-free survival benefits appreciated from DIL boosting, future attention has understandably focused on dose *de-escalation* of the prostate gland itself, as long as DILs (which have been shown to represent the most likely location of radiorecurrent failure) can be successfully dose escalated. The opportunities for studying this dose modification framework with MRIgRT are ripe, with the hope that further gains in both efficacy and toxicity reduction can be realized with MRIgRT through its ability to deliver precise and adaptive radiotherapy.

## On-Table Adaption for Unity ("Unity") and MRIdian ("ViewRay")

The possibility to adapt offers an important theoretical advantage for MRIgRT in prostate cancer. However, it is worth noting that adapting is not essential for achieving improved outcomes in the definitive setting, as none of the patients enrolled on the MIRAGE [60] trial received adaptive treatments. Nevertheless, the adaptive workflow may still offer additional benefits in settings with adequate resources to support regular adaptive workflows. The adaptive workflows for both Unity and VR systems are described below:

### Unity

#### Pretreatment Phase

The Elekta Unity MR-Linac has two adaptive online planning methods: adapt-to-position (ATP) and adapt-to-shape (ATP), with both requiring an initial plan to be generated on a reference image (CT or MRI) [19].

In the pretreatment phase the patient undergoes bladder and rectum preparation to ensure they are imaged and treated with a comfortably full bladder and an empty rectum. The patient then attends a planning CT and/or MRI session which is used to create the reference plan. The reference planning scans are acquired with the patient adequately immobilized. The contours are then outlined on either CT or the MRI depending on the workflow of the treating center. At present, it is permissible to use an MR-only workflow and this is likely to become more common in the future.

The MR-Linac reference plan is generated using the Monaco treatment planning system (TPS) (Elekta AB, Stockholm, Sweden, V5.40.00) using the GPUMCD dose calculation algorithm [68]. The secondary dose calculations are generated with the Raystation TPS (Raysearch Laboratories, Stockholm, Sweden, V8.0.0.61) with a collapsed cone dose calculation algorithm [31]. An example treatment plan is shown in Fig. 16.8.

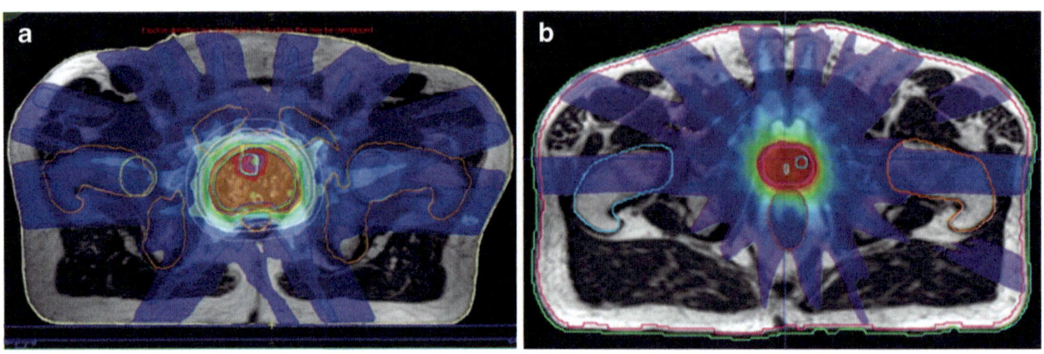


**Fig. 16.8** Radiotherapy plans for prostate only SBRT from (**a**) Unity and (**b**) MRIdian. Both treatments show that the dominant intraprostatic lesion has been contoured

### Treatment Phase

On the day of treatment, the patient will have undergone adequate bowel and bladder preparation. Once on the treatment couch, a session image is acquired with a T2 3D Tra Philips sequence (1-mm slice thickness).

In the "adapt-to-position" workflow, the reference image is registered to the daily MR image and the multileaf collimator (MLC) leaves are adapted to the translation-only rigid registration. The adapted plan is then recalculated on the reference image. This adaption workflow does not account for any changes in the anatomical shape of the target or OARs [31].

In the "adapt-to-shape" workflow, the reference plan is deformably registered to the daily MR image and the contours are propagated either rigidly or deformably. The contours are then edited by the treating physician to match the daily anatomy as seen on the session MRI. Based on the updated contours, Monaco reoptimizes the treatment plan using the optimization constraints derived from the original plan [69].

Once the online daily plan has been generated, checks are performed to ensure the plan parameters are as expected. A second dose calculation may then be performed, and a repeat verification image is acquired. If there has been further motion of the prostate during the ATS planning process, then a further ATP shift may be required prior to dose delivery.

## ViewRay

### Pretreatment Phase

As with conventional IGRT, the pretreatment phase begins with a simulation scan in the desired treatment position. CT and MR images are acquired on a traditional CT scanner and the MR-Linac, respectively, although derivation of synthetic CT scans from MR-only images has been investigated [70] as part of an MRI-only workflow that could reduce systematic errors introduced by image fusion [71]. Registration of the diagnostic MRI and other relevant imaging modalities (including prostate-specific membrane antigen [PSMA] positron emission tomography [PET]) to the simulation MRI is performed. Target volumes and OARs are delineated on the primary simulation MRI. Planning, dose calculations, and QA then proceed prior to treatment.

### Treatment Phase

On the day of treatment, a pilot image is first acquired to confirm appropriate bladder filling and rectal emptying. This is followed by a high-resolution session MRI scan. This higher-resolution scan is fused to the treatment planning MRI based on the prostate CTV. Image verification is then performed and OARs are recontoured if needed. Dose to the new OAR volumes is then predicted, and if it is decided that adaptive planning will be pursued, then the plan is reoptimized to improve dosimetry over the initially predicted dosimetry. The reoptimized plan is then verified and approved and online QA is performed. A repeat MR image is taken to verify positional accuracy, and if positioning is felt to be acceptable, the gating window is set and the treatment is delivered. If adaptive planning is not deployed, the workflow proceeds directly to plan verification and online QA following image verification.

Cine imaging is used for continuous motion monitoring during dose delivery. The prostate and the rectum are monitored in real-time sagittal, axial, and transverse plans. If the prostate moves outside the motion monitoring volume, the treatment is stopped until the prostate returns to the correct position [69].

## Evaluation

### Plan Evaluation

As with CTgRT, a homogenous dose distribution within the prostate gland is preferred, though this can be more challenging if DIL boosting is employed. Planning evaluation goals are discussed in "DVH Constraints" above.

It is worth noting that dosimetry can also be impacted by the technical inability of MRIgRT devices to offer modulated arc therapy, as well as the inability to deploy non-coplanar beam arrangements. While these techniques have both been shown to assist in improving conformality and lower radiation doses to OARs [72], most patients are likely to be adequately served by MRIgRT despite these limitations. Indeed, a direct comparison between MRI-guided IMRT and conventional VMAT planning approaches demonstrated similar dose distributions in the high-dose region and the target volume DVH parameters ($D_{98\%}$, $D_{50\%}$, $D_{2\%}$, and $V_{95\%}$) and similar homogeneity indices (HI). While MRIgRT resulted in a worse conformity index (CI) when compared to VMAT, MRIgRT planning demonstrated similar OAR sparing when normal PTV margins were used, but significantly better sparing once reduced PTV margins were applied [73]. Dosimetric shortfalls, however, may arise in patients with anatomy that renders planning more difficult, for whom fewer planning solutions may be available with the limitations of MRIgRT.

Furthermore, because the primary mechanism of action of RT involves activation of secondary particles which carry an electronic charge, exposure of these secondary charged particles to a strong magnetic field can alter their movement and consequentially alter radiation dosimetry. The electron return effect, also known as the Lorentz force, is an example of such a phenomenon where electrons move in a circular rather than linear pattern in the presence of a magnet [74, 75]. While this effect can, and should, be accounted for during radiation treatment planning, it is predominately a concern at air-tissue interfaces and thus less problematic for the treatment of internal organs such as the prostate, particularly with efforts to reduce rectal gas in place [76].

### Evaluation of Daily Setup

The superior visualization and reduced interobserver interpretation of pelvic anatomy when imaged by MR has already been discussed, but in addition to advantages with target and OAR delineation, MRI also enables more accurate and reproducible matching of the anatomy on the treatment plan, as seen at the time of CT/MR simulation, to the anatomy on the day of treatment in comparison to cone beam CT imaging [77]. Furthermore, since the need for surrogate markers for prostate triangulation is abrogated with MRIgRT, this technology also removes the need for invasive fiducial marker insertion which is unpleasant for patients and introduces the risk of bleeding and infection in a patient population that may harbor an aversion to surgically invasive procedures.

Improved image matching on the day of treatment and more accurate contouring of both treatment targets and organs at risk are estimated to lead to more precise radiotherapy which could result in improved toxicity.

## Outcome

### Definitive Prostate

Superior image quality and the ability to account for interfraction and intrafraction motion is ultimately of clinical relevance because it mitigates the errors that are factored into the CTV to PTV margins, enabling a safe reduction in PTV margin which may consequentially reduce the radiotherapy dose to surrounding normal tissue and in turn reduce acute toxicity. These theoretical benefits have recently been borne out with phase III clinical data from the University of California Los Angeles. The MIRAGE trial is a single-center trial which randomized 156 men receiving SBRT (40 Gy in 5 fractions) to either a CT-guided RT (CTgRT) approach employing a 4-mm CTV to PTV margin versus a MRIgRT approach which employed a 2-mm CTV to PTV margin [60]. Final reporting demonstrated a significant reduction in acute grade ≥2 genitourinary (GU) toxicity with MRI versus CT guidance (24.4% vs 43.4%, $p = 0.01$) as well as a similarly significant reduction in grade ≥2 gastrointestinal (GI) toxicity (0% vs 10.5%, $p = 0.003$). This translated to a 60% reduction in the odds of grade ≥2 GU toxicity with the tighter PTV margins enabled by MRIgRT. Patient-reported urinary and bowel function were also better preserved at the 1-month time point. Daily treatment adaption was not performed on any of the patients enrolled in MIRAGE, suggesting that MR guidance and gating may be enough to significantly reduce the toxicity that would have been seen with CTgRT. MRIgRT may emerge as the preferred treatment modality for men with prostate cancer in centers that have MRIgRT capacity. It remains to be seen whether the gains achieved in the toxicity profile of SBRT could be further improved if daily adaption was incorporated in to the workflow.

EXCALIBUR (NCT04915508) is the phase II nonrandomized successor to SCIMITAR that will exclusively treat with MRIgRT and determine the change in GI and GU reported outcome. MRIgRT is similarly being evaluated on the SHORTER trial (NCT04422132), which will randomize patients to either 36.5 Gy in 5 fractions or 55 Gy in 20 fractions on the MR-Linac with the aim of comparing the toxicity profile between these moderate and ultrahypofractionated arms.

## Post-prostatectomy

While the majority of studies in MRIgRT have focused on targeting and treating intact prostate cancer in the definitive setting, the role of MRIgRT in the post-prostatectomy setting and its potential improvements in convenience and efficacy have recently garnered interest. An increase in understanding of the deformable and dynamic nature of surrounding OARs in the post-prostatectomy setting, including the penile bulb [78], has further motivated interest in the careful implementation of MRI-guided ultrahypofractionated radiotherapy in the post-prostatectomy setting. In the SCIMITAR study, men were treated with 30–34 Gy in 5 fractions to the prostate bed under either CT-guided RT or MRIgRT. With a median follow-up of 29.5 months, patient-reported outcomes showed baseline normalization in EPIC-26 QOL scores by 3 months post SBRT, suggesting that post-radical prostatectomy SBRT is well-tolerated on both platforms [39]. In total, three patients had grade 3 toxicity, yet no patients receiving MRIgRT reported grade 3 GU or grade ≥2 GI toxicity. MRIgRT was also independently associated with improved any-grade GI toxicity and improved bowel QOL. The apparent benefits of MRIgRT are most likely attributed to the reduction in PTV margins with a CTV to PTV margin of 3 mm compared to 5 mm used in CTgRT. This reduction in PTV margins is made possible by improved soft tissue definition and contouring using MRI as well as consistently accurate setup enabled by MR imaging.

## Potential Limitations of MRIgRT

An important and practical limitation of MRIgRT treatment delivery relates to the design of the MR-linear accelerator. As opposed to the "open" nature of CT-guided linear accelerator treatment, the 70-cm closed-bore MRIgRT delivery system requires a much tighter space, resembling that of a traditional MRI. As a result, patients with a large body habitus or patients with claustrophobia or anxiety surrounding treatment may not be well-suited for treatment with MRIgRT. As diagnostic multiparametric MRI is a critical component of the initial workup of prostate cancer, this offers a natural "test" of which patients may be poorly suited to undergo MRIgRT-based therapy, with the option to trial anxiolytics to aid in patient comfort and tolerability.

Treatment time on the MR-Linac is longer than traditional methods of radiotherapy delivery due to the more complex workflow. Adaptive radiotherapy with repeat contouring and replanning in particular takes additional time, with patients on the table for 45–60 min [31, 79]. This is particularly challenging for the treatment of men with prostate cancer as full bladders have been historically required to physically displace small bowel out of the treatment field, but can be difficult for patients to achieve for long periods of time.

In a workflow involving gating, the size of the gating margins as well as the threshold of CTV allowable outside of the gating margin will also influence treatment times, with more rigid gating boundaries contributing to longer treatment times owing to an increased likelihood of prostate CTV extending outside of the allowable margin. However, these concerns about treatment length must be balanced against accuracy of treatment delivery, and individual centers or physicians may adopt different thresholds. Reassuringly, a validated patient-reported treatment questionnaire which collected responses from five centers found that patient-reported experience and satisfaction were consistently high despite the length of treatment [79].

Moreover, a recent series examined motion-compensated dose distributions from 20 patients for all prostate positions and only prostate positions within the gating window and found that beam gating actually had a negligible dosimetric impact in online-adaptive MRIgRT of prostate cancer, albeit in a cohort with low center of mass translations to begin with. They also found that even without gating, reducing the prostate-to-PTV margin from 5 to 3 mm maintained $D95\%$ $\geq 100\%$, suggesting that PTV margins might be successfully reduced without the need for a treatment-prolonging gating approach in the majority patients [80].

To address the dilemma of lengthy treatment times and the requirement for full bladders during treatment, research is ongoing as to whether real-time dose prediction can be deployed to assess whether bladder constraints will be met when bladders are not consistently filled, enabling treatment to potentially move forward with incompletely filled bladders if reasonable dosimetry can be achieved. An alternative strategy for circumventing the requirement of a full bladder would be to rely on online adaption for each fraction—a strategy which has been shown to improve both target and OAR dosimetry in underfilled bladders in a single institution cohort of 22 patients [81].

Given the later-in-life presentation of most prostate cancers, there may be a greater prevalence of relative medical contraindications to treatment on MRI-guided devices in this patient population, thereby theoretically reducing numbers of patients suitable for this technology.

Chief among these relative contraindications include the presence of a pacemaker or implantable defibrillator, as well as metallic prostheses. While treatment of pacemaker-dependent patients with MRIgRT is not an absolute contraindication, careful multidisciplinary collaboration with electrophysiologists, device representatives, and radiation therapists is required to ensure treatment safety and continued pacemaker function. Interrogation of the pacemaker prior to treatment, following each fraction, and after completion of treatment is encouraged. Workflows have been published to ensure safe stereotactic radiation therapy delivery in 1.5 T devices and the importance of daily dose monitoring with thermoluminescent dosimeters is emphasized to ensure dose tolerances for the pacemaker remain well below 2 Gy, even when the pacemaker does not lie within proximity to the treatment field as is often the case for prostate cancer patients [82].

Metal prostheses and vascular stents are other examples of metallic implants that can generate challenges for MRIgRT. These prostheses, particularly when located in the hip and low pelvic regions, can lead to geometric distortions and signal loss in neighboring treatment targets or OARs.

Magnet strength is likely to affect the magnitude of these distortions. A pilot study of three patients from the Netherlands reported their experience with a workflow that employed B0 mapping at the time of simulation to estimate the magnitude of distortions during

MRIgRT and guide the decision on eligibility for MRIgRT of prostate cancer patients with metallic hip implants [83]. Taken together, these points suggest that metallic implants and devices should not be viewed as an absolute contraindication to MRIgRT, though additional protocols and procedures to ensure patient safety and appropriate selection for MRIgRT are paramount.

The maximum craniocaudal field length for the Unity device is 22 cm, whereas the MRIdian device has a maximum length of 24 cm [84]. Many patients receiving definitive treatment for prostate cancer will receive radiation to the prostate alone, and in these scenarios, field length concerns are unlikely to arise.

The selection of patients who benefit from radiotherapy to the prostate and pelvic lymph nodes in the definitive setting is controversial, though men with very-high-risk disease and no evidence of distant metastatic disease on molecular imaging will likely benefit from elective irradiation of the pelvic lymph nodes [85]. In the post-prostatectomy setting, data from RTOG 0534 also supports routine inclusion of pelvic lymph nodes for some patients with a rising PSA following prostatectomy [3].

In such situations where elective pelvic lymph node volumes are included and/or boosting of gross pelvic nodal disease is desired, MRIgRT may not be suitable for all patients due to the need to allow for a PTV margin and penumbra for dose calculation to OARs. Thus, patient selection on the basis of radiation field size may become an important consideration, particularly for the Elekta Unity device.

The lower 0.35 T magnet strength of the MRIdian device limits direct visualization of the DIL during treatment planning. Additionally, initiation of hormone therapy can render challenges for DIL identification with both the MRIdian and the Unity devices. Therefore, offline fusion of diagnostic MRI is often required to enable accurate contouring of DIL(s). This additional step has the potential to introduce further errors into target delineation and hampers the verification of boost volume alignment on the day of treatment. Additionally, functional sequences are more challenging to perform with the lower magnet strength on the MRIdian device.

## Ongoing Trials in MRIgRT for Prostate Cancer

Trials taking place on the Unity MR-Linac are investigating the efficacy and toxicity of treating men with ultrahypofractionated regimes in as little as 2 fractions as well as the ability to visualize and boost the dominant lesion. Furthermore, the ability to avoid dose delivery to OARs is being realized, with the ERECT trial investigating the dose sparing of the neurovascular bundle, internal pudendal artery, corpora cavernosa, and penile bulb [86] (Table 16.4).

Trials investigating MRIgRT with the MRIdian device are similarly interested in reducing total fractions and increasing dose per fraction, with the FORT trial as an example that is assessing the non-inferiority of prostate RT delivery in 2 versus 5 fractions. Other investigational questions center around the safety and feasibility of conventional dose escalation to DILs, as will be investigated in the SIBRT trial, as well as dose escalation in a single fraction employing spatially fractionated GRID therapy, which will be assessed in the LEAD study (Table 16.4).

**Table 16.4** Ongoing trials in prostate cancer MRIgRT

| Trial | NCT | Device | Phase | Planned number of patients | Primary outcome | Dose delivered in treatment arm | Dose delivery in standard arm | Accrual status |
|---|---|---|---|---|---|---|---|---|
| HERMES | 04595019 | Unity | II | 46 | Acute grade 2+ GU toxicity | 24 Gy in 2 fractions to the prostate with an intraprostatic boost to 27 Gy in 2 fractions | 36.25 Gy to the prostate in 5 fractions | Open to recruitment |
| UltraHypo | 05183074 | Unity | II | 50 | Incidence of acute GU and GI toxicity | Not stated | NA | Open to recruitment |
| ERECT | 04861194 | Unity | II | 70 | Erectile dysfunction over 3 years post SBRT | 36.25 Gy in 5 fractions with sparing of the neurovascular bundle, IPA, corpora cavernosa, and penile bulb | NA | Open to recruitment |
| Boost (MSK) | 04997018 | Unity | II | 91 | A reduction in posttreatment biopsy rates at 24 months | 40 Gy in 5 fractions to the prostate with 45 Gy to the dominant lesion | NA | Open to recruitment |
| AFFIRM | 05373316 | Unity | I | 95 | Acute GI and GU toxicity | 35 Gy to the prostate with 50 Gy to the intraprostatic tumor in 5 fractions | NA | Open to recruitment |
| 2SMART | 03588819 | Unity | I | 30 | Quality of life using EPIC | 26 Gy in 2 fractions to the prostate and the DIL dose of up to 32 Gy in 2 fractions delivered 1 week apart | NA | Open to recruitment |
| iSMART | 05600400 | Unity | II | 144 | Change in quality of life function | 27 Gy in 2 fractions to the prostate | Five every other day fractions of 8 Gy | Open to recruitment |
| LEAD | 01411319 | ViewRay | I | 25 | Grade 2 or higher physician-reported treatment-related adverse events | 12–14 Gy in 1 fraction to the mpMRI defined GTV on day 1, followed by standard 38 fraction IMRT | N/A | Completed |
| FORT | 04984343 | ViewRay | II | 136 | Change in patient-reported GI symptoms using EPIC | 37.5 Gy in 5 fractions to the prostate | 25 Gy in 2 fractions to the prostate | Recruiting |
| SIBRT | 03664193 | ViewRay | II | 30 | Feasibility | 35 Gy in 5 fractions to the prostate with prostate lesion SIB to 37.5, 40, 42.5, or 45 Gy | NA | Completed |
| EXCALIBUR | 04915508 | ViewRay | II | 102 | Change in patient-reported GI symptoms using EPIC | 30–34 Gy in 5 fractions | N/A | Recruiting |
| SHORTER | 04422132 | ViewRay | II | 134 | Change in patient-reported GI symptoms using EPIC | 32.5 Gy in 5 fractions | 55 Gy in 20 fractions | Recruiting |

## References


1. Siegel RL, Miller KD, Fuchs HE, Jemal A. Cancer statistics, 2022. CA Cancer J Clin. 2022;72:7–33.
2. Hamdy FC, et al. Fifteen-year outcomes after monitoring, surgery, or radiotherapy for prostate cancer. N Engl J Med. 2023;388:1547. https://doi.org/10.1056/NEJMoa2214122.
3. Pollack A, et al. The addition of androgen deprivation therapy and pelvic lymph node treatment to prostate bed salvage radiotherapy (NRG Oncology/RTOG 0534 SPPORT): an international, multicentre, randomised phase 3 trial. Lancet Lond Engl. 2022;399:1886–901.
4. Valle LF, et al. A systematic review and meta-analysis of local salvage therapies after radiotherapy for prostate cancer (MASTER). Eur Urol. 2020;80:280. https://doi.org/10.1016/j.eururo.2020.11.010.
5. NCC, N. NCCN clinical practice guidelines in oncology: prostate cancer 2020. 2020.
6. Brenner DJ, Hall EJ. Fractionation and protraction for radiotherapy of prostate carcinoma. Int J Radiat Oncol Biol Phys. 1999;43:1095–101.
7. Michalski JM, et al. Effect of standard vs dose-escalated radiation therapy for patients with intermediate-risk prostate cancer: the NRG oncology RTOG 0126 randomized clinical trial. JAMA Oncol. 2018;4:e180039.
8. Dearnaley D, et al. Conventional versus hypofractionated high-dose intensity-modulated radiotherapy for prostate cancer: 5-year outcomes of the randomised, non-inferiority, phase 3 CHHiP trial. Lancet Oncol. 2016;17:1047–60.
9. Lee WR, et al. Randomized phase III noninferiority study comparing two radiotherapy fractionation schedules in patients with low-risk prostate cancer. J Clin Oncol. 2016;34:2325–32.
10. Catton CN, et al. Randomized trial of a hypofractionated radiation regimen for the treatment of localized prostate cancer. J Clin Oncol. 2017;35:1884–90.
11. Widmark A, et al. Ultra-hypofractionated versus conventionally fractionated radiotherapy for prostate cancer: 5-year outcomes of the HYPO-RT-PC randomised, non-inferiority, phase 3 trial. Lancet Lond Engl. 2019;394:385–95.
12. Fransson P, et al. Ultra-hypofractionated versus conventionally fractionated radiotherapy for prostate cancer (HYPO-RT-PC): patient-reported quality-of-life outcomes of a randomised, controlled, non-inferiority, phase 3 trial. Lancet Oncol. 2021;22:235–45.
13. Tree AC, et al. Intensity-modulated radiotherapy versus stereotactic body radiotherapy for prostate cancer (PACE-B): 2-year toxicity results from an open-label, randomised, phase 3, non-inferiority trial. Lancet Oncol. 2022;23:1308–20.
14. Brand DH, et al. Intensity-modulated fractionated radiotherapy versus stereotactic body radiotherapy for prostate cancer (PACE-B): acute toxicity findings from an international, randomised, open-label, phase 3, non-inferiority trial. Lancet Oncol. 2019;20:1531–43.
15. Kupelian P, et al. Stereotactic body radiation therapy for prostate cancer: rational and reasonable. Pract Radiat Oncol. 2015;5:188.
16. Hodges JC, et al. Cost-effectiveness analysis of stereotactic body radiation therapy versus intensity-modulated radiation therapy: an emerging initial radiation treatment option for organ-confined prostate cancer. J Oncol Pract. 2012;8:e31s–7s.
17. Buyyounouski MK, et al. Primary endpoint analysis of a randomized phase III trial of hypofractionated vs. conventional post-prostatectomy radiotherapy: NRG Oncology GU003. Int J Radiat Oncol Biol Phys. 2021;111:S2–3.
18. Tandberg DJ, et al. Postoperative radiation therapy for prostate cancer: comparison of conventional versus hypofractionated radiation regimens. Int J Radiat Oncol Biol Phys. 2018;101:396–405.
19. Winkel D, et al. Adaptive radiotherapy: the Elekta Unity MR-linac concept. Clin Transl Radiat Oncol. 2019;18:54–9.
20. Mutic S, Dempsey JF. The ViewRay system: magnetic resonance-guided and controlled radiotherapy. Semin Radiat Oncol. 2014;24:196–9.
21. Moiseenko V, Liu M, Kristensen S, Gelowitz G, Berthelet E. Effect of bladder filling on doses to prostate and organs at risk: a treatment planning study. J Appl Clin Med Phys. 2006;8:55–68.
22. Zellars RC, et al. Prostate position late in the course of external beam therapy: patterns and predictors. Int J Radiat Oncol Biol Phys. 2000;47:655–60.
23. Brand VJ, Milder MTW, Christianen MEMC, Hoogeman MS, Incrocci L. Seminal vesicle inter- and intra-fraction motion during radiotherapy for prostate cancer: a review. Radiother Oncol. 2022;169:15–24.
24. Alexander SE, et al. Prostate volume changes during extreme and moderately hypofractionated magnetic resonance image-guided radiotherapy. Clin Oncol (R Coll Radiol). 2022;34:e383–91.
25. Ma TM, et al. Dosimetric impact of interfraction prostate and seminal vesicle volume changes and rotation: a post-hoc analysis of a phase III randomized trial of MRI-guided versus CT-guided stereotactic body radiotherapy. Radiother Oncol. 2022;167:203–10.
26. Hötker AM, et al. Prostate cancer: assessing the effects of androgen-deprivation therapy using quantitative diffusion-weighted and dynamic contrast-enhanced MRI. Eur Radiol. 2015;25:2665–72.
27. Kupelian P, et al. Multi-institutional clinical experience with the calypso system in localization and continuous, real-time monitoring of the prostate gland during external radiotherapy. Int J Radiat Oncol Biol Phys. 2007;67:1088–98.
28. Zhu X, et al. Tradeoffs of integrating real-time tracking into IGRT for prostate cancer treatment. Phys Med Biol. 2009;54:N393–401.
29. de Muinck Keizer DM, et al. Prostate intrafraction motion during the preparation and delivery of MR-guided radiotherapy sessions on a 1.5 T MR-Linac. Radiother Oncol. 2020;151:88–94.
30. Kontaxis C, et al. Delivered dose quantification in prostate radiotherapy using online 3D cine imaging

and treatment log files on a combined 1.5 T magnetic resonance imaging and linear accelerator system. Phys Imaging Radiat Oncol. 2020;15:23–9.
31. Dunlop A, et al. Daily adaptive radiotherapy for patients with prostate cancer using a high field MR-linac: initial clinical experiences and assessment of delivered doses compared to a C-arm linac. Clin Transl Radiat Oncol. 2020;23:35–42.
32. Glitzner M, et al. On-line MR imaging for dose validation of abdominal radiotherapy. Phys Med Biol. 2015;60:8869–83.
33. Sritharan K, Tree A. MR-guided radiotherapy for prostate cancer: state of the art and future perspectives. Br J Radiol. 2022;95:20210800.
34. van Schie MA, et al. Quantitative MRI changes during weekly ultra-hypofractionated prostate cancer radiotherapy with integrated boost. Front Oncol. 2019;9:1264.
35. Datta A, Aznar MC, Dubec M, Parker GJM, O'Connor JPB. Delivering functional imaging on the MRI-Linac: current challenges and potential solutions. Clin Oncol (R Coll Radiol). 2018;30:702–10.
36. Kneebone A, et al. Adjuvant radiotherapy versus early salvage radiotherapy following radical prostatectomy (TROG 08.03/ANZUP RAVES): a randomised, controlled, phase 3, non-inferiority trial. Lancet Oncol. 2020;21:1331–40.
37. Parker CC, et al. Timing of radiotherapy after radical prostatectomy (RADICALS-RT): a randomised, controlled phase 3 trial. Lancet Lond Engl. 2020;396:1413–21.
38. Sargos P, et al. Adjuvant radiotherapy versus early salvage radiotherapy plus short-term androgen deprivation therapy in men with localised prostate cancer after radical prostatectomy (GETUG-AFU 17): a randomised, phase 3 trial. Lancet Oncol. 2020;21:1341–52.
39. Ma TM, et al. Quality-of-life outcomes and toxicity profile among patients with localized prostate cancer after radical prostatectomy treated with stereotactic body radiation: the SCIMITAR multicenter phase 2 trial. Int J Radiat Oncol Biol Phys. 2022;115(1):142–52. https://doi.org/10.1016/j.ijrobp.2022.08.041.
40. Jansen BHE, et al. Detection of recurrent prostate cancer using prostate-specific membrane antigen positron emission tomography in patients not meeting the Phoenix criteria for biochemical recurrence after curative radiotherapy. Eur Urol Oncol. 2021;4:821–5.
41. Pfister D, et al. 68 Ga-PSMA-PET/CT helps to select patients for salvage radical prostatectomy with local recurrence after primary radiotherapy for prostate cancer. BJU Int. 2020;126:679–83.
42. Panebianco V, et al. Prostate magnetic resonance imaging for local recurrence reporting (PI-RR): international consensus-based guidelines on multiparametric magnetic resonance imaging for prostate cancer recurrence after radiation therapy and radical prostatectomy. Eur Urol Oncol. 2021;4:868–76.
43. Valle LF, et al. Multiparametric MRI for the detection of local recurrence of prostate cancer in the setting of biochemical recurrence after low dose rate brachytherapy. Diagn Interv Radiol. 2018;24:46–53.
44. Fendler WP, et al. Assessment of 68Ga-PSMA-11 PET accuracy in localizing recurrent prostate cancer: a prospective single-arm clinical trial. JAMA Oncol. 2019;5:856–63.
45. Fuller D, et al. Retreatment for local recurrence of prostatic carcinoma after prior therapeutic irradiation: efficacy and toxicity of HDR-like SBRT. Int J Radiat Oncol Biol Phys. 2020;106:291.
46. Fuller DB, et al. High dose 'HDR-like' prostate SBRT: PSA 10-year results from a mature, multi-institutional clinical trial. Front Oncol. 2022;12:935310.
47. Michalet M, et al. Magnetic resonance-guided reirradiation for local recurrence within the prostate or in the prostate bed: 1-year clinical results of a prospective registry study. Cancers. 2022;14:1943.
48. Barchetti F, Panebianco V. Multiparametric MRI for recurrent prostate cancer post radical prostatectomy and postradiation therapy. Biomed Res Int. 2014;2014:316272.
49. Salembier C, et al. ESTRO ACROP consensus guideline on CT- and MRI-based target volume delineation for primary radiation therapy of localized prostate cancer. Radiother Oncol. 2018;127:49–61.
50. Hall WA, et al. NRG oncology updated international consensus atlas on pelvic lymph node volumes for intact and postoperative prostate cancer. Int J Radiat Oncol Biol Phys. 2021;109:174–85.
51. Harris VA, et al. Consensus guidelines and contouring atlas for pelvic node delineation in prostate and pelvic node intensity modulated radiation therapy. Int J Radiat Oncol Biol Phys. 2015;92:874–83.
52. Michalski JM, et al. Development of RTOG consensus guidelines for the definition of the clinical target volume for postoperative conformal radiation therapy for prostate cancer. Int J Radiat Oncol Biol Phys. 2010;76:361–8.
53. Parker CC, et al. Magnetic resonance imaging in the radiation treatment planning of localized prostate cancer using intra-prostatic fiducial markers for computed tomography co-registration. Radiother Oncol. 2003;66:217–24.
54. Pathmanathan AU, et al. Comparison of prostate delineation on multimodality imaging for MR-guided radiotherapy. Br J Radiol. 2019;92:20180948.
55. White I, et al. Interobserver variability in target volume delineation for CT/MRI simulation and MRI-guided adaptive radiotherapy in rectal cancer. Br J Radiol. 2021;94:20210350.
56. Westley R, Hall E, Tree A. HERMES: delivery of a speedy prostate cancer treatment. Clin Oncol (R Coll Radiol). 2022;34:426–9.
57. Kerkmeijer LGW, et al. Focal boost to the intraprostatic tumor in external beam radiotherapy for patients with localized prostate cancer: results from the FLAME randomized phase III trial. J Clin Oncol. 2021;39:787–96.
58. Draulans C, et al. Primary endpoint analysis of the multicentre phase II hypo-FLAME trial for interme-

diate and high risk prostate cancer. Radiother Oncol. 2020;147:92–8.
59. Tree AC, et al. Standard and hypofractionated dose escalation to intraprostatic tumor nodules in localized prostate cancer: 5-year efficacy and toxicity in the DELINEATE trial. Int J Radiat Oncol Biol Phys. 2023;115:305–16.
60. Kishan AU, et al. Magnetic resonance imaging-guided vs computed tomography-guided stereotactic body radiotherapy for prostate cancer: the MIRAGE randomized clinical trial. JAMA Oncol. 2023;365:e226558. https://doi.org/10.1001/jamaoncol.2022.6558.
61. Rodda S, Morris WJ, Hamm J, Duncan G. ASCENDE-RT: an analysis of health-related quality of life for a randomized trial comparing low-dose-rate brachytherapy boost with dose-escalated external beam boost for high- and intermediate-risk prostate cancer. Int J Radiat Oncol Biol Phys. 2017;98:581–9.
62. Pham J, et al. Evaluation of T2-weighted MRI for visualization and sparing of urethra with MR-guided radiation therapy (MRgRT) on-board MRI. Cancers. 2021;13:3564.
63. Spratt DE, et al. Vessel-sparing radiotherapy for localized prostate cancer to preserve erectile function: a single-arm phase 2 trial. Eur Urol. 2017;72:617–24.
64. Hamdy FC, et al. 10-Year outcomes after monitoring, surgery, or radiotherapy for localized prostate cancer. N Engl J Med. 2016;375:1415–24.
65. Lee JY, Spratt DE, Liss AL, McLaughlin PW. Vessel-sparing radiation and functional anatomy-based preservation for erectile function after prostate radiotherapy. Lancet Oncol. 2016;17:e198–208.
66. Boike TP, et al. Phase I dose-escalation study of stereotactic body radiation therapy for low- and intermediate-risk prostate cancer. J Clin Oncol. 2011;29:2020.
67. Hannan R, et al. Stereotactic body radiation therapy for low and intermediate risk prostate cancer-results from a multi-institutional clinical trial. Eur J Cancer Oxf Engl. 2016;1990(59):142–51.
68. Paudel MR, et al. Experimental evaluation of a GPU-based Monte Carlo dose calculation algorithm in the Monaco treatment planning system. J Appl Clin Med Phys. 2016;17:230–41.
69. Menten MJ, et al. Automatic reconstruction of the delivered dose of the day using MR-linac treatment log files and online MR imaging. Radiother Oncol. 2020;145:88–94.
70. Chen S, Qin A, Zhou D, Yan D. Technical note: U-net-generated synthetic CT images for magnetic resonance imaging-only prostate intensity-modulated radiation therapy treatment planning. Med Phys. 2018;45:5659–65.
71. Tenhunen M, et al. MRI-only based radiation therapy of prostate cancer: workflow and early clinical experience. Acta Oncol. 2018;57:902–7.
72. Ren W, et al. Dosimetric comparison of intensity-modulated radiotherapy and volumetric-modulated arc radiotherapy in patients with prostate cancer: a meta-analysis. J Appl Clin Med Phys. 2016;17:254–62.
73. Da Silva Mendes V, et al. Dosimetric comparison of MR-linac-based IMRT and conventional VMAT treatment plans for prostate cancer. Radiat Oncol Lond Engl. 2021;16:133.
74. Abraham M. Theorie der Elektrizität. Zweiter Band: Elektromagnetische Theorie der Strahlung. Monatshefte Für Math Phys. 1906;17:A39.
75. Hall WA, et al. The transformation of radiation oncology using real-time magnetic resonance guidance: a review. Eur J Cancer. 2019;1990(122):42–52.
76. Yang Y, et al. Longitudinal diffusion MRI for treatment response assessment: preliminary experience using an MRI-guided tri-cobalt 60 radiotherapy system. Med Phys. 2016;43:1369–73.
77. van Herk M, McWilliam A, Dubec M, Faivre-Finn C, Choudhury A. Magnetic resonance imaging-guided radiation therapy: a short strengths, weaknesses, opportunities, and threats analysis. Int J Radiat Oncol Biol Phys. 2018;101:1057–60.
78. Roy A, et al. Assessing inter-fraction changes in the size and position of the penile bulb during daily MR-guided radiation therapy to the prostate bed: do we need to adjust how we plan radiation in the post-radical prostatectomy setting to reduce risk of erectile dysfunction? Clin Genitourin Cancer. 2022;20:e227–32.
79. Barnes H, et al. Development and results of a patient-reported treatment experience questionnaire on a 1.5 T MR-Linac. Clin Transl Radiat Oncol. 2021;30:31–7.
80. Wahlstedt I, et al. Gating has a negligible impact on dose delivered in MRI-guided online adaptive radiotherapy of prostate cancer. Radiother Oncol. 2022;170:205–12.
81. Leeman JE, et al. Magnetic resonance-guided prostate stereotactic body radiation therapy with daily online plan adaptation: results of a prospective phase 1 trial and supplemental cohort. Adv Radiat Oncol. 2022;7:100934.
82. Yang B, et al. Magnetic resonance-guided radiation therapy of patients with cardiovascular implantable electronic device on a 1.5 T magnetic resonance-Linac. Pract Radiat Oncol. 2022;12:e56–61.
83. Keesman R, et al. Clinical workflow for treating patients with a metallic hip prosthesis using magnetic resonance imaging-guided radiotherapy. Phys Imaging Radiat Oncol. 2020;15:85–90.
84. Tocco BR, Kishan AU, Ma TM, Kerkmeijer LGW, Tree AC. MR-guided radiotherapy for prostate cancer. Front Oncol. 2020;10:616291.
85. Murthy V, et al. Prostate-only versus whole-pelvic radiation therapy in high-risk and very high-risk prostate cancer (POP-RT): outcomes from phase III randomized controlled trial. Clin Oncol. 2021;1234:JCO2003282. https://doi.org/10.1200/JCO.20.03282.
86. Verkooijen HM. EREctile Function Preservation for Prostate Cancer Radiation Therapy (ERECT); a prospective phase II trial. 2021. https://clinicaltrials.gov/ct2/show/NCT04861194.

# 17 Breast Radiotherapy Using MR-Linac

Chiara De-Colle, Anna M. Kirby, Desiree van den Bongard, and Maureen Groot Koerkamp

## Introduction

Long-term treatment-related toxicity remains a major concern for patients affected by breast cancer. The favorable prognosis found especially in patients diagnosed with low-risk early-stage tumors directs the current research towards minimizing toxicity while maintaining optimal oncological outcomes. One way to achieve this is by reducing the irradiated volumes. Technical improvements in modern radiotherapy help to achieve this without increasing the risk of missing the target. In breast radiotherapy, partial breast irradiation has been adopted in recent years to reduce the treatment volume. However, correct target visualization and definition is particularly important when performing partial breast irradiation. Neoadjuvant partial breast irradiation could facilitate target definition. With this approach, the volume of irradiated breast tissue can be smaller than in the postoperative setting [1]. In addition, the breast tissue receiving the highest doses could be subsequently surgically removed. Furthermore, MR-guided breast radiotherapy performed on an MR-Linac offers the possibility to better visualize the target and potentially to reduce the margins used to compensate for inter- and intrafraction variation. In this chapter, the rationale and potential advantages as well as specific considerations and practical suggestions for performing breast cancer treatment on an MR-Linac, with a focus on neoadjuvant partial breast irradiation, are discussed.

C. De-Colle (✉)
Department of Radiation Oncology, University Hospital and Medical Faculty, Eberhard Karls University Tübingen, Tübingen, Germany
e-mail: chiara.de-colle@med.uni-tuebingen.de

A. M. Kirby
Department of Radiotherapy, Royal Marsden NHS Foundation Trust and Institute of Cancer Research, Sutton, United Kingdom
e-mail: anna.kirby@rmh.nhs.uk

D. van den Bongard
Department of Radiation Oncology, Amsterdam University Medical Centers, Cancer Center Amsterdam, Cancer Treatment and Quality of Life, Cancer Biology and Immunology, Amsterdam, The Netherlands
e-mail: h.j.vandenbongard@amsterdamumc.nl

M. G. Koerkamp
Department of Radiation Oncology, University Medical Centre Utrecht, Utrecht, The Netherlands
e-mail: M.L.GrootKoerkamp-3@umcutrecht.nl

## Breast Cancer Overview

Breast cancer is the most frequent female cancer, with an incidence of approximately 140 per 100,000 women and a lifetime risk of 1 every 7 women [2, 3]. With more than 2.26 million of new cases of breast cancer diagnosed worldwide in 2020 [4], it accounts for one-third of all cancers diagnosed in women [5].


I. J. Das et al. (eds.), *A Practical Guide to MR-Linac*, https://doi.org/10.1007/978-3-031-48165-9_17

Despite improvements in diagnosis and treatment, breast cancer remains the main cause of death in women below 65 years and the third most common cause of death after cerebral and cardiovascular diseases in women older than 65 years [6]. With 91,826 deaths in 2020 in the EU and 43,250 in 2022 in the USA, the mortality rates are between 20 and 30 per 100,000 women per year [2, 5]. Because of its high incidence and mortality, but at the same time high long-term survival rates and because of predicted further increase in new cases each year [7], breast cancer and long-term treatment-related toxicities represent a major public health burden.

In this context, early diagnosis is essential for better outcome. Thanks to diagnostic improvements, the majority of breast tumors are nowadays detected in early-stage and the 5-year overall survival rate for these patients' subgroup reaches almost 100% [8]. Radiotherapy plays a main role in the treatment of these patients. Attempts to omit radiotherapy in patients with low-risk breast cancer revealed a more than ten-fold higher incidence of relapses in those treated with breast-conserving surgery and hormonal therapy alone [9], indicating that radiotherapy remains an important part of the treatment of patients undergoing a breast-conserving approach.

Given the high survival rates, especially for patients diagnosed with early-stage cancer, much clinical research in this patient group focuses on reducing treatment-related toxicities, without compromising oncological outcomes. Risk of late radiotherapy-induced effects such as heart disease, lung injury, and secondary cancers appears minimal in patients treated with modern techniques, not requiring nodal irradiation and nonsmokers [10–12]. However, side effects in healthy breast tissue remain an issue.

One way to try to maintain the excellent oncological outcomes and reduce the radiotherapy-related breast tissue toxicity is to reduce the irradiated volume. A widely investigated approach is external-beam adjuvant partial breast irradiation. Partial breast irradiation reduces the irradiated breast volume compared to whole breast irradiation and has become a standard approach in selected low-risk breast cancer patients [13–15]. This approach appears to offer better results in terms of acute and late treatment-related toxicity and cosmesis [16, 17]. In the IMPORT LOW trial, significantly better results in terms of changed breast appearance and breast hardness compared to whole breast were reported [17]. However, even though smaller compared to whole breast irradiation, the irradiated breast area when performing external-beam adjuvant partial breast irradiation often represents 1/4 of the breast or more [17, 18], due to the need to cover all the postoperative changes, especially in the era of oncoplastic surgery. Additionally, planning target volume (PTV) margins up to 10 mm are traditionally required to account for the inter- and intrafractional variability. Because both breast irradiation and high-grade subcutaneous fibrosis have been shown to be correlated with the breast volume receiving a high dose [17, 19, 20], small volumes for partial breast irradiation are strongly encouraged.

Another approach, investigated more recently in an increasing number of studies, is neoadjuvant partial breast radiotherapy, the goal of which is to further reduce the irradiated area and consequently minimize treatment-related toxicities and improve cosmetic outcomes. Neoadjuvant partial breast irradiation allows a more precise and less observer-dependent definition of the target, which might translate in lower treatment-related toxicity [21]. With neoadjuvant partial breast irradiation, the possibility to treat the visible tumor with a margin for the clinical target volume (CTV) allows for smaller target volumes compared to the breast CTV in the adjuvant setting.

The new MR-Linac technology improves the tumor visualization and enables for online adaptive radiotherapy, opening the possibility to reduce the CTV-PTV margin and therefore further decrease the irradiated volume.

## Overcoming Challenges with Traditional Approaches

There are two main challenges for target volume reduction for partial breast irradiation that MR-guided breast radiotherapy, compared to traditional breast techniques, might help to overcome. The first challenge is the visualization of the target area in order to precisely define the gross tumor volume (GTV) and CTV. MRI offers a better soft-tissue contrast and detailed view of breast and tumor tissue, which could help to improve target definition in neoadjuvant partial breast irradiation. The second challenge is how to minimize the irradiated volume. The latter could be achieved by visualizing the target before and during dose delivery and by adapting the treatment plan to inter- and intrafraction variabilities to reduce CTV-PTV margins. Hence, the MR-Linac may play a role in overcoming both challenges.

Regarding the first topic, in the adjuvant setting radiation oncologists need to determine the location of the postoperative tumor bed. This could be particularly large after breast tissue mobilization and displacement following oncoplastic surgery. Moreover, the traditional CT-based delineation of the CTV might be challenging when the resection cavity cannot be easily distinguished [22], e.g., because of the absence of a well-defined seroma or when few clips are available (cavity visualization scores 1–3 [23]). MR-based CTV delineation has been associated with lower interobserver variability, improved accuracy, and smaller CTVs in some investigations [24–27]. However, other studies demonstrated no increase in consistency of CTV contouring when MR was used [27–29], sometimes ending up in larger CTVs, mainly due to the inclusion of postoperative breast tissue changes which were not visible on CT [29–31]. In the neoadjuvant scenario, the target delineation of a tumor in situ appears to be more accurate and less affected by interobserver variability compared to the target delineation in the adjuvant setting [1]. Additionally, MR offers clear advantages in breast tumor visualization compared to CT [32]. Therefore, the largest benefit of MR for breast radiotherapy lies in the neoadjuvant setting.

Regarding the second topic, traditional image guidance approaches are characterized by the acquisition of portal images or cone beam CTs before irradiation, which are used to correct variations in patient positioning. However, these image guidance methods cannot correct for changes in the breast contour, e.g., swelling and seroma volume variations, changes in tumor bed location, or differences in breathing pattern or arm position compared to the planning CT. In the setting of standard fractionated adjuvant breast irradiation, a reduction of the lumpectomy cavity volume to less than a half has been described, mainly due to a reduction of the seroma [33]. All such factors contribute to increase in the interfractional variability and have to be included as uncertainties in a PTV margin. In a worst-case scenario, they may lead to a replan being required. Depending on the correction strategy, PTV margins up to 10 mm may be necessary [34]. Treatment on an MR-Linac could facilitate online monitoring and adaptation to interfraction variability [35].

Additionally, the traditional image guidance approaches cannot compensate for intrafraction motion, which therefore also has to be compensated for in the PTV margin. Given the increasing use of hypofractionated (five-fraction) schedules [36], compensation for intrafraction motion becomes increasingly important because the risks of a target missing might have greater consequences due to the lack of the possibility for the inaccuracies in dose delivery to be compensated during other fractions. Also, fewer but longer fractions may require larger PTV margins because intrafraction drift motion increases with longer treatment time [37]. With the MR-Linacs, a precise target visualization before and during irradiation is possible. Treatments performed at the MR-Linac might allow for (almost) no PTV margins, because the interfractional variability is reduced and the intrafractional variability might be tracked and corrected online [37].

After decades of experience with postoperative whole breast irradiation with tangential fields encompassing the whole breast, radiation oncologists performing partial breast irradiation need to switch to a concept where the CTV and, for neoadjuvant breast irradiation, the GTV need to be precisely identified in order to avoid geographical miss. With the excellent soft tissue contrast of MR and the possibility of online daily MR-based plan adaptation, the MR-Linacs offer the tools required for this concept.

In the following sections, practical recommendations for simulation, target and OAR definition, and plan adaptation for treatments performed at the MR-Linac are provided. These recommendations are intended for treatments of patients with low-risk early-stage breast cancer, according to the GEC-ESTRO and ASTRO selection criteria [38, 39].

## Simulation, Contouring, PTV, and OAR

### Simulation: Patient Positioning

Because the MR-Linac bore size is smaller than the CT bore, particular attention needs to be paid to patient positioning for breast radiotherapy at an MR-Linac. When treatments are performed in the classical supine position with arms above the head, an MR-compatible immobilization device less inclined (maximum 5°) compared to a CT-planning setup or a device without inclination is usually required in order to avoid collision between the patient's elbows and the machine. An MR-compatible knee support or feet immobilization device is recommended. The target area should be as close as possible to the isocenter to minimize geometric distortions in the target area, which may be achieved by shifting the patient to the contralateral side of the table. The anterior receiver coil can be placed on coil bridges to avoid breast image deformation [40, 41]. Also, application of flex coils placed over the breasts has been described [42].

Treatments in prone position have been investigated [43–45] and appear to be feasible, but may be challenging, due to the smaller bore size leaving limited space for pendulous breasts to hang freely without touching the table, especially combined with a receiver coil on the back of the patients [41].

When a planning CT is still required, the planning CT and planning MR should be performed with the same immobilization device and, if possible, on the same day in order to improve the consistency of patient and target position and shape between the two imaging modalities.

### Simulation: Image Acquisition

In the adjuvant setting, a T1-weighted sequence without fat suppression is recommended for visualization of the surgical clips, while a T1-weighted scan with fat suppression and especially a T2-weighted scan are preferred for the visualization of seroma [25, 28, 32, 46–49]. In the neoadjuvant setting, a T1-weighted sequence with fat suppression (mDixon) and gadolinium contrast enhancement offers the best visualization of the tumor and its spiculae [50–53]. During treatment at an MR-Linac not all preferred sequences may be available, as only a fixed set of MRI sequences is available on the machines [41]. At the MRIdian, a True-FISP sequence, a balanced steady-state free precession imaging sequence, is used. At the Elekta Unity, a T1-weighted spoiled gradient echo and a T2-weighted fast spin echo sequence are available. To aid tumor delineation, one scan should be acquired before contrast injection, and multiple scans should be acquired at different time points in the first minutes after contrast injection [54]. An example of tumor visualization in a breast cancer patient treated in the neoadjuvant setting at an Elekta Unity MR-Linac is displayed in Fig. 17.1. Diffusion-weighted imaging (DWI) sequences allow for differentiation between malignant and benign lesions, but they do not offer anatomical references, so that in the context of radiotherapy they might be more interesting for evaluation of tumor response after irradiation than for target definition [50]. Regarding the image quality, planning studies demonstrated that images acquired at the MR-Linac are mostly hampered neither by the use of positioning devices nor by breath-related motion [47, 48]. Nevertheless, if needed, MR-navigated or gated sequences might be adopted. To be able to evaluate the impact of the electron stream effect ("Considerations for Breast Treatment Planning at the MR-Linac Global Web Icon"), the simulation scan should cranially include the region of the chin. Caudally, it should include the bottom of the diaphragm. Recommended image slice thickness is 1–2 mm. In the adjuvant setting a 2-mm

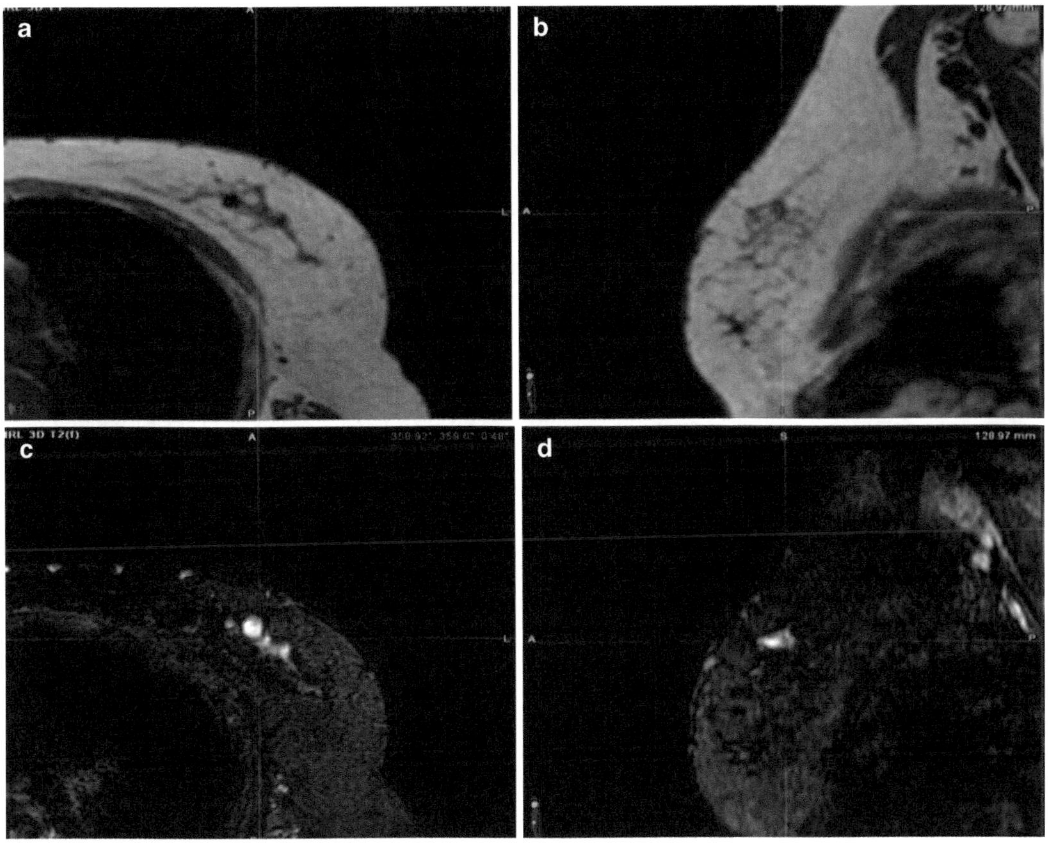


**Fig. 17.1** Example of a patient with a T1b N0 M0 (Stage IA) invasive ductal carcinoma of the left breast treated at an Elekta Unity MR-Linac. The tumor is visible on a 3D T1-weighted sequence (**a**, **b**, axial and sagittal) and a 3D T2-weighted FLAIR sequence (**c**, **d**, axial and sagittal) acquired at the MR-Linac (Courtesy of Adam Currey, Department of Radiation Oncology, Medical College of Wisconsin, Milwaukee, WI, United States)

slice thickness appears appropriate, while in the neoadjuvant setting thinner slices are recommended.

## Contouring: Target Volumes in the Adjuvant Context

As previously discussed, the added value of MR images for target contouring in the context of adjuvant partial breast irradiation is arguable. Nevertheless, in situations where the tumor bed delineation is challenging because few clips are placed, no seroma is visible, and postoperative tissue changes after oncoplastic surgery are difficult to be recognized on CT images, an MR-based contouring might be helpful. The tumor bed should include the lumpectomy cavity, the seroma if present, the clips if present, and the postoperative changes visible on the T1-weighted and T2-weighted images. Based on the IMPORT protocols [17, 55], the CTV is defined as an isotropic expansion of 15 mm of the tumor bed, limited by the anatomical barriers of the ipsilateral breast. Regarding the PTV, as discussed in the previous section, a margin of 5–10 mm is traditionally required for breast treatments. At present, the impact of MR-based plan adaptation on PTV margin reduction has not been clearly quantified yet. In a study conducted at the ViewRay MR $^{60}Co$ system, a PTV reduction of more than 50% could be reached through an MR-based setup correction used to align the surgical cavity [40]. However, other

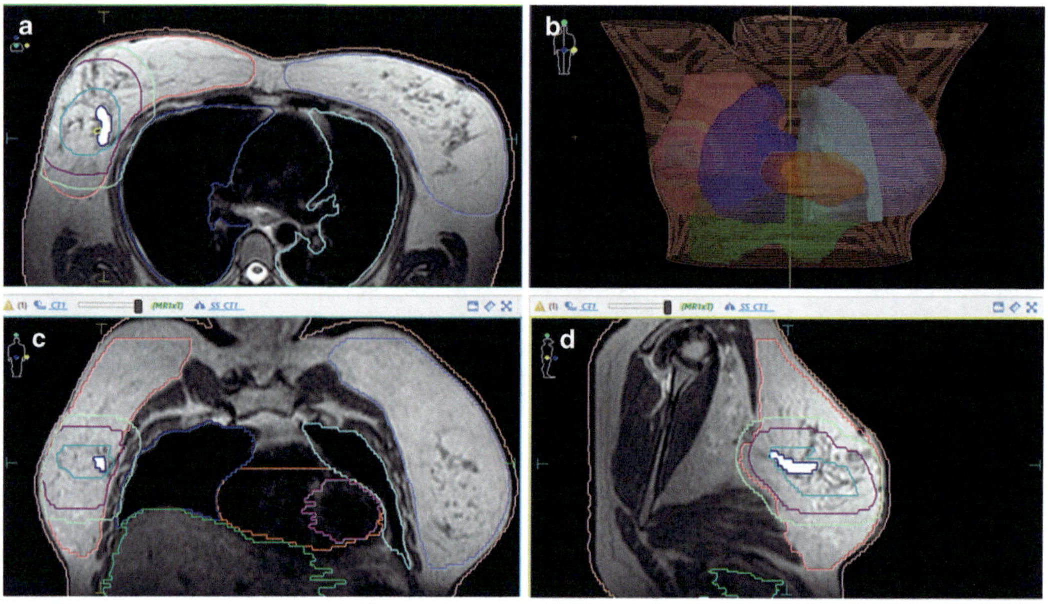

**Fig. 17.2** Example of target contouring for partial breast irradiation on an Elekta Unity MR-Linac in the adjuvant setting. Displayed is a T2-weighted sequence, where the seroma appears bright. The seroma is contoured in dark blue; one clip lays lateral to the seroma (yellow); the tumor bed (emerald green) includes the seroma, the clips, and the postoperative changes; the CTV (violet) is an isotropic expansion of 15 mm, limited to the breast tissue; the PTV (light green) is in this case an isotropic expansion of 10 mm, limited to 5 mm below the patient's surface. (**a**) Axial, (**b**) 3D reconstruction, (**c**) coronal, (**d**) sagittal

investigations suggest that margins of 6–8 mm are still required when performing adjuvant partial breast irradiation at the MR-Linac [56]. Here it should also be mentioned that, compared to traditional treatments, at the MR-Linac the geometric image accuracy represents a potential source of error eventually needed to be compensated within the PTV. Distortions up to 2 mm have been described, especially when the target is distant from the isocenter [57–59]. For breast treatments, this should be considered for targets located in the outer quadrants of big breasts. Finally, a structure for online motion monitoring, e.g., a copy of the PTV, should be defined. An example of target delineation in the adjuvant setting is shown in Fig. 17.2.

## Contouring: Target Volumes in the Neoadjuvant Context

Clear advantages of MR compared to CT have been described in the neoadjuvant setting and contouring guidelines are available [54]. The GTV should include the fiducial marker and the tumor as visible in the T1-weighted gadolinium-enhanced sequence with fat suppression, where it appears hyperintense. Spiculae are included and the GTV should be cross-checked on coronal images. The CTV should be defined as a GTV isotropic expansion of 15–20 mm, limited by the anatomical barriers of the ipsilateral breast [60]. In the setting of neoadjuvant treatments performed at the MR-Linac, smaller PTV margins compared to the adjuvant setting might be used, since it could be assumed that some of the errors contributing to the PTV margins in the adjuvant scenario play here a minor role, as follows. First, due to a better visualization of the target, delineation errors as well as matching errors are minimized. Second, a target deformation, i.e., surgical clip displacement in the adjuvant setting, might be also reduced compared to the adjuvant setting. Other factors contributing to possible errors and vari-

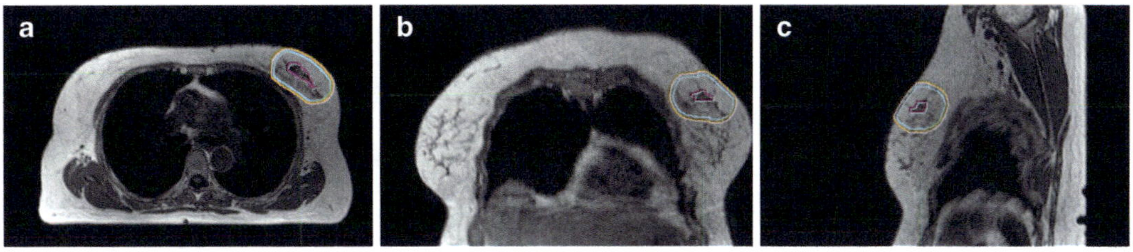


**Fig. 17.3** Example of target contouring for partial breast irradiation on an Elekta Unity MR-Linac in the neoadjuvant setting. The GTV includes the tumor and its spiculae (magenta), the CTV is a 15-mm expansion from the GTV (light blue) and the PTV is a 3-mm expansion from the CTV (orange). (**a**) Axial, (**b**) coronal, (**c**) sagittal (Courtesy of Adam Currey, Department of Radiation Oncology, Medical College of Wisconsin, Milwaukee, WI, United States)

ability, such as intrafraction target motion according to the breathing, geometrical distortion, technical delivery accuracy, and beam penumbra, need to be considered in the neoadjuvant scenario in the same way as in the adjuvant setting. As before, a structure for the motion monitoring, e.g., a copy of the PTV, should be defined. An example of target delineation in the neoadjuvant setting is shown in Fig. 17.3.

## Contouring: Organs at Risk

The following structures are recommended to be identified as organs at risk when performing partial breast irradiation, either neoadjuvant or adjuvant, at the MR-Linac: ipsilateral and contralateral breasts, heart, lungs, and, especially for neoadjuvant partial breast irradiation, chest wall. Additionally, some dummy structures might be required for MR-Linac-specific planning purposes. For the breasts, the breast tissue is best visualized on T1-weighted sequences. T1-weighted mDixon sequences are also preferred for other thoracic organs at risk [61]. To evaluate the electron return effect ("Considerations for Breast Treatment Planning at the MR-Linac Global Web Icon"), two dummy structures might be delineated, namely, a 3–5-mm-thick structure under the skin (between the skin surface and the breast tissue/target volume) and a 3–5-mm-thick structure at the lung/tissue (ribs, intercostal muscles) interface.

## Dose Prescription and DVH Constraints

In the adjuvant setting, there are three preferred fractionation schedules for external beam partial breast irradiation [62]: the 3-week hypofractionated schedule as per IMPORT-LOW, namely, 40 Gy in 15 fractions [17]; the 1-week highly hypofractionated schedule as per FAST-Forward, namely, 26 Gy in 5 fractions [36]; and a 2-week highly hypofractionated schedule based on the Florence protocol, namely, 30 Gy in 5 fractions delivered in nonconsecutive days [63].

In the neoadjuvant setting, different schedules have been applied in clinical trials. Preferred fractionation schedules appear to be 20–24 Gy in one fraction or 30–40 Gy in 3–5 fractions. These fractionations are based on published studies and protocols [51, 52, 60, 64–71] and ongoing trials [64, 70–81]. Depending on the fractionation schedule, different constraints might be used. An example of an MR-Linac treatment plan in the neoadjuvant setting is shown in Fig. 17.4.

A summary of constraints applied in clinical trials of partial breast irradiation, which could be used when performing partial breast irradiation at the MR-Linac, is reported in Table 17.1.

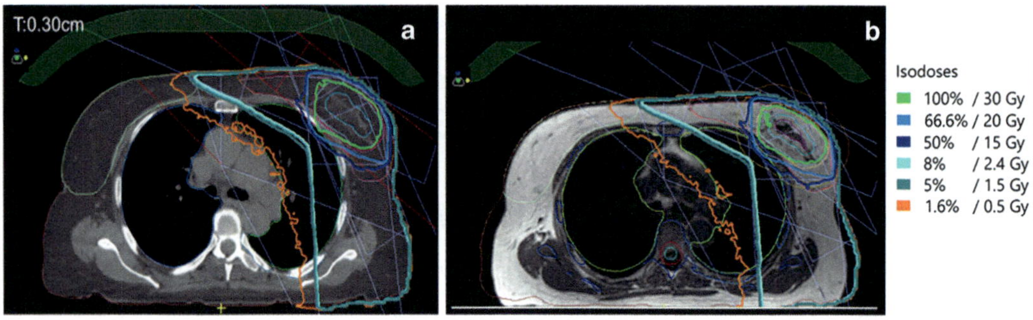


**Fig. 17.4** Treatment plan for a partial breast irradiation on an Elekta Unity MR-Linac in the neoadjuvant setting. The patient was treated with 30 Gy in 5 fractions. (**a**) Base treatment plan, (**b**) the match with the daily MRI at the fifth and last treatment fraction (Courtesy of Adam Currey, Department of Radiation Oncology, Medical College of Wisconsin, Milwaukee, WI, United States)

**Table 17.1** Summary of target coverage and constraints used in different clinical trials in the adjuvant or neoadjuvant setting

| Target/OAR | 30 Gy in 5 fractions [63] | 26 Gy in 5 fractions [36] | Single-dose neoadjuvant APBI [82] |
|---|---|---|---|
| PTV coverage | V28.5 Gy = 100% (100% of PTV covered by 95% of the PD)<br>$D_{max}$ = 31.5 Gy (<105%)<br>$D_{min}$ = 28 Gy | V24.7 Gy = 95% (95% of PTV covered by 95% of the PD)<br>V27.3 Gy <5% (<5% of PTV should receive ≥105% of the PD)<br>V27.8 Gy <2% (<2% of PTV should receive ≥107% of the PD)<br>$D_{max}$ = 28.6 Gy (<110%) | ≥95% dose in ≥97% volume with ≥95% dose in ≥99% of the GTV<br>$D_{max}$ <110% |
| Ipsilateral breast | V15 Gy < 50% | – | $D_{mean}$ ≤4.8 Gy |
| Skin | – | – | $D_{max}$ ≤100% of the PD<br>3 mm: *D* 1 cc ≤13.2 Gy<br>3 mm: *D* 10 cc ≤9.5 Gy<br>5 mm: *D* 1 cc ≤16.3 Gy<br>5 mm: *D* 10 cc ≤12.4 Gy |
| Chest wall | – | – | *D* 20 cc ≤10.2 Gy |
| Heart | V3 Gy <10% | V1.5 Gy <30%<br>V7 Gy <5% | $D_{mean}$ ≤0.6 Gy<br>$D_{max}$ ≤3.4 Gy |
| Ipsilateral lung | V10 Gy <20% | V8 Gy <15% | $D_{mean}$ ≤1.0 Gy |
| Contralateral breast | $D_{max}$ <1 Gy | – | $D_{max}$ ≤1.6 Gy |
| Contralateral lung | V5 Gy <10% | – | $D_{mean}$ ≤0.4 Gy |

*LST* left-sided tumor, *RST* right-sided tumor, *PD* prescribed dose, *Dmean* mean dose, *Dmax* maximal dose, *Dmin* minimal dose

## Considerations for Breast Treatment Planning at the MR-Linac

When performing treatments at the MR-Linac, it must be considered that the presence of the magnetic field impacts the motion of secondary electrons in and around the patient. Compared to the other tumor sites, this is particularly important for patients receiving breast radiotherapy because of the superficial target location. Two effects induced by the magnetic field have to be considered. The electron stream effect may cause an unwanted out-of-field dose deposition to the chin or to the arm of the patient [18, 41, 83–85]. The electron return effect might induce an increased dose to the skin and at the lung/chest wall interface, both air/tissue interfaces [83, 86–88]. It has been shown that both effects are correctly calculated by the modern treatment planning systems [18, 83]. Evaluation in the first patients treated with adjuvant partial breast irradiation at the Elekta Unity MR-Linac showed that the maximal dose to the skin and the interface lung/chest wall was low and that the use of a 1-cm bolus can effectively minimize the dose on the chin or on the arm of the patient [18, 84], as shown in Fig. 17.5. Another possibility might be to shield the upper part of the patient's body with a "bolus curtain" placed approximately in correspondence of the patient's neck (Fig. 17.5). Until now no increased acute skin toxicities have been reported for breast patients treated at the MR-Linac.

Another aspect to consider for breast radiotherapy at the MR-Linac is the possible impact of geometric distortions on treatment planning. Dosimetric variation due to the geometric distortions has been described [58, 89, 90], but this topic needs further investigation in the field of partial breast irradiation. As previously described, the geometric distortions should be taken into account as possible source of errors in the definition of the PTV.

Finally, at the MR-Linacs, volumetric arc techniques are not available and treatments are planned with a step-and-shoot technique. Therefore, the choice of the beam angles is important. It should be considered that some angles are not available or should be avoided in treatment planning at the MR-Linac. Due to technical limitations, angles between 30° and 33° and between 8° and 18° cannot be used at the ViewRay system and at the Elekta system, respectively [91, 92]. Particularly important for

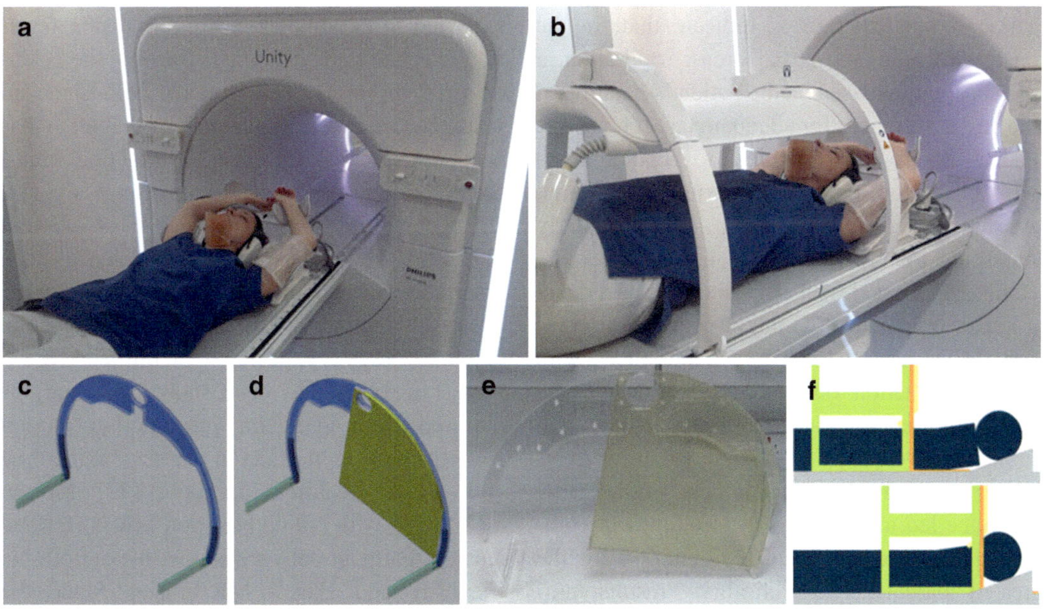


**Fig. 17.5** Setup possibilities to shield the patient from the electron stream effect. (**a**, **b**) A 1-cm bolus can be placed on the patient's arm or chin, depending on the planned out-of-field dose distribution due to the electron stream effect. (**c**–**f**) A "bolus curtain" can be placed between the coil and patient's chin and arms, where a support appositely constructed (**c**: original design) holds a bolus (**d**: original design, **e**: finalized product, **f**: schematic reproduction) (Courtesy of Joan Chick and Jim Sullivan, Royal Marsden NHS Foundation Trust and Institute of Cancer Research, Sutton, United Kingdom)

breast treatments, angles between 130°–150° and 210°–230°, depending on the tumor location and patient's anatomy, should be avoided because of the presence of the couch edges with high-density components [41].

## On-Table Adaptation

The first MR-based breast treatments were performed at the MR $^{60}$Co system, a previous version of the ViewRay MRIdian system which used three $^{60}$Co heads instead of a Linac [40]. In these initial treatments, MR was only used for setup correction; no online adaptation or online motion management was performed. The first experiences with a simple online adaptation based on the patient's position were performed a few years ago at the Elekta MR-Linac [18, 83]. The adapt-to-position method allowed for a quicker adaptation compared to a full adaptation where the target and the organs at risk are recontoured and a completely new plan is generated. These initial experiences indicate that, in the context of adjuvant partial breast irradiation, a plan adaptation based on the patient's position of the day is sufficient for adequate target coverage and allows for the treatment to be completed with an in-room time of approximately 25 min. A short treatment time appears particularly important for breast patients, because they are commonly treated with arms above the head and a longer treatment time might result in discomfort. More recently, the first experiences with online adaptation on the MRIdian MR-Linac were reported for partial breast irradiation in prone position [45]. Use of daily adaptive treatment planning was feasible and online MRI guidance led to smaller target volumes in comparison to conventional CT simulation for the treated patients. The preliminary results indicated that there may be potential to decrease CTV and PTV with the use of online MRI guidance. It should be noticed that no contrast agent was used in aforementioned experiences. In the context of neoadjuvant partial breast irradiation, the use of gadolinium is recommended for target delineation, as described in the previous section. However, currently, contrast agent injection is not recommended for imaging before irradiation, because of potential radiosensitizing properties [93] and the lack of data regarding its use in combination with radiotherapy. In the neoadjuvant setting, as for adjuvant partial breast irradiation, a rigid registration with a simple plan adaptation based on the patient's position of the day appears reasonable, for different reasons. First, tumors are small, and without contrast agent, it might be difficult to exactly identify the lesion edges and its spiculae for precise recontouring. Second, as described in the previous section, neoadjuvant treatments are performed in very few fractions, so that significant changes in the target and organs at risk anatomy (e.g., tumor response, breast swelling) are not to be expected. Nevertheless, in case of significant changes, a full plan adaptation is recommended.

Especially in the context of neoadjuvant partial breast irradiation, the management of the intrafraction motion is particularly important. First, as previously mentioned, the PTVs are expected to be smaller. Second, if using fewer fractions and higher dose per fraction, the beam-on time is longer. Third, at the MR-Linac no volumetric arc techniques are available. It has been documented that the intrafraction breast target motion, mostly due to breathing, is generally regular and limited to a median of 2–3 mm [37, 40, 94, 95]. Larger variations due to deep inspiration or whole-body displacements were observed but rare, and therefore, their impact on target coverage is expected to be limited [40, 94, 95]. However, it has been shown that margins required to account for intrafraction motion without motion mitigation strategies increase with longer treatment durations [37]. Motion monitoring with online beam-on MR performed during treatment delivery is possible at both the ViewRay and Elekta MR-Linacs. Since the recent implementation on the Unity system, gating techniques are available at both systems as well. It should be noted that gating techniques might prolong the delivery time. Cine gating appeared to be an important reason for selecting MR-guided radiotherapy for accelerated partial breast irradiation [96].

## Outcome

Adjuvant partial breast irradiation treatments have been successfully performed at the ViewRay 0.35 T MR $^{60}$Co system [40, 96, 97], at the ViewRay 0.35 T MR-Linac [45], and at the Elekta 1.5 T Unity MR-Linac [18, 83]. At Washington University, almost 200 breast patients were treated at the MR 60Co system [96], but oncological outcomes and toxicity were not described in detail. At Seoul University, 173 patients received adjuvant partial breast irradiation at the MR $^{60}$Co system, with very good results in terms of both toxicity, with no patients with more than grade 3 toxicity, and oncological outcome, with recurrence-free survival rates close to 100% after a 3-year follow-up. A smaller cohort of 11 patients was treated with adjuvant partial breast irradiation at the 1.5 T Elekta MR-Linac with excellent results, i.e., no patients with more than grade 2 toxicity and good or excellent cosmetic results. There were no relapses, but the number of patients was limited and the follow-up short [18, 98]. Also, the small cohort of 11 patients treated with adjuvant partial breast irradiation in prone position at an ViewRay MR-Linac showed no grade 2 or higher toxicities, though reported results only covered a median follow-up of 6 months [45].

Different groups have published their experience with external beam neoadjuvant partial breast radiotherapy performed at a CT-Linac or CyberKnife [51, 52, 64–68, 70]. MR-based radiological and pathological responses ranged between 17–42% [64, 65, 68] and 9–42% [65, 66, 68, 70], respectively. Delayed surgery performed 6–8 months instead of few weeks after radiotherapy appears to correlate with a higher radiological and pathological complete response. These studies could also document the safe profile of high-dose neoadjuvant partial breast irradiation, with only very few cases of grade 3 toxicities, the large majority of patients reporting excellent or good cosmetic results, and very few local relapses. Until now, no reports have been published on neoadjuvant partial breast irradiation performed at an MR-Linac.

Ongoing studies are currently investigating neoadjuvant partial breast irradiation at the MR-Linac [72, 75, 76]. The advantages of MR-guided partial breast radiotherapy have been discussed elsewhere in this chapter and are represented by more precise target definition and margin reduction due to the possibility of a MR-based online plan adaptation and online motion monitoring. Additional advantages may be the use of DCE- and DWI-MR to provide (semi)quantitative parameters, which could be investigated as biomarkers for response assessment after neoadjuvant radiotherapy [21, 64, 99–101].

## Summary and Conclusion

MR-guided radiotherapy for breast cancer patients has no widespread clinical application yet. MR-guided breast radiotherapy may help to minimize the irradiated breast volume in partial breast irradiation by aiding target definition for neoadjuvant partial breast irradiation and by potential margin reduction for both adjuvant and neoadjuvant partial breast irradiation through online adaptation and monitoring when performing treatment on an MR-Linac. MR-Linac-delivered adjuvant partial breast irradiation has been demonstrated feasible and studies evaluating MR-Linac treatment in the neoadjuvant setting are ongoing.

## References


1. van der Leij F, et al. Target volume delineation in external beam partial breast irradiation: less inter-observer variation with preoperative-compared to postoperative delineation. Radiother Oncol. 2014;110(3):467–70.
2. E.E.C.I. System. J.R. Centre, Editor. 2020. https://ecis.jrc.ec.europa.eu/pdf/Breast_cancer_factsheet-Oct_2020.pdf.
3. U.S. Cancer Statistics, T.O.F.C.S.C.f.D.C.a.P. Cancer statistics at a glance, 2019. https://gis.cdc.gov/Cancer/USCS/#/AtAGlance/.
4. World Cancer Research. 2022. https://www.wcrf.org/cancer-trends/breast-cancer-statistics.
5. Siegel RL, et al. Cancer statistics, 2022. CA Cancer J Clin. 2022;72(1):7–33.

6. Eurostat. 2023. https://ec.europa.eu/eurostat/statistics-explained/index.php?title=Causes_of_death_statistics#Major_causes_of_death_in_the_EU_in_2017.
7. Arnold M, et al. Current and future burden of breast cancer: global statistics for 2020 and 2040. Breast. 2022;66:15–23.
8. DeSantis CE, et al. Breast cancer statistics, 2019. CA Cancer J Clin. 2019;69(6):438–51.
9. Kunkler IH, et al. Breast-conserving surgery with or without irradiation in early breast cancer. N Engl J Med. 2023;388(7):585–94.
10. Lee BM, et al. Hypofractionated radiotherapy dose scheme and application of new techniques are associated to a lower incidence of radiation pneumonitis in breast cancer patients. Front Oncol. 2020;10:124.
11. Yeboa DN, Evans SB. Contemporary breast radiotherapy and cardiac toxicity. Semin Radiat Oncol. 2016;26(1):71–8.
12. Taylor C, et al. Estimating the risks of breast cancer radiotherapy: evidence from modern radiation doses to the lungs and heart and from previous randomized trials. J Clin Oncol. 2017;35(15):1641–9.
13. NCCN Clinical Practice Guidelines in Oncology, Breast Cancer, N.C.C. Network, Editor. 2023. p. BINV-I.
14. Interdisziplinäre S3-Leitlinie für die Früherkennung, Diagnostik, Therapie und Nachsorge des Mammakarzinoms. 2021. p. 150.
15. Radiologists TRCo. Postoperative radiotherapy for breast cancer: UK consensus statements. 2016, update 2021.
16. Meattini I, et al. Accelerated partial-breast irradiation compared with whole-breast irradiation for early breast cancer: long-term results of the randomized phase III APBI-IMRT-Florence Trial. J Clin Oncol. 2020;38(35):4175–83.
17. Coles CE, et al. Partial-breast radiotherapy after breast conservation surgery for patients with early breast cancer (UK IMPORT LOW trial): 5-year results from a multicentre, randomised, controlled, phase 3, non-inferiority trial. Lancet. 2017;390(10099):1048–60.
18. De-Colle C, et al. Analysis of the electron-stream effect in patients treated with partial breast irradiation using the 1.5 T MR-linear accelerator. Clin Transl Radiat Oncol. 2021;27:103–8.
19. Hepel JT, et al. Toxicity of three-dimensional conformal radiotherapy for accelerated partial breast irradiation. Int J Radiat Oncol Biol Phys. 2009;75(5):1290–6.
20. Jagsi R, et al. Unacceptable cosmesis in a protocol investigating intensity-modulated radiotherapy with active breathing control for accelerated partial-breast irradiation. Int J Radiat Oncol Biol Phys. 2010;76(1):71–8.
21. Civil YA, et al. Preoperative partial breast irradiation in patients with low-risk breast cancer: a systematic review of literature. Ann Surg Oncol. 2023;30:3263.
22. van Mourik AM, et al. Multiinstitutional study on target volume delineation variation in breast radiotherapy in the presence of guidelines. Radiother Oncol. 2010;94(3):286–91.
23. Landis DM, et al. Variability among breast radiation oncologists in delineation of the postsurgical lumpectomy cavity. Int J Radiat Oncol Biol Phys. 2007;67(5):1299–308.
24. Al-Hammadi N, et al. MRI reduces variation of contouring for boost clinical target volume in breast cancer patients without surgical clips in the tumour bed. Radiol Oncol. 2017;51(2):160–8.
25. Jolicoeur M, et al. Localization of the surgical bed using supine magnetic resonance and computed tomography scan fusion for planification of breast interstitial brachytherapy. Radiother Oncol. 2011;100(3):480–4.
26. Mast M, et al. Target volume delineation in breast conserving radiotherapy: are co-registered CT and MR images of added value? Radiat Oncol. 2014;9:65.
27. Pogson EM, et al. Comparison of magnetic resonance imaging and computed tomography for breast target volume delineation in prone and supine positions. Int J Radiat Oncol Biol Phys. 2016;96(4):905–12.
28. Giezen M, et al. MRI- versus CT-based volume delineation of lumpectomy cavity in supine position in breast-conserving therapy: an exploratory study. Int J Radiat Oncol Biol Phys. 2012;82(4):1332–40.
29. Den Hartogh MD, et al. Post-lumpectomy CT-guided tumor bed delineation for breast boost and partial breast irradiation: can additional pre- and postoperative imaging reduce interobserver variability? Oncol Lett. 2015;10(5):2795–801.
30. Sabine B, Giovanna D, Peter P, Clara J, Bert P, John K. Open low-field magnetic resonance (MR) versus CT scanner (CT) imaging in breast radiotherapy treatment planning. Int J Radiat Oncol Biol Phys. 2005;63:S232–3.
31. Kirby AM, et al. Tumor bed delineation for partial breast and breast boost radiotherapy planned in the prone position: what does MRI add to X-ray CT localization of titanium clips placed in the excision cavity wall? Int J Radiat Oncol Biol Phys. 2009;74(4):1276–82.
32. den Hartogh MD, et al. MRI and CT imaging for preoperative target volume delineation in breast-conserving therapy. Radiat Oncol. 2014;9:63.
33. Tersteeg RJ, et al. Changes in excision cavity volume: prediction of the reduction in absolute volume during breast irradiation. Int J Radiat Oncol Biol Phys. 2009;74(4):1181–5.
34. Topolnjak R, et al. Breast-conserving therapy: radiotherapy margins for breast tumor bed boost. Int J Radiat Oncol Biol Phys. 2008;72(3):941–8.
35. Jeon SH, et al. Seroma change during magnetic resonance imaging-guided partial breast irradiation and its clinical implications. Radiat Oncol. 2017;12(1):103.

36. Murray Brunt A, et al. Hypofractionated breast radiotherapy for 1 week versus 3 weeks (FAST-forward): 5-year efficacy and late normal tissue effects results from a multicentre, non-inferiority, randomised, phase 3 trial. Lancet. 2020;395(10237):1613–26.
37. Hoekstra N, et al. Intrafraction motion during partial breast irradiation depends on treatment time. Radiother Oncol. 2021;159:176–82.
38. Polgár C, et al. Patient selection for accelerated partial-breast irradiation (APBI) after breast-conserving surgery: recommendations of the Groupe Européen de Curiethérapie-European Society for Therapeutic Radiology and Oncology (GEC-ESTRO) breast cancer working group based on clinical evidence (2009). Radiother Oncol. 2010;94(3):264–73.
39. Correa C, et al. Accelerated partial breast irradiation: executive summary for the update of an ASTRO evidence-based consensus statement. Pract Radiat Oncol. 2017;7(2):73–9.
40. Acharya S, et al. Magnetic resonance image guided radiation therapy for external beam accelerated partial-breast irradiation: evaluation of delivered dose and intrafractional cavity motion. Int J Radiat Oncol Biol Phys. 2016;96(4):785–92.
41. Groot Koerkamp ML, et al. Optimizing MR-guided radiotherapy for breast cancer patients. Front Oncol. 2020;10:1107.
42. Moore-Palhares D, et al. Clinical implementation of magnetic resonance imaging simulation for radiation oncology planning: 5 year experience. Radiat Oncol. 2023;18(1):27.
43. Charaghvandi KR, et al. Single dose partial breast irradiation using an MRI linear accelerator in the supine and prone treatment position. Clin Transl Radiat Oncol. 2019;14:1–7.
44. Groot Koerkamp ML, et al. Prone vs. supine accelerated partial breast irradiation on an MR-Linac: a planning study. Radiother Oncol. 2021;165:193–9.
45. Ng J, Pennell R, Formenti SC. The initial experience of MRI-guided precision prone breast irradiation with daily adaptive planning in treating early stage breast cancer patients. Front Oncol. 2022;12:1048512.
46. Paulson ES, et al. Comprehensive MRI simulation methodology using a dedicated MRI scanner in radiation oncology for external beam radiation treatment planning. Med Phys. 2015;42(1):28–39.
47. Ahn KH, et al. MRI guidance for accelerated partial breast irradiation in prone position: imaging protocol design and evaluation. Int J Radiat Oncol Biol Phys. 2009;75(1):285–93.
48. Batumalai V, et al. Assessment of MRI image quality for various setup positions used in breast radiotherapy planning. Radiother Oncol. 2016;119(1):57–60.
49. Liney GP, Moerland MA. Magnetic resonance imaging acquisition techniques for radiotherapy planning. Semin Radiat Oncol. 2014;24(3):160–8.
50. Vasmel JE, Koerkamp G, Kirby A, Russell N, Shaitelman S, Vesprini D, et al. Evaluation of MRI-based guidelines for contouring tumors for preoperative partial breast irradiation. Radiother Oncol. 2019;133:S719.
51. Horton JK, et al. Preoperative single-fraction partial breast radiation therapy: a novel phase 1, dose-escalation protocol with radiation response biomarkers. Int J Radiat Oncol Biol Phys. 2015;92(4):846–55.
52. Guidolin K, et al. Stereotactic image-guided neoadjuvant ablative single-dose radiation, then lumpectomy, for early breast cancer: the SIGNAL prospective single-arm trial of single-dose radiation therapy. Curr Oncol. 2019;26(3):e334–40.
53. Mouawad M, et al. Reducing the dose of gadolinium-based contrast agents for DCE-MRI guided SBRT: the effects on inter and intra observer variability for preoperative target volume delineation in early stage breast cancer patients. Radiother Oncol. 2019;131:60–5.
54. Vasmel JE, et al. Consensus on contouring primary breast tumors on MRI in the setting of neoadjuvant partial breast irradiation in trials. Pract Radiat Oncol. 2020;10(6):e466–74.
55. The Institute of Cancer Research, London. 2013. https://www.icr.ac.uk/media/docs/default-source/default-document-library/import-high-planning-pack-version-4-0-final-23052013-edited.pdf?sfvrsn=6d2c2169_0.
56. Tan IZ, Mitchell A, McNair H, Dunlop A, Herbert T, Nartey J, Lawes R, O'Connell N, Cooper D, De-Colle C, Diaz J, Donovan EK, Han K, Hahn E, Nelms B, Russell N, Kirby AM, on behalf of the MR-Linac Consortium Breast Tumour Working Group. A multicentre study of clinical to planning target volume margins for adjuvant partial breast irradiation delivered on the 1.5 T MR-Linear accelerator. Int J Radiat Oncol Biol Phys. 2023;117(2):e725.
57. Ginn JS, et al. Characterization of spatial distortion in a 0.35 T MRI-guided radiotherapy system. Phys Med Biol. 2017;62(11):4525–40.
58. Fransson A, Andreo P, Pötter R. Aspects of MR image distortions in radiotherapy treatment planning. Strahlenther Onkol. 2001;177(2):59–73.
59. Walker A, et al. MRI distortion: considerations for MRI based radiotherapy treatment planning. Australas Phys Eng Sci Med. 2014;37(1):103–13.
60. Charaghvandi RK, et al. Redefining radiotherapy for early-stage breast cancer with single dose ablative treatment: a study protocol. BMC Cancer. 2017;17(1):181.
61. Whiteside L, et al. To see or not to see: evaluation of magnetic resonance imaging sequences for use in MR-Linac-based radiotherapy treatment. J Med Imaging Radiat Sci. 2022;53(3):362–73.
62. Meattini I, et al. European Society for Radiotherapy and Oncology Advisory Committee in Radiation Oncology Practice consensus recommendations on patient selection and dose and fractionation for external beam radiotherapy in early breast cancer. Lancet Oncol. 2022;23(1):e21–31.

63. Livi L, et al. Accelerated partial breast irradiation using intensity-modulated radiotherapy versus whole breast irradiation: 5-year survival analysis of a phase 3 randomised controlled trial. Eur J Cancer. 2015;51(4):451–63.
64. Weinfurtner RJ, et al. Quantitative changes in intratumoral habitats on MRI correlate with pathologic response in early-stage ER/PR+ HER2− breast cancer treated with preoperative stereotactic ablative body radiotherapy. J Breast Imaging. 2022;4(3):273–84.
65. Bosma SCJ, et al. Five-year results of the preoperative accelerated partial breast irradiation (PAPBI) trial. Int J Radiat Oncol Biol Phys. 2020;106(5):958–67.
66. Nichols E, et al. Preoperative accelerated partial breast irradiation for early-stage breast cancer: preliminary results of a prospective, phase 2 trial. Int J Radiat Oncol Biol Phys. 2017;97(4):747–53.
67. Tiberi D, et al. Tumour response 3 months after neoadjuvant single-fraction radiotherapy for low-risk breast cancer. Curr Oncol. 2020;27(3):155–8.
68. Vasmel JE, et al. Tumor response after neoadjuvant magnetic resonance guided single ablative dose partial breast irradiation. Int J Radiat Oncol Biol Phys. 2020;106(4):821–9.
69. Vanderwalde NA, et al. Phase 2 study of preexcision single-dose intraoperative radiation therapy for early-stage breast cancers: 6-year update with application of the ASTRO accelerated partial breast irradiation consensus statement criteria. Cancer. 2013;119(9):1736–43.
70. Meattini I, et al. Preoperative robotic radiosurgery for early breast cancer: results of the phase II ROCK trial (NCT03520894). Clin Transl Radiat Oncol. 2022;37:94–100.
71. Zerella MA, et al. Single fraction ablative preoperative radiation treatment for early-stage breast cancer: the CRYSTAL study—a phase I/II clinical trial protocol. BMC Cancer. 2022;22(1):358.
72. van den Bongard HJGD. Single-dose preoperative partial breast irradiation in low-risk breast cancer patients (ABLATIVE-2). ClinicalTrials.gov Identifier: NCT05350722.
73. Brackstone M. Stereotactic image-guided neoadjuvant ablative radiation then lumpectomy (SIGNAL 2). ClinicalTrials.gov Identifier: NCT02212860.
74. Yassa M. Single pre-operative radiation therapy—with delayed or no surgery (SPORT-DNS). ClinicalTrials.gov Identifier: NCT05217966.
75. Currey A. MRI-based preoperative accelerated partial breast irradiation. ClinicalTrials.gov Identifier: NCT02728076.
76. Shaitelman SF. Radiation therapy before surgery in treating patients with hormone receptor positive, HER2 negative breast cancer. ClinicalTrials.gov Identifier: NCT03359954.
77. Wapnir I. Phase 2 surgical excision vs neoadjuvant radiotherapy + delayed surgical excision of ductal carcinoma (NORDIS). ClinicalTrials.gov Identifier: NCT03909282.
78. Rahimi A. A phase I dose escalation study of single fraction ablative pre-operative partial breast (S-PBI) for early stage breast cancer. ClinicalTrials.gov Identifier: NCT04040569.
79. Takita C. Study for selected early stage breast cancer (SABER). ClinicalTrials.gov Identifier: NCT04360330.
80. Cedrone L. SBRT in early breast cancer in elderly women. ClinicalTrials.gov Identifier: NCT05071105.
81. Blitzblau R. Preoperative single-fraction radiotherapy in early stage breast cancer. ClinicalTrials.gov Identifier: NCT02482376.
82. Charaghvandi RK, et al. Treatment constraints for single dose external beam preoperative partial breast irradiation in early-stage breast cancer. Clin Transl Radiat Oncol. 2017;6:7–14.
83. Nachbar M, et al. Partial breast irradiation with the 1.5 T MR-Linac: first patient treatment and analysis of electron return and stream effects. Radiother Oncol. 2020;145:30–5.
84. Park JM, et al. Air-electron stream interactions during magnetic resonance IGRT: skin irradiation outside the treatment field during accelerated partial breast irradiation. Strahlenther Onkol. 2018;194(1):50–9.
85. Hackett SL, et al. Spiraling contaminant electrons increase doses to surfaces outside the photon beam of an MRI-linac with a perpendicular magnetic field. Phys Med Biol. 2018;63(9):095001.
86. van Heijst TC, et al. MR-guided breast radiotherapy: feasibility and magnetic-field impact on skin dose. Phys Med Biol. 2013;58(17):5917–30.
87. Raaijmakers AJ, Raaymakers BW, Lagendijk JJ. Integrating a MRI scanner with a 6 MV radiotherapy accelerator: dose increase at tissue-air interfaces in a lateral magnetic field due to returning electrons. Phys Med Biol. 2005;50(7):1363–76.
88. Kim A, et al. Magnetic field dose effects on different radiation beam geometries for hypofractionated partial breast irradiation. J Appl Clin Med Phys. 2017;18(6):62–70.
89. Prott FJ, et al. Effect of distortions and asymmetry in MR images on radiotherapeutic treatment planning. Int J Cancer. 2000;90(1):46–50.
90. Walker A, et al. MRI geometric distortion: impact on tangential whole-breast IMRT. J Appl Clin Med Phys. 2016;17(5):7–19.
91. Klüter S. Technical design and concept of a 0.35 T MR-Linac. Clin Transl Radiat Oncol. 2019;18:98–101.
92. Bijman R, et al. MR-Linac radiotherapy—the beam angle selection problem. Front Oncol. 2021;11:717681.
93. Taupin F, et al. Gadolinium nanoparticles and contrast agent as radiation sensitizers. Phys Med Biol. 2015;60(11):4449–64.
94. van Heijst TC, et al. Quantification of intrafraction motion in breast radiotherapy using supine magnetic resonance imaging. Phys Med Biol. 2016;61(3):1352–70.

95. Groot Koerkamp ML, et al. Intrafraction motion during radiotherapy of breast tumor, breast tumor bed, and individual axillary lymph nodes on cine magnetic resonance imaging. Phys Imaging Radiat Oncol. 2022;23:74–9.
96. Henke LE, et al. Magnetic resonance image-guided radiotherapy (MRIgRT): a 4.5-year clinical experience. Clin Oncol (R Coll Radiol). 2018;30(11):720–7.
97. Lee HI, et al. The acute and late toxicities of MRI-guided external beam partial breast irradiation delivered using a once-per-day regimen. Front Oncol. 2021;11:649301.
98. De-Colle C, et al. Estimation of secondary cancer projected risk after partial breast irradiation at the 1.5 T MR-linac. Strahlenther Onkol. 2022;198(7):622–9.
99. Mouawad M, et al. DCE-MRI assessment of response to neoadjuvant SABR in early stage breast cancer: comparisons of single versus three fraction schemes and two different imaging time delays post-SABR. Clin Transl Radiat Oncol. 2020;21:25–31.
100. Wang C, et al. Assessment of treatment response with diffusion-weighted MRI and dynamic contrast-enhanced MRI in patients with early-stage breast Cancer treated with single-dose preoperative radiotherapy: initial results. Technol Cancer Res Treat. 2016;15(5):651–60.
101. Vasmel JE, et al. Dynamic contrast-enhanced and diffusion-weighted magnetic resonance imaging for response evaluation after single-dose ablative neoadjuvant partial breast irradiation. Adv Radiat Oncol. 2022;7(2):100854.

# 18 Considerations for Using MR Linac for the Treatment of Patients with Gynecologic Cancer: A Practical Guide and Early Clinical Experience

Lorraine Portelance, Ina Jürgenliemk-Schulz, Kyle R. Padgett, Rosa Patricia Castillo, and Astrid van Lier

## Introduction

Radiation therapy (RT) plays a major role in the management of patients with gynecologic cancer. Cervical cancer is the fourth most frequent cancer in women worldwide, with an estimated 604,000 new cases in 2020. However, the incidence of cervical cancer is much higher in low- and middle-income countries [1]. According to the American Cancer Society, cervical cancer incidence in the United States has decreased by more than half since the mid-1970s and other Western countries have observed similar trends [2]. The decline in the incidence rate of cervical cancer seen in developed countries is attributed to successful screening and HPV vaccination programs. The standard of care for the treatment of locally advanced cervical cancer (LACC) is a combination of chemotherapy and radiation therapy (external beam and brachytherapy [BT]). Vaginal cancers are treated with a similar approach. Both pathologies being relatively rare in developed countries, these cancers are ideally treated in centers with expertise in BT. MR-Linac is a technology which allows for the delivery of external beam radiation with higher precision, and patients diagnosed with LACC might benefit from this technology. This chapter will present an overview of these potential opportunities.

Endometrial cancer is by far the most frequent gynecologic malignancy in Western countries. While surgery is the primary treatment for uterine cancer, radiation is key for adjuvant treatment for patients with a higher risk of recurrence (higher-grade disease, deep myometrial invasion, lymphovascular invasion, higher FIGO stage). The dose of radiation delivered in the adjuvant setting, however, is relatively low, and the conformality provided by endocavitary BT and pelvic intensity-modulated radiation therapy (IMRT) is associated with good treatment outcomes. However, the management of patients with inoperable endometrial cancer requires a higher dose of radiation, and MR-Linacs might become an excellent treatment option for these patients.

L. Portelance (✉) · K. R. Padgett
Radiation Oncology Department, Sylvester Comprehensive Cancer Center, Miller School of Medicine, University of Miami, Miami, FL, USA
e-mail: lportelance@med.miami.edu

I. Jürgenliemk-Schulz · A. van Lier
Department of Radiation Oncology, University Medical Center Utrecht, University Utrecht, Utrecht, The Netherlands
e-mail: I.M.Schulz@umcutrecht.nl

R. P. Castillo
Radiology Department, Miller School of Medicine, University of Miami, Miami, FL, USA



I. J. Das et al. (eds.), *A Practical Guide to MR-Linac*, https://doi.org/10.1007/978-3-031-48165-9_18

Radiation is also frequently used as a salvage treatment of recurrent gynecologic cancer, either alone or in combination with chemotherapy or surgery. Additionally, there is a fast-growing literature demonstrating that radiation has a role to play in the management of oligometastatic disease for many disease sites, including gynecologic malignancies [3].

During the last three decades, research in oncology focused on developing systemic agents offering personalized medicine options and adapting the systemic treatment to specific tumor characteristics (targeted agents and immunotherapy, preferably directed to specific genetic signatures). In parallel, RT has significantly benefited from technological advances that lead to improved treatment precision. Breakthroughs in dynamic multi-leaf collimator design and treatment planning systems have improved treatment delivery conformality, translating into decreased treatment toxicity and a better outcome for patients with gynecologic cancer [4].

The introduction of online MRI-guided external beam RT (MRgRT) in 2016 marked a significant milestone and effort to increase the precision of RT delivery. The MR-Linacs provide the advantage of onboard MRI combined with the possibility of performing onboard adaptive planning when necessary. In parallel with the implementation of MRgRT, numerous publications showed the benefits of MRI-guided BT in the same years [5, 6].

The ability to perform online adaptive radiotherapy requires daily imaging datasets with enough quality to delineate both targets and organs at risk (OARs) effectively. These datasets need to be artifact resistant and spatially accurate. Additionally, these high-fidelity onboard imaging systems must be coupled with a fast and robust adaptive treatment planning system and a modern radiation delivery platform. With these requirements in place, onboard imaging and online replanning based on the daily anatomy allows for dose escalation and better sparing of OARs. It is important to note that onboard MRI provides superior soft tissue contrast resolution compared to cone beam computed tomography (CBCT). This increased image quality is essential for online replanning for certain body sites, like the pelvis.

MRI is now commonly used at initial diagnosis for identifying tumors and differentiating them from adjacent organs in the management of patients with gynecologic cancer. Gynecologic cancers often have poor contrast resolution compared to the surrounding soft tissues on CT scans, and consequently, MRI is the diagnostic imaging modality of choice. Beyond its superiority for diagnostic purposes, MRI plays a role throughout all stages of care, including EBRT planning, outcome evaluation, and detecting tumor recurrence.

Cervical, vaginal, endometrial, and oligometastatic gynecologic cancers are being treated on MR-Linacs in increasing numbers worldwide. This chapter provides an overview of the principles of MR imaging for gynecologic cancer, followed by the current state and potential future directions for the clinical use of MR-Linacs in the management of gynecologic cancers.

## Principles of Magnetic Resonance Imaging for Gynecologic Cancer

MRI scanners were first introduced for clinical use just 40 years ago. Significant technology development followed, and today, MRI is a diagnostic tool widely used in oncology. As early as 1990, MRI was described as a promising tool in the management of gynecologic cancers providing superior visualization of tumors and adjacent pelvic organs compared to CT scan [7]. A review published in 1992 by Russell highlighted the potential for MRI guidance to avoid marginal tumor misses in EBRT treatment for gynecologic cancers [8]. In the early 2000s, Potter et al. developed a cervical BT planning approach for patients with LACC, integrating the use of pelvic MRI to delineate tumor targets according to possible remaining cell density at the time of BT planning [9]. Subsequently, efforts of the American Brachytherapy Society (ABS) [10] and the Groupe Europeen de Curietherapie European Society for Radiation Oncology (GEC ESTRO) [11] led to further development of the concepts.

Twenty years later, there is a wealth of literature confirming that MRI-guided adaptive BT leads to better treatment outcomes, particularly in LACC cases [12], paving the way for the use of MRI in treatment planning for EBRT as well.

## Pelvic MRI for Initial Staging of Gynecologic Malignancies

### Initial Staging of Cervical and Vaginal Cancer

MRI plays a vital role in the initial staging of cervix cancer [13], and its findings are now included in the International Federation of Gynecology and Obstetrics (FIGO) and American Joint Committee on Cancer (AJCC) cervical cancer staging system [14]. MRI's high soft tissue contrast resolution and ability to detect disease extent in more than one plane make it the preferred imaging modality for detecting cervical and vaginal cancer [15, 16].

Although gadolinium is not nephrotoxic, a review of recent laboratories should always be done before ordering a contrast-enhanced (CE) MRI; gadolinium should be used cautiously in patients with severe renal insufficiency since these patients are at higher risk of nephrogenic systemic fibrosis. A careful review of creatinine levels should be done when prescribing a CE pelvic MRI for patients with LACC since hydronephrosis and increased creatinine level are common findings in patients with FIGO stage IIIB or IVA LACC.

To minimize bowel motion during pelvic image acquisition, patients may be advised to fast for 4 hours before the examination. Some medical facilities may also administer medication such as scopolamine butyl bromide (Buscopan) or glucagon to decrease peristalsis, although this may cause short-time nausea. Additionally, a low-fiber diet to reduce bowel gas may be recommended to begin the day before the diagnostic MRI.

The two most common field strengths employed when performing diagnostic imaging for cervical cancer are 1.5 tesla and 3.0 tesla. While both field strengths are more than capable of collecting images of sufficient quality to properly diagnose and stage cervical cancer, 3.0 tesla systems have some distinct advantages with a few potential drawbacks as well. The higher field strength of 3.0 tesla systems provides a higher signal-to-noise ratio which is primarily used to increase resolution, improve contrast, and speed up acquisition. On the other hand, the higher field strengths of 3.0 tesla systems come with increased image artifacts when encountering bowel or rectal gas and when hip prostheses or other metal implants are found in the pelvis: surgical clips, etc. These artifacts are more pronounced on certain pulse sequences (gradient recalled echo, echo planar imaging, diffusion-weighted imaging [DWI], among others), many of which are used for functional imaging. To summarize, both 1.5 tesla and 3.0 tesla produce good-quality images, making either choice a suitable option. Hence, 1.5 tesla systems are significantly more common than 3.0 tesla systems and provide images of sufficient quality for diagnosis and staging of cervical cancer, and 3.0 tesla systems generally provide images with superior resolution, albeit with a significantly higher cost, and may produce inferior results in certain patients with excessive bowel gas or with exogenous implanted objects. The key sequences for the evaluation of cervical cancer are the multiplanar fast spin-echo T2-weighted images without fat saturation. Examples of image acquisition protocols used for a diagnostic study and treatment planning simulation pelvic MRI for cervical cancer are presented in Tables 18.1 and 18.2.

During image acquisition for diagnostic purposes, a large field of view is first used to screen the lower abdomen and pelvis for lymphadenopathy, bone metastasis, or any other incidental findings (5–6 mm thickness through the entire low abdomen and pelvis). In the localizer, sagittal T2-weighted images are used to locate the tumor in the cervix and plan the high-spatial-resolution T2-weighted images in axial and coronal oblique planes. The possibility that MRI offers to perform image acquisition in two orthogonal planes along the tumor axis provides important information on disease extent for cervical cancer staging. The cervical tumor size and information showing

**Table 18.1** MRI protocols for image acquisition in the management of cervical cancer: (a) University of Miami: diagnostic image acquisition protocol

| | Planes | Sequences | |
|---|---|---|---|
| Non-contrast (localizer) | Localizer (three planes)<br>Coronal<br>Axial oblique (perpendicular to the cervical canal)<br>Sagittal | T2, TSE, TSE FS<br>T2, TSE<br>T2, TSE<br>T2, TSE | 6 mm (whole pelvis)<br>6 mm (from renal hilum to pubic symphysis)<br>3.5 mm (small FOV)<br>4 mm (small FOV) |
| Contrast gadolinium | Axial<br>Coronal<br>Axial | T1, TSE, TSE FS<br>T1, TSE<br>T1, TSE | 6 mm (whole pelvis)<br>3.5 mm (small FOV)<br>3.5 mm (small FOV) |
| DWI | Axial oblique | *B* value 0–750<br>4 mm | |

Small FOV recommended is 240 × 240 mm

*TSE* turbo spin echo, *FS* fat saturated, *FOV* field of view, *DWI* diffusion-weighted imaging

**Table 18.2** MRI protocols for image acquisition in the management of cervical cancer: (b) University Medical Center Utrecht: MRI protocol for treatment planning simulation

| Sequence | Plane | Details | Bladder filling |
|---|---|---|---|
| T2 TSE | Transverse | 0.48 × 0.48 × 3 mm (whole pelvis) | Medium |
| T2 TSE | 3D (transverse) | 0.63 × 0.63 × 2 $mm^3$ (whole pelvis) | Medium |
| pseudo CT | 3D (transverse) | 1 × 1 × 1 $mm^3$ (whole pelvis) | Medium |
| B0 map[a] | 3D (transverse) | 1.6 × 1.6 × 5 $mm^3$ (whole pelvis) | Medium |
| PD TSE | Transverse | 0.88 × 0.88 × 6.6 (very large FOV including PAO lymph nodes) | Medium |
| DWI SPAIR | Transverse | 1.5 × 1.5 × 5.0 (small FOV) | Medium |
| | | | Wait for extra bladder filling |
| T2 TSE | 3D (transverse) | 0.63 × 0.63 × 2 mm (whole pelvis) | Full |
| | | | Patient empties bladder |
| T2 TSE | 3D (transverse) | 0.63 × 0.63 × 2 mm (whole pelvis) | Empty |

The image acquisition protocol includes specific instructions to assess the impact of bladder filling on the position of the uterus

*PD* proton density

[a]B0 map to verify that the main magnetic field is homogeneous and thus that images are geometrically correct

the relationship between the tumor and adjacent structures (e.g., parametrial tissue involvement and pelvic sidewall invasion) are more accurately assessed on the oblique planes.

In a multicenter study, Zhang et al. [17] demonstrated that the overall accuracy of staging for cervical cancer increased from 61 to 81.4% when MRI findings were combined with clinical findings. Clinical staging particularly underestimates tumor size if there is endocervical tumor growth towards the lower uterine wall. In these situations, MRI measurement of the tumor extent will provide a more appropriate initial evaluation than clinical evaluation.

The cervical tumor is best depicted on T2-weighted images without fat suppression, on which it appears as an expansile or infiltrative mass with increased signal intensity compared with the surrounding normal cervical stroma. The use of intravenous (IV) contrast (gadolinium) in the evaluation of cervical cancer is controversial. Some do not recommend routine administration of IV contrast material for staging protocols, given the various degrees of enhancement of cervical tumors. However, dynamic contrast-enhanced (DCE) MRI is reported to be of value for small lesions, for assessment of residual tumor post-RT, for surgical planning in patients who wish to undergo fertility-sparing surgery, and to distinguish between cervical and endometrial tumors when the primary tumor is a locally extensive adenocarcinoma [18–21].

Although MRI has a high negative predictive value (94–100%) for parametrial invasion, reported sensitivities vary between 38 and 100% [22, 23], and clinical findings will always rule over MRI findings for parametrial tissue involvement. Disruption of the stromal ring, obliteration of the peri-uterine fat planes, encasement of the ureters, and involvement of peri-uterine vasculature are signs consistent with parametrial invasion. Diffusion-weighted image (DWI) sequences add value to pelvic MRI for evaluation of parametrial involvement, increasing the sensitivity and specificity (82% and 97%, respectively) compared with MRI without DWI sequences (72% and 91%, respectively) [24]. Pelvic sidewall invasion is defined as a tumor that is less than 3 mm from the pelvic muscles (internal obturator, levator ani, pyriformis muscle) or the iliac vessels. Gross invasion is easy to recognize and could be associated with an enlargement of the pelvic wall muscles, encasement or narrowing of the iliac vessels, or extension to pelvic bones [25].

To evaluate the extent of vaginal involvement in patients with cervical cancer, combining findings from physical examination with MRI leads to an accuracy as high as 86–93% [26]. Vaginal wall invasion is detected when a high-signal-intensity tumor replaces the low signal intensity of the vaginal wall. Evaluation of vaginal wall integrity may be difficult to assess, with large exophytic cervical tumors protruding into the vagina. The assessment of vaginal wall invasion by cervical tumor can be easier if medical gel used for ultrasound is injected into the vaginal canal before image acquisition [27], as shown in Fig. 18.1.

DWI may improve lesion detection and is obtained in the same two oblique planes (coronal and axial) as tumor extension assessment [28–30]. During DWI, random Brownian motion of water within tissues is detected, which is hindered by the tissue architecture. Information obtained from the DWI study could be used to distinguish malignant from benign tumors by measuring apparent diffusion coefficient (ADC) values. While malignant tumors exhibit a low ADC value, DWI is highly sensitive in differentiating tumors from surrounding tissues.

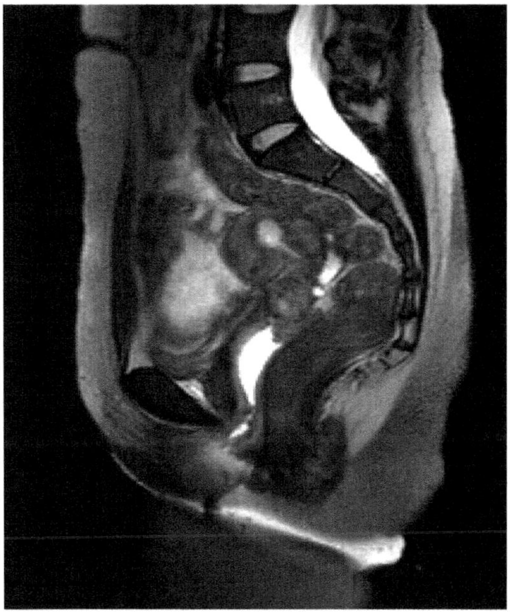

**Fig. 18.1** Siemens 3 T MRI T2 FSE sagittal view of a pelvic MRI of a patient with LACC stage IVA; endovaginal gel was used to fill the vaginal canal, thus allowing for better visualization of disease extent in the upper vaginal wall. Note: Patient had bilateral external nephrostomy tubes; absence of urine in the bladder leads to suboptimal condition to assess bladder involvement

Finally, DWI is obtained in the axial oblique plane with a 3D gradient-echo T1-weighted sequence after IV gadolinium administration. Sequential images are acquired before contrast injection and then during multiple sequential phases of enhancement and a delayed CE T1-weighted image in the axial oblique plane. CE images help with the assessment of invasion of adjacent structures (bladder and rectum) and evaluation of lymph node involvement which is key information for cervical cancer staging. The specificity of MRI for metastatic lymph nodes in cervical cancer has been reported to be as high as 71–100%. MRI criteria indicating lymph node involvement include lymph nodes larger than 1 cm in the short axis, lobulated or spiculated contours, signal intensity like the primary tumor, and heterogeneity of the internal architecture [31]. The presence of lymph nodes with spiculated contours, the same signal intensity as the primary tumor, and heterogeneous signal are also indicative of metastatic disease in lymph nodes that are less than 1 cm in the short axis, even in case of negative PET-CT [32]. Although PET-CT has a role in staging cervical cancer for detecting the location of disease (primary tumor, metastatic nodes, and possible distant metastasis), MRI is superior for precisely defining tumor extension.

Local extension of the primary tumor into the bladder can be evaluated using MRI with a sensitivity of 83% and specificity of 100%. Per FIGO staging, mucosal invasion of the bladder must be present for the tumor to be classified as FIGO stage IVA; mucosal invasion carries a significantly worse prognosis compared with only serosal or muscular invasion [33]. MRI findings that suggest bladder invasion include nodular and irregular bladder wall, tumor protruding into the bladder lumen, and high signal intensity of the anterior aspect of the posterior wall of the bladder on T2-weighted images. The presence of vesicovaginal fistula is diagnostic for bladder wall invasion. Rectal wall invasion by cervical cancer is less frequent than bladder invasion; rectal wall invasion carries a worse prognosis [34]. When bladder or rectal invasion is suspected, on MRI, cystoscopy and proctoscopy can be performed to confirm mucosal involvement (Fig. 18.2).

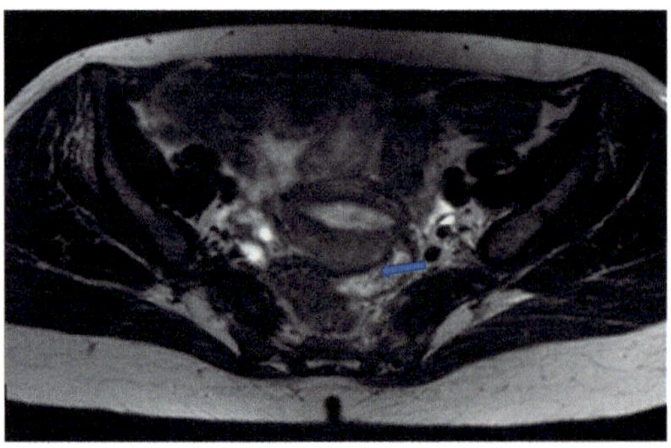

**Fig. 18.2** Siemens 3 T MRI T2 FSE axial view of a pelvic MRI of a patient with LACC stage IVA. The blue arrow indicates the absence of a plane between the cervical gross tumor and the rectal wall, a situation where a rectoscopy could be indicated to rule out rectal involvement

During the EBRT planning process for LACC, the diagnostic MRI should be fused with the planning studies (planning CT or planning MRI). MR-Linacs commercially available do not provide the possibility of image acquisition in an oblique plane perpendicular to the cervical canal, and for now, onboard MRI cannot be used to replace a good-quality diagnostic or treatment planning simulation MRI for the planning process.

## Initial Staging of Endometrial Cancer

Although the staging of endometrial cancer is primarily based on pathologic findings, pelvic MRI at initial diagnosis has real value in certain clinical situations. For patients with inoperable endometrial cancer, pelvic MRI is the best tool to determine the extent of the primary disease (depth of myometrial involvement, cervix stromal invasion, invasion into the surrounding tissue) and the presence of pelvic lymph node metastasis [35]. For patients with inoperable endometrial cancer, when there is no evidence of nodal involvement on imaging, it is common practice to assess the risk of microscopic disease spread in the pelvic nodes by combining the tumor grade with the depth of myometrial invasion seen on MRI using the Creasman rules [36]. When the risk of nodal involvement is low, treating patients with inoperable endometrial cancer with endocavitary BT only is reasonable. In the adjuvant setting for patients who did not undergo a surgical nodal staging procedure, the risk of pelvic node disease added to the tumor grade might be used to formulate individual treatment recommendations. This includes deciding if a plan for postoperative treatment could be limited to endocavitary BT or if pelvic EBRT (for patients with a higher risk of nodal involvement) should be applied with or without additional BT [37].

Pelvic MRI can also provide key information to distinguish between a primary cervical cancer versus a primary endometrial cancer in situations where adenocarcinoma is found on cervical and endometrial biopsies.

## Pelvic MRI During Radiation Treatment

Sequential diagnostic MRIs obtained during EBRT can provide predictive information on tumor response [38–41]. The integration of an MRI in a linear accelerator (MR-Linac) treatment unit (Unity, Elekta, Sweden; MRIdian, View Ray, Cleveland, OH, USA) constitutes a breakthrough in the management of gynecologic cancer. The onboard MRI in the Unity system has a field strength of 1.5 tesla, and the onboard MRI in the MRIdian system has a lower field strength of 0.35 tesla. While a higher-strength magnet might produce better diagnostic image quality, there are several advantages to a low-field-strength magnet in RT. First, low field strength allows the linear accelerator and the MRI to be in closer proximity. This can help increase the dose rate and reduce the penumbra. The second key benefit is the reduction in both image distortion and radiation dose distortion. Finally, for imaging sequence speed, at lower field strength, images can be acquired more quickly, and field strength enables real-time imaging, which MRIdian uses for tissue tracking (providing eight images per second) and beam gating. On board, MRI provides valuable information on changes in tumor volume and position of the tumor and surrounding OAR. Since the onboard image is acquired with the same protocol daily, valuable radio genomics data can be extracted from onboard imaging for further study.

As MRgRT use continues to grow, tools are adapted from diagnostic MRI to the treatment delivery environment. For example, studies have demonstrated that serial ADC measurements during the treatment course can be used as an independent prognostic factor for treatment response, where an increase in ADC values during treatment represents tumor response, thus aiding in identifying good responders [42, 43]. DWI could be used to demonstrate heterogeneity within the tumor, indicating areas of resistant clones and areas of disease regression. Acquiring multiple MRI datasets in the traditional clinical setting is challenging due to patient burden and limited diagnostic MRI resources. The ability to perform serial functional imaging on MR-Linac systems through the course of EBRT brings information that could be used to guide decisions on boosting poorly responding targets or reaching for additional therapies as the treatment course progresses.

Towards the end of a course of pelvic radiation for patients with LACC or vaginal cancer, performing a diagnostic pelvic MRI in preparation for the BT treatment is standard of care. This study provides essential information on residual tumor size and extension, which is taken into consideration in addition to the physical examination findings to guide the BT applicator selection. With improved onboard image quality, this end of EBRT MRI may become obsolete, and the onboard MRI will provide the needed information.

Radiomics is a powerful technique where many imaging features are extracted from the datasets, and automated analysis of these features is performed to correlate imaging findings with impactful clinical endpoints. Following image acquisition, the radiomics process is initiated by segmenting the portion of the image that is to be analyzed (i.e., areas of suspected tumor, cervix, uterus, etc.). Imaging features are then extracted, but the number of features that are extracted can be quite large, up to 5000 [44]. These features are then significantly reduced to a manageable number of features to perform analysis; ideally, features that are useful and reproducible are utilized, and the remainder are excluded. Once the features to be included have been chosen, a statistical framework can be developed with the goal of aiding with diagnosis, treatment response, toxicity, etc. While many of the initial radiomics studies were developed to aid in the diagnostic setting, these powerful techniques have also been applied to the treatment and follow-up phases. Some of the common applications include predicting tumor response and predicting patient outcome, but other applications have been developed or are possible such as dose painting and prediction of side effects [45].

Many of these radiomics studies have been demonstrated on body sites other than the female pelvis. One powerful example of radiomics potential for gynecologic malignancies was provided by Takada et al. [46]. Their research utilized machine learning and radiomics to predict in-field tumor control after definitive RT using two MRI scans, T2-weighted imaging, and ADC datasets. Their research was revealing on several levels; one main finding was that utilizing a volume of interest larger than the GTV resulted in a stronger prediction of in-field tumor control than when the GTV was used without an expansion. Another significant finding was that radiomics analysis on the ADC datasets was more predictive of in-field tumor control than T2-weighted imaging. Other radiomics applications—delta radiomics, dose-painting applications, and prediction of side effects—have not been robustly investigated in the gynecologic cancer setting but are ripe for future investigations.

## Pelvic MRI for Assessment of Treatment Response After Treatment

It is standard of care to use pelvic MRI to assess treatment response after completion of treatment for LACC, vaginal cancer, or inoperable endometrial cancer. It is essential to wait long enough to let the tissue heal from the acute inflammatory reaction seen post RT; therefore, the first follow-up pelvic MRI is usually done about 3 months after treatment completion. In addition to findings from physical examination, important information is obtained from this imaging study: residual tumor size, distinguished between residual tumors and fibrosis. Fibrosis could be distinguished from residual disease, which has an intermediate signal intensity on T2-weighted imaging [47]. Differentiating residual viable tumor or tumor recurrence from posttreatment changes can be challenging; DWI sequences may help to differentiate persistent tumors from fibrosis. Residual viable tumors usually show restriction on DWI sequences. Multiple studies have used ADC as a quantitative metric for treatment response [42, 48]. Interval increase in ADC value (attributed to apoptotic-induced cell death or necrosis) has been seen in patients with a complete response but not in patients with partial or no treatment response [49–52]. Figure 18.3 is an excellent example of a pelvic MRI performed at diagnosis and after treatment, with dramatic changes in the cervical tissue showing an excellent treatment response. FDG positron emission tomography (PET/CT) is frequently used in early follow-up of patients with LACC.

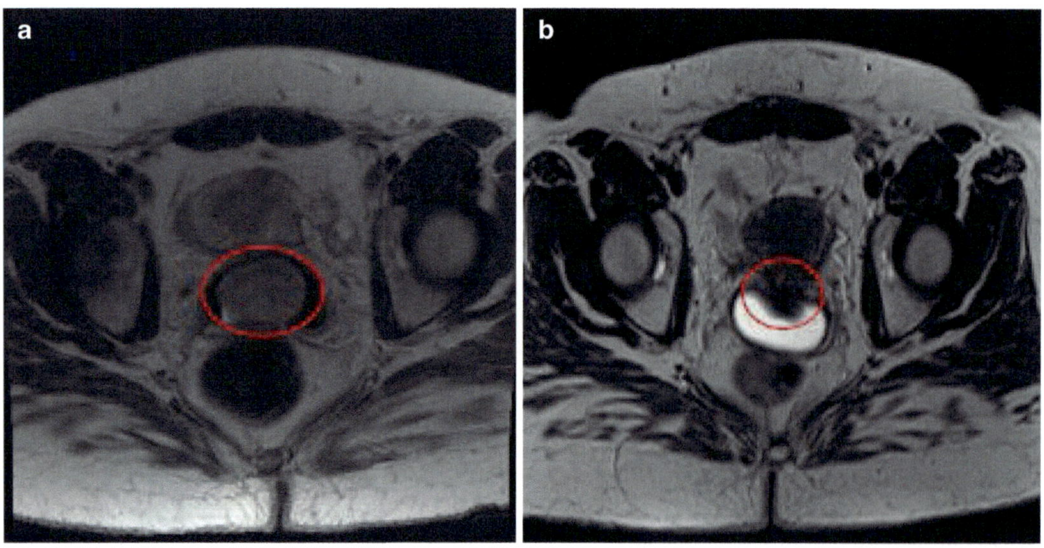


**Fig. 18.3** Siemens 3.0 tesla pelvic MRI, axial T2 view showing the extent of a cervical squamous cell carcinoma at diagnosis (**a**) and after 5 weeks of chemoradiation pre-brachytherapy (**b**)

## Pelvic MRI for the Diagnosis of Disease Recurrence

MRI also plays an essential role in diagnosing pelvic recurrence of gynecologic cancer. The recurrence can be local in the region of the primary tumor, locoregional (in pelvic or para-aortic nodes), or distant. In addition to findings from physical examination and tissue biopsy, a diagnostic MRI should be done to document the extent of a recurrence. While any imaging modality can be used (CT scan, PET/CT, or MRI), pelvic MRI is superior to CT scan, especially in the detection of small recurrences. Locoregional relapse or disease progression for cervical and vaginal cancer usually occurs within the first 2 years after treatment [53]; however, the most frequent sites of recurrence are distant metastases (including distant nodes and visceral metastases) [54]. Tumor recurrence for endometrial cancer usually occurs within 3 years after surgery; the most frequent sites of recurrence are in the vaginal vault and/or in regional pelvic nodes [55].

Disease recurrence should be distinguished from residual disease. The general rule is that disease recurrence is defined as the persistent visualization of a tumor in the same locations as the primary tumor, possible metastatic nodes, or distant metastasis after complete remission and occurring at least 6 months after treatment. Residual disease is defined as incomplete remission or reappearance of the tumor in the original locations in a period shorter than 6 months. For local disease recurrence, a thorough clinical examination and MRI should be performed [56].

The MRI scanning protocol during the follow-up phase of patients with LACC is the same as for the pretreatment MRI. The use of medical gel in the vaginal canal is important to help distinguish vaginal fibrosis from residual/new cervical tumor extension. The sensitivity and specificity to detect recurrent or residual tumors using MRI have been described as 82–100% and 78–100%, respectively [57].

Changes seen in the cervix and surrounding tissues on pelvic MRI after RT are characterized by the recovery of low signal intensity in the cervical stroma and in locations of initial extra-cervical tumor extension, which could be associated with dense fibrosis and decreased cervix volume. DWI and contrast imaging are crucial to distinguish fibrosis from residual tumor; fibrosis will demonstrate delayed enhancement and no restriction in DWI, as opposed to restrictive hypovascular residual masses, which will display restriction in DWI.

For patients treated with surgery (total hysterectomy, modified radical hysterectomy, or radical hysterectomy) at diagnosis, a nodular appearance of the vaginal vault with thickening of the vaginal wall will be best detected by a pelvic MRI.

## Role of MR-Linac in the Management of Cervical Cancer

Surgery is the standard of care for patients diagnosed with early cervical cancer. Very early disease might even be managed by fertility-sparing surgery, while patients with FIGO stage IB1 or IB2 cancer will be treated by a more radical surgical approach (modified radical hysterectomy or radical hysterectomy with bilateral salpingo-oophorectomy and lymph node surgery of various extent—from sentinel node biopsy to full lymphadenectomy). While surgery is the preferred treatment for early-stage disease, the standard of care for patients diagnosed with LACC is a combined modality approach of EBRT (either limited to the pelvis or extended field) combined with weekly cisplatin and intracavitary/interstitial image-guided BT (IGBT). The EBRT volumes cover the central disease (primary tumor, the entire cervix, uterus, parametria, and upper vagina), draining lymphatic regions and nodal boosts to positive nodes. Adding a margin to the central disease to account for inter- and intra-fraction motion is recommended.

Elective para-aortic (PAO) node irradiation may be indicated in selected patients. Treating patients with LACC with prophylactic PAO node irradiation fell out of favor after an intergroup clinical trial showed a survival benefit for pelvic RT combined with chemotherapy over extended field RT in 1999 [58]. However, recent data indicate that patients with high-risk cervical cancer (≥one pathologic node at common iliac or above or ≥three pathologic nodes in the pelvis) might benefit from the inclusion of PAO nodes up to the renal vein even if no suspicious nodes are seen in the lower PAO region below the renal vein on initial workup imaging [59, 60].

MR-Linacs (Unity, ViewRay) allow physicians to perform online adaptive radiation therapy (ART) based on the onboard MRIs of the day. Onboard MRIs provide valuable information on changes in tumor position or volume and the daily position of the surrounding OAR.

Based on the improvement in treatment outcomes observed with the development of image-guided adaptive BT (IGABT) led by the GEC ESTRO gynecology working group, we have evidence that integrating MRI information in treatment planning adds value to the management of patients with LACC. There is now a bulk of literature demonstrating that IGABT leads to better tumor control, increased survival, and decreased treatment toxicity [5, 12, 61–64]. When performing MRI-guided (MRg) BT, the treatment volume accounts for the topography of the primary tumor at diagnosis and the regression observed during EBRT [11] instead of using a fixed-point A for treatment planning. IGABT is supported by both the GEC ESTRO and the ABS and several guidelines have been published [11, 12, 65, 66].

It is well-established that both IGABT and EBRT benefit from MRI guidance [67]. Since the 1990s, EBRT treatment delivery has evolved from the use of port films and skin marks for position verification to the use of cone beam CT (CBCT) with or without fiducial markers for more precise targeting of soft tissue lesions translating in a reduction in treatment margins and irradiated volumes. For the curative treatment of most gynecologic cancers, the use of daily onboard image guidance is the standard of care, but the suboptimal soft tissue contrast provided by CBCT makes it challenging at times to distinguish soft tissue tumors from surrounding normal tissues. Given the fact that MRI provides superior soft tissue contrast than CT, it is an imaging modality of choice to verify the daily delivery of radiation. Another advantage of onboard imaging using MRI is that it is not associated with additional ionizing radiation exposure.

Online MRgRT is a strategy that allows for increased EBRT conformality and more individualized treatment plans. Treating cervix cancer with an MRgRT approach can allow for a further reduction in treated volume, which should lead to reduced treatment toxicity. MRgRT could also allow for dose escalation where needed, thus improving tumor control since IGABT results clearly demonstrate the existence of a dose-response curve in patients with LACC [5, 61, 64].

Patient selection for MRgRT depends on patients and disease characteristics. The presence of non-MRI-compatible implantable cardiac electronic devices or other types of incompatible metallic implants/foreign bodies constitutes contraindication to MRgRT. Nephrostomy catheters (internal or external) are usually MRI-compatible, but verifying on the manufacturer's website is always safer before bringing a patient to the MR-Linac room. Clinical conditions which constitute a formal or relative contraindication to MRgRT also include claustrophobia, severe anxiety, and pain preventing the patient from holding the same position for a long time, given the fact that the whole replanning and treatment delivery process might take up to 60 min, and morbid obesity.

In the curative treatment of LACC, MRgRT may also be used for nodal boosts or primary tumor boosts if first-line BT is not feasible, and MRg stereotactic body radiation therapy (SBRT) could be applied to both nodal and soft tissue metastasis to achieve target tumor control with limited morbidity. The use of MRgRT for patients who cannot undergo BT is discussed in further detail at the end of this section. Strategies for the use of MRgRT for the delivery of pelvic radiation are described in the next section.

## Adaptive MRgRT in the Management of LACC

In the management of patients diagnosed with LACC, it is well known that large inter-fraction motion occurs in the target volume localized in the center of the pelvis due to variations in the volume and shape of the surrounding pelvic organs (bladder, rectum, and sigmoid) seen during the delivery of pelvic EBRT. Haripotepornkul et al. [68] calculated the inter-fraction movement of the cervix during IMRT in different axes and demonstrated that during a course of treatment, the cervix could move: 1.9 mm in the lateral direction, 4.1 in the vertical direction, and 4.2 mm in the anterior-posterior axis. The simplest strategy to deal with target inter-fraction and intra-fraction motion has been to add an internal target motion margin of 1.5–2.0 cm into the planning target volume (PTV) [69]. This security margin is necessary to ensure a full dose is delivered to the target during every fraction. The cost of this approach is that a large volume of the surrounding normal organs receives the dose prescribed to the target volume. Inter-fraction motion uncertainties are eliminated when onboard adaptive radiation is used during MRgRT. However, a margin for intra-fraction motion should always be used as treatment time per fraction increases with MRgRT compared to traditional RT fractions, and longer treatment time could lead to changes in bladder filling and central PTV position during the delivery process.

When using an online adaptive approach, the contours of the central target volume could be adjusted daily to correct for inter-fraction motion during EBRT, resulting in reduced central PTV margins and decreased volume of OARs that receive the prescribed dose. Studies on the use of MRgRT demonstrated that daily MRI permits the adaptation of EBRT plans to daily tumor and OAR positions [40]. The use of ART potentially leads to a considerable reduction in OAR dose by facilitating improved accuracy of treatment delivery and enabling margin reduction. A recent study comparing various ART techniques using CBCT with standard margins, reduced margins, and MRgRT found that significant OAR sparing can be achieved using MRgRT [70].

Consensus contouring guidelines for IMRT for cervical cancer advise including the whole uterus [69] in the target volume. Since, when using CT scan, it is challenging to outline the extent of a cervical tumor in the uterine wall, the consensus was to include the whole uterus in the CTV and to add a large margin on this volume to account for inter- and intra-fraction motion which could be significant when attempting to cover potential daily fundus motion. The ability of MRI to depict tumor extension in the uterine wall introduces the possibility of targeting only the primary tumor, whole cervix parametrial tissue, with a security margin above the disease extension in the uterine wall rather than including the entire uterus in the central tumor volume. Preliminary modeling suggests this is a feasible approach that could reduce OAR doses. Based on CTV to PTV margins of 3–5 mm, an online adaptive planning strategy can reduce the dose to the rectum V40Gy by 36–47%, the dose to the bladder V40Gy by 43–59%, and the dose to the bowel V40Gy by 13–30% compared to a nonadaptive approach [70].

Kozak published a single-institution retrospective study of 53 patients with LACC treated per institution policy with less than whole uterus irradiation volume. The study found that the patients had comparable locoregional control and a reduced bowel dose (V40 and D200cc) compared to historical series [71]. These results are encouraging since the incidence of cervical cancer in developed countries where this technology is available is decreasing, and preliminary results of MRgRT should be tested in a large multicenter prospective study to investigate the safety of reducing treatment volumes as described above (Fig. 18.4).

Another caveat associated with the MRgRT approach for a course of pelvic radiation is that the online adaptive process is time-consuming. Until the MR-Linac technology offers a faster replanning process with robust auto-segmentation tools, ART will remain resource intensive and would be challenging to integrate into the clinic in nonacademic centers. Since the first clinical use of MR-Linac technology in 2014, significant improvements have been made in the treatment replanning process. The latest MRIdian upgrade (Ai3) enables the completion of multiple tasks happening in parallel (for example, target contouring could be done by the

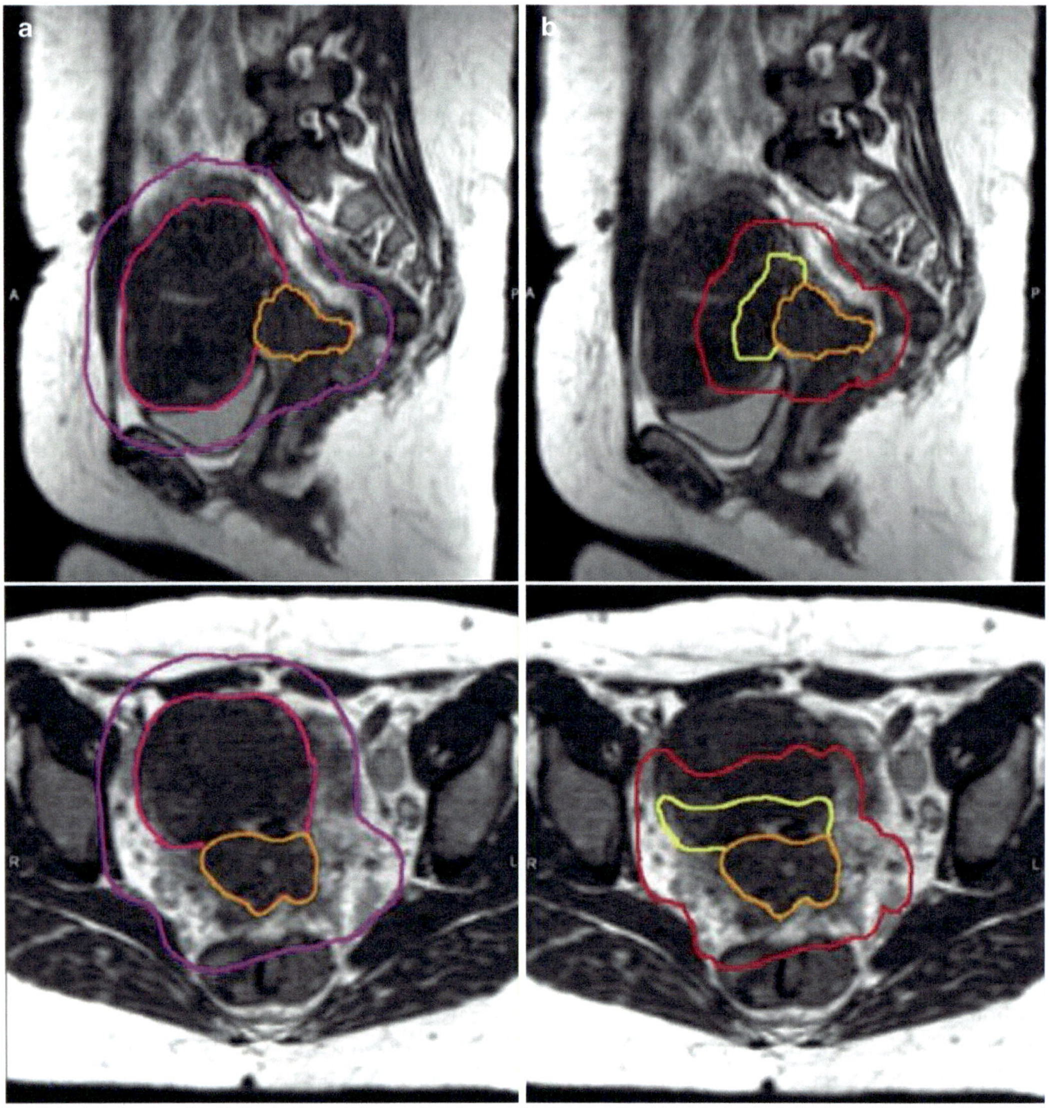


**Fig. 18.4** Sagittal and axial view of a 0.35 T MRI showing the PTV in a situation where the whole uterus is being included in the treatment volume (**a**) versus an approach where a 2-cm margin within the uterine wall is added to the GTV (**b**)

treating physician while a trained technologist simultaneously contours the OARs, and the physicist performs the electron density calculation), which translates into a reduction in the time required to complete the whole on-board adaptive process. As MRgRT and auto-segmentation technology continue to improve, the burden of daily adaptive planning will likely be significantly reduced, and daily adaptive treatment for patients with LACC will become accessible.

A practical approach to mitigate the large treatment margins necessary for accounting for inter-fraction motion without spending time on a daily adaptive plan is to utilize a plan-of-the-day (POTD) approach [72]. The POTD technique uses an individualized IMRT plan library that is selected based on the patient's internal anatomy at the time of the daily setup. The POTD technique has the potential to reduce the treatment margins compared to conventional treatments, and it has a more manageable workload and

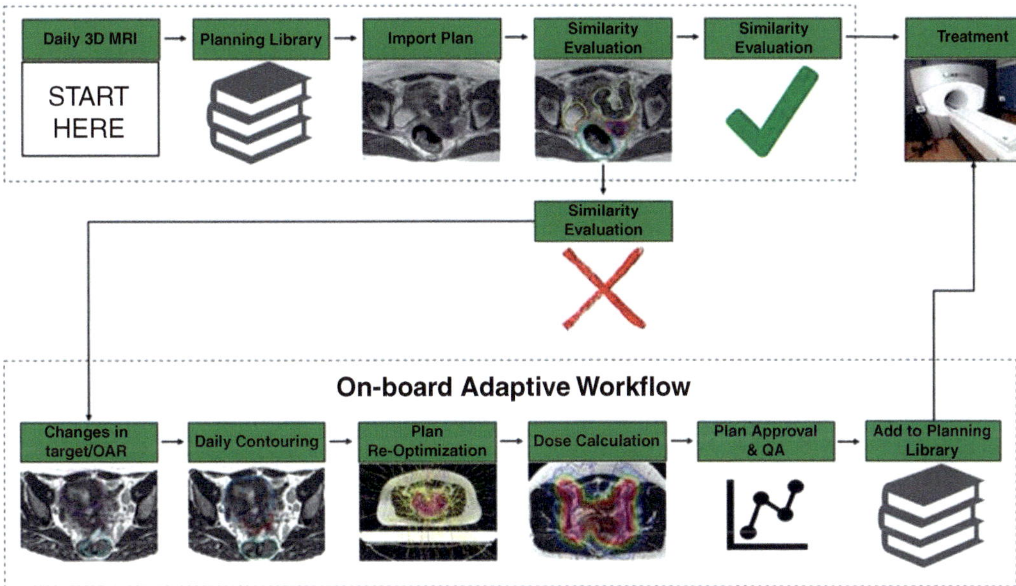


**Fig. 18.5** Illustration of the plan-of-the-day approach

faster treatment time compared to daily ART. Buschmann et al. [72] published their experience with 16 patients using a volumetric modulated arc therapy (VMAT) plan library for bladder full, bladder empty, and a motion robust backup plan, where plan selection is based upon daily setup CBCT. MR-Linac systems could use a similar methodology. MR-Linac offers the added benefit of allowing the users to create a plan library as needed for fractions that do not have a suitable match in the current library for the patient, resulting in an adapted plan for the day and an additional entry into the plan library. This process used in the University of Miami for selected cases is described in Fig. 18.5.

The improved image quality of onboard MRI compared to CBCT eases plan selection for those fractions where a predefined plan will suffice. Although using MRgRT for LACC has been seen as one of the key examples for using the ART approach with MR-Linacs, to date, clinical implementation of this treatment is limited. In addition to the time constraint, another drawback of the currently available MR-Linac systems is the limited treatment field size (feet/head extent: 22 cm [Unity/Elekta], 24.1 cm [MRIdian]). With this field length limitation, only patients with low-risk LACC could be treated with MRgRT since extending the treatment volume to include the common iliac nodes or PAO will lead to a volume that would not be compatible with the field constraints of the MR-Linac. A multiple isocenter approach may solve this; however, this would translate into long treatment times, added complexity, and the risk of irradiating the same bowel volume twice. Until the MR-Linac systems can treat longer volumes, patients with intermediate- or high-risk LACC are not good candidates for this technology.

Studies exploring the benefits of integrated MR-Linacs for enhanced target and OAR visualization and online adaptation to treat LACC with hypofractionated schedules are in progress [73]. If successful, these studies will facilitate the broader adoption of daily replanning for cervical cancer.

## MRgRT Boost When Brachytherapy Is Not Feasible

MRgRT can be used to substitute the final BT boost for patients who cannot be treated with BT for different reasons (e.g., patients with comorbidities limiting their capacity to undergo invasive procedures, patients unfit for any type of anesthesia, patients receiving therapeutic anticoagulation therapy, BT implantation technically not feasible). The first preliminary experience with this novel treatment approach was presented at the ESTRO annual meeting in 2020 [74], and a single-institution series has been published [75].

Many series on the use of SBRT to substitute for BT have been published [76–78]. When using SBRT to deliver the final boost, instead of BT, the target dose is more limited since the dose to OAR must be maintained within safe constraints. Focus on the OAR constraints is important for finding a balance between treatment-related morbidity and quality of life against tumor control. The high toxicity reported in a single-institution phase II clinical trial on the use of SBRT for patients with LACC clearly demonstrates that the use of SBRT boost as a replacement for BT has been associated with increased toxicity and decreased efficacy and should be used with caution in the context of clinical trials [78].

Using the recommended BT, OAR dose constraints from ABS or GEC ESTRO guidelines, and an isotoxic approach seems the way to move forward when using SBRT to deliver the boost. In such an approach, the OAR dose drives the choices in treatment planning. It can be expected that daily online replanning with MR-Linacs may allow for the safe delivery of a higher dose to the target while maintaining D2cc to the bladder, rectum, and sigmoid within recommended guidelines. Using MR-Linacs could allow for the safe delivery of a higher dose to the target compared to nonadaptive CBCT-guided treatments [74]. It was demonstrated in the BT literature that adhering to high-dose levels to the HR-CTV increases the chance of obtaining local control [12, 79]. Recent preliminary studies show that the target dose is reduced using SBRT schedules on both MR-Linacs and standard CBCT Linacs compared to BT [74, 78].

Until proven otherwise, BT should remain the first choice to deliver the final boost to the tumor. A SEER review published by Han et al. has clearly demonstrated that the use of BT in the management of patients with LACC is associated with improved survival [80]. The Society of Gynecologic Oncology and the ABS have recently published a review to reinforce that BT is a critical component of primary radiation therapy for LACC [81].

MR-Linac treatments are not considered replacements for BT but should be seen as an alternative for selected cases where BT is not feasible. In these cases, MRgRT might be preferable over CBCT-guided SBRT as the better soft tissue contrast of MRI allows to better tailor the dose to the tumor according to the daily online anatomy and away from the surrounding organs. Currently, the clinical experience with MR-Linac treatments in the above-described situation is slowly growing, as is the number of centers that use them. We regularly notice that the delivered target dose is somewhat less than the planned one. This is a notable difference from the BT approaches. The reason is that during BT delivery, the applicator and, therefore, the radiation source positions are inside or near the tumor and therefore fixed in relation to the treatment targets during dose delivery. If the target moves, the applicator will move with it; OARs, however, can move independently. This fixed relation of target and radiation source is not reproducible when delivering EBRT with a steady source at a distance and independently changing tumor and OAR position during radiation delivery. MR-Linac centers are still in their learning curves for these specific treatments, experience technical challenges, and continuously adapt approaches according to the slowly growing experience. Centers are collaborating with others to discuss treatment planning approaches, fractionation schedules, dose-volume parameters, and uncertainties during dose delivery. The feasibility of the new treatment approaches can be investigated in longitudinal prospective cohort studies. One of them is the Momentum cohort within the MR-Linac consortium (The MOMENTUM Study: The Multiple Outcome Evaluation of Radiation Therapy Using the MR-Linac Study—[ClinicalTrials.gov]), which provides the possibility of cooperation for prospective data registration including treatment parameters and clinical outcome data in terms of tumor control, morbidity, and quality of life.

Finally, the availability of MR-Linac treatment units in the radiation oncology clinic has the additional benefit of providing easy access to MRI datasets with applicators in place to aid in MRg BT planning. This can significantly simplify the logistics of doing IGABT for many institutions that until now had relied on the limited availability of MRI scanners in the diagnostic radiology department [82].

**Case Example 1 Primary cervix cancer, stage IV; BT boost to primary tumor not applicable; alternatively, MR-Linac boost is applied, and treatment intention is pelvic control.**

A 54-year-old patient presents with hemodynamic-relevant vaginal bleeding, abdominal pain, and bladder cramps. The diagnostic workup reveals a FIGO stage IVA cervical cancer with infiltration into distant parametrial tissue bilaterally, into the upper vagina, rectovaginal septum, and rectal wall. In addition, there is infiltration of the bladder wall and both ureters, associated with bilateral hydronephrosis. The patient's history includes brainstem infarction with persistent hemiplegia and the need for anticoagulation. The multidisciplinary recommendation is to offer a curative treatment approach. EBRT is delivered with VMAT (45 Gy elective dose in 25 fractions to tumor and lymphatic drainage) with bilateral simultaneous integrated boosts (2.35 Gy per fraction) for two metastatic obturator nodes. Concurrent chemotherapy cannot be administered due to severely impaired kidney function and comorbidity. The initial plan includes a BT boost (four HDR fractions, aiming at a total D90 HR-CTV of 85–90 Gy EQD2$\alpha/\beta$ = 10).

MRI is repeated after 18 EBRT fractions but unfortunately shows only minor tumor regression with persistent invasion of the rectum and bladder. Tumor volume is reduced from 174 to 118 mm$^3$, including a decrease of the largest tumor extension from 93 to 80 mm. BT is not considered feasible, and therefore, a boost to the primary tumor is delivered on a 1.5 T MR-Linac. Four IMRT fractions are given, using an 11-field beam arrangement. Planning is done using an isotoxic approach with priority given to hard OAR dose constraints over target coverage. The HR-CTV and relevant OAR are being re-contoured for the pretreatment online plan and before each fraction based on the actual anatomy. For each fraction, daily MRI and online ART are performed. The pelvic EBRT dose (45 Gy or 44.25 Gy, EQD2$\alpha/\beta$ = 10) is added to the dose from the four online adaptive plans to calculate the cumulative dose, which sums up as follows: D90 HR-CTV, 76.4 Gy EQD2$\alpha/\beta$ = 10 (i.e., 6.0 Gy, or 8.1 Gy, EQD2$\alpha/\beta$ = 10 per fraction). OAR doses are for bladder D2cc, 90.9; rectum D2cc, 70.0; sigmoid D2cc, 47.3; and bowel D2cc, 74.9 Gy EQD2$\alpha/\beta$ = 3. Although D90 HR-CTV is below the recommended dose (D90 $\geq$85–90 Gy EQD2$\alpha/\beta$ = 10), using this stereotactic planning approach at least part of the HR-CTV receives this dose aim with V90 Gy EQD2$\alpha/\beta$ = 10 = 19% and V85 Gy EQD2$\alpha/\beta$ = 10 = 64%. Our institutional approach (UMCU) for CBCT Linacs would allow for a total D90 HR-CTV of 70 Gy EQD2$\alpha/\beta$ = 10 using VMAT with uniform target dose distribution.

Treatment is overall well tolerated without unexpected early toxicity. At first follow-up 3 months after treatment, ongoing tumor regression is visible on MRI. Vaginal bleeding and pain have resolved completely. The patient experiences no major side effects, especially no organ fistulation (Fig. 18.6).

## Role of MR-Linac in the Management of Recurrent Cervical Cancer

Cervical cancer will most frequently recur in the pelvis, either in the vaginal vault for patients who were initially treated by a surgical approach or in the cervix, parametrium, pelvic, and PAO nodes for patients treated with chemoradiation [83]. After ruling out the presence of any metastatic spread outside of the pelvis, patients who did not receive a prior course of radiation should be treated with pelvic radiation combined with chemotherapy and BT in case of local relapse. For the reasons mentioned above, delivering a whole

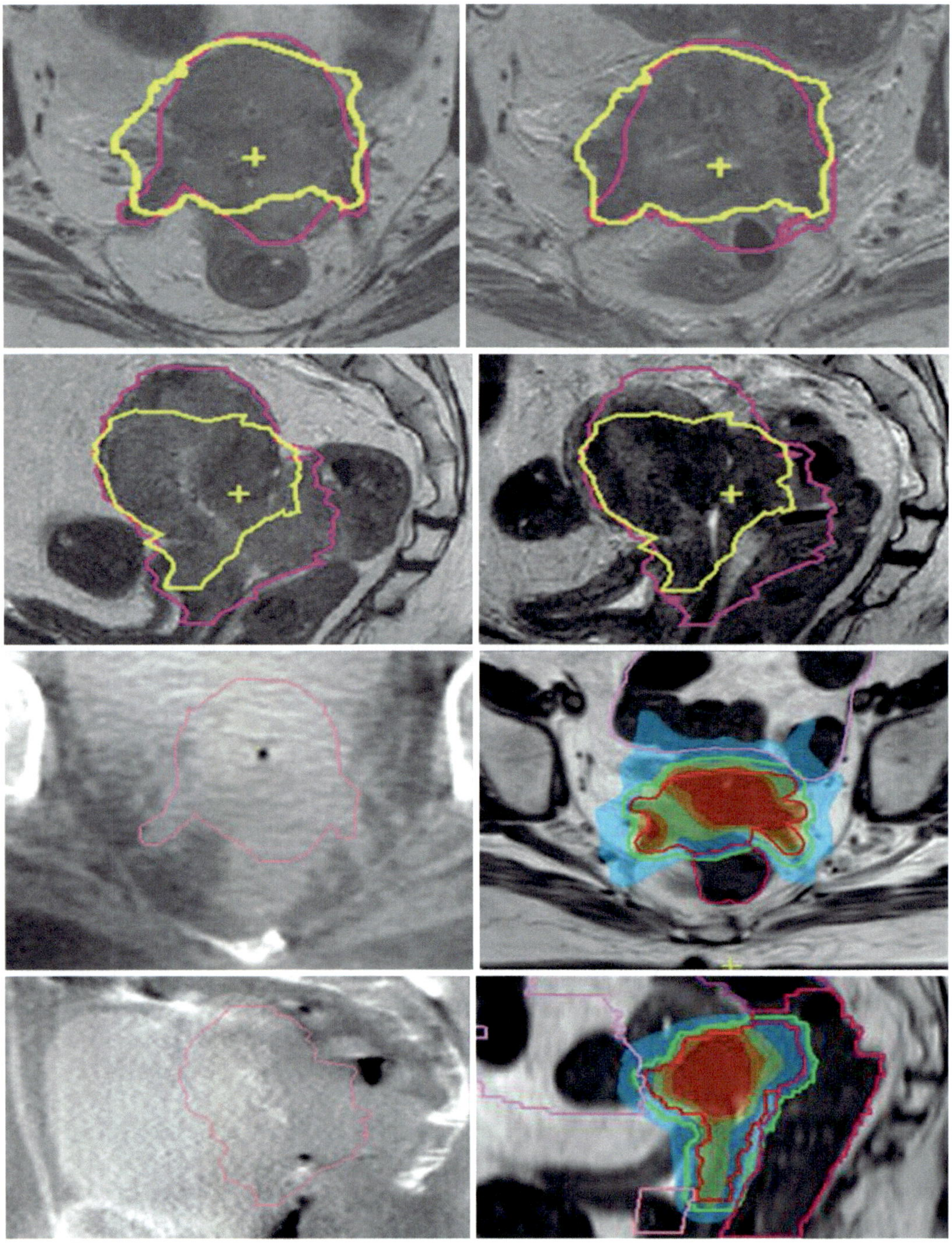

**Fig. 18.6** Cervix cancer; MR-Linac boost of CTV-HR after elective radiotherapy. Cervix cancer; MR-Linac boost to HR-CTV after pelvic EBRT. Left column (top to bottom), transversal and sagittal T2-weighted MRI at the time of treatment planning and onboard CBCT scans in the first week of elective EBRT. Right column (top to bottom), MRI scans after 32.4 Gy of elective EBRT and first. MR-Linac boost plans. For comparison, the initial and boost HR-CTV contours are shown on the MRI scans (pink at the time of treatment planning and yellow after 32.4 Gy EBRT). On the CBCT scans, initial HR-CTV is shown. For the MR-Linac boost plan, the online delineation of the first fraction for HR-CTV and rectum is shown. The images show the improved visualization of MRI compared to CBCT for online position verification

course of pelvic radiation on the MR-Linac system might be tedious. However, a boost to the site of recurrence using MRgRT to bring the tumor volume to an ablative dose might be of interest if BT is not possible or in case of nodal disease recurrence. Hadi et al. published early clinical results of MR-g SBRT boost for patients with locally advanced or recurrent gynecological cancers ineligible for BT. Five patients in this series had recurrent cervical cancer and were successfully treated to a median cumulative dose of 73.6 Gy with a combination of pelvic IMRT and MRgRT boost of 21 Gy in 4 fractions [75].

For patients treated primarily with combined chemoradiation as a curative intent who present a central pelvic recurrence, surgery might be a treatment option [84]. For all other pelvic recurrence or pelvic progression, chemotherapy will generally be offered. For patients who reach a plateau on systemic treatment, MRgRT is a promising technology to deliver an ablative dose to a site of recurrent disease. When using MRgRT in this clinical situation, typically a hypofractionation approach will be used (40–50 Gy in 5 fractions), and the advantage of performing ART will allow for safer dose delivery, sparing the surrounding organs from receiving a high dose. In a phase II prospective clinical trial of treating inoperable pancreatic cancer, Parikh et al. have demonstrated that using MRgRT and an isotoxic approach, a dose of 50 Gy in 5 fractions could be delivered to a target volume adjacent to surrounding organs without any grade 3 or higher acute toxicity [85]. These results achieved treating upper abdomen targets are likely to be reproducible when treating any pelvic or abdominal target near luminal organs.

**Case Example 2 Recurrent cervical cancer after complicated course of previous curative treatment approaches, second salvage approach with curative intention including MR-Linac boost to the recurrent tumor in the vaginal vault.**

A 47-year-old patient with recurrent cervical cancer in the vaginal vault. Two years ago, the patient is diagnosed with FIGO stage IB2 adenocarcinoma of the cervix, curative surgery is intended, but intraoperatively, nodal pathology is detected in the right obturator region. Further surgery is omitted, leaving the uterus and cervix in situ. Decision for chemoradiation is made and started 3 weeks after surgery. After four EBRT fractions and one cycle of cisplatin, patient develops a massive thromboembolic event in the superior mesenteric artery A (outside irradiated volume). Emergency surgery follows, including removal of necrotic bowel parts (nearly the entire small bowel and part of the right hemicolon) and colostomy. The patient survives with dependency on total parenteral nutrition. Two months later, cervical cancer restaging is performed, showing only minor progression. Therefore, radical hysterectomy and lymphadenectomy are performed with curative intent. The pathology report reveals adenocarcinoma of the cervix with lymphovascular space invasion component, tumor size 27 mm, R0 resection, and no lymph node involvement. The patient recovers slowly but remains dependent on parenteral nutrition. Fifteen months later, a solitary vaginal vault recurrence is detected during a routine follow-up. On MRI and PET-CT, there was no indication of nodal or distant spread, and the multidisciplinary tumor board advised for second salvage treatment using radiotherapy consisting of limited elective field EBRT and boost to the vaginal tumor. BT is considered not feasible due to tumor extension into the bladder wall, its close relation to remaining large bowel loops, and the need for anticoagulation treatment. Alternatively, radiotherapy salvage treatment is considered, including small-field elective irradiation and MR-Linac boost. The elective EBRT field includes some margin into the surrounding of the vaginal tumor in all directions (respecting boundaries of surrounding, not affected organs) and the entire vagina, with 45 Gy being administered in 25 fractions. This is immediately followed by an SBRT boost using a 1.5 T MR-Linac. A schedule of 6 fractions is delivered with 3 fractions per week and respecting an overall treatment time of 6 weeks aiming at a total dose of 85–90 Gy (EQD2$\alpha/\beta$ 10) to the vaginal tumor remnant without violation of the OAR constraints. For the MR-Linac boost, an isotoxic approach is used, giving priority to OAR dose constraints during the planning process. For the offline pretreatment planning situation, the prescribed dose to the vault recurrence (D90 HR-CTV), calculated as the sum of elective EBRT and the six boost fractions, is 98.7 Gy (EQD2$\alpha/\beta$ 10) with all OAR constraints being tolerated. In the online and actual treatment situation, the total D90 HR-CTV based on the sum of daily pre-fraction plans is somewhat lower with 94.0 Gy (EQD2$\alpha/\beta$ 10) but well in the range of the intended target dose. The total D90 HR-CTV, as calculated on the anatomy on post-fraction MRI, is 91.7 Gy (EQD2$\alpha/\beta$ 10). For details regarding OAR dose levels and differences calculated for moments in time, see the table below. Differences are mainly caused by changes in intra-fraction bladder filling, causing a minor position shift of the target.

Three months after salvage treatment, clinical investigation and MRI reveals complete remission of the vaginal vault tumor and no signs of progression elsewhere. The patient is recovering well without signs of additional damage to the remaining bowel parts. She reports G2 urinary urgency and frequency and G1 vaginal stenosis and mucositis (Fig. 18.7, Table 18.3).

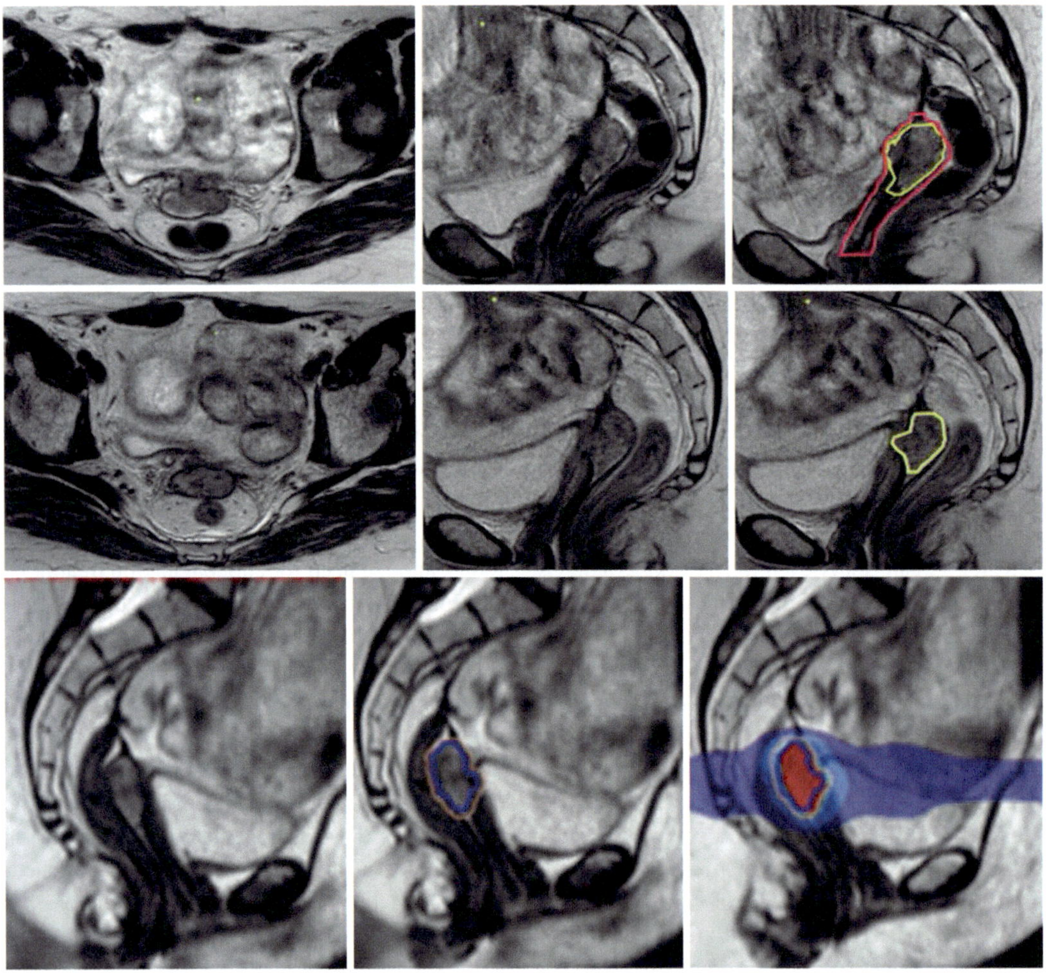

**Fig. 18.7** Recurrent cervical cancer after complicated course of previous curative treatment approaches, second salvage approach with curative intent including MR-Linac boost to the vaginal vault recurrence. Upper row shows images acquired on 1.5 T MR sim for elective EBRT treatment planning, CTV including entire vagina in red, GTV of recurrent tumor in vagina top in yellow; middle row shows images acquired on 1.5 T MR sim for offline pre-treatment planning of the MRL boost showing only minor tumor regression; lower row shows images acquired on 1.5 T MR-Linac for online contouring and treatment planning, treatment targets are contoured with GTV tumor dark blue and PTV 3 mm orange, and dose distribution of the online plan is shown with D98% in red, D90% in yellow, D50% in light blue, and D20% in dark blue

**Table 18.3** Dose level differences based on changing anatomical relations of target and organs at risk during MR-Linac boost delivery

| Dose in Gy | | | | | | | |
|---|---|---|---|---|---|---|---|
| | | Dose | | | Constraints | | Differences |
| | | Offline | Online | Post-recontoured | | | Post-recontoured—online |
| HR-CTV | **D98%** | 95.6 | 91.0 | 88.1 | | | −3.0 |
| | **D90%** | 98.7 | 94.0 | 91.7 | >90 | | −2.2 |
| PTV+3 mm | **D98%** | 85.4 | 83.0 | 78.4 | <80 | | −4.5 |
| OAR | | | | | **Soft** | **Hard** | |
| Bladder | **D2cc** | 78.3 | 79.4 | 85.7 | <80 | <90 | 6.2 |
| | **D10cc** | 67.6 | 66.0 | 68.6 | | <75 | 2.6 |
| Rectum | **D2cc** | 74.6 | 73.7 | 70.1 | <65 | <75 | −3.7 |
| | **D5cc** | 62.7 | 61.1 | 58.8 | | <66 | −2.3 |
| Sigmoid | **D2cc** | 59.5 | 55.8 | 56.5 | <70 | <75 | 0.7 |
| | **D5cc** | 54.3 | 52.1 | 52.2 | | | 0.1 |
| Bowel | **D2cc** | 47.7 | 47.9 | 46.8 | <70 | <75 | −1.0 |
| | **D5cc** | 45.7 | 46.2 | 45.4 | | | −0.8 |
| Urethra | **D0.01cc** | 44.0 | 44.0 | 44.0 | | <90 | 0.0 |
| Ureter right | **D0.01cc** | 102.3 | 98.4 | 102.2 | | <90 | 3.8 |
| Ureter left | **D0.01cc** | 100.1 | 99.0 | 101.1 | | <90 | 2.1 |

In the management of cervical cancer, an isolated PAO node recurrence is considered a locoregional recurrence and should be treated with a curative approach. If the patient did not receive prior radiation to PAO nodes, the whole PAO area (from the superior border of the previous field to 2–3 cm above the involved node) should be irradiated using fractionated RT with chemotherapy with a boost (to 55–59 Gy) to the involved node. Here again, MRgRT is the technology of choice to deliver the boost since surrounding bowel loops could be spared from high doses. An alternative would be resecting the involved node; however, radiation to an ablative dose is a nice noninvasive alternative.

The dose levels recommended to achieve disease control are 45–50 Gy in 25–28 fractions for the elective nodal volume and a dose range of 50–65 Gy for macroscopic disease [86]. The proximity of these nodes to the bowel is a dose-limiting factor. Severe duodenal morbidity was reported after PAO irradiation using simultaneous integrated boost to nodes in the upper abdomen [85, 87]. MR-Linacs might be the technology of choice in these situations as daily visualization of the anatomy, together with the possibility for online ART, allows for a broader therapeutic window with better tailoring of the dose to the metastatic nodes and away from the surrounding bowel (duodenum) loops.

The management of cervical cancer recurrence to metastatic sites is discussed in the last section of this chapter on the management of oligometastatic disease. It can also be seen in Chap. 12 of this book.

## Role of MR-Linac in the Management of Endometrial Cancer

### Treatment of Inoperable Endometrial Cancer

The standard treatment for nonmetastatic endometrial cancer is surgery consisting of hysterectomy with bilateral salpingo-oophorectomy with or without regional lymph node dissection or sentinel lymph node mapping. Surgery may need to be followed by radiotherapy and/or systemic treatment depending on histopathologic risk factors.

A minority of patients diagnosed with early endometrial cancer cannot undergo surgery due to advanced age, poor performance status, or medical contraindications to anesthesia. These patients can be treated with definitive radiotherapy consisting of BT with or without EBRT (+/− chemotherapy). Depending on the tumor stage, good disease control and long-term survival could be achieved [86–89]. In a cohort of 1322 patients with endometrial cancer treated with radiotherapy alone (EBRT and/or BT) for various reasons, the disease-specific survival at 5 years was 78.5%, and severe late morbidity (grade ≥ or 3) was as low as 3.7% for the combined treatment approach [90]. In a smaller retrospective study, 74 patients with stage I and II endometrial cancer received a combination of pelvic EBRT and BT with curative intent, resulting in a 3-year progression-free survival of 68% with a median interval of 43.5 months [87]. In selected patients, BT alone could be used with curative intent, with excellent LC up to 100% in well-selected patients [37, 90, 91].

In this population of women who are not surgical candidates, invasive BT procedures requiring anesthesia might not be an option either, and treatment alternatives to deliver an ablative dose to the tumor are valuable. For patients not able to undergo surgery and BT, MRgRT can be an alternative treatment technique to deliver an ablative dose to the tumor.

Daily changes in the position of the uterus due to bladder or rectal filling can be observed and adjusted for. To save time in the replanning process, a POTD method could be implemented. As described earlier, for patients with LACC, the MR-Linac treatment fields are limited in a cranial-caudal direction. The currently available field lengths (Unity/Elekta, 22 cm; MRIdian/ViewRay, 24.1 cm) can be too limited for pelvic fields in tall patients. If MR-Linac field length limits its use for the entire course of treatment, MR-Linacs constitute a nice option for boosting the uterus and any metastatic nodes in addition to elective EBRT when a BT boost to the uterus is not feasible. Daily MRI and replanning allow for better targeting of the HRCTV while adapting for the variable positions of the sigmoid, small bowel, and bladder.

**Case Example 3 Primary endometrial cancer; coronary bypass surgery is indicated prior to uterine surgery, MR-Linac boost is applied to control vaginal bleeding prior to any surgery, and treatment intent is curative.**

A 68-year-old patient presents with postmenopausal bleeding. The diagnostic workup (including a CT scan) is suggestive of low-stage, high-grade endometrial cancer (endometroid adenocarcinoma). Gynecological surgery with curative intent is indicated, consisting of laparoscopic staging with hysterectomy, bilateral salpingo-ovariectomy, and pelvic/para-aortic nodal dissection. During diagnostic workup, severe coronary artery disease is diagnosed with an indication for coronary artery bypass, resulting in an absolute contraindication for the intended gynecological surgery. However, clinically relevant vaginal bleeding persists and interferes with the necessary cardiac interventions. Therefore, the decision to offer a short course of hypofractionated radiotherapy prior to any other treatment is taken. The treatment plan in this situation is (1) short-course SBRT to the entire uterus to control vaginal bleeding, (2) coronary artery bypass, (3) and radical uterine surgery.

The patient is treated on a 1.5 T MR-Linac with a total of ten online adaptive SBRT fractions (4 fractions per week). Total planning dose aim of ≥30 Gy (32.5 Gy EQD2) to the entire uterus, including the cervix, with at least ≥40 Gy (46.7 Gy EQD2) to the GTV as depicted on MRI is easily achievable and not violating the clinical OAR dose constraints. The final online delivered total dose levels are within the planning aims for both GTV and the entire uterus with doses as high as 43 Gy (51.2 Gy EQD2) reached in GTV.

The entire RT series is delivered as intended. Vaginal bleeding resolves entirely, and treatment-related morbidity is mild with G1 diarrhea and urinary frequency at the end of treatment. Three weeks after finalizing RT, a cardiac artery bypass is performed as intended. A radical hysterectomy is done 6 weeks later revealing necrosis but no remaining vital tumor on pathology. At her 2-year follow-up, the patient is alive without signs of tumor relapse (Fig. 18.8).

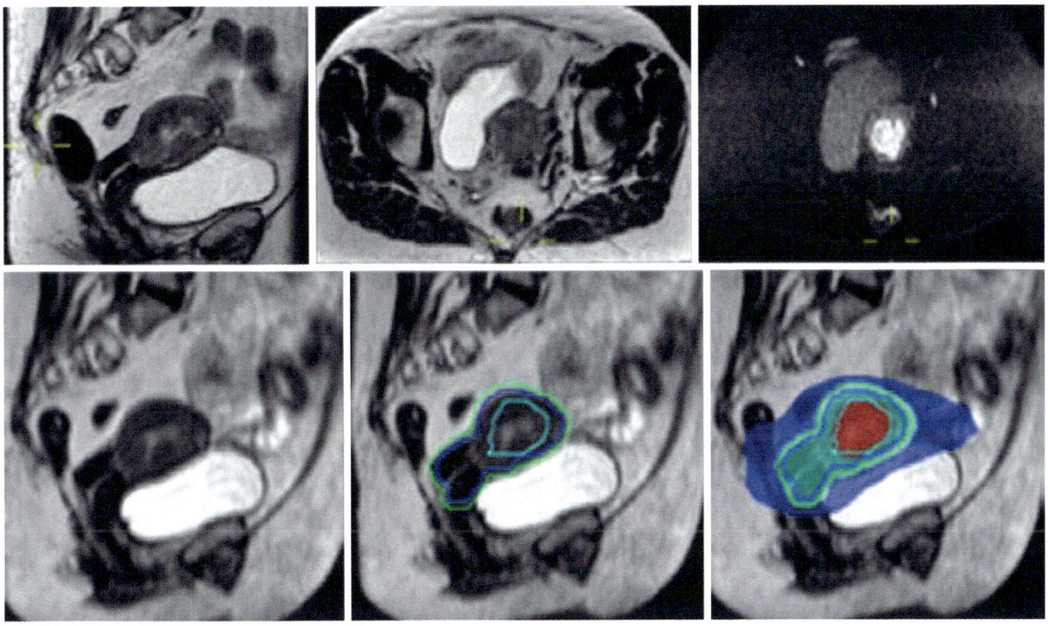

**Fig. 18.8** Endometrial cancer; MR-Linac treatment to prevent vaginal bleeding and further tumor progression while waiting for surgery. Upper row shows images acquired on 1.5 T MR sim for offline pretreatment planning; sagittal and transversal views on T2 3D image and transversal DWI spair. Lower row shows images acquired on 1.5 T MR-Linac for online contouring and treatment planning; treatment targets are contoured with GTV tumor light blue, CTV entire uterus dark blue, and PTV 3 mm green; dose distribution of the online plan is shown with GTV D98% in red, CTV D98% in green, and D50% in blue

## Treatment of Recurrent Endometrial Cancer

### Vaginal Vault Recurrences

Depending on the initial management, treatment of vaginal recurrence of endometrial cancer may consist of surgery or RT [92, 93]. In patients who did not receive a prior course of pelvic radiation, EBRT combined with BT is usually the preferred treatment option. While BT is the modality of choice to deliver a boost to vaginal recurrence, this approach might be challenging at times.

Endocavitary BT using single-channel vaginal applicators can only be used if the thickness of the tumor volume to be treated is less than 5 mm [94]. Multichannel endocavitary vaginal applicators can be used for focal lesions up to 1 cm thick, but interstitial BT is required for a larger volume of disease. After a hysterectomy, small bowel loops and other pelvic organs are near the vaginal vault. Unless real-time image guidance or a laparoscopic approach is used, there is a risk of causing damage to the surrounding organs during BT needle insertion.

Although BT offers the most definitive boost treatment and high LC rates in this clinical situation [95, 96], it is associated with significant risk. In an overview of 28 patients described by Fokdal et al. the 2-year LC rate was 92% [96]. In addition, not all recurrences are amenable to BT. An example may be a rectovaginal septum recurrence in close proximity to the rectal wall. In many situations, interstitial implants might be needed but are not achievable or are associated with a high risk of fistula and other complications. An external beam boost using MRgRT and adjusting the plan to the daily anatomy using an isotoxic approach may provide a good alternative.

Using the BT recommended dose constraints to the OARs (D2cc to bladder <75 Gy, rectum and sigmoid <65 Gy EQD2), isotoxic ART allows to keep the dose away from the uninvolved rectum and other surrounding organs. Avoiding extremely high doses around interstitial BT needles might result in less normal organ damage and necrosis.

**Case Example 4 Primary endometrial cancer; limited early recurrence in the vaginal vault and abdominal wall; salvage radiotherapy is intended to include BT boost to the vaginal tumor; BT however, is not applicable, and alternatively, MR-Linac boost is applied.**

A 68-year-old patient presents with vaginal bleeding. She is diagnosed with low-grade/high-stage cancer of the uterus and is treated with laparoscopic hysterectomy and bilateral salpingo-oophorectomy. The final pathological report reveals a FIGO stage IIIB grade 2 endometroid adenocarcinoma, which qualifies for postoperative EBRT. On treatment planning CT scan and MRI, an early recurrent tumor is detected in the vaginal vault, with a second lesion seen in the anterior abdominal wall (most likely a laparoscopic port site recurrence). A biopsy of both lesions confirms metastases from endometrial cancer. The patient is considered ineligible for additional surgery due to comorbidities, including obesity. Definitive radiotherapy is planned, consisting of sequentially 1.5 T MR-Linac SBRT for the abdominal wall metastasis (35 Gy in five fractions), followed by pelvic EBRT with VMAT on a conventional Linac (45 Gy in 25 fractions) and finally a sequential MR-Linac boost to the vaginal vault recurrence (28 Gy in four fractions). Vaginal vault BT is considered but not deemed feasible since a complex interstitial approach under anesthesia would be necessary but not feasible due to comorbidity. A composite approach is done for each of the three radiotherapy courses (see Fig. 18.9 below).

For the MR-Linac boosts to the vaginal vault recurrence, an isotoxic approach is used, giving priority to hard OAR dose constraints during the planning process, calculating the sum of elective EBRT and the four boost fractions. For the offline pretreatment planning situation, the total dose to the vault recurrence (D90 HR-CTV) is calculated being 91.9 Gy (EQD2$\alpha/\beta$ 10). In the online and actual treatment situation, however, the total D90 HR-CTV based on the sum of daily online planning is 82 Gy (EQD2$\alpha/\beta$ 10). For D2 cc bladder and rectum, the pretreatment and online doses are 84.9 Gy/73.5 Gy and 68.3 Gy/73.7 Gy (EQD2, $\alpha/\beta$ 3), respectively. The differences in pretreatment and online doses are mainly caused by variations in rectum positions and filling status, which in our isotoxic planning approach results in a reduced target dose. For the abdominal wall metastasis, the SBRT planning aim is 35 Gy in five fractions to 95% of the target. For both the pretreatment and online plan, the GTV35 Gy (EQD2$\alpha/\beta$ 10 = 50 Gy) has a median value of 100%.

MRI performed at 3 months after treatment shows no residual tumor in both locations and no evidence of disease progression. At this moment patient does not report any unexpected or grade ≥3 treatment toxicity (Fig. 18.9).

**Case Example 5 Vaginal vault recurrence; MR-Linac boost after a course of pelvic IMRT curative treatment intent.**

A 60-year-old patient is diagnosed with a vaginal vault recurrence of endometrial cancer. The tumor extends to the right pelvic wall causing hydronephrosis (A). After completing a course of pelvic IMRT to a dose of 45 Gy in 25 fractions, the patient receives a boost of 30 Gy in 6 fractions to the PTV using MRgRT. The HRCTV is expanded by 3 mm to generate the PTV. Then the optimized PTV (PTVopt) is obtained by subtracting 3 mm from the surrounding OARs (B). The dose constraints for the composite (pelvic IMRT and MR-Linac boost) are the following: dose to D2cc to bladder, 77 Gy; rectum, 65 Gy; sigmoid, 65 Gy; and small bowel, 65 Gy. An isotoxic plan is generated, meeting all these dose constraints, and a daily adaptive approach is used. More than 2 years posttreatment, imaging shows a complete response, and the patient has no treatment toxicity (Fig. 18.10, Table 18.4).

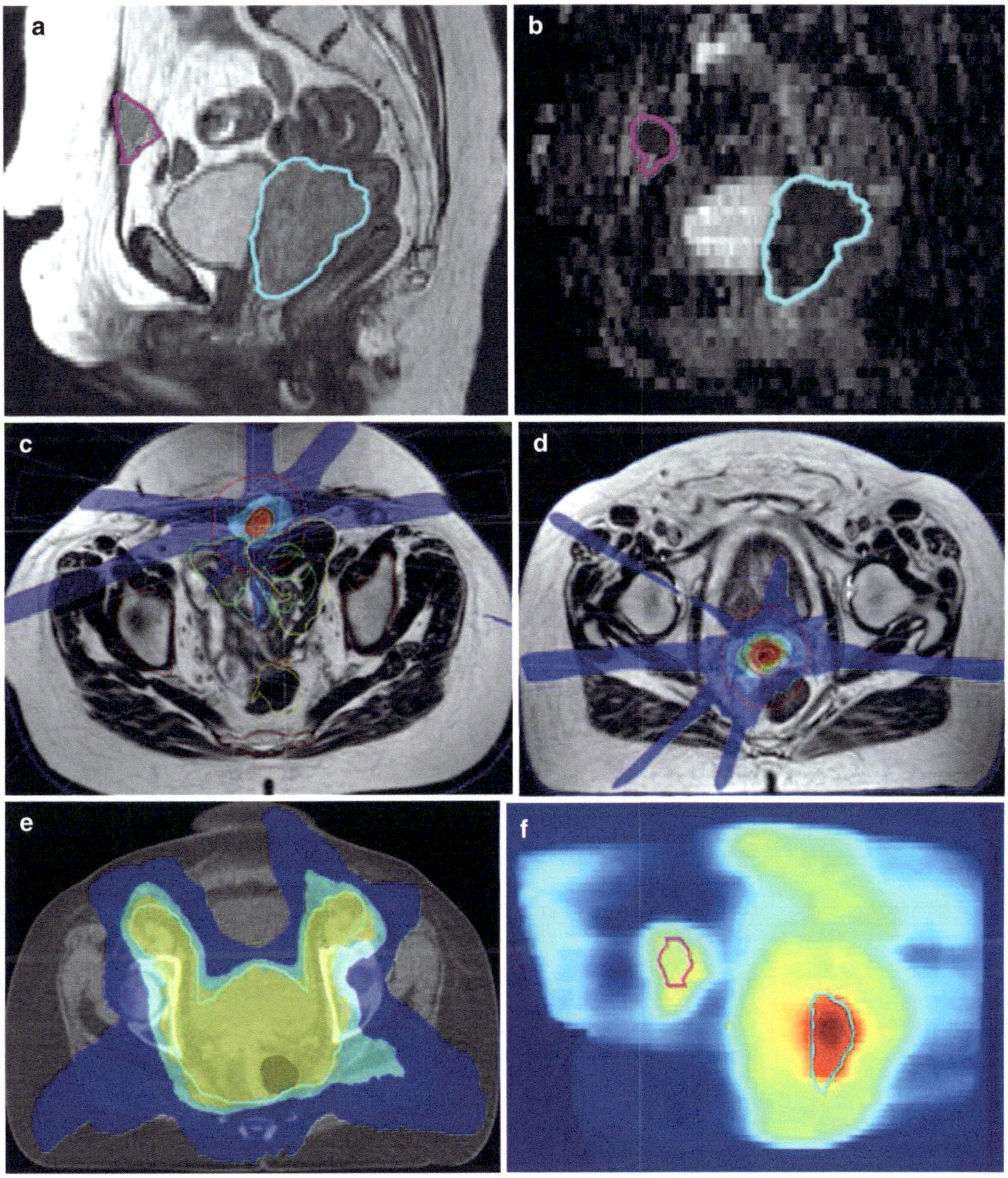


**Fig. 18.9** Endometrial cancer; MR-Linac SBRT for concurrent local recurrent abdominal wall metastasis. Endometrial cancer; MR-Linac SBRT for concurrent vaginal vault recurrence and anterior abdominal wall metastasis. (**a**) T2 sagittal, (**b**) ADC map derived from diffusion-weighted image acquired with MR-Linac. Abdominal metastasis in pink and vaginal vault recurrence in blue. Target and OAR are clearly visualized on the MR images allowing for daily adaptation. (**c**, **d**) Typical daily MR-Linac plans for both lesions (isodoses red 110%, orange 100%, blue 25%). (**e**) Elective EBRT plan on planning CT (yellow 95%, green 82%, blue 52%). (**f**) Overlay of elective and boost plans (range 0–70 Gy physical dose)

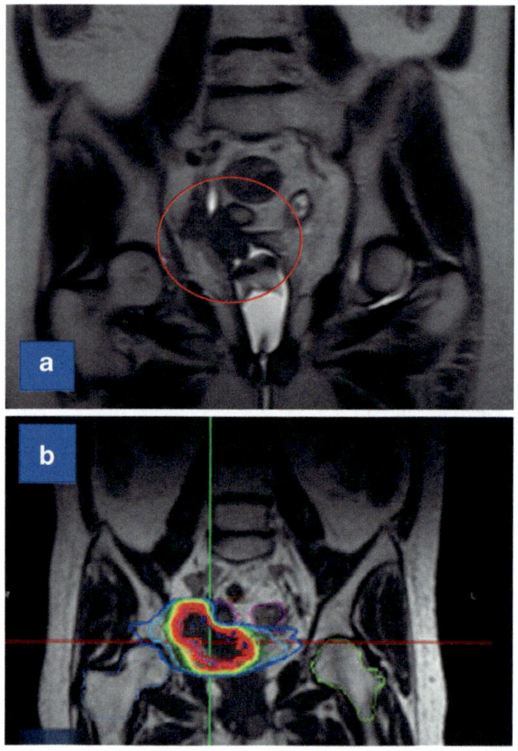


**Fig. 18.10** A coronal view of 3.0 tesla diagnostic MRI showing the residual disease at the vaginal vault extending to the right pelvic wall and causing a hydroureter. (**b**) Isodose line achieved for the MR-Linac boost, the red isodose line being the 30 Gy isodose

**Table 18.4** Prescription of the MR-Linac boost for a vaginal vault recurrence of endometrial cancer

| | | | | | | | | | | | | | |
|---|---|---|---|---|---|---|---|---|---|---|---|---|---|
| **EBRT** | | | | | | | | | | | | | |
| Dose/fx | 1.8 | | Tumor α/β | | | 10 | | | | | | | |
| fx w/o shield | 25 | | OAR α/β | | | 3 | | | | | | | |
| fx w/ shield | 0 | | | | | | | | | | | | |
| **HDR** | | **fx1** | **fx2** | **fx3** | **fx4** | **fx5** | **fx6** | **EBRT** | **HDR** | **Total** | **Goal** | | |
| HRCTV | RX | 5.0 | 5.0 | 5.0 | 5.0 | 5.0 | 5.0 | 45.0 | 30.0 | | | | |
| | $EQD2_{10}$ | 6.3 | 6.3 | 6.3 | 6.3 | 6.3 | 6.3 | 44.3 | 37.6 | 81.9 | $80 \le x \le 90$ | | |
| Bladder | $D_{2cc}$ | 4.0 | 4.0 | 4.0 | 4.0 | 4.0 | 4.0 | 45.0 | 24.0 | | | | |
| | $EQD2_3$ | 5.6 | 5.6 | 5.6 | 5.6 | 5.6 | 5.6 | 43.2 | 33.6 | 76.8 | < 75 | 75 - 90 | > 90* |
| Rectum | $D_{2cc}$ | 3.0 | 3.0 | 3.0 | 3.0 | 3.0 | 3.0 | 45.0 | 18.0 | | | | |
| | $EQD2_3$ | 3.6 | 3.6 | 3.6 | 3.6 | 3.6 | 3.6 | 43.2 | 21.6 | 64.8 | < 65 | 65 - 75 | > 75* |
| Sigmoid | $D_{2cc}$ | 3.0 | 3.0 | 3.0 | 3.0 | 3.0 | 3.0 | 45.0 | 18.0 | | | | |
| | $EQD2_3$ | 3.6 | 3.6 | 3.6 | 3.6 | 3.6 | 3.6 | 43.2 | 21.6 | 64.8 | < 65 | 65 - 75 | > 75* |
| Bowel | $D_{2cc}$ | 3.0 | 3.0 | 3.0 | 3.0 | 3.0 | 3.0 | 45.0 | 18.0 | | | | |
| | $EQD2_3$ | 3.6 | 3.6 | 3.6 | 3.6 | 3.6 | 3.6 | 43.2 | 21.6 | 64.8 | < 65 | 65 - 75 | > 75* |

## Treatment of Oligometastatic Gynecologic Cancer

The concept of oligometastatic disease was first introduced in 1995 by Hellman and Weiselbaum [97], who described a subcategory of patients with metastatic disease whose prognosis is better than widespread metastatic disease and might even be approached with curative intent. SBRT of oligometastatic disease has been reported to increase survival while preserving the quality of life. Chapter 12 reviews the rationale for using MRgRT to deliver an ablative dose to oligometastatic disease, and the reader should refer to this section of the book to get a full scope of the subject. This section will briefly present a review of

**Table 18.5** Published series on the use of SBRT for oligometastatic gynecologic cancer

| | Patients | Dose | LC | Toxicity |
|---|---|---|---|---|
| Mixed Gyn cancer | | | | |
| Kunos et al. (2012) [98] | 50 pts.<br>Metastatic GYN ca all sites | 24 Gy/3 fx<br>(Rx at 70% isodose)<br>$BED_{10}$: 43 | LC 100%<br>(CR + PR)<br>At 2 years<br>Only 1 CR | Gde 1 and 2: 76%<br>Gde 3: 4%<br>Gde 4: 2% |
| Mesko et al. (2017) [99] | 28 pts.<br>(47 lesions)<br>Metastatic GYN Ca all sites | Median total dose (Gy) 40<br>Median Fx: 5<br>$BED_{10}$: >72 associated with better outcome | LC 83%<br>34% (SD)<br>32% (PR)<br>17% (CR)<br>17% (PD)<br>Median FU: 12 months | Gde 3: 1 pt. |
| Ovarian cancer | | | | |
| Lazzari et al. (2018) [100] | 82 pts<br>Ovarian cancer | 24 Gy/3 fx<br>$BED_{10}$: 43 | LC 70% all sites<br>(81% if LN)<br>At 17.4 months | No Gde 3–4 toxicity |
| Macchia et al. (2020) [101] | 449 pts<br>Ovarian cancer | Median total dose (Gy): 25 (5–75)<br>Median no. of fx 4 (1–13)<br>Median dose/fx, (Gy): 8 (3–30)<br>Note: BED 10 >70 Gy = higher CR in MVA | LC 81.9%<br>Median FU: 22 months | No Gde 3–4 toxicity<br>Gde 1–2: 20.7% |

the literature (Table 18.5) supporting the use of SBRT for oligometastatic gynecologic cancer with a discussion on guidelines to select an appropriate dose fractionation for oligometastatic gynecologic malignancies.

Surgery is rarely possible in the management of oligometastatic gynecologic cancer, either due to unfavorable locations and/or patients unfit for surgery [102]. Combined with systemic treatment, salvage irradiation can be used to treat these disease sites. The fact that SBRT is delivered in only a few treatment sessions and is a noninvasive ablative treatment alternative makes it a great local therapy approach to integrate with systemic treatment with minimal interference with the systemic treatment schedule. Radiation can also be offered to patients with oligometastatic gynecologic cancer to allow for a pause in systemic treatment or postpone resuming systemic treatment [103]. MRI guidance and ART offer the advantage of better tumor localization and high precision in dose delivery, which is especially important when planning to deliver a high dose per fraction for a target surrounded by luminal organs. Another important feature of MR-Linac systems is the advantage of real-time tracking and automatic beam stopping. For oligometastatic targets localized in the chest or upper abdomen, intra-fraction target motion is important to consider, and having the ability to perform real-time tracking with automatic beam stopping when motion is detected allows treatment with reduced margins and increased safety.

Patients are often asymptomatic from the site of oligometastatic disease when the disease burden is low. However, the paradigm that radiation should only be offered to palliate symptoms in the metastatic status is now being challenged, and data support that there is a potential survival benefit in treating these small sites of metastatic deposits [3]. SBRT for small target volumes provides good tumor control rates with a low toxicity profile [104–106].

Focusing only on the site of recurrence carries the risk of out-of-field local progression in other regional lymph nodes. Therefore, it is essential to carefully assess the clinical situation at the time of disease recurrence. Treatment of a patient with cervical cancer presenting a PAO recurrence should be managed as a locoregional progression. These patients should be offered a curative treatment with a comprehensive chemoradiation approach, as discussed in this chapter in the section on the management of cervical cancer recurrence. Out-of-field progression could be prevented by using larger EBRT fields; however, in some situations, overlapping volumes of treatment fields with previous radiotherapy areas lead to higher morbidity rates [104], and in some situations, patients would not tolerate longer fractionation and larger treatment volumes.

There is a clear dose-response correlation between higher biologically equivalent doses (BEDs) and LC rates [98, 99, 107, 108]. For lymph node metastases, 5-year LC rates of uterine, cervical, and ovarian cancer range from 70 to 97%, and good rates of disease-free survival and overall survival are reported in retrospective series [100, 101, 105, 108–113]. Lymph node metastases of gynecologic malignancies are often located in the pelvis or abdomen, where conventional SBRT using CBCT provides poor soft tissue contrast. When delivering treatment on standard Linac, dose escalation will be limited in these clinical situations since the dose needs to be maintained below the dose tolerance of the surrounding bowel. In this clinical situation, MRgRT offers significant advantages, as it allows direct visualization of the metastases on MRI and enables margin reduction or dose escalation strategies by using online ART and automated gating systems [114].

Lazzari reported on the outcome of SBRT in oligometastatic ovarian cancer [100]. In this series, SBRT was used for oligorecurrent or oligoprogressive disease in heavily pretreated patients (median of three prior systemic therapy regimens). Excellent LC was achieved without any $\geq$ grade 3 acute or late toxicity. The median systemic treatment-free interval after SBRT in this cohort was 7.4 months, and over one-third of patients were still disease-free 1 year after SBRT. SBRT was able to postpone the need to reinitiate systemic therapy allowing patients to recover from systemic treatment toxicity for many months before resuming treatment. If focused treatment is being used, when needed, multiple SBRT courses could be used in case of subsequent recurrence.

A large retrospective multicenter analysis of SBRT of 449 ovarian cancer lesions found that patient's age $\leq$60-year-old, a PTV size of $\leq$18 $cm^3$, lymph node disease, and a BED (a/b10) of >70 Gy were independent predictive factors of complete response on multivariate analysis [101]. This analysis led to the development of the ongoing MITO-RT3/RAD prospective clinical trial, a multicenter phase II study aimed at identifying potential predictors of response and clinical outcomes after SBRT treatment for patients with oligometastatic/persistent/recurrent ovarian cancer.

The study hypothesis is that SBRT delivered by predefined treatment schedules and shared constraints could improve the complete response rate, and the primary endpoint is the complete clinical response rate. All patients accrued will be treated with a radiotherapy dose in the range of 30–50 Gy by 1, 3, or 5 fractions to all sites of active metastatic disease. The best response for each lesion basis will be evaluated by CT scan, positron emission tomography/CT, or MRI in case of brain lesions every 3 months. Approximately 205 lesions will be treated (90 lymph nodes and 115 parenchyma lesions). The investigators plan to complete enrollment by spring 2023, and the analysis will be completed in the same year [115].

The fractionation of SBRT varies widely based on tumor-related parameters (size, proximity to vulnerable OAR, tumor motion) and if the target lies in a previously irradiated field. Breathing motion for oligometastasis located in the chest or upper abdomen and changes in the filling status of surrounding OARs in the pelvis can present a real challenge. MRgRT can improve the feasibility of dose escalation to sites of oligometastatic in the abdomen and pelvis. In addition, the fact that the MR-Linac system can automatically gate the beam by using real-time tracking eliminates the need for invasive implantation of fiducial markers and the addition of an ITV to account for inter-fraction motion, leading to reduced healthy surrounding tissue irradiation.

For many tumor sites, the combination of SBRT with immunotherapy has been shown to enhance the immune response. With the emergent use of immunotherapy in managing gynecologic cancer, MRg SBRT might play a dominant role in the multidisciplinary treatment strategy for oligometastatic gynecologic cancer.

## Conclusion

MR-Linac is an emerging technology that broadens the radiation treatment options for patients with gynecologic cancers. This promising technology is still very recent, and tremendous work needs to be done to explore its clinical relevance and possible superiority over treatment systems with CT scanners on board. Gynecologic cancers develop in organs associated with large interfraction motion and significant changes in shape and extension through the course of external beam treatment. Therefore, MRgRT, with the potential of daily online treatment planning according to the actual anatomical situation, could significantly improve the treatment outcome. For boosts to primary tumor lesions in LACC and vaginal cancer, image-guided BT remains the treatment of choice, and MRgRT should only be considered if BT is not feasible. Through joint efforts and collaboration, clinical evidence to explore the full potential of this new technology for the management of women with gynecologic cancer should be built during the next decade.

## References

1. Sung H, Ferlay J, Siegel RL, Jemal A, et al. Global cancer statistics 2020; GLOBOCAN estimates of incidence and mortality worldwide for 36 cancers in 185 countries. CA Cancer J Clin. 2021;71(3):209–49. https://doi.org/10.3322/caac.21660.
2. Siegel RL, Miller KD, Wagle NS, Jemal A. Cancer statistics 2023. CA Cancer J Clin. 2023;73(1):17–48. https://doi.org/10.3322/caac.21763.
3. Palma DA, Olson R, Harrow S, et al. Stereotactic ablative radiotherapy for the comprehensive treatment of oligometastatic cancers: long-term results of the SABR-COMET phase II randomized trial. J Clin Oncol. 2020;38(25):2038–838. https://doi.org/10.1200/JCO.20.00818.
4. Gregoire V, Guckenberger M, Haustermans K, et al. Image guidance in radiation therapy for better cure of cancer. Mol Oncol. 2020;14:1470–91.
5. Sturdza A, Pötter R, Fokdal LU, Haie-Meder C, et al. Image-guided brachytherapy in locally advanced cervical cancer: improved pelvic control and survival in RetroEMBRACE, a multicenter cohort study. Radiother Oncol. 2016;120(3):428–33. https://doi.org/10.1016/j.radonc.2016.03.011.
6. Westerveld H, Schmid MP, Nout RA, et al. Image-guided adaptive brachytherapy (IGABT) for primary vaginal cancer: results of the international multicenter RetroEMBRAVE cohort study. Cancers (Basel). 2021;13(6):1459. https://doi.org/10.3390/cancers13061459.
7. Scoutt LM, McCarthy SM. Applications of magnetic resonance imaging to gynecology. Top Magn Reson Imaging. 1990;2(3):37–49.
8. Russell AH, Anderson M, Walter J, Kinney W, Smith L, Scudder S. The integration of computed tomography and magnetic resonance imaging in treatment planning for gynecologic cancer. Clin Obstet Gynecol. 1992;35(1):55–72. https://doi.org/10.1097/00003081-199203000-0001.
9. Pötter R, Dimopoulos J, Georg P, et al. Clinical impact of MRI assisted dose volume adaptation and dose escalation in brachytherapy of locally advanced cervix cancer. Radiother Oncol. 2007;83(2):148–55. https://doi.org/10.1016/j.radonc.2007.04.012.
10. Nag S, Chao C, Erickson B, et al. The American Brachytherapy Society recommendations for low-dose-rate brachytherapy for carcinoma of the cervix. Int J Radiat Oncol Biol Phys. 2002;52(1):33–48. https://doi.org/10.1016/s0360-3016(01)01755-2.
11. Haie-Meder C, Pötter R, Van Limbergen E, et al. Gynaecological (GYN) GEC-ESTRO Working Group Recommendations from Gynaecological (GYN) GEC-ESTRO Working Group (I): concepts and terms in 3D image-based 3D treatment planning in cervix cancer brachytherapy with emphasis on MRI assessment of GTV and CTV. Radiother Oncol. 2005;74(3):235–45. https://doi.org/10.1016/j.radonc.2004.12.015.
12. Pötter R, Tanderup K, Schmid MP, EMBRACE Collaborative Group, et al. MRI-guided adaptive brachytherapy in locally advanced cervical cancer (EMBRACE-I): a multicenter prospective cohort study. Lancet Oncol. 2021;22(4):538–47. https://doi.org/10.1016/S1470-2045(20)30753-1.
13. Balleyguier C, Sala E, Da Cunha T, et al. Staging of uterine cervical cancer with MRI: guidelines of the European Society of Urogenital Radiology. Eur Radiol. 2011;21(5):1102–10. https://doi.org/10.1007/s00330-010-1998-x.
14. AJCC cancer staging manual. 8th ed. Cervical cancer staging. Springer; 2017.
15. Sala E, Rockall A, Rangarajan D, Kubik-Huch RA. The role of dynamic contrast enhanced and diffusion-weighted magnetic resonance imaging in

the female pelvis. Eur J Radiol. 2010;76(3):367–85. https://doi.org/10.1016/j.ejrad.2010.01.026.
16. Otero-Garcia MM, Mesa-Alvarez A, Nikolic O, et al. Role of MRI in staging and follow up of endometrial and cervical cancer: pitfalls and mimickers. Insights Imaging. 2019;10(19):1–22. https://doi.org/10.1186/s13244-019-0696-8.
17. Zhang W, Chen C, Liu P, et al. Impact of pelvic MRI in routine clinical practice on staging of IB1-IIA2 cervical cancer. Cancer Manag Res. 2019;11:3603–9. https://doi.org/10.2147/CMAR.S197496.
18. Akita A, Shinmoto H, Hayashi S, et al. Comparison of T2-weighted and contrast-enhanced T1-weighted MR imaging at 1.5 T for assessing the local extent of cervical carcinoma. Eur Radiol. 2011;21(9):1850–7. https://doi.org/10.1007/s00330-011-2122-6.
19. Meissnitzer M, Forstner R. MRI of endometrium cancer—how we do it. Cancer Imaging. 2016;16:11. https://doi.org/10.1186/s40644-016-0069-1.
20. Seki H, Azumi R, Kimura M, Sakai K. Stromal invasion by carcinoma of the cervix: assessment with dynamic MR imaging. Am J Roentgenol. 1997;168(6):1579–85. https://doi.org/10.2214/ajr.168.6.9168730.
21. Vargas HA, Akin O, Zheng J, et al. The value of MR imaging when the site of uterine cancer origin is uncertain. Radiology. 2011;258(3):785–92. https://doi.org/10.1148/radiol.10101147.
22. Shweel MA, Abdel-Gawad EA, Abdel-Gawad EA, Abdelghany HS, Abdel-Rahman AM, Ibrahim EM. Uterine cervical malignancy: diagnostic accuracy of MRI with histopathologic correlation. J Clin Imaging Sci. 2012;2:42. https://doi.org/10.4103/2156-7514.99175.
23. Thomeer MG, Gerestein C, Spronk S, van Doorn HC, van der Ham E, Hunink MG. Clinical examination versus magnetic resonance imaging in the pretreatment staging of cervical carcinoma: systematic review and meta-analysis. Eur Radiol. 2013;23(7):2005–18. https://doi.org/10.1007/s00330-013-2783-4.
24. Woo S, Suh CH, Kim SY, Cho JY, Kim SH. Magnetic resonance imaging for detection of parametrial invasion in cervical cancer: an updated systematic review and meta-analysis of the literature between 2012 and 2016. Eur Radiol. 2018;28(2):530–41. https://doi.org/10.1007/s00330-017-4958-x.
25. Bourgioti C, Chatoupis K, Moulopoulos LA. Current imaging strategies for the evaluation of uterine cervical cancer. World J Radiol. 2016;8(4):342–54. https://doi.org/10.4329/wjr.v8.i4.342.
26. Sala E, Wakely S, Senior E, Lomas D. MRI of malignant neoplasms of the uterine corpus and cervix. Am J Roentgenol. 2007;188(6):1577–87. https://doi.org/10.2214/AJR.06.1196.
27. Atcı N, Özgür T, Öztürk F, Dolapçıoğlu KS. Utility of intravaginal ultrasound gel for local staging of cervical carcinoma on MRI. Clin Imaging. 2016;40(6):1104–7. https://doi.org/10.1016/j.clinimag.2016.07.004.
28. Lin G, Ho KC, Wang JJ, et al. Detection of lymph node metastasis in cervical and uterine cancers by diffusion-weighted magnetic resonance imaging at 3T. J Magn Reson Imaging. 2008;28(1):128–35. https://doi.org/10.1002/jmri.21412.
29. Liu Y. Benign ovarian and endometrial uptake on FDG PET-CT: patterns and pitfalls. Ann Nucl Med. 2009;23(2):107–12. https://doi.org/10.1007/s12149-008-0227-z.
30. Nakai G, Matsuki M, Inada Y, et al. Detection and evaluation of pelvic lymph nodes in patients with gynecologic malignancies using body diffusion-weighted magnetic resonance imaging. J Comput Assist Tomogr. 2008;32(5):764–8. https://doi.org/10.1097/RCT.0b013e318153fd43.
31. Choi HJ, Kim SH, Seo SS, et al. MRI for pretreatment lymph node staging in uterine cervical cancer. Am J Roentgenol. 2006;187(5):W538–43. https://doi.org/10.2214/AJR.05.0263.
32. Bates DB, Homsi ME, Chang KJ, et al. MRI for rectal cancer: staging, MR-CRM, EMVI, lymph node staging and post-treatment response. Clin Colorectal Cancer. 2022;21(1):10–8. https://doi.org/10.1016/j.clcc.2021.10.007.
33. Nam H, Huh SJ, Park W, et al. Prognostic significance of MRI-detected bladder muscle and/or serosal invasion in patients with cervical cancer treated with radiotherapy. Br J Radiol. 2010;83:868–73. https://doi.org/10.1259/bjr/66646798.
34. Wakatsuki M, Kato S, Kiyohara H, et al. The prognostic value of rectal invasion for stage IVA uterine cervical cancer treated with radiation therapy. BMC Cancer. 2016;16:244. https://doi.org/10.1186/s12885-016-2268-3.
35. Nougaret S, Horta M, Sala E, et al. Endometrial cancer MRI staging: update guidelines of the European Society of Urogenital Radiology. Eur Radiol. 2019;29(2):792–805. https://doi.org/10.1007/s00330-018-5515.
36. Creasman WT, Morrow CP, Bundy BN, Homesly HD, Graham JE, Heller PB. Surgical pathologic spread patterns of endometrial cancer. A Gynecologic Oncology Group Study. Cancer. 1987;60(8):2035–41. https://doi.org/10.1002/1097-0142(19901015)60:8.
37. Schwarz JK, Beriwal S, Esthappan J, et al. Consensus statement for brachytherapy for the treatment of medically inoperable endometrial cancer. Brachytherapy. 2015;14(5):587–99. https://doi.org/10.1016/j.brachy.2015.06.002.
38. van de Bunt L, van der Heide UA, Ketelaars M, de Kort GAP, Jugenliemk-Schulz IM. Conventional, conformal, and intensity-modulated radiation therapy treatment planning of external beam radiotherapy for cervical cancer: the impact of tumor regression. Int J Radiat Oncol Biol Phys. 2006;64(1):189–96. https://doi.org/10.1016/j.ijrobp.2005.04.025.
39. van de Bunt L, Jürgenliemk-Schulz IM, de Kort GAP, Roesink JM, Tersteeg RJHA, van der Heide UA. Motion and deformation of the target volumes

during IMRT for cervical cancer: what margins do we need? Radiother Oncol. 2008;88(2):233–40. https://doi.org/10.1016/j.radonc.2007.12.017.

40. Kerkhof EM, Raaymakers BW, van der Heide UA, van de Bunt L, Jürgenliemk-Schulz IM, Lagendijk JJW. Online MRI guidance for healthy tissue sparing in patients with cervical cancer: an IMRT planning study. Radiother Oncol. 2008;88(2):241–9. https://doi.org/10.1016/j.radonc.2008.04.009.
41. Bowen SR, Yuh WTC, Hippe DS, et al. Tumor radiomic heterogeneity: multiparametric functional imaging to characterize variability and predict response following cervical cancer radiation therapy. J Magn Reson Imaging. 2018;47(5):1388–96. https://doi.org/10.1002/jmri.25874.
42. De Boer P, Mandija S, Werensteijn-Honingh AM, van den Berg CAT, de Leeuw AAC, Jürgenliemk Schulz IM. Cervical cancer apparent diffusion coefficient values during external beam radiotherapy. Phys Imaging Radiat Oncol. 2019;9:77–82. https://doi.org/10.1016/j.phro.2019.03.001.
43. Han K, Croke J, Foltz W, et al. A prospective study of DWI, DCE-MRI and FDG PET imaging for target delineation in brachytherapy for cervical cancer. Radiother Oncol. 2016;120(3):519–25. https://doi.org/10.1016/j.radonc.2016.08.002.
44. Nougaret S, McCague C, Tibermacine H, Vargas HA, Rizzo S, Sala E. Radiomics and radiogenomics in ovarian cancer: a literature review. Abdom Radiol (NY). 2021;46(6):2308–22. https://doi.org/10.1007/s00261-020-02820-z.
45. Michalet M, Azria D, Tardieu M, Tibermacine H, Nougaret S. Radiomics in radiation oncology for gynecological malignancies: a review of literature. Br J Radiol. 2021;94(1125):20210032.
46. Takada A, Yokota H, Watanabe Nemoto M, Horikoshi T, Matsushima J, Uno T. A multi-scanner study of MRI radiomics in uterine cervical cancer: prediction of infield tumor control after definitive radiotherapy based on a machine learning method including peritumoral regions. Jpn J Radiol. 2020;38(3):265–73. https://doi.org/10.1007/s11604-019-00917-0.
47. Engin G. Cervical cancer: MR imaging findings before, during, and after radiation therapy. Eur Radiol. 2006;16(2):313–24. https://doi.org/10.1007/s00330-005-2804-z.
48. Dappa E, Elger T, Hasenburg A, Düber C, Battista MJ, Hötker AM. The value of advanced MRI techniques in the assessment of cervical cancer: a review. Insights Imaging. 2017;8:471–81.
49. Fu C, Feng X, Bian D, et al. Simultaneous changes of magnetic resonance diffusion-weighted imaging and pathological microstructure in locally advanced cervical cancer caused by neoadjuvant chemotherapy. J Magn Reson Imaging. 2015;42:427–35.
50. Kuang F, Yan Z, Wang J, Rao Z. The value of diffusion-weighted MRI to evaluate the response to radiochemotherapy for cervical cancer. Magn Reson Imaging. 2014;32:342–9.
51. Liu Y, Sun H, Bai R, Ye Z. Time-window of early detection of response to concurrent chemoradiation in cervical cancer by using diffusion-weighted MR imaging: a pilot study. Radiat Oncol. 2015;10:185.
52. Somoye G, Harry V, Semple S, et al. Early diffusion-weighted magnetic resonance imaging can predict survival in women with locally advanced cancer of the cervix treated with combined chemo-radiation. Eur Radiol. 2012;22:2319–27.
53. de Foucher T, Bendifallah S, Ouldamer L, et al. Patterns of recurrence and prognosis in locally advanced FIGO stage IB2 to IIB cervical cancer: retrospective multicenter study from the FRANCOGYN group France. Eur J Surg Oncol. 2019;45(4):659–65. https://doi.org/10.1016/j.ejso.2018.11.014.
54. Schmid MP, Franckena M, Kirchheiner K, et al. Distant metastasis in patients with cervical cancer after primary radiotherapy with or without chemotherapy and image-guided adaptive brachytherapy. Gynecol Oncol. 2014;133(2):256–62. https://doi.org/10.1016/j.ygyno.2014.02.004.
55. Sohaib SA, Houghton SL, Meroni R, Rockall AG, Blake P, Reznek RH. Recurrent endometrial cancer: patterns of recurrent disease and assessment of prognosis. Clin Radiol. 2007;62(1):28–34. https://doi.org/10.1016/j.crad.2006.06.015.
56. Lucas R, Dias JL, Cunha TM. Added value of diffusion-weighted MRI in detection of cervical cancer recurrence: comparison with morphologic and dynamic contrast-enhanced MRI sequences. Diagn Interv Radiol. 2015;21(5):368–75. https://doi.org/10.5152/dir.2015.14427.
57. Meads C, Davenport C, Małysiak S, et al. Evaluating PET-CT in the detection and management of recurrence cervical cancer: systematic reviews of diagnostic accuracy and subjective elicitation. BJOG. 2014;121(4):398–407. https://doi.org/10.1111/1471-0528.12488.
58. Morris M, Eifel PJ, Lu J, et al. Pelvic radiation with concurrent chemotherapy compared with pelvic and para-aortic radiation for high-risk cervical cancer. N Engl J Med. 1999;340(15):1137–43. https://doi.org/10.1056/NEJM199904153401501.
59. Peters M, de Leeuw AAC, Nomden CN, et al. Risk factors for nodal failure after radiochemotherapy and image-guided brachytherapy in locally advanced cervical cancer: an EMBRACE analysis. Radiother Oncol. 2021;163:150–8. https://doi.org/10.1016/j.radonc.2021.08.020.
60. Bukkems LJH, Jurgenliemk-Schulz IM, van der Leij F, Peters M, Gerestein CG, Zweemer RP. The impact of para-aortic lymph node irradiation on disease-free survival in patients with cervical cancer: a systematic review and meta-analysis. Clin Transl Radiat Oncol. 2022;35:97–103. https://doi.org/10.1016/j.ctro.2022.05.006.
61. Pötter R, Tanderup K, Kirisits C, The EMBRACE Collaborative Group, et al. The EMBRACE II study: the outcome and prospect of two decades of evolu-

tion within the GEC-ESTRO GYN working group and the EMBRACE studies. Clin Transl Radiat Oncol. 2018;11(9):48–60. https://doi.org/10.1016/j.ctro.2018.01.001.
62. Tanderup K, Lindegaard JC, Kirisits C, et al. Image guided adaptive brachytherapy in cervix cancer: a new paradigm changing clinical practice and outcome. Radiother Oncol. 2016;120(3):365–9. https://doi.org/10.1016/j.radonc.2016.08.007.
63. Grover S, Harkenrider MM, Cho LP, et al. Image guided cervical brachytherapy: 2014 survey of the American Brachytherapy Society. Int J Radiat Oncol Biol Phys. 2014;94:598–60. https://doi.org/10.1016/j.ijrobp.2015.11.024.
64. Pötter R, Georg P, Dimopoulos JCA, et al. Clinical outcome of protocol-based image (MRI) guided adaptive brachytherapy combined with 3D conformal radiotherapy with or without chemotherapy in patients with locally advanced cervical cancer. Radiother Oncol. 2011;100:116–23. https://doi.org/10.1016/j.radonc.2011.07.012.
65. Potter R, Haie-Meder C, Van Limbergen E, et al. GEC ESTRO Working Group. Recommendations from Gynaecological (GYN) GEC ESTO Working Group (II): concepts and terms in 3D image-based treatment planning in cervix cancer brachytherapy-3D dose volume parameters and aspects of 3D image-based anatomy, radiation physics, radiobiology. Radiother Oncol. 2006;78(1):67–77. https://doi.org/10.1016/j.radonc.2005.11.014.
66. Nag S, Cardenes H, Chang S, et al. Image-guided brachytherapy working group. Proposed guidelines for image-based intracavitary brachytherapy for cervical carcinoma: report from image-guided brachytherapy working group. Int J Radiat Oncol Biol Phys. 2004;60(4):1160–72. https://doi.org/10.1016/j.ijrobp.2004.04.032.
67. Berger T, Seppenwoolde Y, Pötter R, EMBRACE Collaborative Group, et al. Importance of technique, target selection, contouring, dose prescription, and dose-planning in external beam radiation therapy for cervical cancer: evolution of practice from EMBRACE-I to II. Int J Radiat Oncol Biol Phys. 2019;104(4):885–94. https://doi.org/10.1016/j.ijrobp.2019.03.020.
68. Hariteportunkel NH, Nath SK, Scanderbeg D, Saenz C, Yashar CM. Evaluation of intra-and interfraction movement of the cervix during intensity modulated radiation therapy. Radiother Oncol. 2011;98(3):347–51. https://doi.org/10.1016/j.radonc.2010.11.015.
69. Lim K, Small W, Portelance L, et al. For the GYN IMRT consortium consensus guidelines for delineation of clinical target volume for intensity-modulated pelvic radiotherapy for the definitive treatment of cervix cancer. Int J Rad Oncol Biol Phys. 2011;79(2):348–55. https://doi.org/10.1016/ijrobp.2009.10.075.
70. White I, Mcquaid E, Godwin E, et al. Image-guided adaptive external beam radiotherapy (EBRT) planning for cervical cancer—a comparison of offline, online and MRI-guided techniques to reduce organ at risk (OAR) dose using a dose accumulation model. Int J Radiat Oncol Biol Phys. 2019;105(1):S250–1.
71. Kozak M, Koenig JL, von Eyben R, Kidd EA. Less than whole uterus irradiation for locally advanced cervical cancer maintains locoregional control and decreases radiation dose to bowel. Pract Oncol Radiother. 2018;9(2):e164–71. https://doi.org/10.1016/j.prro.2018.10.009.
72. Buschmann M, Majercakova K, Sturdza A, et al. Image guided adaptive external beam radiation therapy for cervix cancer: evaluation of a clinically implemented plan-of-the-day technique. Med Phys. 2018;28(3):184–95. https://doi.org/10.1016/j.zemedi.2017.09.004.
73. Mendez LC, Raziee H, Davidson M, et al. Should we embrace hypofractionated radiotherapy for cervical cancer? A technical note on management during COVID-19 pandemic. Radiother Oncol. 2020;148:270–3. https://doi.org/10.1016/j.radonc.2020.05.032.
74. Hes J, Van Lier A, De Leeuw A, Kroon P, DeGroot-van Breugel E, Jurgenkiemk-Schulz I. OC-0713: MR-Linac boosts for patients with cervix cancer ineligible for brachytherapy; preliminary experience. Radiother Oncol. 2020;152:S402–3. https://doi.org/10.1016/S0167-8140(21)00735-09.
75. Hadi I, Eze C, Schönecker S, von Bestenbostel R, et al. MR-guided SBRT boost for patients with locally advanced or recurrent gynecological cancers ineligible for brachytherapy: feasibility and early clinical experience. Radiat Oncol. 2022;17(8):1–9. https://doi.org/10.1186/s13014-022-01981-z.
76. Mahmoud O, Kilic S, Khan AJ, Beriwal S, Small W. External beam techniques to boost cervical cancer when brachytherapy is not an option—theories and applications. Ann Transl Med. 2017;5(10):5. https://doi.org/10.21037/atm.2017.03.102.
77. Kilic S, Cracchiolo B, Mahmoud O. Non-brachytherapy alternatives in cervical cancer radiotherapy: why not? Appl Radiat Oncol. 2015;4:11–7.
78. Albuquerque K, Tumati V, Lea J, Ahn C, Richardson D, Miller D, Timmerman R. A phase II trial of stereotactic ablative radiation therapy as a boost for locally advanced cervical cancer. Int J Radiat Oncol Biol Phys. 2020;106(3):464–71. https://doi.org/10.1016/j.ijrobp.2019.10.042.
79. Tanderup K, Fokdal LY, Sturdza A, et al. Effect of tumor dose, volume and overall treatment time on local control after radiochemotherapy including MRI guided brachytherapy of locally advanced cervical cancer. Radiother Oncol. 2016;120(3):441–6. https://doi.org/10.1016/j.radonc.2017.01.022.
80. Han K, Milosevic M, Fyles A, Pintilie M, Viswanathan AN. Trends in the utilization of brachytherapy in cervical cancer in the United States. Int J Radiat Oncol Biol Phys. 2013;87(1):111–9. https://doi.org/10.1016/j.ijrobp.2013.05.033.

81. Holschneider CH, Petereit DG, Chu C, et al. Brachytherapy: a critical component of primary radiation therapy for cervical cancer: from the Society of Gynecologic Oncology (SGO) and the American Brachytherapy Society (ABS). Gynecol Oncol. 2019;152(3):540–7. https://doi.org/10.1016/j.ygyno.2018.10.016.
82. Ko HC, Huang JY, Miller JR, et al. Novel use of ViewRay MRI guidance for high-dose-rate brachytherapy in the treatment of cervical cancer. Brachytherapy. 2018;17:680–8. https://doi.org/10.1016/j.brachy.2018.04.005.
83. Babar S, Rockall A, Goode A, Shepherd J, Reznek R. Magnetic resonance imaging appearances of recurrent cervical carcinoma. Int J Gynecol Cancer. 2007;17(3):637–45. https://doi.org/10.1111/j.1525-1438.2007.00849.x.
84. Pervin S, Ruma FI, Rahman K, et al. Adjuvant hysterectomy in patients with residual disease after radiation for locally advanced cervical cancer: a prospective longitudinal study. J Glob Oncol. 2019;4:157. https://doi.org/10.1200/JGO.18.00157.
85. Parikh PJ, Lee P, Low DA, et al. Multi-institutional phase 2 trial of ablative 5-fraction stereotactic MR-guided on 2 table adaptive radiation therapy (SMART) for inoperable pancreatic cancer. Int J Radiat Biol Oncol Phys. 2023;117:799–808. https://doi.org/10.1016/j.ijrobp.2023.05.023.
86. Yoo S, Hegarty SE, Mishra MV, Patel N, Cantrell LA, Showalter TN. Definitive radiation therapy for stage I–II endometrial cancer: an observational study of nonoperative management. Am J Clin Oncol. 2017;40(6):582–9. https://doi.org/10.1097/COC.0000000000000204.
87. Podzielinski I, Randall ME, Breheny PJ, et al. Primary radiation therapy for medically inoperable patients with clinical stage I and II endometrial carcinoma. Gynecol Oncol. 2012;124(1):36–41. https://doi.org/10.1016/j.ygyno.2011.09.022.
88. Nag S, Erickson B, Parikh S, Gupta N, Varia M, Glasgow G. The American Brachytherapy Society recommendations for high-dose-rate brachytherapy for carcinoma of the endometrium. Int J Radiat Oncol Biol Phys. 2000;48(3):779–90. https://doi.org/10.1016/s0360-3016(00)00689-1.
89. van der Steen-Banasik E, Christiaens M, Shash E, et al. Systemic review: radiation therapy alone in medical non-operable endometrial carcinoma. Eur J Cancer. 2016;65:172–81. https://doi.org/10.1016/j.ejca.2016.07.005.
90. van der Steen-Banasik E. Primary brachytherapy as a radical treatment for endometrial carcinoma. J Contemp Brachytherapy. 2014;6(1):106–12. https://doi.org/10.5114/jcb.2014.42028.
91. Weitmann HD, Pötter R, Waldhäusl C, Nechvile E, Kirisits C, Knocke TH. Pilot study in the treatment of endometrial carcinoma with 3D image-based high-dose-rate brachytherapy using modified Heyman packing: clinical experience and dose-volume histogram analysis. Int J Radiat Oncol Biol Phys. 2005;62(2):468–78. https://doi.org/10.1016/j.ijrobp.2004.10.013.
92. Hardarson HA, Heidemann LN, dePont CR, Mogensen O, Jochumsen KM. Vaginal vault recurrences of endometrial cancer in non-irradiated patients—radiotherapy or surgery. Gynecol Oncol Rep. 2015;11:26–30. https://doi.org/10.1016/j.gore.2015.01.002.
93. Pai HH, Souhami L, Clark BG, Roman T. Isolated vaginal recurrences in endometrial carcinoma: treatment results using high-dose-rate intracavitary brachytherapy and external beam radiotherapy. Gynecol Oncol. 1997;66(2):300–7. https://doi.org/10.1006/gyno.1997.4752.
94. Kamrava M, Sushil Beriwal S, Erickson B, et al. American Brachytherapy Society recurrent carcinoma of the endometrium task force patterns of care and review of the literature. Brachytherapy. 2017;16(6):1129–43. https://doi.org/10.1016/j.brachy.2017.07.012.
95. Weitmann HD, Knocke TH, Waldhäusl C, Pötter R. Ultrasound-guided interstitial brachytherapy in the treatment of advanced vaginal recurrences from cervical and endometrial carcinoma. Strahlenther Onkol. 2006;182(2):86–95. https://doi.org/10.1007/s00066-006-1420-4.
96. Fokdal L, Tanderup K, Nielsen SK, et al. Image and laparoscopic guided interstitial brachytherapy for locally advanced primary or recurrent gynaecological cancer using the adaptive GEC ESTRO target concept. Radiother Oncol. 2011;100(3):473–9. https://doi.org/10.1016/j.radonc.2011.08.016.
97. Hellman S, Weichselbaum RR. Oligometastasis. J Clin Oncol. 1995;13:8–10. https://doi.org/10.1200/JCO.1995.13.1.8.
98. Kunos CA, Brindle J, Waggoner S, et al. Phase II clinical trial of robotic stereotactic body radiosurgery for metastatic gynecologic malignancies. Front Oncol. 2012;2(181):1–6. https://doi.org/10.3389/fonc.2012.00181.
99. Mesko S, Sandler K, Cohen J, Konecny G, Steinberg M, Kamrava M. Clinical outcomes for stereotactic ablative radiotherapy in oligometastatic and oligoprogressive gynecological malignancies. Int J Gynecol Cancer. 2017;27:403–8. https://doi.org/10.1097/IGC.0000000000000869.
100. Lazzari R, Ronchi S, Gandini S, et al. Stereotactic body radiation therapy for oligometastatic ovarian cancer: a step toward a drug holiday. Int J Radiat Oncol Biol Phys. 2018;101:650–60. https://doi.org/10.1016/j.ijrobp.2018.03.058.
101. Macchia G, Lazzari R, Colombo N, et al. A large, multicenter, retrospective study on efficacy and safety of stereotactic body radiotherapy (SBRT) in oligometastatic ovarian cancer (MITO RT1 study): a collaboration of MITO, AIRO GYN, and MaNGO groups. Oncologist. 2020;25(2):e311–20. https://doi.org/10.1634/theoncologist.2019-0309.
102. Chiva LM, Lapuente F, González-Cortijo L, et al. Surgical treatment of recurrent cervical cancer:

state of the art and new achievements. Gynecol Oncol. 2008;110(3 Suppl 2):S60–6. https://doi.org/10.1016/j.ygyno.2008.05.024.
103. Kubota H, Tsujino K, Sulaiman NS, et al. Comparison of salvage therapies for isolated para-aortic lymph node recurrence in patients with uterine cervical cancer after definitive treatment. Radiat Oncol. 2019;14:236. https://doi.org/10.1186/s13014-019-1442-6.
104. Jereczek-Fossa BA, Ronchi S, Orecchia R. Is stereotactic body radiotherapy (SBRT) in lymph node oligometastatic patients feasible and effective? Rep Pract Oncol Radiother. 2015;20:472–83. https://doi.org/10.1016/j.rpor.2014.10.004.
105. Ning MS, Ahobila V, Jhingran A, et al. Outcomes and patterns of relapse after definitive radiation therapy for oligometastatic cervical cancer. Gynecol Oncol. 2018;148:132–8. https://doi.org/10.1016/j.ygyno.2017.10.017.
106. Jadon R, Higgins E, Hanna L, Evans M, Coles B, Staffurth J. A systematic review of dose-volume predictors and constraints for late bowel toxicity following pelvic radiotherapy. Radiat Oncol. 2019;14(1):57. https://doi.org/10.1186/s13014-019-1262-8.
107. Onderdonk BE, Chmura SJ. The Yin and Yang of cytoreductive SBRT in oligometastases and beyond. Front Oncol. 2019;9:1–6. https://doi.org/10.3389/fonc.2019.00706.
108. Park HJ, Chang AR, Seo Y, et al. Stereotactic body radiotherapy for recurrent or oligometastatic uterine cervix cancer: a cooperative study of the Korean Radiation Oncology Group (KROG 14-11). Anticancer Res. 2015;35(9):5103–10.
109. Matsushita H, Jingu K, Umezawa R, et al. Stereotactic radiotherapy for oligometastases in lymph nodes—a review. Technol Cancer Res Treat. 2018;17:1–8. https://doi.org/10.1177/1533033818803597.
110. Loi M, Frelinghuysen M, Klass ND, et al. Locoregional control and survival after lymph node SBRT in oligometastatic disease. Clin Exp Metastasis. 2018;35:625–33. https://doi.org/10.1007/s10585-018-9922-x.
111. Iftode C, D'Agostino GR, Tozzi A, et al. Stereotactic body radiation therapy in oligometastatic ovarian cancer: a promising therapeutic approach. Int J Gynecol Cancer. 2018;28:1507–13. https://doi.org/10.1097/IGC.0000000000001324.
112. Kowalchuk RO, Waters MR, Richardson KM, et al. Stereotactic body radiation therapy in the treatment of ovarian cancer. Radiat Oncol. 2020;15:1–10. https://doi.org/10.1186/s13014-020-01564-w.
113. Trippa F, Draghini L, di Marzo A, et al. Long-term palliation of lymph node oligometastatic ovarian carcinoma after repeated stereotactic body radiotherapy. Case Rep. 2020;106:6. https://doi.org/10.1177/0300891620952852.
114. Heike L, Kashani R, Robinson C, et al. Phase I trial of stereotactic MR-guided online adaptive radiation therapy (SMART) for the treatment of oligometastatic or unresectable primary malignancies of the abdomen. Radiother Oncol. 2018;126(3):519–26. https://doi.org/10.1016/j.radonc.2017.11.032.
115. Macchia G, Jereczek-Fossa BA, et al. Efficacy and safety of stereotactic body radiotherapy (SBRT) in oligometastatic/persistent/recurrent ovarian cancer: a prospective, multicenter phase II study (MITO-RT3/RAD). Int J Gynecol Cancer. 2022;32(7):939–43. https://doi.org/10.1136/ijgc-2021-002709.

# 19 MR-Guided Radiotherapy in the Pediatric and Adolescent Patient

Margaret M. Kozak, Sean Mahase, Bryan Traughber, Mitchell Machtay, and John Buatti

## Introduction

### Disease Overview

In 2022 in the United States, an estimated 10,470 children aged 0–14 years and 5480 adolescents aged 15–19 years were diagnosed with cancer [1]. The most common types of cancer in children are leukemias, followed by brain and other central nervous system (CNS) tumors and lymphomas. Based on Surveillance, Epidemiology, and End Results (SEER) data, the 5-year survival for all children diagnosed with brain and other CNS cancers is about 75%, compared to about 86% for children diagnosed with leukemia [2]. This relatively high cure rate when compared to adult cancers underscores the importance of tailoring cancer therapy with the goal of avoiding adverse effects that may not become apparent until much later in life [3, 4]. In this chapter, we will focus primarily on pediatric solid tumors, which comprise approximately 38% of cancer diagnoses in children aged 0–14 years and 30% of diagnoses in adolescents aged 15–19 years; they are also the predominant type of pediatric malignancy seen in radiation oncology clinics around the world [5]. These include but are not limited to pediatric brain and CNS tumors, soft-tissue and bone sarcomas, neuroblastoma, and Wilms tumor.

### Modality Approach Based on Tumor Type

The key tenet of highly conformal radiotherapy in children is to treat the tumor to a therapeutic dose while sparing as much surrounding normal tissue from radiation exposure as possible. A number of critical components need to be considered in order for an appropriate radiation plan to be created and safely delivered. These include patient-, target-, and technique-specific factors that allow for highly conformal RT delivery:

- Patient immobilization, localization, and motion are critical to the design of a sophisticated RT plan. Basic patient positioning, including location of the arms and legs, patient comfort especially for treatments that are expected to take >20 min, and an understanding

M. M. Kozak (✉)
Department of Radiation Oncology, University of Iowa Hospitals and Clinics, Iowa City, IA, USA

Department of Radiation Oncology, UT Southwestern Medical Center, Dallas, TX, USA
e-mail: malgorzata.kozak@utsouthwestern.edu

S. Mahase · B. Traughber · M. Machtay
Department of Radiation Oncology, Penn State Cancer Institute, Hershey, PA, USA
e-mail: smahase@pennstatehealth.psu.edu; btraughber@pennstatehealth.psu.edu; mmachtay@pennstatehealth.psu.edu

J. Buatti
Department of Radiation Oncology, University of Iowa Hospitals and Clinics, Iowa City, IA, USA
e-mail: john-buatti@uiowa.edu



I. J. Das et al. (eds.), *A Practical Guide to MR-Linac*, https://doi.org/10.1007/978-3-031-48165-9_19

of the beam directions are all taken into consideration at the time of radiation simulation.

- Target and organ-at-risk (OAR) motion are both influenced by respiration, setup error, and/or involuntary movement during RT delivery. Obtaining a 4D-CT during the simulation process is critical for targets in the lungs and abdomen. The use of a face or head and shoulders mask is important to immobilize targets located in the brain and head and neck regions. Instructing the patient about appropriate bladder and rectal filling can have an impact on doses to critical structure located in the pelvis.
- The development of onboard imaging with cone beam CT allows for daily assessment of patient and target positioning and verification of appropriate dose distribution.
- The ability to modulate and sculpt the radiation beam was enabled by the development of intensity-modulated radiotherapy (IMRT), which uses static gantry positions to deliver radiation from multiple angles. VMAT is an enhanced version of IMRT in which the gantry continually rotates around the patient, which allows for both faster RT delivery and increased conformality.
- Certain treatment techniques such as stereotactic body radiotherapy (SBRT) allow for highly conformal dose distributions but are limited by adjacent OAR dose constraints. For example, a lung nodule near the mainstem bronchus may need to be underdosed to meet the bronchial dose constraint, while SBRT of a liver metastasis is determined by whether sufficient normal liver can be spared from therapeutic radiation. Currently, SBRT for pediatric malignancies is primarily used to treat metastatic sites.
- Minimizing the planning target volume (PTV), defined as the margin required for setup error, decreases the overall treatment volume. For this to be feasible, patient immobilization and setup accuracy are of utmost importance. Setup accuracy can be increased by daily image-guided RT (IGRT) and can also be improved by daily adaptation of the treatment plan.
- Finally, proton therapy is commonly used in the treatment of pediatric malignancies. The potential benefit of protons rests in their physical properties, namely, the Bragg peak, which yields a steep dose falloff with the added ability to decrease integral dose and exit dose. However, there persist certain uncertainties with respect to proton dose deposition and biologic effects that also must be considered when choosing the most appropriate treatment modality for a pediatric or adolescent cancer patient.

The aforementioned considerations do not account for any positional or physiological changes that occur after the pre-treatment images are obtained. Additionally, intrafraction changes due to respiratory motion, peristalsis, and bladder filling are also not corrected for, which may alter the intended dose distribution. Potential consequences include increased dose to OARs and decreased dose to the target. Methods to mitigate these issues include treating to a more conservative dose or using larger planning volumes to account for these uncertainties. Another solution entails placing fiducials in the target, requiring the patient to undergo an invasive procedure. Finally, major changes in anatomy or tumor volume during a treatment course may require pausing treatment while a new simulation and treatment plan are performed, thus lengthening the total treatment package time.

The advent of MRI linear accelerator (MR-linac)-based RT has revolutionized radiation delivery by offering the ability for daily plan adaptation based on changes in both target and OAR anatomy. MRI-guided radiation therapy (MRgRT) utilizes a linear accelerator with MRI capabilities, enabling real-time MRI-based detection of the treatment target and organs at risk. MR-based imaging provides superior visualization of soft-tissue structures over CT-based imaging [6, 7]. In contrast to conventional radiation-delivery machines reliant on static imaging, MRgRT platforms possess intrafraction motion monitoring capabilities to account for tumor changes, internal anatomy movement, and changes in bowel gas or bladder fill, providing a truer representation of the dose distribution. This is utilized for gating, in which radiation is delivered in parallel with a specific parameter, such as the target volume being fully encompassed within a predefined boundary. Additionally, substantial changes in target or normal structures can be accounted for in real time, recontoured, and replanned using same-day imaging (adaptive planning) without treatment breaks. Gating and adaptive planning are distinct abilities, with gating accounting for intrafraction motion while adaptive planning adjusts for interfraction changes. This allows for both reduced PTV volumes and optimal OAR sparing. It also allows for daily assessment of patient response to treatment, with the eventual goal of deriving biomarkers from quantitative MRI techniques to better understand tumor biology [8–10]. Finally, an inherent benefit of MRgRT for pediatric patients is eliminating radiation exposure from pre-treatment KV X-rays or cone beam CT scans. MRgRT platforms have only been in routine use since 2014. Data on this emerging technology is primarily retrospective, focusing on feasibility, safety, and dosimetry, with the first phase three prospective clinical trials reported recently [11]. While the ideal candidates for MRgRT are still under investigation, it likely provides the greatest benefits in regions subject to physiologic motion, such as the thorax, abdomen, and pelvis. Thus far, the clinical experience in MRgRT has been primarily in the adult population. However, in certain scenarios, pediatric patient may also significantly benefit from this approach. The decision of whether to offer a pediatric patient photon versus protons, VMAT versus MRgRT, is nuanced and integrates a variety of factors. In this chapter, we will review some common pediatric diagnoses, and propose ways in which MRgRT can be used to optimize radiation delivery in this clinical setting.

## MR-Guided Linear Accelerator Treatment Workflow

The MRIdian linear accelerator (ViewRay Technologies, Inc.), approved in 2017, consists of a 0.35-T MRI, a 70-cm bore diameter, and 6-MV photons [12]. The MRIdian possesses adaptive planning and real-time automated gating functions. A true fast imaging with steady-state free precession (TRUFI) sequence is obtained for both volumetric clinical imaging and 2D cine MRI for gating [13]. The MRIdian is currently pending FDA clearance to treat pediatric malignancies.

The Elekta Unity combines a wide-bore 1.5-Tesla MR imaging system and a next-generation Elekta 7-MV linear accelerator [14]. Like the MRIdian, the Unity has the ability for online adaptive planning and real-time automated gating in the near future. It contains a Monaco TPS capable of two types of adaptation—adapt-to-position (ATP) and adapt-to-shape (ATS) [15]. The ATP workflow adapts the plan according to changes in daily patient positioning as compared to the reference plan. For this workflow, the contours remain unchanged. The ATS workflow is employed when significant changes in patient anatomy require recontouring by the treating physician, and a new plan is generated in real time with the goal of meeting dose constraints that were set at the time of initial RT planning.

Many components of MRgRT planning are similar to conventional radiation planning, including simulation and patient immobilization, imaging (and adjunct imaging fusions) for target and organ-at-risk contouring, and treatment planning. MRgRT patients undergo an MR simulation for treatment planning as well as a CT simulation to obtain electron density data. The CT scan can be deformably registered to the daily MR scan for adaptive dose calculations. Target and organ-at-risk delineation are performed on the MRI dataset for plan generation, which obviates uncertainties involved with using fused image sets which may not have been obtained in the treatment position [16]. However, adjunct diagnostic scans can be fused to the reference MRI planning scan to aid in target delineation. Certain structures with a predefined expansion (e.g., using a 3-mm CTV expansion to create a PTV) or Boolean operations (e.g., removing overlap between two contoured structures) can be set as "rules" automatically applied during adaptive treatments to reduce table time. Treatment planning is via TPS software with Monte Carlo-based dose calculations accounting for the magnetic field influence on dose distribution. Treatment plans are limited to step-and-shoot IMRT with coplanar beam arrangements. In spite of these limitations, treatment plans are comparable to conventional linac plans [17–19]. Additionally, the initial MRgRT plan represents a reference point that can be utilized for the entire treatment course, or further improved through adaptive planning on interfraction changes.

Each treatment entails screening the patient for MR compatibility. The coils must be placed ideally centered along imaging isocenter while covering the target area. It is important to ensure the coils aren't in direct contact with the skin. Once the patient is appropriately positioned, image acquisition and registration are performed. Upon evaluating the original target volumes registered to the new MRI, the patient can be treated using the original plan. The MRIdian also allows for couch shifts [16]. A predicted plan can also be created, entailing recalculating the original plan with contours deformed with respect to same-day anatomy [13, 20]. After reviewing the treatment plan, the provider may elect to treat with the original plan or create an adapted plan. The MRIdian offered two adaptive planning options. The first entails reoptimizing segment weighting while retaining their respective silhouettes, which can be performed relatively quickly [21]. The alternative is complete reoptimization which includes changing priorities for treatment targets and normal tissues in the planning algorithm. Given the time constraints involved in replanning while the patient is in the treatment position, it is critical to perform adequate quality assurance, including comparing recontoured structures to those created for the initial plan, and verifying electron density. During treatment, the 2D cine MR acquisition functions in a single sagittal plane with images up to 5–20 mm thick continuously obtained at 4–8 frames per second, enabling target volume position verification [13]. The beam turns off when the target moves outside the defined boundary. A 3-mm gating margin is commonly utilized for many tumors treated on the MRIdian [22].

## Rationale for MRgRT Based on Tumor Type and Location

### Supratentorial Brain Tumors

Primary brain and other central nervous system (CNS) tumors are the most common solid tumors of childhood. There are a wide range of diagnoses in this group, with heterogeneous tumor types and variation in tumor location. Over half of pediatric CNS tumors arise in the supratentorial brain, the most common of which are low-grade astrocytomas. Other types of primary brain tumors include high-grade gliomas, embryonal CNS tumors, germ cell tumors, and craniopharyngiomas. Craniopharyngiomas have the lowest 5-year overall survival rates among nonmalignant CNS tumors [23]. These tumors require special attention as they frequently have cystic components that can change in size and shape over the course of radiation delivery. Typically, repeat MRI every 1–2 weeks is indicated for these patients with cyst aspiration either through indwelling Ommaya reservoirs or via a stereotactic approach for symptomatic patients. The use of daily MRgRT for this patient population can allow for significantly improved visualization of the target, daily adaptation of the treatment volume as needed, and a reduction in PTV volume. Typical PTV margins for craniopharyngiomas range from 0.5 to 1 cm to account for changes in tumor size over the course of treatment. In small children, the use of a 2- to 3-mm PTV would significantly decrease the dose to adjacent critical structures such as the pituitary gland, optic chiasm, globes, and surrounding normal brain. This is of utmost importance when considering that the long-term control of craniopharyngiomas following subtotal resection and radiotherapy is excellent and long-term sequelae of treatment can be severe. A study from St. Jude Children's Hospital using smaller PTV margins of 3–5 mm did not show any decrement in tumor control [24].

For other tumors in the supratentorial brain that do not change in size and shape as significantly as a craniopharyngioma, MRgRT may be useful if there are significant changes in the size of the resection cavity or FLAIR signal surrounding the tumor. However, the primary research focus in this area in adult patients with glioblastoma has been quantitative imaging to discover functional response-based adaptation during radiation treatment. This has not yet been done in the pediatric population. At this time, we do not routinely treat pediatric supratentorial brain tumors using MRgRT; however, there may be select scenarios in which this would be beneficial.

### Tumors of the Posterior Fossa

Tumors of the posterior fossa also form a heterogeneous group, the most common of which are low-grade astrocytic tumors, medulloblastoma, and ependymoma. The primary treatment approach for these tumors is maximal safe surgical resection. In patients with low-grade astrocytic tumors, observation may be an appropriate next step especially in the setting of a gross total resection. Medulloblastoma requires craniospinal irradiation (CSI), which is not easily done using MRgRT due to multiple factors including large field sizes and long daily treatment delivery. Many of these patients will benefit from proton CSI, especially if they are of young age. The tumor bed or posterior fossa boost in patients with medulloblastoma is also very well defined, with CTV expansions of 1.0–1.5 cm. Providing the patient does not require a posterior fossa boost, the cochlea can be relatively well spared from high radiation doses with modern RT planning techniques. Therefore, we do not recommend that medulloblastoma patients be treated with MRgRT for reasons of target delineation or OAR avoidance.

Intracranial ependymomas comprise approximately 5–8% of brain tumors in children, most of which present in the posterior fossa, but can be supratentorial as well. The most important determinant of outcome in ependymoma is extent of surgical resection. Complicating this is the fact that these tumors arise from the floor of the fourth ventricle and can extend along cranial nerves and into the foramina of Luschka and Magendie making complete resection difficult and at times leaving patients with permanent cranial nerve deficits. These tumors do not usually invade the brainstem, but due to their location, residual disease can be left at the brainstem interface. The most recent Children's Oncology Group trial ACNS0121 included adjuvant RT for all patients with posterior fossa ependymoma as well as all patients who underwent subtotal resection [25]. The CTV is comprised of the tumor bed, any residual disease, and a 5- to 10-mm margin [26]. The PTV is typically 3–5 mm. CSI is not needed unless the patient presents with neuraxis dissemination at diagnosis. Though the posterior fossa is a rigid structure, patients with residual disease may benefit from improved alignment using MRgRT, smaller treatment volumes, and adaptive replanning to help keep hot spots away from the brainstem and minimize potential for radiation necrosis [27–29]. However, a large multi-institutional study of pediatric patients receiving photon irradiation for posterior fossa tumors demonstrated extremely low rates of brainstem toxicity with conventionally fractionated RT [27]. In this series of 107 patients, the median prescribed dose was 55.8 Gy with maximum doses to the brainstem ranging from 50.4 to 93.7 Gy. Only 2 patients developed late brainstem toxicity, and those patients received Dmax to the brainstem of 58.0 Gy and 57.6 Gy.

Figure 19.1 depicts a case of a brainstem glioma with a posterior cystic component that fluctuated in size during RT. The patient experienced worsening of bulbar symptoms at the time of cyst enlargement and therefore was transitioned from a regular Versa linear accelerator to the Unity MR-linac for the remainder of his treatment. Weekly adaptive replanning was performed on an as-needed basis. The patient also received Avastin infusions to help with tumor swelling as well.

### Spinal Cord Tumors

Spinal tumors comprise a heterogeneous group, with intramedullary tumors frequently arising from the spinal cord, intradural/extramedullary tumors arising from the dura, and extradural tumors originating from outside central nervous system. Intramedullary tumors commonly include ependymomas and low-grade astrocytomas. Meningiomas and nerve sheath tumors are the most prevalent intradural/extramedullary tumors. Metastases account for the overwhelming majority of extradural spinal tumors. Given the rarity of many of these tumors, multidisciplinary management is tantamount to determining the optimal treatment paradigm which chiefly entails resection.

Malignant primary spinal tumors and spinal metastases can significantly impact quality of life due to pain, pathological fracture, and spinal cord compression [30, 31]. Postoperative radiotherapy

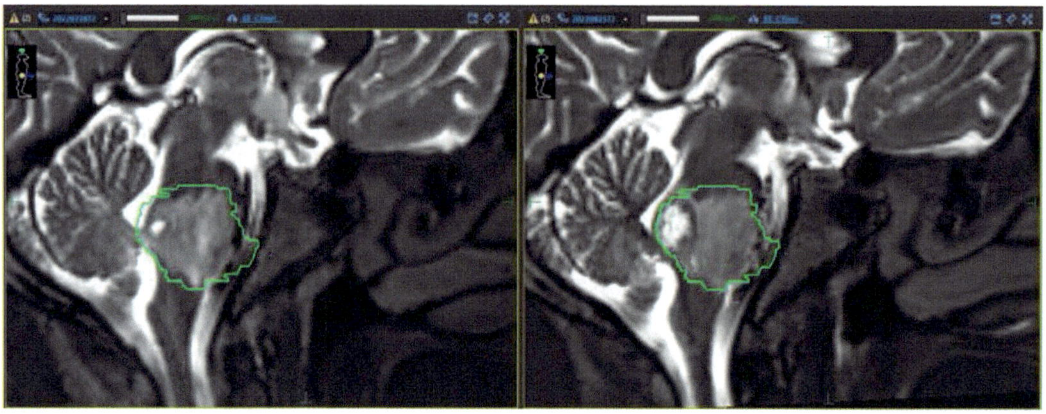

**Fig. 19.1** MR-linac images of a patient with a brainstem glioma with cystic component that fluctuated in size over the course of the patient's treatment, causing intermittently worsening symptoms. The patient was initially planned for standard linear accelerator treatment, however complained of worsening symptoms 1 week after initiating treatment. A diagnostic MRI demonstrated interval enlargement of the cystic component of the tumor. The patient was transitioned to the MR-linac for adaptive replanning as needed. He eventually required Avastin infusion due to worsening cyst enlargement and severe, unrelenting symptoms

is considered standard of care following resection of malignant spinal tumors to minimize local recurrence [32, 33]. Definitive radiotherapy is also considered for unresectable patients. In particular, the frequency and complexity of spinal metastasis radiotherapy has increased in parallel with systemic therapy advances conferring longer overall survival to metastatic cancer patients. Historically, low-dose palliative radiotherapy was the sole option for symptom management, but with disappointing local control rates [31, 34, 35]. Advances in image guidance, technology, and planning software promoted use of SBRT for spinal metastases to safely deliver a higher biologically equivalent dose. While prospective trials utilizing SBRT for spinal tumors are ongoing, it's utilization is increasing [36].

When treating a patient with definitive or postoperative radiotherapy, radiation dose is chiefly limited by spinal cord or thecal sac constraints. Visualization of the spinal cord and treatment volume are essential to successful and safe radiation delivery. This often entails obtaining and fusing MRI or myelogram sequences to CT-based planning scans. Limitations to radiotherapy utilizing conventional linac-based platforms include uncertainties regarding spinal cord location necessitating the use of PRVs, as well as setup uncertainties, which can decrease effective tumor coverage [37–39]. MRgRT platforms enable better soft-tissue delineation for planning, and real-time visualization of the spinal cord to reduce setup uncertainties and improve dose coverage to the PTV [40]. Prospective studies in adults are under development, with initial experiences demonstrating feasibility and safety of this approach [41]. These studies will be imperative to determining whether these putative benefits of MRgRT translate clinically in adults before applying this technology in pediatric populations.

### Head and Neck Cancers

Head and neck cancers comprise about 12% of all childhood malignancies. The most common histologies include lymphomas (~40%), rhabdomyosarcoma (~20%), and nasopharyngeal carcinomas (~15%) [42–44]. The benefit of MRgRT in patients with head and neck malignancies has primarily been studied in the adult population. Due to the significantly improved soft-tissue contrast, MRgRT allows for more precise tumor delineation and margin reduction [45] and affords the added benefit of assessing tumor response over the course of treatment. Another problem that is often encountered in the head and neck adult population are alterations in patient and tumor anatomy due to weight loss [46] caused by oral toxicity secondary to treatment. Weekly

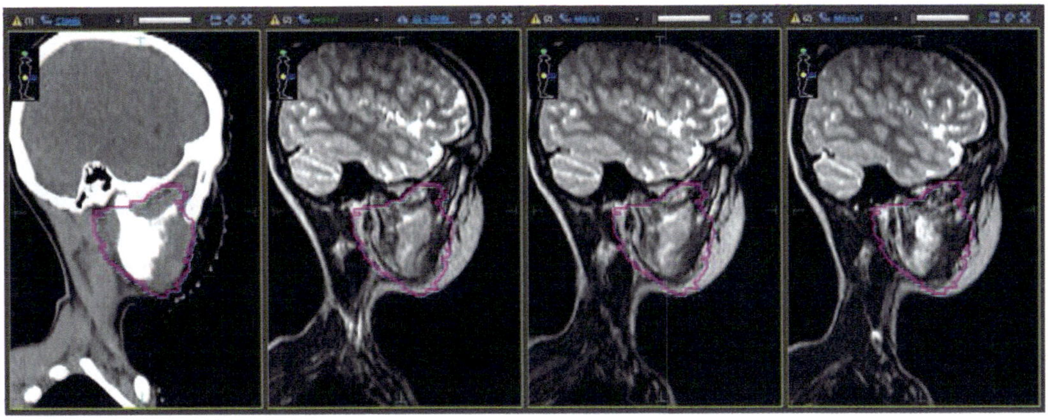

**Fig. 19.2** Example of a palliative treatment plan for a neuroblastoma mandibular metastasis. The patient was initially planned for standard linear accelerator treatment, however complained of significant swelling of the mass a few days into treatment. He was subsequently transferred to the MR-linac and put under anesthesia to finish out his course. The tumor initially enlarged, but after 5 days of treatment began to decrease in size. The purple line shows the initial GTV, and, from left to right, the T2 images show decreasing size of the GTV (T2 bright mass) over time. The treatment volume was adapated as needed with the goals of decreasing acute oral toxicity during the course of radiotherapy

offline adaptation may be sufficient for addressing these issues; however, preliminary studies to date report mixed results [47–49]. Additionally, some tumors may fluctuate in size more quickly than others necessitating daily adaptive replanning. This is not feasible with conventional treatment approaches as re-simulating and replanning require routine imaging that places a strain on busy clinic schedules [47]. Especially for young children who may require anesthesia for imaging purposes, daily online adaptation can significantly streamline the process (Fig. 19.2).

Further improvements in outcomes using functional MRI technologies are being explored, including the use of quantitative MRI biomarkers such as diffusion-weighted imaging (DWI), dynamic contrast-enhanced (DCE) MRI, MR relaxometry, and radiomics [50–54]. Adaptation to shrinking tumor volumes continues to be an active area of investigation in adult head and neck malignancies with long-term outcomes results not yet available [55–57]. Given the uncertainty in microscopic tumor extension, indistinct tumor borders, and tumor infiltration into adjacent tissues, the treating physician can consider the use of a CTV margin to account for microscopic disease depending on the clinical scenario. There may be a significant advantage to tumor alignment and OAR sparing with the use of smaller PTV margins; however, this should be evaluated on a case-by-case basis.

### Pediatric Leukemias

Acute lymphoblastic leukemia (ALL) is the most common malignancy in children accounting for ~80% of pediatric leukemia cases. The next most common is acute myeloblastic leukemia (AML) comprising ~20% of cases. There are about 2500–3000 children diagnosed with ALL each year, with a peak incidence between ages 2 and 5. Treatment of ALL is heavily based on intensive chemotherapy regimens, as well as intrathecal chemotherapy and potentially cranial irradiation depending on the risk grouping of the patient [58–60]. In recent years, cranial irradiation has fallen out of favor and been replaced by more intensive chemotherapeutic regimens [61, 62]. Similarly, AML is primarily treated with chemotherapy with little role for RT unless certain clinical scenarios arise.

One of such clinical scenarios includes the formation of solid tumors known as myeloid sarcomas, previously called chloromas; they are infrequent but have been seen to arise in patients with relapsed or refractory AML [63, 64]. The most commonly involved sites include the bone,

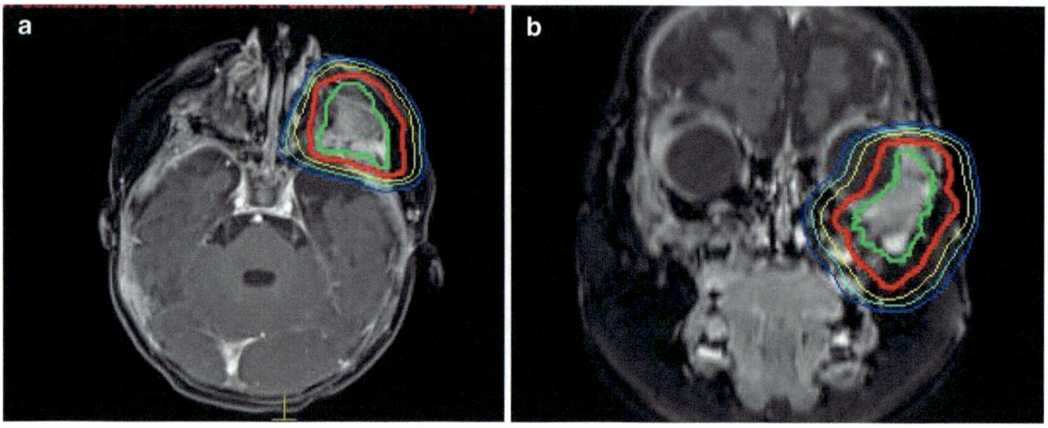


**Fig. 19.3** MR-guided radiation plan for a 6-month-old infant with relapsed refractory AML and a painful orbital chloroma. The patient was planned for MR-guided treatment due to superior daily alignment and ability to use a small PTV expansion to limit dose to surrounding tissues as much as possible. The GTV was defined on T1 post-contrast imaging. The PTV was a 2-mm expansion on the GTV. Panel **a** shows the axial slice and panel **b** shows the coronal slice. Green = GTV; red = 6 Gy; yellow = 5.4 Gy; teal = 4.8 Gy; blue = 4.2 Gy

skin, orbit, lymph nodes, GI tract, and CNS. In children, the skin and orbit are most commonly described [65]. Prior data has not shown a benefit to the addition of RT for the treatment of chloromas; however, radiation is commonly used to palliate chloroma-related vision loss or cord compression [66, 67]. Due to their inherent radiosensitivity and a potentially rapid local response, the use of adaptive treatment to treat myeloid sarcomas in children is particularly appealing (Fig. 19.3). The proposed dose is 20–24 Gy in 10–12 fractions, which has been shown to result in excellent local control and palliation of symptoms [68] and is also recommended by the ILROG guidelines [69]. However, for palliation of symptoms, doses as low as 6 Gy have been used [70]. These lesions can be well visualized using FDG-PET imaging and can also be seen well on MRI [71–76]. The use of daily MRgRT may be especially useful for myeloid sarcomas of the orbit, where the use of small target volumes and adaptive replanning may help minimize treatment toxicities. Especially in very young children with poor prognosis, the use of lower doses for palliation should be considered.

## Pediatric Lymphoma

Lymphomas in children are the third most common type of cancer after ALL and CNS tumors. Non-Hodgkin's lymphoma (NHL) comprises about 60% of lymphoma cases in children under the age of 15 years. In adolescents aged 15–18 years, Hodgkin's lymphoma predominates. The three main types of NHL in children include Burkitt's lymphoma, lymphoblastic lymphoma, and large cell lymphoma [77]. Treatment outcomes for NHL are superior to those seen in adults. The role of radiation is largely limited due to advances in chemotherapy regimens but can be considered on a case-by-case basis in patients with relapsed or refractory disease [78–80]. At our institution, we treated a case of relapsed refractory DLBCL of the jejunum using MR-guided radiotherapy. Daily MRI allowed for improved visualization and alignment to the tumor, as well as assessment of the tumor's relationship to nearby critical structures, such as the stomach. The patient required adaptive replanning over the course of treatment due to fluctuations in bowel and stomach position in order to meet maximum dose constraints.

Pediatric Hodgkin's lymphoma accounts for 5–6% of childhood cancers and is more common in adolescents rather than young children. Patients most commonly present with painless cervical adenopathy or large mediastinal masses in up to 66% of patients. Treatment paradigms have evolved from risk-adapted therapy to response-adapted therapy based on PET scan evaluations over the course of chemotherapy [81, 82]. Radiation volumes have also evolved from involved field to involved site and now to involved nodal RT. Radiation was historically administered to all sites of initial disease involvement and now is selectively administered to sites of bulky disease and sites of either slow early response (SER) or partial response (PR). In the SWOG 1826 trial, only sites of residual disease after completion of systemic therapy are irradiated (NCT03907488). To date, no reports of MRgRT for Hodgkin's lymphoma exist. However, a survey of pediatric oncology practitioners ($n$ = 12) indicated an expected benefit to MR-guided treatment for patients with mediastinal lymphomas [83]. Seventy-five percent of respondents indicated a benefit of PTV margin reduction, and 67% indicated a benefit of daily adaptation to a shifting treatment volume, the use of online replanning for OAR avoidance, and compensation for intrafraction motion, and 50% felt there might be a benefit to the use of functional imaging. Furthermore, a large portion of respondents felt there would be a benefit to the use of MR guidance to the treatment of lymph nodes located in the mediastinum, upper abdomen including peri-hilar and splenic lymph nodes, and those located in the pelvis.

## Abdominal Tumors

### Wilms Tumor

Wilms tumor is the most common kidney tumor in children and accounts for ~6% of childhood malignancies. There are approximately 500–600 new cases per year in the United States [84]. More than two-thirds of cases arise in children under the age of 5 years, and 10–13% of cases are associated with genetic predisposition syndromes including WAGR, Denys-Drash, and Beckwith-Wiedemann. Patients are typically treated with upfront surgical resection followed by chemotherapy with or without adjuvant RT. In general, indications for flank or whole abdomen RT are as follows: stage III favorable histology, stages I–III focal anaplasia, and stages I–III diffuse anaplasia. Patients with lung metastases may or may not receive whole lung irradiation based on their tumor molecular findings and/or their response to 6 weeks of chemotherapy [85–87]. Further definitions of staging for Wilms tumor and what constitutes anaplasia are beyond the scope of this book.

With respect to radiation volumes, the standard of care in the United States for flank and whole abdomen RT remains parallel-opposed (AP-PA) treatment fields using 4-MV or 6-MV photons [88]. In Europe, there has been increasing interest in the use of more streamlined radiation fields for postoperative flank treatment. Mul et al. [89] defined conformal postoperative RT volumes using VMAT. Using 20 consecutive cases from their institution, they evaluated dose to OARs and total body volume (TBV) for both conventional AP-PA and VMAT plans. Dose constraints included tail of the pancreas to <10 Gy, heart D50 <5 Gy, and mammary bud mean <10 Gy. Their results showed an estimated dose reduction to OARs by 60%. VMAT increased the TBV receiving 10% prescription dose but decreased the irradiated TBV for all other doses. If this technique were to be more widely adopted, utilization of MR-guided technologies would likely prove effective in further reducing doses to adjacent OARS such as the heart and stomach.

One unique clinical scenario in which we have found a benefit to MR guidance for Wilms tumor is that of an unresectable inferior vena cava (IVC) tumor thrombus [90]. IVC tumor thrombus is found in 4–10% of Wilms tumor cases. Removal of the tumor is required at the time of surgery—in a large SIOP/GPOH study, thrombus was completely removed in 85%, incompletely in 12%, and not removed in 2% [91]. For patients with incompletely removed thrombus, the recommendation is to deliver a boost dose of radiation targeting residual disease to a total of 21.6 Gy. The thrombus region is included in the initial flank treatment field, and then an additional 10.6 Gy in 1.8 Gy/fraction is delivered to a volume encompassing the tumor thrombus. We have been treating the tumor thrombus using MRgRT as the IVC is well visualized on T2 MRI and this affords superior alignment to the region and allows for the use of significantly smaller treatment volumes (Fig. 19.4). Our standard protocol is to obtain a 4D CT simulation scan with contrast as well as an MRI with and without contrast for RT planning. The IVC and residual thrombus are identified, and an internal target volume (ITV) is created, if necessary. We then apply a 3-mm PTV. We have found that with this approach, mean doses to OARs including the heart, bowel, liver, and stomach are all decreased when compared to conventional linear accelerator treatment [90]. If the primary tumor was on the contralateral side, we also aggressively constrain the remaining kidney to a composite mean of <10–11 Gy. We ensure that the dose gradient across the adjacent vertebral bodies is less than 5 Gy to avoid future risk of scoliosis [92].

Finally, we would be remiss without mentioning the potential for whole abdomen RT using IMRT or VMAT, which has been shown to better spare OARs and provide superior PTV coverage to standard AP-PA treatment fields [93]. The potential benefit of adaptive therapy in this setting would include superior visualization and avoidance of the remaining kidney, as well as adaptive replanning for patients whose abdominal circumference changes over the course of treatment. Some potential limitations of this approach using MRgRT would be the larger field size and longer time under anesthesia as most patients with Wilms tumor will be under the age of 5 and unable to lay still on their own for the duration of treatment.

### Neuroblastoma

Neuroblastoma is the most common malignancy in infants under the age of 12 months, and 90% of cases are diagnosed prior to the age of 5 [94]. Neuroblastoma cells derive from embryonic neural crest cells of the peripheral nervous system, and tumors are comprised of several different histologic subtypes including ganglioneuroma, ganglioneuroblastoma, and neuroblastoma. Neuroblastomas can arise anywhere along the peripheral nervous system with most located in the adrenal gland (35%), abdomen (30%), and posterior mediastinum (20%) [95]. Treatment involves surgical resection and chemotherapy, and patients with high-risk neuroblastoma will also receive RT. Other considerations for RT include Pepper syndrome, in which infants experience respiratory failure due to liver enlargement from tumor.

Radiation treatment planning should be done in accordance with standard guidelines. The use of 4D CT or 4D MRI will aid in the delineation of treatment volumes near the diaphragm. Target volumes in neuroblastoma are well defined, and we instruct our US readers to review the most contemporary neuroblastoma protocols for further details. The GTV should encompass the

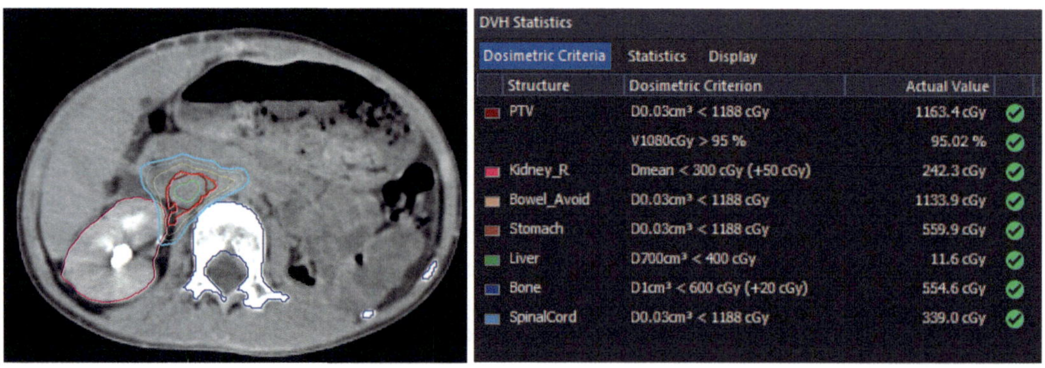


**Fig. 19.4** MR-linac radiation plan for a child with Wilms tumor and unresectable IVC thrombus remaining after neoadjuvant chemotherapy. The patient received conventional flank radiation using AP-PA fields on a standard linear accelerator to a dose of 10.8 Gy. The tumor thrombus was subsequently boosted to another 10.8 Gy (total dose to thrombus was 21.6 Gy) using MR-guided IMRT. The main objectives were to avoid a gradient across the vertebral body of >5 Gy, limit the dose to the remaining adjacent kidney to a mean of <3 Gy, and keep coverage of the PTV at >95%

region of the primary tumor and any involved lymph nodes at diagnosis but should not include fixed OARs including the kidney, bones, liver, etc. unless the tumor invaded into these regions. The CTV is an anatomically confined 1.0-cm expansion, and its superior and inferior extent should encompass the initial extent of the presurgical tumor volume. Adjacent vertebral bodies are traditionally treated to 18 Gy to purposefully cause symmetric growth stunting and prevent scoliosis. The PTV is 0.3- to 0.8-cm expansion to account for physiologic motion. The most recent ANBL1531 protocol allows for the use of a 0.3-cm PTV for most tumor sites provided the patient undergoes daily cone beam CT.

Typical beam arrangements are AP-PA; however, over the past two decades, other approaches including 3D-CRT, IMRT, and protons have been investigated. Paulino et al. compared the use of traditional AP-PA plans to IMRT for the treatment of abdominal neuroblastoma and evaluated doses to adjacent OARs [96]. They found that dose to kidneys was decreased with the use of IMRT at the expense of increased doses to the liver, stomach, and spleen. IMRT was not found to be beneficial for well-lateralized tumors. The authors concluded that treatment recommendations should be made on a case-by-case basis.

At our institution, we have selectively employed MR-guided radiotherapy for patients with neuroblastoma tumors located in the abdomen and for neuroblastoma metastases. We have found that based on changes in kidney position, daily plan adaptation is beneficial for abdominal tumors near that structure. In addition, we have found that neuroblastoma metastases can change rapidly in size over the course of treatment—from tumor swelling to shrinkage—making daily adaptive therapy an excellent option in select clinical scenarios.

## Pelvic Tumors in Children and Adolescents

Pelvic tumors in children are a rare, heterogeneous group that includes urogenital tract rhabdomyosarcoma (bladder, prostate, vagina), neuroblastoma, Ewing sarcoma, and germ cell tumors, among others. Some of these tumors can be surgically addressed, and others will require radiotherapy [97–99].

Genitourinary rhabdomyosarcoma of the bladder and prostate is typically treated with chemotherapy and radiation to enable organ preservation. Certain cases are amenable to brachytherapy, which provides optimal sparing of adjacent OARs such as the rectum and can significantly minimize late rectal toxicity and decrease rates of secondary malignancy [100]. However, for some patients, brachytherapy is either not accessible or feasible due to tumor location, and external beam radiation is recommended.

For bladder rhabdomyosarcoma treated with external beam radiation, tumor targeting can prove complex due to daily changes in bladder size and shape. In addition, the rectum frequently changes in its filling, and this is more difficult to control in small children than in adults. In these respects, daily adaptive MRgRT can be especially useful in ensuring adequate target coverage while minimizing doses to adjacent critical structures such as the rectum (Fig. 19.5). The bladder and rectum can be recontoured daily, and the PTV margin can be kept to 3 mm to account for any potential changes during the adaptation process. Even with the use of protons, margins of 5–10 mm for aperture opening and additional range to account for setup uncertainties and tissue heterogeneity are required. A study by Cotter et al. showed that protons were superior to IMRT in sparing dose to bladder, testes, femoral heads, and pelvic bones [101]. However, there was no difference with respect to dose to the rectum, prostate, penile bulb, or bowel. To date, no comparative study of the use of MRgRT and either proton radiation or IMRT has been done. We have successfully treated bladder/prostate rhabdomyosarcoma using MRgRT and felt that this approach successfully minimized dose to adjacent critical structures through the ability to use daily adapt-to-shape treatment planning.

Other instances of pelvic disease in the pediatric and adolescent population include Ewing sarcoma. The most common pelvic subsite for Ewing sarcoma is the sacrum (44%), followed by iliac/buttock region (40%) and the ischiopubic-

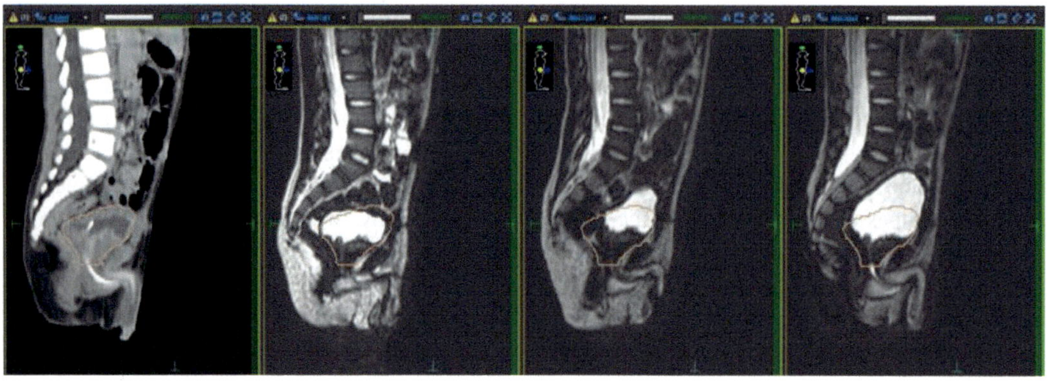

**Fig. 19.5** Panel on the left shows the original treatment planning CT scan for a 3-year-old patient with bladder rhabdomyosarcoma (light orange = CTV). Due to the patient's age and need for anesthesia, it was not possible to give bladder filling instructions. Because we expected bladder filling to vary significantly over the course of radiotherapy, we elected to treat the tumor using MR guidance for adaptive planning as needed. The patient required daily adapt-to-shape (ATS) to account for changes in bladder filling. The panels show the wide range of changes that occurred over the course of the patient's treatment. MR guidance allowed us to employ a small PTV margin of 3 mm and therefore to significantly limit the dose to surrounding normal structures. The rectum (not shown) was also recontoured as needed

acetabular region (16%). Historically, pelvic Ewing sarcoma has been associated with inferior outcomes due to part to adjacent pelvic anatomy such as bowel, nerves, etc. that limit the ability to achieve high radiation doses to the region [102]. However, the most recent data from AEWS1031 challenge that paradigm and show that RT is an effective modality for local control. The cumulative incidence of local failure by tumor site was 8.6% for pelvis, 12.5% for extremities, 6.7% for spine/chest wall/skull, and 3.6% for extraosseous tumors ($p = 0.61$) [103]. That being said, local failure was higher for tumors ≥200 mL (14.3%) versus those <200 mL (5.9%, $p = 0.04$), and Ewing sarcoma of the pelvis tends to present with larger tumors overall compared to other subsites.

Over the past several years, radiation oncologists have pursued dose escalation in an effort to improve local control outcomes for patients with Ewing sarcoma. A retrospective study of 48 patients with pelvic Ewing sarcoma from Mayo Clinic demonstrated improved local control with definitive radiation doses of ≥56 Gy [104]. A phase II study from St. Jude documented zero local failures with dose escalation to 64.8 Gy for tumors >8 cm in size [105]. In addition, that study employed limited margin RT using conformal or intensity-modulated techniques. The treatment volume incorporated a 1-cm constrained margin around the gross tumor. The estimated cumulative incidence of local failure was 4.4 ± 3.1%. The most common acute toxicity was radiation dermatitis. Late grade 3 toxicities were related to bone, joint, skin, or subcutaneous tissue injury in the region of the high-dose radiation field. A more recent analysis from the University of Florida comprised of large tumors with a median size and volume of 10.5 cm and 472 $cm^3$ showed no local progression in five patients who received doses of ≥59.4 Gy RBE [106]. Only two patients treated with proton radiation alone developed local recurrence, and both those patients received doses to tumor of <59.4 Gy. The overall local control was 91% for patients receiving radiation alone. There were five late grade 2+ toxicities including premature ovarian failure after 12-Gy RBE to the ovary on the affected side, hip dislocation after treatment for a left ischium/pubic primary, bladder fibrosis in the setting of bladder Dmax of 54-Gy RBE, grade 3 wound infection after preoperative RT, grade 2 pelvic pain of unclear etiology, and grade 2 limb-length discrepancy.

For pelvic primaries, MR-guided treatment could be beneficial in decreasing treatment margins, allowing for accurate patient positioning, and adapting to the changing shape of nearby OARs such as the bowel, bladder, and rectum. It could potentially allow for the daily identification of ovarian position, depending on patient age. In general, we recommend approaching these complex clinical scenarios on a case-by-case basis.

## Patient Selection and Technical Challenges

Potential pediatric MRgRT candidates are subject to many limitations as adults. They must pass the same screening criteria [107], entailing metallic implants that may be ferromagnetic or cause image distortion, and patient girth. The respective MRgRT platform's magnetic field strength's impact on particle trajectory, inability to perform couch kicks, and utilizing of noncoplanar beams must be accounted for during treatment planning [108–110]. MRgRT utilizing adaptive planning is resource- and time-intensive: a radiation oncologist, physicist, and dosimetrist (and possibly nursing and anesthesia for cases requiring sedation) must be present for each treatment. Patient setup, imaging, modification of volumes requiring familiarity with MRI-based anatomy, and recalculating a treatment plan may necessitate 30–60 min per fraction. This also translates to fewer patients treated daily compared to traditional linac-based plans averaging 15–30 min per fraction [111–115]. These aforementioned factors may prove to be financially limiting to a practice, curbing widespread adoption of MRgRT, especially in community settings. Finally, prospective studies are necessitated to validate potential significant therapeutic benefits conferred by dosimetric advantages as proven improvements in toxicity and quality-of-life may ultimately prove to be cost-effective. This endeavor may require several years for trial development, implementation, accrual and data maturation.

There are several planning considerations unique to the pediatric population. The coil weight must be safely tolerable; custom setups may be necessitated. It is also important to monitor patient body temperature during treatment. Diapers should be checked prior to treatment to minimize the risk of thermal injury secondary to moisture. Hearing protection should be utilized every treatment to minimize auditory toxicity. Pediatric and young adults not requiring sedation may not tolerate MRgRT due to claustrophobia. They must also remain still for longer periods compared to conventional linear accelerator treatments as positional intrafraction changes will undermine adaptive planning. These considerations may be addressed with a combination of anxiolytics, patient education, parental presence, and child life services [116]. Anesthesiology should be informed of pending cases requiring sedation to ensure they have MRI zone 4-compatible equipment [117]. The MRgRT suite must be equipped to monitor vitals during treatment. The distance from the vault to console may require longer oxygen tubes. A dry run with anesthesiology is recommended for the first sedation case. This can determine the need for additional time-outs throughout the adaptive process and points of no return. A volunteer adult patient can wear patient-monitoring devices for treatment to test radiation interference with monitoring equipment, as well as imaging artifacts. Finally, the increased time under anesthesia due to adaptive planning must be weight against the benefits of MRgRT.

## Future Directions

Improving the therapeutic ratio by reducing target volumes, tracking intratumoral motion in real time, and daily adaption of OAR changes promote consideration of several concepts currently under investigation for adult malignancies. Historically radioresistant malignancies, such as Ewing sarcoma and rhabdomyosarcoma, may benefit from dose escalation to the gross tumor. Similarly, inoperable gross disease in the pelvic and abdominal regions can be safely dose-escalated. The MR-linac adaptive planning features can be combined with functional imaging (e.g., FDG, MIBG) to selectively modify dose distribution based on therapeutic response.

Tumor regions demonstrating rapid regression and/or decreased metabolic activity may be amenable to dose reduction to spare organs at risk while resistant regions will benefit from dose escalation. Integrating artificial intelligence may mitigate resource demands through facilitating adaptive planning and quality assurance [118]. Finally, development of a MR-guided particle therapy platform is under investigation [119].

Pediatric malignancies are heterogeneous and relatively rare compared to typical adult malignancies. Prospective, multi-institutional clinical trials are necessitated to sufficiently power studies showing whether the putative dosimetric benefits of MRgRT correlate with clinical outcomes. Equally important is determining which tumor histologies or target sites derive benefits from MRgRT versus proton therapy or conventional linac-based treatment. This requires selection of appropriate clinical endpoints for evaluation. Radiation oncologists must be adept in MRI-based imaging, as well as the benefits and limitations of MRgRT to appropriately select ideal patients who will benefit from this emerging approach.

## Conclusions

Photon- and proton-based radiation therapies are planned and delivered through CT-based image guidance. MRgRT is a state-of-the-art technology with the potential to revolutionize pediatric cancer treatment. Its real-time imaging and adaptive planning capabilities allow high precision radiation delivery which may improve the therapeutic ratio compared to traditional radiotherapy platforms. As technology continues to advance, it is expected that MRgRT will become an even more crucial component of radiation therapy in the treatment of pediatric cancers.

## References


1. Siegel RL, Miller KD, Fuchs HE, Jemal A. Cancer statistics, 2022. CA Cancer J Clin. 2022;72(1):7–33.
2. National Cancer Institute. Cancer in children and adolescents. 2021. https://www.cancer.gov/types/childhood-cancers/child-adolescent-cancers-fact-sheet.
3. Fidler-Benaoudia MM, Oeffinger KC, Yasui Y, Robison LL, Winter DL, Reulen RC, et al. A comparison of late mortality among survivors of childhood cancer in the United States and United Kingdom. J Natl Cancer Inst. 2021;113(5):562–71.
4. Phillips SM, Padgett LS, Leisenring WM, Stratton KK, Bishop K, Krull KR, et al. Survivors of childhood cancer in the United States: prevalence and burden of morbidity. Cancer Epidemiol Biomark Prev. 2015;24(4):653–63.
5. Steliarova-Foucher E, Colombet M, Ries LAG, Moreno F, Dolya A, Bray F, et al. International incidence of childhood cancer, 2001-10: a population-based registry study. Lancet Oncol. 2017;18(6):719–31.
6. Fischer-Valuck BW, Henke L, Green O, Kashani R, Acharya S, Bradley JD, et al. Two-and-a-half-year clinical experience with the world's first magnetic resonance image guided radiation therapy system. Adv Radiat Oncol. 2017;2(3):485–93.
7. Metcalfe P, Liney GP, Holloway L, Walker A, Barton M, Delaney GP, et al. The potential for an enhanced role for MRI in radiation-therapy treatment planning. Technol Cancer Res Treat. 2013;12(5):429–46.
8. Dunlop A, Mitchell A, Tree A, Barnes H, Bower L, Chick J, et al. Daily adaptive radiotherapy for patients with prostate cancer using a high field MR-linac: initial clinical experiences and assessment of delivered doses compared to a C-arm linac. Clin Transl Radiat Oncol. 2020;23:35–42.
9. Henke L, Kashani R, Yang D, Zhao T, Green O, Olsen L, et al. Simulated online adaptive magnetic resonance-guided stereotactic body radiation therapy for the treatment of oligometastatic disease of the abdomen and central thorax: characterization of potential advantages. Int J Radiat Oncol Biol Phys. 2016;96(5):1078–86.
10. Winkel D, Bol GH, Werensteijn-Honingh AM, Intven MPW, Eppinga WSC, Hes J, et al. Target coverage and dose criteria based evaluation of the first clinical 1.5T MR-linac SBRT treatments of lymph node oligometastases compared with conventional CBCT-linac treatment. Radiother Oncol. 2020;146:118–25.
11. Kishan AU, Ma TM, Lamb JM, Casado M, Wilhalme H, Low DA, et al. Magnetic resonance imaging-guided vs computed tomography-guided stereotactic body radiotherapy for prostate cancer: the MIRAGE randomized clinical trial. JAMA Oncol. 2023;9(3):365–73.
12. Lagendijk JJ, Raaymakers BW, Raaijmakers AJ, Overweg J, Brown KJ, Kerkhof EM, et al. MRI/linac integration. Radiother Oncol. 2008;86(1):25–9.
13. Klüter S. Technical design and concept of a 0.35 T MR-Linac. Clin Transl Radiat Oncol. 2019;18:98–101.

14. Dunkerley DAP, et al. Clinical implementational and site-specific workflows for a 1.5 T MR-Linac. J Clin Med. 2022;11(6):1662.
15. Rammohan N, et al. History of technological advancements towards MR-Linac: the future of image-guided radiotherapy. J Clin Med. 2022;11(16):4730.
16. Christiansen RL, Dysager L, Bertelsen AS, Hansen O, Brink C, Bernchou U. Accuracy of automatic deformable structure propagation for high-field MRI guided prostate radiotherapy. Radiat Oncol. 2020;15(1):32.
17. Yadav P, Musunuru HB, Witt JS, Bassetti M, Bayouth J, Baschnagel AM. Dosimetric study for spine stereotactic body radiation therapy: magnetic resonance guided linear accelerator versus volumetric modulated arc therapy. Radiol Oncol. 2019;53(3):362–8.
18. Chamberlain M, Krayenbuehl J, van Timmeren JE, Wilke L, Andratschke N, Garcia Schüler H, et al. Head and neck radiotherapy on the MR linac: a multicenter planning challenge amongst MRIdian platform users. Strahlenther Onkol. 2021;197(12):1093–103.
19. Musunuru HB, Yadav P, Olson SJ, Anderson BM. Improved ipsilateral breast and Chest Wall sparing with MR-guided 3-fraction accelerated partial breast irradiation: a dosimetric study comparing MR-Linac and CT-Linac plans. Adv Radiat Oncol. 2021;6(3):100654.
20. Acharya S, Fischer-Valuck BW, Kashani R, Parikh P, Yang D, Zhao T, et al. Online magnetic resonance image guided adaptive radiation therapy: first clinical applications. Int J Radiat Oncol Biol Phys. 2016;94(2):394–403.
21. van Timmeren JE, Chamberlain M, Krayenbuehl J, Wilke L, Ehrbar S, Bogowicz M, et al. Comparison of beam segment versus full plan re-optimization in daily magnetic resonance imaging-guided online-adaptive radiotherapy. Phys Imaging Radiat Oncol. 2021;17:43–6.
22. van Sörnsen de Koste JR, Palacios MA, Bruynzeel AME, Slotman BJ, Senan S, Lagerwaard FJ. MR-guided gated stereotactic radiation therapy delivery for lung, adrenal, and pancreatic tumors: a geometric analysis. Int J Radiat Oncol Biol Phys. 2018;102(4):858–66.
23. Ostrom QT, Cioffi G, Waite K, Kruchko C, Barnholtz-Sloan JS. CBTRUS statistical report: primary brain and other central nervous system tumors diagnosed in the United States in 2014-2018. Neuro Oncol. 2021;23(12 Suppl 2):iii1–iii105.
24. Merchant TE, Kun LE, Hua CH, Wu S, Xiong X, Sanford RA, et al. Disease control after reduced volume conformal and intensity modulated radiation therapy for childhood craniopharyngioma. Int J Radiat Oncol Biol Phys. 2013;85(4):e187–92.
25. Merchant TE, Bendel AE, Sabin ND, Burger PC, Shaw DW, Chang E, et al. Conformal radiation therapy for pediatric ependymoma, chemotherapy for incompletely resected Ependymoma, and observation for completely resected, supratentorial ependymoma. J Clin Oncol. 2019;37(12):974–83.
26. Peters S, Merta J, Schmidt L, Jazmati D, Kramer PH, Blase C, et al. Evaluation of dose, volume, and outcome in children with localized, intracranial ependymoma treated with proton therapy within the prospective KiProReg study. Neuro Oncol. 2022;24(7):1193–202.
27. Devine CA, Liu KX, Ioakeim-Ioannidou M, Susko M, Poussaint TY, Huisman T, et al. Brainstem injury in pediatric patients receiving posterior fossa photon radiation. Int J Radiat Oncol Biol Phys. 2019;105(5):1034–42.
28. Orukari I, Perkins S, Zhao T, Huang J, Caruthers DF, Duriseti S. Brainstem toxicity in pediatric patients treated with protons using a single-vault synchrocyclotron system. Int J Part Ther. 2022;9(1):12–7.
29. Yock TI, Constine LS, Mahajan A. Protons, the brainstem, and toxicity: ingredients for an emerging dialectic. Acta Oncol. 2014;53(10):1279–82.
30. Rades D, Veninga T, Stalpers LJ, Basic H, Rudat V, Karstens JH, et al. Outcome after radiotherapy alone for metastatic spinal cord compression in patients with oligometastases. J Clin Oncol. 2007;25(1):50–6.
31. Rades D, Schild SE, Abrahm JL. Treatment of painful bone metastases. Nat Rev Clin Oncol. 2010;7(4):220–9.
32. Pica A, Miller R, Villà S, Kadish SP, Anacak Y, Abusaris H, et al. The results of surgery, with or without radiotherapy, for primary spinal myxopapillary ependymoma: a retrospective study from the rare cancer network. Int J Radiat Oncol Biol Phys. 2009;74(4):1114–20.
33. Kotecha R, Tom MC, Naik M, Angelov L, Benzel EC, Reddy CA, et al. Analyzing the role of adjuvant or salvage radiotherapy for spinal myxopapillary ependymomas. J Neurosurg Spine. 2020:1–6.
34. Zeng L, Chow E, Bedard G, Zhang L, Fairchild A, Vassiliou V, et al. Quality of life after palliative radiation therapy for patients with painful bone metastases: results of an international study validating the EORTC QLQ-BM22. Int J Radiat Oncol Biol Phys. 2012;84(3):e337–42.
35. Mizumoto M, Harada H, Asakura H, Hashimoto T, Furutani K, Hashii H, et al. Radiotherapy for patients with metastases to the spinal column: a review of 603 patients at Shizuoka cancer center hospital. Int J Radiat Oncol Biol Phys. 2011;79(1):208–13.
36. Pan H, Simpson DR, Mell LK, Mundt AJ, Lawson JD. A survey of stereotactic body radiotherapy use in the United States. Cancer. 2011;117(19):4566–72.
37. Redmond KJ, Robertson S, Lo SS, Soltys SG, Ryu S, McNutt T, et al. Consensus contouring guidelines for postoperative stereotactic body radiation therapy for metastatic solid tumor malignancies to the spine. Int J Radiat Oncol Biol Phys. 2017;97(1):64–74.

38. Ryu SI, Chang SD, Kim DH, Murphy MJ, Le QT, Martin DP, et al. Image-guided hypo-fractionated stereotactic radiosurgery to spinal lesions. Neurosurgery. 2001;49(4):838–46.
39. Gerszten PC, Ozhasoglu C, Burton SA, Vogel WJ, Atkins BA, Kalnicki S, et al. CyberKnife frameless stereotactic radiosurgery for spinal lesions: clinical experience in 125 cases. Neurosurgery. 2004;55(1):89–98; discussion 98–9.
40. Han EY, Aima M, Hughes N, Briere TM, Yeboa DN, Castillo P, et al. Feasibility of spinal stereotactic body radiotherapy in Elekta Unity. J Radiosurg SBRT. 2020;7(2):127–34.
41. Han EY, Yeboa DN, Briere TM, Yang J, Wang H. Dosimetric analysis of MR-LINAC treatment plans for salvage spine SBRT re-irradiation. J Appl Clin Med Phys. 2022;23(10):e13752.
42. Albright JT, Topham AK, Reilly JS. Pediatric head and neck malignancies: US incidence and trends over 2 decades. Arch Otolaryngol Head Neck Surg. 2002;128(6):655–9.
43. Arboleda LP, Perez-de-Oliveira ME, Hoffmann IL, Cardinalli IA, Gallagher KP, Santos-Silva AR, et al. Clinical manifestations of head and neck cancer in pediatric patients, an analysis of 253 cases in a single Brazilian center. Med Oral Patol Oral Cir Bucal. 2022;27(3):e285–e93.
44. Sengupta S, Pal R, Saha S, Bera SP, Pal I, Tuli IP. Spectrum of head and neck cancer in children. J Indian Assoc Pediatr Surg. 2009;14(4):200–3.
45. Boeke S, Monnich D, van Timmeren JE, Balermpas P. MR-guided radiotherapy for head and neck cancer: current developments, perspectives, and challenges. Front Oncol. 2021;11:616156.
46. Gurney-Champion OJ, McQuaid D, Dunlop A, Wong KH, Welsh LC, Riddell AM, et al. MRI-based assessment of 3D intrafractional motion of head and neck cancer for radiation therapy. Int J Radiat Oncol Biol Phys. 2018;100(2):306–16.
47. Chatterjee S, Maulik S, Prasath S, Arun B, Das A, Chakrabarty S, et al. Xerostomia quality of life and resource requirements following parotid sparing adaptive radiotherapy in head and neck cancers: results of a prospective cohort study (study ID CTRI/2017/11/010683). Radiother Oncol. 2022;168:250–5.
48. Castelli J, Simon A, Lafond C, Perichon N, Rigaud B, Chajon E, et al. Adaptive radiotherapy for head and neck cancer. Acta Oncol. 2018;57(10):1284–92.
49. Zhang P, Simon A, Rigaud B, Castelli J, Ospina Arango JD, Nassef M, et al. Optimal adaptive IMRT strategy to spare the parotid glands in oropharyngeal cancer. Radiother Oncol. 2016;120(1):41–7.
50. Bernstein JM, Homer JJ, West CM. Dynamic contrast-enhanced magnetic resonance imaging biomarkers in head and neck cancer: potential to guide treatment? A systematic review. Oral Oncol. 2014;50(10):963–70.
51. Fatima K, Dasgupta A, DiCenzo D, Kolios C, Quiaoit K, Saifuddin M, et al. Ultrasound delta-radiomics during radiotherapy to predict recurrence in patients with head and neck squamous cell carcinoma. Clin Transl Radiat Oncol. 2021;28:62–70.
52. McGee KP, Hwang KP, Sullivan DC, Kurhanewicz J, Hu Y, Wang J, et al. Magnetic resonance biomarkers in radiation oncology: the report of AAPM task group 294. Med Phys. 2021;48(7):e697–732.
53. Shukla-Dave A, Obuchowski NA, Chenevert TL, Jambawalikar S, Schwartz LH, Malyarenko D, et al. Quantitative imaging biomarkers alliance (QIBA) recommendations for improved precision of DWI and DCE-MRI derived biomarkers in multicenter oncology trials. J Magn Reson Imaging. 2019;49(7):e101–e21.
54. Tran WT, Suraweera H, Quaioit K, Cardenas D, Leong KX, Karam I, et al. Predictive quantitative ultrasound radiomic markers associated with treatment response in head and neck cancer. Future Sci OA. 2019;6(1):FSO433.
55. Bahig H, Yuan Y, Mohamed ASR, Brock KK, Ng SP, Wang J, et al. Magnetic resonance-based response assessment and dose adaptation in human papilloma virus positive tumors of the oropharynx treated with radiotherapy (MR-ADAPTOR): an R-IDEAL stage 2a-2b/Bayesian phase II trial. Clin Transl Radiat Oncol. 2018;13:19–23.
56. McDonald BA, Vedam S, Yang J, Wang J, Castillo P, Lee B, et al. Initial feasibility and clinical implementation of daily MR-guided adaptive head and neck cancer radiation therapy on a 1.5T MR-Linac system: prospective R-IDEAL 2a/2b systematic clinical evaluation of technical innovation. Int J Radiat Oncol Biol Phys. 2021;109(5):1606–18.
57. Subesinghe M, Scarsbrook AF, Sourbron S, Wilson DJ, McDermott G, Speight R, et al. Alterations in anatomic and functional imaging parameters with repeated FDG PET-CT and MRI during radiotherapy for head and neck cancer: a pilot study. BMC Cancer. 2015;15:137.
58. Laningham FH, Kun LE, Reddick WE, Ogg RJ, Morris EB, Pui CH. Childhood central nervous system leukemia: historical perspectives, current therapy, and acute neurological sequelae. Neuroradiology. 2007;49(11):873–88.
59. Nachman J, Sather HN, Cherlow JM, Sensel MG, Gaynon PS, Lukens JN, et al. Response of children with high-risk acute lymphoblastic leukemia treated with and without cranial irradiation: a report from the Children's cancer group. J Clin Oncol. 1998;16(3):920–30.
60. Waber DP, Turek J, Catania L, Stevenson K, Robaey P, Romero I, et al. Neuropsychological outcomes from a randomized trial of triple intrathecal chemotherapy compared with 18 Gy cranial radiation as CNS treatment in acute lymphoblastic leukemia: findings from Dana-Farber Cancer Institute ALL consortium protocol 95-01. J Clin Oncol. 2007;25(31):4914–21.
61. McNeer JL, Schmiegelow K. Management of CNS disease in pediatric acute lymphoblastic leukemia. Curr Hematol Malig Rep. 2022;17(1):1–14.

62. Thastrup M, Duguid A, Mirian C, Schmiegelow K, Halsey C. Central nervous system involvement in childhood acute lymphoblastic leukemia: challenges and solutions. Leukemia. 2022;36(12):2751–68.
63. Brooks HW, Evans AE, Glass RM, Pang EM. Chloromas of the head and neck in childhood: the initial manifestation of myeloid leukemia in three patients. Arch Otolaryngol. 1974;100(4):306–8.
64. Lusher JM. Chloroma as a presenting feature of acute leukemia: a report of two cases in children. Am J Dis Child. 1964;108(1):62–6.
65. Samborska M, Derwich K, Skalska-Sadowska J, Kurzawa P, Wachowiak J. Myeloid sarcoma in children—diagnostic and therapeutic difficulties. Contemp Oncol (Pozn). 2016;20(6):444–8.
66. Dusenbery KE, Howells WB, Arthur DC, Alonzo T, Lee JW, Kobrinsky N, et al. Extramedullary leukemia in children with newly diagnosed acute myeloid leukemia: a report from the Children's cancer group. J Pediatr Hematol Oncol. 2003;25(10):760–8.
67. Majdoul S, Colson-Durand L, To NH, Belkacemi Y. Adaptive radiotherapy for an uncommon chloroma. Case Rep Oncol. 2016;9(3):593–8.
68. Bakst R, Wolden S, Yahalom J. Radiation therapy for chloroma (granulocytic sarcoma). Int J Radiat Oncol Biol Phys. 2012;82(5):1816–22.
69. Bakst RL, Dabaja BS, Specht LK, Yahalom J. Use of radiation in extramedullary leukemia/chloroma: guidelines from the International Lymphoma Radiation Oncology Group. Int J Radiat Oncol Biol Phys. 2018;102(2):314–9.
70. Lee DY, Baron J, Wright CM, Plastaras JP, Perl AE, Paydar I. Radiation therapy for chemotherapy refractory gingival myeloid sarcoma. Front Oncol. 2021;11:671514.
71. AlSemari MA, Perrotta M, Russo C, Alkatan HM, Maktabi A, Elkhamary S, et al. Orbital myeloid sarcoma (chloroma): report of 2 cases and literature review. Am J Ophthalmol Case Rep. 2020;19:100806.
72. Matsui M, Yamanaka J, Shichino H, Sato N, Kubota K, Matsushita T. FDG-PET/CT for detection of extramedullary disease in 2 pediatric patients with AML. J Pediatr Hematol Oncol. 2016;38(5):398–401.
73. Shinagare AB, Krajewski KM, Hornick JL, Zukotynski K, Kurra V, Jagannathan JP, et al. MRI for evaluation of myeloid sarcoma in adults: a single-institution 10-year experience. AJR Am J Roentgenol. 2012;199(6):1193–8.
74. Singh A, Kumar P, Chandrashekhara SH, Kumar A. Unravelling chloroma: review of imaging findings. Br J Radiol. 2017;90(1075):20160710.
75. Wu J, Jiang G, Wu J, Ou L, Zhang C. 18F-FDG PET/CT imaging of testicular myeloid sarcoma in a pediatric patient. Clin Nucl Med. 2021;46(1):84–5.
76. Zhang M, Yang X, Si Y, Kan Y, Yang J. 18 F-FDG PET/CT findings of myeloid sarcoma involving the left ventricle and kidney in a pediatric patient. Clin Nucl Med. 2023;48(1):103–5.
77. Sandlund JT, Martin MG. Non-Hodgkin lymphoma across the pediatric and adolescent and young adult age spectrum. Hematology. 2016;2016(1):589–97.
78. Burkhardt B, Woessmann W, Zimmermann M, Kontny U, Vormoor J, Doerffel W, et al. Impact of cranial radiotherapy on central nervous system prophylaxis in children and adolescents with central nervous system-negative stage III or IV lymphoblastic lymphoma. J Clin Oncol. 2006;24(3):491–9.
79. Link MP, Shuster JJ, Donaldson SS, Berard CW, Murphy SB. Treatment of children and young adults with early-stage non-Hodgkin's lymphoma. N Engl J Med. 1997;337(18):1259–66.
80. Seidemann K, Tiemann M, Schrappe M, Yakisan E, Simonitsch I, Janka-Schaub G, et al. Short-pulse B-non-Hodgkin lymphoma-type chemotherapy is efficacious treatment for pediatric anaplastic large cell lymphoma: a report of the Berlin-Frankfurt-Munster group trial NHL-BFM 90. Blood. 2001;97(12):3699–706.
81. Bair SM, Svoboda J. Response-adapted treatment strategies in Hodgkin lymphoma using PET imaging. PET Clin. 2019;14(3):353–68.
82. Hall MD, Terezakis SA, Lucas JT, Gallop-Evans E, Dieckmann K, Constine LS, et al. Radiation therapy across pediatric Hodgkin lymphoma research group protocols: a report from the staging, evaluation, and response criteria harmonization (SEARCH) for childhood, adolescent, and young adult Hodgkin lymphoma (CAYAHL) group. Int J Radiat Oncol Biol Phys. 2022;112(2):317–34.
83. Seravalli E, Kroon PS, Buatti JM, Hall MD, Mandeville HC, Marcus KJ, et al. The potential role of MR-guided adaptive radiotherapy in pediatric oncology: results from a SIOPE-COG survey. Clin Transl Radiat Oncol. 2021;29:71–8.
84. National Cancer Institute. Childhood Cancer Data Initiative, National Childhood Cancer Registry Explorer. 2023. https://nccrexplorer.ccdi.cancer.gov/.
85. Gadd S, Huff V, Skol AD, Renfro LA, Fernandez CV, Mullen EA, et al. Genetic changes associated with relapse in favorable histology Wilms tumor: a Children's oncology group AREN03B2 study. Cell Rep Med. 2022;3(6):100644.
86. Gratias EJ, Dome JS, Jennings LJ, Chi YY, Tian J, Anderson J, et al. Association of chromosome 1q gain with inferior survival in favorable-histology Wilms tumor: a report from the Children's oncology group. J Clin Oncol. 2016;34(26):3189–94.
87. Grundy PE, Breslow NE, Li S, Perlman E, Beckwith JB, Ritchey ML, et al. Loss of heterozygosity for chromosomes 1p and 16q is an adverse prognostic factor in favorable-histology Wilms tumor: a report from the National Wilms Tumor Study Group. J Clin Oncol. 2005;23(29):7312–21.
88. Wilm's tumor. In: Gunderson LL, editor. Gunderson and Tepper's clinical radiation oncology. 4th ed. Philadelphia: Elsevier; 2015.

89. Mul J, Seravalli E, Bosman ME, van de Ven CP, Littooij AS, van Grotel M, et al. Estimated clinical benefit of combining highly conformal target volumes with volumetric-modulated arc therapy (VMAT) versus conventional flank irradiation in pediatric renal tumors. Clin Transl Radiat Oncol. 2021;29:20–6.
90. Kozak MM, Crompton D, Gross BA, Harshman L, Dickens D, Snyder J, et al. Initial clinical applications treating pediatric and adolescent patients using MR-guided radiotherapy. Front Oncol. 2022;12:962926.
91. Meier CM, Furtwangler R, von Schweinitz D, Stein R, Welter N, Wagenpfeil S, et al. Vena cava thrombus in patients with Wilms tumor. Cancers (Basel). 2022;14(16):3924.
92. Hoeben BA, Carrie C, Timmermann B, Mandeville HC, Gandola L, Dieckmann K, et al. Management of vertebral radiotherapy dose in paediatric patients with cancer: consensus recommendations from the SIOPE radiotherapy working group. Lancet Oncol. 2019;20(3):e155–e66.
93. Chen MJ, Leao CR, Simoes RCP, Belletti FS, Figueiredo MLS, Cypriano MS. Kidney-sparing whole abdominal irradiation in Wilms tumor: potential advantages of VMAT technique. Pediatr Blood Cancer. 2020;67(5):e28223.
94. Ward E, DeSantis C, Robbins A, Kohler B, Jemal A. Childhood and adolescent cancer statistics, 2014. CA Cancer J Clin. 2014;64(2):83–103.
95. Vo KT, Matthay KK, Neuhaus J, London WB, Hero B, Ambros PF, et al. Clinical, biologic, and prognostic differences on the basis of primary tumor site in neuroblastoma: a report from the international neuroblastoma risk group project. J Clin Oncol. 2014;32(28):3169–76.
96. Paulino AC, Ferenci MS, Chiang KY, Nowlan AW, Marcus RB Jr. Comparison of conventional to intensity modulated radiation therapy for abdominal neuroblastoma. Pediatr Blood Cancer. 2006;46(7):739–44.
97. Fleming ID, Etcubanas E, Patterson R, Rao B, Pratt C, Hustu O, et al. The role of surgical resection when combined with chemotherapy and radiation in the management of pelvic rhabdomyosarcoma. Ann Surg. 1984;199(5):509–14.
98. Groff DB. Pelvic neoplasms in children. J Surg Oncol. 2001;77(1):65–71.
99. Zobel M, Zamora A, Sura A, Wang L, Stein J, Marachelian A, et al. The clinical management and outcomes of pelvic neuroblastic tumors. J Surg Res. 2020;249:8–12.
100. Romano E, Simon R, Minard-Colin V, Martin V, Bockel S, Espenel S, et al. Analysis of radiation dose/volume effect relationship for anorectal morbidity in children treated for pelvic malignancies. Int J Radiat Oncol Biol Phys. 2021;109(1):231–41.
101. Cotter SE, Herrup DA, Friedmann A, Macdonald SM, Pieretti RV, Robinson G, et al. Proton radiotherapy for pediatric bladder/prostate rhabdomyosarcoma: clinical outcomes and dosimetry compared to intensity-modulated radiation therapy. Int J Radiat Oncol Biol Phys. 2011;81(5):1367–73.
102. Ahmed SK, Witten BG, Harmsen WS, Rose PS, Krailo M, Marcus KJ, et al. Analysis of local control outcomes and clinical prognostic factors in localized pelvic Ewing sarcoma patients treated with radiation therapy: a report from the Children's oncology group. Int J Radiat Oncol Biol Phys. 2023;115(2):337–46.
103. Ahmed SK, Indelicato DJ, Chuba PJ, Krailo M, Buxton A, Randall RL, et al. Local failure in non-metastatic Ewing sarcoma patients treated with definitive radiation therapy on AEWS1031: a report from the Children's oncology group. Int J Radiat Oncol Biol Phys. 2022;114(3 Suppl):S69–70.
104. Ahmed SK, Robinson SI, Arndt CAS, Petersen IA, Haddock MG, Rose PS, et al. Pelvis Ewing sarcoma: local control and survival in the modern era. Pediatr Blood Cancer. 2017;64(9).
105. Talleur AC, Navid F, Spunt SL, McCarville MB, Wu J, Mao S, et al. Limited margin radiation therapy for children and young adults with Ewing sarcoma achieves high rates of local tumor control. Int J Radiat Oncol Biol Phys. 2016;96(1):119–26.
106. Uezono H, Indelicato DJ, Rotondo RL, Mailhot Vega RB, Bradfield SM, Morris CG, et al. Treatment outcomes after proton therapy for Ewing sarcoma of the pelvis. Int J Radiat Oncol Biol Phys. 2020;107(5):974–81.
107. Adaptive radiation therapy. Med Phys. 2011;38(7):4467–4468.
108. Prior P, Chen X, Botros M, Paulson ES, Lawton C, Erickson B, et al. MRI-based IMRT planning for MR-linac: comparison between CT- and MRI-based plans for pancreatic and prostate cancers. Phys Med Biol. 2016;61(10):3819–42.
109. Chen X, Prior P, Chen GP, Schultz CJ, Li XA. Technical note: dose effects of 1.5 T transverse magnetic field on tissue interfaces in MRI-guided radiotherapy. Med Phys. 2016;43(8):4797.
110. Raaijmakers AJ, Raaymakers BW, Lagendijk JJ. Magnetic-field-induced dose effects in MR-guided radiotherapy systems: dependence on the magnetic field strength. Phys Med Biol. 2008;53(4):909–23.
111. Li XA, Liu F, Tai A, Ahunbay E, Chen G, Kelly T, et al. Development of an online adaptive solution to account for inter- and intra-fractional variations. Radiother Oncol. 2011;100(3):370–4.
112. Ahunbay EE, Li XA. Gradient maintenance: a new algorithm for fast online replanning. Med Phys. 2015;42(6):2863–76.
113. Zhang Y, Paulson E, Lim S, Hall WA, Ahunbay E, Mickevicius NJ, et al. A patient-specific autosegmentation strategy using multi-input deformable image registration for magnetic resonance imaging-guided online adaptive radiation therapy: a feasibility study. Adv Radiat Oncol. 2020;5(6):1350–8.
114. Ates O, Ahunbay EE, Moreau M, Li XA. Technical note: a fast online adaptive replanning method for

VMAT using flattening filter free beams. Med Phys. 2016;43(6):2756–64.
115. Zhang J, Ahunbay E, Li XA. Technical note: acceleration of online adaptive replanning with automation and parallel operations. Med Phys. 2018;45(10):4370–6.
116. Boik N, Hall MD. Psychosocial support for pediatric patients at proton therapy institutions. Int J Part Ther. 2020;7(1):28–33.
117. Salerno S, Granata C, Trapenese M, Cannata V, Curione D, Rossi Espagnet MC, et al. Is MRI imaging in pediatric age totally safe? A critical reprisal. Radiol Med. 2018;123(9):695–702.
118. Huynh E, Hosny A, Guthier C, Bitterman DS, Petit SF, Haas-Kogan DA, et al. Artificial intelligence in radiation oncology. Nat Rev Clin Oncol. 2020;17(12):771–81.
119. Fuchs H, Padilla-Cabal F, Oborn BM, Georg D. Commissioning a beam line for MR-guided particle therapy assisted by in silico methods. Med Phys. 2023;50(2):1019–28.

# 20 MR-Linac-Guided Adaptive Radiotherapy for Brain Tumors

Tugce Kutuk, Jay Detsky, Arjun Sahgal, and Rupesh Kotecha

## Disease Overview

The term "brain tumors" refers to malignancies that originate in the brain (primary brain tumors) as well as those that metastasize to the central nervous system (CNS) with a primary site of disease origin arising from an extracranial site. Primary brain tumors consist of various histologic types with different tumor growth rates, patterns of disease progression, and treatment paradigms [1]. Approximately 30% of primary brain tumors in the USA are malignant, while the remainder are benign. Meningiomas and glial tumors (such as glioblastoma multiforme [GBM], astrocytoma, oligodendroglioma, etc.) account for approximately two-thirds of all primary brain tumors in adults, with GBM as the most common primary malignant brain tumor [2]. On the other hand, metastases to the CNS most commonly arise from lung, melanoma, and breast cancers, and are treated in a similar manner despite the different primary tumor histologies and biologic behaviors. Metastases significantly outpace the incidence of primary brain tumors with a 10:1 ratio and an annual estimated incidence exceeding 150,000 [3]. Primary and metastatic brain tumors can cause a wide range of symptoms depending on their size, location, and rate of growth, as well as the functional characteristics and eloquence of the substructures of the brain affected. Comprehensive history, neurologic examination, and diagnostic neuroimaging studies are necessary in the initial evaluation of the patient suspected to have a brain tumor.

GBM is a highly malignant tumor that grows quickly and infiltrates into the surrounding brain tissue [4]. The most common symptoms of GBM include headaches, focal neurologic deficits, seizures, nausea, and vomiting. As measured from initial diagnosis, the median survival is less than 2 years, and the overall survival remains poor [5]. Despite advancements in radiotherapy technology, developments of novel systemic therapies, and improvements in surgical techniques, only 5–10% of patients survive up to 5 years [6]. For patients with newly diagnosed GBM, the current standard of care includes maximal safe resection, followed by concurrent chemoradiotherapy and adjuvant temozolomide. Tumor-treating fields (TTFields) can also be applied adjuvantly [7]. Conventionally fractionated radiotherapy courses (i.e., 60 Gy in 30 fractions) are typically used for younger patients with good performance status or

T. Kutuk · R. Kotecha (✉)
Department of Radiation Oncology, Miami Cancer Institute, Baptist Health South Florida, Miami, FL, USA
e-mail: TugceK@baptisthealth.net; RupeshK@baptisthealth.net

J. Detsky · A. Sahgal
Department of Radiation Oncology, Odette Cancer Centre, Sunnybrook Health Sciences Centre, University of Toronto, Toronto, ON, Canada
e-mail: Jay.Detsky@sunnybrook.ca; Arjun.Sahgal@sunnybrook.ca



I. J. Das et al. (eds.), *A Practical Guide to MR-Linac*, https://doi.org/10.1007/978-3-031-48165-9_20

those treated in clinical trials, whereas hypofractionated regimens (i.e., 40 Gy in 15 fractions) are used for elderly patients [8–10]. Extreme hypofractionation (i.e., 25 Gy in 5 fractions) can also be used as a palliative measure to stabilize symptomatology over medical management alone [11].

Brain metastases occur in approximately 15–30% of all cancer patients at some point during their disease course [12]. The incidence of brain metastases is increasing due to a combination of factors such as patients living longer with their systemic disease, increased screening and surveillance for asymptomatic patients, and the improved sensitivity of imaging techniques such as magnetic resonance imaging (MRI) [13]. Treatment for brain metastases typically involves a combination of approaches including surgery, radiotherapy, chemotherapy, targeted agents, and immunotherapies [14]. The specific treatment plan depends on factors such as the number and location of the brain metastases, the patient's overall performance status, and the type of primary cancer.

Whole-brain radiation therapy (WBRT) has been the historical standard of care for treating brain metastases. However, the development of stereotactic radiosurgery (SRS) has revolutionized the treatment of brain metastases, due to improved disease control, reduced detriment to neurocognitive function, and improved quality of life compared to WBRT [15–19]. For patients with small intact brain metastasis, SRS is typically delivered in a single fraction [20]. Given the inferior local control rates with large intact brain metastasis (>2 cm), fractionated SRS (FSRS), delivered in 3–5 fractions, is frequently recommended [21]. SRS or FSRS can also be delivered preoperatively or postoperatively to improve local control over surgery alone and as an alternative to WBRT [22, 23].

As radiotherapy represents a primary treatment modality used in the management of GBM and brain metastases, the focus of this chapter is to review the potential applications of magnetic resonance-guided radiation therapy (MRgRT) for these two categories of CNS malignancies.

## Overcoming Challenges with Traditional Approaches

Computed tomography (CT) is a widely used imaging technique for simulation, target volume delineation, treatment planning, and image guidance during treatment. However, CT alone may not be sufficient to achieve the ultimate goal of delivering radiation dose efficiently and precisely to the target volume while sparing surrounding normal tissues and organs at risk (OARs). In the treatment of brain tumors such as glioblastoma (GBM) or brain metastases, MRI provides superior resolution of the tumor extent and OARs and is the standard of care for treatment planning. Despite these benefits, MRI has a limited role in image guidance during treatment. To address this, cone-beam CT is routinely applied for positional accuracy. Nonetheless, the technology's capabilities with respect to image guidance are not yet sufficient for the brain (Fig. 20.1).

The integration of MRI technologies into the radiotherapy workflow has been growing rapidly over the past decade, as there has been greater availability and access to diagnostic and dedicated radiotherapy simulation MRI scanners. In addition to the use of MRI for anatomic visualization at the time of treatment planning, specific MRI sequences can be applied to investigate the physiology and metabolism of CNS tumors (i.e., diffusion, perfusion, chemical exchange saturation transfer, magnetic transferase, and spectroscopy sequences) [24]. Diffusion-weighted imaging (DWI), for example, can detect changes in water mobility within the tumor, which represents changes in the tumor's cellularity and necrosis, and this information can be used to predict the tumor's response to chemotherapy and radiotherapy, as well as to predict overall survival rates [25–28]. Alternatively, perfusion imaging can measure blood flow within the tumor, which can provide information about the tumor's blood supply and metabolism. This can also be used to predict the tumor's response to therapy and overall survival [29]. The metabolic profile of a tumor can also be determined by MRI spectroscopy, chemical exchange saturation transfer, and magnetic transferase imaging as potential imaging

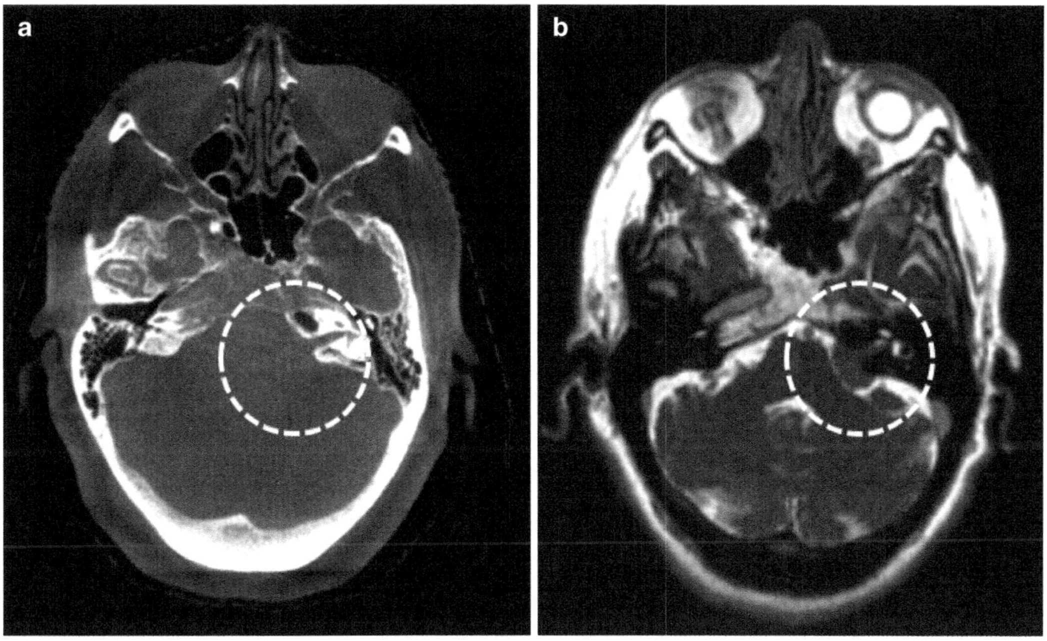


**Fig. 20.1** Imaging comparison of a cerebellopontine angle meningioma using (**a**) cone-beam computed tomography (CBCT) and (**b**) ViewRay MRIdian™ TrueFISP magnetic resonance imaging (MRI). It was difficult to delineate the tumor and adjacent brainstem on the CBCT images, whereas both were clearly visualized on the MRIdian image. This demonstrates the superior resolution and sensitivity of MRIdian MRI for detecting and localizing such tumors

surrogates of tumor aggressiveness [30–36]. The time course of MRI acquisition has been limited to the time of treatment and then approximately 1–3 months after the completion of chemoradiotherapy. However, recent developments in these sequences are significant, and parameters derived from these sequences are being integrated into the radiotherapy process to guide boost volumes in clinical trials [37].

MRgRT also provides the unique advantage of visualizing tumors on a daily basis, both during treatment (intra-treatment) and in between treatments (inter-treatment), using advanced imaging sequences that can provide additional information about the tumor's physiology during therapy which opens up the possibility of personalizing the response of the tumor to therapy [38–41]. MRI-guided linear accelerator (MR-Linac) devices have unique advantages for CNS tumors to frequently image tumors during a course of radiotherapy using both anatomically guided and physiologically focused sequences [32], hence unlocking the potential to use this real-time information during the treatment course to potentially affect patient outcomes [42, 43].

One of the key benefits of MR-Linac devices is their ability to overcome the challenge of involuntary movements, such as respiration and digestion, which can cause changes in the position of tumors during the course of radiotherapy. This is not an issue in the brain, making functional interfraction sequences more robust and clinically applicable. Therefore, the focus has been on understanding tumor dynamics during radiotherapy; a recent example of this is a deeper understanding of the migration of glioma during a course of radiotherapy and the potential to adapt [44].

There are other potential benefits from MRgRT as certain targets are sensitive to changes that can occur with little external motion. Although not common for brain metastasis or GBMs, optic pathway meningiomas or metastases to the orbit or retina are known to be affected by optic nerve motion [45, 46]. In fact, the displacement of the optic nerves in certain positions can even exceed 1 cm [46]. Even if intrafraction motion is not a concern, interfraction changes in the tumor size and shape can still occur between treatment sessions, which can be problematic. These changes can be caused by factors such as edema, inflammation, or changes in the tumor's blood supply [44]. To address this issue, it is common practice to perform diagnostic MRI scans weekly or biweekly to ensure adequate target dose coverage and to adapt the radiotherapy plan accordingly, if necessary, based on anatomical changes. However, with MRgRT, the diagnostic and treatment planning can be done simultaneously in the same session, thus avoiding the need for additional diagnostic MRI scans, which can save time and resources.

Two commercial systems are currently available for MRgRT: the Elekta Unity™ (Elekta AB, Stockholm, Sweden) and the ViewRay MRIdian™ A3i (ViewRay Inc., Cleveland, OH, USA) systems. The Elekta Unity™ utilizes a 1.5-Tesla (T) magnet, the most powerful magnet commercially available for MRgRT. In the Unity™ system, the photon beam is placed on a rotating gantry, which passes through a superconducting cryostat with a configuration that avoids angles that may interfere with imaging quality, in order to overcome interference from the magnet. Also, the electron gun is mounted in a zone with zero field, and specific settings are used for shims to minimize image degradation [47]. The ViewRay MRIdian™ system uses similar design characteristics to the Elekta Unity™. However, it uses a less powerful magnet (0.35 T) which leads to a weak magnetic field that places less stress on the linear accelerator yet allows for sufficient image quality for image guidance and treatment adaptation [47]. The image quality is not the same as a 1.5 T MRI and that has to be considered as functional sequences cannot be obtained at low-field-strength MRI. Additionally, the ViewRay MRIdian™ system uses a balanced steady-state free precession sequence, also known as true fast imaging with steady-state precession (TrueFISP). This steady-state free precession (SSFP) sequence provides the highest signal-to-noise ratio per unit of time to yield rapidly acquired, high-quality images. With TrueFISP, the resulting signal is related to the ratio of T2 to T1, but ultimately yields a predominantly T2-weighted sequence [48–50]. This is because the SSFP sequence produces a bright signal from tissues with short T2 relaxation times, such as water while suppressing signals from tissues with longer T2 relaxation times. Therefore, TrueFISP is particularly useful for imaging tissues with high water content, such as the brain [51, 52].

## Glioblastoma

Although limited in long-term follow-up, recent studies have demonstrated evolving tumor dynamics, geographic shifts, and changes in enhancement and edema throughout a fractionated radiotherapy course for GBM. In a prospective study that evaluated sequential MRI scans conducted during chemoradiotherapy for patients with GBM, target dynamics were evaluated [44]. In that study, the gross tumor volume (GTV) and clinical target volume (CTV), along with tumor migration, were evaluated at fraction 0, fraction 10, fraction 20, and 1 month after treatment. Among the majority of patients who were enrolled, the GTV decreased from a median of 18.4 $cm^3$ at fraction 0, 14.7 $cm^3$ at fraction 10, 13.7 $cm^3$ at fraction 20, and 13 $cm^3$ at 1 month after treatment. Moreover, in their report, tumor shrinkage was associated with tumor migration. The median GTV migrations were greater than 5 mm in 54% of patients at fraction 10, 58% of patients at fraction 20, and 57% of patients 1 month after treatment. This amount of tumor migration during chemoradiotherapy exposes patients to the possibility of geographic tumor miss, in particular, if applying margin reduction strategies. Prior studies have also demonstrated the potential geographic miss due to tumor migration, although to a smaller degree [53, 54]. Consistent findings of tumor migration during treatment further reinforce the need to evaluate the role of MRgRT with on-table adaptation for patients undergoing chemoradiotherapy for GBM.

In addition to tumor migration, changes in the primary tumor and cavity are also observed and reported in MRgRT series. In a case series, Mehta et al. evaluated the daily MRgRT images of three GBM patients treated on the MRIdian™ system to determine whether changes in resection cavity volume and cerebral edema could be observed on daily MRIs [55]. There was a general trend of daily decrease in the resection cavity volume in these three patients analyzed. The edema volume of 1 patient initially increased during the first 13 fractions, but then began to decrease until the end of the treatment. In a recent case report of a patient treated with the new MRIdian® A3i BrainTx™, imaging changes were also observed during a course of conventionally fractionated chemoradiotherapy [56]. While starting radiotherapy, the surgical cavity volume on T1 was 6.4 cc (representing 100%); on weekly repeat full MRI, the postoperative cavity relative decrease was 21.9%, 29.7%, 50.0%, 58.6%, and 67.5% from baseline. Additionally, the initial T2-weighted-fluid-attenuated inversion recovery (T2/FLAIR) signal abnormality volume at baseline was 60.3 cc (representing 100%) and showed a relative decrease in volume of 79.6%, 80.0%, 81.5%, 81.8%, and 86.5% on weekly MRIs (Fig. 20.2a and b). It is likely that regular interval imaging with MRgRT, based on the gradual decrease in cavity size and edema volume, can assist in adaptive planning which results in reduced irradiated volume and a better sparing of normal brain tissue. In an alternative scenario, a brain tumor may grow during treatment in some cases, and it can be necessary to enlarge the target volume. Additionally, daily imaging can also allow for the monitoring of cerebral edema, which may facilitate the correlation of clinical symptoms and the follow-up of corticosteroid response. Future studies between both MRgRT platforms are needed to understand which timings are critical for tumor and edema evaluation.

In another study from Jones et al., 14 GBM patients underwent tri-Cobalt-60 MRgRT in 30 fractions over 6 weeks delivered with concurrent temozolomide, and the tumor volume was delineated on MRI before each of the 30 fractions [57]. In their study, 4 of 14 patients demonstrated increases of at least 25% in T2 hyperintense volume through treatment. This volume expansion correlated with both T2/FLAIR and contrast-enhanced volume expansion on post-radiotherapy diagnostic MRIs. The observed growth in early edema was suggestive of early pseudoprogression, and these patients exhibited improved outcomes. In three of these four cases, a significant volume increase started at week 3, with the most prominent changes occurring during weeks 4 and 5. In addition to visualization of these structural changes, advanced imaging techniques can add potential information during a radiotherapy course. For example, in one series, it was detected that patients who developed progressive disease had significantly lower relative cerebral blood volume (rCBV) during their third week of treatment [58]. Apparent diffusion coefficient (ADC) values were also shown to increase during treatment of tumors responding to radiotherapy [59]. Therefore, the potential for these advanced imaging features to be integrated into treatment plan delivery review is key to unlocking the full power of MRgRT systems.

## Brain Metastasis

Significant volume changes can also occur with brain metastases during treatment or from the time of simulation to delivery of radiotherapy; adaptation may therefore be useful for improving treatment accuracy, especially if limited additional planning margins are employed. Salkeld and colleagues demonstrated that a change in management was required in 46% of lesions with 7-day interscan intervals, increasing to 62% of lesions at intervals longer than 7 days in their study. In their analysis, the overall mean change was 0.15 $cm^3$ for brain metastases in approximately 1 week [60]. In a study with 531 brain metastases in 101 consecutive patients at our center, 20.5% of the lesions would require a change in management within 7 days, and 33.3% of lesions would require replanning if the interval exceeded 7 days [61]. Additionally, the

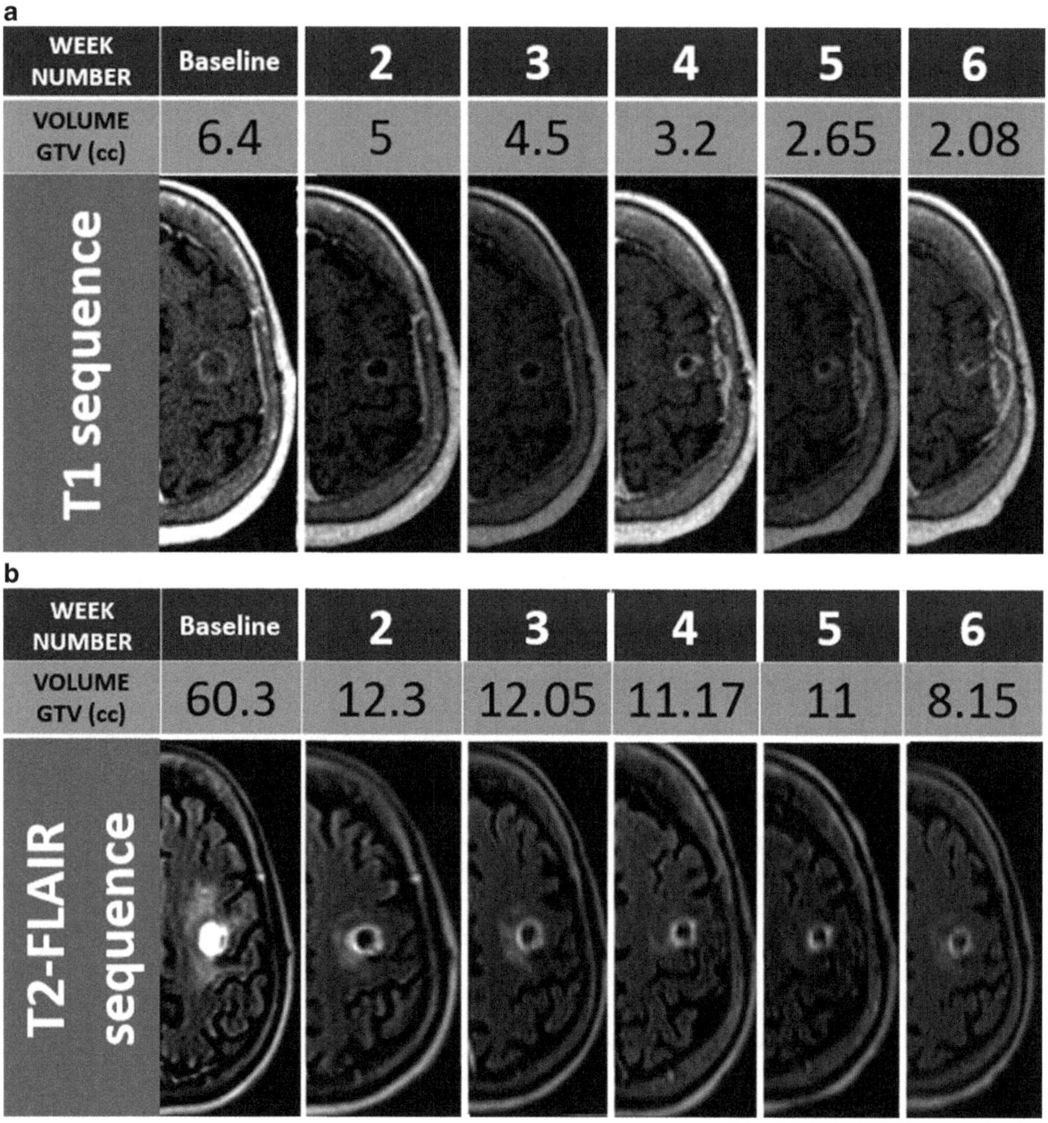


**Fig. 20.2** Axial T1 post-contrast (**a**) and T2/FLAIR (**b**) MRI sequences obtained weekly during radiotherapy. On sequential imaging, both the resection cavity and surrounding enhancement and the peri-tumoral FLAIR volumes decreased during the chemoradiotherapy course. (**a**) Axial T1 post-contrast images. (**b**) Axial T2/FLAIR images

displacement of the tumor centroid was observed in 66.8% of lesions, and in 12.4% of cases, the magnitude exceeded 1 mm. Ultimately, each day between treatment planning MRI acquisition and treatment was associated with an increased tumor margin expansion beyond 1 mm by 5%. Similarly, tumor dynamic changes have also been reported for cavity volumes after resection of brain metastases [62]. This suggests that in some situations, anatomic adaptation might be useful for brain metastasis treatment. Even though these short courses only allow for a limited amount of time for adaptation, studies have shown that changes in MRI from 1 day to 1 week are correlated with treatment responses [63, 64]. It may be possible to adjust the treatment plan or to predict early response in cases of SRS by daily monitoring with MRgRT. An example brain metastasis case treated with FSRS is demonstrated in Fig. 20.3 to show the tumor volume change during the treat-

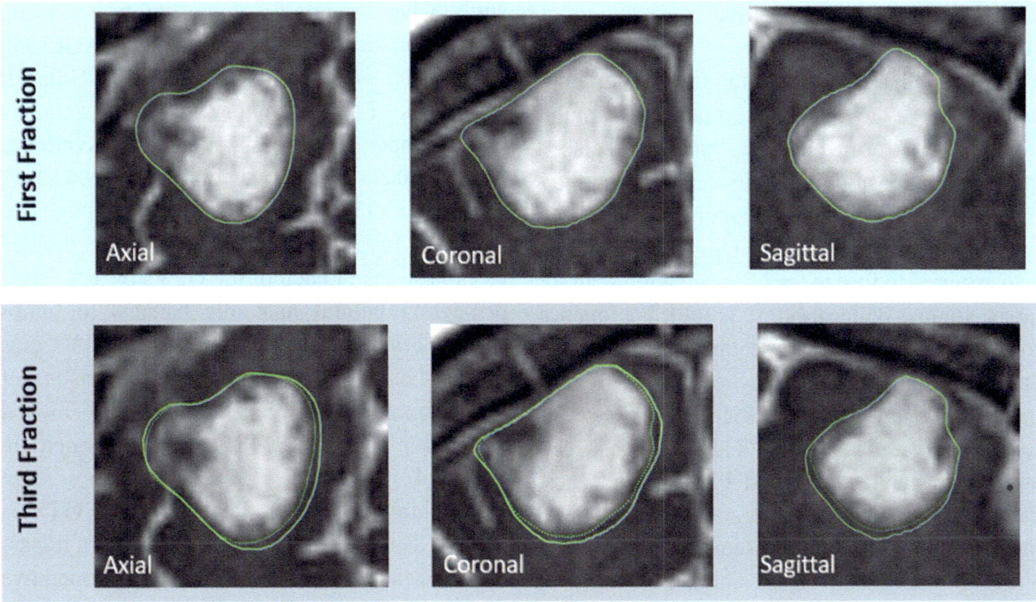


**Fig. 20.3** An example brain metastasis resection cavity case treated with fractionated stereotactic radiosurgery (FSRS). The patient was a 54-year-old male with a history of lung cancer who presented with a single brain metastasis and underwent gross total resection. He underwent FSRS to the resection cavity to a dose of 27 Gy in three fractions. The MRIs were performed at the time of each fraction to assess the target volume change. The interval tumor shrinkage observed between the first (volume, 6.5 cc) and third (volume, 5.0 cc) fractions is demonstrated

ment course. A recent study conducted by Tan et al. evaluated the changes in the CTV during five-fraction FSRS for resected brain metastases to quantify the interfraction cavity dynamic changes [65]. Their findings showed that there were significant cavity dynamic changes, with 80% of patients experiencing a reduction in T1 post-contrast CTV volume at third fraction compared to simulation MRI. Additionally, the observed variability in CTV contouring between T1 post-contrast and T2/FLAIR sequences in the study indicated that T1 post-contrast sequence should be the preferred choice for precise target delineation.

Based on the findings of the aforementioned studies, it can be inferred that personalized treatment adaptation through MRgRT facilitates the detection of significant changes in the tumor dynamics and utilizing MR-Linacs for treatment adaptation would enhance CTV coverage while minimizing radiation exposure to healthy brain tissue and reducing the risk of radiation necrosis. However, the implementation of SRS on MR-Linac devices is a significant challenge. SRS requires highly conformal planning and delivery, accurate patient setup, and position confirmation, and management of intrafraction motion and several different SRS platforms handle these issues in different ways. Previous feasibility studies have demonstrated that MR-Linac is capable of generating SRS plans that meet clinical requirements for brain metastases [65–67]. However, these studies still demonstrated that plan quality is still limited in comparison with other dedicated SRS systems.

In conclusion, MRgRT is a new and evolving technology, and it is still being researched to determine its optimal use in the treatment of GBM and brain metastases. The advantages of MRgRT include improved targeting accuracy which will ultimately translate into improved disease control and reduced risk of side effects.

## Contouring and Planning

To deliver radiation to targets in the brain, a non-contrast CT (1- to 3-mm slice thickness) simulation scan is performed initially. Patients are placed in an optimal position utilizing a customized thermoplastic mask with head support which is used for daily positioning. An MRI simulation scan is also acquired to fuse to the CT simulation scan for target volume delineation; a TrueFISP sequence for ViewRay MRIdian™ treatments; and a standard T1-weighted gradient echo series with contrast (T1 + c) and a volumetric T2/FLAIR series for Unity treatments. Additional sequences, such as T1 post-contrast, T2, and FLAIR sequences, can also be obtained and coregistered to the treatment planning TrueFISP sequence. With the ViewRay MRIdian™, the simulation study is exported to the planning software for coregistration with the previous diagnostic MRI sequences (i.e., T1-magnetization-prepared rapid gradient echo [T1-MPRAGE] and T2/FLAIR), using a rigid registration to the area of interest. MR-Linac-acquired brain imaging sequences for MRIdian™ and Unity™ are shown in Table 20.1.

The standard Radiation Therapy Oncology Group (RTOG) planning approach for GBM involves a two-phase approach, with the GTVs being defined using diagnostic postoperative MRIs. The GTV46 Gy is defined as the T2/FLAIR contrast-enhanced volume, and the GTV14 Gy (boost) is defined as the MRI T1 contrast-enhanced resection cavity. In our recent study, when the target volumes delineated using the diagnostic MRI sequences (MPRAGE and T2/FLAIR) were compared to the MRIdian™ sequences (T1 post-contrast and T2/FLAIR) in one example GBM case, a good concordance between the volumes was observed: 68.4 cc vs 68.3 cc for GTV 46 Gy and 6.2 cc versus 6.4 cc for GTV 14 Gy, respectively. CTVs are defined as expansions of each of the GTVs by 1.5–2 cm, respecting natural anatomic barriers to tumor spread. Finally, PTVs are defined as CTVs plus uniform expansions of 3 mm. Comparisons of image qualities between diagnostic (3 T) and simulation (0.5 T) MRIs are illustrated in Fig. 20.4.

For brain metastasis treatments, the GTV is defined as the lesion itself for intact brain metastasis and the resection cavity in the postoperative setting. No additional CTV margin is added for intact brain metastasis, and a 2-mm margin is added for resection cavities [68, 69]. These targets are delineated using the T1 post-contrast imaging study. Finally, PTVs are defined as CTV plus uniform expansions of 1 mm.

In the MRIdian™ treatment planning system (TPS), the density from the CT is rigidly registered to the planning MRI for dose calculation purposes. To simplify the treatment planning process, bilateral structures such as lens left + right (L + R) and optic nerve L + R, which have a maximum point dose constraint, are merged into a single structure. Furthermore, to ensure that the dose to the OARs is within acceptable limits, a

**Table 20.1** MRgRT-acquired brain imaging sequences

| | ViewRay MRIdian | Elekta Unity | |
|---|---|---|---|
| Scan | Details | Scan | Details |
| Daily imaging | | | |
| TrueFISP | 1-mm slice thickness; FOV 25 × 25 × 25 cm³ | T1-weighted GRE | 1.1 × 1.1 × 2.2 mm voxels; FOV 27 × 20 × 20 cm³ |
| | | T2/FLAIR | 1.1 × 1.1 × 2.2 mm voxels; FOV 27 × 20 × 20 cm³ |
| Weekly imaging—IV contrast | | | |
| T1 | 1-mm slice thickness; FOV 20 × 20 × 20 cm³ | | |
| T2 | 1.5-mm slice thickness; FOV 20 × 20 × 20 cm³ | | |
| FLAIR | 2.5-mm slice thickness; FOV 15 × 15 × 25 cm³ (in-plane 1.5 mm) | | |

*TrueFISP* True fast imaging with steady state precession, *FOV* Field of view, *GRE* Gradient recalled echo, *FLAIR* Fluid-attenuated inversion recovery

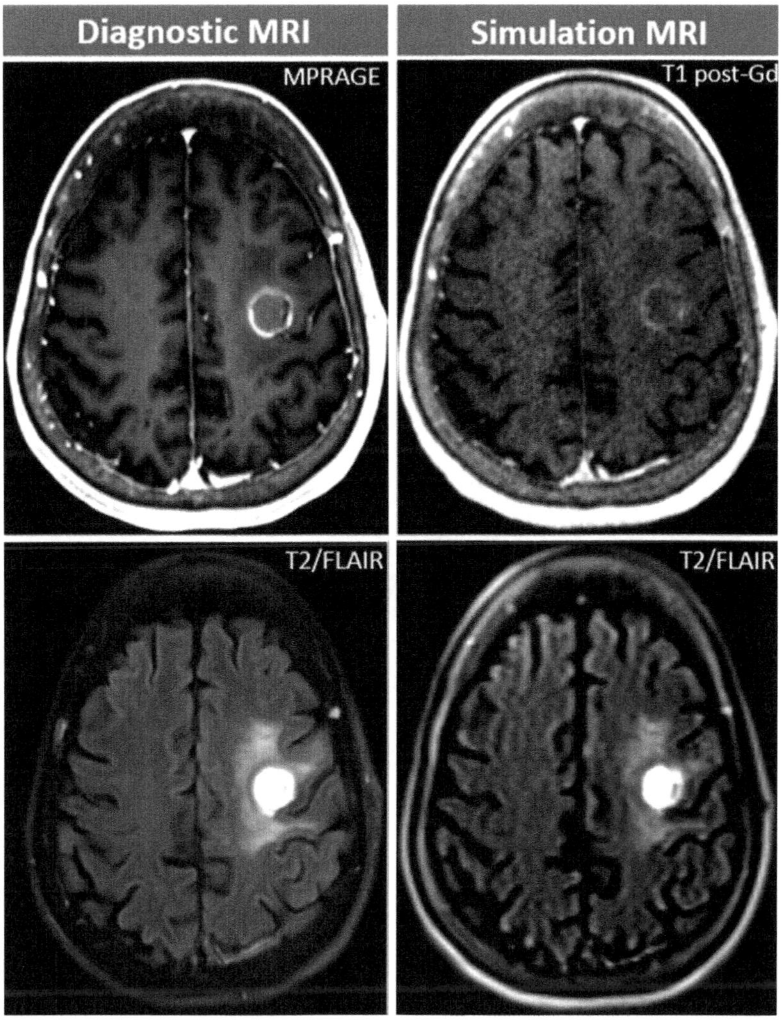


**Fig. 20.4** Comparison between diagnostic (3 T) and simulation (0.5 T) MRIs for a gross totally resected glioblastoma. Represented in the left side are two axial slices (T1-MPRAGE and T2/FLAIR sequences). On the right side are the corresponding simulation MRIs (T1-post Gd and T2/FLAIR sequences)

planning organ-at-risk volume (OAR_PRV) is created, which encompasses any OARs that have a maximum point dose constraint and overlap with the target volumes. In order to optimize the radiation dose, the planning target volume optimized structure (PTV_opt) is generated by subtracting the OAR_PRV from the PTV. To control low- and high-dose spills, two rings are created. A high-dose ring is used to control prescription dose conformality, which is created as a function of rings PTV + 4 mm minus PTV + 2 mm. Low-dose spilling is controlled by creating a low-dose ring which is a function of the skin minus the PTV + 4 mm ring. To help with daily localization matching, a GTV_TrueFISP is delineated as the enhanced region of the GTV on the TrueFISP MRI. The selection of the radiation beam positions in the treatment plan is meticulously determined based on the size and location of the target volume. For GBM cases, for example, a pseudo-half-arc beam arrangement is typically used, which involves delivering radiation from multiple angles around the involved hemisphere of the brain. In order to ensure that the total radiation dose between the initial and boost plans achieves the desired constraints for each OAR, the cumulative constraints of OARs are adjusted based on the constraint limitations in each plan. For instance, if the optic chiasm constraint is $D_{0.03cc} < 54$ Gy and the prescribed dose to the target volume is 60 Gy, each plan takes into account the overlapping structures with a lower tolerance. To meet the maximum dose constraint, the chiasm is limited to 44 Gy in the primary plan and 10 Gy in the boost plan. Thus, this guarantees that the $D_{0.03cc} < 54$ Gy requirement is satisfied in the composite of the primary and boost plans. A composite dose distribution is utilized at the end of planning to assess the summation of the initial and boost plans. An example GBM radiotherapy plan planned for ViewRay MRIdian is demonstrated in Fig. 20.5. For the comparison purpose, the isodose distributions for a right frontal brain metastasis planned with Gamma Knife® (Elekta, Stockholm, Sweden), CyberKnife® (Accuray, Wisconsin, USA), and TrueBeam® Linac (Varian, Palo Alto, California, USA) SRS are illustrated in Fig. 20.6.

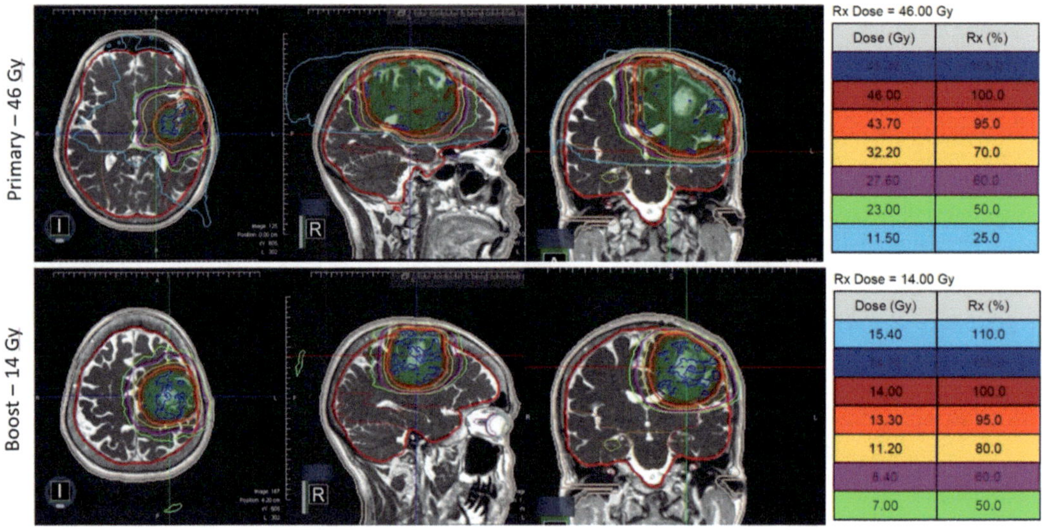


**Fig. 20.5** An example glioblastoma radiotherapy plan and corresponding isodose distribution with ViewRay MRIdian. A 2-phase technique was employed in this case, 46 Gy in 23 fractions with a boost to 60 Gy in 30 fractions

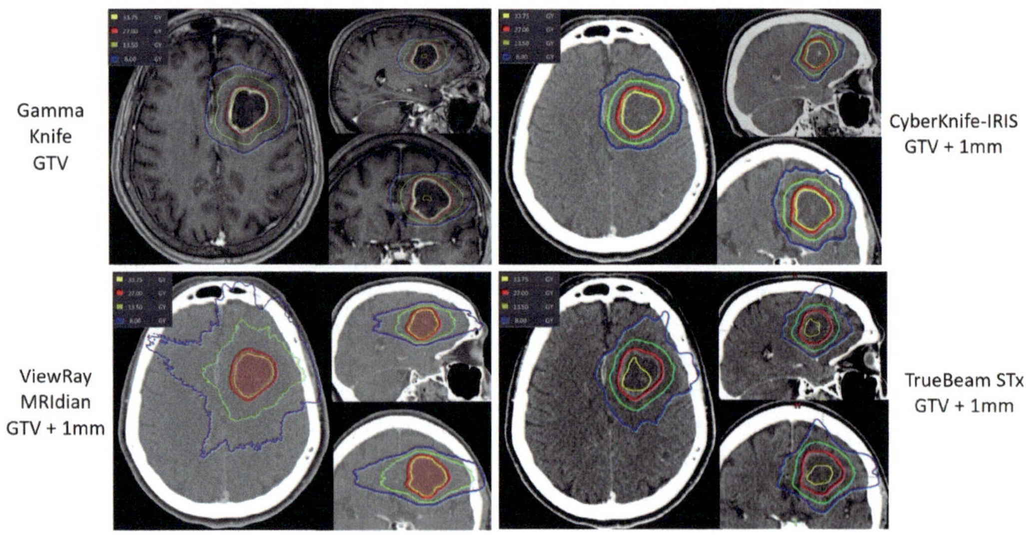


**Fig. 20.6** Comparison of MRgRT to other radiation modalities. Isodose distributions for a left frontal brain metastasis planned with Gamma Knife radiosurgery, CyberKnife, MRIdian, and TrueBeam Linac radiosurgery are illustrated to demonstrate the differences in isodose distribution among the radiotherapy modalities. All plans were planned 27 Gy in 3 fractions

Monaco (Elekta CMS, Maryland Heights, MO, USA) is the TPS used for the Unity™ system. Similar to the MRIdian™, the CT simulation is rigidly registered to the MRI simulation to use electron densities to calculate delivered radiation dose. PRVs for OARs in close proximity to (or overlapping with) the PTV are used specifically for the brainstem, optic chiasm, and optic nerves. Reference OARs using a 2-mm margin on the globes and skull (split into left/right) are created for verification during onboard treatment to ensure proper patient positioning since rotations cannot be corrected on the Unity™ treatment bed. An IMRT plan with a default 9 beam (which can be adjusted to between 7 and 11) is created with a 2-mm dose grid. Dose calculation is performed with Monaco TPS Monte Carlo algorithm coined GPUMCD. A stoplight system is used with "ideal" and "tolerance" constraints for both CTV and PTV coverage and OAR limits based on institutional guidelines. A green light demonstrates passing the ideal constraint; yellow is for beyond ideal but still within tolerance; and red light indicates an unacceptable level of target coverage or overdosing of an OAR. This same system is used for rapid plan evaluation during daily onboard replanning. Further technical details on treatment planning for the Unity™ system are described by Ruschin et al. [70].

## Dose Constraints

Institutional dosimetric constraints for the nearby CNS OARs and treatment planning directives for glioblastoma and brain metastases are presented in Tables 20.2 and 20.3.

**Table 20.2** Institutional glioblastoma dose constraints for organ-at-risk (OAR) and target volume coverage directives

| Glioblastoma dose constraints | | | |
|---|---|---|---|
| OAR | | | Target planning dose |
| Uninvolved brain | $D_{mean}$ | ≤30 Gy | ≥99% of the GTV receives ≥100% of PD |
| Total brain | $D_{0.03cc}$ | <66 Gy | ≥99% of the CTV receives ≥100% of PD |
| Brainstem | $D_{0.03cc}$ | ≤55 Gy | ≥95% of the PTV receives ≥100% of PD |
| Cochlea | $D_{5\%}$ | 55 Gy | |
| | $D_{mean}$ | ≤35 Gy | |
| Spinal cord | $D_{0.03cc}$ | ≤45 Gy | |
| Ocular globe | $D_{0.03cc}$ | ≤45 Gy | |
| | $D_{15\%}$ | ≤10 Gy | |
| Lens | $D_{0.03cc}$ | ≤7 Gy | |
| Optic nerves | $D_{0.03cc}$ | ≤55 Gy | |
| Chiasm | $D_{0.03cc}$ | ≤56 Gy | |
| Retina | $D_{0.03cc}$ | ≤45 Gy | |
| Hippocampus | $D_{mean}$ | ≤20 Gy | |
| | $D_{40\%}$ | ≤7.3 Gy | |
| Pituitary gland | $D_{mean}$ | ≤35 Gy | |
| Lacrimal gland | $D_{mean}$ | ≤25 Gy | |
| Temporal lobe | $D_{mean}$ | ≤20 Gy | |

*OARs* organs at risk

**Table 20.3** Institutional brain metastasis dose constraints for organ-at-risk (OAR) and target volume coverage directives

| Brain metastasis dose constraints for SRS and FSRS | | | | | | | | |
|---|---|---|---|---|---|---|---|---|
| One fraction | | | Three fractions | | | Five fractions | | |
| OAR | | | OAR | | | OAR | | |
| Uninvolved brain | $D_{mean}$ | ≤8 Gy | Uninvolved brain | $D_{mean}$ | ≤18 Gy | Uninvolved brain | $D_{mean}$ | ≤18 Gy |
| | V12Gy | ≤10 cc | | V20Gy | ≤20 cc | | V20Gy | ≤20 cc |
| Brainstem | $D_{0.03cc}$ | ≤15 Gy | Brainstem | $D_{0.03cc}$ | ≤23.1 Gy | Brainstem | $D_{0.03cc}$ | ≤31 Gy |
| | $D_{0.5cc}$ | ≤10 Gy | | $D_{0.5cc}$ | ≤18 Gy | | $D_{0.5cc}$ | ≤23 Gy |
| Optic pathway | $D_{0.03cc}$ | ≤10 Gy | Optic pathway | $D_{0.03cc}$ | ≤17.4 Gy | Optic pathway | $D_{0.03cc}$ | ≤25 Gy |
| | $D_{0.2cc}$ | ≤8 Gy | | $D_{0.2cc}$ | ≤15.3 Gy | | $D_{0.2cc}$ | ≤23 Gy |
| Cochlea | $D_{0.03cc}$ | ≤9 Gy | Cochlea | $D_{0.03cc}$ | ≤17.1 Gy | Cochlea | $D_{0.03cc}$ | ≤25 Gy |
| Spinal cord and medulla | $D_{0.03cc}$ | ≤14 Gy | Spinal cord and medulla | $D_{0.03cc}$ | ≤21.9 Gy | Spinal cord and medulla | $D_{0.03cc}$ | ≤30 Gy |
| | $D_{0.35cc}$ | ≤10 Gy | | $D_{0.35cc}$ | ≤18 Gy | | $D_{0.35cc}$ | ≤23 Gy |
| | | | **Target planning dose** | | | | | |
| ≥99.5% of the PTV receives ≥100% of PD | | | | | | | | |

*OARs* organs at risk, *PD* prescribed dose, *FSRS* fractionated stereotactic radiosurgery

## Dose Escalation

The median survival for GBM remains poor at approximately 15–16 months in contemporary series, despite surgical intervention, conventional radiotherapy, chemotherapy, and in some cases with TTFields [6–8, 71, 72]. Radiotherapy is typically aimed at treating the surgical bed, enhancing residual tumor on MRI, and a margin of 1–2 cm around the surgical site [73]. In addition to prolonging the time to tumor progression and increasing survival, the addition of temozolomide to radiotherapy enhances its therapeutic efficacy with a synergistic effect [6]. Despite radiotherapy and temozolomide improving survival, most of the recurrences are local, and locoregional recurrences occur where the highest radiotherapy dose is delivered (e.g., >57 Gy or the 95% isodose line of prescribed 60 Gy) [74, 75].

Increasing the dose of radiotherapy, either through concomitant escalation of the total dose or dose per fraction, represents approaches that have been tested to overcome hypoxia-related resistance of GBM. Standard radiotherapy in postoperative setting currently employs a total dose of 60 Gy in 30 fractions [6]. Dose-escalated radiotherapy has not proven to be beneficial for GBM patients in previous prospective studies [76, 77]. The most significant aspect of these trials is that they were conducted either without chemotherapy or with nitrosourea-based chemotherapy instead of the use of temozolomide, as a radiosensitizer. In a historical joint study of the RTOG and Eastern Cooperative Oncology Group (ECOG), no survival benefit was observed with a 10-Gy focal boost to 60-Gy WBRT [77]. However, a recent systematic review and meta-analysis of 22 prospective trials demonstrated that dose-escalated radiotherapy (defined as the equivalent dose in 2-Gy fractions) >60 Gy without chemotherapy was more effective in terms of progression-free and overall survival compared to 60-Gy standard-dose radiotherapy without chemotherapy [78]. In the contemporary era of concurrent and adjuvant temozolomide, a prospective phase I/II study was conducted to evaluate the efficacy of dose escalation using an IMRT technique [79]. They reported that temozolomide was tolerated to a maximum of 75 Gy in 30 fractions. Interestingly, the probability of local failure decreased as dose escalation increased, and the median survival was 20.1 months, suggesting a higher degree of efficacy compared to previous studies. These findings paralleled results from prior studies of radiotherapy dose escalation without temozolomide, where local control improved with radiotherapy dose escalation [80,

81]. Most recently, the NRG Oncology BN001 study was designed to test the hypothesis that radiotherapy intensification with temozolomide can prolong survival in patients with newly diagnosed GBM based on the promising results of radiotherapy dose escalation combined with concurrent and adjuvant temozolomide. In this trial, 75 Gy was administered to the surgical bed, along with residual enhancing tumor plus a volumetric margin of 5 mm, with a differential dose painting of 50 Gy in 30 fractions to the non-enhancing tumor and a volumetric margin of 2 cm, along with concurrent and adjuvant temozolomide [82]. Although no benefit to dose escalation was observed in the photon group (and proton therapy-specific results are pending at this time), a detailed pattern of failure analysis is needed to fully understand the potential and limitations of that dose-escalation approach.

Instead of dose escalation to the postoperative tumor bed, MRgRT offers the ability to deliver a more personalized form of dose escalation. If, as an example, MRI indicates that a tumor becomes increasingly cellular, metabolic, or angiogenic as it goes through treatment, it may be appropriate to consider RT dose escalation as a method of treatment intensification. One way to answer this question is to use MRgRT to obtain multidimensional MRIs with high temporal frequency with MR-Linac devices throughout treatment to reveal changes in tumor physiology that may be useful in identifying treatment decisions that are adaptive to tumor physiology. MRI-based physiologic metrics that are used to predict patient outcomes should be examined in future studies to determine which changes and at what time points are most predictive of treatment outcome. Obtaining and evaluating these images during radiotherapy will allow for early adaptation of treatment in an effort to improve survival outcomes. By utilizing the nuclear magnetic resonance phenomenon in conjunction with magnetic resonance spectroscopic imaging (MRSI), it is possible to identify the chemical composition of tissues noninvasively [83]. Rapidly proliferating cells are found to be enriched with choline-containing compounds (Cho), which is a structure in the phospholipid bilayer of the cell membrane. The amount of N-acetyl aspartate (NAA), a protein found in neurons, is diminished when the neuronal environment is destroyed in tissue samples. The ratio of Cho to NAA (Cho/NAA) has been previously shown to be elevated in patients with GBM and can serve as a biomarker to differentiate regions of tumor from normal healthy brain tissue [84, 85]. Based on these results, studies have assessed the potential of MRSI for radiotherapy planning in GBM patients [85–87]. However, very few clinical trials have managed to integrate MRSI for treatment of brain tumors prospectively. SPECTRO-GLIO is one such trial where regions of the brain with a ratio of Cho/NAA >2 received an escalated radiation dose of 72 Gy; however, the results have not been reported yet [88]. In a pilot trial at 3 institutions with 30 patients, the authors assessed the feasibility, safety, and efficacy of MRI spectroscopy-guided RT dose escalation to 75 Gy for newly diagnosed GBMs [89]. For radiotherapy, standard tumor volumes based on T2/FLAIR and T1 post-contrast MRIs were treated in 30 fractions to 50.1 and 60 Gy, respectively. An additional high-risk volume based on residual tumor and Cho/NAA ≥2 was treated to 75 Gy. At a median follow-up of 21.4 months, the median overall survival and progression-free survival were 23.0 and 16.6 months, respectively. Eleven of 30 patients experienced grade 3 or greater toxicities that were at least possibly due to their treatment. Of the seven patients who experienced these by 9 months post-radiotherapy, most were attributable to temozolomide, and only one case could potentially be ascribed to radiotherapy. The integration of MRgRT imaging into the treatment workflow may help to facilitate this process for radiotherapy planning.

SRS for brain metastases can be delivered as either a single fraction (generally 18–24 Gy) or as FSRS ranging from 24 to 30 Gy in 3 fractions, or 25–35 Gy in 5 fractions [90]. The RTOG 90-05 dose-escalation trial was the initial prospective study to understand the maximum tolerated single-fraction doses for SRS for recurrent primary brain tumors and brain metastases, although the relationship between dose and local control was not reported [20]. Large retrospective series have supported the RTOG 90-05 dose prescriptions, for example, in one series of 3034 brain metastases ≤2 cm treated with SRS, a dose of 24 was associated with reduced local failure compared to 19–23 Gy (dose de-escalation) [91]. Further dose-escalation efforts with single-fraction SRS, however, are limited due to the perceived increased risk of radiation necrosis. A recent systematic review including 10 studies with 720 brain metastases treated with FSRS showed that the tumor control probability increases with the higher $BED_{10}$. Local control at 1 and 2 years was 84% and 73%, respectively, for lesions treated with a $BED_{10}$ of at least 57.6 Gy compared to 72% and 60% for those treated with $BED_{10}$ less than 57.6 Gy [92]. In one of the few randomized trials, Kim et al. compared the local control and radiation necrosis rate with 24, 27, or 30 Gy in 3 fractions (8, 9, or 10 Gy per fraction, respectively) for brain metastases larger than 3 cm treated on the Gamma Knife® [93]. In their dose-escalation study, 27 Gy in 3 fractions appeared to be the best regimen of FSRS because 30 Gy in 3 fractions was associated with an increased risk of radiation necrosis (hazard ratio, 7.2, $p = 0.047$), and 24 Gy in 3 fractions was associated with an increased risk of local failure (hazard ratio, 2.5, $p = 0.037$).

Despite the early postoperative SRS clinical trials utilizing single-fraction SRS, which yielded modest disease recurrence rates (15–39%), various subsequent studies have explored the potential of postoperative FSRS to enhance local control rates and reduce the risk of radiation necrosis [23, 94–96]. The consensus guidelines of the International Society of Stereotactic Radiosurgery recommend postoperative dose and fractionation schedules ranging from approximately 30 to 50 Gy $EQD2_{10}$, 50 to 70 $EQD2_5$, and 70 to 90 $EQD2_2$ [97]. The A071801 trial (NCT04114981) recently completed accrual and involved randomizing patients with resected brain metastasis larger than 2 cm to either single-fraction SRS or FSRS. The trial's primary endpoint was the time to local recurrence, and it is expected to generate further prospective comparative data to aid in guiding clinical practice.

## On-Table Adaptation

### ViewRay MRIdian™

During daily setup for treatment, a volumetric MRI is obtained to visualize the target volume and OARs. Tumor target volumes are registered rigidly from the simulation MRI to the daily localization MRI scan, and OAR contours are propagated to the MRI scan of the day using deformable image registration. GTV and OAR contours within 3 cm radially of the PTV and 2 cm cranial or caudal to the PTV are edited to reflect the anatomy of the day. Online plans are reoptimized with the TPS, using the same beam parameters and optimization objectives. Following this, patient-specific quality assurance is performed with the patient in treatment position, using an independent Monte Carlo dose calculation followed by a gamma analysis comparison with dose distribution from the MRIdian™ planning system [38, 98]. After plan approval, treatment delivery occurs using visual feedback provided to the patient on a monitor projecting the real-time MRI.

### Elekta Unity™

There are two primary modes for daily adaptation on the Unity™ system: adapt-to-position (ATP) and adapt-to-shape (ATS). ATP uses a single MRI sequence (typically a T1-weighted 3D non-contrast scan, but a volumetric FLAIR scan can also be used) that is fused to the MRI simulation, providing rigid deformations for the target and OARs. A virtual couch shift is applied to adjust for differences in patient positioning [70]. The same IMRT beam angles are used, reoptimization is performed, and the plan is evaluated using the green/yellow/red stoplight system described in section "Contouring and Planning". An example online adaptive interface for a GBM plan on Unity system is shown in Fig. 20.7. Target and OAR contours are not manually adjusted during an ATP fraction. An independent dose check by medical physics is performed to ensure the dose difference between

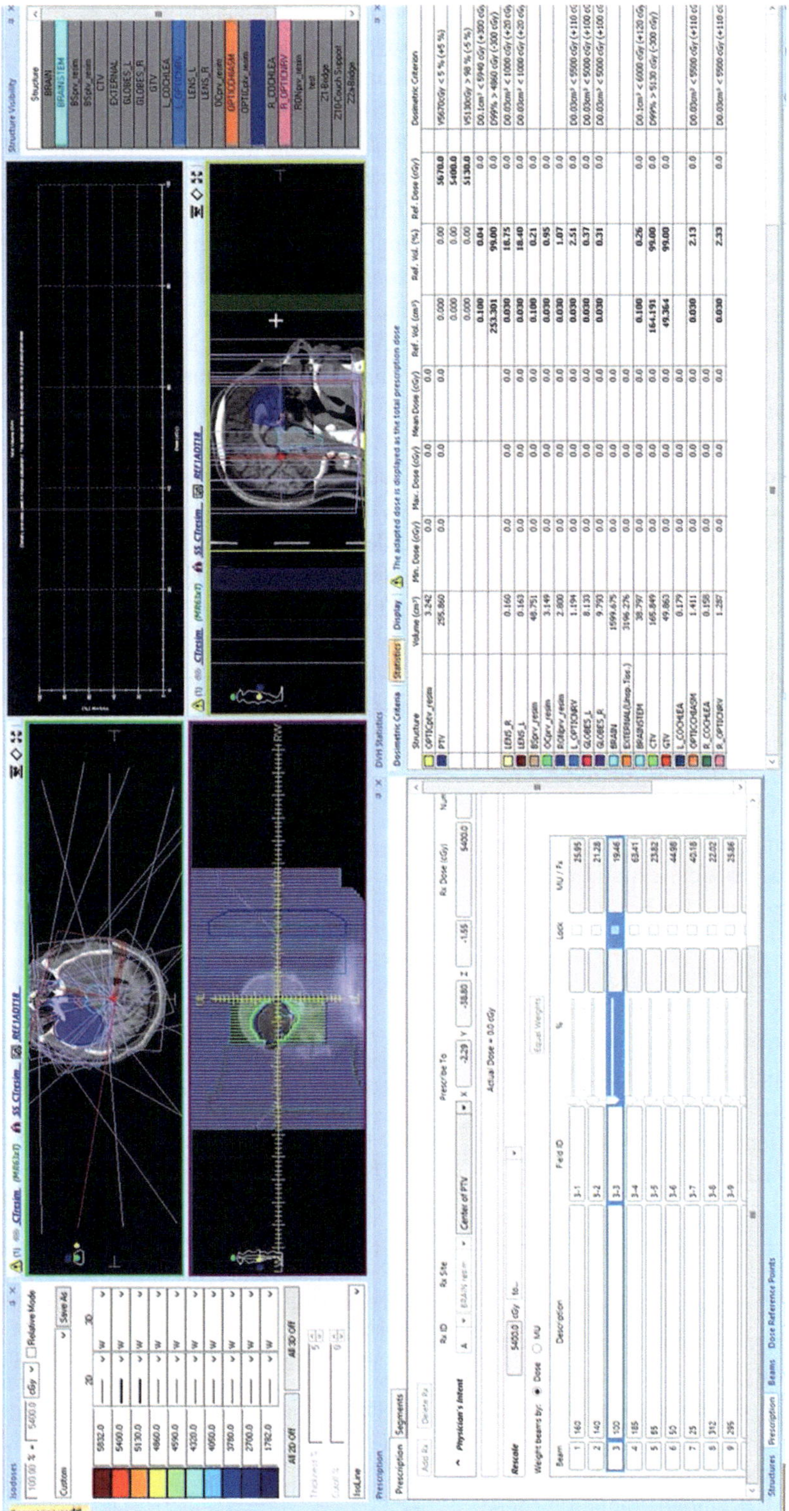

**Fig. 20.7** The figure illustrates the online adaptive interface for a glioblastoma multiforme (GBM) treatment plan for Unity™ system. The top panel shows the beam arrangement over the CTsim, along with the collimator arrangement for one beam. The top-right panel displays the dose-volume histogram (DVH), which is currently empty. The bottom-left panel shows a sagittal magnetic resonance imaging (MRI) reconstructed from the day's MRI scan. The bottom-middle panel displays the optimization window, while the bottom-right panel shows the target and organ-at-risk (OAR) constraints. Once the plan is ready, the stoplight system (green/yellow/red) will indicate the plan's quality based on ideal, tolerance, and unacceptable numbers

the treatment planning software and the physics-independent verification software is within 4% [99]. During the contouring, plan optimization, and beam-on phases, additional MRI sequences such as diffusion, perfusion, magnetization transfer, quantitative relaxometry, and functional MRI can be acquired. Cine MRI during beam-on can also be acquired to monitor for motion but has been not found to be helpful due to the lack of motion with proper immobilization. The initial experience using ATP in ten glioma patients on the Unity™ reported a mean fraction time of 37.3 min, and all fractions met the independent physics tolerance check [99].

The ATS mode for intracranial radiotherapy was developed for the UNITED (NCT04726397) clinical trial. Once a week during a course of fractionated radiotherapy for GBM, gadolinium contrast is injected (outside the Unity™), the patient is positioned, and a 3D T1 + c sequence, as well as a volumetric FLAIR sequence, is acquired. The MRI images are fused to the MRI simulation images, and either rigid or deformable registrations can be used for the targets and OARs. The treating radiation oncologist is present and edits the GTV (on the T1 + c images) and OARs; a 5-mm volumetric expansion is applied respecting anatomical boundaries, and the CTV can then be manually expanded beyond 5 mm to incorporate involved areas of FLAIR hyperintense signal. A new radiation plan is optimized (using the same IMRT beam angles as used during the reference plan optimization) and reviewed by the radiation oncologist. This new ATS plan then becomes the reference plan for the subsequent four ATP fractions until the next ATS fraction is developed a week later.

## Other

There are currently several clinical trials investigating the use of MRgRT for the treatment of GBM. Some examples of ongoing trials include:

(a) A multi-institutional pilot clinical trial of spectroscopic MRI-guided radiation dose escalation for newly diagnosed GBM (NCT03137888)—This trial is investigating the use of MR spectroscopy to guide radiation dose escalation in patients with newly diagnosed GBM.

(b) UNIty-Based MR-Linac Guided AdapTive RadiothErapy for High-GraDe Glioma: A Phase 2 Trial (UNITED) (NCT04726397)—This trial is studying the use of a small-margin (5 mm from GTV to CTV), daily adaptive MRgRT approach for GBM using the Unity™ system.

(c) UNITy-BasED MR-Linac Adaptive SimultaneousIntegratedHypofractionationed Boost Trial for High-Grade Glioma in the Elderly (UNITED2) (NCT05565521)—This study is using a small-margin approach as well when treating to 40 Gy/15 fractions for elderly GBM, but adding a simultaneously integrated boost to 52.5 in 15 fractions to the GTV only (with a PTV expansion).

## References

1. Lah TT, Novak M, Breznik B. Brain malignancies: glioblastoma and brain metastases. Semin Cancer Biol. 2020;60:262–73.
2. Ostrom QT, Price M, Neff C, Cioffi G, Waite KA, Kruchko C, et al. CBTRUS statistical report: primary brain and other central nervous system tumors diagnosed in the United States in 2015-2019. Neuro Oncol. 2022;24(Suppl 5):v1–v95.
3. Sacks P, Rahman M. Epidemiology of brain metastases. Neurosurg Clin N Am. 2020;31(4):481–8.
4. Seker-Polat F, Pinarbasi Degirmenci N, Solaroglu I, Bagci-Onder T. Tumor cell infiltration into the brain in glioblastoma: from mechanisms to clinical perspectives. Cancers. 2022;14(2):443.
5. Bell EH, Pugh SL, McElroy JP, Gilbert MR, Mehta M, Klimowicz AC, et al. Molecular-based recursive partitioning analysis model for glioblastoma in the Temozolomide era: a correlative analysis based on NRG oncology RTOG 0525. JAMA Oncol. 2017;3(6):784–92.
6. Stupp R, Hegi ME, Mason WP, van den Bent MJ, Taphoorn MJ, Janzer RC, et al. Effects of radiotherapy with concomitant and adjuvant temozolomide versus radiotherapy alone on survival in glioblastoma in a randomised phase III study: 5-year analysis of the EORTC-NCIC trial. Lancet Oncol. 2009;10(5):459–66.
7. Stupp R, Taillibert S, Kanner A, Read W, Steinberg D, Lhermitte B, et al. Effect of tumor-treating fields plus maintenance Temozolomide vs maintenance

Temozolomide alone on survival in patients with glioblastoma: a randomized clinical trial. JAMA. 2017;318(23):2306–16.
8. Stupp R, Mason WP, van den Bent MJ, Weller M, Fisher B, Taphoorn MJ, et al. Radiotherapy plus concomitant and adjuvant temozolomide for glioblastoma. N Engl J Med. 2005;352(10):987–96.
9. Roa W, Brasher PM, Bauman G, Anthes M, Bruera E, Chan A, et al. Abbreviated course of radiation therapy in older patients with glioblastoma multiforme: a prospective randomized clinical trial. J Clin Oncol. 2004;22(9):1583–8.
10. Malmström A, Grønberg BH, Marosi C, Stupp R, Frappaz D, Schultz H, et al. Temozolomide versus standard 6-week radiotherapy versus hypofractionated radiotherapy in patients older than 60 years with glioblastoma: the Nordic randomised, phase 3 trial. Lancet Oncol. 2012;13(9):916–26.
11. Roa W, Kepka L, Kumar N, Sinaika V, Matiello J, Lomidze D, et al. International Atomic Energy Agency randomized phase III study of radiation therapy in elderly and/or frail patients with newly diagnosed glioblastoma multiforme. J Clin Oncol. 2015;33(35):4145–50.
12. Kotecha R, Gondi V, Ahluwalia MS, Brastianos PK, Mehta MP. Recent advances in managing brain metastasis. F1000Res. 2018;7:F1000 Faculty Rev-1772.
13. Suh JH, Kotecha R, Chao ST, Ahluwalia MS, Sahgal A, Chang EL. Current approaches to the management of brain metastases. Nat Rev Clin Oncol. 2020;17(5):279–99.
14. Vogelbaum MA, Brown PD, Messersmith H, Brastianos PK, Burri S, Cahill D, et al. Treatment for brain metastases: ASCO-SNO-ASTRO guideline. J Clin Oncol. 2022;40(5):492–516.
15. Gondi V, Bauman G, Bradfield L, Burri SH, Cabrera AR, Cunningham DA, et al. Radiation therapy for brain metastases: an ASTRO clinical practice guideline. Pract Radiat Oncol. 2022;12(4):265–82.
16. Chang EL, Wefel JS, Hess KR, Allen PK, Lang FF, Kornguth DG, et al. Neurocognition in patients with brain metastases treated with radiosurgery or radiosurgery plus whole-brain irradiation: a randomised controlled trial. Lancet Oncol. 2009;10(11):1037–44.
17. Aoyama H, Shirato H, Tago M, Nakagawa K, Toyoda T, Hatano K, et al. Stereotactic radiosurgery plus whole-brain radiation therapy vs stereotactic radiosurgery alone for treatment of brain metastases: a randomized controlled trial. JAMA. 2006;295(21):2483–91.
18. Shinde A, Akhavan D, Sedrak M, Glaser S, Amini A. Shifting paradigms: whole brain radiation therapy versus stereotactic radiosurgery for brain metastases. CNS. CNS Oncol. 2019;8(1):CNS27.
19. Kocher M, Soffietti R, Abacioglu U, Villa S, Fauchon F, Baumert BG, et al. Adjuvant whole-brain radiotherapy versus observation after radiosurgery or surgical resection of one to three cerebral metastases: results of the EORTC 22952-26001 study. J Clin Oncol. 2011;29(2):134–41.
20. Shaw E, Scott C, Souhami L, Dinapoli R, Kline R, Loeffler J, et al. Single dose radiosurgical treatment of recurrent previously irradiated primary brain tumors and brain metastases: final report of RTOG protocol 90-05. Int J Radiat Oncol Biol Phys. 2000;47(2):291–8.
21. Masucci GL. Hypofractionated radiation therapy for large brain metastases. Front Oncol. 2018;8:379.
22. Prabhu RS, Dhakal R, Vaslow ZK, Dan T, Mishra MV, Murphy ES, et al. Preoperative radiosurgery for resected brain metastases: the PROPS-BM multicenter cohort study. Int J Radiat Oncol Biol Phys. 2021;111(3):764–72.
23. Mahajan A, Ahmed S, McAleer MF, Weinberg JS, Li J, Brown P, et al. Post-operative stereotactic radiosurgery versus observation for completely resected brain metastases: a single-centre, randomised, controlled, phase 3 trial. Lancet Oncol. 2017;18(8):1040–8.
24. Padhani AR, Liu G, Koh DM, Chenevert TL, Thoeny HC, Takahara T, et al. Diffusion-weighted magnetic resonance imaging as a cancer biomarker: consensus and recommendations. Neoplasia. 2009;11(2):102–25.
25. Ellingson BM, Malkin MG, Rand SD, Connelly JM, Quinsey C, LaViolette PS, et al. Validation of functional diffusion maps (fDMs) as a biomarker for human glioma cellularity. J Magn Reson Imaging. 2010;31(3):538–48.
26. Hein PA, Eskey CJ, Dunn JF, Hug EB. Diffusion-weighted imaging in the follow-up of treated high-grade gliomas: tumor recurrence versus radiation injury. AJNR Am J Neuroradiol. 2004;25(2):201–9.
27. Hamstra DA, Rehemtulla A, Ross BD. Diffusion magnetic resonance imaging: a biomarker for treatment response in oncology. J Clin Oncol. 2007;25(26):4104–9.
28. Jabehdar Maralani P, Myrehaug S, Mehrabian H, Chan AKM, Wintermark M, Heyn C, et al. Intravoxel incoherent motion (IVIM) modeling of diffusion MRI during chemoradiation predicts therapeutic response in IDH wildtype glioblastoma. Radiother Oncol. 2021;156:258–65.
29. Larsson C, Groote I, Vardal J, Kleppesto M, Odland A, Brandal P, et al. Prediction of survival and progression in glioblastoma patients using temporal perfusion changes during radiochemotherapy. Magn Reson Imaging. 2020;68:106–12.
30. Pirzkall A, McKnight TR, Graves EE, Carol MP, Sneed PK, Wara WW, et al. MR-spectroscopy guided target delineation for high-grade gliomas. Int J Radiat Oncol Biol Phys. 2001;50(4):915–28.
31. Mehrabian H, Chan RW, Sahgal A, Chen H, Theriault A, Lam WW, et al. Chemical exchange saturation transfer MRI for differentiating radiation necrosis from tumor progression in brain metastasis-application in a clinical setting. J Magn Reson Imaging. 2022;57(6):1713–25.
32. Chan RW, Lawrence LSP, Oglesby RT, Chen H, Stewart J, Theriault A, et al. Chemical exchange saturation transfer MRI in central nervous system

tumours on a 1.5 T MR-Linac. Radiother Oncol. 2021;162:140–9.
33. Maralani PJ, Myrehaug S, Mehrabian H, Chan AK, Wintermark M, Heyn C, et al. ADC, D, f dataset calculated through the simplified IVIM model, with MGMT promoter methylation, age, and ECOG, in 38 patients with wildtype IDH glioblastoma. Data Brief. 2021;35:106950.
34. Chan RW, Chen H, Myrehaug S, Atenafu EG, Stanisz GJ, Stewart J, et al. Quantitative CEST and MT at 1.5T for monitoring treatment response in glioblastoma: early and late tumor progression during chemoradiation. J Neurooncol. 2021;151(2):267–78.
35. Karami E, Soliman H, Ruschin M, Sahgal A, Myrehaug S, Tseng CL, et al. Quantitative MRI biomarkers of stereotactic radiotherapy outcome in brain metastasis. Sci Rep. 2019;9(1):19830.
36. Mehrabian H, Detsky J, Soliman H, Sahgal A, Stanisz GJ. Advanced magnetic resonance imaging techniques in management of brain metastases. Front Oncol. 2019;9:440.
37. Kim MM, Sun Y, Aryal MP, Parmar HA, Piert M, Rosen B, et al. A phase 2 study of dose-intensified chemoradiation using biologically based target volume definition in patients with newly diagnosed glioblastoma. Int J Radiat Oncol Biol Phys. 2021;110(3):792–803.
38. Chuong MD, Bryant J, Mittauer KE, Hall M, Kotecha R, Alvarez D, et al. Ablative 5-fraction stereotactic magnetic resonance-guided radiation therapy with on-table adaptive replanning and elective nodal irradiation for inoperable pancreas cancer. Pract Radiat Oncol. 2021;11(2):134–47.
39. Kutuk T, Herrera R, Mustafayev TZ, Gungor G, Ugurluer G, Atalar B, et al. Multi-institutional outcomes of stereotactic magnetic resonance image guided adaptive radiation therapy with a median biologically effective dose of 100 Gy(10) for non-bone oligometastases. Adv Radiat Oncol. 2022;7(6):100978.
40. Kutuk T, McCulloch J, Mittauer KE, Romaguera T, Alvarez D, Gutierrez AN, et al. Daily online adaptive magnetic resonance image (MRI) guided stereotactic body radiation therapy for primary renal cell cancer. Med Dosim. 2021;46(3):289–94.
41. Ugurluer G, Mustafayev TZ, Gungor G, Atalar B, Abacioglu U, Sengoz M, et al. Stereotactic MR-guided online adaptive radiation therapy (SMART) for the treatment of liver metastases in oligometastatic patients: initial clinical experience. Radiat Oncol J. 2021;39(1):33–40.
42. Cao Y, Tseng CL, Balter JM, Teng F, Parmar HA, Sahgal A. MR-guided radiation therapy: transformative technology and its role in the central nervous system. Neuro Oncol. 2017;19(suppl_2):ii16–29.
43. Maziero D, Straza MW, Ford JC, Bovi JA, Diwanji T, Stoyanova R, et al. MR-guided radiotherapy for brain and spine tumors. Front Oncol. 2021;11:626100.
44. Stewart J, Sahgal A, Lee Y, Soliman H, Tseng CL, Detsky J, et al. Quantitating interfraction target dynamics during concurrent chemoradiation for glioblastoma: a prospective serial imaging study. Int J Radiat Oncol Biol Phys. 2021;109(3):736–46.
45. Mesny E, Jacob J, Culot F, Calugaru V, Jenny C, Fonti B, et al. Optic nerve motion and gaze direction: their impact on intraorbital tumor radiotherapy. Cancer Radiother. 2022;26(5):678–83.
46. Qing K, Nie K, Liu B, Feng X, Stone JR, Cui T, et al. The impact of optic nerve movement on intracranial radiation treatment. Front Oncol. 2022;12:803329.
47. Liney GP, Whelan B, Oborn B, Barton M, Keall P. MRI-linear accelerator radiotherapy systems. Clin Oncol (R Coll Radiol). 2018;30(11):686–91.
48. Bonert M, Schneider M, Solyanik O, Hellbach K, Bondesson D, Gaass T, et al. Diagnostic accuracy of magnetic resonance imaging for the detection of pulmonary nodules simulated in a dedicated porcine chest phantom. PLoS One. 2020;15(12):e0244382.
49. Kluter S. Technical design and concept of a 0.35 T MR-Linac. Clin Transl Radiat Oncol. 2019;18:98–101.
50. Dajani S, Hill VB, Kalapurakal JA, Horbinski CM, Nesbit EG, Sachdev S, et al. Imaging of GBM in the age of molecular markers and MRI guided adaptive radiation therapy. J Clin Med. 2022;11(19):5961.
51. Huang TY, Huang IJ, Chen CY, Scheffler K, Chung HW, Cheng HC. Are TrueFISP images T2/T1-weighted? Magn Reson Med. 2002;48(4):684–8.
52. Schmitt P, Griswold MA, Jakob PM, Kotas M, Gulani V, Flentje M, et al. Inversion recovery TrueFISP: quantification of T(1), T(2), and spin density. Magn Reson Med. 2004;51(4):661–7.
53. Manon R, Hui S, Chinnaiyan P, Suh J, Chang E, Timmerman R, et al. The impact of mid-treatment MRI on defining boost volumes in the radiation treatment of glioblastoma multiforme. Technol Cancer Res Treat. 2004;3(3):303–7.
54. Tsien C, Gomez-Hassan D, Ten Haken RK, Tatro D, Junck L, Chenevert TL, et al. Evaluating changes in tumor volume using magnetic resonance imaging during the course of radiotherapy treatment of high-grade gliomas: implications for conformal dose-escalation studies. Int J Radiat Oncol Biol Phys. 2005;62(2):328–32.
55. Mehta S, Gajjar SR, Padgett KR, Asher D, Stoyanova R, Ford JC, et al. Daily tracking of glioblastoma resection cavity, cerebral edema, and tumor volume with MRI-guided radiation therapy. Cureus. 2018;10(3):e2346.
56. La Rosa A, Mittauer KE, Rzepczynski AE, Chuong MD, Kutuk T, Bassiri N, et al. Treatment of glioblastoma using MRIdian(R) A3i BrainTx: imaging and treatment workflow demonstration. Med Dosim. 2023;48(3):127–33.
57. Kolton K Jones SD, Maziero D, et al. MRI-guided radiotherapy identifies early pseudoprogression of glioblastoma. Preprint (Version 1) available at Research Square 2020.
58. Tsien C, Galbán CJ, Chenevert TL, Johnson TD, Hamstra DA, Sundgren PC, et al. Parametric response map as an imaging biomarker to distinguish progres-

sion from pseudoprogression in high-grade glioma. J Clin Oncol. 2010;28(13):2293–9.
59. Yang Y, Cao M, Sheng K, Gao Y, Chen A, Kamrava M, et al. Longitudinal diffusion MRI for treatment response assessment: preliminary experience using an MRI-guided tri-cobalt 60 radiotherapy system. Med Phys. 2016;43(3):1369–73.
60. Salkeld AL, Hau EKC, Nahar N, Sykes JR, Wang W, Thwaites DI. Changes in brain metastasis during radiosurgical planning. Int J Radiat Oncol Biol Phys. 2018;102(4):727–33.
61. Kutuk T, Tolakanahalli R, Williams A, Tom MC, Vadhan JD, Appel H, et al. Impact of MRI timing on tumor volume and anatomic displacement for brain metastases undergoing stereotactic radiosurgery. Neurooncol Pract. 2021;8(6):674–83.
62. Atalar B, Choi CY, Harsh GR 4th, Chang SD, Gibbs IC, Adler JR, et al. Cavity volume dynamics after resection of brain metastases and timing of postresection cavity stereotactic radiosurgery. Neurosurgery. 2013;72(2):180–5; discussion 5.
63. Taunk NK, Oh JH, Shukla-Dave A, Beal K, Vachha B, Holodny A, et al. Early posttreatment assessment of MRI perfusion biomarkers can predict long-term response of lung cancer brain metastases to stereotactic radiosurgery. Neuro Oncol. 2018;20(4):567–75.
64. Jakubovic R, Sahgal A, Soliman H, Milwid R, Zhang L, Eilaghi A, et al. Magnetic resonance imaging-based tumour perfusion parameters are biomarkers predicting response after radiation to brain metastases. Clin Oncol (R Coll Radiol). 2014;26(11):704–12.
65. Tan H, Stewart J, Ruschin M, Wang MH, Myrehaug S, Tseng CL, et al. Inter-fraction dynamics during post-operative 5 fraction cavity hypofractionated stereotactic radiotherapy with a MR LINAC: a prospective serial imaging study. J Neurooncol. 2022;156(3):569–77.
66. Tseng CL, Eppinga W, Seravalli E, Hackett S, Brand E, Ruschin M, et al. Dosimetric feasibility of the hybrid magnetic resonance imaging (MRI)-linac system (MRL) for brain metastases: the impact of the magnetic field. Radiother Oncol. 2017;125(2):273–9.
67. Wen N, Kim J, Doemer A, Glide-Hurst C, Chetty IJ, Liu C, et al. Evaluation of a magnetic resonance guided linear accelerator for stereotactic radiosurgery treatment. Radiother Oncol. 2018;127(3):460–6.
68. Choi CY, Chang SD, Gibbs IC, Adler JR, Harsh GR 4th, Lieberson RE, et al. Stereotactic radiosurgery of the postoperative resection cavity for brain metastases: prospective evaluation of target margin on tumor control. Int J Radiat Oncol Biol Phys. 2012;84(2):336–42.
69. Kutuk T, Kotecha R, Tolakanahalli R, Wieczorek DJJ, Lee YC, Ahluwalia MS, et al. Zero setup margin mask versus frame immobilization during gamma knife((R)) icon stereotactic radiosurgery for brain metastases. Cancers (Basel). 2022;14(14):3392.
70. Ruschin M, Sahgal A, Tseng CL, Sonier M, Keller B, Lee Y. Dosimetric impact of using a virtual couch shift for online correction of setup errors for brain patients on an integrated high-field magnetic resonance imaging linear accelerator. Int J Radiat Oncol Biol Phys. 2017;98(3):699–708.
71. Fabian D, Guillermo Prieto Eibl MDP, Alnahhas I, Sebastian N, Giglio P, Puduvalli V, et al. Treatment of glioblastoma (GBM) with the addition of tumor-treating fields (TTF): a review. Cancers (Basel). 2019;11(2):174.
72. Kotecha R, Odia Y, Khosla AA, Ahluwalia MS. Key clinical principles in the management of glioblastoma. JCO Oncol Pract. 2023;19:180–9.
73. Tseng CL, Stewart J, Whitfield G, Verhoeff JJC, Bovi J, Soliman H, et al. Glioma consensus contouring recommendations from a MR-Linac International Consortium Research Group and evaluation of a CT-MRI and MRI-only workflow. J Neurooncol. 2020;149(2):305–14.
74. Chan JL, Lee SW, Fraass BA, Normolle DP, Greenberg HS, Junck LR, et al. Survival and failure patterns of high-grade gliomas after three-dimensional conformal radiotherapy. J Clin Oncol. 2002;20(6):1635–42.
75. Milano MT, Okunieff P, Donatello RS, Mohile NA, Sul J, Walter KA, et al. Patterns and timing of recurrence after temozolomide-based chemoradiation for glioblastoma. Int J Radiat Oncol Biol Phys. 2010;78(4):1147–55.
76. Mallick S, Kunhiparambath H, Gupta S, Benson R, Sharma S, Laviraj MA, et al. Hypofractionated accelerated radiotherapy (HART) with concurrent and adjuvant temozolomide in newly diagnosed glioblastoma: a phase II randomized trial (HART-GBM trial). J Neurooncol. 2018;140(1):75–82.
77. Nelson DF, Diener-West M, Horton J, Chang CH, Schoenfeld D, Nelson JS. Combined modality approach to treatment of malignant gliomas—re-evaluation of RTOG 7401/ECOG 1374 with long-term follow-up: a joint study of the Radiation Therapy Oncology Group and the Eastern Cooperative Oncology Group. NCI Monogr. 1988;6:279–84.
78. Singh R, Lehrer EJ, Wang M, Perlow HK, Zaorsky NG, Trifiletti DM, et al. Dose escalated radiation therapy for glioblastoma multiforme: an international systematic review and meta-analysis of 22 prospective trials. Int J Radiat Oncol Biol Phys. 2021;111(2):371–84.
79. Tsien CI, Brown D, Normolle D, Schipper M, Piert M, Junck L, et al. Concurrent temozolomide and dose-escalated intensity-modulated radiation therapy in newly diagnosed glioblastoma. Clin Cancer Res. 2012;18(1):273–9.
80. Tanaka M, Ino Y, Nakagawa K, Tago M, Todo T. High-dose conformal radiotherapy for supratentorial malignant glioma: a historical comparison. Lancet Oncol. 2005;6(12):953–60.
81. Nakagawa K, Aoki Y, Fujimaki T, Tago M, Terahara A, Karasawa K, et al. High-dose conformal radiotherapy influenced the pattern of failure but did not improve survival in glioblastoma multiforme. Int J Radiat Oncol Biol Phys. 1998;40(5):1141–9.

82. Gondi V, Pugh S, Tsien C, Chenevert T, Gilbert M, Omuro A, et al. Radiotherapy (RT) dose-intensification (DI) using intensity-modulated RT (IMRT) versus standard-dose (SD) RT with Temozolomide (TMZ) in newly diagnosed glioblastoma (GBM): preliminary results of NRG oncology BN001. Int J Radiat Oncol Biol Phys. 2020;108(3):S22–3.
83. Nelson SJ, Vigneron DB, Dillon WP. Serial evaluation of patients with brain tumors using volume MRI and 3D 1H MRSI. NMR Biomed. 1999;12(3):123–38.
84. Narayana A, Chang J, Thakur S, Huang W, Karimi S, Hou B, et al. Use of MR spectroscopy and functional imaging in the treatment planning of gliomas. Br J Radiol. 2007;80(953):347–54.
85. Pirzkall A, Li X, Oh J, Chang S, Berger MS, Larson DA, et al. 3D MRSI for resected high-grade gliomas before RT: tumor extent according to metabolic activity in relation to MRI. Int J Radiat Oncol Biol Phys. 2004;59(1):126–37.
86. Chang J, Thakur SB, Huang W, Narayana A. Magnetic resonance spectroscopy imaging (MRSI) and brain functional magnetic resonance imaging (fMRI) for radiotherapy treatment planning of glioma. Technol Cancer Res Treat. 2008;7(5):349–62.
87. Parra NA, Maudsley AA, Gupta RK, Ishkanian F, Huang K, Walker GR, et al. Volumetric spectroscopic imaging of glioblastoma multiforme radiation treatment volumes. Int J Radiat Oncol Biol Phys. 2014;90(2):376–84.
88. Laprie A, Ken S, Filleron T, Lubrano V, Vieillevigne L, Tensaouti F, et al. Dose-painting multicenter phase III trial in newly diagnosed glioblastoma: the SPECTRO-GLIO trial comparing arm A standard radiochemotherapy to arm B radiochemotherapy with simultaneous integrated boost guided by MR spectroscopic imaging. BMC Cancer. 2019;19(1):167.
89. Ramesh K, Mellon EA, Gurbani SS, Weinberg BD, Schreibmann E, Sheriff SA, et al. A multi-institutional pilot clinical trial of spectroscopic MRI-guided radiation dose escalation for newly diagnosed glioblastoma. Neurooncol Adv. 2022;4(1):vdac006.
90. Singh K, Saxena S, Khosla AA, McDermott MW, Kotecha RR, Ahluwalia MS. Update on the management of brain metastasis. Neurotherapeutics. 2022;19(6):1772–81.
91. Mohammadi AM, Schroeder JL, Angelov L, Chao ST, Murphy ES, Yu JS, et al. Impact of the radiosurgery prescription dose on the local control of small (2 cm or smaller) brain metastases. J Neurosurg. 2017;126(3):735–43.
92. Baliga S, Garg MK, Fox J, Kalnicki S, Lasala PA, Welch MR, et al. Fractionated stereotactic radiation therapy for brain metastases: a systematic review with tumour control probability modelling. Br J Radiol. 2017;90(1070):20160666.
93. Kim KH, Kong DS, Cho KR, Lee MH, Choi JW, Seol HJ, et al. Outcome evaluation of patients treated with fractionated gamma knife radiosurgery for large (>3 cm) brain metastases: a dose-escalation study. J Neurosurg. 2019:1–10.
94. Brown PD, Ballman KV, Cerhan JH, Anderson SK, Carrero XW, Whitton AC, et al. Postoperative stereotactic radiosurgery compared with whole brain radiotherapy for resected metastatic brain disease (NCCTG N107C/CEC.3): a multicentre, randomised, controlled, phase 3 trial. Lancet Oncol. 2017;18(8):1049–60.
95. Brennan C, Yang TJ, Hilden P, Zhang Z, Chan K, Yamada Y, et al. A phase 2 trial of stereotactic radiosurgery boost after surgical resection for brain metastases. Int J Radiat Oncol Biol Phys. 2014;88(1):130–6.
96. Minniti G, Niyazi M, Andratschke N, Guckenberger M, Palmer JD, Shih HA, et al. Current status and recent advances in resection cavity irradiation of brain metastases. Radiat Oncol. 2021;16(1):73.
97. Redmond KJ, De Salles AAF, Fariselli L, Levivier M, Ma L, Paddick I, et al. Stereotactic radiosurgery for postoperative metastatic surgical cavities: a critical review and international stereotactic radiosurgery society (ISRS) practice guidelines. Int J Radiat Oncol Biol Phys. 2021;111(1):68–80.
98. Li HH, Rodriguez VL, Green OL, Hu Y, Kashani R, Wooten HO, et al. Patient-specific quality assurance for the delivery of (60)Co intensity modulated radiation therapy subject to a 0.35-T lateral magnetic field. Int J Radiat Oncol Biol Phys. 2015;91(1):65–72.
99. Tseng CL, Chen H, Stewart J, Lau AZ, Chan RW, Lawrence LSP, et al. High grade glioma radiation therapy on a high field 1.5 Tesla MR-Linac—workflow and initial experience with daily adapt-to-position (ATP) MR guidance: a first report. Front Oncol. 2022;12:1060098.

# Treatment of Head and Neck Cancers with MR-Linac

# 21

Julia Khriguian, Laila Gharzai, Jolien Heukelom, Brigid McDonald, and Clifton D. Fuller

## Introduction

Advances in magnetic resonance imaging (MRI), image-guided radiotherapy (IGRT), and radiation delivery, such as with MR-Linac (MRL) technology, have revolutionized the field of radiation oncology by providing superior soft-tissue contrast imaging for better tumor and surrounding normal structure delineation and real-time imaging during radiotherapy for monitoring and treatment adaptation. Treatment of head and neck cancers poses a unique challenge to clinicians as tumours are in close proximity to multiple critical organs. Precise radiation delivery to the tumor while sparing healthy tissues is thus paramount to avoid excess treatment toxicity. This chapter will explore the current state of MRL use for the treatment of head and neck cancers and review the advantages, challenges, and potential impact on patient outcomes of this technology as well as future directions.

J. Khriguian · B. McDonald · C. D. Fuller (✉)
Department of Radiation Oncology, The University of Texas MD Anderson Cancer Center, Houston, TX, USA
e-mail: julia.khriguian.med@ssss.gouv.qc.ca; bmcdonald@mdanderson.org; cdfuller@mdanderson.org

L. Gharzai
Department of Radiation Oncology, Northwestern Memorial Hospital, Northwestern University Feinberg School of Medicine, Chicago, IL, USA
e-mail: laila.gharzai@northwestern.edu

J. Heukelom
Department of Radiation Oncology (Maastro), GROW School for Oncology and Reproduction, Maastricht University Medical Centre, Maastricht, Netherlands
e-mail: jolien.heukelom@maastro.nl

## Disease Overview

### Epidemiology and Risk Factors

Head and neck squamous cell carcinomas (HNSCC) account for 90% of all head and neck cancers. HNSCC includes cancers that affect the oral cavity, pharynx, larynx, and sinuses. In this chapter, we will primarily discuss pharyngeal and laryngeal HNSCC as these comprise the majority of patients treated with definitive radiotherapy (RT) with or without chemotherapy. The estimated incidence of HNSCC varies between regions of the world and populations, with higher rates seen in developing countries [1]. In the United States, the estimated incidence of oral cavity and pharyngeal HNSCC is 54,000 new cases per year with a mortality rate of 11,000 deaths per year [2]. The incidence of oral cavity and pharyngeal HNSCC has been increasing by 1% annually, primarily driven by the increased incidence of oral human papillomavirus (HPV) infection [2]. HNSCC is also a male-predominant disease with a 2:1 to 4:1 ratio [2] with the largest difference in incidence being for oropharyngeal



I. J. Das et al. (eds.), *A Practical Guide to MR-Linac*, https://doi.org/10.1007/978-3-031-48165-9_21

SCC (OPSCC). The average age at diagnosis of patients with HPV-related OPSCC is also lower than that for non-HPV-related (mean of 56 versus 63 years old) [3]. In the United States, a 2% per year increase in HPV-related HNSCC incidence was seen in men and 1% in women from 2010 to 2019 [2]. Conversely, there was a 0.8% per year decline in smoking-associated HNSCC [2].

In the United States, it is estimated that 70% of OPSCC are caused by HPV infection [4] with the majority being from HPV16, which often leads to an overexpression of the p16 protein. HPV-related OPSCC is distinct from the rest of HNSCC with better treatment response and prognosis, and this has led to changes in OPSCC staging accounting for HPV status in the new American Joint Committee on Cancer eighth edition. Recent evidence also shows that nearly 20% of hypopharyngeal HNSCC are HPV-associated and may carry a better prognosis [5], although further investigation is warranted. In fact, non-OPSCC p16-positive tumors in the head and neck may also, in general, carry a better prognosis [6–8], although further investigation is warranted. In the setting of nasopharyngeal HNSCC, Epstein-Barr virus (EBV) infection is a major risk factor, and is seen more commonly in endemic countries. The most important risk factors for non-HPV-related HNSCC are tobacco smoking and alcohol consumption. This also has led to the theory of field cancerization in HNSCC [9] due to these common etiologic factors, and high rates of development of secondary HNSCC. Many other environmental and lifestyle risk factors exist as well, depending on the site of HNSCC, but are beyond the scope of this chapter.

## Current Treatment Paradigms for HNSCC

The treatment of HNSCC typically involves a multidisciplinary approach that includes surgery, RT, and/or chemotherapy. The choice of treatment depends on the stage and location of the tumor, the patient's performance status and overall health, and their preferences as different treatment modalities may carry different toxicity profiles. Surgery is more commonly used for the treatment of early-stage HNSCC and may involve the removal of the tumor and surrounding tissue including elective nodal dissection of nodal basins at high risk of harboring micrometastatic nodal disease. In advanced disease, surgery may be more extensive and combined with RT and/or chemotherapy. Due to the morbidity that may be associated with extensive surgeries needed to treat advanced disease, definitive treatment with RT and chemotherapy alone may be preferred.

The most commonly used definitive RT dose regimen for HNSCC is 69.96–70 Gy in 33–35 fractions delivered daily 5 days a week. In early-stage cancers, patients are typically treated with RT alone; in more advanced diseases, patients typically receive concomitant platinum-based chemotherapy (usually cisplatin) [10, 11]. Induction chemotherapy may also be used for stage III/IV nasopharyngeal cancer [12], hypopharyngeal cancer [13], or when tumor shrinkage may be needed prior to local therapy (i.e., decreasing treatment volumes if very extensive disease or near critical organs at risk (OAR)). In cisplatin-ineligible patients, carboplatin [14], docetaxel [15], cetuximab [16], or accelerating RT to six treatments per week instead of five may be considered [17]. In the postoperative setting, 60–66 Gy in 30–33 fractions is given depending on pathologic risk factors and margin status. In cases with positive margins or extracapsular extension (ECE), cisplatin-based chemotherapy is added concomitantly to adjuvant RT. The mainstay therapy for oral cavity and sinus SCC is surgery with or without postoperative RT or chemoRT depending on high-risk pathological features such as advanced T stage, N stage, presence of ECE, close or positive margins, perineural invasion (PNI), and high-grade disease. In cases of laryngeal SCC with extensive thyroid cartilage invasion, up-front total laryngectomy and neck dissection may be preferred prior to chemoRT due to high risk of local failures with chemoRT.

## Changing Landscape of HNSCC Treatments

The landscape of HNSCC treatments is changing as a result of a better understanding of the molecular and genetic mechanisms underlying this heterogeneous disease. One of the key discoveries in HNSCC has been the recognition that a significant proportion of these tumors, particularly arising from the oropharynx, are caused by HPV infection. This has led to a better risk stratification of the disease as these tumors respond better to therapy and carry a better prognosis [18]. Another important discovery is the recognition of the role of EBV in the development of nasopharyngeal cancer and the use of plasma EBV DNA titers to follow treatment response and potentially drive treatment decisions. Both HPV- and EBV-derived biomarkers are currently being investigated in treatment de-escalation studies as these tumors may be more responsive to standard therapy. Additionally, the identification of genetic mutations and other molecular changes as well as the discovery of the tumor microenvironment (TME) has led to the investigation of targeted therapies and immunotherapy in the setting of HNSCC. These discoveries, along with the technologic advances with less invasive surgeries such as transoral robotic surgery and more precise RT with 3D planning with CT and MRI, intensity-modulated RT (IMRT), and volumetric modulated arc therapy (VMAT), as well as IGRT with daily kV imaging, cone beam CT (CBCT), and now MRL, have led to the development of more effective and personalized treatments for HNSCC.

## Overcoming Challenges

The head and neck region is a small anatomic area containing many structures responsible for vital functions such as breathing, swallowing, and phonation. Limiting dose to OARs is particularly challenging due to the close proximity of targets and high-dose areas to these structures. Consequently, RT for HNSCC requires a strong knowledge of anatomy and pattern of disease spread as well as understanding of its unique RT planning complexities. HNSCC subtypes are highly diverse with different molecular profiles and response to treatment. Although HNSCC RT doses and volumes have been relatively stable and very similar throughout most head and neck sites for the past few decades, there is an increasing body of evidence and potentially an upcoming paradigm shift in favor of dose and/or volume adaptation, particularly in the setting of HPV-related oropharyngeal cancers. There is also an increasing awareness of the need for highly conformal and precise RT techniques in order to decrease dose to OARs and limit RT-related toxicities and their consequent impact on quality of life.

The use of MRL in the setting of HNSCC has not yet been well-established and data is still maturing. However, two major areas of improvement in HNSCC treatment may be satisfied by this technique—the ability to provide high-quality anatomic and functional imaging during treatment allowing for a better understanding of this heterogeneous group of tumors and the opportunity for volume and/or dose adaptation based on tumor response allowing for an increase in therapeutic ratio.

## Heterogeneity of HNSCC

Many discoveries in the past two decades have led us to understand that HNSCC constitute a heterogeneous group of tumors with varying genetic and molecular features, locations, and radiosensitivity. Although current standards of treatment have not yet changed to account for these features, many trials are currently underway to investigate treatment personalization by either revisiting the role of different systemic therapies or adapting RT doses and/or volumes. One way to better understand the different behaviors of HNSCC is through pre-, intra-, and post-therapy imaging—an endeavor that may be achieved through MR-guided RT (MRgRT). Treatment personalization may also be facilitated with MRgRT as tumor shrinkage may lead to volume adaptation and functional sequences such as diffusion-weighted (DWI), and perfusion imaging may provide additional information on tumor kinetics allowing for dose de-escalation.

## Complex Anatomy

The head and neck region has a complex anatomy and contains many radiosensitive OARs, such as the salivary glands, mandible, pharyngeal constrictor muscles, brachial plexus, spinal cord, brainstem, and brain, in close proximity to high-dose RT target volumes. High contrast imaging with crisp anatomic resolution is needed for adequate soft-tissue and tumor delineation and primary tumor staging. MRI in particular may be superior to CT in primary tumor delineation in certain instances such as laryngeal cartilage invasion [19] and perineural spread [20]. Retropharyngeal node detection may also be superior with MRI and fludeoxyglucose (FDG)-positron emission tomography (PET/CT) [21]. Any nodal disease may also be better assessed by PET/CT [22, 23]. Ideally, all three imaging modalities should be used for accurate target volume and OAR delineation [24].

## Complex Planning

As the head and neck region has a small surface area, changes in weight and/or tumor size may lead to changes in RT dose distribution warranting RT replanning and volume adaptation. With IGRT, these changes can be monitored more readily. Adaptive RT (ART) is also used when tumors are close to or abutting critical OARs such as the base of skull where the brain, brainstem, and optic chiasm may not tolerate the prescription RT dose. Volume adaptation as the tumor shrinks may thus be particularly helpful in limiting dose to these structures. The optimal time for RT replanning is not well-established, but is frequently done at around fraction 15 [25] as tumors start to have apparent shrinkage or weight loss becomes significant as toxicities such as mucositis, dysphagia, xerostomia, and dysgeusia develop and limit food intake.

Another particularity of HNSCC RT planning is the use of different prescription doses depending on risk of disease involvement. In general, the gross primary and nodal disease receive 70 Gy in 35 fractions, and elective nodal basins receive 56 Gy in 35 fractions (or equivalent dose) using either a simultaneous boost (SIB) or sequential cone-down method to deliver higher tumoricidal doses to gross disease or areas at increased risk of involvement. Some may also use an intermediate dose level of 59.5–63 Gy in 35 fractions (or equivalent dose) to cover areas at intermediate risk of harboring disease or containing suspicious nodes. When using the SIB technique, different areas receive different doses in the same amount of fractions (e.g., 2 Gy/fraction for high-risk volumes, 1.8 Gy/fraction for intermediate-risk volumes, and 1.6 Gy/fraction for low-risk/elective volumes). This was possible with the advent of IMRT and, particularly, VMAT. Conformality may also be better with SIB technique [26]. As VMAT is not possible with MRL, however, a sequential cone-down method is more often employed with MRL treatments. Where there may be an advantage for MRL is for dose-painting based on functional multiparametric MRI, but further investigation is needed to determine feasibility and time points during the course of RT to obtain this imaging.

## Patterns of Failure of HNSCC

Unfortunately, the most common type of failure in HNSCC is local and/or regional with 65% of recurrences arising in the primary/local location and 36% in the nodal/regional location [27]. Most recurrences also occur within the first 3 years post-therapy [27]. Another common occurrence in patients with HNSCC is the development of second primary tumors, with a 20-year cumulative risk of as high as 36% [28]. This is attributed to the relative high incidence of alcohol consumption and smokers among patients with HNSCC in the pre-HPV era. These tumors may also initially be mistaken for recurrences until further investigation is undertaken (e.g., specific pathological assessment of the clonal relation of different tumors).

There is increasing evidence for and use of reirradiation (re-RT) in HNSCC due to the high numbers of local/locoregional recurrences and second primaries in the head and neck region. This is important as it stresses the up-front need to minimize dose to OARs as much as possible and have conformal RT plans as some patients may require additional RT in the future. Altered fractionated RT [29, 30], stereotactic body RT (SBRT) [31], proton therapy [32], and brachytherapy [33] are all options that may be used in cases of re-RT as well in order to limit dose to OARs.

## The Role of Hypofractionation and SBRT in HNSCC

Hypofractionation refers to the use of fraction sizes of >2 Gy/fraction, and SBRT refers to a highly precise technique delivering high doses of RT in typically five or fewer fractions. Moderate hypofractionation has the strongest data in the setting of nasopharyngeal and early glottic HNSCC where fraction sizes of 2.12–2.75 Gy/fraction have been commonly used [34, 35]. The use of SBRT has mostly been reserved for cases of re-RT. The SOAR-HN (NCT03164460) phase II trial is currently accruing patients to compare SBRT to IMRT/proton therapy for recurrent HNSCC in patients having received previous RT. There are also ongoing trials evaluating dose-escalated hypofractionation and incorporation of SBRT boost in the definitive RT setting. The DEHART (NCT04477759) phase I trial is currently investigating the feasibility of hypofractionated dose escalation to target areas with residual tumor and dose de-escalation to areas of tumor regression using MRgRT and concurrent atezolizumab. The MRL-02 (NCT05160714) dose-finding phase I trial is assessing SBRT boost to high-risk subvolumes based on apparent diffusion coefficient (ADC) changes, and the SHORT-OPC (NCT04178174) phase II trial is comparing de-escalated chemoRT with SBRT boost to GTV to standard chemoRT in HPV-positive OPSCC. Lastly, SBRT for early-stage glottic SCC has also shown promise in well-selected patients, with early results showing feasibility [36–38]. However, potential toxicity concerns caused a phase I study to close early due to severe toxicity with vocal cord ulceration and arytenoid cartilage necrosis in 2 patients out of 13 enrolled [38]. The University of Texas Southwestern Medical Center, however, has published encouraging preliminary results [36, 37] and continues to recruit patients to the LT-SABR trial (NCT03548285).

## Current Literature on MRL Use in HNSCC

The current literature on MRL use in HNSCC is limited. Prior to widespread implementation, it should be determined whether local and locoregional control outcomes can be maintained or improved with this new technology and treatment adaptation. Additionally, confirmation is needed to show that dose to OARs can be decreased and that this ultimately results in less RT-related toxicities. Establishing these benefits and the non-inferiority of MRL will justify the use of this technique that is currently resource- and time-intensive. The standardization of IGRT schedules and best MRI sequences for HNSCC would also be beneficial.

Evidence is accumulating for the feasibility of MRgRT for HNSCC, and a few ongoing trials are accruing patients globally, with the majority being for OPSCC. The MD Anderson MRL Development Working Group conducted a prospective in silico study assessing the feasibility and potential dosimetric advantages of MR-guided dose adaptation in HPV-associated OPSCC [39]. Two plans were generated—a standard one and an adaptive one with serial imaging done every 2 weeks. The prescription dose to the initial planning target volume (PTV) was 69.96 Gy in 33 fractions. For the adaptive plan, a PTVadaptive was created (if detectable tumor shrinkage) encompassing the residual tumor, and the previously involved volume received a minimum of 50.16 Gy, which was the same dose used for the elective nodal volumes. This adaptation resulted in a decrease of 12.2 Gy to 95% of the PTVinitial volume (70.7 Gy for the standard plan and 58.5 Gy for the adaptive plan) [39]. Dose to OARs was also reduced, and this translated into a decreased risk of RT-related toxicity, particularly dysphagia, use of feeding tube 6 months post-RT, and hypothyroidism when using the normal tissue complication probability (NTCP) models [39]. The next step is to confirm these findings in vivo in the setting of a trial. McDonald et al. did an initial feasibility study with ten patients and described a clinical workflow for HNSCC MRL [40] with the Unity system, which will be described later in this chapter. Chen et al. also reported their experience with HNSCC MRL with the ViewRay system in both primary and recurrent/second primary disease re-RT setting [41, 42].

The MR-ADAPTOR (NCT03224000), ART-OPC (NCT04901234), MARTHA (NCT03972072), and INSIGHT-2 (NCT04242459) trials are currently accruing and evaluating MRL adapted RT for HNSCC. MR-ADAPTOR is a phase II non-inferiority trial comparing standard IMRT with 69.96 Gy in 33 fractions versus weekly MR-guided dose adaptation and elective volume de-escalation in low-risk HPV-associated OPSCC being treated with RT alone, with the primary endpoint being locoregional control at 6 months [43]. The ART-OPC trial is a phase II trial comparing standard IMRT to MR-guided RT with a single mid-treatment volume adaptation at week 3 in patients being treated with chemoRT with AJCC eighth edition T3-4N0-three tumors. The primary endpoint of this study is patient-reported dysphagia at 6 months post-RT. The MARTHA trial is a single-arm study evaluating RT-induced xerostomia in the setting of daily MR-IGRT with once-weekly offline volume adaptation in patients with stage II–IVB HNSCC requiring bilateral neck RT either definitively or postoperatively. Lastly, the INSIGHT-2 trial is a volume and dose adaptation phase I/II study with three arms (HPV-associated OPSCC, HPV-negative OPSCC, and base of skull HNSCC) evaluating MR-guided ART with volume adaptation at weeks 2 and 4 in all three arms, and dose escalation to HPV-negative disease with suboptimal response to therapy. In the latter arm, treatment effect will be evaluated after fraction 10 based on ADC response with dose escalation being done to the non-responding clinical target volume (CTV).

## Contouring and PTV

Knowledge in anatomy and radiology are crucial for accurate target and OAR contouring in HNSCC. CT, MRI, and positron emission technology CT (PET/CT) are often used to help guide treatment and target delineation. CT is useful to evaluate bone involvement, and contrast helps clearly visualize the tumor and any surrounding tissue changes. PET/CT may be used to confirm or locate gross tumor volumes (GTV), including suspicious nodes [44–46]. MRI is particularly useful for better visualization of soft tissues, bone marrow involvement, retropharyngeal lymph node involvement, and presence and extent of perineural spread [47, 48], all of which affect staging and/or RT planning.

### MRI Sequences

The most commonly used MRI sequences for target delineation are T1-weighted (T1w) and T1 contrast-enhanced (T1wC+), T2-weighted (T2w), DWI, and dynamic contrast-enhanced (DCE). T1w MRI sequences provide anatomical information and are typically used to evaluate the tumor's size and location, particularly when combined with contrast administration and with fat suppression [49], while T2w sequences can be used to evaluate the extent of tumor infiltration and associated edema (i.e., ECE in nearby muscles such as the sternocleidomastoid). DWI sequences provide information on tumor aggressiveness and potential for lymph node involvement by assessing hypercellularity via diffusion restriction with low ADC values, and DCE is used to evaluate the tumor blood flow and vascularity which may correlate with histological grade [47]. Additionally, steady-state free precession (SSFP) sequences provide better assessment of cranial nerves, particularly at the base of skull [48]. Lastly, the true fast imaging with steady-state free precession (TRUFI) sequence, which will be further explained later in this chapter and is a mixture of T1w/T2w images, is often used with the ViewRay system and provides excellent soft-tissue contrast by suppressing background tissue signal while enhancing the signal from tumor.

For MRL treatments, anatomical non-fat-suppressed T2w or T1w/T2w TRUFI scans are used for setup imaging and plan adaptation in the online MRL workflows for HNSCC [40–42, 50]. However, depending on clinical needs, other sequences with improved tumor contrast may be of interest. Salzillo et al. optimized a fat-suppressed T2w MRI sequence for HNSCC imaging on the 1.5T Unity MRL and found that the optimized sequence improved conspicuity of the GTV, lymph nodes, and parotid glands compared to the standard non-fat-suppressed T2w sequence used on the Unity system [51]. Currently, the ViewRay system only offers the TRUFI sequence for anatomical imaging, but other MRI sequences may be useful for disease site-specific applications and should be explored with this system.

T1wC + MRI is another technique that can improve lesion contrast in HNSCC [52], but the safety and feasibility of gadolinium contrast agents used during the daily online adaptive workflow is still being explored. The gadolinium in commonly used contrast agents has been shown in two separate studies to be stable when irradiated to doses up to 100 Gy with 7 MV photons on the Unity MRL [53, 54]. In both of these studies, no un-chelated aqueous gadolinium(III) was found in the irradiated samples, which is toxic and accumulates in the kidneys [55]. Petronek et al. investigated the effects of gadolinium irradiation in mouse glioblastoma xenograft models by injecting gadolinium contrast and irradiating the tumors up to 18 Gy with a 220 kVp small animal irradiator [56]. Based on measured blood urea nitrogen and creatinine levels, no evidence of acute kidney toxicity was found. In addition to the safety of gadolinium contrast during RT, it is also crucial to understand how the presence of gadolinium impacts the radiation dose. In a study investigating dosimetric differences between HNSCC treatment plans generated on CT images without contrast and with simulated gadolinium contrast, the presence of gadolinium resulted in median dose deviations of less than 1.2% for all reported dosimetric parameters [57]. These results are encouraging for the safe and feasible use of gadolinium contrast for daily MRL treatments for HNSCC, but further in vivo studies are needed.

In addition to anatomical sequences, functional MRI techniques such as DWI are of interest for both treatment planning and response assessment using MR-Linacs. Changes in the associated quantitative imaging biomarker, the ADC, during RT have been shown to predict ultimate oncological outcomes in HNSCC [58], so DWI acquired on MR-Linacs may potentially be used for treatment plan adaptation based on mid-therapy response. DWI has been acquired in HNSCC for both the Unity 1.5T MR-Linac [59, 60] and the ViewRay 0.35T MR-Linac [61], although the low-field strength of the ViewRay system limits the signal-to-noise ratio (SNR) and may pose difficulties for accurate and robust ADC measurements. While the 1.5T field strength of the Unity system matches that of many diagnostic MRI scanners, differences in the gradient and receive coil hardware also limit SNR and have major implications for acquisition of DWI on this system [62]. However, McDonald et al. showed that DWI sequences on the MR-Linac had similar in vivo repeatability/reproducibility and phantom SNR values to a 1.5T MR sim [60]. Furthermore, geometric distortion and susceptibility artifacts are major concerns with echo planar imaging, the most common type of DWI sequence. Other types of DWI sequences such as turbo spin echo-based acquisitions may improve spatial accuracy [63] and, when properly optimized, have been shown to have comparable or improved repeatability/reproducibility and SNR compared to echo planar imaging on the 1.5T MR-Linac [60].

## Contouring

In order to be able to safely adapt RT volumes during therapy without compromising tumor control, one must be able to confidently delineate residual tumors. Strong knowledge for patterns of recurrence of HNSCC with nodal drainage based on disease site involvement is also crucial. As opposed to many cancer sites treated with RT, HNSCC treatments often encompass elective coverage of nodal areas at risk of harboring microscopic disease in the neck. Weissmann et al. have proposed an elective nodal volume reduction contouring approach, but validation via incorporation into a prospective randomized trial is still pending [64]. For this chapter, the focus lies on standard and widely used contouring approaches based on published guidelines [65–71]. Table 21.1 describes target volume contours based on disease site. As the main benefits of

**Table 21.1** HNSCC Target Volumes

| Disease site | Target volumes | Notes |
|---|---|---|
| Oropharynx [66, 67] | GTV = GTVp + GTVn<br>CTVhigh = GTV + 5 mm<br>Optional: CTVint = CTVhigh + at least involved lymph node levels<br>CTVlow = CTVhigh + 5 mm and bilateral levels II–IV (include level Ib if base of the tongue or level II node involved, include level V ± retropharyngeal nodes if N1 disease or level IV involved, include retrostyloid nodes is level II node involved)<br>CTV = CTVhigh + CTVint + CTVlow<br>PTV = CTV + 3–5 mm depending on institutional standards | • Contralateral LN RT may be omitted in well-lateralized tonsillar SCC with ipsilateral nodes only and limited to the tonsillar fossa or anterior pillar without base of tongue invasion or soft palate invasion and >1 cm from midline [72–75] |
| Nasopharynx [71] | GTV = GTVp + GTVn<br>CTVhigh = GTV + 5 mm<br>CTVint = CTVhigh + 5 mm and high-risk regions including the entire nasopharynx, posterior 1/3 of the nasal cavity, pterygopalatine fossa, parapharyngeal space, anterior 1/2 of the clivus (or entirety if involved), foramen ovale and rotundum, 1/2 of sphenoid sinus (or entirety if T3–T4 disease), posterior 1/2 of ethmoid sinus, 1/3 of maxillary sinus and cavernous sinus if T3–T4 disease ± at least involved lymph node levels<br>CTVlow = CTVint + bilateral retropharyngeal nodes and levels Ib–V (include retrostyloid nodes if nasal cavity/hard palate/maxillary sinus and/or level II involved—may omit level Ib in node-negative disease)<br>CTV = CTVhigh + CTVint + CTVlow<br>PTV = CTV + 3–5 mm depending on institutional standards | • Unlike in the other disease sites, the use of CTVint is commonly used for nasopharyngeal SCC<br>• In the post-induction chemotherapy setting, the pre-induction volume is included in CTVint [76] |
| Larynx and hypopharynx [66, 67] | GTV = GTVp + GTVn<br>CTVhigh = GTV + 5 mm and including the entire larynx from the top of the thyroid cartilage to the bottom of the cricoid cartilage<br>Optional: CTVint = CTVhigh + at least involved lymph node levels<br>CTVlow = CTVhigh + 5 mm and bilateral levels II–IV (include retropharyngeal nodes, levels Ib, V, and VI if hypopharyngeal SCC, ipsilateral level Ib if base of the tongue or level II involved, level VI if subglottic extension and/or anterior soft-tissue extension and/or if non-oncologic emergency tracheostomy was performed)<br>CTV = CTVhigh + CTVint + CTVlow<br>* CTVint and CTVlow are not applicable for early glottic SCC where elective nodal RT is not performed<br>PTV = CTV + 3–5 mm for nodal volumes and up to 10 mm for primary tumor volume depending on institutional standards (may also consider non-isotropic expansion for laryngeal SCC primary, accounting for glottic movement with deglutition and breathing [77, 78] | • Although guidelines have been published including smaller volumes for early glottic SCC not including the entire larynx, this has yet to be validated in a prospective study. The VOCAL-RT is currently open and addressing this question of vocal cord versus complete larynx RT (NCT03759431)<br>• In the post-induction chemotherapy setting, the pre-induction volume may be included in CTVint |

*GTVp* primary gross tumor volume, *GTVn* nodal gross tumor volume, *CTVhigh* high-risk clinical target volume, *CTVint* intermediate-risk clinical target volume, *CTVlow* low-risk clinical target volume, *PTV* planning target volume

MR-guided ART are primarily suspected in the definitive setting, postoperative and oral cavity SCC volumes will not be discussed in this chapter. In certain centers, CTV intermediate risk may be omitted in non-nasopharyngeal HNSCC. Further, as VMAT is not currently possible with either of the commercially available MRLs, it may be simpler to use two prescription dose levels only.

## PTV Margins

PTV margins are often specific to the treatment machine, IGRT schedule/modality, and immobilization devices and consequently can vary widely from one institution to another. In HNSCC, tightening PTV margins is particularly important in order to minimize dose to OARs in close proximity to high-dose regions and decrease treatment-related toxicity [79]. This is achieved by limiting inter- and intra-fraction motion with the use of rigid immobilization and daily onboard image verification. MRL machines and daily MR-guided IGRT provide an opportunity for PTV margin reduction as daily images provide superior soft-tissue contrast and inter-fraction uncertainty is decreased. However, prior to establishing an optimal PTV margin expansion, MRL-compatible devices and appropriate IGRT protocols/schedules must be standardized, which is an area of ongoing investigation.

## OAR/Dose-Volume Histogram (DVH) Constraints

OARs are a unique challenge in the head/neck region due to the number of structures important for vital function such as breathing, phonation, and swallowing. HNSCC treatment planning has typically consisted of defining critical structures such as the brainstem and optic nerves, which are prioritized over tumor coverage, as well as lower priority structures such as swallowing muscles and/or salivary glands for which tumor coverage is prioritized.

## OAR Contouring

OAR contouring follows standard computed tomography (CT)-based guidelines [69], and all OAR structures seen on CT are readily identifiable with MRI (either 0.35 T or 1.5 T). CT-based consensus guidelines for OAR contouring exist [69] (Figs. 21.1 and 21.2) with concise descriptions of the various OARs within the head/neck region. Critical OARs in the head/neck region include the spinal cord, brainstem, brain, optic nerves, and chiasm. Important noncritical OARs include the parotid glands, submandibular glands, oral cavity, lips, mandible, pharyngeal constrictors, larynx, esophagus, thyroid gland, and swallowing muscles. No MRI-based contouring guidelines exist, and the nuances of identifying structures on MRI as compared to CT have not yet been defined.

## Constraints

The Quantitative Analysis of Normal Tissue Effects in the Clinic (QUANTEC) [69, 80] review updated the classic paper by Emami et al. [81] and provided more modern dose constraints in the three-dimensional era. While this provided important dose constraints relevant for treatment planning the head/neck region, the QUANTEC review did not provide a comprehensive overview of OAR constraints specific to the head/neck region, including lacking guidance on structures such as the mandible, brachial plexus, oral cavity, and submandibular glands. Ongoing head/neck clinical trials by NRG/RTOG provide additional input on modern dose constraints. The three-arm NRG HN005 trial (NCT03952585) examining IMRT dose reduction and altering systemic therapy offers additional guidance on critical structures (Table 21.2).

More recent studies in the IMRT era have suggested that as compared to traditional 3D-conformal treatment planning, tighter dose constraints may be met with acceptable target coverage and may minimize long-term toxicities. A single-institution retrospective review from the

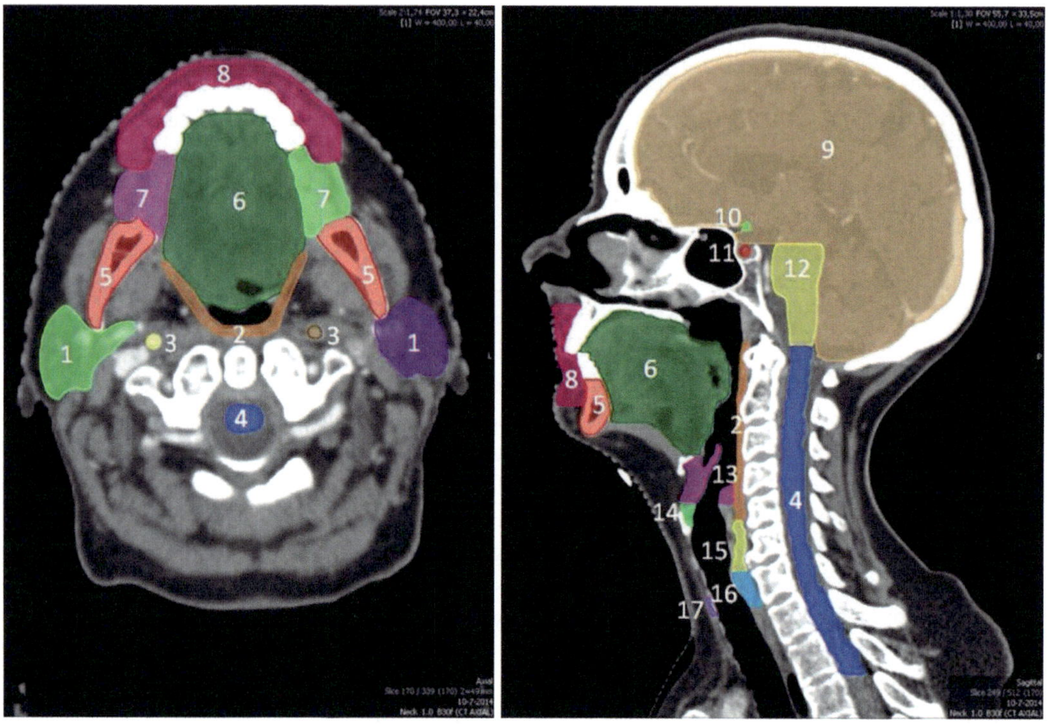


**Fig. 21.1** Axial (left) and sagittal (right) view of the consensus delineations of the parotid glands (1), pharyngeal constrictor muscles (2), carotid arteries (3), spinal cord (4), mandible (5), extended oral cavity (6), buccal mucosa (7), lips (8), brain (9), chiasm (10), pituitary gland (11), brainstem (12), supraglottic larynx (13), glottic area (14), cricopharyngeal inlet (15), cervical esophagus (16), and thyroid (17) (Reprinted from Radiotherapy and Oncology, Vol. 117, Charlotte L. Brouwer, Roel J.H.M. Steenbakkers, Jean Bourhis, Wilfried Budach, Cai Grau, Vincent Grégoire, Marcel van Herk, Anne Lee, Philippe Maingon, Chris Nutting, Brian O'Sullivan, Sandro V. Porceddu, David I. Rosenthal, Nanna M. Sijtsema et al., CT-based delineation of organs at risk in the head and neck region: DAHANCA, EORTC, GORTEC, HKNPCSG, NCIC CTG, NCRI, NRG Oncology and TROG consensus guidelines, p. 8, Copyright (2015), with permission from Elsevier [69])

University of Pennsylvania [82] examining treatment plans in postoperative OPSCC found that dose to various organs could be consistently reduced, offering additional tighter dose constraints. For example, mean dose to the larynx can usually be reduced to a mean of 20 Gy, down from 40 Gy as suggested by NRG and 50 Gy as suggested by QUANTEC (Table 21.2).

Despite these constraints, there remain a number of acute toxicities which are challenging to manage or prevent during treatment, such as dysgeusia [87]. With improvements in outcomes for patients with head/neck cancer in the modern era, particularly in the setting of increased incidence of HPV-related OPSCC, the effect of longer-term toxicities on patients' quality of life is becoming increasingly important. There remain a number of longer-term toxicities for which dose constraints are lacking, such as minimizing the effect of radiation-induced lymphedema or trismus through contouring structures such as the anterior neck or pterygoid muscles, respectively. The improved soft-tissue delineation of the MRI may allow for improvements in understanding of dose constraints to prevent these acute and late toxicities.

## Planning Challenges

The novel techniques utilized to integrate an MRI into a linear accelerator present challenges for the creation of optimal radiation plans during treatment planning, which is already difficult due

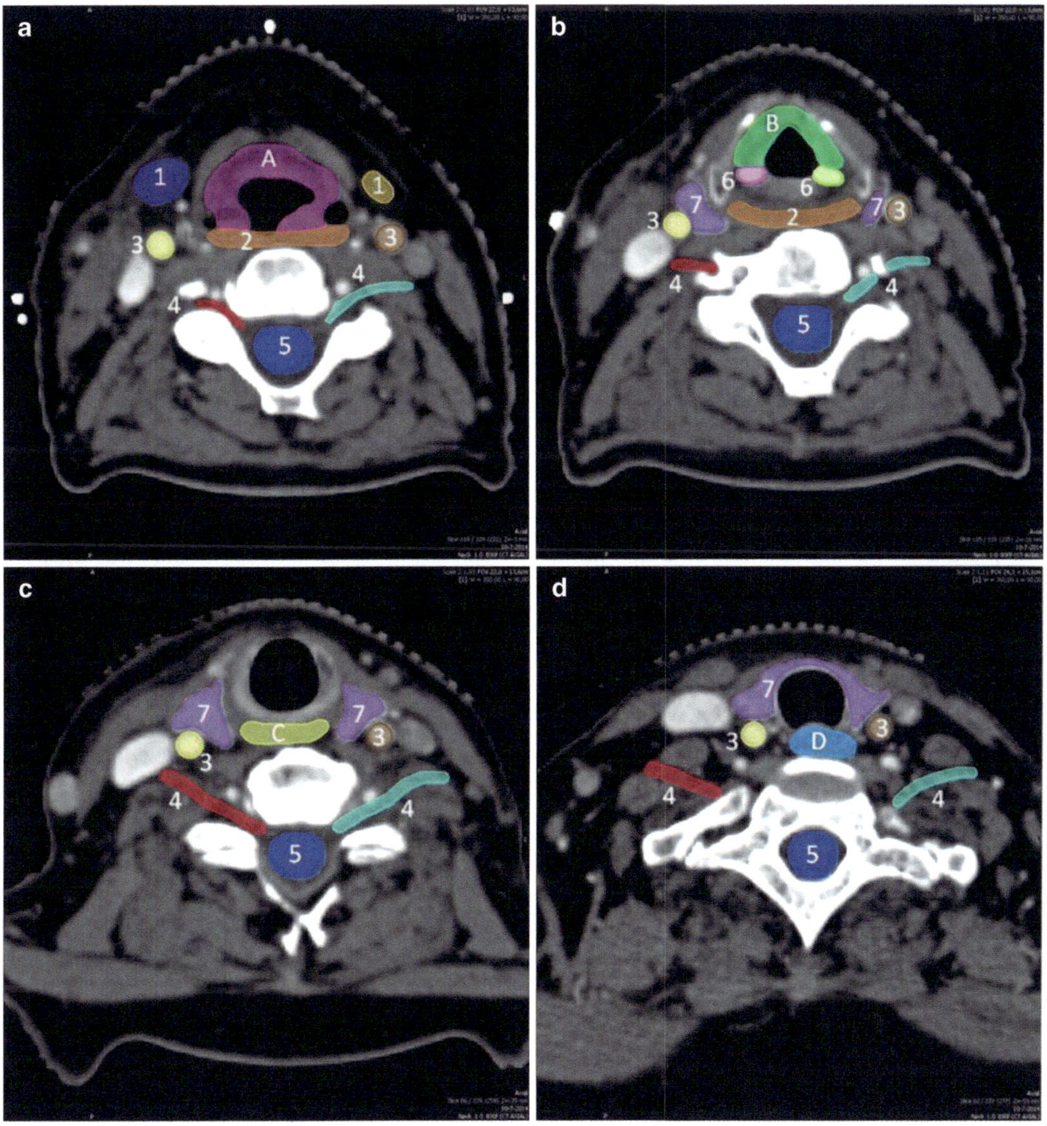


**Fig. 21.2** Axial CT slices showing the delineation of the supraglottic larynx (A) (**a**), glottic area (B) (**b**), cricopharyngeal inlet muscle (C) (**c**), and cervical esophagus (D) (**d**). Other organs at risks visible are the submandibular glands (1), pharyngeal constrictor muscles (2), carotid arteries (3), brachial plexus (4), spinal cord (5), arytenoids (6), and thyroid (7). (Reprinted from Radiotherapy and Oncology, Vol. 117, Charlotte L. Brouwer, Roel J.H.M. Steenbakkers, Jean Bourhis, Wilfried Budach, Cai Grau, Vincent Grégoire, Marcel van Herk, Anne Lee, Philippe Maingon, Chris Nutting, Brian O'Sullivan, Sandro V. Porceddu, David I. Rosenthal, Nanna M. Sijtsema et al., CT-based delineation of organs at risk in the head and neck region: DAHANCA, EORTC, GORTEC, HKNPCSG, NCIC CTG, NCRI, NRG Oncology and TROG consensus guidelines, p. 8, Copyright (2015), with permission from Elsevier [69])

to the number of critical structures in the head and neck region. Treatment planning typically requires balancing competing priorities of tumor coverage and sparing nearby OAR. Many standard head and neck treatment plans now utilize VMAT to balance these competing priorities, which offers an optimal method for sparing critical organs with faster treatment times [88] as compared to IMRT. However, VMAT cannot be delivered with an MRL due to the machine's

**Table 21.2** OAR/DVH constraints

| Structure | QUANTEC [80] | NRG HN005 (NCT03952585) | Single institution |
|---|---|---|---|
| Brainstem | Max ≤54 Gy | Max ≤50 Gy | N/A |
| Spinal cord | Max ≤50 Gy | Max ≤45 Gy | N/A |
| Brachial plexus | N/A | Max ≤66 Gy | N/A |
| Parotid | Mean ≤ 25 Gy | Mean ≤ 26 Gy | Mean ≤ 20 Gy [82] |
| Larynx | Mean ≤ 50 Gy | Mean ≤ 40 Gy | Mean ≤ 20 Gy [82] |
| Pharynx/constrictors | Mean ≤ 50 Gy | Mean ≤ 45 Gy | Mean ≤ 40 Gy [82] |
| Oral cavity | N/A | Mean ≤ 35 Gy | Mean ≤ 20 Gy [82] |
| Esophagus | Mean ≤ 34 Gy | Mean ≤ 30 Gy | Mean ≤ 20 Gy [82] |
| Submandibular gland | N/A | Mean ≤ 39 Gy | Mean ≤ 30 Gy [82] |
| Cochlea | Mean ≤ 45 Gy | Mean ≤ 35 Gy | Mean ≤ 35 Gy [83] |
| Mandible | N/A | Max ≤73.5 Gy | V35 Gy < 30% [84] |
| Thyroid | N/A | Mean ≤ 50 Gy | Mean ≤ 42 Gy [85]<br>V50 Gy ≤ 24% [86] |

inability to rotate from the weight of the shielding present on the machine. Despite this limitation, early work on treatment plan comparison with IMRT on an MRL suggests that high-quality treatment plans are feasible [89].

The number of critical organs present in the head and neck region presents a significant challenge. Most treatment plans require multiple target volumes, nearby OARs, and numerous optimization structures, and the balancing of so many structure constraints has implications for plan quality. For typical VMAT planning, patients typically have 2–3 target volumes, approximately 10 OARs, 5–6 optimization structures (such as rings), and approximately 10 substructures for overlapping areas (e.g., left parotid minus a nearby target area). Finally, additional structures are often drawn to account for areas of hot spots and cold spots during planning, for a total of over 30 structures for a typical plan.

For the ViewRay MRL system, in particular, treatment planning is currently limited to 32 structures. Additionally, optimization structures within the ViewRay planning system must be rule-based, and the rule-based operations do not allow for complex operations. For example, to create a 1 cm ring around a target volume, three structures must be created: a planning volume plus 0.2 cm, a planning volume plus 1.2 cm, and then a third structure created by the subtraction of the previous two. For multiple dose levels, this rule-based requirement creates a need for multiple optimization structures for each dose level. Typically at least four additional structures are often drawn during treatment planning: two structures of air and water to allow for density overrides and an optional tracking structure and accompanying boundary structure. Thus, a relatively simple head and neck plan often rapidly approaches the structure limits available for the ViewRay system. Newer software updates on the ViewRay system may allow for modest increases in these structure limits. Additionally, treatment plans utilizing the ViewRay system are typically heterogeneous, with higher hot spots than seen in VMAT plans. Thus, numerous planning challenges exist that may degrade the quality of treatment plans available utilizing the MRL system.

## Dose Escalation

Locoregional recurrences are the most common type of treatment failure for HNSCC, with distant metastases seen less often as the first site of relapse [90]. The locoregional failures predominantly occur inside the high-dose volumes [91]. It has been stipulated that high-risk subvolumes within the high-dose/GTV volume might be the cause of the recurrences, as primary tumors are known to be heterogeneous. Identification and subsequent targeting of these subvolumes might lead to increased tumor control, although this is challenging and remains to be elucidated. The determination of the exact location of these subvolumes is critical but difficult. To achieve this, identification of several possible high-risk subvolumes has been proposed using PET/CT scans. Examples of these are hypoxic and/or FDG-avid regions in which a tumor may be more likely to recur [91, 92]. In the last decade, the increased use of MRI has led to improved knowledge of tumor characteristics and behavior. Therefore, MR-guided RT trials may lead to the discovery of other high-risk subvolumes. Dose escalation to such subvolumes (or to the whole primary tumor) is the topic of several ongoing clinical trials. Both the boost subvolume and the optimal boost dose are currently being investigated.

## Optimal Dose

To date, there is no consensus on the optimal escalation dose and is a field of active investigation. A pioneering trial in the field of dose escalation in search of an optimal dose was conducted by Madani et al. [93]. In this phase I study, FDG-PET dose painting by number approach was used, and the maximum median tolerated dose was determined to be 80.9 Gy [93]. Higher dose was tolerated in the acute phase, but led to ulcers 4–10 months after treatment for several patients receiving the higher doses. Long-term follow-up did not show improved survival after treatment with higher doses [94].

Another FDG-PET-based dose escalation study is the ARTFORCE trial (NCT01504815). In this trial, dose to a high FDG uptake subvolume is escalated to a maximum total dose of 84 Gy to 2% of the subvolume, with a minimum dose of 70 Gy and a mean dose of approximately 77 Gy [95]. Interestingly, this study does not boost the entire GTV but a subvolume defined by high FDG uptake and does not randomize between a boost directed at the area of high FDG uptake and an alternate subvolume determined otherwise. However, this study design does not allow us to differentiate whether the effect of a random boost might achieve the same results as specifically boosting the area of higher FDG uptake (i.e., determining if the increased dose is more important than the location of the boost).

Several years after the start of accrual to the aforementioned trials, the MRL-02 study opened (NCT05160714). This was a novel trial because it is the first HNSCC dose escalation trial done on the MRL. It aims to define the maximal tolerated dose in an ADC-based subvolume using an adaptive SBRT boost. A real-time adaptive boost to the ADC-high-risk volume is delivered once a week in this study. This boost is gradually increased using four dose levels of 5 × 5 Gy, 5 × 5.5 Gy, 5 × 6 Gy, and 5 × 6.5 Gy in a classical 3 + 3 dose-finding design where three patients per radiation dose level are enrolled. The primary endpoint is the dose-limiting toxicity, and regional tumor control is a secondary objective.

Another MRI-based phase I dose escalation study is PIRATES study which investigates dose escalation with particle therapy [96]. The study is a safety and feasibility trial of MRI-guided adaptive particle RT [96]. Interestingly, this trial includes a boost regime (MRI guided at week 5–7) as well as a hyperfractionation regime with twice-daily fractions. To what extent results from MRI-directed dose escalation with proton therapy are applicable to photon therapy on a MRL device is unclear. It is to be noted that this trial did not use MRL but instead used high-frequency offline MRI simulation studies for adaptation guidance.

Lastly, the University of Michigan published a multicenter randomized phase II study assessing physiologic MRI-based dose escalation in patients with poor prognosis HNSCC and showed decreased locoregional failure with this technique (40% versus 18% at 2 years) [97]. However, no significant difference was seen in disease-free survival, overall survival, or toxicity. The boost subvolume was based on low blood volume/ADC maps at fraction 10 of RT and was escalated to 80 Gy [97].

The implicit assumption of many dose escalation trials is that optimizing tumor control by dose escalation does not require increasing the dose uniformly across the primary tumor. Partly driven by the attempt to spare surrounding OARs, these trials will oftentimes only escalate dose to a predefined subvolume assumed to harbor an increased risk for tumor recurrence. Finding the optimal subvolume to boost is an important topic of investigation, as well as which imaging modality/sequences to use in order to determine such subvolumes. Dose escalation trials, particularly those in search of an optimal boost subvolume, have thus developed alongside the adoption of increasingly advanced imaging techniques.

For a long time, a PET/CT scan was the most informative scan regarding biological behavior of HNSCC. FDG has been and remains the most commonly used tracer for this disease. Studies have shown that tumor recurrences occur in the FDG uptake areas [91]. Ensuingly, the investigation of a boost to this area seems logical. Both the trial by Madani et al. [93] and the ARTFORCE trial direct the high dose to the high FDG uptake region. Moreover, in the original protocol, the ARTFORCE study included a PET/CT scan with zirconium-labeled cetuximab (89Zr-Cetuximab) to investigate whether high uptake on that scan was predictive of tumor response to cetuximab, and to assess whether low uptake might indicate patients with tumors requiring higher RT dose for disease control. This intervention, however, was taken out of the protocol after it was amended. Other PET/CT tracers that have been coined as possibly indicative of high-risk areas for tumor recurrence and include areas of low Zr89-cetuximab uptake [98], FAZA [99], and F-miso [100, 101].

Recent advancements and the increased availability and adoption of MRI scanners have created new opportunities for MRI-guided dose escalation. Examples of MRI-guided dose escalation trials include the MRL-02 and INSIGHT trials for photon therapy and the PIRATES study for proton therapy. The MRL-02 and PIRATES studies have been described above. The INSIGHT-2 study (NCT04242459) includes patients with OPSCC and base of skull tumors. It is a phase I/II trial consisting of three cohorts (HPV-associated OPSCC, HPV-negative OPSCC, and base of skull HNSCC). Prescribed doses are 54 Gy in 30 fractions to nodal areas at risk of harboring microscopic disease and 65 Gy in 30 fractions to primary and nodal tumor volume. Dose to HPV-associated OPSCC and base of skull HNSCC is to be adapted based on anatomical and/or volumetric changes during therapy. Of note, this is not quite dose escalation for these tumors. The dose to HPV-negative tumors is, however, escalated based on a classification as responders vs. nonresponders, which is based on ADC response at week 2 of chemoRT. Responders get target volumes adapted at weeks 2 and 4 based on anatomical changes, and nonresponders get an increase in dose/fraction to the CTV for fractions 11–30. The INSIGHT-2 trial is also a good example of how dose escalation to volumes based on pre-treatment images can be substituted by ART to strategically increase dose to the primary tumor. This strategy can be used to direct dose with online (real-time) adaptive RT treatments, as done with MRL devices [102]. Interestingly, these same devices have also opened up possibilities for ART with the purpose of dose de-escalation for tumors with a good prognosis.

## Dose De-escalation

Dose de-escalation for HNSCC may be done on two levels: to the nodal volumes and to the primary tumor based on imaging and pathologic risk factors. Dose de-escalation to elective nodal fields is an area of active investigation. The general idea is that increased accuracy of diagnostic imaging and RT techniques may allow safe dose de-escalation to the elective nodal volumes without compromising tumor control, potentially reducing treatment-related toxicity. Many of the elective doses that are being used currently are based either on historical experience or on trials that relied on physical examination to determine the nodal status in the head and neck region. Because clinical practice and available imaging are more accurate nowadays, a node-negative neck in the current era is thought to harbor a lower risk of involvement than node-negative necks from the era of trials done decades ago, and, therefore, may be treated effectively with less dose than was used at the time. Examples of trials that investigated this hypothesis include the UPGRADE-RT trial [103] and a study by Deschuymer et al. [104]. Additionally, for patients with HNSCC with a relatively good prognosis (i.e., HPV-associated OPSCC [18]), dose de-escalation to the primary tumor might be feasible. This can be achieved by an up-front dose reduction to the target volumes or by implementing ART regimens where the treatment volumes are cropped and adjusted according to anatomic and volumetric changes during RT with resulting dose reduction to surrounding areas in consequence (e.g., ARTreduco trial [102]).

Based on retrospective studies, dose de-escalation to the primary tumor seems feasible and safe for specific low-risk patient cohorts [105]. Clinical trials are currently ongoing. An example is the pilot DELPHI study (NCT05544136), in which dose to the high-risk volume is lowered stepwise from 66 to 55 Gy. Another example of a dose de-escalation trial is from Memorial Sloan Kettering (NCT05544136) where positive nodes receive an experimental, lower-than-standard dose of 50 Gy, 44 Gy, or 40 Gy if 18F-FMISO PET is also hypoxia-negative. Hypoxic patients are taken off protocol and receive the standard 70 Gy. Depending on the results, these studies may prelude online real-time adaptive dose de-escalation strategies guided by functional imaging.

In conclusion, the employment of dose escalation as well as dose de-escalation for HNSCC is an area of active investigation and may potentially offer important disease control and toxicity benefits. We are currently awaiting more data on dose (de-)escalation with multiple trials open for enrollment. These trials will establish which HNSCC (sub)volumes need to be boosted or treated less aggressively, the sequences to be used for their identification as well as the optimal dose to which they should be treated.

## On-Table Adaptation

Both currently available commercial MRL platforms support online treatment plan adaptation for HNSCC patients based on both positional and anatomical changes throughout RT. The exact workflow differs between the Unity and ViewRay MRIdian systems and may vary among different users of the same device, but many considerations remain the same. Site-specific workflows for HNSCC have been described by McDonald et al. [40] and Gupta et al. [50] for the Elekta Unity and in two separate studies by Chen et al. [41, 42] for the ViewRay MRIdian cobalt-60 unit. For both systems, patients undergo CT and/or MR simulation for treatment planning purposes prior to the start of treatment. The treatment plan created on the pre-treatment image serves as the reference plan for future adaptive plans created during the online workflow. After the first treatment, any prior treatment plan may be selected as the reference plan, including the initial (pre-treatment) plan.

### Elekta Unity Online Workflow

First, the patient is positioned on the treatment table with a custom thermoplastic head and neck immobilization device. The treatment area is imaged using a T2w MRI sequence. Next, the daily MRI is rigidly registered to the reference image in Monaco, the associated treatment planning system, to determine the isocenter shift to account for daily setup variations and any positional changes. Because the Unity table can only move in a single direction (in and out of the bore), the isocenter shift must be incorporated into the adaptive plan (i.e., "virtual isocenter shift").

Next, the user decides between the Adapt-to-Position (ATP) or Adapt-to-Shape (ATS) workflows [106]. ATP applies the virtual isocenter shift to the reference plan and either recalculates the dose from the reference plan or reoptimizes the reference plan based on the new target position. However, ATP does not allow segmentation on the daily setup image and uses the reference image for dose calculation. In ATS, the reference image and daily image are deformably registered and the contours propagated onto the daily image. The target volume and OAR contours may be edited to account for anatomical changes throughout RT. Electron density information for dose calculation is transferred to the contours on the MRI using bulk density assignment [106]. The treatment plan is then reoptimized on the current anatomy. The segmentation editing step tends to be a major bottleneck of the online ATS workflow due to the large number of structures in the head and neck region and the poor performance of the current deformable image registration algorithm within Monaco [40, 50]. For this reason, some sites have adopted an offline ATS/online ATP strategy in which a prior online image is used to create an ATS plan offline that is used as the reference plan for ATP during the next treatment [40]. An alternative "ATS-Lite" approach was proposed by Gupta et al. [50] and involves daily online ATS using deformable image registration for the external patient contour but rigid registration for all other structures with weekly contour review by a physician.

After the treatment plan has been reoptimized and/or recalculated, the dose distribution and DVHs are reviewed by the clinician. The dosimetric criteria and DVHs for both the reference plan and the adaptive plan can be viewed side by side in Monaco for easy comparison. Factors considered for plan evaluation are discussed in the following section ("Evaluation"). Once the plan has been approved by the clinician, the beam is delivered. A two-dimensional balanced fast-field echo MRI sequence in three orthogonal directions is used for motion monitoring of the target volume or a reference OAR during beam delivery [107]. The Unity system is currently not capable of motion gating, but the beam can be stopped manually if gross motion is observed. However, that is less of a concern for HNSCC patients in an immobilization mask unless severe weight loss during RT causes the mask not to fit properly [108].

The median (range) total treatment time was 46 (31–85) min in the online ATP workflow reported by McDonald et al. [40]. However, these treatment times were for the first ten HNSCC patients treated on the MRL at their institution, and treatments may now be streamlined with more experience. For the first two patients treated with the ATS-Lite approach reported by Gupta et al. [50], the median (range) treatment times were 40 (28–47) and 32 (25–44) min.

## ViewRay MRIdian Online Workflow

Similar to the Unity workflow, patients are first set up on the table with a custom immobilization mask and imaged using a true fast imaging with steady-state free precession (TRUFI) sequence, which has a combined T1 and T2 contrast weighting [41, 42]. The daily image is fused with the reference image. If any isocenter shifts need to be performed, the table can be physically moved in all three dimensions. Next, contours are propagated to the daily image with either rigid or deformable registration, and the contours may be edited as necessary. Electron densities are transferred using bulk density assignment. The dose distribution from the reference plan is then recalculated on the current anatomy, and the dose distribution and DVHs are reviewed and compared to the reference plan. The clinician chooses to treat with the reference plan or to proceed with treatment plan adaptation. If they decide to adapt the plan, it undergoes reoptimization and is subsequently reviewed by the clinicians and delivered after final approval [109]. During beam delivery, motion is tracked with a two-dimensional TRUFI sequence for beam gating. The beam will automatically shut off if the target falls outside of a pre-specified margin for a given percentage of time [110].

In the reports by Chen et al. [41, 42], the majority of patients did not undergo any adaptive replanning and were treated with the reference plan during every fraction. Although adaptation was generally not performed, treatment with online MR-guided RT was particularly advantageous for patients undergoing re-RT of recurrent and new primary HNSCC [41]. The enhanced soft-tissue visualization during treatment setup allowed clinicians to reduce PTV margins, limiting dose to previously irradiated normal tissues, and improving confidence in the precision of each delivered treatment. However, Chen et al.

stated that online ART for HNSCC is an area of active investigation within their institution [42]. Furthermore, in these studies, motion was tracked for observational purposes during treatments, but no margin was set for automatic beam termination based on target motion.

## Additional Considerations for On-Table Adaptation

### Treatment Plan Adaptation Strategies

Beyond the general workflows for MR-guided ART for HNSCC described above, the actual treatment plan adaptation paradigm can vary considerably depending on clinical needs. Many treatment factors are not currently standardized, such as the frequency of adaptive replanning and whether target volumes and/or prescription doses should be modified throughout treatment. Heukelom and Fuller have introduced a nomenclature for describing various clinical ART intentions including serial plan verification, dose escalation to tumors or tumor subvolumes, dose de-escalation for OAR sparing, and dose modification based on shrinking tumor volumes [102]. They also describe implementation techniques such as fixed-interval adaptation and adaptation triggered by plan deviation from a pre-specified threshold such as tumor volume, target dose coverage, or accumulated dose. As MR-guided ART becomes increasingly utilized for HNSCC, it is important to report specific details about plan adaptation in any studies and to prospectively evaluate outcomes of each adaptation strategy with clinical trials. A number of recently completed and current clinical trials for MR-guided ART have been described previously in this chapter.

### Use of CT Vs. MRI for Reference Plan

On both MRL platforms, either CT or MRI can be used as the planning image for the initial reference plan. The primary advantage of CT is that the pixel values contain quantitative information that can be converted to electron densities for dose calculation. Also, since CT simulation is universally used for standard RT workflows, many clinics have implemented autosegmentation tools such as deep learning models or atlas-based autosegmentation for contouring OARs on CT, which speed up the treatment planning process considerably [111, 112]. A number of research-grade autosegmentation tools have shown promise for use on MRI, but a large volume of high-quality training data/atlases and thorough validation are needed for these tools to be clinically implemented [113, 114].

MRI can be used for the reference plan instead of CT as done currently and may potentially streamline the online adaptive process for HNSCC. MRI offers better soft-tissue visualization compared to CT, which can potentially improve segmentation accuracy. Also, one of the major challenges in online plan adaptation for HNSCC is that extensive manual contour correction is often necessary after propagating contours from the reference image to the daily image. McDonald et al. [40] compared the accuracy of OAR contours propagated via deformable image registration in Monaco between the pre-treatment CT or MRI and daily MRI images throughout RT. The results showed that MR-to-MR outperformed CT-to-MR on average for all structures, but that some individual outlier cases demonstrated very poor registration accuracy with MR-to-MR. Nonetheless, using MRI for the initial reference plan may help improve contour propagation accuracy for many patients, thus reducing the time spent manually correcting contours while the patient is on the table.

One practical consideration of MRI-based treatment planning is that it requires electron density mapping for dose calculation. In both MRL treatment planning systems, bulk density assignment is used, which involves assigning a single electron density value to all voxels within a structure [106, 109]. The density value can be the mean value from the patient's CT simulation if it is available, or it can be provided from a population-based look-up table. While no studies have specifically evaluated bulk density assignment in MRI using structure-specific density values, studies that assigned bulk density values to water (soft tissue), bone, and air resulted in dose differences within 2.5% of CT-based dose calculation across all target structures [115, 116]. Chin et al. noted that dental implants were not adequately visualized on MRI and would require additional density corrections for accurate dose calculation [115]. Still, if standardized electron density values are used for each structure type, the MRI simulation can be used for the reference plan without the need to acquire a CT simulation, which reduces the time burden on both patients and clinical staff.

### Intrafraction Motion

The Elekta Unity system is currently capable of real-time motion monitoring during treatment delivery [107], while the ViewRay MRIdian system has both motion monitoring and beam gating capabilities (although only in one plane currently) [110]. Unlike Unity, in which the user would have to manually terminate the beam if gross motion is observed, the ViewRay beam can be set to automatically turn off if the target falls outside of a specific margin for a certain percentage of time. The current implementation of the ViewRay allows tracking/gating in a single plane, but the next software release will enable two-plane tracking/gating. For both systems, the ability for the beam to move [42] in real time with the moving target are is an area of active research and development [117–121]. While these features may be especially beneficial for lung and abdominal tumors subject to large displacements from respiratory motion, their utility for HNSCC may be more limited because patients are well-immobilized in thermoplastic head and neck masks.

However, tumors are not completely static in the head and neck; they may be displaced by respiratory, swallowing, and/or tongue motion [42, 108, 122]. Bruijnen et al. quantified the patterns and extent of motion in nasopharyngeal, oropharyngeal, and laryngeal tumors and found that laryngeal tumors had the largest overall maximum displacements and nasopharyngeal tumors the smallest [108]. They used the displacements quantified among a large cohort of patients to calculate modified PTV margins, but they noted that tumor motion varies considerably among individual patients and that the population-based PTV expansions may not be suitable for every patient. Bradley et al. assessed the impact of swallowing motion and duration on delivered dose in HNSCC RT [123]. For most patients, swallowing was infrequent enough to have minimal impact on delivered dose, but for patients with long and frequent swallows, they estimated that delivered dose to the PTV could be reduced by up to 10% in the worst-case scenario. Based on these results, real-time motion tracking and adaptation could be beneficial for head and neck cancer patients with large tumor displacements and long swallowing durations to ensure that the full dose is delivered during each treatment.

### MRI Coils

The MRI coils used for imaging have important implications for both radiation treatment planning and image quality. The Unity system uses a 2 × 4-channel radiotranslucent coil system, with a posterior coil segment beneath the table and an anterior coil segment that is mounted on a ring and suspended above the patient [124]. Any radiosensitive and highly attenuating coil components are positioned on the sides of the coil, producing a rectangular radiotranslucent window that allows a maximum radiation field size of 22 cm in the superior-inferior direction. In the ViewRay system, users can choose from two types of radiotranslucent flexible foam coils, a 2 × 6 channel torso coil or a 2 × 5 channel head and neck coil, which can both be positioned posteriorly and anteriorly [125]. The maximum super-inferior field size is 24 cm. The limited field sizes of the systems restrict the selection of HNSCC patients who can be treated on the devices, potentially excluding patients with multi-level lymph node involvement or patients with cancers of the nasopharynx/paranasal sinuses who would require larger field sizes [126]. In studies at two different institutions, the treatment plans of HNSCC patients who were previously treated with conventional RT were analyzed, and the studies found that 66% and 95% of previous patients had field sizes small enough to be treated on the Unity MRL [127, 128].

In addition to field size considerations, the design of the coils also impacts image quality.

The flexible coils of the ViewRay system can be positioned very close to and conformed around the patient's surface to maximize signal-to-noise ratio. However, reproducibility of the coil setup can be challenging, which is important during the online adaptive workflow so that the dose attenuation of the coils can be properly accounted for during treatment planning. The Unity system uses a rigid anterior coil, which is helpful for setup reproducibility but limits the signal-to-noise ratio for HNSCC imaging because the coil cannot be conformed to the patient's contour. Although the image quality of T1w and T2w images using this coil is adequate for MR simulation and online plan adaptation [40], signal-to-noise ratio is slightly hindered compared to conventional MRI scanners and may have implications for advanced quantitative MRI sequences such as DWI. Thus, the development of dedicated head and neck coils for the Unity system is of great interest to the community.

## Evaluation

Treating HNSCC on MRL introduces several new considerations for evaluating treatment plan quality. Treatment plans must not only be evaluated before the start of treatment as with conventional RT, but adaptive plans must also be reviewed during each treatment prior to delivery. Patient-specific IMRT quality assurance (QA) is typically performed for each reference plan before treatment, but it cannot be performed on adaptive plans before each treatment because the patient is on the treatment table. Therefore, it is typically done after each treatment as a sanity check, but if QA were to fail, it may indicate that a new reference plan should be created to ensure that future adaptive plans meet quality control standards. In a report of the first 10 head and neck patients treated on the Unity MRL at the MD Anderson Cancer Center, all 253 adaptive plans passed QA, and the median gamma pass rate was 99.9% [40]. High pass rates have been reported for other disease sites as well, leading to studies using statistical methods to determine whether QA actually needs to be performed for every adaptive plan [129–131]. Kim et al.'s statistical analysis of brain cancer treatments indicated that the gamma pass rates of the reference plan and first fraction were sufficient to calculate a 99.9% confidence interval that accurately encompassed the mean gamma pass rate of all remaining fractions [131]. Strand et al. found that ATS and SBRT plans had lower overall gamma pass rates but still passed QA [130]. Their results also indicated that if reference plans pass QA, then adaptive plans may also be expected to pass. While these results are promising, similar analyses should be performed specifically for HNSCC treatment plans due to the complexity of these plans.

In addition to IMRT QA, adaptive plans are evaluated prior to treatment delivery based on whether pre-specified dosimetric criteria are met. Head and neck treatment plans can be very complex and often involve a large number of target volumes and OARs. Dosimetric constraint violations occur frequently with adaptive plans generated with Monaco for the Unity MRL, although the frequency of violation depends on the optimization strategy used [40, 50]. No data is currently available in the literature for ViewRay. In Monaco, users are able to specify both optimal and mandatory dosimetric constraints, but there is currently no easy way to track whether each constraint was met in previous plans during the online adaptation workflow. Furthermore, there is no way to accumulate the delivered dose from previous fractions during online plan adaptation or to see if each constraint is being met in the sum of all delivered plans [60]. If we could see this information in the treatment planning system when adapting plans, it would help clinicians decide whether to accept a constraint failure for a single adaptive plan or whether replanning needs to be done. Some clinics export treatment data for dose accumulation in third-party software, but this process is time-consuming and introduces uncertainties and variability between centers using different platforms.

## Outcome

To assess the true benefit of the introduction of MRL devices, standardization of outcome measures across the field of head and neck radiation oncology is paramount. This is important for both pillars of successful treatment: tumor control and toxicity reporting.

### Toxicity Reporting

Currently, toxicity reporting is done by physicians and/or patients, typically using different questionnaires or scales. Well-known toxicity reporting scales include the Common Terminology Criteria for Adverse Events (CTCAE) and Radiation Therapy Oncology Group (RTOG)/European Organisation for Research and Treatment of Cancer (EORTC) scoring systems that allow scoring of general complaints such as weight loss and fatigue, as well as site-specific toxicities such as xerostomia and dysphagia. The Late Effects Normal Tissue Task Force-Subjective, Objective, Management, Analytic (LENT-SOMA) scale is similar but not used as regularly. In addition to these general scoring systems for toxicity reporting, there are more focused scoring systems for specific toxicities available, such as the Edinburgh Dysphagia Score (EDS) [132] for dysphagia or Functional Oral Intake Scale (FOIS) [133] for oral intake. The FOIS scale has been validated using data from several different countries such as Iran [134], Italy [135], and Germany [136]. The Danish Head and Neck Cancer (DAHANCA) group has developed its own scoring system, the DAHANCA toxicity scale. Equally important compared to physician-rated toxicity are patient-reported outcome measures. For these, similar to physician-rated toxicity, scoring can be quite general or HNSCC-specific. Examples of these include MDASI-HN [70], EORTC QLQ-HN35 [137]/QLQ-C30 [138]/EQ-D5-L5 [137], FACT-HN [139], MDADI for dysphagia [140], and VHI-10 for voice quality [141].

Unfortunately, physician-rated and patient-rated outcome measures regarding toxicity can differ substantially. For example, Kaae et al. showed that, for xerostomia, one of the most frequent late toxicities after HNSCC treatment, patient-, and physician-rated toxicity showed weak correlation. Moreover, patients tended to report more severe toxicity compared to physicians [142].

It is hypothesized that the use of MRL devices may lead to a reduction in the toxicity profile for HNSCC due to increased accuracy of contouring combined with innumerable ART options. Yet, this can only be measured if standardization of toxicity reporting as well as tumor control outcome reporting has been achieved.

### Tumor Control Outcome Reporting

Traditionally, outcomes of clinical studies include overall survival, distant metastases, local control, and locoregional control. However, with increased options for treatment modulation, the definition of subvolumes of tumor to (de-)escalate dose to outcome reporting has become increasingly complex. Moreover, the traditional reporting simply does not provide sufficient detail for assessment of treatment strategy efficacy. A standardized means of reporting outcome measures in the era of IMRT was coined in 2016 by Mohamed et al. [143], but this strategy has not yet been widely implemented. In (real-time, online) ART with changing boost volumes, what is the dose to report? Accurate dose reporting might only be possible if an accurate approximation of the actual dose delivered to the tumor/studied subvolume/voxel can be documented. However, as mentioned above, dose accumulation for MRL devices does not currently allow differentiation of dose to that detail.

### Future Perspectives on Outcome Measurement

MRL devices allow for the collection of immense amounts of data, either pre-treatment, post-treatment, or even with on-beam imaging using several different sequences. Big data in HNSCC may further drive technological advancements such as the identification of cellular mechanisms driving HNSCC pathogenesis, new prognostic and predictive factors, or prediction of the effectiveness and tolerance of RT [144]. These prediction models, in turn, can aid physicians in personalized treatment choices, patient education (e.g., CITOR calculator for comprehensive individual toxicity risk assessment [145]), and/or selection of treatment modality (e.g., proton vs. photon treatment [146]).

## Challenges and Future Directions

### Technological Challenges

General MRL challenges are particularly acute in the head and neck region. There remain a number of technological challenges that need to be solved for wider use and implementation of MRL in HNSCC. As discussed in previous sections, the inability to deliver VMAT with the MRL leads to challenges in creating treatment plans that offer optimal balance of tumor coverage and organ sparing. Given machine constraints, patients must be treated with step-and-shoot IMRT. As discussed previously, this may lead to less optimal organ sparing and heterogeneity or hotter plans [88]. This leads to important patient- and plan-selection considerations, where the benefits of a treatment plan on an MRL with better soft-tissue imaging and ability to adapt treatment may be outweighed by better planning techniques available on a conventional linear accelerator.

The lower magnet strength of the two currently commercially available MRL systems (0.35T ViewRay and 1.5T Unity) as compared to standard diagnostic quality magnet strength (3T) may lead to lower-quality images utilizing the MRL system [147]. Identification of optimal MRI sequences for these lower-strength magnets that offer the best delineation of head and neck structures is needed. Ongoing development of additional sequences that can be deployed on the MRL, such as DWI sequences [148], could provide additional tumor or treatment response data that allows for more personalized treatment adaptation.

MRI is less frequently used than CT and FDG-PET in the diagnosis and surveillance of HNSCC. When MRI is used in treatment planning, interobserver variability in tumor delineation exists [149], which raises concern for tumor coverage and appropriate dosing to microscopic margins. With the better soft-tissue delineation available on the MRL, as well as potential real-time gating as in the ViewRay system, opportunity exists to create smaller margins on tumor, which further requires confidence in treatment volume delineation and minimization of interobserver variability. Furthermore, MRI-specific contouring guidelines to minimize interobserver variability in contouring OARs are needed.

MRI acquisition requires patient coils for optimal images. Given the lower magnet strength, coils must be placed close to the area of interest on the patient in order to optimize image quality. The immobilization required to treat HNSCC patients, which typically includes a mask to minimize motion, adds to additional devices that must be placed upon the patient in addition to the required coils; given the small size of the MRI bore, this can present patient setup challenges.

Finally, current approaches to treatment planning require obtaining a standard CT simulation image to obtain electron density information to allow dose calculation. Ongoing efforts to develop synthetic CTs from MR images [150] to eliminate this additional CT-based step would allow for streamlining of treatment planning pathways. Different development of algorithms for both the Unity and the ViewRay MRL are needed, as well as evaluation of ability to clinically implement these synthetic CTs.

### Staffing and Time-Related Challenges

Standard HNSCC RT typically consists of conventionally fractionated RT given over a period of 6–7 weeks. The potential incorporation of MRL into these long treatment courses raises significant time-based and staffing-based challenges.

The average time for a single head and neck treatment on a CT-based linear accelerator is less than 10 min. CBCTs for IGRT can be obtained in a fraction of a minute, and patients can be aligned quickly for treatment delivery. For MRL treatments, due to the requirement of utilizing step-and-shoot IMRT, patients typically require a minimum of 25 min of table time. This includes a need to obtain MR images, which is a minimum of 2 min. For patients undergoing adaptation, additional editing of contours adds a minimum of 10–15 additional minutes.

For long treatment courses, such as standard HNSCC plans, this difference in treatment delivery time equates to a large difference in table time needed for each patient. For example, a patient undergoing a definitive course of 35 fractions of radiation would require less than 6 h of total treatment time for a CT-based linear accelerator, and 14.5 h for an MRL treatment (assuming no adaptation). This total treatment time and required table time represent a significant burden on staffing and limitations for treating additional patients. In contrast to the typical several-week timeframe of treatment of HNSCC, RT for other areas such as early-stage lung cancers can be treated in as few as 3–5 treatments; thus, 7–10 early-stage lung cancer patients could be treated in the same timeframe as a single conventionally treated HNSCC patient utilizing the MRL platform. The use of hypofractionation in HNSCC remains limited, as mentioned above, with typical uses including SBRT for small re-RT cases.

Replanning or treatment adaptation during a course of head and neck RT is common; patients frequently experience anatomical changes from weight loss or tumor response [151]. In standard treatment planning workflows, patients who are identified as requiring replanning are scheduled for repeat CT simulation. Treatment volumes from the prior plan are transferred and potentially deformed from the original simulation scan, and then updated by the physician. Once this has been completed, the plan is reoptimized, typically with some prior knowledge of achievable objectives for this particular plan. The whole process takes approximately 24–48 h.

In contrast, reoptimization on the MRL platform once contours have been edited can take less than 5 min for a complex treatment plan. Physician input remains necessary for volume adaptation, which can variably take 10–20 min depending on the treatment volumes, but the reoptimization process is significantly shorter utilizing the MRL platform. This replanning process can occur online or offline, after the patient has completed treatment and with plans to initiate a new plan on the next fraction. This faster reoptimization can allow for more frequent plan adaptations allowing for rapid response to patient changes, although with unclear dosimetric benefits of frequent adaptation [152] and unknown optimal timing or frequency of replanning. Of note, for patients who experience very significant anatomic shifts such as significant weight loss, reoptimization on the MRL does require repeat simulation in order to ensure accurate electron density calculations, potentially decreasing the time-saving advantage of rapid replanning.

Physician manual input on contouring structures remains critical and, given the number of structures that need to be contoured in the head and neck region, can require a significant amount of physician time. Autocontouring options exist for CT-based images [153], but there are currently no commercially available autocontouring software options available for the MRL. To increase the feasibility of widespread use of the MRL, in addition to creating MRI-based contouring guidelines, autocontouring software would increase accessibility and reduce time burden for patient treatment on the MRL.

## Patient-Specific Challenges

General MRI safety precautions must be followed when treating patients on an MRL, including monitoring for implantable metal devices such as pacemakers. However, some challenges may be more likely to occur for HNSCC patients as compared to others.

Patient setup for head and neck treatment planning often requires custom devices for optimal patient anatomy. Reproducible shoulder positioning is important for patients getting neck treatment [154]; shoulder pulls can be used as an additional treatment device to monitor this. Bite blocks can be used to displace adjacent mucosa and increase dose falloff. A bolus may also be needed for tumors near the skin, including patients with a thin neck receiving laryngeal RT or having undergone emergency tracheostomy with tumor cut-through. Current commercially available devices are intended for use with CT simulations and CT-based linear accelerators. Use of such additional patient-specific devices needs to be MRI compatible for use with the MRL.

Screening for claustrophobia and ensuring patient comfort during treatment are critical for HNSCC treatment on the MRL, given both the longer treatment times and the patient devices required for setup. Patients require a treatment mask for immobilization and placement of coils for optimal image quality, and the MRL treatment bore is small (70 cm for both Unity and ViewRay systems). Some patients may be more at risk of experiencing claustrophobia. Anecdotally, our experience has been that patients tolerate the mask/coil placement in the smaller bore without issues when appropriately counseled beforehand.

Older patients are at risk of hearing loss, and a proportion of patients may require hearing aids as assistive devices during treatment. Within treatment of HNSCC, concurrent chemotherapy agents also carry the risk of inducing and further exacerbating ototoxicity. Patients with pre-existing hearing loss may not be candidates for treatment on the MRL given the need to be able to communicate with patients without these metal-containing assistive devices during the course of MRL treatment (setup, image acquisition, potential reoptimization, and treatment).

## Future Directions

MRL is a promising new technology for the treatment of HNSCC, but will require additional work to optimally identify the best patients, best diseases, and best anatomy that will lend itself favorably to the benefits of this technology.

As noted above, ongoing research is needed to overcome challenges associated with this technology, including further refinement of the machine to allow for faster gantry rotation and potential VMAT treatment. Imaging-based refinements will continue to add to the ability to define target volumes and potentially allow for treatment adaptation based on novel sequences such as DWI. This may allow for a future of functional adaptation, allowing for treatment changes that prevent acute toxicities or minimize the risk of late toxicities.

Further clinical trial development for the role and safety of hypofractionation in HNSCC may further increase the appeal of the use of the MRL in HNSCC treatment, and may help overcome the staffing-/time-related challenges of conventionally fractionated treatment. Novel work on hypofractionated larynx treatment [37] lends itself particularly well to the MRL platform, and the ViewRay platform's ability to allow for real-time tracking may allow for additional developments in this space. MRL may also be interesting in the setting of re-RT where SBRT and hypofractionation may be used more commonly.

## Summary

The use of MRL in HNSCC has potential for treatment personalization and may offer an opportunity to study treatment changes and response through the collection of serial MRIs with functional sequences throughout therapy. Future works assessing treatment and toxicity outcomes as well as ways to optimize workflow and address current challenges with MRL for HNSCC are warranted and under way with multiple trials open internationally.

## References

1. Warnakulasuriya S. Global epidemiology of oral and oropharyngeal cancer. Oral Oncol. 2009;45(4–5):309–16.
2. Siegel RL, Miller KD, Fuchs HE, Jemal A. Cancer statistics, 2022. CA Cancer J Clin. 2022;72(1):7–33.
3. Rettig EM, Fakhry C, Khararjian A, Westra WH. Age profile of patients with oropharyngeal squamous cell

carcinoma. JAMA Otolaryngol Head Neck Surg. 2018;144(6):538–9.
4. Chaturvedi AK, Engels EA, Pfeiffer RM, Hernandez BY, Xiao W, Kim E, et al. Human papillomavirus and rising oropharyngeal cancer incidence in the United States. J Clin Oncol. 2011;29(32):4294–301.
5. Patel EJ, Oliver JR, Jacobson AS, Li Z, Hu KS, Tam M, et al. Human papillomavirus in patients with hypopharyngeal squamous cell carcinoma. Otolaryngol Head Neck Surg. 2022;166(1):109–17.
6. Ohno S, Arnheiter H, Dubois-Dalcq M, Lazzarini RA. Immunocytochemical localization of vesicular stomatitis virus proteins N and NS with monoclonal antibodies. Histochemistry. 1985;82(2):185–96.
7. Chung CH, Zhang Q, Kong CS, Harris J, Fertig EJ, Harari PM, et al. p16 protein expression and human papillomavirus status as prognostic biomarkers of nonoropharyngeal head and neck squamous cell carcinoma. J Clin Oncol. 2014;32(35):3930–8.
8. Stephen JK, Divine G, Chen KM, Chitale D, Havard S, Worsham MJ. Significance of p16 in site-specific HPV positive and HPV negative head and neck squamous cell carcinoma. Cancer Clin Oncol. 2013;2(1):51–61.
9. Slaughter DP. Multicentric origin of intraoral carcinoma. Surgery. 1946;20(1):133–46.
10. Adelstein DJ, Li Y, Adams GL, Wagner H Jr, Kish JA, Ensley JF, et al. An intergroup phase III comparison of standard radiation therapy and two schedules of concurrent chemoradiotherapy in patients with unresectable squamous cell head and neck cancer. J Clin Oncol. 2003;21(1):92–8.
11. Pignon JP, le Maitre A, Maillard E, Bourhis J, MACH-NC Collaborative Group. Meta-analysis of chemotherapy in head and neck cancer (MACH-NC): an update on 93 randomised trials and 17,346 patients. Radiother Oncol. 2009;92(1):4–14.
12. Zhang Y, Chen L, Hu GQ, Zhang N, Zhu XD, Yang KY, et al. Gemcitabine and cisplatin induction chemotherapy in nasopharyngeal carcinoma. N Engl J Med. 2019;381(12):1124–35.
13. Lefebvre JL, Andry G, Chevalier D, Luboinski B, Collette L, Traissac L, et al. Laryngeal preservation with induction chemotherapy for hypopharyngeal squamous cell carcinoma: 10-year results of EORTC trial 24891. Ann Oncol. 2012;23(10):2708–14.
14. Sun L, Candelieri-Surette D, Anglin-Foote T, Lynch JA, Maxwell KN, D'Avella C, et al. Cetuximab-based vs carboplatin-based chemoradiotherapy for patients with head and neck cancer. JAMA Otolaryngol Head Neck Surg. 2022;148(11):1022–8.
15. Patil VM, Noronha V, Menon N, Singh A, Ghosh-Laskar S, Budrukkar A, et al. Results of phase III randomized trial for use of docetaxel as a radiosensitizer in patients with head and neck cancer, unsuitable for cisplatin-based chemoradiation. J Clin Oncol. 2023;41(13):2350–61.
16. Bonner JA, Harari PM, Giralt J, Cohen RB, Jones CU, Sur RK, et al. Radiotherapy plus cetuximab for locoregionally advanced head and neck cancer: 5-year survival data from a phase 3 randomised trial, and relation between cetuximab-induced rash and survival. Lancet Oncol. 2010;11(1):21–8.
17. Overgaard J, Hansen HS, Specht L, Overgaard M, Grau C, Andersen E, et al. Five compared with six fractions per week of conventional radiotherapy of squamous-cell carcinoma of head and neck: DAHANCA 6 and 7 randomised controlled trial. Lancet. 2003;362(9388):933–40.
18. Ang KK, Harris J, Wheeler R, Weber R, Rosenthal DI, Nguyen-Tan PF, et al. Human papillomavirus and survival of patients with oropharyngeal cancer. N Engl J Med. 2010;363(1):24–35.
19. Becker M, Zbaren P, Casselman JW, Kohler R, Dulguerov P, Becker CD. Neoplastic invasion of laryngeal cartilage: reassessment of criteria for diagnosis at MR imaging. Radiology. 2008;249(2):551–9.
20. Ong CK, Chong VF. Imaging of perineural spread in head and neck tumours. Cancer Imaging. 2010;10 Spec no A(1A):S92–8.
21. Kim JH, Choi KY, Lee SH, Lee DJ, Park BJ, Yoon DY, et al. The value of CT, MRI, and PET-CT in detecting retropharyngeal lymph node metastasis of head and neck squamous cell carcinoma. BMC Med Imaging. 2020;20(1):88.
22. Adams S, Baum RP, Stuckensen T, Bitter K, Hor G. Prospective comparison of 18F-FDG PET with conventional imaging modalities (CT, MRI, US) in lymph node staging of head and neck cancer. Eur J Nucl Med. 1998;25(9):1255–60.
23. Schmid DT, Stoeckli SJ, Bandhauer F, Huguenin P, Schmid S, von Schulthess GK, et al. Impact of positron emission tomography on the initial staging and therapy in locoregional advanced squamous cell carcinoma of the head and neck. Laryngoscope. 2003;113(5):888–91.
24. Chauhan D, Rawat S, Sharma MK, Ahlawat P, Pal M, Gupta G, et al. Improving the accuracy of target volume delineation by combined use of computed tomography, magnetic resonance imaging and positron emission tomography in head and neck carcinomas. J Cancer Res Ther. 2015;11(4):746–51.
25. Figen M, Colpan Oksuz D, Duman E, Prestwich R, Dyker K, Cardale K, et al. Radiotherapy for head and neck cancer: evaluation of triggered adaptive replanning in routine practice. Front Oncol. 2020;10:579917.
26. Mohan R, Wu Q, Manning M, Schmidt-Ullrich R. Radiobiological considerations in the design of fractionation strategies for intensity-modulated radiation therapy of head and neck cancers. Int J Radiat Oncol Biol Phys. 2000;46(3):619–30.
27. Pagh A, Grau C, Overgaard J. Failure pattern and salvage treatment after radical treatment of head and neck cancer. Acta Oncol. 2016;55(5):625–32.
28. Chuang SC, Scelo G, Tonita JM, Tamaro S, Jonasson JG, Kliewer EV, et al. Risk of second primary cancer among patients with head and neck cancers: a pooled analysis of 13 cancer registries. Int J Cancer. 2008;123(10):2390–6.

29. Kim YS. Reirradiation of head and neck cancer in the era of intensity-modulated radiotherapy: patient selection, practical aspects, and current evidence. Radiat Oncol J. 2017;35(1):1–15.
30. You R, Liu YP, Xie YL, Lin C, Duan CY, Chen DP, et al. Hyperfractionation compared with standard fractionation in intensity-modulated radiotherapy for patients with locally advanced recurrent nasopharyngeal carcinoma: a multicentre, randomised, open-label, phase 3 trial. Lancet. 2023;401:917.
31. Yamazaki H, Kodani N, Ogita M, Sato K, Himei K. Reirradiation of head and neck cancer focusing on hypofractionated stereotactic body radiation therapy. Radiat Oncol. 2011;6:98.
32. Phan J, Sio TT, Nguyen TP, Takiar V, Gunn GB, Garden AS, et al. Reirradiation of head and neck cancers with proton therapy: outcomes and analyses. Int J Radiat Oncol Biol Phys. 2016;96(1):30–41.
33. Rodin J, Bar-Ad V, Cognetti D, Curry J, Johnson J, Zender C, et al. A systematic review of treating recurrent head and neck cancer: a reintroduction of brachytherapy with or without surgery. J Contemp Brachytherapy. 2018;10(5):454–62.
34. Le QT, Fu KK, Kroll S, Ryu JK, Quivey JM, Meyler TS, et al. Influence of fraction size, total dose, and overall time on local control of T1-T2 glottic carcinoma. Int J Radiat Oncol Biol Phys. 1997;39(1):115–26.
35. Kachhwaha A, Jakhar SL, Syiem T, Sharma N, Kumar HS, Sharma A. Hypofractionated radiotherapy versus conventional radiotherapy in early glottic cancer T1-2N0M0: a randomized study. J Cancer Res Ther. 2021;17(6):1499–502.
36. Schwartz DL, Sosa A, Chun SG, Ding C, Xie XJ, Nedzi LA, et al. SBRT for early-stage glottic larynx cancer-initial clinical outcomes from a phase I clinical trial. PLoS One. 2017;12(3):e0172055.
37. Sher DJ, Timmerman RD, Nedzi L, Ding C, Pham NL, Zhao B, et al. Phase 1 fractional dose-escalation study of equipotent stereotactic radiation therapy regimens for early-stage Glottic larynx cancer. Int J Radiat Oncol Biol Phys. 2019;105(1):110–8.
38. Kang BH, Yu T, Kim JH, Park JM, Kim JI, Chung EJ, et al. Early closure of a phase 1 clinical trial for SABR in early-stage glottic cancer. Int J Radiat Oncol Biol Phys. 2019;105(1):104–9.
39. Mohamed ASR, Bahig H, Aristophanous M, Blanchard P, Kamal M, Ding Y, et al. Prospective in silico study of the feasibility and dosimetric advantages of MRI-guided dose adaptation for human papillomavirus positive oropharyngeal cancer patients compared with standard IMRT. Clin Transl Radiat Oncol. 2018;11:11–8.
40. McDonald BA, Vedam S, Yang J, Wang J, Castillo P, Lee B, et al. Initial feasibility and clinical implementation of daily MR-guided adaptive head and neck cancer radiation therapy on a 1.5T MR-Linac system: prospective R-IDEAL 2a/2b systematic clinical evaluation of technical innovation. Int J Radiat Oncol Biol Phys. 2021;109(5):1606–18.
41. Chen AM, Cao M, Hsu S, Lamb J, Mikaeilian A, Yang Y, et al. Magnetic resonance imaging guided reirradiation of recurrent and second primary head and neck cancer. Adv Radiat Oncol. 2017;2(2):167–75.
42. Chen AM, Hsu S, Lamb J, Yang Y, Agazaryan N, Steinberg ML, et al. MRI-guided radiotherapy for head and neck cancer: initial clinical experience. Clin Transl Oncol. 2018;20(2):160–8.
43. Bahig H, Yuan Y, Mohamed ASR, Brock KK, Ng SP, Wang J, et al. Magnetic resonance-based response assessment and dose adaptation in human papilloma virus positive tumors of the oropharynx treated with radiotherapy (MR-ADAPTOR): an R-IDEAL stage 2a-2b/Bayesian phase II trial. Clin Transl Radiat Oncol. 2018;13:19–23.
44. Kyzas PA, Evangelou E, Denaxa-Kyza D, Ioannidis JP. 18F-fluorodeoxyglucose positron emission tomography to evaluate cervical node metastases in patients with head and neck squamous cell carcinoma: a meta-analysis. J Natl Cancer Inst. 2008;100(10):712–20.
45. Kim SY, Kim JS, Doo H, Lee H, Lee JH, Cho KJ, et al. Combined [18F]fluorodeoxyglucose positron emission tomography and computed tomography for detecting contralateral neck metastases in patients with head and neck squamous cell carcinoma. Oral Oncol. 2011;47(5):376–80.
46. Nguyen A, Luginbuhl A, Cognetti D, Van Abel K, Bar-Ad V, Intenzo C, et al. Effectiveness of PET/CT in the preoperative evaluation of neck disease. Laryngoscope. 2014;124(1):159–64.
47. Vishwanath V, Jafarieh S, Rembielak A. The role of imaging in head and neck cancer: an overview of different imaging modalities in primary diagnosis and staging of the disease. J Contemp Brachytherapy. 2020;12(5):512–8.
48. Widmann G, Henninger B, Kremser C, Jaschke W. MRI sequences in head & neck radiology—state of the art. Rofo. 2017;189(5):413–22.
49. Ross MR, Schomer DF, Chappell P, Enzmann DR. MR imaging of head and neck tumors: comparison of T1-weighted contrast-enhanced fat-suppressed images with conventional T2-weighted and fast spin-echo T2-weighted images. AJR Am J Roentgenol. 1994;163(1):173–8.
50. Gupta A, Dunlop A, Mitchell A, McQuaid D, Nill S, Barnes H, et al. Online adaptive radiotherapy for head and neck cancers on the MR linear accelerator: introducing a novel modified adapt-to-shape approach. Clin Transl Radiat Oncol. 2022;32:48–51.
51. Salzillo TC, Dresner AM, Way A, Wahid KA, McDonald BA, Mulder S, et al. Development and implementation of optimized endogenous contrast sequences for delineation in adaptive radiotherapy on a 1.5T MR-Linear-accelerator (MR-Linac): a prospective R-IDEAL Stage 0-2a quantitative/qualitative evaluation of *in vivo* site-specific quality-assurance using a 3D T2 fat-suppressed

platform for head and neck cancer. medRxiv. 2022:2022.06.24.22276839.
52. Phillips CD, Gay SB, Newton RL, Levine PA. Gadolinium-enhanced MRI of tumors of the head and neck. Head Neck. 1990;12(4):308–15.
53. Wang J, Salzillo T, Jiang Y, Mackeyev Y, David Fuller C, Chung C, et al. Stability of MRI contrast agents in high-energy radiation of a 1.5T MR-Linac. Radiother Oncol. 2021;161:55–64.
54. Mahmood F, Nielsen UG, Jorgensen CB, Brink C, Thomsen HS, Hansen RH. Safety of gadolinium based contrast agents in magnetic resonance imaging-guided radiotherapy—an investigation of chelate stability using relaxometry. Phys Imaging Radiat Oncol. 2022;21:96–100.
55. Ersoy H, Rybicki FJ. Biochemical safety profiles of gadolinium-based extracellular contrast agents and nephrogenic systemic fibrosis. J Magn Reson Imaging. 2007;26(5):1190–7.
56. Petronek MS, Steinbach EJ, Kalen AL, Builta ZJ, Callaghan CM, Hyer DE, et al. Assessment of Gadobutrol safety in combination with ionizing radiation using a preclinical MRI-guided radiotherapy model. Radiat Res. 2021;195(3):230–4.
57. Hales RB, Chuter R, McWilliam A, Salah A, Dubec M, Freear L, et al. The impact of gadolinium-based MR contrast on radiotherapy planning for oropharyngeal treatment on the MR Linac. Med Phys. 2022;49(1):510–20.
58. Payabvash S. Quantitative diffusion magnetic resonance imaging in head and neck tumors. Quant Imaging Med Surg. 2018;8(10):1052–65.
59. Habrich J, Boeke S, Nachbar M, Nikolaou K, Schick F, Gani C, et al. Repeatability of diffusion-weighted magnetic resonance imaging in head and neck cancer at a 1.5 T MR-Linac. Radiother Oncol. 2022;174:141–8.
60. McDonald BA, Zachiu C, Christodouleas J, Naser MA, Ruschin M, Sonke JJ, et al. Dose accumulation for MR-guided adaptive radiotherapy: from practical considerations to state-of-the-art clinical implementation. Front Oncol. 2022;12:1086258.
61. Yang Y, Cao M, Sheng K, Gao Y, Chen A, Kamrava M, et al. Longitudinal diffusion MRI for treatment response assessment: preliminary experience using an MRI-guided tri-cobalt 60 radiotherapy system. Med Phys. 2016;43(3):1369–73.
62. Kooreman ES, van Houdt PJ, Keesman R, Pos FJ, van Pelt VWJ, Nowee ME, et al. ADC measurements on the Unity MR-linac - a recommendation on behalf of the Elekta Unity MR-linac consortium. Radiother Oncol. 2020;153:106–13.
63. Schakel T, Hoogduin JM, Terhaard CHJ, Philippens MEP. Technical note: diffusion-weighted MRI with minimal distortion in head-and-neck radiotherapy using a turbo spin echo acquisition method. Med Phys. 2017;44(8):4188–93.
64. Weissmann T, Speer S, Putz F, Lettmaier S, Schubert P, Shariff M, et al. Reduction of elective radiotherapy treatment volume in definitive treatment of locally advanced head and neck cancer-comparison of a prospective trial with a revised simulated contouring approach. J Clin Med. 2021;10(20):4653.
65. Gregoire V, Ang K, Budach W, Grau C, Hamoir M, Langendijk JA, et al. Delineation of the neck node levels for head and neck tumors: a 2013 update. DAHANCA, EORTC, HKNPCSG, NCIC CTG, NCRI, RTOG, TROG consensus guidelines. Radiother Oncol. 2014;110(1):172–81.
66. Gregoire V, Evans M, Le QT, Bourhis J, Budach V, Chen A, et al. Delineation of the primary tumour clinical target volumes (CTV-P) in laryngeal, hypopharyngeal, oropharyngeal and oral cavity squamous cell carcinoma: AIRO, CACA, DAHANCA, EORTC, GEORCC, GORTEC, HKNPCSG, HNCIG, IAG-KHT, LPRHHT, NCIC CTG, NCRI, NRG Oncology, PHNS, SBRT, SOMERA, SRO, SSHNO, TROG consensus guidelines. Radiother Oncol. 2018;126(1):3–24.
67. Gregoire V, Grau C, Lapeyre M, Maingon P. Target volume selection and delineation (T and N) for primary radiation treatment of oral cavity, oropharyngeal, hypopharyngeal and laryngeal squamous cell carcinoma. Oral Oncol. 2018;87:131–7.
68. Machiels JP, Rene Leemans C, Golusinski W, Grau C, Licitra L, Gregoire V, et al. Squamous cell carcinoma of the oral cavity, larynx, oropharynx and hypopharynx: EHNS-ESMO-ESTRO clinical practice guidelines for diagnosis, treatment and follow-up. Ann Oncol. 2020;31(11):1462–75.
69. Brouwer CL, Steenbakkers RJ, Bourhis J, Budach W, Grau C, Gregoire V, et al. CT-based delineation of organs at risk in the head and neck region: DAHANCA, EORTC, GORTEC, HKNPCSG, NCIC CTG, NCRI, NRG oncology and TROG consensus guidelines. Radiother Oncol. 2015;117(1):83–90.
70. Cleeland CS, Mendoza TR, Wang XS, Chou C, Harle MT, Morrissey M, et al. Assessing symptom distress in cancer patients: the M.D. Anderson symptom inventory. Cancer. 2000;89(7):1634–46.
71. Lee AW, Ng WT, Pan JJ, Poh SS, Ahn YC, AlHussain H, et al. International guideline for the delineation of the clinical target volumes (CTV) for nasopharyngeal carcinoma. Radiother Oncol. 2018;126(1):25–36.
72. O'Sullivan B, Warde P, Grice B, Goh C, Payne D, Liu FF, et al. The benefits and pitfalls of ipsilateral radiotherapy in carcinoma of the tonsillar region. Int J Radiat Oncol Biol Phys. 2001;51(2):332–43.
73. Chronowski GM, Garden AS, Morrison WH, Frank SJ, Schwartz DL, Shah SJ, et al. Unilateral radiotherapy for the treatment of tonsil cancer. Int J Radiat Oncol Biol Phys. 2012;83(1):204–9.
74. Dan TD, Raben D, Schneider CJ, Hockstein NG, Witt RL, Dzeda M, et al. Freedom from local and regional failure of contralateral neck with ipsilateral neck radiotherapy for node-positive tonsil cancer: updated results of an institutional clinical management approach. Oral Oncol. 2015;51(6):616–21.
75. Kennedy WR, Herman MP, Deraniyagala RL, Amdur RJ, Werning JW, Dziegielewski PT, et al.

Ipsilateral radiotherapy for squamous cell carcinoma of the tonsil. Eur Arch Otorhinolaryngol. 2016;273(8):2151–6.
76. Xue F, Hu C, He X. Induction chemotherapy followed by intensity-modulated radiotherapy with reduced gross tumor volume delineation for stage T3-4 nasopharyngeal carcinoma. Onco Targets Ther. 2017;10:3329–36.
77. Perillo A, Landoni V, Farneti A, Sanguineti G. Organ motion in linac-based SBRT for glottic cancer. Radiat Oncol. 2021;16(1):106.
78. Kwa SL, Al-Mamgani A, Osman SO, Gangsaas A, Levendag PC, Heijmen BJ. Inter- and Intrafraction target motion in highly focused single vocal cord irradiation of T1a larynx cancer patients. Int J Radiat Oncol Biol Phys. 2015;93(1):190–5.
79. Navran A, Heemsbergen W, Janssen T, Hamming-Vrieze O, Jonker M, Zuur C, et al. The impact of margin reduction on outcome and toxicity in head and neck cancer patients treated with image-guided volumetric modulated arc therapy (VMAT). Radiother Oncol. 2019;130:25–31.
80. Marks LB, Yorke ED, Jackson A, Ten Haken RK, Constine LS, Eisbruch A, et al. Use of normal tissue complication probability models in the clinic. Int J Radiat Oncol Biol Phys. 2010;76(3 Suppl):S10–9.
81. Emami B, Lyman J, Brown A, Coia L, Goitein M, Munzenrider JE, et al. Tolerance of normal tissue to therapeutic irradiation. Int J Radiat Oncol Biol Phys. 1991;21(1):109–22.
82. Su W, Wright CM, Lee DY, Kim M, Anstadt EJ, Teo BK, et al. Stricter postoperative oropharyngeal cancer radiation therapy normal tissue dose constraints are feasible. Pract Radiat Oncol. 2022;12(4):e282–5.
83. Yock TI, Murphy B, Bass JK, Ronckers CM, Kremer L, Baliga S, et al. Modeling the risk of hearing loss from radiotherapy in childhood cancer survivors: initial results from the pediatric normal tissue effects in the clinic (PENTEC) hearing loss task force. Int J Radiat Oncol Biol Phys. 2019;105(1):S191.
84. van Dijk LV, Abusaif AA, Rigert J, Naser MA, Hutcheson KA, Lai SY, et al. Normal tissue complication probability (NTCP) prediction model for osteoradionecrosis of the mandible in patients with head and neck cancer after radiation therapy: large-scale observational cohort. Int J Radiat Oncol Biol Phys. 2021;111(2):549–58.
85. Kamal M, Peeler CR, Yepes P, Mohamed ASR, Blanchard P, Frank S, et al. Radiation-induced hypothyroidism after radical intensity modulated radiation therapy for oropharyngeal carcinoma. Adv Radiat Oncol. 2020;5(1):111–9.
86. Zhou L, Chen J, Shen W, Chen ZL, Huang S, Tao CJ, et al. Thyroid V(50) is a risk factor for hypothyroidism in patients with nasopharyngeal carcinoma treated with intensity-modulated radiation therapy: a retrospective study. Radiat Oncol. 2020;15(1):68.
87. Chen WC, Hsu CM, Tsai YT, Lin MH, Tsai MS, Chang GH, et al. Prospective evaluation of taste function in patients with head and neck cancer receiving intensity-modulated radiotherapy. JAMA Otolaryngol Head Neck Surg. 2022;148(7):604–11.
88. Ouyang Z, Liu Shen Z, Murray E, Kolar M, LaHurd D, Yu N, et al. Evaluation of auto-planning in IMRT and VMAT for head and neck cancer. J Appl Clin Med Phys. 2019;20(7):39–47.
89. Chamberlain M, Krayenbuehl J, van Timmeren JE, Wilke L, Andratschke N, Garcia Schüler H, et al. Head and neck radiotherapy on the MR linac: a multicenter planning challenge amongst MRIdian platform users. Strahlenther Onkol. 2021;197(12):1093–103.
90. Lacas B, Carmel A, Landais C, Wong SJ, Licitra L, Tobias JS, et al. Meta-analysis of chemotherapy in head and neck cancer (MACH-NC): an update on 107 randomized trials and 19,805 patients, on behalf of MACH-NC group. Radiother Oncol. 2021;156:281–93.
91. Soto DE, Kessler ML, Piert M, Eisbruch A. Correlation between pretreatment FDG-PET biological target volume and anatomical location of failure after radiation therapy for head and neck cancers. Radiother Oncol. 2008;89(1):13–8.
92. Mohamed ASR, Cardenas CE, Garden AS, Awan MJ, Rock CD, Westergaard SA, et al. Patterns-of-failure guided biological target volume definition for head and neck cancer patients: FDG-PET and dosimetric analysis of dose escalation candidate subregions. Radiother Oncol. 2017;124(2):248–55.
93. Madani I, Duprez F, Boterberg T, Van de Wiele C, Bonte K, Deron P, et al. Maximum tolerated dose in a phase I trial on adaptive dose painting by numbers for head and neck cancer. Radiother Oncol. 2011;101(3):351–5.
94. Berwouts D, Madani I, Duprez F, Olteanu AL, Vercauteren T, Boterberg T, et al. Long-term outcome of (18) F-fluorodeoxyglucose-positron emission tomography-guided dose painting for head and neck cancer: matched case-control study. Head Neck. 2017;39(11):2264–75.
95. Heukelom J, Hamming O, Bartelink H, Hoebers F, Giralt J, Herlestam T, et al. Adaptive and innovative radiation treatment FOR improving cancer treatment outcomE (ARTFORCE); a randomized controlled phase II trial for individualized treatment of head and neck cancer. BMC Cancer. 2013;13:84.
96. van Dijk LV, Frank SJ, Yuan Y, Gunn B, Moreno AC, Mohamed ASR, et al. Proton image-guided radiation assignment for therapeutic escalation via selection of locally advanced head and neck cancer patients [PIRATES]: a phase I safety and feasibility trial of MRI-guided adaptive particle radiotherapy. Clin Transl Radiat Oncol. 2022;32:35–40.
97. Mierzwa ML, Aryal M, Lee C, Schipper M, VanTil M, Morales K, et al. Randomized phase II study of physiologic MRI-directed adaptive radiation boost in poor prognosis head and neck cancer. Clin Cancer Res. 2022;28(23):5049–57.
98. Zhang Y, Hong H, Cai W. PET tracers based on Zirconium-89. Curr Radiopharm. 2011;4(2):131–9.

99. Grosu AL, Souvatzoglou M, Roper B, Dobritz M, Wiedenmann N, Jacob V, et al. Hypoxia imaging with FAZA-PET and theoretical considerations with regard to dose painting for individualization of radiotherapy in patients with head and neck cancer. Int J Radiat Oncol Biol Phys. 2007;69(2):541–51.
100. Bittner MI, Wiedenmann N, Bucher S, Hentschel M, Mix M, Weber WA, et al. Exploratory geographical analysis of hypoxic subvolumes using 18F-MISO-PET imaging in patients with head and neck cancer in the course of primary chemoradiotherapy. Radiother Oncol. 2013;108(3):511–6.
101. Zschaeck S, Haase R, Abolmaali N, Perrin R, Stutzer K, Appold S, et al. Spatial distribution of FMISO in head and neck squamous cell carcinomas during radio-chemotherapy and its correlation to pattern of failure. Acta Oncol. 2015;54(9):1355–63.
102. Heukelom J, Fuller CD. Head and neck cancer adaptive radiation therapy (ART): conceptual considerations for the informed clinician. Semin Radiat Oncol. 2019;29(3):258–73.
103. van den Bosch S, Dijkema T, Kunze-Busch MC, Terhaard CH, Raaijmakers CP, Doornaert PA, et al. Uniform FDG-PET guided GRAdient Dose prEscription to reduce late Radiation Toxicity (UPGRADE-RT): study protocol for a randomized clinical trial with dose reduction to the elective neck in head and neck squamous cell carcinoma. BMC Cancer. 2017;17(1):208.
104. Deschuymer S, Nevens D, Duprez F, Daisne JF, Dok R, Laenen A, et al. Randomized clinical trial on reduction of radiotherapy dose to the elective neck in head and neck squamous cell carcinoma; update of the long-term tumor outcome. Radiother Oncol. 2020;143:24–9.
105. Gabani P, Lin AJ, Barnes J, Oppelt P, Adkins DR, Rich JT, et al. Radiation therapy dose de-escalation compared to standard dose radiation therapy in definitive treatment of HPV-positive oropharyngeal squamous cell carcinoma. Radiother Oncol. 2019;134:81–8.
106. Winkel D, Bol GH, Kroon PS, van Asselen B, Hackett SS, Werensteijn-Honingh AM, et al. Adaptive radiotherapy: the Elekta Unity MR-linac concept. Clin Transl Radiat Oncol. 2019;18:54–9.
107. de Muinck Keizer DM, Kerkmeijer LGW, Willigenburg T, van Lier A, Hartogh MDD, van der Voort van Zyp JRN, et al. Prostate intrafraction motion during the preparation and delivery of MR-guided radiotherapy sessions on a 1.5T MR-Linac. Radiother Oncol. 2020;151:88–94.
108. Bruijnen T, Stemkens B, Terhaard CHJ, Lagendijk JJW, Raaijmakers CPJ, Tijssen RHN. Intrafraction motion quantification and planning target volume margin determination of head-and-neck tumors using cine magnetic resonance imaging. Radiother Oncol. 2019;130:82–8.
109. Kluter S. Technical design and concept of a 0.35 T MR-Linac. Clin Transl Radiat Oncol. 2019;18:98–101.
110. Green OL, Rankine LJ, Cai B, Curcuru A, Kashani R, Rodriguez V, et al. First clinical implementation of real-time, real anatomy tracking and radiation beam control. Med Phys. 2018;45:3728.
111. Zhong Y, Yang Y, Fang Y, Wang J, Hu W. A preliminary experience of implementing deep-learning based auto-segmentation in head and neck cancer: a study on real-world clinical cases. Front Oncol. 2021;11:638197.
112. Nikolov S, Blackwell S, Zverovitch A, Mendes R, Livne M, De Fauw J, et al. Clinically applicable segmentation of head and neck anatomy for radiotherapy: deep learning algorithm development and validation study. J Med Internet Res. 2021;23(7):e26151.
113. Kieselmann JP, Kamerling CP, Burgos N, Menten MJ, Fuller CD, Nill S, et al. Geometric and dosimetric evaluations of atlas-based segmentation methods of MR images in the head and neck region. Phys Med Biol. 2018;63(14):145007.
114. Korte JC, Hardcastle N, Ng SP, Clark B, Kron T, Jackson P. Cascaded deep learning-based auto-segmentation for head and neck cancer patients: organs at risk on T2-weighted magnetic resonance imaging. Med Phys. 2021;48(12):7757–72.
115. Chin AL, Lin A, Anamalayil S, Teo BK. Feasibility and limitations of bulk density assignment in MRI for head and neck IMRT treatment planning. J Appl Clin Med Phys. 2014;15(5):4851.
116. Karotki A, Mah K, Meijer G, Meltsner M. Comparison of bulk electron density and voxel-based electron density treatment planning. J Appl Clin Med Phys. 2011;12(4):3522.
117. Crijns SP, Kok JG, Lagendijk JJ, Raaymakers BW. Towards MRI-guided linear accelerator control: gating on an MRI accelerator. Phys Med Biol. 2011;56(15):4815–25.
118. Kontaxis C, Bol GH, Lagendijk JJ, Raaymakers BW. A new methodology for inter- and intrafraction plan adaptation for the MR-linac. Phys Med Biol. 2015;60(19):7485–97.
119. Zachiu C, Papadakis N, Ries M, Moonen C, Denis de Senneville B. An improved optical flow tracking technique for real-time MR-guided beam therapies in moving organs. Phys Med Biol. 2015;60(23):9003–29.
120. Willigenburg T, Zachiu C, Lagendijk JJW, van der Voort van Zyp JRN, de Boer HCJ, Raaymakers BW. Fast and accurate deformable contour propagation for intra-fraction adaptive magnetic resonance-guided prostate radiotherapy. Phys Imaging Radiat Oncol. 2022;21:62–5.
121. Uijtewaal P, Borman PTS, Woodhead PL, Kontaxis C, Hackett SL, Verhoeff J, et al. First experimental demonstration of VMAT combined with MLC tracking for single and multi fraction lung SBRT on an MR-linac. Radiother Oncol. 2022;174:149–57.
122. Yang J, Mohamed ASR, Bahig H, Ding Y, Wang J, Ng SP, et al. Automatic registration of 2D MR cine

images for swallowing motion estimation. PLoS One. 2020;15(2):e0228652.
123. Bradley JA, Paulson ES, Ahunbay E, Schultz C, Li XA, Wang D. Dynamic MRI analysis of tumor and organ motion during rest and deglutition and margin assessment for radiotherapy of head-and-neck cancer. Int J Radiat Oncol Biol Phys. 2011;81(5):e803–12.
124. Hoogcarspel SJ, Zijlema SE, Tijssen RHN, Kerkmeijer LGW, Jurgenliemk-Schulz IM, Lagendijk JJW, et al. Characterization of the first RF coil dedicated to 1.5 T MR guided radiotherapy. Phys Med Biol. 2018;63(2):025014.
125. Cuccia F, Alongi F, Belka C, Boldrini L, Hörner-Rieber J, McNair H, et al. Patient positioning and immobilization procedures for hybrid MR-Linac systems. Radiat Oncol. 2021;16(1):183.
126. Boeke S, Monnich D, van Timmeren JE, Balermpas P. MR-guided radiotherapy for head and neck cancer: current developments, perspectives, and challenges. Front Oncol. 2021;11:616156.
127. Chuter RW, Whitehurst P, Choudhury A, van Herk M, McWilliam A. Technical note: investigating the impact of field size on patient selection for the 1.5T MR-Linac. Med Phys. 2017;44(11):5667–71.
128. Ng-Cheng-Hin B, Nutting C, Newbold K, Bhide S, McQuaid D, Dunlop A, et al. The impact of restricted length of treatment field and anthropometric factors on selection of head and neck cancer patients for treatment on the MR-Linac. Br J Radiol. 2020;93(1111):20200023.
129. Monnich D, Winter J, Nachbar M, Kunzel L, Boeke S, Gani C, et al. Quality assurance of IMRT treatment plans for a 1.5 T MR-linac using a 2D ionization chamber array and a static solid phantom. Phys Med Biol. 2020;65(16):16NT01.
130. Strand S, Boczkowski A, Smith B, Snyder JE, Hyer DE, Yaddanapudi S, et al. Analysis of patient-specific quality assurance for Elekta Unity adaptive plans using statistical process control methodology. J Appl Clin Med Phys. 2021;22(4):99–107.
131. Kim A, Ruschin ME, McCann C, Au P, Singh A, Sahgal A, et al. Streamlining patient-specific quality assurance for the 1.5T MR-Linac using a classic statistics tool. Int J Radiat Oncol Biol Phys. 2020;108(3):e350–1.
132. Rhatigan E, Tyrmpas I, Murray G, Plevris JN. Scoring system to identify patients at high risk of oesophageal cancer. Br J Surg. 2010;97(12):1831–7.
133. Crary MA, Mann GD, Groher ME. Initial psychometric assessment of a functional oral intake scale for dysphagia in stroke patients. Arch Phys Med Rehabil. 2005;86(8):1516–20.
134. Bakhtiyari J, Tohidast SA, Mansuri B, Azimi H, Ebadi A. The Persian version of the Functional Oral Intake Scale (FOIS-P): a validation study on stroke patients with dysphagia. Logoped Phoniatr Vocol. 2022;47(2):133–8.
135. Ninfa A, Pizzorni N, Eplite A, Moltisanti C, Schindler A. Validation of the Italian version of the Functional Oral Intake Scale (FOIS-It) against fiber-optic endoscopic evaluation of swallowing and nutritional status. Dysphagia. 2022;37(1):137–47.
136. Hamzic S, Braun T, Juenemann M, Butz M, Voswinckel R, Belly M, et al. Validation of the German version of Functional Oral Intake Scale (FOIS-G) for flexible endoscopic evaluation of swallowing (FEES). Dysphagia. 2021;36(1):130–9.
137. Bjordal K, Hammerlid E, Ahlner-Elmqvist M, de Graeff A, Boysen M, Evensen JF, et al. Quality of life in head and neck cancer patients: validation of the European Organization for Research and Treatment of cancer quality of life questionnaire-H&N35. J Clin Oncol. 1999;17(3):1008–19.
138. Kaasa S, Bjordal K, Aaronson N, Moum T, Wist E, Hagen S, et al. The EORTC core quality of life questionnaire (QLQ-C30): validity and reliability when analysed with patients treated with palliative radiotherapy. Eur J Cancer. 1995;31A(13–14):2260–3.
139. List MA, D'Antonio LL, Cella DF, Siston A, Mumby P, Haraf D, et al. The performance status scale for head and neck cancer patients and the functional assessment of cancer therapy-head and neck scale. A study of utility and validity. Cancer. 1996;77(11):2294–301.
140. Chen AY, Frankowski R, Bishop-Leone J, Hebert T, Leyk S, Lewin J, et al. The development and validation of a dysphagia-specific quality-of-life questionnaire for patients with head and neck cancer: the M. D. Anderson dysphagia inventory. Arch Otolaryngol Head Neck Surg. 2001;127(7):870–6.
141. Rosen CA, Lee AS, Osborne J, Zullo T, Murry T. Development and validation of the voice handicap index-10. Laryngoscope. 2004;114(9):1549–56.
142. Kaae JK, Johnsen L, Hansen CR, Kristensen MH, Brink C, Eriksen JG. Relationship between patient and physician-rated xerostomia and dose distribution to the oral cavity and salivary glands for head and neck cancer patients after radiotherapy. Acta Oncol. 2019;58(10):1366–72.
143. Mohamed AS, Rosenthal DI, Awan MJ, Garden AS, Kocak-Uzel E, Belal AM, et al. Methodology for analysis and reporting patterns of failure in the era of IMRT: head and neck cancer applications. Radiat Oncol. 2016;11(1):95.
144. Resteghini C, Trama A, Borgonovi E, Hosni H, Corrao G, Orlandi E, et al. Big data in head and neck cancer. Curr Treat Options Oncol. 2018;19(12):62.
145. Van den Bosch L, van der Schaaf A, van der Laan HP, Hoebers FJP, Wijers OB, van den Hoek JGM, et al. Comprehensive toxicity risk profiling in radiation therapy for head and neck cancer: a new concept for individually optimised treatment. Radiother Oncol. 2021;157:147–54.
146. Langendijk JA, Hoebers FJP, de Jong MA, Doornaert P, Terhaard CHJ, Steenbakkers R, et al. National protocol for model-based selection for proton therapy in head and neck cancer. Int J Part Ther. 2021;8(1):354–65.
147. Lewis BC, Shin J, Maraghechi B, Quinn B, Cole M, Barberi E, et al. Assessment of a novel commercial

large field of view phantom for comprehensive MR imaging quality assurance of a 0.35T MRgRT system. J Appl Clin Med Phys. 2022;23(4):e13535.
148. Ng SP, Cardenas CE, Bahig H, Elgohari B, Wang J, Johnson JM, et al. Changes in apparent diffusion coefficient (ADC) in serial weekly MRI during radiotherapy in patients with head and neck cancer: results from the PREDICT-HN study. Curr Oncol. 2022;29(9):6303–13.
149. Anderson CM, Sun W, Buatti JM, Maley JE, Policeni B, Mott SL, et al. Interobserver and intermodality variability in GTV delineation on simulation CT, FDG-PET, and MR images of head and neck cancer. Jacobs J Radiat Oncol. 2014;1(1):006.
150. Li X, Yadav P, McMillan AB. Synthetic computed tomography generation from 0.35T magnetic resonance images for magnetic resonance-only radiation therapy planning using perceptual loss models. Pract Radiat Oncol. 2022;12(1):e40–8.
151. Morgan HE, Sher DJ. Adaptive radiotherapy for head and neck cancer. Cancers Head Neck. 2020;5:1.
152. Ahn PH, Chen CC, Ahn AI, Hong L, Scripes PG, Shen J, et al. Adaptive planning in intensity-modulated radiation therapy for head and neck cancers: single-institution experience and clinical implications. Int J Radiat Oncol Biol Phys. 2011;80(3):677–85.
153. Oktay O, Nanavati J, Schwaighofer A, Carter D, Bristow M, Tanno R, et al. Evaluation of deep learning to augment image-guided radiotherapy for head and neck and prostate cancers. JAMA Netw Open. 2020;3(11):e2027426.
154. Neubauer E, Dong L, Followill DS, Garden AS, Court LE, White RA, et al. Assessment of shoulder position variation and its impact on IMRT and VMAT doses for head and neck cancer. Radiat Oncol. 2012;7:19.

# 22 The Australian MRI-Linac: A 10-Year Journey

Gary Liney

## Introduction

Having been asked many times to contribute to papers and book chapters on MRI-Linac technology over the years, it becomes increasingly difficult to provide the reader with new material—even in such an evolving field. In this case, serendipity means this latest invitation comes at the end of my contract and association with the Australian MRI-Linac (AMRL) project. This has given me the opportunity to look back on my last 10 years' work on this system. What follows is a personal reflection of the progress that was made as well as some of the issues that were not always possible to overcome. I will talk about some of the distinguishing features and technical specifications only briefly, and for a more detailed discussion, the reader is directed elsewhere [1, 2]. The points made are my own and not necessarily the opinion of others on the project. However, as someone that led the scientific development from an empty bunker way back in 2012 to a fully functioning system, I feel I am uniquely qualified to do this. Furthermore, while acknowledging there were (and still are) countless people associated with the project, I will instead be focusing on the work undertaken by the small team I led as part of the development of the AMRL, which ultimately delivered the working system.

My involvement in the project began in mid-2012 when I applied to the position of MRI physicist at Ingham Institute in southwest Sydney. This was to lead the scientific input into two key programs of research, namely, the dedicated MRI simulator at Liverpool cancer therapy center and the prototype MRI-Linear accelerator being developed in a research bunker within that same department. I was leaving a job-for-life in the UK's national health service for a 3-year contract, which on the face of it seemed rather strange. However, I was also swapping a position of a lead imaging physicist in a UK radiotherapy department that had a perpetual "5-year plan" to acquire an MRI scanner—that still has not materialized—to one in Australia that would have access to dedicated MRI equipment on both a clinical and research basis. Or so I thought. Imagine my horror when I arrived to see that neither of these pieces of MRI equipment actually even existed: The magnet for the MRI-Linac had encountered serious delays, and the business case for the simulator had not started. And so began what turned into my decade-long involvement.

## The Australian MRL: A 10-Year Journey

The Australian MRI-Linac (AMRL) was a collaboration across multiple institutions including the Universities of NSW, Sydney, Queensland,

G. Liney (✉)
Sydney, NSW, Australia
e-mail: glmriconsultant@gmail.com



I. J. Das et al. (eds.), *A Practical Guide to MR-Linac*, https://doi.org/10.1007/978-3-031-48165-9_22

and Wollongong. It was operated by the Ingham Institute and housed in a dedicated research bunker which was completed in August 2012 within Liverpool hospital in southwest Sydney. The project was the brain child of the "three amigos" Paul Keall, Michael Barton, and Stuart Crozier who were the lead CIs that had procured a significant amount of NHMRC funding [3].

The original intention of the Australian MRL had been to provide a system that allowed for a degree of flexibility in its operation with the potential for both inline and perpendicular orientations. At the time there was still much debate as to which beam-magnetic field configuration would win the day, with both having advantages and disadvantages in terms of interactions and engineering complexities. The AMRL system was designed around a large open bore that could be accessed by the patient or linear accelerator in either orientation. In order to achieve this degree of flexibility, the Linac gantry was to be fixed during treatment, and instead the patient would be moved and rotated in front of the beam.

Similar-looking split-bore designs had been used in the 1990s for surgical interventions. However, at 1 Tesla, the field strength selected for the AMRL was double that previously achieved; this ambitious figure was a best guess compromise between optimal image quality and practicality. Furthermore, a 30-cm working imaging volume and a near-zero field region 1.4 m from the isocenter had also been stipulated in the blueprint. The homogeneity and active shielding requirements of such a design required extreme precession engineering of four superconducting formers, and these took nearly 2 years to successfully manufacture. I entered my first meeting of the project in November 2012 and left an hour or so later (armed with a photograph of the "high level" marker board plan) having being told I was expected to lead the entire project through to its completion! Looking back, I could not have imagined just how much work and responsibility was falling on my shoulders, but it would certainly be *challenging*.

The first period of my tenure was however spent on producing a business case for the aforementioned MRI simulator and its ensuing tender process. After a very nervous few months, the decision to finally go ahead was granted at New South Wales state level, and this ultimately led to the installation of a 3-T Siemens Skyra midway through 2013. The implementation of this scanner into clinical practice was a principal focus of the first couple of years—especially with the ongoing delays on the AMRL—but in so doing there were clearly many indirect benefits to the AMRL program. A good example of this was the testing of a manually driven rotating couch prototype that fitted inside the 70-cm-wide bore system (built by the late Rob Wilkins) and allowed us to replicate and study the effects and imaging performance of patient rotation [4]. Beyond the protocol development and training on this scanner, I was asked to negotiate a research agreement with Siemens. This also led to us to access many early preclinical sequences such as ultrashort TE and 4D with these being relevant to MR-only planning and motion management.

Around the end of 2013, we recruited our chief engineer Bin Dong—initially having little experience of radiotherapy—he went on to become a vital cog in the operation of the AMRL and was a fantastic support over the years. By mid-2014, with the projected 12-month delay to the magnet already slipping, it was decided that I would return to England on a fact-finding mission. In a combined road trip of 1000 miles, I visited the factory producing the gradient coils on the south coast; the Agilent factory providing the magnet, near Oxford; and the manufacturer of the remaining fourth magnet former in the northeast. At this point only three of the four formers had been successfully made with cracks having appeared in the last of the original set. The added time pressure was that Agilent were getting out of the bespoke magnet business and instead concentrating on small-bore systems going forward, meaning that ours was to be the last magnet to leave their factory. I returned to Australia with many reassurances and some trepidation concerning the completion of the magnet.

In the meantime, there was plenty of activity proceeding in the bunker and the adjacent control and equipment rooms. The portable linear accelerator (a Linatron from Varian and now Varex) was delivered and over the next few months was installed into a fixed position in the middle of the bunker. The gradient cabinets were also installed at this stage, composed from a tailor-made list of Siemens Avanto components. I became concerned that no thought was being given to the proximity of the Linatron and committing to a fixed position seemed too risky. At this point in the MRL story, interactions between components was still very much trial and error. Out of an abundance of caution, I began exploring a rail system that could position the Linac at varying distances. Initially a design was considered that moved it on an arc between the two proposed orientations, but this was quickly abandoned for a design that allowed it to more simply translate backwards and forwards at the end of the bore. We had taken the first stage of committing the MRI-Linac to an inline orientation—but it also meant we now had the ability to vary the interacting field strength by virtue of its proximity to the magnet. A couple of photographs from the early stages of the project are shown in Fig. 22.1.

The research relationship with Siemens would pay dividends from the MR-Linac perspective in one other major way. As the delays to magnet production dragged on, I approached Siemens to see whether we could make use of an ex-clinical scanner that was being removed from Royal Perth Hospital and potentially scrapped. And so in November 2015, this 1.5-Tesla Sonata was driven 2000 miles cross country to be installed in our empty bunker. The agreement would see us use this MRI for the next 5 or 6 months allowing us to do anything other than human imaging. Both myself and Bin worked over the Christmas period to gain as much experimental data of the beam-field interactions as we could [5]. This included our first ever beam-on image which, with a nod to the Utrecht group's first pork chop image, was a kangaroo steak! As it turned out, we only got about half of the expected time with the magnet owing to a water leak and an unintended quench. The prohibitive cost of refilling the helium put a sudden end to "Phase I" of the project. Nevertheless, in the intervening period, we had managed to achieve a fair amount of progress. This included a firsthand practical demonstration of the electron focusing effect and its potential mitigation by shooting the beam off axis. We set up radiochromic films at various points through and out the other side of the bore and could see the primary photon divergence while the electrons were first pulled and then repelled from opposite sides of the isocenter. These films generated a lot of excitement, although we struggled to convince some of our colleagues as to the explanation! We had also demonstrated a dosimetric application for the radiation induction we were observing in some receiver coils, which would lead to a US patent

**Fig. 22.1** Early photographs from the Australian MRI-Linac project. (Left) The Varian Linatron being installed on fixed footplates toward the end of 2014. This would eventually be mounted onto a sliding rail system. (Right) The RF cage nearing completion; the quench pipe awaits connection to the magnet in this photo taken in mid-2015

being granted [6]. Sadly this was never exploited further and in fact Ingham's first and only technology patent has since been allowed to lapse.

The Agilent magnet was finally installed in April 2017 to a flurry of media interest. The 12-ton magnet looked immense as it was being craned overhead. I spoke to a local reporter and jokingly said that the only thing that could go wrong was if it was too big to fit in the room. And of course, as if to tempt fate, that is exactly what happened! Errors on the factory blueprint had shown a shorter scaffold overhang above the magnet meaning it would not fit through the bunker door in the way intended. Some last-minute robust adjustments were made to the entrance of the bunker, and it eventually went through "lengthways." Once inside the room, it then took many hours of iterations to move the magnet with specialist lifting equipment into its final position. This had to be as close as possible to the cage wall—the thickness of a cable—and exactly square on, to allow the MLC to benefit from the zero field just 1.4 m from the middle of the magnet.

To complete the hardware, a dedicated split gradient module and RF transmit receive coil were designed to fit inside the opening with as little intrusion into the gap as possible as can be seen in Fig. 22.2. The gradient coils were manufactured as a single modular unit for each magnet half by Tesla, UK. Each half of the module, which contains a series of passive shim trays, had been test fitted in the UK, shipped separately, and was then reinserted by a fork lift once the magnet was inside the bunker. Also shown in Fig. 22.2 are two of the four rungs that make up the birdcage design of the integrated transceiver coil. This was designed (Magnetica, Brisbane) and tested using a full-scale mock-up of the magnet gap and was the last of the MRI hardware to be installed [7]. The four legs of the coil run along the support pillars in the magnet gap to maximize the patient space. In maintaining this opening, the primary beam is unobstructed creating no radiation induction effects, and image quality is unaffected [8].

Much of the next few months were spent on the hardware integration chiefly by our engineer Bin, as he grappled with third party and homebuilt apparatus to link up the manufactured components. Having completed this, attention turned to software integration led by Kevin Zhang with some assistance from Siemens. All things relevant to a 1.5-Tesla system had to be reconfigured to the new (actually just shy of 1.0 Tesla) field strength. In addition, the reconstruction engine was modified so images could be sent to control the MLC. Our first ever test image of a very distorted QA phantom was acquired via remote help from Erlangen in Germany. A few tweaks to the gradient parameters later and we were ready to image our first subject. I was more than happy to volunteer (admittedly the queue to do so was very short!) and climbed in between the magnet pillars keeping 40 tons of attractive force apart. The first images are shown in Fig. 22.3, and seeing these at the end of that test session will rank as a most memorable moment of my career. In the May of that year, we had got the MLC tracking to work just in time for our planned bunker tour on the final day of the MRI in RT symposium being held in Sydney. A demonstration involved Bin moving a bottle of water backward and forward manually in front of the MLC whose leaf positions dutifully followed the sub-second images. This managed to impress all of our visitors even the more seasoned MRI-Linac experts among the contingent.

**Fig. 22.2** Photograph of the bore of the system showing the patient opening circa 2017. A glass RF window was installed at this time to further increase the light and the feeling of space. The patient opening is 78 × 50 cm, and the table can be removed to permit imaging in the standing position

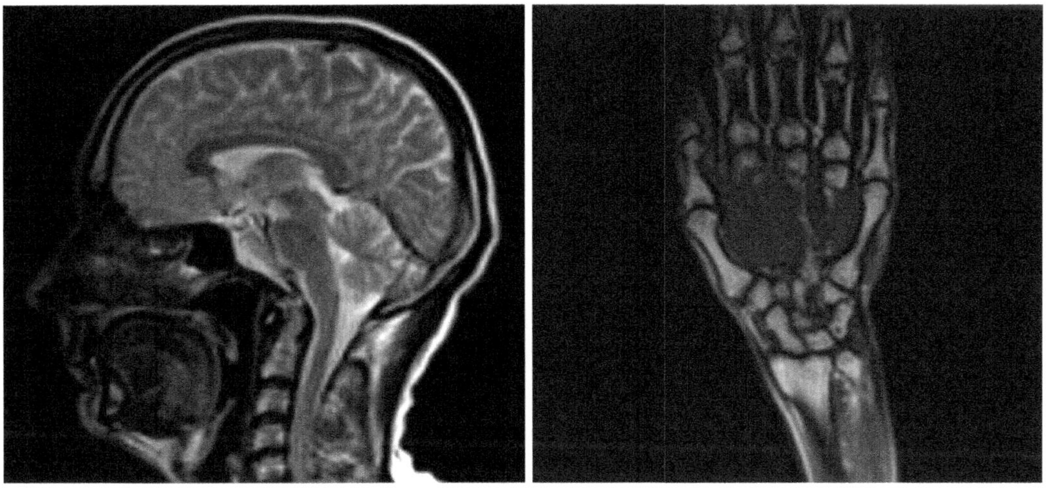

**Fig. 22.3** (Left) The official first image on the Australian MRI-Linac. This is a 5-min T2-weighted TSE image of the author taken in the "beams' eye view" using the open transceiver coil. (Right) The actual first human image (also of the author) which was acquired indirectly as part of initial RF heating measurements

Attention turned to completing the commissioning of the beam and dosimetry. Urszula Jelen, a medical physicist with experience working on particle accelerators, was brought on board to complete this work [9]. At this moment in 2017, no one had yet been treated on an MRL anywhere in the world, and although we were clearly many years behind the commercial developments, it really did feel like we could make some impact. We started to examine various planning and treatment strategies including how we could determine patient position from rapid imaging as well as demonstrating MR-MR registration possibilities between MRL and high-resolution data from the 3 T.

Once the AMRL was up and running a number of PhD projects from the various collaborating institutions were able to begin in earnest. One of the more fruitful of these was a study investigating the effects of gadolinium dose enhancement in rodents [10, 11]. This actually led to the first treatment use of an inline MRL anywhere in the world. It also enabled skin dose mitigation in prospective patient trials via a simple bolus to be studied. Ongoing solutions to this issue are still being examined including off-axis irradiation to keep electrons down the isocenter and out of the beam, work which builds directly on earlier tests performed on the Sonata magnet during Phase I [12].

By the end of 2018, the commercial systems were up and running in other parts of the world, and a first installation in Australia was scheduled for Townsville, northern Queensland, in 2019. At this point I was loudly advocating the need for us to adopt a more general MRI guidance approach in response to these developments. Clearly there were a lot of things we could achieve with the system from a patient perspective alone. It was extremely open and well tolerated, and at one point we inserted an RF window to maximize this feeling of space even further. We'd already started using this space to great effect to perform vertical imaging which even attracted the attention of the Australian defense force who were interested in upright imaging as a way of designing tailored body armor. The magnet had other possibilities too, in terms of adjunct therapeutic equipment that could fit in and around the bore enabling MRI guidance of various complementary treatments.

The final component of the AMRL was the rotating patient couch which was developed over three prototypes and over several years working closely with Craig Andrews and his colleagues at Design Momentum as shown in Fig. 22.4. After a number of attempts at trying to accommodate motors inside the scan room, I got them to completely rethink the approach and instead we came up with a solution that worked from outside the

**Fig. 22.4** (Left) Early days of imaging using a wooden bench top that had to be lifted into place and climbed onto by subjects. By mid-2017 this had been replaced with a moveable table. In this photo chief engineer Bin Dong is about to pass earplugs to Gary Liney prior to some imaging tests. (Right) The final version of the rotating couch shown in situ taken in 2022...after all that, I hear this is now broken!

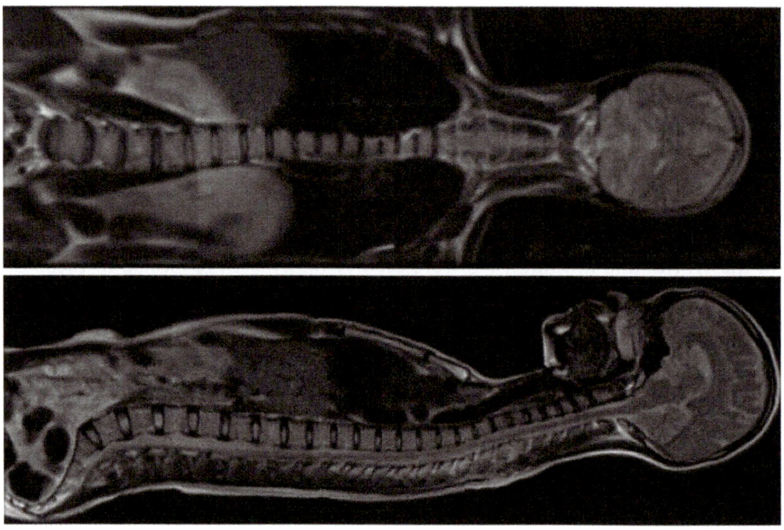

**Fig. 22.5** Illustration of the extended FOV possible by a moving table acquisition. In total an 80-cm length is possible by concatenating smaller sections to provide a final distortion-free image

room. The final design is now driven from external motors that connect to the bed via shafts that pass through a waveguide; this reduces artifacts and makes for a more accurate and quiet performance. The patient rotation provides a unique solution to the fixed gantry design and is potentially very attractive in situations where the beam line may be much harder to move as in the case of proton therapy. The couch is easily docked and retracted and once removed still allows clear access to the bore for vertical imaging providing an alternative treatment position. The large couch translation from the isocenter also allows extended FOV imaging which effectively increases the DSV in the x-direction by a factor of three. By concatenating smaller image sections with simultaneous incremental table movements, we could provide distortion-free images of up to 80 cm in length with no need for gradient correction, B0 shimming, or additional RF coils as shown in Fig. 22.5. A very similar implementation now exists on some clinical scanners. I had described this method in our first group meeting in 2012, and we had actually tried this with the manual table previously, but it wasn't until the accuracy of the new couch could we achieve it properly. Example images shown in Fig. 22.5 show a longer FOV than what is even possible on clinical scanners, with greater uniformity and no need for surface coils on the subject. This may have benefits in the spine and as the couch can also move in two other directions could be applied to other anatomy. And with that last project, and the couch installation now complete, marking just over 10 years in post, my involvement was no longer required.

## Summary

The Australian MRI-Linac was certainly a significant part of my life. With neither the financial or personnel resources of the other big MRL developments, the scale of the task cannot be overstated. Engagement of wider clinical staff in a department unused to this type of work was always tentative. The development ultimately depended on a very small number of enthusiastic and dedicated people. Many careers have benefitted enormously from their association with the AMRL, although sadly those of us directly involved have had to move on. The failure to retain our engineer—so key to its continued operation—was a particularly significant blow.

So what of the future of the AMRL? Well as part of the proposed clinical trial, the Ingham Institute handed over temporary ownership of its once flagship research program to the cancer therapy team in 2020. This team was charged with delivering the first patient treatment. The trial is a basic parallel pair radiation treatment for palliative brain cases and is now the sole focus of the AMRL. The acquisition of the MRI plays no part in the therapy, a move that looks increasingly out of step with the adaptive MR-guided treatments being performed at other hospitals. As of time of writing, the animal study remains the only treatment yet performed on the system, and while this could be seen as a failure with respect to the competing technologies, it should be remembered that the original intention had never been to treat (or even image) a patient. In my opinion it would have been much better to concentrate on the unique aspects of the system—the magnet—and develop a more general MRI guidance approach. It remains to be seen whether there are any plans to change direction in the future.

A much more tangible legacy has been the impetus that the project provided generally to the MRI in RT field within Australia. It was a focal point of numerous meetings and conferences over the years, including the international MRI in RT symposium, which I was very excited to chair in Sydney in 2017. This remains the only time the event has taken pace outside of Europe or the USA. This conference led to a national spin-off which is still very active. The landscape of MRI in RT in Australia has certainly changed in the last decade. When I arrived in 2012, there was no dedicated MRI use in radiotherapy and of course no MRI-Linacs. As of today, there are four MRI simulators and four installations of the Elekta Unity MRI-Linac system. These figures stand up very well against other countries when you consider they serve a combined population of only 26 million. Interestingly only one of these departments has both an MRL and MRI scanner in situ. Three of the MRI-simulator sites operate in isolation and have no access to MRI-Linac. These are quite different working models, and it will be interesting to see how this plays out in the next few years. The dedicated MRI simulator at Liverpool hospital—after a 10-year lifespan—has still never exceeded 40% patient throughput; Access to a nearby MRI-guided treatment system would clearly be an advantage, and those centers lucky enough to have both would seem to be in the ideal position to exploit MRI use to its maximum. In any case, there is clearly a strong future for MRI in RT in Australia, and although I am no longer included in it, I am proud to have contributed to this in some way.

## References

1. Whelan B, Liney GP, Oborn B, Keall P. MRI linac systems. In: Liney GP, van der Heide U, editors. MRI for radiotherapy. Springer. 2019. pp. 155–168.
2. Liney GP, Whelan B, Oborn B, Barton M, Keall P. MRI-linear accelerator radiotherapy systems. Clin Oncol (R Coll Radiol). 2018;30(11):686–91. https://doi.org/10.1016/j.clon.2018.08.003.

3. Keall PJ, Barton M, Crozier S. The Australian magnetic resonance imaging linac program. Semin Radiat Oncol. 2014;24:203–6.
4. Whelan BM, Liney GP, Dowling J, et al. An MRI-compatible patient rotation system design, construction, and first organ deformation results. Med Phys. 2017;44(2):581–8. https://doi.org/10.1002/mp.12065.
5. Liney GP, Dong B, Begg J, et al. Experimental results from a prototype high-field inline MRI-linac. Med Phys. 2016;43:5188. https://doi.org/10.1118/1.4961395.
6. Liney G, Dong B. Radio-frequency coil for simultaneous imaging and dosimetry on an MRI-Linac. Patent no. 10391335.
7. Destruel A, Weber E, Hughes I, Li Y, Liu F, Crozier S. Design of a whole-body radiofrequency coil for image-guided radiotherapy treatment in a MRI-Linac system 3076 Proc. of ISMRM. 2015.
8. Liney GP, Dong B, Begg J, et al. Imaging performance of a dedicated radiation transparent RF coil on a 1.0 Tesla inline MRI-linac. Phys Med Biol. 2018;63:135005.
9. Jelen U, Dong B, Begg J, et al. Dosimetric optimization and commissioning of a high-field inline MRI-Linac. Front Oncol. 2020;10:136. https://doi.org/10.3389/fonc.2020.00136.
10. Liney GP, Jelen U, Byrne H, et al. Technical note: the first live treatment on a 1.0 Tesla inline MRI-linac. Med Phys. 2019;46(7):3254–8. https://doi.org/10.1002/mp.13556.
11. Byrne HL, et al. Enhanced MRI-guided radiotherapy with gadolinium-based nanoparticles: preclinical evaluation with an MRI-linac. Cancer Nano. 2020;11:9. https://doi.org/10.1186/s12645-020-00065-5.
12. Roberts NF, Patterson E, Jelen U, et al. Experimental characterization of magnetically focused electron contamination at the surface of a high-field inline MRI-linac. Med Phys. 2019;46(12):5780–9. https://doi.org/10.1002/mp.13847. Epub 2019 Oct 21.

# Low-Field MR Imaging

# 23

Yingli Yang, Shaolei Li, Daniel A. Low, Zhimin Li, and Peng Hu

## Introduction

Magnetic resonance imaging (MRI) is a mainstay diagnosis tool in modern medicine, providing significant value in both scientific and clinical applications. MRI offers excellent soft-tissue contrast and the capability of probing a wealth of physiological and physical parameters. The development of MRI systems in the early days started with low-field systems and permanent magnets. These permanent magnet designs had an upper limit in achievable field strength and relatively poor $B_0$ field uniformity. In the late 1980s and early 1990s, superconducting magnets were developed which allowed higher field strengths and better field homogeneity. Since then, MRI has been on a continuous trajectory of higher field strengths in pursuit of greater signal to noise ratio (SNR), with 3-T scanners becoming a clinical mainstay and 7-T and greater systems being more widely adopted.

However, in recent years, there has been a renaissance in low-field MRI systems because of economic, technological, and scientific factors. From an economic perspective, MRI is widely used in high-income countries (HICs) with 0.19 scans/persons annually in the USA [1, 2] but has limited adoption in low-income and middle-income countries (LMICs) due to high costs and many technical barriers. Even inside HICs, MRI is usually unavailable to rural areas or for patients with disability or device constraints [3]. Worldwide, only 10 percent of the population has access to MRI, and even within LMICs an uneven distribution of this important diagnostic tool persists [4, 5]. There are two major factors responsible for this situation: (1) the cost of installation and post-installation maintenance and (2) the complexity of operating MR systems and (3) technical know-how. Given the extraordinary imbalance in healthcare availability in LMICs, there has been increasing interest in revisiting low-field MRI, which, while not producing the highest-quality images, should be able to provide diagnostically useful information. From a tech-

Y. Yang (✉) · S. Li
Institute for Medical Imaging Technology, Ruijin Hospital, Shanghai Jiaotong University School of Medicine, Shanghai, China

Department of Radiology, Ruijin Hospital, Shanghai Jiaotong University, School of Medicine, Shanghai, China
e-mail: sli135uiuc@126.com

D. A. Low
Department of Radiation Oncology, University of California, Los Angeles, CA, USA
e-mail: Dlow@mednet.ucla.edu

Z. Li
Department of Radiology, Northwestern University, Chicago, IL, USA
e-mail: zhimin.li@nm.org

P. Hu
School of Biomedical Engineering, ShanghaiTech University, Shanghai, China



I. J. Das et al. (eds.), *A Practical Guide to MR-Linac*, https://doi.org/10.1007/978-3-031-48165-9_23

nological perspective, the past 30 years have seen tremendous advances in relevant technology, from magnet design to gradient and radiofrequency system performance. Many technologies developed after the era of lower-performance low-field systems, such as active shielding, parallel imaging [6], and compressed sensing [7], can now be adopted by modern low-field systems, greatly boosting their performance. More importantly, such a setup could potentially allow for interesting applications that would be challenging for conventional higher-field strength systems. For example, MRI-guided catheter interventions, where the patient undergoes a catheterization procedure under continuous or intermittent MRI guidance and real-time feedback, are potentially more feasible at lower field strengths due to reduced heating of various devices [8, 9] in the MRI environment. Lung imaging is another example where low-field systems could provide superior imaging due to the reduction of magnetic susceptibility effects [10].

Low-field MRI is usually defined as systems with <1.0-T main magnetic fields [2]. In Fig. 23.1, low-field MRI systems are shown with a variety of main magnetic fields. Originally, the performance of low-field MRI systems was rated lower than their actual clinical usefulness, and they were viewed as low-cost but poorly performing systems. The older type of MRI systems had limited spatial resolution associated with poor image quality, limited available receiver coils, limited types of imaging sequences and parameters, and insufficient temporal resolution associated with low SNR. In the early phase of MRI development, it was generally accepted that higher static field strength would inevitably lead to improved performance. This was based on the notion that the MR signal amplitude is a quadratic function of the static magnetic field strength while the noise is only linearly proportional to the field strength. According to this principle, a 3-T MRI system would theoretically have 15 times the SNR compared to a 0.2-T MRI system [11, 12].

The concept of linear SNR dependence on magnetic field strength has since been determined to be not entirely true [13, 14]. T1 and T2 changes at higher field strengths could partially negate those SNR gains due to higher magnetic polarization at higher field strengths. The progression in both software, including pulse sequences and hardware, including gradient and imaging coils, has enabled relatively high-performance imaging techniques even at lower field strengths. In the past, imaging sequences for low-field MRI systems were relatively limited and basic, but now they have become quite sophisticated and are comparable to the imaging sequences of 1.5-T MRI systems that are widely used clinically [8]. In addition, the low SNR inherent in low-field MRI systems has been improved by image reconstruction methods using deep learning and de-noising techniques. This has further improved image quality, and studies have shown sufficient imaging performance for various clinical applications [15, 16]. Low field also provides benefits from a patient safety and comfort perspective, in terms of reduced risks of projectiles, decreased implanted device heating, and ability to image tissues closer to implants due

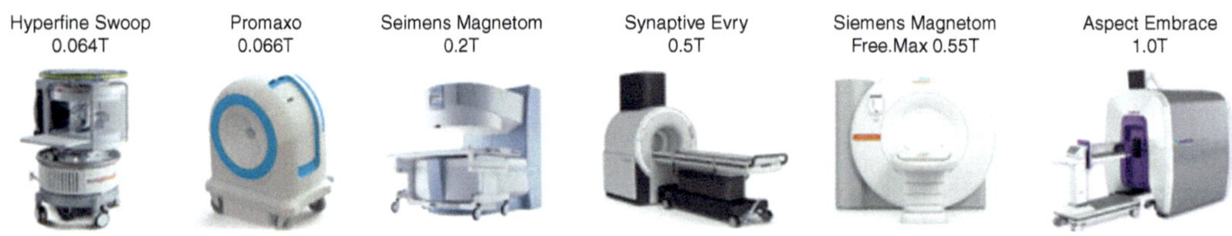


**Fig. 23.1** Defining "low-field" MRI. This figure defines how scanners at difference field strengths will be categorized in the article, with the following boundaries illustrated on the bottom: low-field (LF) ≤ 1.0 T < high-field (HF) ≤ 3 T < very-high-field (VHF) < 7 T ≤ ultra-high-field (UHF). Select commercially available scanners with a field strength of 1 Tesla or lower are illustrated in this figure. Scanners are not to scale. Scanner images are copyright of the respective manufacturers. (Images used with permission or in accordance with manufacturer pol [2])

to smaller magnetic susceptibility, reduced specific absorption rate (SAR) for high-SAR pulse sequences, and reduced acoustic noise because of lowered forces on the gradient coil windings with a given current amplitude [12]. The other attractive points of low-field systems are mobility due to the lower weight, which could take MRI to the point-of-care, similar to an ultrasound unit.

Lower-field MRI still faces significant challenges, and it remains to be seen how newer devices will be deployed clinically. In this chapter, we discuss the advantages of and challenges for low-field MRI systems. In addition to diagnostic imaging, we will also further illustrate the potential application of low-field MRI to radiotherapy in the setting of MRI-guided linear accelerators.

## Physical Properties and Engineering Challenges of Low-Field MRI

Since the invention of MRI, researchers have made efforts to pursue higher magnetic fields. Worldwide, there have been almost the same number of 1.5-T and 3-T systems sold [17]. The efforts of increasing static field already focused on 5-T or even 7-T whole-body systems, some of which have already advanced to the stage of targeted clinical and clinical research studies. Although they were not developed for low-field MRI, many advances in hardware and post-processing algorithms can benefit low-field imaging. In this section, we review the current trends and developments in the various hardware components and the engineering challenges of low-field MRI scanners.

### Physical Properties of Low-Field MRI

One of the key properties of MRI systems is the static magnetic field. MRI needs a very strong magnetic field strength, and conventional MRI uses a superconducting coil that requires a cryogen system to remain superconducting. Permanent and/or resistive electromagnets had been the mainstay for low-field systems until the past 3–5 years. The homogeneity of the magnetic field should be on the order of parts per million (ppm) over an ellipsoidal imaging volume, with fluctuations during the scan period of less than 100 nT [12]. For a low-field system operating at between 0.25 and 0.5 T, there are two basic choices of the magnet, one using a permanent neodymium-iron-boron magnet or one that combines a permanent magnet with an additional electromagnet. A more recent system, the Siemens Free.Max, uses a 0.55-T superconducting magnet.

### Designing Coils for Low-Field MRI

An MRI system as shown in Fig. 23.2 is a mix of several subsystems which provide the necessary functions for imaging.

The general consideration for designing coils (especially RF coils) includes the direction of the main magnetic field, bore diameter, size of the homogeneous volume, and the relation between electric current and magnetic field. The coils are designed as resonant structures to maximize their efficiency and sensitivity. The resonant frequency is given by the formula:

$$f = \frac{1}{2\pi\sqrt{LC}} \tag{23.1}$$

where $L$ is the inductance in Henries and $C$ is the capacitance in Faradays. The orientation of the static ($B_0$) field in most conventional low-field systems is vertical, which means that a solenoid coil can be used, which generally provides severalfold greater efficiency than a transverse resonator.

The main factor for designing a coil is to maximize SNR, which is proportional to the static magnetic field. The major source of noise in the MRI signal is different between low-field and high-field systems. At a lower field of $<0.5$ T, the MRI signal noise in general is dominated by noise from the RF signal processing chain, such as A/D converters, amplifiers, etc. However, at higher field strengths, noise from the magnetizations in the subject becomes more significant. One of the major advances in the past two decades in clinical systems has been the incorporation of multiple receiver elements (receiver arrays) both

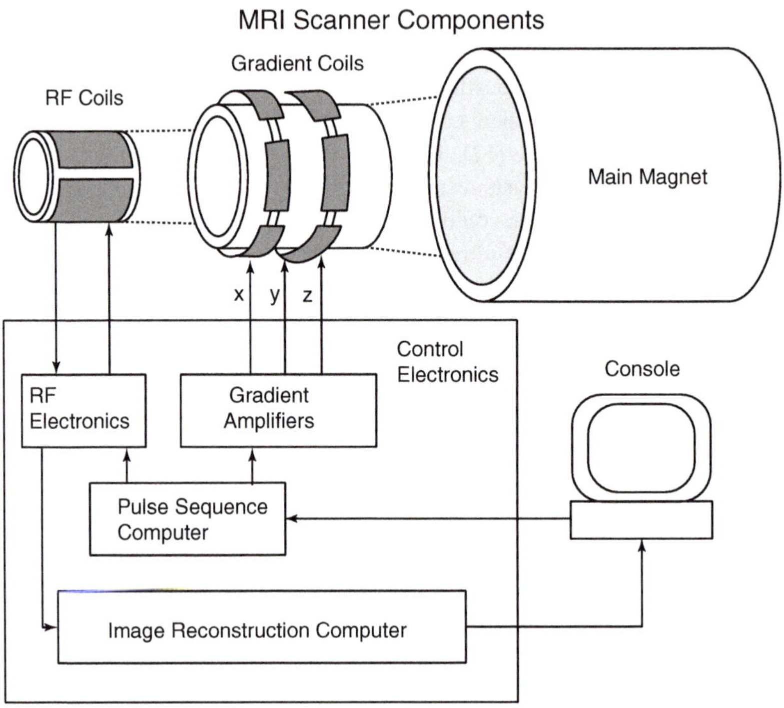


**Fig. 23.2** The basic components of any MRI system: the main magnet produces the $B_0$ field, necessary to align the spins and achieve equilibrium. Gradient coils enable image encoding in the x, y, and z directions (i.e., the frequency, phase, and slice-encoding directions). The RF coil is the part of the MRI system that excites the aligned spins and receives an RF signal back from the sample. All the components are controlled and interfaced with the user via a console (Gruber et al. JMRI 2018) [18]

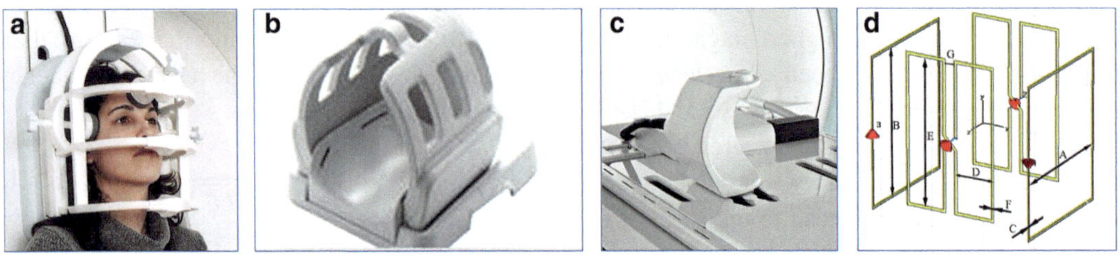


**Fig. 23.3** Examples of RF coils used on low-field MRI systems. (**a**) Quadrature transmit/receive coil on the vertical 0.6-T Fonar system. (**b**) Four-element head array on the 0.25-T Esaote. (**c**) Shoulder-phased array for the Siemens 0.35-T Magnetom. (**d**) One example of a research-phased array designed for 0.25 T with a loop/butterfly coil arrangement. (**a**–**c** courtesy of FONAR Corporation, Esaote and Siemens Healthineers, respectively) (J.P Marques JMRI 2019) [12]

for higher SNR and reduced imaging time when using parallel imaging techniques (examples are shown in Fig. 23.3). In this way, there is an SNR gain mainly in the periphery of the image, and the use of multiple receive elements also enables faster scanning times through sparse sampling of k-space [19, 20]; see details of k-space in Chap. 5 of this book .

With a very large inductance at low field, the decoupling of receive coil elements is challenging because instead of being well coupled to the body, they tend to couple to each other. The inter-coil coupling is calculated as

$$X_L = \omega L \qquad (23.2)$$

where $L$ is the inductance and ω is the angular frequency. Two decoupling strategies, capacitive decoupling and geometric decoupling, are usually considered solutions, but they both have drawbacks. Preamp decoupling is another solution. Based on the factors from the previous paragraphs, some designs, such as optimized wire patterns and a separate transmitter/receiver including quadrature coil and array coils, are engendering interest. In this case, many low-field systems start to use receive arrays with their elements being of relatively large size so that coil noise does not dominate. Additional research in this area is being done that might provide a flexible coil [21].

## Permanent Magnet System

The earliest low-field MRI systems were constructed using various forms of electromagnets and/or permanent magnets. Nowadays, a static field of 1.5 T has become the clinical standard in HICs, almost completely replacing the older lower-field strength (0.2–1 T) systems that had an important role during the 1980s [17]. The early designs of low-field MRI used permanent magnet arrays (PMAs), with and without iron yokes. A PMA has no power consumption and is low-cost. Compared to a superconducting magnet, a PMA does not require sophisticated cooling systems to operate, although the field strength is relatively low. Also, compared to a resistive electromagnet, it has no heat dissipation resulting from providing the main static field. The materials used in PMAs vary, as does the maximum achievable field strength. For example, a magnet based on neodymium can generate a magnetic field of around 1.4 T. The equivalent surface current parallel to the magnetization, if being considered as dipole moment as shown in Fig. 23.4, can be as high as the order of $10^4$ A. Even though this is a simplified example, it highlights how powerful these magnets are compared to electromagnets.

One category of permanent magnet systems that is popular in both research and commercial applications are yoked magnets. An example [22] of a yoked magnet that is shown in Fig. 23.5 is able to generate a strong static magnetic field with good homogeneity.

Despite the achievements of yoked magnets, they are still limited by their heavy weight, requiring a supporting structure usually made of steel and iron. Conversely, the category termed yokeless magnets is more diverse and can broadly be categorized into two subtypes: homogeneous magnets and single-sided magnets. The homogeneous magnets are typically cylindrical and bipolar shaped. Efforts have been made to reduce field inhomogeneity by increasing the diameter of the cylinder or decreasing the space between the rings; it is also possible to use inhomogeneity for spatial encoding as an example shown in Fig. 23.6 from Vogel's work [23].

The same principle can be used for single-sided magnets as well even though it is harder to control the field. Single-sided magnets usually have small sizes and are also easier to position. The cylindrical Halbach array is used for generating a homogeneous encoding field as shown in Fig. 23.7.

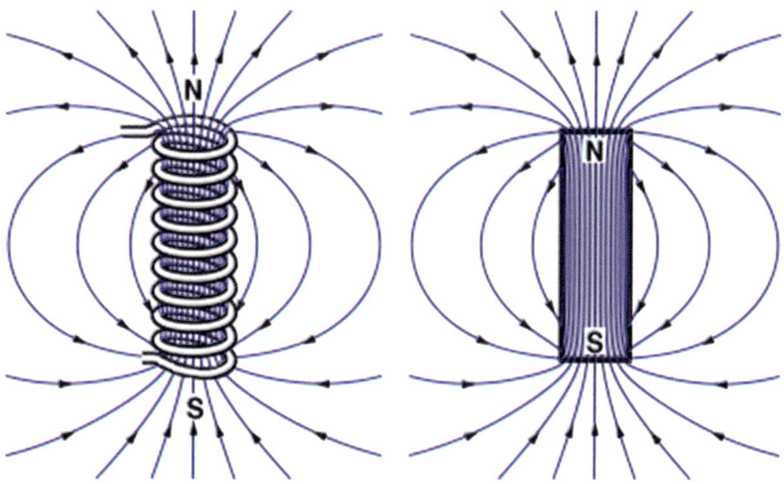


**Fig. 23.4** Permanent magnets generate a magnetic field similar to that of a coil with an equivalent current. (From Wikipedia)

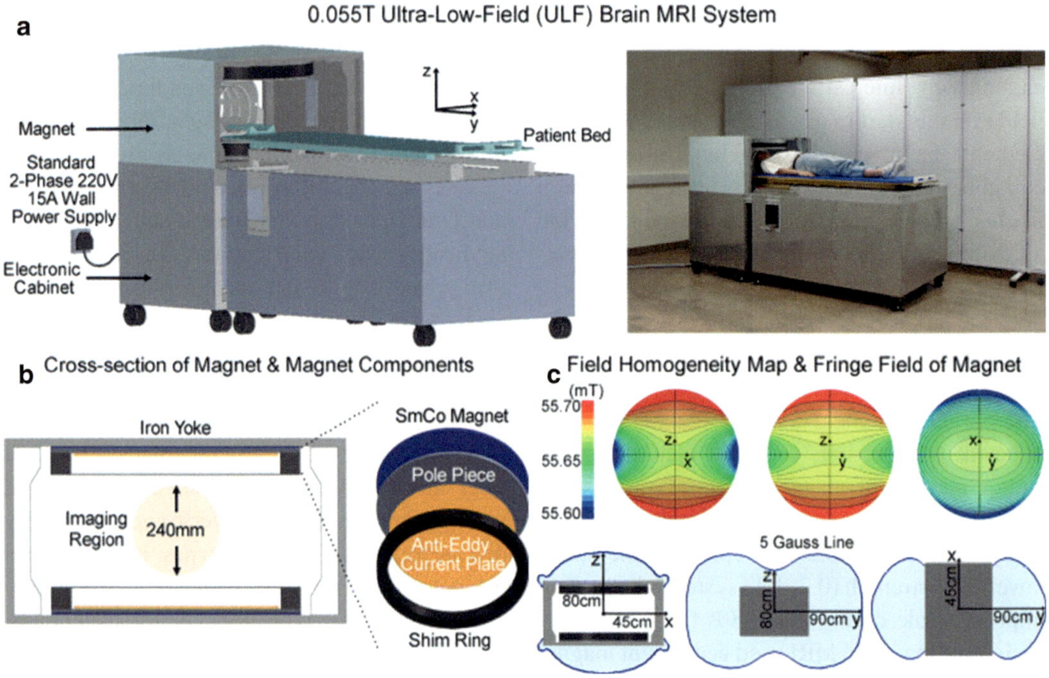


**Fig. 23.5** Prototype of a low-cost low-power and shielding-free brain ultra-low-field (ULF) magnetic resonance imaging (MRI) scanner with homogenous 0.055-Tesla magnetic field and small 5-Gauss fringe field (Liu et al. [22])

A series of applications for using Halbach magnets with external shimming for portable low-field scanners has shown optimal results, and researchers are continuously optimizing the field pattern of cylindrical Halbach arrays in a way that retains construction simplicity.

## Hardware Requirement for Point-of-Care and Portable MRI

A major advantage of low-field MRI systems is their accessibility, which makes them ideal as portable MRI scanners. We can even consider a single-handed device in the future. A group of representative MRI systems with different field strengths are shown in Fig. 23.8.

Regardless of the field strength, safety is always of critical importance in clinical workflow. The major hazards during MRI scanning are projectile accidents and injury to patients with implanted devices. Hence, a safety zone is necessary for conventional MRI. The American College of Radiology has published an MRI safety manual in which four zones with increasing levels of danger are defined.

As shown in Fig. 23.9, only zone I is considered completely safe to the public where the magnetic field is 5 Gauss (0.5 mT) or less [26]. However, a locked-off safety is feasible for portable devices. One solution is to use low-field magnets with a small magnetic footprint and fringe field as mentioned in the previous section. A physical barrier or at least an indicator has also been used in some commercial products. Hyperfine [27] uses a barrier for the 5-Gauss zone. Another tactic is to use an on/off magnet. Synaptive [28] 0.5-T product adopts this solution allowing a relatively high magnetic field since the magnetic field can be shut down.

The infrastructure and logistics are other important aspects of designing portable MRI scanners. To increase accessibility, infrastructure requirements, specifically cooling and power, need to be minimized. Portable scanners also need to be able to operate from standard wall outlets. Several groups and companies have demonstrated the achievability of this requirement

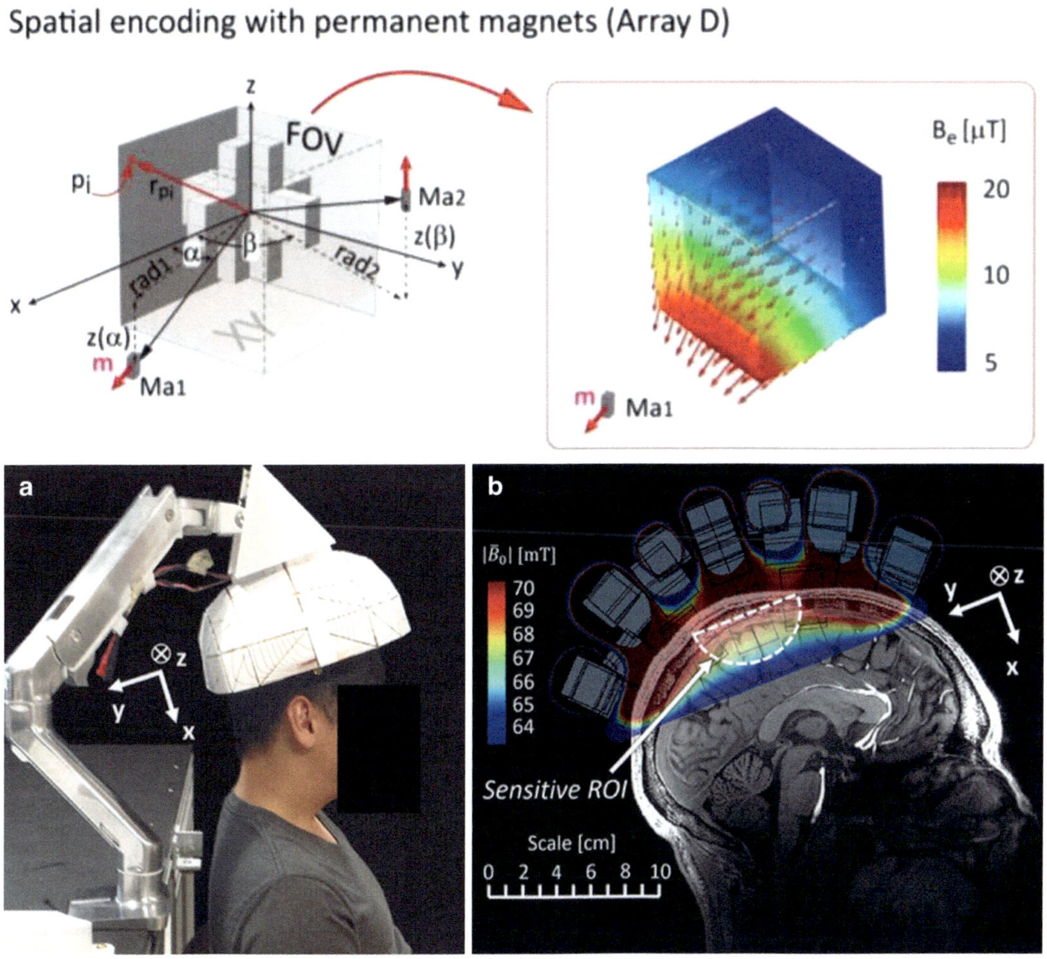


**Fig. 23.6** (Top) Spatial encoding with magnet arrays [23]. (Bottom) Single-sided magnets from McDaniel et al. [24]

(e.g., Fig. 23.5). Pushing even further, it is conceivable to power low-field MRI systems using batteries, a feature that exists for some portable CT scanners.

The portable MRI also needs a different cooling system than conventional scanners. Since a cryogenic environment is not practical, a water-cooling system is becoming a popular choice. Most of the portable MRI permanent magnets and the water-cooling systems are used in conjunction with compact unshielded gradient coils.

For portable systems, a critical logistical consideration is the integration of the patient table with the scanner. For example, Hyperfine uses a fold-down patient table and integrates it with a standard patient bed for scanning. In general, the design of a portable MRI needs to be lightweight and compact in order to be practically transportable.

A portable MRI scanner operates in locations that typically lack the RF shielding that conventional high-field systems use to prevent external artifact-generating RF interference. As a result, portable scanners operate without shielding or retrospectively removing interference from the images. Also, compact magnet coils can generate nonlinear gradient fields and field inhomogeneities that cause image distortion. Nonlinearity corrections and shimming are necessary to reduce these effects, and retrospective corrections during the image reconstruction process are likely necessary. One approach is shown in Fig. 23.10.

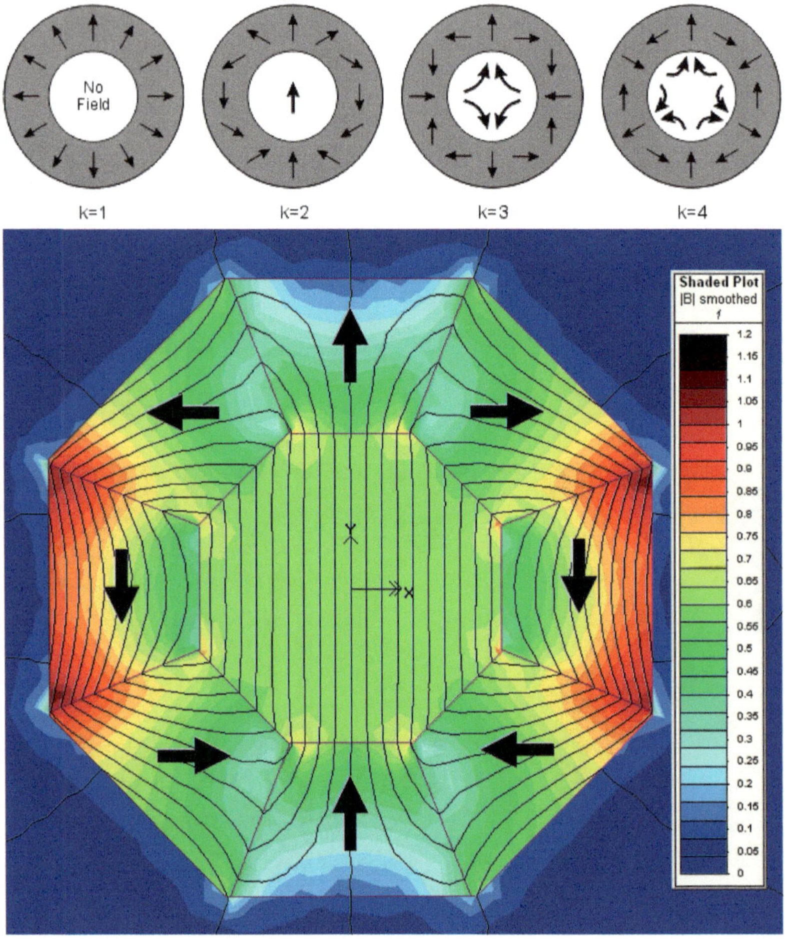


**Fig. 23.7** Field direction and simulation of higher-order circular Halbach array (from Wikipedia)

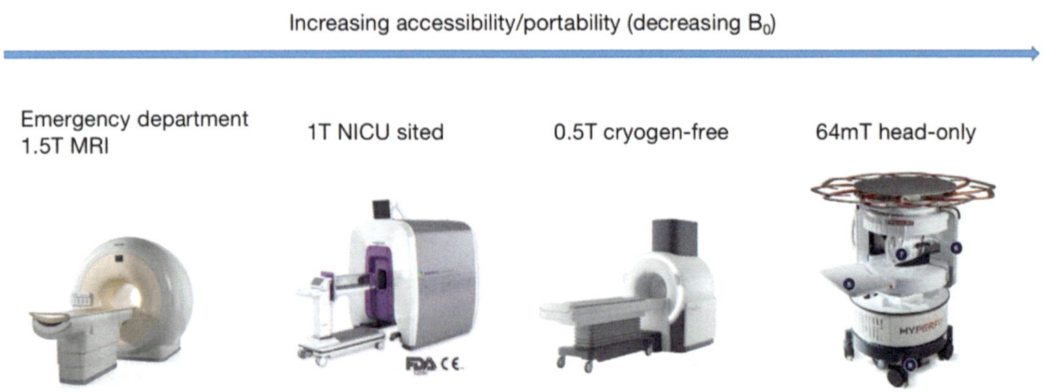


**Fig. 23.8** The MRI systems with different magnitude of static magnetic fields. The hardware requirements for these systems are different

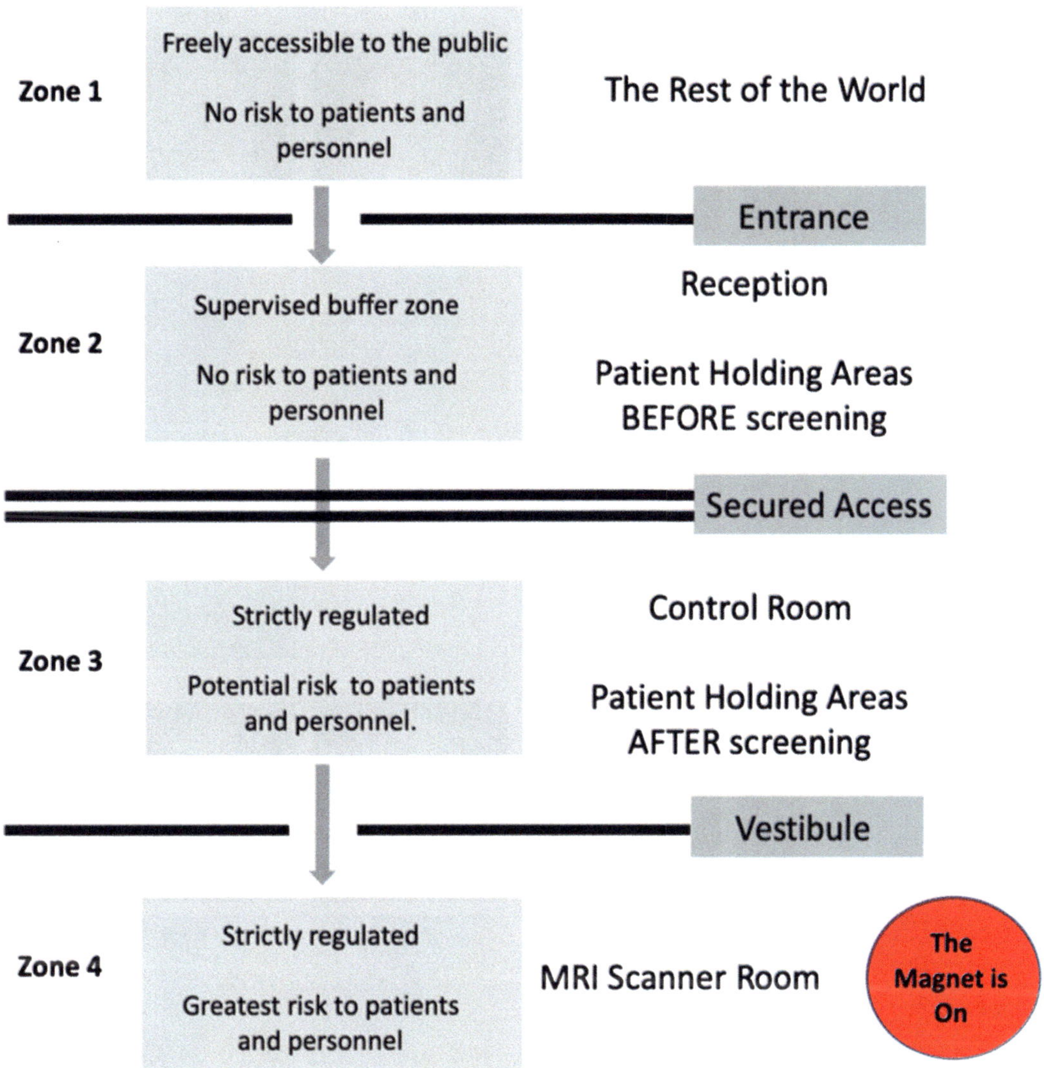


**Fig. 23.9** The MRI zone sign set from the American College of Radiology MRI safety manual [25]

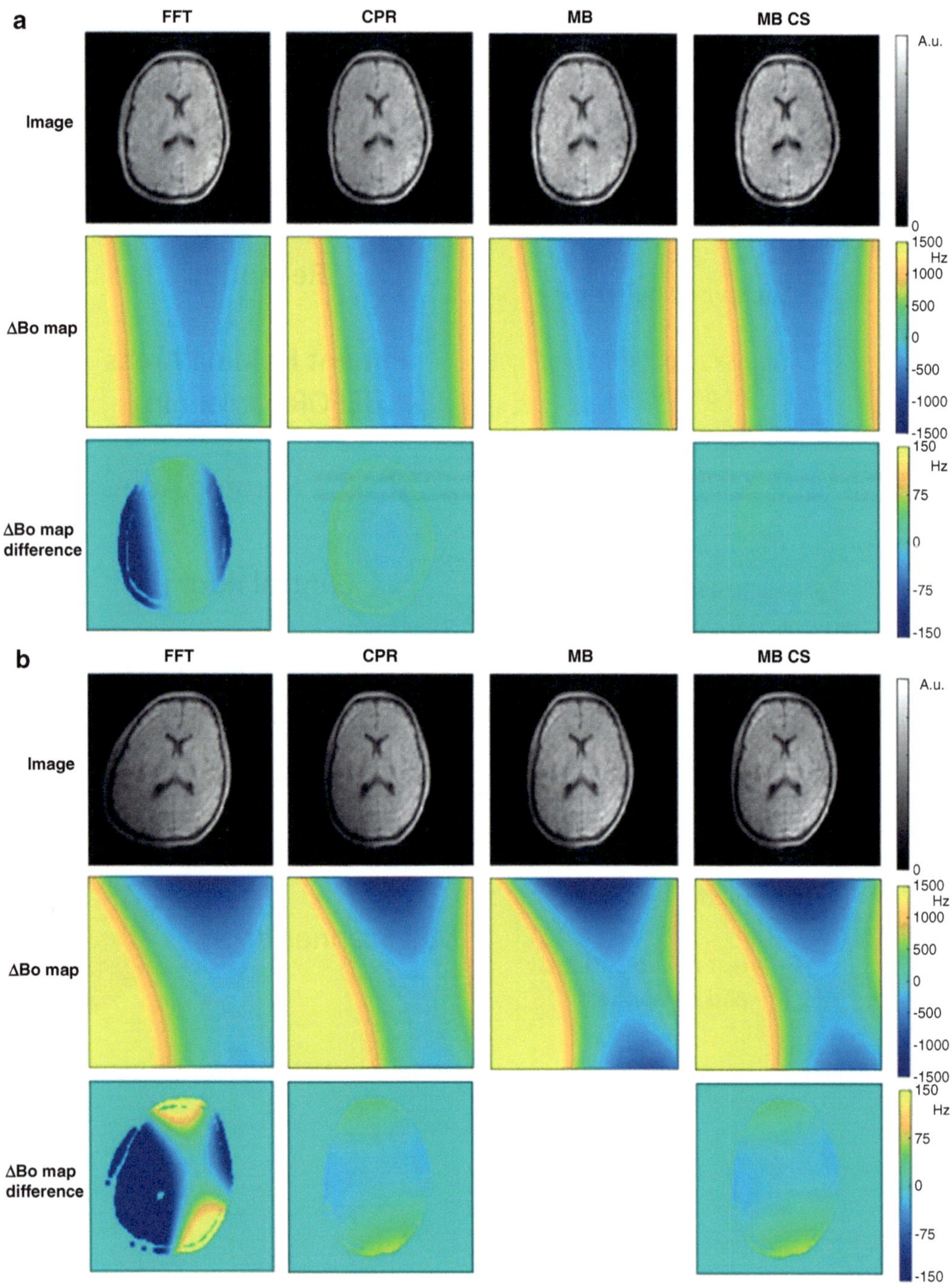


**Fig. 23.10** Comparison of FFT, iterative MB, and iterative CPR approach for an in vivo brain scan. (**a**) Mild B0 inhomogeneities (~ 600 Hz). (Top row) The FFT reconstruction results in a stretch of the brain along the readout direction (left to right). Both iterative CPR and iterative MB image reconstruction result in low errors in the estimated field map, and correct for the corresponding image deformations. The estimated $\Delta B_0$ maps (middle row) and the differences with respect to that of the FFT approach (bottom row) show a similar improvement in the accuracy of the field map estimation when using the iterative CPR and MB approaches. The MB CS reconstruction for an undersampling factor of 2 shows the flexibility of the MB approach toward undersampling. (**b**) For strong $B_0$ inhomogeneities (~ 1500 Hz), here obtained by attaching an external magnet to the Faraday cage, the iterative MB approach still results in a uniform signal intensity, while the iterative CPR approach shows more signal shading in the lower-left corner of the brain [29]

## Advantages of Low-Field MRI

Conventional MRI, although widely used in most clinical areas, still has limitations. For some patients with metal implants, conventional MRI can cause safety risks, and the metal implants can cause artifacts as well. Low-field MRI has advantages in resolving these issues. Low-field MRI is also good at imaging near air-tissue interfaces (e.g., lung, airway). In this section, a review of the advantages and limitations of low-field MRI in these cases is presented.

### Imaging of Patients with Implants

When preparing an MRI scan, the MRI system operator needs to follow their institutional MRI safety protocols, which typically involve a safety questionnaire about whether the patient has metallic implants. In cases where patients do have implants, the operator may need to consult resources, such as calling the device manufacturer, reading books, or online resources by Shellock et al. [26] to determine whether the patient can be scanned.

There are several issues with metallic implants and devices. The most apparent one is the magnetic field interaction in form of both force and torque exerted on the devices, which can cause them to dislodge. Another significant concern is device heating caused by the RF field. At a higher field with correspondingly higher Larmor frequencies, the fast-changing electromagnetic field of the RF pulse can substantially raise the device surface temperature, which can cause tissue damage. This heating issue is generally mitigated at lower field strengths, resulting in overall better safety profiles. Another issue is the artifact caused by metal implants. Metal implants disturb the main field because the implant's magnetic susceptibility differs dramatically from human tissue. These susceptibility differences cause a large-scale main field variation that extends well beyond the physical boundary of the devices. As a result, tissues at a short distance from these implants (often within centimeters) are not imaged well due to off-resonance artifacts. Numerous metal implants cause artifacts on MRI images. Metal causes spatial variations in the static magnetic field ($B_0$) and T2* dephasing as well as the frequency shift $\Delta f$, resulting in slice shifts by the magnitude ($\Delta f/BW_{RF}$ * slice width) and readout displacements of ($\Delta f/BW_{pix}$ * pixel width). It can also cause background gradients and a $B_1$ shading effect. Examples of $B_0$-inhomogeneity metal artifacts are shown in Fig. 23.11. Figure 23.12 shows the dependence of signal void artifacts on the field strength. The device-dependent signal void tends to be larger at higher field strengths. Although the artifacts cannot be completely removed, low-field scanners inherently have smaller artifacts than higher-field scanners due to their reduced device-dependent off-resonance.

Many methods have been proposed to correct distortion artifacts caused by metal implants. For example, to remove in-plane distortion artifacts, a sequence named view-angle tilting (VAT) can be used. VAT corrects for in-plane distortions by adding a gradient on the slice-select axis during readout. The sequence does not require additional scanning time than the normal spin-echo sequence. Figure 23.13 illustrates the use of VAT.

VAT does not fix through-slice distortion, and to fix this problem, another sequence called multispectral imaging (MSI) is considered. This method uses a combination of spin echo and VAT to generate 3D images of each slice. After adding up the slices, the cumulative distortion-free images are reconstructed.

Table 23.1 shows a comparison of 0.55 T vs. 3 T in terms of several parameters related to MRI safety and image quality.

At the lower field, the SNR is lower, but all the other parameters from Table 23.1 are more desirable. The lower SAR means better implant safety, and high-SAR sequences are easier to perform at a lower field; the lower off-resonance means the smaller range of frequency to correct for; the longer T2* enables longer readout; a higher peak $B_1$ means a higher bandwidth and a more efficient encoding.

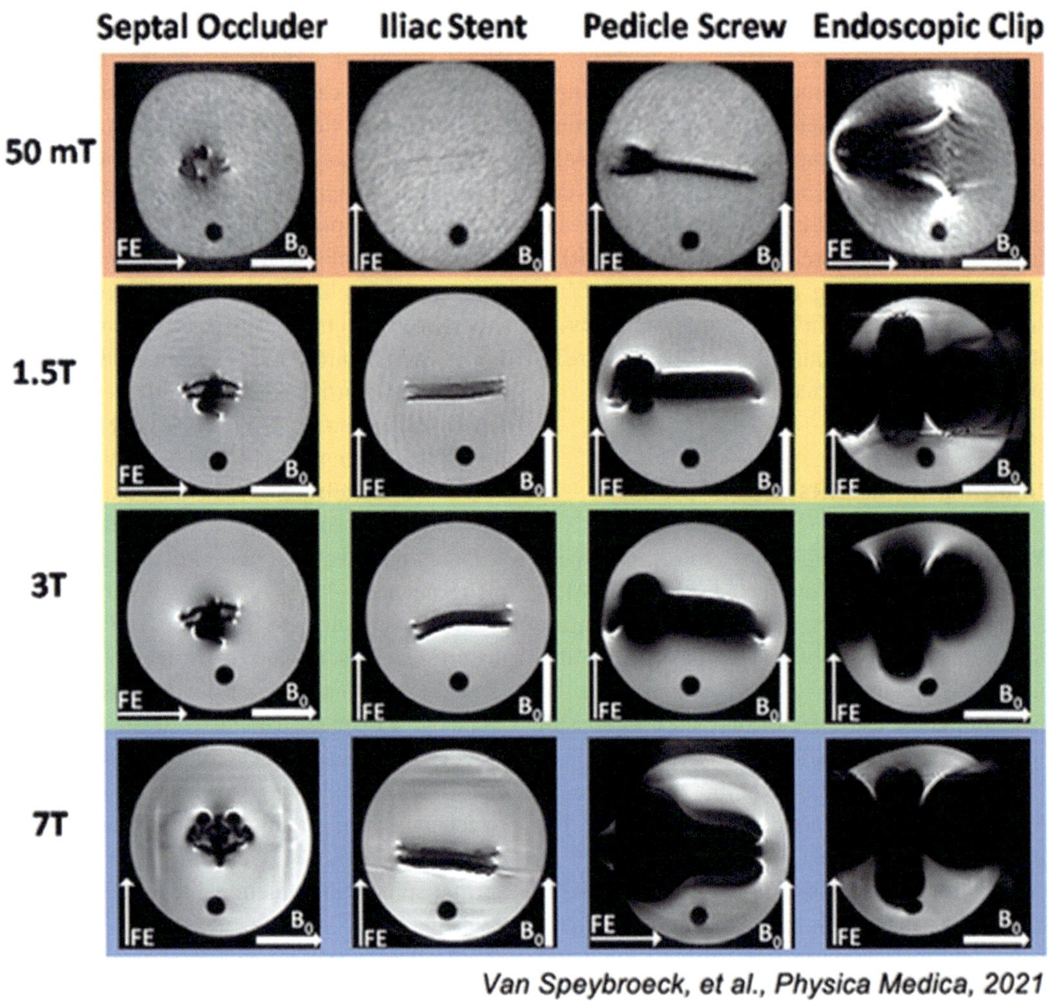


**Fig. 23.11** Metal artifact from 50 mT to 7 T. Here we show Fig. 7 from Van Speybroek et al. [44] where the "worst-case" scenario images for four types of metal implant are demonstrated. (Images reused via open-access license) [30]

## Imaging of Organs with Air-Tissue Interfaces

MRI signal strength depends on water proton density. This causes an issue for imaging organs with air-tissue interfaces, such as the lungs. The lungs have much lower proton density and diminished intrinsic MRI signal. Usually, there is a relatively low SNR and long acquisition speed when imaging these organs. Low-field imaging can improve morphological imaging because, at low field, there is a relatively longer T2* value and susceptibility also decreases. In this section, we will review the role of low-field imaging in the lungs.

The lung has long been neglected by conventional proton MR imaging methods. Numerous studies have described lung MRI methods using hyperpolarized gases [33]. This is due to unique intrinsic difficulties such as the lower proton density and susceptibility artifacts caused by multiple air-tissue interfaces. Clinically speaking, CT is commonly used for lung imaging. However, CT provides excellent spatial morphological information but limited functional information. Pulmonary function tests (PETs) are global measures of lung function but lack highly resolved spatial information and are inadequate in assessing disease heterogeneity. Then, the question becomes whether

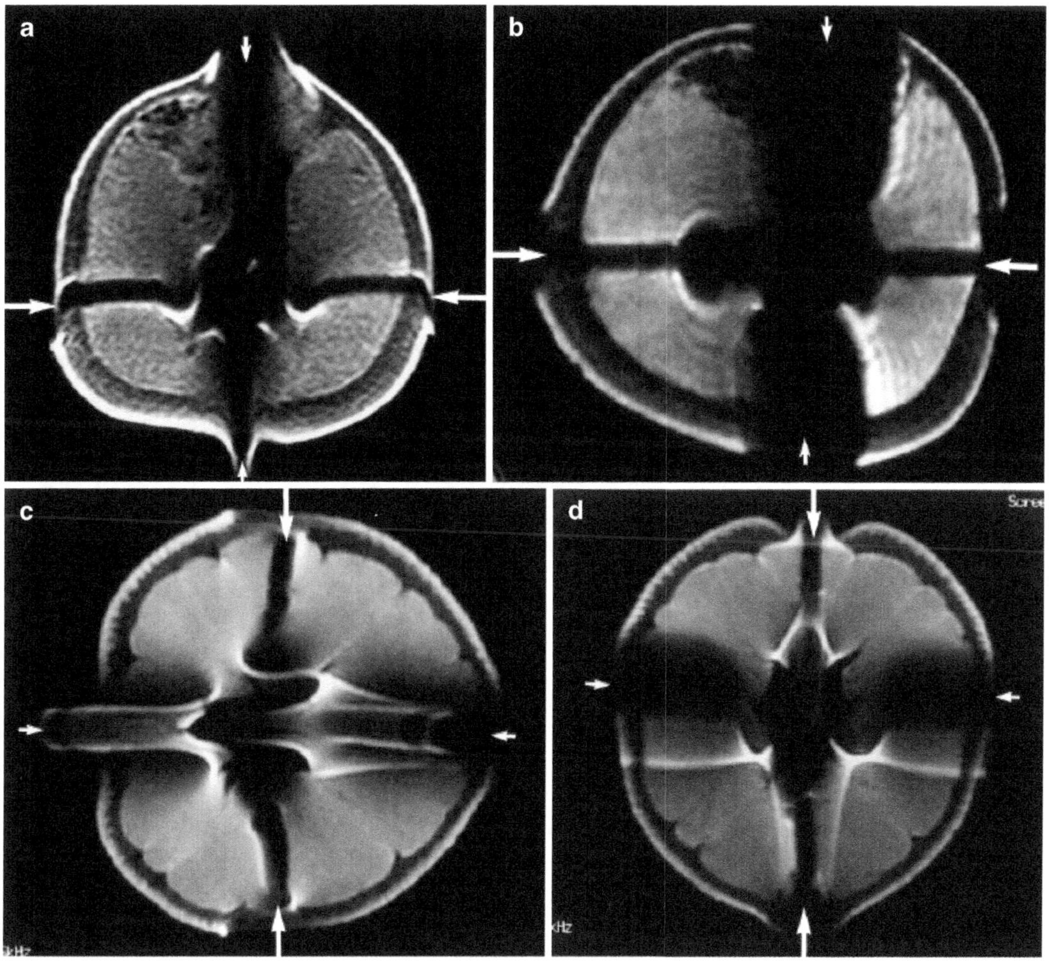


**Fig. 23.12** An orange containing six ferromagnetic metallic surgical screws (5 mm in diameter) is imaged at 0.2 T in the axial plane using (**a**) spin-echo T1-weighted and (**b**) gradient-echo T1-weighted sequences. The same phantom is imaged at 1.5 T in the coronal plane (**c**, **d**), using spin-echo T1-weighted sequences, the frequency-encoding axis being switched between the two series. Note that the screws parallel to the direction of B0 (long arrows) produce less artifacts than those perpendicular to the direction of B0 (short arrows). The artifacts are more important on (**b**) gradient-echo than on (**a**) spin-echo sequences because of loss of signal intensity due to intravoxel dephasing (T2* effect). There is less image distortion at (**a**) 0.2 T than at (**c**) 1.5 T for comparable screws (those parallel to the direction of B0; long arrows), the frequency-encoding gradient being perpendicular to the direction of B0. Note that when the phase and frequency-encoding axes are switched (**c**, **d**), the image distortion in the screws parallel to the direction of B0 (long arrows) is less marked when the frequency encoding axis is parallel to the direction of B0 (**d**) and the distortion increases in the screws perpendicular to the direction of B0 (short arrows). Interchange of the frequency and phase-encoding axes also changes the shape and orientation of the artifacts in the direction of the frequency-encoding axis [31]

there exists a one-stop shop for morphologic and spatially resolved functional imaging.

As we have mentioned in the beginning, conventional MRI suffers from low SNR, and it cannot provide enough details of the structure of the lung. It can be boosted by techniques using ultrashort echo (UTE) [34]. A study [35] by Campbell-Washburn comparing T2-weighed imaging with CT using a 0.55-T MRI shows a good correlation between them.

Another example is shown in Fig. 23.14 from a COVID-19 pneumonia patient.

Other than morphologic imaging, functional imaging is also an important aspect to consider.

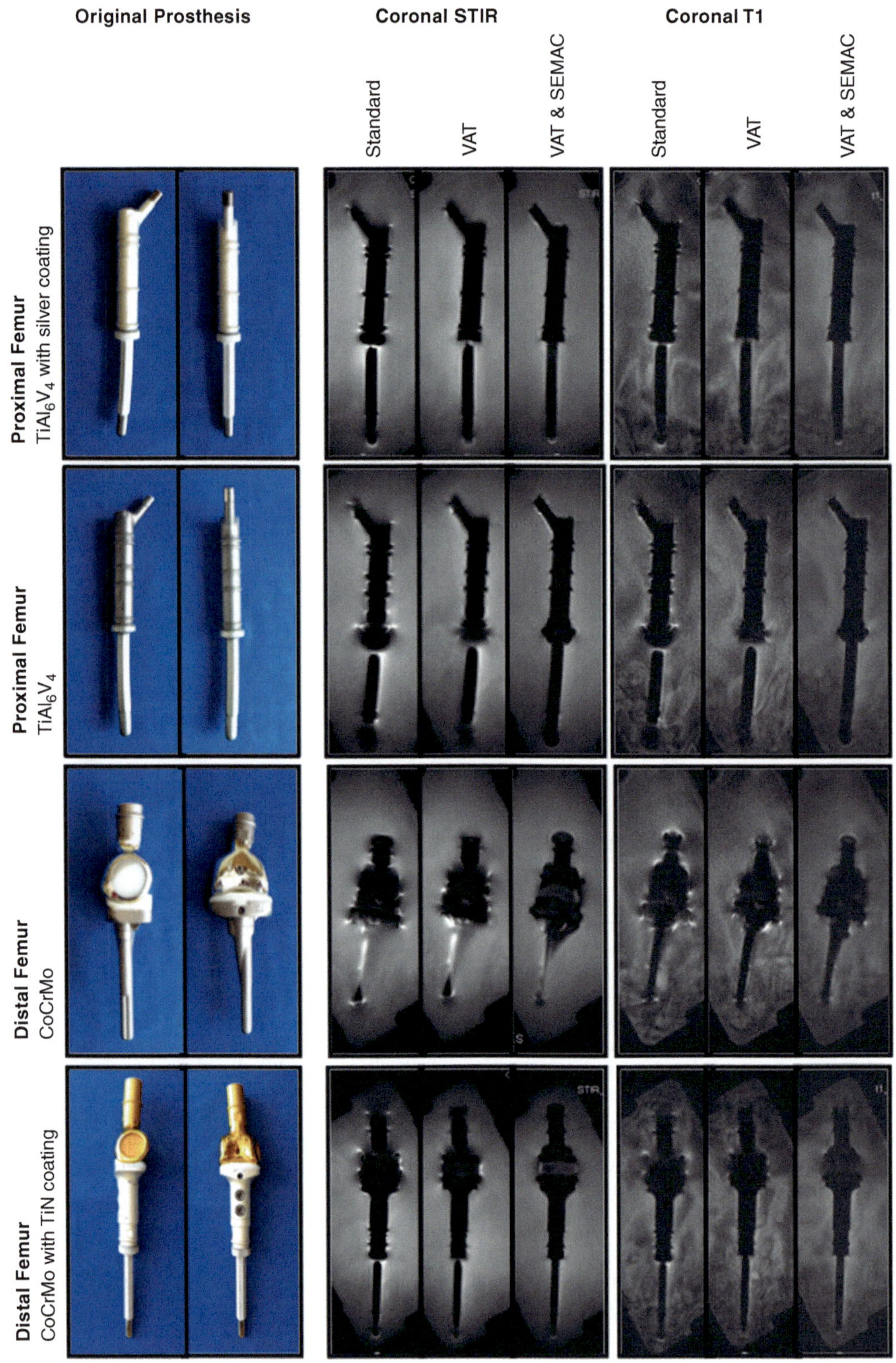


**Fig. 23.13** Experimental model. Tumor endoprostheses were embedded in water. Specifications and materials are indicated on the left. Central slices of coronal STIR and coronal T1-w MR pulse sequences are presented for each technique (standard; VAT, view-angle tilting; VAT&SEMAC, view-angle tilting and slice-encoding metal artifact correction) [32]

**Table 23.1** Comparing low-field and high-field parameters from Siemens. Bold, advantages; italics, disadvantages

| Parameter | Relation to $B_0$ | 0.55 vs 3 (T) | Comment |
|---|---|---|---|
| SNR efficiency | $\propto B_0$ | *Overall ~4× lower* | *A proportional loss of SNR* |
| T1 | $\propto B_0^{0.3}$ | **TR 1.7× shorter** | **Small SNR efficiency gain** |
| SAR | $\propto B_0^2$ | **30× lower** | **Reduced SAR** |
| Off-resonance ($\Delta B_0$) | $\propto B_0$ | **5.5× lower** | **Reduced $\Delta f$** |
| T2* | $\approx 1/B_0$ | **Much longer** | **Longer readout** |
| Peak $B_1$ | – | **3× higher** | **High BW, short RF** |

Fortunately, low-field MRI is still useful to provide information about lung function such as lung compliance, ventilation, and perfusion. One example is a 4D imaging technique called the free-breathing acquisition scheme that records respiration. By recording spirometry measures of forced expiratory volume (FEV) and forced vital capacity (FVC) normal breathing and deep breathing, we notice a good correlation between MR and traditional measurements. The MRI information can tell a better story than spirometry measures. For instance, MR can help to calculate the flow volume loop for the left/right lung separately, which is impossible to get from spirometry measures. Pushing even further, even the information about subregions of the

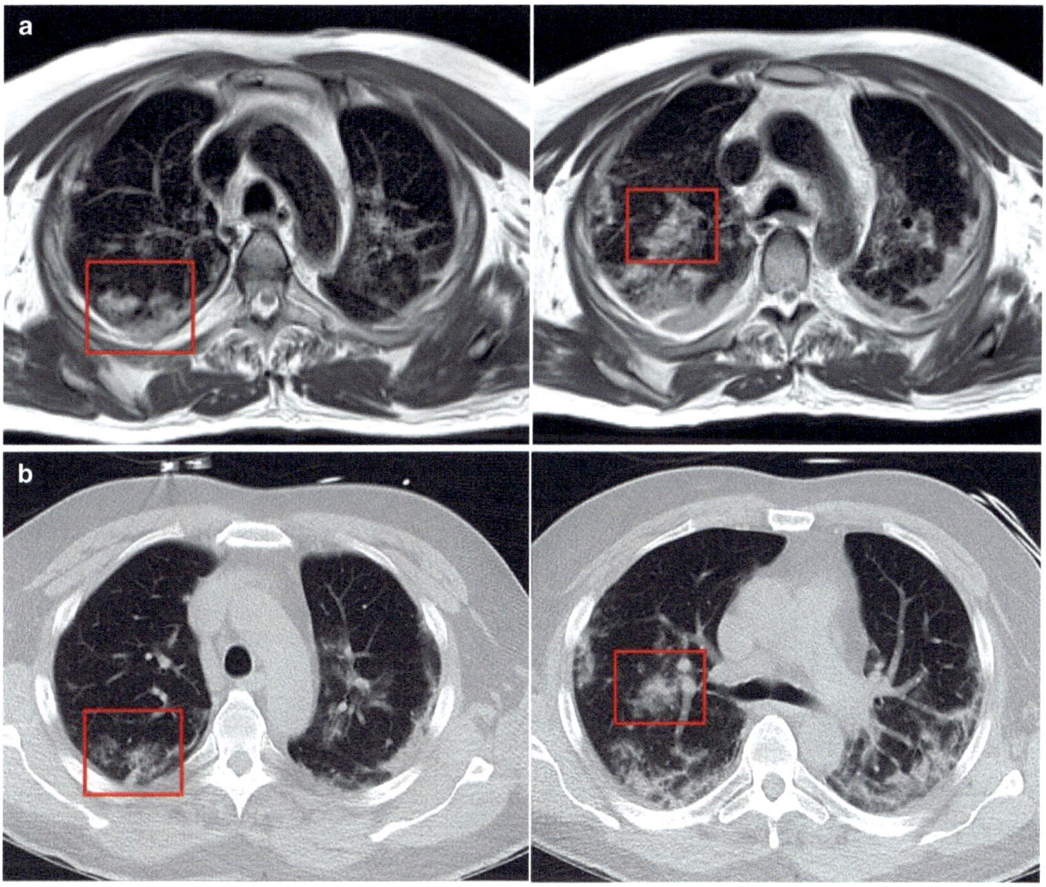


**Fig. 23.14** Images in a 61-year-old man with coronavirus disease 2019 (COVID-19) infection. Shown are two axial sections from, A, MRI scan obtained with prototype 0.55-T Magnetom Aera scanner (Siemens Healthineers) with spatial resolution of 1.2 mm × 1.2 mm × 6 mm (respiratory-gated fast-BLADE turbo spin-echo acquisition. Multifocal lesions equivalent to ground-glass opacities on CT image and peripheral distribution of consolidative opacities characteristic of COVID-2019 are observed with both imaging modalities. In some regions, severity of abnormality appears worse on MRI scan, as depicted by red rectangles. (Campbell-Washburn et al. [36])

lung is available. What's more, with the help of Fourier decomposition, we can even obtain microscopic information such as ventilation and perfusion maps as shown in Fig. 23.15.

Oxygen-enhanced MRI is another option of promising functional MRI in low field as 100% oxygen can be a T1-shortening MR contrast agent and T1 relaxivity of oxygen increases at low field. One could measure the percentage of signal enhancement from oxygen inhalation, an interesting and innovative method for probing lung ventilation. A low oxygen enhancement can be used to identify areas of lung function deficit. Figure 23.16 gives an example.

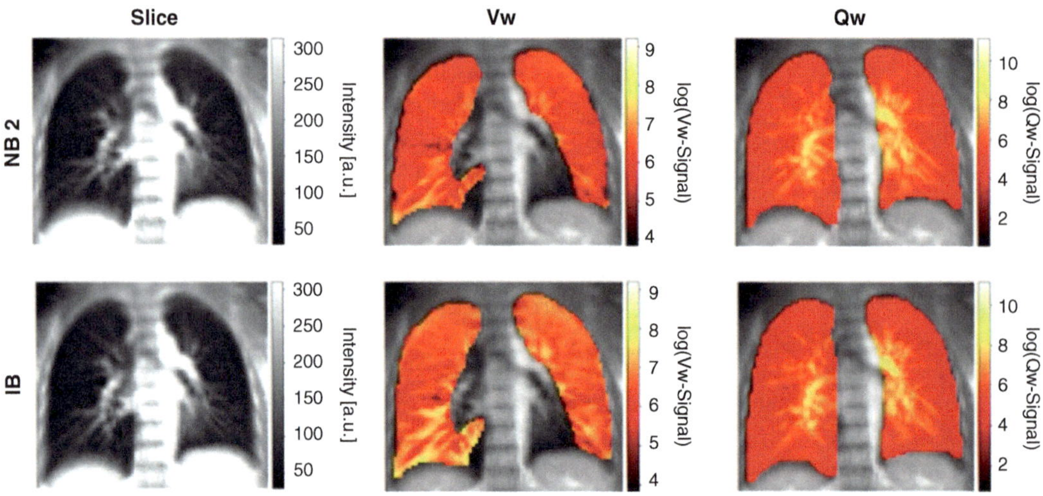


**Fig. 23.15** Exemplary ventilation-(Vw) and perfusion-weighted (Qw) maps. The unnormalized V- and Q-maps for the normal (NB 2) and the first irregular breathing scan (IB, deep breathing) exemplarily presented for the aorta slice of Volunteer 4. The heart and large vessels were excluded from the segmentation of the Vw-maps. The maps were Gaussian filtered and logarithmically plotted [37]

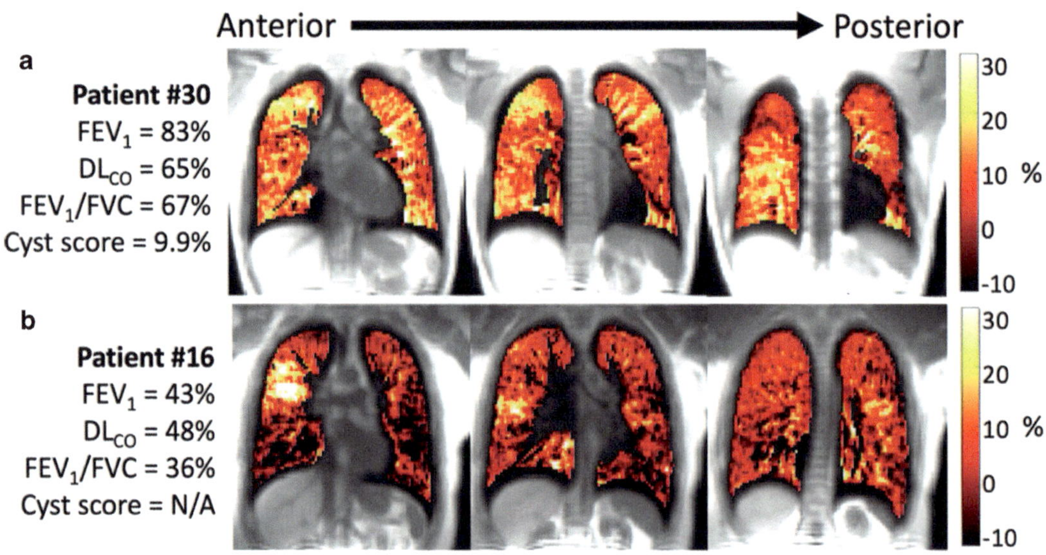


**Fig. 23.16** Percent signal-enhancement maps generated by oxygen-enhanced functional lung MRI. The color scale represents the percent signal enhancement within the lung tissue from inhalation of 100% oxygen. Three slices are presented for two different patients classified moderate (**a**) and severe (**b**) by FEV1, DLCO, and CT images. Regional variation in enhancement is indicative of local functional deficits (Bhattacharya et al. [38])

There are also techniques that can offer tissue characterization such as MR fingerprinting [39] which can provide parameter maps such as T1/T2 maps. MR fingerprinting obtains quantitative images by template matching of the MRI signal evolution over time, rather than conventional Fourier encoding methods. Therefore, it is conceivable that MR fingerprinting may be less sensitive to the intrinsic low SNR at a lower field.

In conclusion, there exists a need for one-stop shop spatially resolved morphologic and functional imaging of air-tissue interface organs like lungs. Low-field MRI coupled with advances in image acquisition and reconstruction can be an enabling technology to deliver on this promise.

## Clinical Applications of Low-Field MRI in Radiotherapy

In the past decade, low-field MRI products have proliferated and are quickly gaining popularity in a variety of clinical fields including radiotherapy. Reviewing the history of MRI, the differences between high-field and low-field MRI scanners appeared during the early phase of MRI history in the 1980s [40], and the gap expanded after 1985 when the first 1.5-T scanner was introduced [41]. The high-field scanners became more popular and gained a dominant market share because of their higher SNR per unit time, which permits faster imaging, higher resolution, greater contrast sensitivity, and more advanced sequences. With the technical advances mentioned earlier, some commercial systems have received FDA clearance and been put on the market since 2018. The commercial products include 0.064-T Hyperfine Swoop head scanner, 0.066-T Promaxo prostate scanner, 0.5-T Synaptive Evry intraoperative scanner, and 0.55-T Siemens Magnetom FreeMax. While high-field devices still dominate the market because of higher image quality, two main drivers support this recent lower-field revitalization: (1) lower costs and (2) recent technological developments leading to improved image quality. These advantages also result in MRI scanners' accessibility in all the other areas beyond the diagnostic imaging field [42].

In radiotherapy, the MR-guided radiotherapy (MRgRT) is playing a more important role. Conventional image-guided radiotherapy which is more commonly based on CT images is not able to fully address the challenges that lead to errors in treatment such as differentiating tumor from organ or patient position changes and motions. The soft-tissue contrast is often low, and it lacks the ability to monitor the treatment responses during radiation delivery. Last few years, MR-guided linear accelerators (MR-Linacs) such as ViewRay MRIdian (0.35 T) and Elekta Unity (1.5 T) have helped researchers expand the application boundary of MR-Linac. In this section, we review current clinical applications of MRI with great potential in radiotherapy, especially low-field scanners.

### MR-Linac Design with Low-Field Magnet

Hybrid magnetic resonance (MR)-guided linear accelerators represent a new development in the field of radiation oncology. By harnessing the favorable combination of onboard MR imaging with the possibility of daily recalculating the treatment plan based on real-time anatomy, the accuracy in target and organs-at-risk identification is expected to improve, with the aim of providing the best-tailored treatment. To date, two main MR-Linac hybrid machines are available, Elekta Unity (1.5 T) and ViewRay MRIdian (0.35 T). Given the relative novelty of this technology, these devices still raise practical issues and require groups of researchers to address the issues.

It is not a simple job to combine MR and RT. MR receives very weak RF signals from patients, and even small noises can contaminate the image. The strong magnetic field of MR is also able to influence the function of the linear accelerator. The linear accelerator is also a significant source of RF interference. Given the current situation, the lower field MR-Linac offers the benefits of reduced engineering and shielding challenges in designing these systems. Theoretically, it may also offer favorable charac-

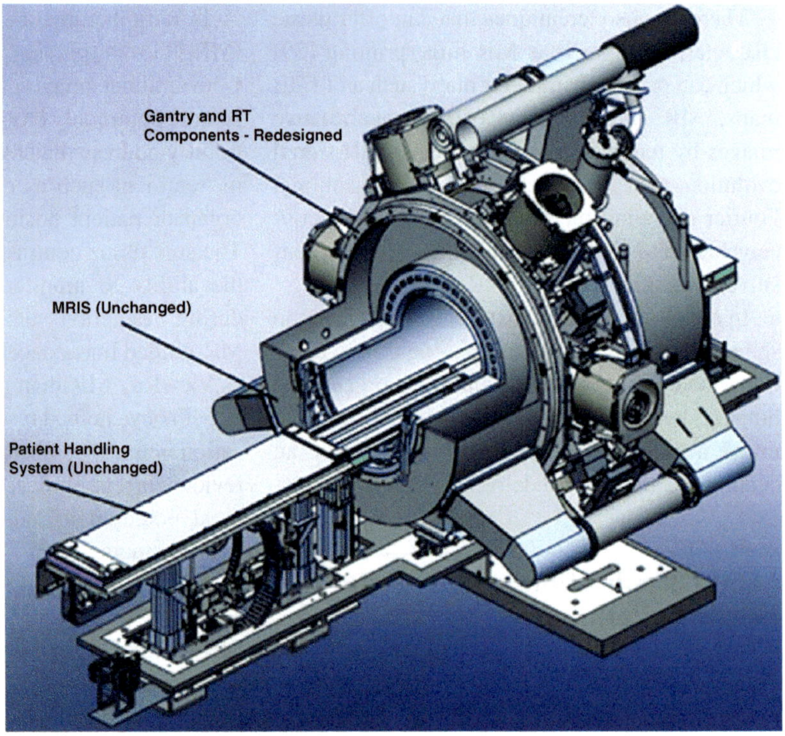


**Fig. 23.17** ViewRay MRIdian MRL model that replaced the initial tri-cobalt design in 2017. (Rammohan et al. [43])

teristics regarding the impact of the electron return effect on the radiation dose distribution. Currently, the ViewRay MRIdian, as shown in Fig. 23.17, is the only FDA-approved low-field MR-Linac, and in this section, we will discuss more details about the design of this system.

The details of MRIdian systems are described as follows [43]. The system consists of two hardware components: an MRI system and an RT delivery system [43]. The MRI scanner is double donut shaped with a 28-cm gap in between, generating a 0.35-T static magnetic field and a 50-cm-diameter spherical field of view. The isocenter of the MRI matches that of the RT system. The gradient coils have an inner diameter of 80 cm. The coils are self-shielded to help against thrust and torque. The maximum magnetic field gradient and slew rate can reach 10 mT/m and 200 T/m/s on each axis, respectively. The transmit coil can cover the whole body. There is also an integrated RT shield to span the magnet gap while maintaining uniform radiation beam attenuation to prevent beam heterogeneities. Two sets of receive coils, the torso coil and the head- and neck coil, consist of six- and five-element phase arrays, respectively, in each half. Both coils are made with flexible foam. The system shields from leaking RF using copper lined with carbon fiber to absorb internally reflecting RF. The MRI system is able to apply intra-fraction soft-tissue-based tracking by acquiring *ciné* planar images simultaneously with radiation delivery.

The first generation of MRIdian contained a ring gantry housing three cobalt-60 sources, providing a nominal dose rate of 550 cGy/min with fresh sources [43]. The system was later upgraded by replacing the cobalt with a linear accelerator which is able to generate a 6-MV flattening filter-free photon beam, as shown in Fig. 23.17. The beam has a nominal dose rate of 600 MU per minute at the 90-cm isocenter. The linear accelerator and its subcomponents are shielded from the magnetic field by five ferromagnetic steel concentric cylinders as well as three additional metal shields on the electron gun.

Thanks to great soft-tissue contrast, MR-Linacs have shown their potential in adaptive RT and intra-factional motion management. MR-Linacs have advantages in treating tumors that grow in areas prone to deformation and motion such as the pancreas and liver. This system is able to minimize the target area by tracking in real-time the tumor motion, saving as much normal tissue as possible.

Unlike CT, MRI scanning is quite versatile, given multiple choices of functional MRI sequences such as diffusion weight imaging or contrast-enhanced imaging. MRIdian has installed these functional MRI sequences which are able to provide additional physiological and biological information during the whole treatment. Combining all this information, it is promising to imagine a personalized RT treatment plan that maximizes the treatment effectiveness.

## Potentials of Low-Field MRI in Radiotherapy

### High-Speed MRI

There are already low-field MRI scanners being used in point-of-care testing. Both commercial and experimental products have reached a reasonable image resolution using conventional sequences such as FSE, GRE, etc. The scanner from Hyperfine is able to operate T1-/T2-weighted, FLAIR, and DWI sequences, while the other experimental groups also have the ability to offer comparable options. We are even impressed by the image quality with noise cancellation techniques which are often combined with deep learning reconstruction. The whole scan usually takes around 30 min. It is not difficult to conclude that since the image quality of the low-field scanner is already comparable with the state-of-the-art techniques, a fast acquisition scheme or high-speed MRI with low field is going to be helpful to show the clinical value of the low-field MRI. For MRgRT, increasing the images per unit time is considered very valuable, especially for quantitative MRI which is already quite time-consuming.

The approaches of fast acquisition need to harness low-field physics changes. For example, at low field, T1 and T2 decrease differently, with T1 approaching T2. In this case, when people use balanced steady-state free precession (bSSFP) approaches, the steady-state magnetization, $M_{ss}$, depends on the ratio $T_2/T_1$ as [44]:

$$M_{ss} = \frac{1}{2} M_0 \sqrt{T_2 / T_1}. \tag{23.3}$$

where $M_0$ is the initial magnetic moment. Hence, at a low-field limit, this steady-state magnetization can reach up to 50% of the total magnetization, making it one of the most efficient sequences.

The bSSFP sequence also needs good magnetic field homogeneity which is incompatible with the small footprint scanners. Nevertheless, some most recent initiatives to resolve the field homogeneity with passive shimming and first-order gradient DC shimming show very promising results, enabling bSSFP imaging without artifacts with reasonable TR on the order of 7–10 ms [22].

In the MR fingerprinting approach [39], the signal's pattern gives a unique set of magnetic properties. The images of a series of bSSFP with varied flip angles (FA) and TR are generated, and the signals are matched to the precomputed dictionary, as shown in Fig. 23.18. From this approach, several quantitative maps can be extracted at once. For instance, an initiative of low-field MRF like OPTIMUM is able to generate six quantified parametric maps in a small footprint scanner of 0.1 T (Fig. 23.19) [45]. Furthermore, with bSSFP sequences, OPTIMUM generates 270 2D images in around 8.5 min.

We can even go faster by using non-Cartesian sampling such as spiral or radial samplings. There are already proof-of-concept works from AMT center at the University of Basel, which used radial k-space sampling of 100 spokes with an image matrix in size of 129*129*15. The total acquisition time is around 2.5 min. In conclusion, it is optimal to improve treatment quality by applying high-speed low-field MRI techniques, and there still exist many more to be explored in

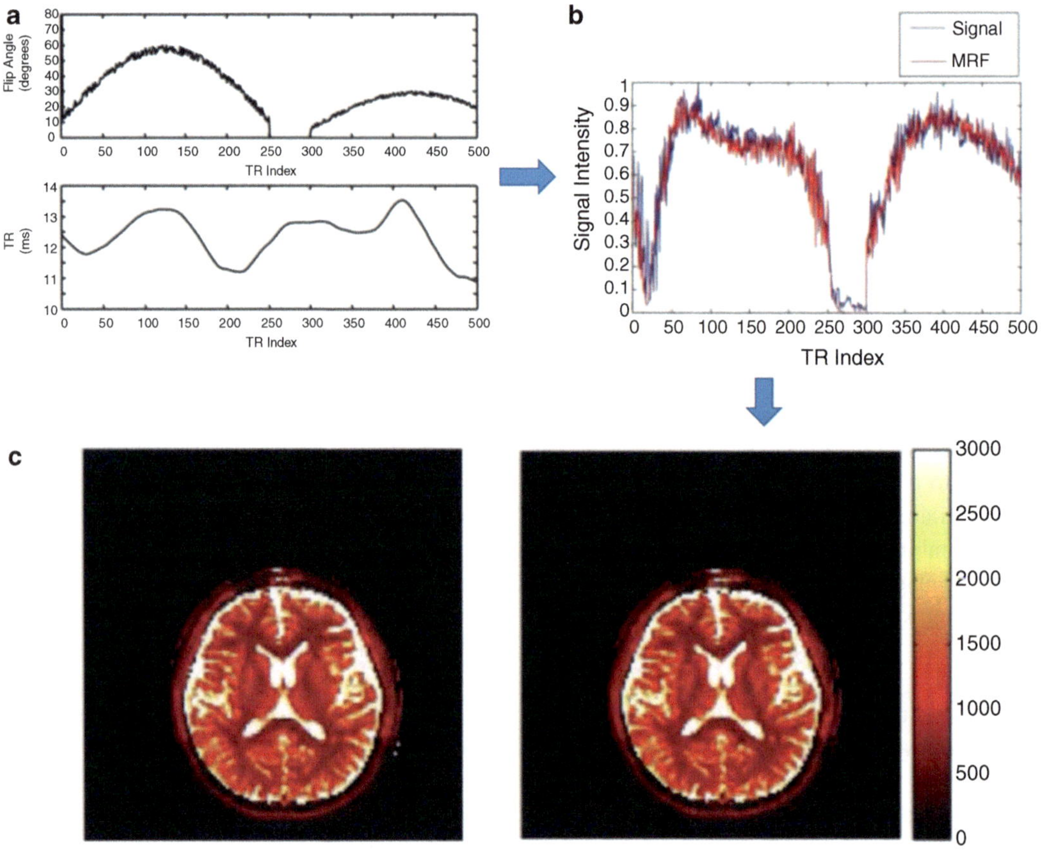


**Fig. 23.18** The paradigm of MRF from acquisition (**a**) to its match to the dictionary (**b**) and the quantitative maps (**c**) (Ma et al. 2013—with editing [39])

the future. In radiotherapy, such a technique can be used to improve the efficiency of workflow.

## 4DMRI

4D MRI is a quantified assessment of the breathing motion of tumors and organs. In radiotherapy of abdominal tumors, this motion needs to be considered when calculating the delivered dose accurately. Currently, 4D CT is conventionally used in radiotherapy for motion evaluation with high spatial resolution. However, it has several limitations, such as low temporal resolution, poor soft-tissue contrast, and motion-induced artifacts. These problems undermine tumor visualization and accurate motion quantification. To overcome these disadvantages, 4D MRI is a promising alternative due to its good soft-tissue contrast and lack of ionizing radiation. In this section, we discuss the potential of 4D MRI applications in radiotherapy.

Many efforts have already been made on this topic. These include the 2D-CINE imaging and 2D navigators which both have disadvantages like "stitching" artifacts or low scan efficiency. More recently, 3D k-space binning-based techniques such as SG-KS-4D-MRI and ROCK-4D-MRI have been used to replace 2D image sorting methods. The ROCK-4D-MRI uses a 4D Rotating Cartesian K-space (ROCK) sampling trajectory which improves scan efficiency so that the high-resolution 4D images of the entire torso may be acquired within 5 min using a 1.5-T diagnostic MRI scanner [46].

The same techniques were recently tested on the 0.35-T MRI-guided radiotherapy (MRgRT) system [47]. The motion quantification based on

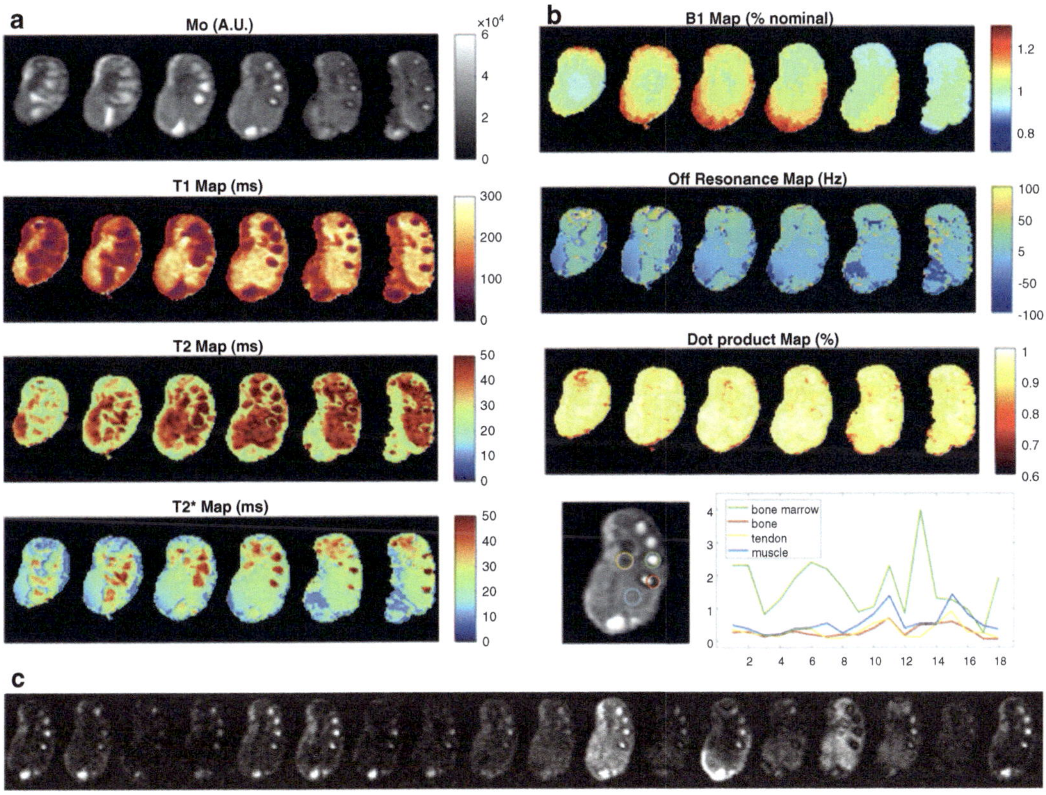


**Fig. 23.19** LF OPTIMUM results in vivo at 0.1 T. (**a**) Quantified maps in the human hand/wrist that includes proton density M0, T1, T2, T2*, B0, FA homogeneity B1+, and matching efficiency (dot product map) for six central slices. Matrix size is 90 × 81 × 15; corresponding voxel size is [1.5 × 1.5 × 6.5] mm$^3$. Total imaging time is 8.5 min with number of averages = 2. (**b**) Signal evolution (i.e., MR fingerprints) reported for different tissue types in a central slice. (**c**) Corresponding MR image series (Sarracanie et al. [45])

4D MRI was compared with those based on 2D-CINE, and the motion measurements based on 4D MRI matched well with 2D-CINE. Then, the image quality of 4D MRI was evaluated against 4D CT, and the image quality scores of 4D MRI were significantly higher than 4D CT, with better tumor contrast and fewer motion artifacts, as shown in Figs. 23.20 and 23.21.

These researchers have shown the clinical feasibility of using 4D MRI to acquire high-quality, respiratory motion-resolved 4D MRI in a low-field MRgRT system.

### Diffusion-Weighted MRI

The diffusion-weighted MRI (DWI) is a commonly used functional MRI technique for probing tissue cellularity and tumor response to radiotherapy and chemotherapy. This technique was originally developed for the diagnosis of stroke and later for cancer diagnosis. In radiotherapy, DWI is considered a noninvasive method to assess the tissue microenvironment before, during, and after cancer therapy. The apparent diffusion coefficient (ADC) calculated from DWI is closely related to tissue diffusivity, with low ADC values corresponding to more restricted diffusion and vice versa. An increase in ADC values after treatment reflects a decrease in tumor cellularity because of tumor cell death, indicating a good response. This parameter offers a useful and early tool to predict treatment response with a diagnosis MRI [48].

Researchers have shown the feasibility of performing DWI on the 0.35-T MRIdian MR-Linac system [49] and continued studies about DWI technical development as well as clinical applica-

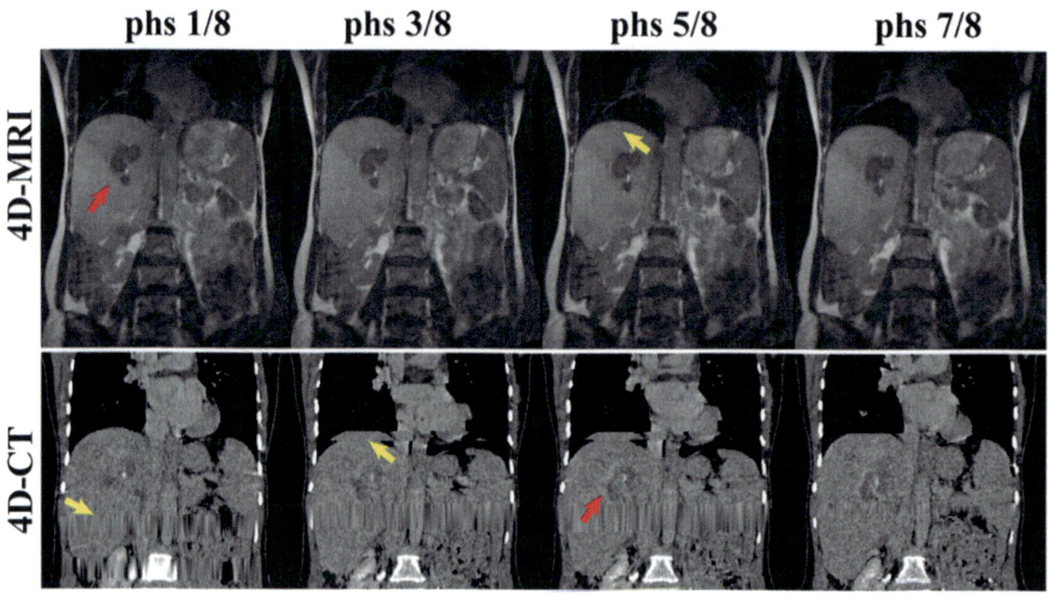


**Fig. 23.20** The 4D-MRI images (top row) and 4D-CT images (bottom row) on a patient with liver metastasis. The 4D-MRI images have better soft-tissue contrast and are free of the stitching artifacts. The boundary of the tumor is much better defined in 4D MRI than in 4D-CT (Han et al. [47])

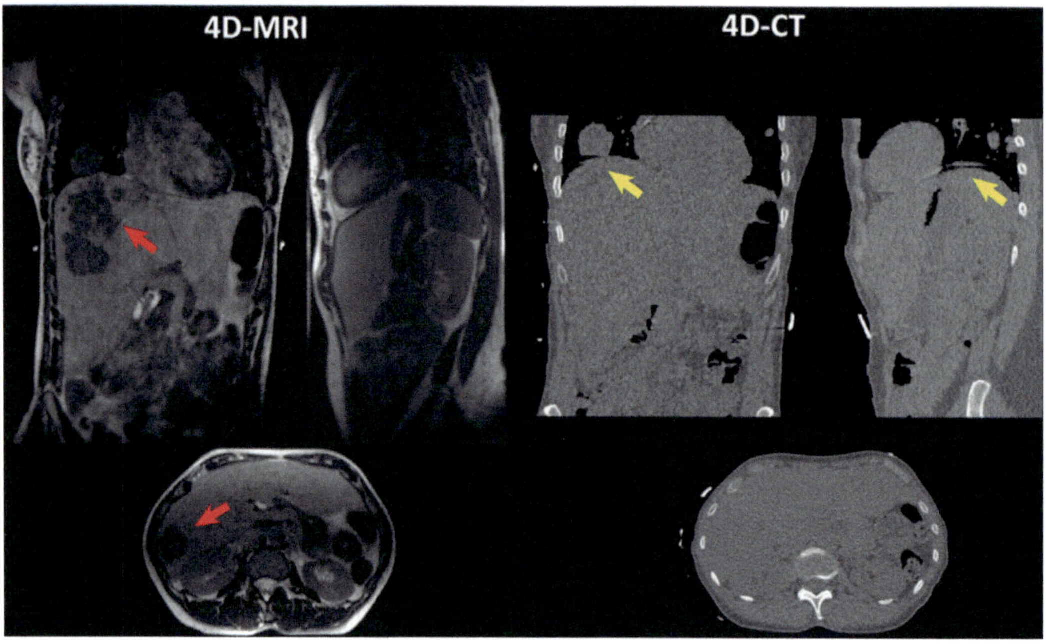


**Fig. 23.21** End-expiration phase of 4D-MRI and 4D-CT images acquired on a patient with liver cancer, reformatted to coronal, sagittal, and transversal slice orientations (Han et al. [47])

tions at 0.35 T. Some groups tried to compensate for image distortion by using improved sequences with special designs or better positioning. Researchers from UCLA [50] have developed a specially designed DWI sequence for the low-field 0.35-T scanner called diffusion-prepared turbo spin-echo (DP-TSE) readout to generate distortion-free DWI images. Another group from Washington University [51] discovered that results showed that variation of the ADC quantification accuracy and geometric fidelity on the 0.35-T MR-Linac system was dependent on the gantry angle. Following this direction, another group from *Fondazione Policlinico Universitario* in Italy focused on optimizing the SE-EPI-based DWI protocol on the low-field MR-Linac aiming to maximize imaging quality and minimize acquisition time [52]. The value of using longitudinal DWI acquired on the low-field MR-Linac for treatment response prediction was also demonstrated for both rectal cancer and sarcoma [53–55]. DWI has already been confirmed as a quite practical tool on low-field MRI with broad application prospects and can potentially bring more clinical value when combined with other advanced techniques such as advanced diffusion models, intravoxel incoherent motion (IVIM) [56, 57], and continuous-time random walk (CTRW) [58, 59].

### Optimized Protocols for Specific Clinical Purpose: Urethra MRI

With the low-field MR-guided RT introduced in RT, there are new possibilities and needs to optimize MR protocols for specific clinical purposes, such as to visualize small critical structures for better protection with improved patient positioning and online adaptive RT. One example is the genitourinary (GU) toxicity of stereotactic body radiotherapy (SBRT). SBRT is now a widely accepted standard-of-care option for localized prostate cancer, but its late grade GU toxicity remains a significant challenge. Typically, bladder sparring is a common method to reduce toxicity, but it was discovered that urethral injury is also a contributor. In the past, delineating the urethra on CT images was not easy due to the lack of contrast between the prostatic urethra and prostate tissue. On MRIdian, researchers optimized a T2-weighted MRI sequence for delineating the prostatic urethra. The sequence they used was a 3D half-Fourier acquisition single-shot turbo spin-echo (3D HASTE). The UCLA group first demonstrated the clinical value of acquiring MR urethra imaging on an MR-Linac [60]. Onboard urethra MRI was also acquired at two time points during the entire radiation course: after MR simulation and after the final treatment fraction. Their results show that good visualization of the prostatic urethra (Fig. 23.22) can be a valuable tool to help better protect the urethra through patient setup or online adaptive RT, given the large variation visualized using the proposed MR technique [61].

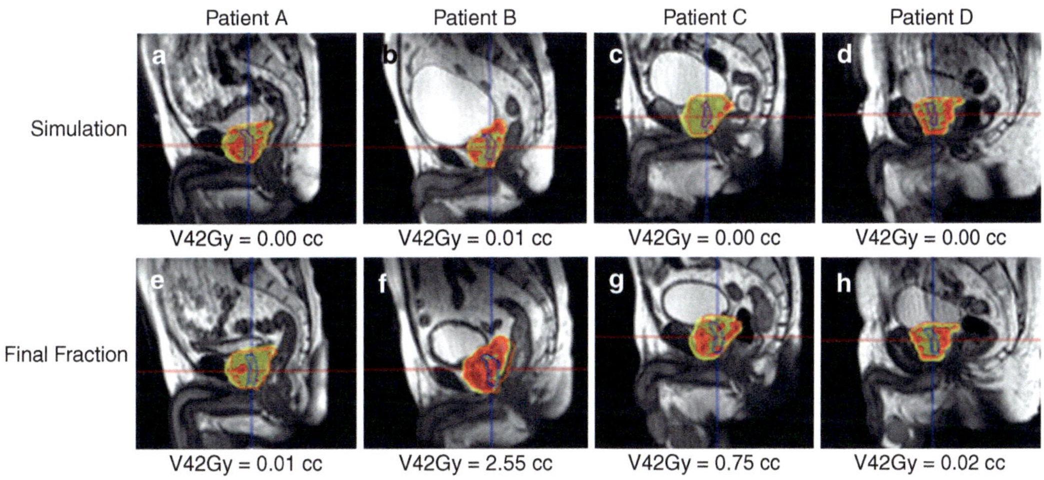


**Fig. 23.22** A radiation oncologist's calculated and estimated dose and urethral V42Gy for patient A–D's simulation (**a–d**) and final fraction (**e–h**) bSSFP. RO1 simulation/final HASTE urethra contour, purple/blue. Orange contour, PTV. Red, 105% (42 Gy) isodose region; yellow, 95% (38 Gy) isodose region (Pham et al. [61])

## Conclusion

Great advances in hardware and imaging techniques of low-field MRI systems beyond common perception have happened. Since the 1990s, it has been possible to apply advanced techniques such as DWI and contrast-enhanced imaging. Compared with conventional MRI, low-field MRI is more economic and more convenient. Moreover, researchers have a better understanding of the physical characteristics of low-field systems which can be superior to high-field MRI in many ways. For instance, T1-shortening effects and low magnetic susceptibility artifacts are associated with static field strength. Recent developments in hardware such as the use of permanent magnet arrays, sophisticated radiofrequency coils, and optimized sequences have shown that the static field strength itself does not necessarily need to be high. Numerous products in both academia and industry have given optimistic results, some of which are better than those from conventional MRI for certain tasks. In addition, with the recent remarkable developments in low-field MRgRT, the MRIdian system, the onboard low-field MRI as one of the solutions for MR and RT integration, shows great potential. Clinically, the fundamental value of MRI is superior tissue contrast information. For this reason, the low-field MRI can surely help in optimizing the treatment workflow, combined with advanced imaging techniques including those we have mentioned in the previous section. Overall, we are enthusiastic about low-field MRI systems and believe that they will continue to develop and gain more implementations in the future.

## References


1. Hong AS, Levin D, Parker L, Rao VM, Ross-Degnan D, Wharam JF. Trends in diagnostic imaging utilization among Medicare and commercially insured adults from 2003 through 2016. Radiology. 2020;294(2):342–50.
2. Arnold TC, Freeman CW, Litt B, Stein JM. Low-field MRI: clinical promise and challenges. J Magn Reson Imaging. 2023;57(1):25–44.
3. Ginde AA, Foianini A, Renner DM, Valley M, Camargo CA. Availability and quality of computed tomography and magnetic resonance imaging equipment in U.S. emergency departments. Acad Emerg Med. 2008;15(8):780–3.
4. Health equipment—magnetic resonance imaging (MRI) units—OECD data. [Internet]. OECD; 2018. https://data.oecd.org/healtheqt/magnetic-resonance-imaging-mri-units.htm.
5. WHO|Atlas: multiple sclerosis resources in the world 2008. [Internet]. WHO; 2018. http://www.who.int/mental_health/neurology/atlas_multiple_sclerosis_resources_2008/en/.
6. Niendorf T, Sodickson DK. Parallel imaging in cardiovascular MRI: methods and applications. NMR Biomed. 2006;19(3):325–41.
7. Lustig M, Donoho D, Pauly JM. Sparse MRI: the application of compressed sensing for rapid MR imaging. Magn Reson Med. 2007;58(6):1182–95.
8. Campbell-Washburn AE, Ramasawmy R, Restivo MC, Bhattacharya I, Basar B, Herzka DA, et al. Opportunities in interventional and diagnostic imaging by using high-performance low-field-strength MRI. Radiology. 2019;293(2):384–93.
9. Basar B, Sonmez M, Yildirim DK, Paul R, Herzka DA, Kocaturk O, et al. Susceptibility artifacts from metallic markers and cardiac catheterization devices on a high-performance 0.55 T MRI system. Magn Reson Imaging. 2021;77:14–20.
10. Hori M, Hagiwara A, Goto M, Wada A, Aoki S. Low-field magnetic resonance imaging: its history and renaissance. Invest Radiol. 2021;56(11):669–79.
11. Hayashi N, Watanabe Y, Masumoto T, Mori H, Aoki S, Ohtomo K, et al. Utilization of low-field MR scanners. Magn Reson Med Sci. 2004;3(1):27–38.
12. Marques JP, Simonis FFJ, Webb AG. Low-field MRI: an MR physics perspective. J Magn Reson Imaging. 2019;49(6):1528–42.
13. Ejbjerg BJ, Narvestad E, Jacobsen S, Thomsen HS, Østergaard M. Optimised, low cost, low field dedicated extremity MRI is highly specific and sensitive for synovitis and bone erosions in rheumatoid arthritis wrist and finger joints: comparison with conventional high field MRI and radiography. Ann Rheum Dis. 2005;64(9):1280–7.
14. Yamada K, Miyahara K, Sato M, Hirose T, Yasugi Y, Matsuda Y, et al. Optimizing technical conditions for magnetic resonance imaging of the rat brain and abdomen in a low magnetic field. Vet Radiol Ultrasound. 1995;36(6):523–7.
15. Iturri-Clavero F, Galbarriatu-Gutierrez L, Gonzalez-Uriarte A, Tamayo-Medel G, de Orte K, Martinez-Ruiz A, et al. "Low-field" intraoperative MRI: a new scenario, a new adaptation. Clin Radiol. 2016;71(11):1193–8.
16. Klein HM. Clinical low-field strength magnetic resonance imaging: a practical guide to accessible MRI. Springer; 2016.
17. Magnetic Resonance Imaging Systems Market by Architecture (Open MRI Systems and Closed MRI Systems (Standard Bore and Wide Bore)), Field Strength (Low-to-mid Field, High-field (1.5T and

3T), and Very-high Field)—Global Forecast to 2023 [Internet]. Markets and Markets. https://www.marketsandmarkets.com/requestsampleNew.asp?id=99.
18. Gruber B, Froeling M, Leiner T, Klomp DWJ. RF coils: a practical guide for nonphysicists. J Magn Reson Imaging. 2018;48(3):590–604.
19. Pruessmann KP, Weiger M, Scheidegger MB, Boesiger P. SENSE: sensitivity encoding for fast MRI. Magn Reson Med. 1999;42(5):952–62.
20. Griswold MA, Jakob PM, Heidemann RM, Nittka M, Jellus V, Wang J, et al. Generalized autocalibrating partially parallel acquisitions (GRAPPA). Magn Reson Med. 2002;47(6):1202–10.
21. McGee KP, Campeau NG, Witte RJ, Rossman PJ, Christopherson JA, Tryggestad EJ, et al. Evaluation of a new, highly flexible radiofrequency coil for MR simulation of patients undergoing external beam radiation therapy. J Clin Med. 2022;11(20):5984.
22. Liu Y, Leong ATL, Zhao Y, Xiao L, Mak HKF, Tsang ACO, et al. A low-cost and shielding-free ultra-low-field brain MRI scanner. Nat Commun. 2021;12(1):7238.
23. Vogel MW, Guridi RP, Su J, Vegh V, Reutens DC. 3D-spatial encoding with permanent magnets for ultra-low field magnetic resonance imaging. Sci Rep. 2019;9(1):1522.
24. McDaniel PC, Cooley CZ, Stockmann JP, Wald LL. The MR cap: a single-sided MRI system designed for potential point-of-care limited field-of-view brain imaging. Magn Reson Med. 2019;82(5):1946–60.
25. Kodzwa R. ACR manual on contrast media: 2018 updates. Radiol Technol. 2019;91(1):97–100.
26. Sammet S. Magnetic resonance safety. Abdom Radiol (NY). 2016;41(3):444–51.
27. Swoop® Portable MR Imaging System™ Details and specifications [Internet]. Hyperfine Inc. 2023. https://hyperfine.io/swoop/details-and-specifications.
28. SYNAPTIVE MRI Everywhere for everyone [Internet]. [42] Synaptive Medical Inc. 2023. https://www.synaptivemedical.com/products/mri/.
29. Koolstra K, O'Reilly T, Börnert P, Webb A. Image distortion correction for MRI in low field permanent magnet systems with strong B0 inhomogeneity and gradient field nonlinearities. Magn Reson Mater Phys Biol Med. 2021;34(4):631–42.
30. Haskell MW, Nielsen JF, Noll DC. Off-resonance artifact correction for MRI: a review. NMR Biomed. 2023;36(5):e4867.
31. Guermazi A, Miaux Y, Zaim S, Peterfy CG, White D, Genant HK. Metallic artefacts in MR imaging: effects of main field orientation and strength. Clin Radiol. 2003;58(4):322–8.
32. Jungmann PM, Ganter C, Schaeffeler CJ, Bauer JS, Baum T, Meier R, et al. View-angle tilting and slice-encoding metal artifact correction for artifact reduction in MRI: experimental sequence optimization for orthopaedic tumor Endoprostheses and clinical application. PLoS One. 2015;10(4):e0124922.
33. Mugler JP, Altes TA. Hyperpolarized 129ssMRI of the human lung. J Magn Reson Imaging. 2013;37(2):313–31.
34. Togao O, Tsuji R, Ohno Y, Dimitrov I, Takahashi M. Ultrashort echo time (UTE) MRI of the lung: assessment of tissue density in the lung parenchyma. Magn Reson Med. 2010;64(5):1491–8.
35. Campbell-Washburn AE, Malayeri AA, Jones EC, Moss J, Fennelly KP, Olivier KN, et al. T2-weighted lung imaging using a 0.55-T MRI system. Radiol Cardiothorac Imaging. 2021;3(3):e200611.
36. Campbell-Washburn AE, Suffredini AF, Chen MY. High-performance 0.55-T lung MRI in patient with COVID-19 infection. Radiology. 2021;299(2):E246–7.
37. Klaar R, Rabe M, Gaass T, Schneider MJ, Benlala I, Eze C, et al. Ventilation and perfusion MRI at a 0.35 T MR-Linac: feasibility and reproducibility study. Radiat Oncol. 2023;18(1):58.
38. Bhattacharya I, Ramasawmy R, Javed A, Lowery M, Henry J, Mancini C, et al. Assessment of lung structure and regional function using 0.55 T MRI in patients with lymphangioleiomyomatosis. Invest Radiol. 2022;57(3):178–86.
39. Ma D, Gulani V, Seiberlich N, Liu K, Sunshine JL, Duerk JL, et al. Magnetic resonance fingerprinting. Nature. 2013;495(7440):187–92.
40. PubMed by Year—The Finest Worklog [Internet]. https://esperr.github.io/pubmed-by-year/.
41. Edelman RR. The history of MR imaging as seen through the pages of radiology. Radiology. 2014;273(2 Suppl):S181–200.
42. Sarracanie M, Salameh N. Low-field MRI: how low can we go? A fresh view on an old debate. Front Phys. 2020;8:172.
43. Rammohan N, Randall JW, Yadav P. History of technological advancements towards MR-Linac: the future of image-guided radiotherapy. J Clin Med. 2022;11(16):4730.
44. Scheffler K, Lehnhardt S. Principles and applications of balanced SSFP techniques. Eur Radiol. 2003;13(11):2409–18.
45. Sarracanie M. Fast quantitative low-field magnetic resonance imaging with OPTIMUM-optimized magnetic resonance fingerprinting using a stationary steady-state Cartesian approach and accelerated acquisition schedules. Invest Radiol. 2022;57(4):263–71.
46. Han F, Zhou Z, Han E, Gao Y, Nguyen KL, Finn JP, et al. Self-gated 4D multiphase, steady-state imaging with contrast enhancement (MUSIC) using rotating cartesian K-space (ROCK): validation in children with congenital heart disease. Magn Reson Med. 2017;78(2):472–83.
47. Han F, Zhou Z, Du D, Gao Y, Rashid S, Cao M, et al. Respiratory motion-resolved, self-gated 4D-MRI using rotating Cartesian K-space (ROCK): initial clinical experience on an MRI-guided radiotherapy system. Radiother Oncol. 2018;127(3):467–73.

48. Huang WY, Wen JB, Wu G, Yin B, Li JJ, Geng DY. Diffusion-weighted imaging for predicting and monitoring primary central nervous system lymphoma treatment response. AJNR Am J Neuroradiol. 2016;37(11):2010–8.
49. Yang Y, Cao M, Sheng K, Gao Y, Chen A, Kamrava M, et al. Longitudinal diffusion MRI for treatment response assessment: preliminary experience using an MRI-guided tri-cobalt 60 radiotherapy system. Med Phys. 2016;43(3):1369–73.
50. Gao Y, Han F, Zhou Z, Cao M, Kaprealian T, Kamrava M, et al. Distortion-free diffusion MRI using an MRI-guided Tri-Cobalt 60 radiotherapy system: sequence verification and preliminary clinical experience. Med Phys. 2017;44(10):5357–66.
51. Lewis B, Guta A, Mackey S, Gach HM, Mutic S, Green O, et al. Evaluation of diffusion-weighted MRI and geometric distortion on a 0.35T MR-LINAC at multiple gantry angles. J Appl Clin Med Phys. 2021;22(2):118–25.
52. Nardini M, Capotosti A, Mazzoni LN, Cusumano D, Boldrini L, Chiloiro G, et al. Tuning the optimal diffusion-weighted MRI parameters on a 0.35-T MR-Linac for clinical implementation: a phantom study. Front Oncol. 2022;12:867792.
53. Gao Y, Ghodrati V, Kalbasi A, Fu J, Ruan D, Cao M, et al. Prediction of soft tissue sarcoma response to radiotherapy using longitudinal diffusion MRI and a deep neural network with generative adversarial network-based data augmentation. Med Phys. 2021;48(6):3262–372.
54. Gao Y, Kalbasi A, Hsu W, Ruan D, Fu J, Shao J, et al. Treatment effect prediction for sarcoma patients treated with preoperative radiotherapy using radiomics features from longitudinal diffusion-weighted MRIs. Phys Med Biol. 2020;65(17):175006.
55. Shaverdian N, Yang Y, Hu P, Hart S, Sheng K, Lamb J, et al. Feasibility evaluation of diffusion-weighted imaging using an integrated MRI-radiotherapy system for response assessment to neoadjuvant therapy in rectal cancer. Br J Radiol. 2017;90(1071):20160739.
56. Le Bihan D, Breton E, Lallemand D, Grenier P, Cabanis E, Laval-Jeantet M. MR imaging of intravoxel incoherent motions: application to diffusion and perfusion in neurologic disorders. Radiology. 1986;161(2):401–7.
57. Iima M, Le Bihan D. Clinical Intravoxel incoherent motion and diffusion MR imaging: past, present, and future. Radiology. 2016;278(1):13–32.
58. Ingo C, Magin RL, Colon-Perez L, Triplett W, Mareci TH. On random walks and entropy in diffusion-weighted magnetic resonance imaging studies of neural tissue: random walks and entropy in diffusion-weighted MRI. Magn Reson Med. 2014;71(2):617–27.
59. Karaman MM, Sui Y, Wang H, Magin RL, Li Y, Zhou XJ. Differentiating low- and high-grade pediatric brain tumors using a continuous-time random-walk diffusion model at high *b* -values: pediatric brain tumors using a CTRW diffusion model. Magn Reson Med. 2016;76(4):1149–57.
60. Pham J, Savjani RR, Gao Y, Cao M, Hu P, Sheng K, et al. Evaluation of T2-weighted MRI for visualization and sparing of urethra with MR-guided radiation therapy (MRgRT) on-board MRI. Cancers. 2021;13(14):3564.
61. Pham J, Savjani RR, Yoon SM, Yang T, Gao Y, Cao M, et al. Urethral interfractional geometric and dosimetric variations of prostate cancer patients: a study using an onboard MRI. Front Oncol. 2022;12:916254.

# Image-Based Biomarkers in Magnetic Resonance-Guided Radiotherapy (MRgRT)

24

Andrea D'Aviero, Haidy Nasief, Charlotte Robert, and Davide Cusumano

## Introduction

Artifical intelligence (AI) is a scientific field with the main aim to explore relationships and connections in datasets with different data formats to extract meaningful information. Using complex algorithms, AI is able of performing specific tasks, reducing workload for humans, and playing a relevant role in the world of automation [1] but also providing new knowledge that is at the core of healthcare decision support tools. The application of AI to healthcare has increased in last years in different settings, from diagnosis to outcome prediction [2, 3]. Radiotherapy (RT) in particular has been consistently improving by the introduction of AI, providing new tools to accelerate the patient workflow [1], supporting clinicians and physicists in the manual steps such as contouring and treatment planning [4–8], improving workflow process and outcome evaluations.

In this context, radiomics represents one of the most advanced applications in radiotherapy, consisting in extracting numerical parameters from images and then using such parameters to predict treatment responses or treatment toxicities [9, 10]. Radiomic features aim to describe tumors from a genetic and phenotypical point of view through image intensity analysis [11]. In other words, in the era of omics-guided medicine, radiomics is intended to support the definition of image-based biomarkers to be used in treatment decisions, alone or in conjunction with markers extracted by other "omics," such as genomics, metabolomics, transcriptomics, or proteomics [3, 12, 13]. Radiomics can be separated into two categories: conventional radiomics, which relies on handcrafted features, and deep-learning-based radiomics, for which the characteristics of interest are defined by the algorithm itself. Conventional radiomics consists in a complex process of extraction of quantitative image features from a region of interest (ROI) followed by

A. D'Aviero
UO Radioterapia Oncologica, Mater Olbia Hospital, Olbia, Italy
e-mail: andrea.daviero@materolbia.com

H. G. Nasief
Department of Radiation Oncology, Medical College of Wisconsin, Milwaukee, WI, USA
e-mail: hnaseif@mcw.edu

C. Robert (✉)
UMR 1030 Radiothérapie Moléculaire et Innovation Thérapeutique, ImmunoRadAI, Université Paris-Saclay, Institut Gustave Roussy, Inserm, Villejuif, France

Department of Radiation Oncology, Gustave Roussy, Villejuif, France
e-mail: CH.ROBERT@gustaveroussy.fr

D. Cusumano
Fondazione Policlinico Universitario "Agostino Gemelli" IRCCS, Rome, Italy

UO Fisica Medica e Radioprotezione, Mater Olbia Hospital, Olbia, Italy
e-mail: davide.cusumano@materolbia.com



I. J. Das et al. (eds.), *A Practical Guide to MR-Linac*, https://doi.org/10.1007/978-3-031-48165-9_24

a model creation correlated to clinical endpoints [9, 12, 14–16].

As a subdomain of radiomics, delta-radiomic strategies aim in particular to quantify the patient sensitivity to the treatment using longitudinal assessment to study the temporal variation of the radiomic features during the course of treatment: the potentialities of this approach are very relevant, and delta radiomics seems able to overcome the performance obtainable by radiomic features extracted in single time frames, as demonstrated by some experiences reported in more details in the next paragraphs [15, 17, 18]. Within imaging technologies, magnetic resonance imaging (MRI) represents one of the most interesting fields of application of radiomics due to the increased soft tissue differentiation and availability of complementary image contrasts. The implementation in RT of hybrid MR-Linac systems represents a unique opportunity in the development and application of delta-radiomic approaches [17].

The aim of this chapter is to define the efforts of radiomics in providing reliable biomarkers for RT treatment optimization in MRgRT.

## Radiomic Workflow in the Context of MRgRT

In the context of MRgRT, only conventional handcrafted radiomics and its delta-radiomics variant have been implemented to date. The delta-radiomic process can be summarized into the following major steps as shown in Fig. 24.1:

1. Acquiring longitudinal images during treatment.
2. ROI delineation—At this stage, registering images acquired at the different time frames and populating ROI to subsequent fractions can be an option for time saving.
3. Contour quality assurance checks.
4. MRI preprocessing (including in particular bias correction, spatial resampling, intensity standardization)—Use of a posteriori radiomic features harmonization strategies can be also an option.
5. Extracting radiomic features from reference and daily images, and calculating delta-radiomic features as the relative net change of radiomic features as compared to the reference image.
6. Usage of feature selection strategies when necessary.
7. Building of the model for the appropriate clinical task.
8. Evaluation of the model's generalizability performances on an internal test set (suboptimal) or an independent external test set [17].

Unlike computed tomography (CT) or positron-emission tomography (PET), MR imaging has the disadvantage of being a nonquantitative imaging modality for conventional T1- or T2-weighted anatomical sequences, resulting in high intra- and inter-image intensity variability. MR image intensities are therefore highly dependent on manufacturer, sequence type, and acquisition parameters such as magnetic field, repetition and echo times, and voxel size, all of which having a significant impact on the radiomic indices extracted [19–21]. This observation makes the development of robust radiomic signatures that can be generalized to a multicenter scale even more complex. To solve this problem, several methods have been proposed, ranging from simple image preprocessing methods applied in the image domain to methods applied in the feature domain itself, notably a posteriori statistical method ComBat and its variants [22]. In MRI, although preprocessing methods depend on the tumor location, they conventionally include steps for correcting magnetic field inhomogeneities, spatial resampling, and intensity harmonization [10]. More recently, deep-learning methods have been proposed as an interesting alternative for image harmonization [23]. A final step before feature extraction involves gray-level discretization, which consists of grouping similar intensities into bins, with the aim of minimizing the impact of noise on texture features and reducing computation time.

Traditional radiomics then consists of extracting hand-created features calculated using predefined mathematical formulae [24]. The features include semantic and agnostic features. Semantic

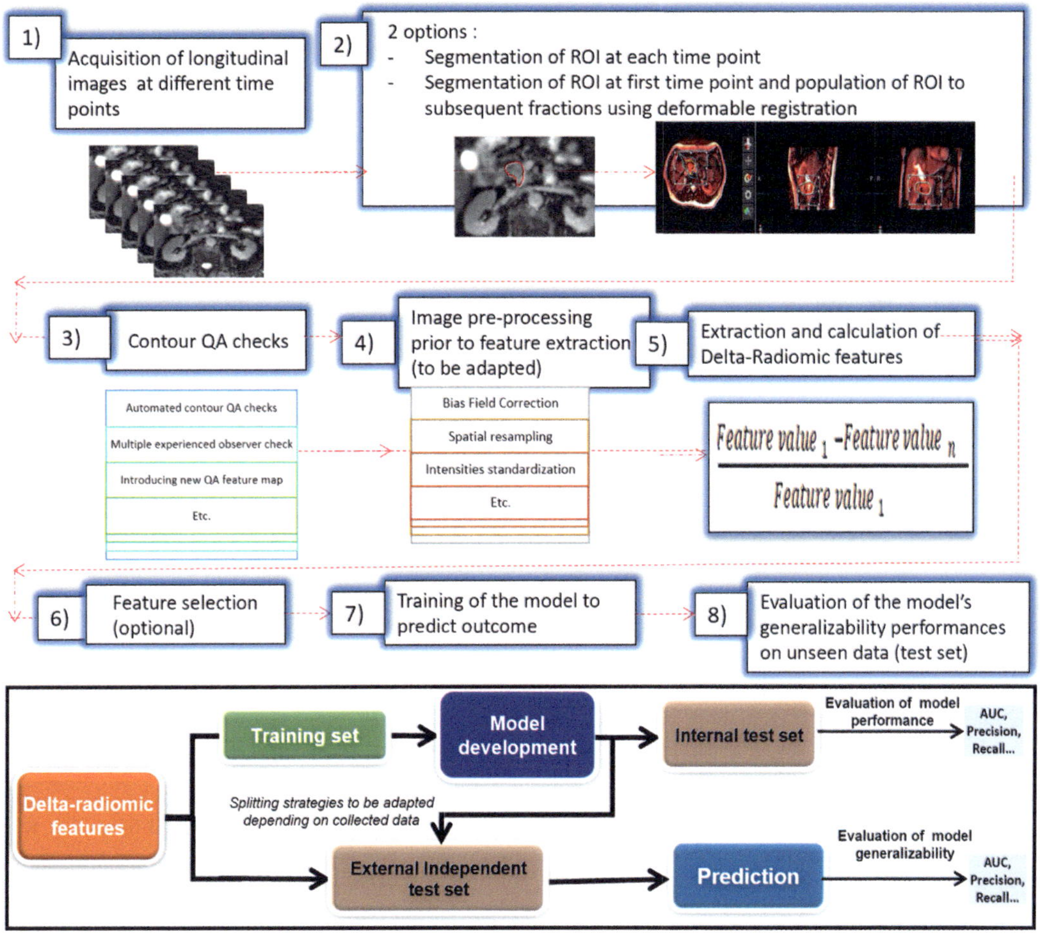


**Fig. 24.1** Major steps for delta-radiomic feature extraction and model building in MRgRT as described in the text

features refer to features conventionally used in the vocabulary of radiologists to describe lesions. Agnostic features (Fig. 24.2) can be divided into four subgroups:

1. Shape-based features describing the 3D geometry of the segmented structure
2. First-order features quantifying the distribution of voxel intensities
3. Second-order features, also known as textural features, providing information on the arrangement of grayscale values in the region of interest
4. Higher-order features extracted from filtered images such as wavelets, Laplacian of Gaussian (LoG) filter, square, square root, logarithmic, and exponential filters

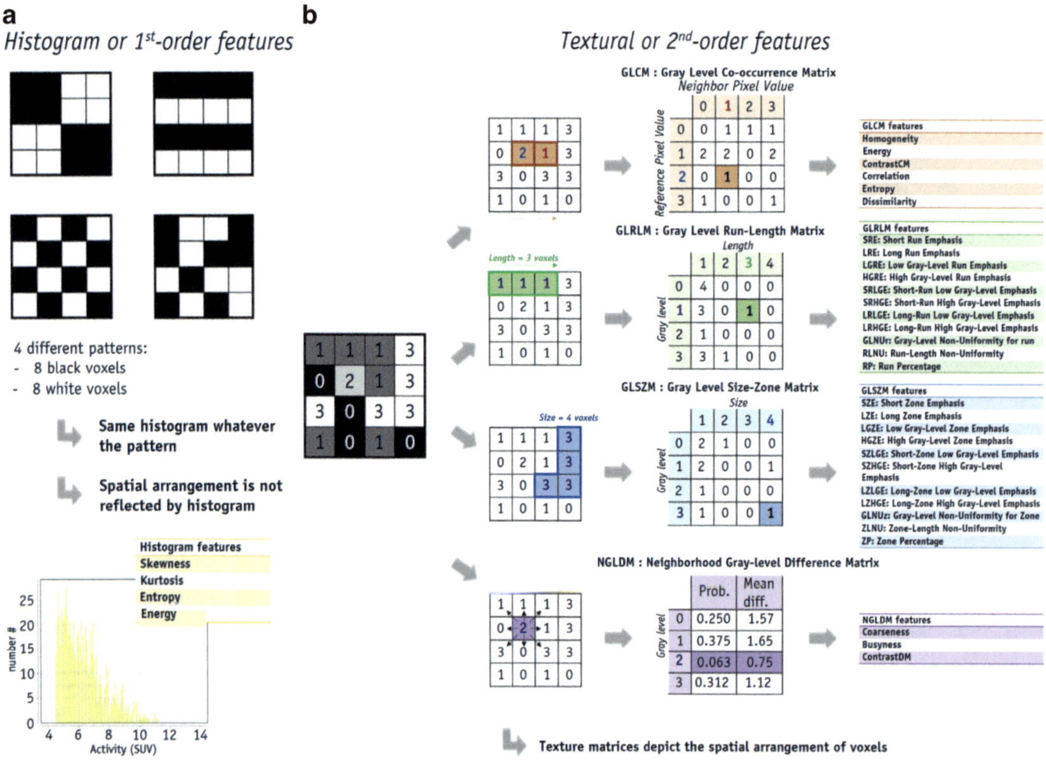


**Fig. 24.2** Histogram or first-order features reflect voxel intensity distribution only (**a**). Unlike histogram-based features, textural or second-order features (**b**) derived from texture matrices (e.g., co-occurrence, run length, zone size matrices) reflect the complex and unique spatial arrangement of voxels [24]. (Reprinted from [24], with permission from Elsevier)

## Imaging Biomarkers in MRgRT

The development of MRgRT techniques has been made possible by the spread of hybrid MR-Linac units, based on the implementation of radiotherapy delivery systems with different onboard MR scanners. A summary of the radiomic signatures related to these units is presented below.

### Elekta's MR-Linac System (Unity)

Elekta's hybrid MR-Linac system (Unity) has a field strength of 1.5 T associated with a 7 MV flattening filter-free (FFF) linear accelerator and is defined as high-field system by the scientific community as described in Chap. 10. Evidence in the field of radiomics of such a system is limited to date, given its recent clinical implementation.

A preliminary experience focused on prostate cancer has recently compared radiomic features extracted from MR simulator with those extracted from MR-Linac. For a total of 26 patients, the authors acquired 3 MR images: 1 using an MR-simulator and 2 using the 1.5 T MR-Linac before the delivery of the first treatment fraction. They used a 3D T2-w-TSE sequence considering the gross tumor volume (GTV) and planning target volume (PTV) as regions of interest for radiomic analysis and extracted 1023 radiomic features. The results of this study reported that only approximately 9% of the radiomic features extracted showed high repeatability and reproducibility [25].

Very recently, a study was performed using T2-weighted MRI scans acquired on a 1.5 T MR-Linac to investigate the ability of MRI radiomics to correctly classify patients according to their D'Amico score (high risk vs. low/intermediate risk). A total of 111 patients were enrolled for model building (76 patients) and testing (35 patients), resulting in ROC-AUC values of 0.718 and 0.750 for the CTV- and PTV-based models in the test cohort [26], respectively.

Another study was performed using combination of wavelet delta-radiomic (WDRFs) textures extracted from 90 daily motion-averaged MRI sets derived from 4D MRIs acquired during MR-guided adaptive radiation therapy along with CA19-9, CEA, metastatic status, and follow-up data. They developed and demonstrated the method using 90 datasets extracted from 18 patients treated in 5 stereotactic body radiation therapy fractions with a 1.5 T MR-Linac and showed that incorporating clinical biomarkers with the significant WDRFs increased the concordance statistics equal to 0.65 using clinical biomarkers alone to 0.96 using the combination of delta radiomics with the clinical biomarkers [27].

Delta radiomics has also been utilized recently to aid in the treatment adaptation process to predict necessity of online adaptive replanning based on wavelet multiscale delta-radiomic textures extracted from 119 daily MRIs acquired during MRgART for 24 pancreatic cancer patients treated on a 1.5 T MR-Linac [28]. The system offers two adaptive planning techniques to address the inter-fractional variations based on the daily MRI: an adapt-to-position (ATP) workflow where an adaptive plan is generated and optimized using the original contours from the reference images by shifting the multi-leaf collimators (MLC) based on the co-registration of the daily and the reference image and an adapt-to-shape (ATS) workflow (online adaptive replanning), in which a new plan is generated based on the anatomy of the day requiring recontouring on the daily MRI to fully correct for inter-fractional variations. The latter however is labor intensive and time-consuming and might not be necessary for each treatment; thus, there is a necessity to determine when it is needed during online treatment adaptation process. They developed a model based on multiscale wavelet radiomic features extracted from reference and daily MRIs to predict the necessity of online adaptation automatically and objectively in MRgRT. The correlation between the features extracted from the reference and the daily MRIs calculated in a ROI can indicate the presence of anatomical differences. A high correlation indicated small anatomic difference between the reference and daily images, implying that ATS is not necessary and that ATP is sufficient for that fraction, while a low correlation would indicate substantial anatomic difference between the daily and reference images, suggesting that ATS is necessary for the fraction. To select the optimum ROI for feature extraction, isodose surface (IDS) from the reference plan was converted into structures, and the region enclosing the 100–50% IDS was selected as the optimal ROI. Typically, the 100% IDS covers the tumor volume, while the region between 100 and 50% IDSs contains the dose-limiting OAR(s) with high-dose gradient (Fig. 24.3). Features extracted from this ROI can detect anatomical changes that can affect the dose-limiting OARs and be used to guide the decision toward online ATS versus ATP workflow [28].

## ViewRay's MR-Linac System (MRIdian)

Current evidence in the field of radiomics based on onboard MRI is mainly based on the deployment of hybrid MR-Linac units equipped with low-field MR scanner (0.35 T). Such systems have in fact been widely used in different centers, showing very promising results in terms of feasibility and clinical effectiveness [6, 29]. The low-field MR-Linac system joins a radiation therapy delivery system consisting of a 6MV FFF linear accelerator (MRIdian® ViewRay Inc. system, Mountain View, California, USA) with a 0.35 T MR scanner [30, 31]. The low-field MR system is described in Chap. 9, and low-field imaging is presented in Chap. 23 of this book.

One of the earliest experiences in this field was provided by Boldrini et al. who conducted a feasibility study of delta-radiomic analysis in locally advanced rectal cancer (LARC) [15]. In this cluster of patients, pathological complete

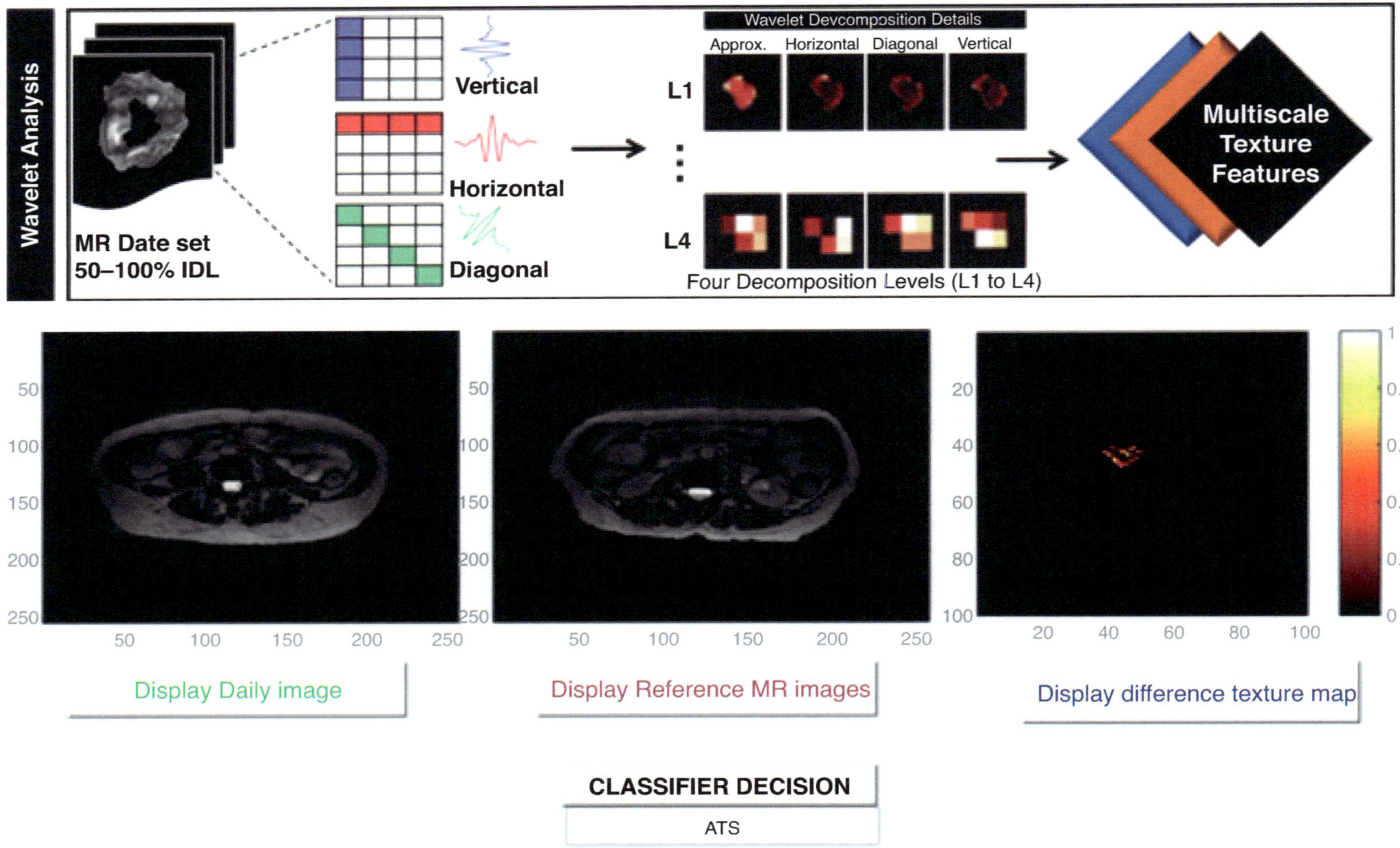


**Fig. 24.3** Wavelet multiscale delta-radiomic textures extracted from different decomposition levels from high-dose gradient regions enclosing the 100–50% IDS and classifier building to automatically determine the necessity of online adaptive replanning using multiscale wavelet-based radiomic features.

response (pCR) is generally associated with better local control (LC), metastasis-free survival (MFS), and overall survival (OS) rates [32]. The authors explored the possibility of establishing predictive models for clinical complete response (cCR), with the aim of being able to consider conservative approaches in the clinical management of patients with LARC. This predictivity was analyzed considering delta-radiomic features in a cohort of 16 patients with LARC undergoing neoadjuvant MRgRT.

Radiomic feature extraction was performed starting from GTV at different treatment time points, analyzing the change from baseline in terms of ratio, corresponding to the treatment simulation. The main result of this hypothesis-generating study was the validity of the delta-radiomic features, which showed higher results in terms of predictive performance with respect to radiomic features extracted at the single time points. In particular, in the study two parameters were shown to be able to fully identify complete responders, describing the variation of a morphological parameter (delta variation of least length, $\Delta L_{least}$) and a textural one (variation of gray-level non-uniformity, $\Delta_{glnu}$) between the second week of treatment and simulation [33], as associated with cCR.

The validity of these parameters was then subsequently evaluated in a multicenter setting, with data from 43 LARC patients treated with the MRgRT technique in 2 different institutions. To obtain a homogeneous analysis of the extracted data, the radiomic features $\Delta L_{least}$ and $\Delta_{glnu}$ were extracted from MR images acquired at identical biologically effective doses (BED) corresponding to 26.8 Gy. The prediction accuracies of cCR in the validation study were 81% and 63% for $\Delta L_{least}$ and $\Delta_{glnu}$, respectively [33]. The study reported also promising results of $\Delta L_{least}$ in predicting pCR, with an accuracy of 79%.

With a similar design, a logistic regression model was built to predict 2-year disease-free survival (2yDFS) from multi-institutional data including clinical and radiomic variables extracted from the GTV. In this study, 48 LARC patients were retrospectively enrolled, 8 of them (16.7%) having developed distant metastases at a 2-year follow-up. The best performing model was based on a delta-radiomic feature, analyzing the variation in the area/surface ratio between the 54 Gy BED and the simulation time points (AUC of 0.92) [34].

Recent evidence has also demonstrated the validity of an additional biomarker of treatment response in LARCs called early regression index (ERITCP). This index models the radiobiological behavior of the tumor by considering its volumetric regression, from simulated treatment ($V_{pre}$) to mid-therapy ($V_{mid}$) [35]. The first theorization of ERITCP was performed on T2-weighted MR images at 1.5 T and was found to be adequate in identifying patients with and without pCR with a good sensitivity (86%) and negative predictive value (90%) and an area under the receiver operating characteristic (ROC) curve (AUC) of 0.81 (0.69–0.89 as 95% confidence interval, CI). The identified cutoff value was found to be 13.1. The predictive power of ERITCP was also externally validated on low-field (0.35 T) MR-Linac system, using a different image contrast (T2*/T1) and identifying the BED value of 25 Gy as the optimal time point to make predictions [36]. The impact of ERITCP has been translated into clinical practice through a clinical trial focused on a dose escalation protocol based on pCR prediction [37]. Patients enrolled in the clinical trial start their MRIgRT treatment with a prescribed dose of 55 Gy in 25 fractions on the GTV plus the corresponding mesorectum and 45 Gy in 25 fractions on the drainage nodes. ERITCP is calculated at fraction 10: patients showing good prognosis (ERITCP <13.1) continue their initial treatment; the others change the treatment schedule, intensifying the dose on residual GTV through an additional boost, scaling up to 60.1 Gy.

Additionally, the potentialities of ERITCP have been explored also in the context of cervical cancer. In a preliminary experience carried out on 16 patients affected by locally advanced cervical cancer (LACC) and treated using 0.35 T MRgRT, Cusumano et al. showed that ERITCP was able to predict pCR with an area under ROC curve of 0.84 [38]. In another experience performed on 88 LACC patients, ERITCP showed significant performances in predicting not only pCR but also local control, disease-free survival, and overall survival after 2 years from the end of treatment [39]. Interestingly, in this experience, ERITCP was calculated also starting from diffusion-weighted (DW) MRI and not only starting from T2-weighted, and the authors observed that the parameter calculated on DW images showed higher performance with respect to T2 (AUC of 0.77 and 0.75 for DW and T2-w, respectively, in predicting pCR).

Another interesting site where MRgRT reported promising results was the treatment of locally advanced pancreatic cancer (LAPC). The availability of high soft tissue contrast images and the possibility of online adapting every single fraction have led to different dose escalation studies with excellent results in terms of local control and overall survival [40]. Pancreatic neoplasms represent a further tumor site where radiomic analysis has yielded interesting results, with Simpson et al.'s study being the first to evaluate the possibility of extracting radiomic features [41]. Analyzing features extracted from the GTV of 20 patients with LAPCs, machine learning models based on LASSO regression and random forest were proposed, with an AUC of 0.81 (95% CI 0.59–1) for the best model. A similar study was conducted on 26 patients with locally advanced and borderline resectable pancreatic adenocarcinoma treated with MRgRT of 50 Gy in 5 fractions. The results showed that linear normalization of the image intensities to the median kidney value improved the reproducibility of the radiomic features, with histogram skewness being significantly associated with progression-free survival ($p = 0.005$, HR = 2.75) [42].

In LAPC, the 1-year local control rate (1y-LC) has also been the focus of studies using radiomic analyses to define predictive models. A retrospective study of 35 patients treated with adaptive MRgRT identified a textural feature capable of predicting 1y-LC [43]. This feature was the variation of cluster shade with respect to the simulation, calculated when a BED value of 40 Gy was reached ($p = 0.005$, AUC = 0.78). The results of these studies would seem to hypothesize the possibility of characterizing the heterogeneity of pancreatic neoplasms through extracted radiological features, capable of early identifying treatment responders.

## Future Developments and Perspectives

The development of delta-radiomics algorithms and related evidence is opening a new perspective in the radiation treatment of oncological diseases. The variation of radiomic parameters during RT treatment as a presumptive index of tumor response could lead to the development of response prediction models to be used for accurate adaptation of ongoing treatment. This perspective is indeed the focus of randomized clinical trials that will assess the appropriateness of dose escalation in patients identified by predictive response models as poor responders [44].

In view of the progressive increase in the number of centers equipped with hybrid MR-Linac units, the definition of multicenter collaborative networks for large-scale validation of predictive models is desirable [45]. The integration of image standardization methods into the radiomic workflow, which have so far been little applied in this MR-Linac context, could become a point of attention to ensure the generalization of the models developed in a multicenter setting [46, 47]. A further relevant perspective in defining the future role of AI and radiomics in MRgRT treatment is that of integrating biomarkers derived from other omics sciences (proteomics, genomics, metabolomics) for the personalization of each treatment.

This chapter has reviewed the current literature about radiomics applied to MRgRT, focusing on delta-radiomics approach, which is made possible by the acquisition of longitudinal images at several points during the course of radiotherapy treatment. Current knowledge is essentially based on low-field MRI, due to its earlier implementation, with promising results for several pathologies, i.e., rectal, cervical, and pancreatic cancers. The analysis of regression kinetics, which are ultimately very simple to extract and the opposite of the black box concept, seems promising for the introduction of personalized treatments and is still being evaluated in clinical trials. Preliminary analyses on the 1.5 T MR-Linac are very promising and can play a critical role in online treatment adaptation workflows. The integration of delta radiomics with clinical biomarkers is another interesting extension that can increase the ability to create interpretable omics-based models to be developed into invaluable tool to predict treatment outcomes.

## References


1. Vandewinckele L, Claessens M, Dinkla A, et al. Overview of artificial intelligence-based applications in radiotherapy: recommendations for implementation and quality assurance. Radiother Oncol. 2020;153:55–66.
2. Ramesh AN, Kambhampati C, Monson JRT, et al. Artificial intelligence in medicine. Ann R Coll Surg Engl. 2004;86:334–8.
3. Limkin EJ, Sun R, Dercle L, et al. Promises and challenges for the implementation of computational medical imaging (radiomics) in oncology. Ann Oncol. 2017;28:1191–206.
4. Casà C, Piras A, D'Aviero A, et al. The impact of radiomics in diagnosis and staging of pancreatic cancer. Ther Adv Gastrointest Endosc. 2022;15:26317745221081596.
5. Boulanger M, Nunes J-C, Chourak H, et al. Deep learning methods to generate synthetic CT from MRI in radiotherapy: a literature review. Phys Med. 2021;89:265–81.
6. Cusumano D, Boldrini L, Dhont J, et al. Artificial intelligence in magnetic resonance guided radiotherapy: medical and physical considerations on state of art and future perspectives. Phys Med. 2021;85:175–91.
7. Robert C, Munoz A, Moreau D, et al. Clinical implementation of deep-learning based auto-contouring tools—experience of three French radiotherapy centers. Cancer Radiother. 2021;25:607–16.
8. McIntosh C, Conroy L, Tjong MC, et al. Clinical integration of machine learning for curative-intent radiation treatment of patients with prostate cancer. Nat Med. 2021;27:999–1005.
9. Lambin P, Rios-Velazquez E, Leijenaar R, et al. Radiomics: extracting more information from medical images using advanced feature analysis. Eur J Cancer. 2012;48:441–6.
10. Dercle L, Henry T, Carré A, et al. Reinventing radiation therapy with machine learning and imaging biomarkers (radiomics): state-of-the-art, challenges and perspectives. Methods. 2021;188:44–60.
11. Kumar V, Gu Y, Basu S, et al. Radiomics: the process and the challenges. Magn Reson Imaging. 2012;30:1234–48.
12. Lambin P, Leijenaar RTH, Deist TM, et al. Radiomics: the bridge between medical imaging and personalized medicine. Nat Rev Clin Oncol. 2017;14:749–62.
13. Bodalal Z, Trebeschi S, Beets-Tan R. Radiomics: a critical step towards integrated healthcare. Insights Imaging. 2018;9:911–4.
14. Chiloiro G, Rodriguez-Carnero P, Lenkowicz J, et al. Delta Radiomics can predict distant metastasis in locally advanced rectal cancer: the challenge to personalize the cure. Front Oncol. 2020;10:595012.
15. Boldrini L, Cusumano D, Chiloiro G, et al. Delta radiomics for rectal cancer response prediction with hybrid 0.35 T magnetic resonance-guided radiotherapy (MRgRT): a hypothesis-generating study for an innovative personalized medicine approach. Radiol Med. 2019;124:145–53.
16. Gardin I, Grégoire V, Gibon D, et al. Radiomics: principles and radiotherapy applications. Crit Rev Oncol Hematol. 2019;138:44–50.
17. Nardone V, Reginelli A, Grassi R, et al. Delta radiomics: a systematic review. Radiol Med. 2021;126:1571–83.
18. Cusumano D, Boldrini L, Yadav P, et al. Delta Radiomics analysis for local control prediction in pancreatic cancer patients treated using magnetic resonance guided radiotherapy. Diagnostics. 2021;11:72.
19. Ford J, Dogan N, Young L, et al. Quantitative radiomics: impact of pulse sequence parameter selection on MRI-based textural features of the brain. Contrast Media Mol Imaging. 2018;2018:1729071.
20. Buch K, Kuno H, Qureshi MM, et al. Quantitative variations in texture analysis features dependent on MRI scanning parameters: a phantom model. J Appl Clin Med Phys. 2018;19:253–64.
21. Mayerhoefer ME, Szomolanyi P, Jirak D, et al. Effects of MRI acquisition parameter variations and protocol heterogeneity on the results of texture analysis and pattern discrimination: an application-oriented study. Med Phys. 2009;36:1236–43.
22. Da-Ano R, Visvikis D, Hatt M. Harmonization strategies for multicenter radiomics investigations. Phys Med Biol. 2020;65:24TR02.
23. Tixier F, Jaouen V, Hognon C, et al. Evaluation of conventional and deep learning based image harmonization methods in radiomics studies. Phys Med Biol. 2021;66:245009.
24. Reuzé S, Schernberg A, Orlhac F, et al. Radiomics in nuclear medicine applied to radiation therapy: methods, pitfalls, and challenges. Int J Radiat Oncol Biol Phys. 2018;102:1117–42.
25. Xue C, Yuan J, Poon DM, et al. Reliability of MRI radiomics features in MR-guided radiotherapy for prostate cancer: repeatability, reproducibility, and within-subject agreement. Med Phys. 2021;48:6976–86.
26. Zhou Y, Yuan J, Xue C, et al. A pilot study of MRI radiomics for high-risk prostate cancer stratification in 1.5 T MR-guided radiotherapy. Magn Reson Med. 2023;89:2088–99.
27. Nasief HG, Chen X, Erickson BA, et al. Improving outcomes prediction using a combination of MRI

Wavelet Delta-Radiomic textures and clinical biomarkers for MR-guided adaptive radiation therapy of pancreatic cancer. Int J Radiat Oncol Biol Phys. 2022;114:S79–80.
28. Nasief HG, Parchur AK, Omari E, et al. Predicting necessity of daily online adaptive replanning based on wavelet image features for MRI guided adaptive radiation therapy. Radiother Oncol. 2022;176:165–71.
29. Kurz C, Buizza G, Landry G, et al. Medical physics challenges in clinical MR-guided radiotherapy. Radiat Oncol. 2020;15:93.
30. Klüter S. Technical design and concept of a 0.35 T MR-Linac. Clin Transl Radiat Oncol. 2019;18:98–101.
31. Mutic S, Dempsey JF. The ViewRay system: magnetic resonance-guided and controlled radiotherapy. Semin Radiat Oncol. 2014;24:196–9.
32. Martin ST, Heneghan HM, Winter DC. Systematic review and meta-analysis of outcomes following pathological complete response to neoadjuvant chemoradiotherapy for rectal cancer. Br J Surg. 2012;99:918–28.
33. Cusumano D, Boldrini L, Yadav P, et al. Delta radiomics for rectal cancer response prediction using low field magnetic resonance guided radiotherapy: an external validation. Phys Med. 2021;84:186–91.
34. Chiloiro G, Boldrini L, Preziosi F, et al. A predictive model of 2yDFS during MR-guided RT neoadjuvant chemoradiotherapy in locally advanced rectal cancer patients. Front Oncol. 2022;12:831712.
35. Fiorino C, Gumina C, Passoni P, et al. A TCP-based early regression index predicts the pathological response in neo-adjuvant radio-chemotherapy of rectal cancer. Radiother Oncol. 2018;128:564–8.
36. Cusumano D, Boldrini L, Yadav P, et al. External validation of early regression index (ERITCP) as predictor of pathologic complete response in rectal cancer using magnetic resonance-guided radiation therapy. Int J Radiat Oncol Biol Phys. 2020;108:1347–56.
37. Chiloiro G. THeragnostic Utilities for Neoplastic DisEases of the rectum by MRI guided radiotherapy. clinicaltrials.gov; 2021.
38. Cusumano D, Catucci F, Romano A, et al. Evaluation of an early regression index (ERITCP) as predictor of pathological complete response in cervical cancer: a pilot-study. Appl Sci. 2020;10:8001.
39. Cusumano D, Russo L, Gui B, et al. Evaluation of early regression index as response predictor in cervical cancer: a retrospective study on T2 and DWI MR images. Radiother Oncol. 2022;174:30–6.
40. Rudra S, Jiang N, Rosenberg SA, et al. Using adaptive magnetic resonance image-guided radiation therapy for treatment of inoperable pancreatic cancer. Cancer Med. 2019;8:2123–32.
41. Simpson G, Spieler B, Dogan N, et al. Predictive value of 0.35 T magnetic resonance imaging radiomic features in stereotactic ablative body radiotherapy of pancreatic cancer: a pilot study. Med Phys. 2020;47:3682–90.
42. Tomaszewski MR, Latifi K, Boyer E, et al. Delta radiomics analysis of magnetic resonance guided radiotherapy imaging data can enable treatment response prediction in pancreatic cancer. Radiat Oncol. 2021;16:237.
43. Cusumano D, Boldrini L, Yadav P, et al. Delta Radiomics analysis for local control prediction in pancreatic cancer patients treated using magnetic resonance guided radiotherapy. Diagnostics (Basel). 2021;11:72.
44. Chiloiro G, Cusumano D, Boldrini L, et al. THUNDER 2: THeragnostic Utilities for Neoplastic DisEases of the rectum by MRI guided radiotherapy. BMC Cancer. 2022;22:67.
45. de Mol van Otterloo SR, Christodouleas JP, Blezer ELA, et al. The MOMENTUM study: an international registry for the evidence-based introduction of MR-guided adaptive therapy. Front Oncol. 2020;10:1328.
46. Orlhac F, Lecler A, Savatovski J, et al. How can we combat multicenter variability in MR radiomics? Validation of a correction procedure. Eur Radiol. 2021;31:2272–80.
47. Carré A, Klausner G, Edjlali M, et al. Standardization of brain MR images across machines and protocols: bridging the gap for MRI-based radiomics. Sci Rep. 2020;10:12340.

# Epilogue and Outlook

# 25

Filippo Alongi, Bharat B. Mittal, Indra J. Das, and Poonam Yadav

## Introduction

MR-Linac is a nascent technology that allows for simultaneous real-time imaging and precise radiation treatment. It has great potential to impact clinical practice and clinical research in radiation oncology. There have been several review articles describing its usage and potential [1, 2]. Additionally, there is momentum behind MRI-only-based radiation treatment [3–5]. There was a need to gather and summarize the knowledge and literature on this technology, and this was the starting point of our thoughts for a book, which we undertook beginning in the Spring of 2022. The 25 chapters collected here are organized in 4 major sections: General, MR-Linac systems, Clinical, and Miscellaneous futuristic topics. Our view was that chapters should be written by experts in each field; hence, our authors represent a vast number of researchers and clinicians around the world, and the chapters cover every system available for patient treatment.

The general section of the book provides an Introduction (Chap. 1) and then addresses the Role of MRI (Chap. 2), Clinical Necessities (Chap. 3), Education and Training (Chap. 4), Imaging Sequence (Chap. 5), Motion Management (Chap. 6), Synthetic CT (Chap. 7), and Treatment Planning (Chap. 8). The next section describes the existing MR-Linac systems which are being used to treat patients. These include the ViewRay MRIdian (Chap. 9), Elekta Unity (Chap. 10), and Aurora Magneton (Chap. 11).

Since MR-Linac is used for many disease sites, it was decided that the coverage should be given to the most common disease sites, which would benefit readers looking to glean from the experience of expert authors and adapt their treatment options and policies accordingly. Thus, in the Clinical section, we included chapters on Oligometastatic Diseases (Chap. 12), Pancreas (Chap. 13), Liver (Chap. 14), Lung (Chap. 15), Prostate (Chap. 16), Breast (Chap. 17), Gynecological Diseases (Chap. 18), Pediatrics (Chap. 19), Brain (Chap. 20), and Head and Neck (Chap. 21).

The final section of the book includes chapters which are more futuristic. These chapters cover the Australian Sydney Project on MR-Linac (Chap. 22), Low-Field Imaging (Chap. 23), and

F. Alongi (✉)
University of Brescia, Brescia, Italy

Advanced Radiation Oncology Department, IRCCS Sacro Cuore Don Calabria Hospital, Negrar di Valpolicella, Italy

B. B. Mittal · I. J. Das · P. Yadav
Department of Radiation Oncology, Northwestern Memorial Hospital, Northwestern University Feinberg School of Medicine, Chicago, IL, USA
e-mail: b-mittal@northwestern.edu; indra.das@nm.org; poonam.yadav@northwestern.edu



I. J. Das et al. (eds.), *A Practical Guide to MR-Linac*, https://doi.org/10.1007/978-3-031-48165-9_25

Artificial Intelligence and Image Texture Analysis (Chap. 24).

A wealth of information about MR-Linac technology and its emerging clinical applications has been collated here, with up-to-date knowledge and insights. Future editions of this book will include important technological advances, especially autosegmentation and gating tools integrated into the MR-Linac workflow, but more importantly the clinical outcome data from many ongoing trials.

## Wish List and Future of MR-Linac

The future of MR-Linac depends on technological advances but more importantly on the clinical outcome, which is described in detail by Rammohan et al. [6]. Currently, a large amount of clinical data has been reported, but most of it with a short follow-up and a limited population size, as shown in the various clinical trials and references [5, 7–27].

Some of the potential improvements of this innovative technology are elaborated below. At present, most MR-Linac setups can only accommodate 6–8 patients/day, which is not optimal considering the number of waiting lists and treatment requests. To increase the number of treatable patients, several aspects have to be incorporated:

- Patient comfort procedures should be streamlined.
- Gantry size of the MR-Linac and field of view need to be increased.
- Autosegmentation needs to be faster and more reliable.
- Dose rate should be modulated in order to minimize treatment time.
- Volumetric modulated arc therapy (VMAT) should be standard to improve dose distribution and shorten treatment time for each session.
- Better image sequencing to enhance morphological and functional aspects.
- Online cumulative dose option, to increase the quality of the more complex treatment plans.
- 4D MRI option to help manage anatomic motion.
- Accurate synthetic CT option to reduce dependence on CT imaging.
- Simplified imaging quality assurance (QA) option for low-field MRI.
- 3D couch motion (missing in the current Unity system).
- Curate an image library, to increase clinical use in challenging cases.
- Add an option to check the robustness of the original plan.
- Technical challenges in the breast due to electron return effect [28, 29].
- Choice of multiple-beam energy.
- Develop methods for biomarkers and radiomics [30–36].

## Summary

To conclude, MR-Linac represents one of the most advanced medical devices, combining MRI and linear accelerator technology. It unifies the capabilities of an MRI scanner and a linear accelerator into one system, enabling real-time imaging and improving radiotherapy delivery during tumor treatment.

The MR-Linac system allows for precise and personalized radiotherapy by providing, as was not possible before, high-quality real-time images of the tumor and surrounding tissue. Moreover, it allows physicians to track tumor location; account for changes in shape, size, and position during treatment; and adjust radiation therapy accordingly. These opportunities and challenges have been described by Hehakaya et al. [37].

MR-Linac can offer several advantages by combining MRI and radiotherapy in one machine. It can improve the precision of treatment and reduce potential damage to healthy tissue around the tumor. Real-time imaging also enables adaptive treatment planning, allowing radiation to be precisely delivered to target areas while minimizing stress on healthy tissue.

However, MR-Linac technology is relatively new, and its evolution is still ongoing. Research and development work is being continuously carried out to further improve its capabilities and clinical applications. Promising improvements include:

- Improved targeting, to minimize healthy tissue damage and increase effectiveness of the treatment
- Adaptive treatment, especially for lesions that move or change shape during the treatment sessions, including lung tumors or upper abdominal sites, affected directly or indirectly by breathing
- Soft tissue visualization, allowing users to identify and precisely target lesions difficult to define with standard imaging

Indeed, MR-Linac technology has the potential capability to advance radiotherapy techniques and definitively improve outcomes for cancer patients. Nevertheless, while MR-Linac represents one of the most innovative steps in modern radiation therapy, its clinical availability and application in clinical practice may vary across different countries and facilities. The impact of MR-Linac on clinical outcomes still remains an unresolved question, which will become clearer with the many clinical trials that are ongoing. Future studies and technological improvements are awaited.

## References


1. Pollard JM, Wen Z, Sadagopan R, Wang J, et al. The future of image-guided radiotherapy will be MR guided. Br J Radiol. 2017;90:20160667.
2. Das IJ, McGee KP, Tyagi N, Wang H. Role and future of MRI in radiation oncology. Br J Radiol. 2019;92:20180505.
3. Owrangi AM, Greer PB, Glide-Hurst CK. MRI-only treatment planning: benefits and challenges. Phys Med Biol. 2018;63:05tr01.
4. Srinivasan S, Dasgupta A, Chatterjee A, Baheti A, et al. The promise of magnetic resonance imaging in radiation oncology practice in the management of brain, prostate, and GI malignancies. JCO Glob Oncol. 2022;8:e2100366.
5. Chandarana H, Wang H, Tijssen RHN, Das IJ. Emerging role of MRI in radiation therapy. J Magn Reson Imaging. 2018;48:1468–78.
6. Rammohan N, Randall JW, Yadav P. History of technological advancements towards MR-Linac: the future of image-guided radiotherapy. J Clin Med. 2022;11:4730.
7. Kishan AU, Ma TM, Lamb JM, Casado M, et al. Magnetic resonance imaging-guided vs computed tomography-guided stereotactic body radiotherapy for prostate cancer: the MIRAGE randomized clinical trial. JAMA Oncol. 2023;9:365–73.
8. Michalet M, Bettaïeb O, Khalfi S, Ghorbel A, et al. Stereotactic MR-guided radiotherapy for adrenal gland metastases: first clinical results. J Clin Med. 2022;12:291.
9. Alongi F, Rigo M, Figlia V, Nicosia L, et al. 1.5T MR-guided daily-adaptive SBRT for prostate cancer: preliminary report of toxicity and quality of life of the first 100 patients. J Pers Med. 2022;12:1982.
10. Nierer L, Eze C, da Silva MV, Braun J, et al. Dosimetric benefit of MR-guided online adaptive radiotherapy in different tumor entities: liver, lung, abdominal lymph nodes, pancreas and prostate. Radiat Oncol. 2022;17:53.
11. Nicosia L, Trapani G, Rigo M, Giaj-Levra N, et al. 1.5 T MR-guided daily adapted SBRT on lymph node oligometastases from prostate cancer. J Clin Med. 2022;11:6658.
12. Chuong MD, Herrera R, Kaiser A, Rubens M, et al. Induction chemotherapy and ablative stereotactic magnetic resonance image-guided adaptive radiation therapy for inoperable pancreas cancer. Front Oncol. 2022;12:888462.
13. Cuccia F, Rigo M, Figlia V, Giaj-Levra N, et al. 1.5T MR-guided daily adaptive stereotactic body radiotherapy for prostate re-irradiation: a preliminary report of toxicity and clinical outcomes. Front Oncol. 2022;12:858740.
14. Versteeg AL, Sahgal A, Kawahara N, Rhines LD, et al. Patient satisfaction with treatment outcomes after surgery and/or radiotherapy for spinal metastases. Cancer. 2019;125:4269–77.
15. Sayan M, Serbez I, Teymur B, Gur G, et al. Patient-reported tolerance of magnetic resonance-guided radiation therapy. Front Oncol. 2020;10:1782.
16. Tan H, Cheung P, Louie AV, Myrehaug S, et al. Outcomes of extra-cranial stereotactic body radiotherapy for metastatic breast cancer: treatment indication matters. Radiother Oncol. 2021;161:159–65.
17. Weykamp F, Hoegen P, Klüter S, Spindeldreier CK, et al. Magnetic resonance-guided stereotactic body radiotherapy of liver tumors: initial clinical experience and patient-reported outcomes. Front Oncol. 2021;11:610637.
18. Hehakaya C, van der Voort van Zyp JRN, Vanneste BGL, Grutters JPC, et al. Early health economic analysis of 1.5 T MRI-guided radiotherapy for localized prostate cancer: decision analytic modelling. Radiother Oncol. 2021;161:74–82.
19. de Mol van Otterloo SR, Christodouleas JP, Blezer ELA, Akhiat H, et al. The MOMENTUM study: an international registry for the evidence-based introduc-

tion of MR-guided adaptive therapy. Front Oncol. 2020;10:1328.
20. Christiansen RL, Dysager L, Hansen CR, Jensen HR, et al. Online adaptive radiotherapy potentially reduces toxicity for high-risk prostate cancer treatment. Radiother Oncol. 2022;167:165–71.
21. Chuong MD, Bryant J, Mittauer KE, Hall M, et al. Ablative 5-fraction stereotactic magnetic resonance-guided radiation therapy with on-table adaptive replanning and elective nodal irradiation for inoperable pancreas cancer. Pract Radiat Oncol. 2021;11:134–47.
22. Krishnan S, Chadha AS, Suh Y, Chen HC, et al. Focal radiation therapy dose escalation improves overall survival in locally advanced pancreatic cancer patients receiving induction chemotherapy and consolidative chemoradiation. Int J Radiat Oncol Biol Phys. 2016;94:755–65.
23. Luterstein E, Cao M, Lamb J, Raldow AC, et al. Stereotactic MRI-guided adaptive radiation therapy (SMART) for locally advanced pancreatic cancer: a promising approach. Cureus. 2018;10:e2324.
24. Finazzi T, Haasbeek CJA, Spoelstra FOB, Palacios MA, et al. Clinical outcomes of stereotactic MR-guided adaptive radiation therapy for high-risk lung tumors. Int J Radiat Oncol Biol Phys. 2020;107:270–8.
25. Alongi F, Rigo M, Figlia V, Cuccia F, et al. 1.5T MR-guided and daily adapted SBRT for prostate cancer: feasibility, preliminary clinical tolerability, quality of life and patient-reported outcomes during treatment. Radiat Oncol. 2020;15:69.
26. Ugurluer G, Atalar B, Zoto Mustafayev T, Gungor G, et al. Magnetic resonance image-guided adaptive stereotactic body radiotherapy for prostate cancer: preliminary results of outcome and toxicity. Br J Radiol. 2021;94:20200696.
27. Bruynzeel AME, Tetar SU, Oei SS, Senan S, et al. A prospective single-arm phase 2 study of stereotactic magnetic resonance guided adaptive radiation therapy for prostate cancer: early toxicity results. Int J Radiat Oncol Biol Phys. 2019;105:1086–94.
28. De-Colle C, Nachbar M, Mönnich D, Boeke S, et al. Analysis of the electron-stream effect in patients treated with partial breast irradiation using the 1.5 T MR-linear accelerator. Clin Transl Radiat Oncol. 2021;27:103–8.
29. Nachbar M, Mönnich D, Boeke S, Gani C, et al. Partial breast irradiation with the 1.5 T MR-Linac: first patient treatment and analysis of electron return and stream effects. Radiother Oncol. 2020;145:30–5.
30. McGee KP, Hwang KP, Sullivan DC, Kurhanewicz J, et al. Magnetic resonance biomarkers in radiation oncology: the report of AAPM task group 294. Med Phys. 2021;48:e697–732.
31. Otazo R, Lambin P, Pignol JP, Ladd ME, et al. MRI-guided radiation therapy: an emerging paradigm in adaptive radiation oncology. Radiology. 2021;298:248–60.
32. Mata LA, Retamero JA, Gupta RT, García Figueras R, et al. Artificial intelligence-assisted prostate cancer diagnosis: radiologic-pathologic correlation. Radiographics. 2021;41:1676–97.
33. Granzier RWY, Verbakel NMH, Ibrahim A, van Timmeren JE, et al. MRI-based radiomics in breast cancer: feature robustness with respect to inter-observer segmentation variability. Sci Rep. 2020;10:14163.
34. Cusumano D, Boldrini L, Dhont J, Fiorino C, et al. Artificial intelligence in magnetic resonance guided radiotherapy: medical and physical considerations on state of art and future perspectives. Phys Med. 2021;85:175–91.
35. Gui B, Autorino R, Miccò M, Nardangeli A, et al. Pretreatment MRI Radiomics based response prediction model in locally advanced cervical cancer. Diagnostics (Basel). 2021;11:631.
36. Cusumano D, Russo L, Gui B, Autorino R, et al. Evaluation of early regression index as response predictor in cervical cancer: a retrospective study on T2 and DWI MR images. Radiother Oncol. 2022;174:30–6.
37. Hehakaya C, Sharma AM, van der Voort Van Zijp JRN, Grobbee DE, et al. Implementation of magnetic resonance imaging-guided radiation therapy in routine care: opportunities and challenges in the United States. Adv Radiat Oncol. 2022;7:100953.

# Index


**A**
Abdominal organs, 264
Adaptive radiotherapy (ART), 2, 5, 8, 21, 149, 170, 227, 398
Adapt-to-position (ATP) workflow, 279, 463
Adapt-to-shape (ATS) workflow, 279, 463
Adjust radiation therapy, 470
Agilent magnet, 428
Alberta Linac-MR, 205–211
Alkaline phosphatase (ALP), 258
Alpha-fetoprotein (AFP), 258
American Association for the Study of Liver Diseases (AASLD), 258
American Brachytherapy Society, 324
American College of Radiology, 438
American Joint Committee on Cancer (AJCC) cervical cancer staging system, 325
Androgen deprivation therapy (ADT), 285, 286
Apparent diffusion coefficient (ADC), 278, 399
Atlas (multi-atlas) approach, 96, 97
Aurora-RT, historical milestone and development, 193–195
Australian MRI-Linac (AMRL)
  adjunct therapeutic equipment, 429
  advantages and disadvantages, 426
  beam and dosimetry, 429
  beam-field interactions, 427
  final component of, 429
  gadolinium dose enhancement, 429
  manufactured components, 428
  patient rotation, 430
  planning and treatment strategies, 429
  portable linear accelerator, 426
  protocol development and training, 426
  radiation induction, 427
  surgical interventions, 426

**B**
Balanced steady-state free precession (bSSFP) sequence, 201
Beams geometries, 132
Biochemical recurrence (BCR), 286
Biologically effective dose (BED), 226, 278
Bladder rhabdomyosarcoma, 365
Bo field homogeneity, 160
Boolean operations, 357
Borderline resectable pancreatic cancer, 245
Borderline resectable tumors, 244
Brain metastases, 376, 379–381
Brain tumors
  brain metastases, 379–381
  contouring and planning, 382–385
  dose constraints, 385–388
  glioblastoma, 378–380
  on-table adaptation
    Elekta UnityTM, 388, 390
    Viewray MRIdianTM, 388
  primary and metastatic brain tumors, 375
  traditional approaches, 376, 377
  in United States, 375
  whole brain radiation therapy (WBRT), 376
Breast radiotherapy
  contouring, 311–313
  diagnosis and treatment, 308
  dose prescription and DVH constraints, 313, 314
  early diagnosis, 308
  image acquisition, 310, 311
  incidence of, 307
  at MR-Linac, 314–316
  oncological outcomes, 308
  on-table adaptation, 316
  outcome, 317
  overcoming challenges, 308, 309
  patient positioning, 310
Breath-hold techniques, 72, 73, 78
Breath-hold treatments, 71
Breathing fatigue, 30
Bulk density assignment (BDA), 92, 96

**C**
Carcinoembryonic antigen (CEA), 258
Cardiac motion, 70
Central nervous system (CNS), 375
Central pelvic recurrence, 339



I. J. Das et al. (eds.), *A Practical Guide to MR-Linac*, https://doi.org/10.1007/978-3-031-48165-9

Child-Pugh (CP) scoring system, 259
Children's Oncology Group trial ACNS0121, 359
Ciné imaging, 14, 15
Cine MRI with breath-hold technique and target tracking, 73, 74
Claustrophobia, 27
Clinical target volume (CTV), 308
Compressed Sensing and Principal Component Analysis (CS-PCA), 201, 202
Computed tomography (CT), 22, 41, 46, 376
  in radiation therapy, 89
Conditional GAN (cGAN), 100
Cone beam computed tomography (CBCT), 21–23, 89, 324, 332
Constraint violations, 275
Contrast enhancement, 56–58
Conventional fractionation, 286
Conventional image-guided radiotherapy, 449
Conventional radiomics, 459
Conventional treatment approaches, 361
Conventionally fractionated radiotherapy, 375
Craniospinal irradiation (CSI), 359
Cyberknife™, 287, 288

**D**
Daily dose accumulation, 279
Deep inspiratory breath hold (DIBH) technique, 275
Deep-learning methods, 434, 460
Delta-radiomic process, 460, 463, 466
De-noising techniques, 434
Dice's Similarity Index (DSI), 200
Diffusion weighted imaging (DWI), 53, 54, 60, 278
Diffusion-prepared turbo spin-echo readout (DP-TSE), 455
Diffusion-weighted imaging (DWI), 310, 361, 453, 455, 466
Dixon method, 6
Dose calculation, 14, 134, 135
  electron return effect, 135
  heterogeneity corrections, 136
Dose constraints, 263
Dose escalation, 262, 263
Dose-volume-histograms (DVHs), 224, 225, 261, 262
Dynamic contrast-enhanced (DCE) MRI, 25, 278, 361

**E**
Early regression index (ERITCP), 465
Elective para-aortic (PAO) node irradiation, 331
Electron density (ED), 128
  CT primary, 128
  MR primary, 129, 130, 132
Electron density data, 275
Electron return effect (ERE), 135, 150, 151, 177
Electron streaming effect (ESE), 177
Elekta Unity MR-Linac, 59
  4D-MRI, 73
Elekta Unity system, 36, 79, 175, 252, 260, 378, 388, 390
  beam characteristics, 176
  transverse magnetic field, 176
Endometrial cancer, 323
  inoperable endometrial cancer, 341–343
European GEC ESTRO, 324
External beam radiotherapy (EBRT), 258, 285, 291

**F**
Fast Field Echo (FFE) sequences, 280
Fat reduction techniques, 55
Fiducials, 259
Fixed conformal, 153
Forced expiratory volume (FEV), 447
Forced shallow breathing with abdominal compression, 71
Fourier transformation, 49
Full width at half maximum (FWHM), 160
Functional biomarkers, 139
Fuzzy c-means (FCM) algorithms, 98

**G**
Gadolinium contrast enhancement, 310
Gadoxetate, 260
Gantry isocenter tolerance, 164
Gaussian mixture models (GMM), 97
Genitourinary rhabdomyosarcoma, 365
Geometric fidelity, 103
Gibbs/truncation artifact, 63
Glioblastoma, 378–380
Gross tumor volume (GTV), 248, 309
Gynecological cancer
  MR Linacs, role of
    adaptive MRgRT, 332–335
    endometrial cancer (*see* Endometrial cancer)
    MRgRT, 335–338
    recurrent cervical cancer, 337, 339–341
  MRI, 324
    cervical and vaginal cancer, 325–328
    endometrial cancer, 328
    pelvic MRI, 329–331
  oligometastatic disease, 346–349
  recurrent endometrial cancer (*see* Recurrent endometrial cancer)

**H**
Halbach magnets, 438
Head and neck neoplasms
  complex anatomy, 398
  complex planning, 398
  constraints, 403, 404
  contouring and PTV, 402
    MRI sequences, 400, 401
  current treatment paradigms for, 396
  dose de-escalation, 408, 409
  dose escalation
    optimal dose, 407, 408
  epidemiology and risk factors, 395, 396

heterogeneity of, 397, 398
hypofractionation and SBRT in, 399
MRL use in, 399, 400
OAR contouring, 403–405
on-table adaptation
CT vs. MRI for reference plan, 411, 412
Elekta unity online workflow, 409, 410
evaluation, 413
intrafraction motion, 412
MRI coils, 412, 413
treatment plan adaptation strategies, 411
ViewRay MRIdian online workflow, 410, 411
outcomes
future perspectives on outcome measurement, 414
toxicity reporting, 414
tumor control outcome reporting, 414
patient-specific challenges, 416, 417
patterns of failure of, 398, 399
planning challenges, 404, 406
PTV margins, 403
staffing and time-related challenges, 415, 416
technological challenges, 415
treatment, changing landscape of, 397
Hepatic iron deposition, 55
Hepatocellular carcinoma (HCC), 257
Heterogeneity corrections, 136
High-Z artifacts, 11
Homogeneous magnets, 437
Hyperfine, 438, 439

I
Image acquisition, 325
Image-guided adaptive BT (IGABT), 332
Image-guided radiation therapy (IGRT), 5, 6, 21, 35, 149, 287
Imaging artifacts, 65
Immobilization, planning considerations for, 84, 85
Immunotherapy, 349
Implanted electromagnetic transponders, 287
Innovative technology, 470
Intensity modulated radiation therapy (IMRT), 133, 167, 323
Inter- and intra-observer variability, 24
Internal target volumes (ITV), 274
International Federation of Gynecology and Obstetrics (FIGO), 325
Interobserver contouring variation, 40
Intracranial ependymomas, 359
Intrafractional planar imaging, 290
Intrahepatic cholangiocarcinoma (ICC), 257
Intravoxel incoherent motion (IVIM), 455
Ion chambers, 195
Irradiation margin, 7
Isocenter placement, 132

K
Kinetic energy released in matter (KERMA), 176

L
Larmor frequency, 104
Liver malignancies, 258
Locally advanced cervical cancer (LACC), 323, 465
Locally advanced pancreatic cancer (LAPC), 246, 466
Locally advanced rectal cancer (LARC), 463
Lorentz force, 134, 151, 195
Low-field MR imaging
advantages of, 443–449
clinical applications of, 449
clinical purpose, 455
designing coils for, 435–437
diffusion-weighted MRI (DWI), 453, 455
engineering challenges of, 435
4D MRI, 452–454
high-income countries (HICs), 433
high-speed MRI, 451, 453
linear SNR dependence, 434, 435
low-field magnet, 449–451
magnetic fields, 434
MR signal amplitude, 434
permanent magnet system, 437, 438
physical properties of, 435
physiological and physical parameters, 433
point-of-care and portable MRI, 438, 439, 442

M
MAESTRO trial, 262
Magnetic resonance fingerprinting, 449, 451
Magnetic resonance for calculating attenuation (MRCAT) prostate, 109
Magnetic resonance imaging (MRI), 6, 24–26, 45, 46, 286, 460
advantages, 6
artifacts, 63, 64
autocontouring and segmentation, 13, 14
ciné imaging, 14, 15
contouring, 13
contrast enhancement, 56–58
diffusion weighted imaging (DWI), 53, 54
dose calculation, 14
fat reduction techniques, 55
high-Z artifacts, 11, 12
imaging protocols, 59
margin, 6–8
MR education and training, 14, 15
MR simulation, 12, 13
multimodality imaging
superiority, 9, 10
x-ray imaging, 8, 9
physics, 47–49
problem with, 6
proton density (PD), 53
safety, 26, 27
safety considerations, 65
spin echo and gradient echo, variation in, 52, 53
T1-weighted imaging, T2-weighted imaging, and pulse sequences, 49–52

Magnetic resonance imaging with linear accelerators (MR-Linac), 2, 21, 35, 39, 106, 149, 471
acceptance testing for, 163, 164
architectural and structural planning, 178, 179
breath-hold techniques, 72, 73
breathing fatigue, 30
cine MRI with breath-hold technique and target tracking, 73, 74
claustrophobia, 27
commissioning, and routine QA, 184–190
construction and installation timeline, 179, 180
economic analysis and cost, 177, 178
4D computed tomography, 23
4D-MRI, 72
historical perspective, 1, 2
limitation, 73
logistics and economics, 157
magnetic resonance imaging (MRI), 24–26
safety, 26, 27
magnetic susceptibility, 30
metallic implants, 27–29
in oligometastatic disease (*see* Oligometastatic disease (OMD))
on-board imaging systems, 21–23
patient comfort, 29
patient selection, 26
planning and dosimetry challenges for, 152
reference dosimetry, 165
sequence in, 59
Elekta Unity MR-Linac, 59
Navigator, 60–63
ViewRay MRIdian MR-Linac, 59
volumetric TRUFI sequence, 60
staffing and training, 181
team member roles and responsibilities, 180, 181
treatment planning systems (TPS) (*see* Treatment planning systems (TPS))
thermal effects, 30
wishlist and future of, 470
workflow setup and considerations
fall back scenarios, 184
implementation of new workflows, 184
pre-treatment preparations and online workflow, 182, 183
Magnetic resonance-guided radiation therapy (MRgRT), 6, 35, 36, 123, 178, 449
advantages, 259, 260
for brain tumors (*see* Brain tumors)
challenges, 280, 281
contouring, 260, 276–278
dose escalation, 262, 263, 278
DVH constraints, 261, 262
DVH metrics/evaluation, 279
evaluation, 264, 265
imaging biomarkers in
Elekta's MR-Linac system (Unity), 462–464
future developments and perspectives, 466, 467
ViewRay's MR-Linac system (MRIdian), 463, 465, 466
OARs, 260
outcomes, 265–267, 279, 280
MRI-specific training and equipment, 37, 38, 40
pediatric and adolescent patient
abdominal tumors, 363, 364
head and neck cancers, 360, 361
modality approach, 355–357
MR-guided linear accelerator treatment workflow, 357, 358
neuroblastoma, 364, 365
patient selection and technical challenges, 367
pediatric leukemias, 361, 362
pediatric lymphoma, 362, 363
pelvic tumors in, 365–367
posterior fossa, 359, 360
spinal tumors, 359, 360
supratentorial brain tumors, 358
potential limitations of, 299–301
preparation, 36, 37
for prostate cancer (*see* Prostate cancer)
PTV, and OAR, 276–278
radiomic workflow, 460–462
sample patient plans, 267, 268
on table adaptation, 278, 279
on-table adaptive radiotherapy (On-ART), 263, 264
training and education during the delivery phase, 39
training and education procedures, 40, 41
traditional approaches, 274–276
Magnetic resonance relaxometry, 361
Magnetic susceptibility, 30
Magnetic susceptibility artefacts, 106
Maximum tolerated dose (MTD), 273
Mayneord F factor, 165
Mean absolute error (MAE), 102
Mean centroid displacement (MCD), 200
Medical linear accelerator, 2
Medulloblastoma, 359
Metachronous, 218
Metal implants, 443
Metal prostheses, 300
Metallic implants, 27–29
Metastasis-free interval (MFI), 222
MIRAGE trial, 298
Monte Carlo codes, 134–135
Monte Carlo dose calculation algorithms, 135, 357
Monte Carlo (MC) methods, 152
Motion
cardiac, 70
encompassing methods, 71
management, 69
physiological, 70
respiratory, 69
MRIdian machine
acceptance testing for MRI, 160
commissioning of, 169–172
couch transmission, 165
dosimetry consistency with gating, 168
IMRT functionality, 167
in-house courses and hands-on training sessions, 158
MLC linearity and positional accuracy, 166
MLC transmission and leakage, 166
MR safety for, 155, 157
MR tracking latency, 168
practicing contouring and workflow, 158, 159

radiation field size, shape, symmetry, and penumbra accuracy, 167
receiver coils, SNR and uniformity, 161–163
system design of, 152–155
treatment planning system (TPS), 382
ViewRay training, 157
Multichannel endocavitary vaginal applicators, 343
Multileaf collimator (MLC), 297
Multimodality approach, 5
Multiparametric (mp) MRI, 139
Multiple image acquisitions with x-rays, 21
Multispectral imaging (MSI), 443

N
Nabothian cysts, 45
Neoadjuvant partial breast irradiation, 307
Neoadjuvant partial breast radiotherapy, 308
Neoadjuvant therapy, 245
Net magnetization vector, 48
Neuroblastoma, 364, 365
Non-alcoholic steatohepatitis (NASH), 257
Non-Cartesian sampling, 451
Non-Hodgkin's lymphoma (NHL), 362
Non-Invasive Intra-Fractional Tumor-Tracked Radiotherapy (NifteRT), 205–210
Nonlinearity corrections, 439
Non-small cell lung cancer (NSCLC), 220, 273

O
Offline (inter-fraction) adaptive planning, 137
Oligometastatic disease (OMD), 346–349
classification, 219
contouring, PTV, and OAR, 223, 224
definition of, 217
dose escalation, 226, 227
DVH constraints, 224, 225
evaluation, 233
on table adaptation, 227
outcome, 233, 234
overcoming challenges with traditional approaches, 222, 223
patient selection characteristics, 221, 222
SBRT Treatment for, 218, 220, 221
subgroups, 218
treatment of, 234
Oligopersistence, 218
Oligorecurrence, 218
On-board imaging systems, 21–23
Online adaptive approach, 138, 278, 279, 333
Online-adaptive MRgSBRT, 226
Online MRgRT, 332
On-table adaptive radiotherapy (On-ART), 263, 264
Optimization process, 134
Optimized conformal, 153
Organs at risk (OARs), 5, 69, 259, 273, 324
Osteosarcoma, 54
Out-of-field progression, 348
Oxygen-enhanced MRI, 448

P
Pancreas, 244
Pancreatic cancer
Elekta Unity
contouring, PTV, and OARs for, 249, 250
on-table adaptation for, 252
overcoming challenges of radiotherapy in, 246, 247
plan evaluation, 254
symptoms and signs of, 243
treatment, 244–246
ViewRay MRIdian
contouring, PTV, and OARs for, 248
on-table adaptation for, 252
Pancreatic ductal adenocarcinoma (PDAC), 243
Pancreatic neoplasms, 466
Partial breast irradiation, 308
Pathological complete response (pCR), 463–465
Patient coaching
feedback mechanism, 85
patient education, 85
Pediatric Hodgkin's lymphoma, 362
Percent depth dose (PDD), 151, 165, 166
Permanent magnet arrays (PMAs), 437
Permanent magnet system, 437, 438
Physiological motion, 70
Planning target volume (PTV), 308, 333, 356, 462
Planning treatment volume (PTV), 273, 274
Portable scanners, 438
Positron emission tomography (PET), 8
Post-operative radiotherapy, 359
Post-prostatectomy, 299
Post-prostatectomy RT, 289
Preamp decoupling, 436
Primary liver cancer, 257
Primary sclerosing cholangitis (PSC), 257
Prior Data Assisted Compressed Sensing (PDACS), 199, 200
Process failure mode and effect analysis (P-FMEA), 37
Prospective acquisition correction (PACE) technique, 60
Prostate cancer
dose accumulation, 288
dose escalation, 295, 296
DVH constraints, 292, 295
evaluation of daily setup, 298
exposure to ionizing radiation, 289
functional response, 289
motion management, 287, 288
natural history of, 285
OAR, 291, 292
ongoing trials in MRIgRT, 301, 303
outcome, 298, 299
plan evaluation, 297, 298
PTV, 291
target volumes, 290, 291
ultrahypofractionation in, 289, 290
Unity
pre-treatment phase, 296
treatment phase, 297
ViewRay
pretreatment phase, 297
treatment phase, 297

Prostate specific membrane antigen (PSMA)-PET, 218
Prosthetic devices, 11
Proton density (PD), 53
Proton therapy, 356
Pseudo CT, 91
Pulmonary function tests (PETs), 444
Pulse-coupled neural networks (PCNN), 208
Pulse sequences, 49, 50, 325

**Q**
Quality assurance, 141
Quantitative Analysis of Normal Tissue Effects in the Clinic (QUANTEC), 403

**R**
Radiation-induced liver disease (RILD), 258
Radiation oncology, 5
Radiation therapy, 1, 324
Radiation Therapy Oncology Group (RTOG) planning approach, 382
Radical prostatectomy (RP), 285
Radiomics, 361, 459
Radiotherapy (RT), 28, 431, 459
Real-time (dynamic) adaptive planning, 138
Real time motion management, 199–203
  accelerated MR acquisition and compressed sensing MRI, 199
  CNN performance, 204
  coherent versus incoherent undersampling for CNN reconstruction, 204
  CS-PCA, 201, 202
  PDACS, 199, 200
  real-time CNN implementation, 203
  time-domain PCA-based real-time reconstruction for dynamic undersampled MRI, 202, 203
Real-time tumor tracking, 278
Recurrent endometrial cancer, vaginal vault recurrences, 343, 344, 346
Region of interest (ROI), 161, 459
Residual gradient non-linearity-induced distortions, 104, 105
Resistive electromagnet, 437
Respiratory-gated radiation therapy with X-ray approaches, 71
Respiratory gating methods, 23
Respiratory motion, 69, 76–78, 275
Respiratory motion management
  MRI-guided linear accelerator techniques for
    breath-hold techniques, 72, 73
    cine MRI with breath-hold technique and target tracking, 73, 74
    4D-MRI, 72
    limitation, 73
  tracking and gating, 79
    considerations for, 81, 82
    multi-target tracking, 82, 83
    single plane vs. multiplane tracking, 79, 81
  X-ray-and CT-based techniques for, 71
    breath-hold treatments, 71
    forced shallow breathing with abdominal compression, 71
    motion encompassing methods, 71
    respiratory-gated radiation therapy with X-ray approaches, 71
    tracking, 72
Rotating bi-planar Linac-MR, RE coils for, 197, 198

**S**
Secondary malignancies, 258
Seminal trials, 285
Shimming, 439
Simulation
  patient positioning and immobilization, 75, 76
  respiratory motion, 76–78
Single-shot fast spin echo (SSFSE), 52
Single-sided magnets, 437
Soft tissue differentiation, 290
Soft tissue visualization, 278
Solid state detectors, 196
Sørensen-Dice index, 102
Spatial integrity, 162
Spiculae, 312
Spinal tumors, 359
Standard fractionated adjuvant breast irradiation, 309
Stereotactic ablative body radiation (SABR), 219–220
Stereotactic body radiotherapy (SBRT), 222, 259, 273, 286, 332, 356, 455
  for oligometastatic disease, 218, 220, 221
Structural similarity metric (SSIM), 102
Superconducting magnet, 437
Superiority of MRI, 9, 10
Supratentorial brain tumors, 358
Susceptibility artefacts, 105
Sydney project MR-Linac, 3
Synthetic CT scans (sCT)
  applications of, 91
  generation from MRI for radiotherapy dose calculation, 92
    atlas (multi-atlas) approach, 96, 97
    bulk density assignment (BDA), 92, 96
    voxel-based methods, 97–101
  implementation in MR radiotherapy workflows, 108–114
  metrics for evaluation, 95, 101
    accuracy of patient repositioning, 103
    dosimetric impact, 102, 103
    geometric fidelity, 102
    intensity-based metrics, 102
    MRI-only workflow, 103
  MRI-QA for distortion/artefact-free synthetic CT, 103–106

**T**
Tesla MRI scanner, 260
Thermal effects, 30
3D conformal radiation therapy (3DCRT), 153

Three-dimensional conformal treatment planning, 132
3D gradient-echo (GRE) sequence, 60
3D half-Fourier acquisition single-shot turbo spin-echo (3D HASTE), 455
Tracheal implant, 28
Traditional radiomics, 460
Transient charged particle equilibrium (TCPE), 151
Treatment delivery system (TDS), 172
Treatment planning systems (TPS)
  adaptive planning considerations, 136
    anatomical changes, 138
    functional changes, 139
    reference plans, 136
    timescales, 137
  beams geometries, 132
  commissioning and QA, 124
  contouring, 125
    imaging datasets, 125, 126
    margins, 126
    motion management with cine imaging, 127
    online adaptive considerations, 127, 128
  dose calculation, 134, 135
    electron return effect, 135
    heterogeneity corrections, 136
  electron density (ED), 128
    CT primary, 128
    MR primary, 129, 130
  hardware and software requirements, 123
  IMRT, 133
  integration with treatment delivery software, 124
  isocenter placement, 132
  optimization process, 134
  patient database, 124
  plan evaluation
    dose calculation accuracy, 140
    metrics, 140
    quality assurance, 141
    summation, 140
  three-dimensional conformal treatment planning, 132
True fast imaging with steady state free precession (TRUFI) sequence, 248, 310, 357
Tumor microenvironment (TME), 397
Tumor-treating fields (TTFields), 375
Turbo Field Echo (TFE), 280
T1-weighted imaging, 49–51
T2-weighted imaging, 49–53
2D cine, 60
2D U-Net architecture, 99

**U**

UCLA SMART constraints, 262
Ultra-hypofractionation, 228
Ultrasound (US) surveillance, 258
University of California, Los Angeles (UCLA), 267
Urethral avoidance, 291

**V**

VacQfix Vacuum Cushions (Qfix), 75
Vaginal vault recurrences, 343, 344, 346
Vascular stents, 300
Vascular structures, 291
View-angle tilting (VAT), 443
ViewRay MR $^{60}$Co system, 311
ViewRaw MRIdian system, 260
ViewRay MRIdian MR-Linac, 59
ViewRay MRIdian system, 149, 156, 248, 252, 378, 388
ViewRay training, 157
Volume adaptation, 398
Volumetric imaging, 288
Volumetric modulated arc therapy (VMAT) capabilities, 280, 335, 470
Volumetric TRUFI sequence, 60
Voxel-based methods, 97–101

**W**

Wavelet delta-radiomic (WDRFs) textures, 463
Whole brain radiation therapy (WBRT), 376
Wilms tumor, 363, 364

**X**

X-ray-based respiratory motion management, limitation of, 72
X-ray imaging, 8, 9
X-ray tube technology, 1

GPSR Compliance

*The European Union's (EU) General Product Safety Regulation (GPSR) is a set of rules that requires consumer products to be safe and our obligations to ensure this.*

*If you have any concerns about our products, you can contact us on ProductSafety@springernature.com*

In case Publisher is established outside the EU, the EU authorized representative is:

Springer Nature Customer Service Center GmbH
Europaplatz 3
69115 Heidelberg, Germany

**Batch number: 10732807**

Printed by Printforce, the Netherlands